FLUID MECHANICS AND THE SPH METHOD

Fluid Mechanics and the SPH Method

Theory and Applications

D. VIOLEAU

EDF R&D

OXFORD
UNIVERSITY PRESS

OXFORD
UNIVERSITY PRESS

Great Clarendon Street, Oxford, OX2 6DP,
United Kingdom

Oxford University Press is a department of the University of Oxford.
It furthers the University's objective of excellence in research, scholarship,
and education by publishing worldwide. Oxford is a registered trade mark of
Oxford University Press in the UK and in certain other countries

First published 2012
First published in paperback 2015

Impression: 1

Published in the United States of America by Oxford University Press
198 Madison Avenue, New York, NY 10016, United States of America

British Library Cataloguing in Publication Data
Data available

Library of Congress Cataloging in Publication Data
Data available

ISBN 978–0–19–965552–6 (Hbk.)
ISBN 978–0–19–874423–8 (Pbk.)

Printed and bound in Great Britain by
Clays Ltd, St Ives plc

Preface

Among all the books which I have read in the field of physics, especially fluid mechanics, only the unrivalled volumes of the L.D. Landau and E. Lifshitz's course left me fully unfrustrated. This is presumably an effect of my irrepressible rigour, which was undoubtedly inherited from my education in which mathematics prevailed over physics. As I became more knowledgeable in fluid mechanics, I found that the gaps which had been left unfilled in my general approach to that construction were more and more flagrant, giving a feeling of incompleteness. Scientific books and papers too often referred to a sector in physics which I was not familiar with. Intuitively, however, they allowed me to consider answers which, ultimately, I managed to come across in some other book, at the cost of painstaking effort. Since I became more aware of that issue when I worked as a teacher, I contemplated writing myself a text that would meet my own expectations; thus, I had to gain basic knowledge of these missing elements, which resulted in the first part of this book. It occurs in the simple, sometimes rather ingenuous form, of a theoretical fluid flow physics course in which the admitted principles have been as much reduced as possible.

I began working on the SPH (smoothed particle hydrodynamics) method at a complete loss. As I discovered its leading principles, I luckily noticed, as I browsed through literature, that such a numerical approach had close links to theoretical mechanics; that made me even keener to understand the latter's fundamentals, while becoming gradually more confident in the former. The desire to gather my skills in SPH in a pedagogic text slowly gained ground; thus, I had to make constant references to the theoretical laws governing fluids, which it is based on. Since the idea of a book on hydraulics was simultaneously germinating, it became obvious to me that I had to deal with both aspects of my research work as jointly as possible. That idea gave birth to this book. Nevertheless, because of the huge corpus of fluid mechanics literature, along with the ever increasing number of publications about the SPH method, I had to make a number of choices as regards the topics being dealt with herein. Predictably enough, I have decided to give prevalence to those which I am familiar with. That is why I will only discuss the weakly compressible or incompressible, possibly turbulent flows.

This book comprises two parts that refer to each other. The first, which deals with the fundamentals of hydraulics, is composed of four chapters. It is based on the elementary principles of both Lagrangian and Hamiltonian mechanics. The specific laws governing a system of macroscopic particles are built, then the major systems (statistical physics) involving the dissipative processes are explained. The continua are then discussed; lastly, a fairly exhaustive account of turbulence is given. The second part comprises the last four chapters. It discloses the bases of the Lagrangian numerical method SPH from the continuous equations which are introduced in the first part, as well as from discrete variational principles, setting out the method's specific properties of conservativity and invariance. Various numerical schemes are compared, permanently referring to physics as dealt with in the first part. Applications to schematic instances are then discussed; ultimately, practical applications to the dimensioning of coastal and fluvial structures are considered.

This book is intended for scientists, doctoral students, teachers and engineers who want to enjoy a rather unified approach to the theoretical bases of hydraulics or who want to get more skills in the SPH method in the light of these principles. In this form, it will probably look too complex to those who fully rely on empirism; on the other hand, it is perhaps insufficiently anchored to mathematical formalism for the hard core of theoreticians. I think it will nevertheless inspire many readers with a feeling of unity, answering many questions without any detrimental abstruse formalism. I hope it will be helpful to young researchers and doctoral students involved in the increasing development of interest about the SPH method.

February 2012 D.V.

Acknowledgements

Writing a book on one's own is a strenuous and awkward task, and is misleading, since the author is actually indebted to an endless list of people. Such is the case with this textbook; I am only afraid I might ignore some of the indirect contributors.

I would like to express my gratitude first to the students and researchers who have taken part in some of my works that this project has stemmed from. I must emphasize two names: Réza Issa and Eun-Sug Lee; I enjoyed so much supervising their doctoral works in the SPH method. The discussions developed in the second part of this book owe quite a lot to them, as well as to several trainees from high schools, namely Dr Sylvain Faure, Hélène Flament, Adrien Foucart, Asven Gariah, Antoine Joly, Christophe Magnin-Decugis, Dr Jilal Hammouch, Bastien Pelliccioli and Sandrine Piccon.

The first part of this book has been greatly influenced both by my teaching activities and my confident colleagues Prof. Pierre-Louis Viollet and Prof. Michel Benoit.

A number of scientists and engineers from various industry or university organizations provided me with effective assistance in my research work about the SPH method. I must therefore mention the names of Dr John Biddiscombe, Dr Clément Buvat, David Latino, Prof. Dominique Laurence, Dr Jean-Christophe Marongiu, Dr Charles Moulinec, Dr Etienne Parkinson, Dr Damien Pham Van Bang, Stéphane Ploix, Dr Ben Rogers and Prof. Peter Stansby. I also want to express my thanks to Dr Kamel Abed-Meraïm, Dr Sofiane Benhamadouche, Jacques Chorda, Dr Laurent David, Jean-Michel Hervouet, Dr Marie-Madeleine Maubourguet, Dr Emmanuel de Nanteuil, Khaled Saleh, Dr Juan Uribe and Dr Pascal Vezolles.

I want to express particular thanks to the valiant proofreaders: Martin Ferrand, Prof. Francis Lebœuf, Prof. Pierre Ferrant, Prof. Stefano Sibilla and Omar Mahmood. I am also grateful to Prof. Philippe Fraunié, Jean-Paul Chabard and Prof. Mathieu Mory, who assisted them assessing this work, within the frame of an official application for the allocation of an accreditation to supervise research, as well as Thierry Pain, who took on the translation from French to English.

This book was written in the course of my research activities at EDF R&D, with the support of the European Community through the Marie Curie project *Transfer of Knowledge Agreement* ESPHI (*European SPH Initiative*). A large part of that work was done at Manchester University (United Kingdom). Owing to the fruitful exchanges which I used to enjoy with many researchers, I am also much indebted to the SPHERIC working group (*SPH European Research Interest Community*).

Contents

Global introduction

Such particle-oriented methods as SPH (smoothed particle hydrodynamics) have become more and more popular since the advent of the intensive numerical calculation. Being relevant for modelling complex and rapidly varying processes, they have been increasingly acknowledged, since the late 1990s, as a valid alternative to grid methods for the modelling of possibly combined motions of fluids and solids. Among the assets which contribute to their success are their simplicity and smartness, their user-friendly programming and the wide range of their possible applications. Their proximity to the practical and intuitive reality of everyday mechanics provides them with an additional advantage in the eyes of their most active promoters who see them as the most 'natural' way to model continuous media. A close analysis of these methods, however, makes it possible to credit them with deeper qualities stemming from their close relationship to the theoretical bases of point mechanics. Throughout this book, we will endeavour to highlight these affinities, by constructing as general equations as possible for the motion of mutually interacting particle systems, before deriving the equations of the SPH method and their main properties.

The works conducted by J.-L. Lagrange, then by W.R. Hamilton, laid rigorous and smart foundations for mechanics. That formalism, which is the subject matter of the first chapter of this book, provides a rigorous framework for writing the equations of motion of a system which may be considered as a material point, or as a group of material points, having masses, positions, velocities and energies, and being subject to mutual forces. The variational least action principle, which is at the core of these discussions, subsequently leads to the basic conservation properties. One may also consider that each of these points, which is then referred to as 'particle', in turn consists of subsystems (molecules), which makes it possible to include the notions of density, internal energy and pressure into that formalism. The particular case of systems comprising a very large number of degrees of freedom, that is macroscopic sets of molecules, may be treated likewise from statistical considerations, which were first contemplated by L. Boltzmann. We establish their principles in Chapter 2, which can be used to introduce the notion of irreversibility, resulting in particular in the consideration of particle friction forces. The concepts of entropy, kinetic coefficients and energy dissipation rate then appear in the particle formalism. The discrete equations achieved through this procedure already give an early indication of those equations as solved by the SPH method, with very close notations and similar properties.

Among the main issues to be settled by a theoricist is the treatment of the particle density and internal forces (pressure and friction), the evolution of which can only be investigated through the formalism of continuous media. On the basis of the arguments of the first two chapters, Chapter 3 sets out the suitable tools for establishing the formulation of these quantities in the particular case of the viscous fluids, especially when the density hardly varies in both space and time. Several approaches leading to the Euler and Navier–Stokes equations are highlighted; they are initially based on the abovementioned discrete equations, then on the Boltzmann equation as described in Chapter 2, lastly on the variational principles extended to the continuous media, whereupon well defined continuous forms of the conservation laws can be obtained. The notions of stress tensor, molecular viscosity, then surface tension coefficient appear naturally through the concept of local strain. Furthermore, seeking solutions to these continuous equations leads us to consider the question of turbulence throughout Chapter 4. After a phenomenological study, the main turbulent closure models are established, giving rise to further mechanical quantities, namely Reynolds stresses, turbulent kinetic energy, eddy viscosity, and so on.

At the end of the first section, all the tools of point mechanics and conventional fluid mechanics are available to work out numerical—especially Lagrangian—solution techniques. It seems, however, that the discrete (Chapters 1 and 2) and continuous (Chapters 3 and 4) approaches have diverged. In the second section of the book, they will be lumped together again through the general principles of the SPH method, which are set out in Chapter 5. From the notion of the interpolation kernel, the basic equations of this approach are established in two ways for the treatment of the viscosity-free fluids, by discretizing the continuous equations on the one hand and applying a least action principle on the other. The discrete conservation laws are then naturally satisfied, although an important reservation should be mentioned: the formalism of Chapter 1 only contemplates a discrete nature for the space variables, whereas time remains a continuous variable, which makes the theoretical treatment of the notion of conservation easier. With a numerical approach, since time has to be discretized as well, the conservation laws should be re-examined according to the selected temporal scheme, which in turn affects the model's numerical stability properties. Chapter 6 extends SPH formalism to the notions of friction, surface tension and turbulence. In addition, it put forwards solutions both for the treatment of the wall boundary conditions and the calculation of pressure and it briefly discusses the treatment of the rigid bodies moving about within a fluid.

The implementation of the SPH method with respect to schematic application cases has been amply dealt with in the literature, which is evidence of its predictive capacities, but also of its drawbacks, particularly as regards the computational time and the required memory capacity. The applications to practical cases are therefore less numerous, but have become more common since the 2000s. Chapters 7 and 8 are dedicated to this kind of analysis. The former is restricted to schematic cases in order to ascertain the validity of the numerical method through a comparison with other methods, with laboratory experiments or even with semi-analytical solutions as derived from the first section of this book. The last chapter addresses instances which are close to practice in hydraulic engineering, in the cases of complex and fairly varied flows, along with several quantitative validations. Readers will find some assurance that the SPH method is suitable for effectively tackling many problems related to the dimensioning of water works.

At first sight, readers may be puzzled by the numerous subject matters being tackled. Actually, no important result is left to chance, as might be expected from the first few lines of this introduction. For example, Boltzmann's equation is demonstrated in Chapter 2 before explanations are provided in Chapter 3 for using it when constructing the Navier–Stokes equations, though the latter are previously derived from simpler principles. The reasons for such a seeming complexity lie, in my opinion, in the required grasp of the various links between the fundamental fluid mechanics equations. This relatively complete review, however, does not imply any thorough reading; readers in a hurry may miss out some sections without prejudice to understanding. Yet, in order to provide a main thread, a concern for coherence is systematically put forward; thus, each section, when it is necessary, will refer to the useful notions as set out in the previous chapters. For example, the time integration schemes for solving the discrete equations in the SPH method, as dealt with in Chapter 5, are selected according to the properties of the Hamilton equations as discussed in Chapter 1.

As can be seen, this book provides an overview of particle theory for fluid modelling. It is particle-oriented, first of all, due to the immediate utilization of this formalism for establishing the fundamental equations of physics which it treats, then only through the SPH method considered as a way of numerically solving these equations. Reading it will, indeed, require prior skills in elementary mathematics: particularly matrix algebra, integration and differentiation. Those unfamiliar with the tensorial formalism or the notion of Fourier transform will find brief information in the appendices.

Taking into account the great variety of processes being considered in this book, a high number of notations, some of which are redundant, will inevitably be employed. For instance, notation $\tilde{\mathbf{u}}$ denotes, in Chapters 2 and 3, the velocity of a molecule, whereas in Chapter 4 it refers to the filtered velocity of a fluid with a turbulent flow regime. Likewise, the generalized momenta p_i should not be likened to pressure p, and notation T refers sometimes to temperature and sometimes to a time scale. Actual ambiguities, however, are rare and only related to specific quantities from different chapters. Thus, we hope no confusion will be possible; footnotes will sometimes highlight distinctions which the reader will probably have made himself.

Physics of weakly compressible fluids

Part I

Lagrangian and Hamiltonian mechanics

1

1.1 Introduction

The scope of this chapter consists in defining, from simple concepts, the mechanical laws governing the systems of material points, that is which can be considered as a more or less extensive collection of entities liable to be likened to point systems. The purpose of this is to prepare for the next chapter, which will be dedicated to the mechanics of large systems (statistical mechanics, or thermodynamics).

The laws of classical mechanics, as they are usually taught, are often perceived as arbitrary or excessively founded on results. Now it is comparatively easy to prepare a theoretical framework, based upon a small number of intuitive assumptions consistent with observation, where these laws (Newton's second law, conservation laws) occur naturally and take a clear physical significance. The keystone to that theory of mechanics is the *least action principle*, the single really constraining hypothesis, since it looks unwarranted at first sight. It happens, however, that this single principle, accompanied by natural assumptions and in keeping with the daily experience, is sufficient for simply and neatly building the foundations of mechanics and highlighting their links to the properties of regularity in both time and space.

The least action principle, which was initially conceived by P.-L. Moreau de Maupertuis in 1748, had been faintly recognized by Fermat in this 'least time principle' (1661); the latter had derived, out of a minimum principle, the laws of light refraction which had initially been empirically laid by Descartes. This principle, which was explored more deeply and operated by Lagrange at the end of the18th century, then by Hamilton in the subsequent decades, proved to be quite fruitful, especially in physics, since it is far from being only applicable to mechanics. Historically, indeed, Einstein did manage to work out his theory of special relativity by thoroughly reviewing the foundations of classical mechanics and the least action principle. As far back as 1916, Eddington suggested a similar approach to derive the laws of general relativity, which had been built through very intuitive means by Einstein one year before. Electromagnetism, as well as geometrical optics which is a particular case of it, can also be founded on that principle against a relativistic background (Landau and Lifshitz, 1975). Feynman, as long ago as the 1940s, could outline a conceptual framework (based upon the notion of trajectory integrals) allowing the provision non-relativistic quantum mechanics with foundations based upon the least action principle, before it found applications in the relativistic quantum theory (Landau et al., 1982). More recently, the physics of water waves found a theoretical spectral framework resting on that formalism (Bretherton and Garret, 1969), yielding outstanding applications. Among the attempts to quantify gravitation, string theory is also based on similar concepts (see e.g. Zwiebach, 2009).

This chapter, apart from giving suitable expertise in the actual meanings of the equations of mechanics, will bring some intellectual satisfaction to all those who are reluctant about extensive empirism. It can also be used to lay the foundations of variational mechanics, which is implemented by the end of Chapter 3 to discuss the continua, as well as for a rigorous introduction to the SPH numerical method in the second part of this book. That theoretical framework, however, will only be basically set out; more solid—but drier—foundations can be built (for instance, refer to Swaters, 2000). We have elected to refer primarily to Landau & Lifshitz, 1982a); nevertheless, we have preferred to give explanations in a slightly different way; thus, the least action principle will come after the principle of inertia, with reference to Boudenot (1989).

First of all, we will set out the Lagrangian mechanics formalism for any discrete mechanical system, based upon the Least action principle yielding the Lagrange equations. The case of a system of interacting particles (either microscopic or macroscopic) will then be thoroughly discussed; afterwards, we will come to Newton's second law. The notion of energy (Hamiltonian) will be used to establish the Hamilton equations before switching to the concept of discrete conversation laws. We will then present a general case which will be taken up throughout the book, namely the simple oscillating pendulum. Lastly, an account of the systems consisting of sub-systems (macroscopic particles) will enable us to deal with the cases of rigid bodies, then fluids, resting upon the notion of pressure and internal energy. The Hamilton equations will be reformalized against this new background.

1.2 Least action principle

In this section, we state the basics of Lagrangian Mechanics. After defining the concepts of Galilean frame and Galileo's relativity principle, we introduce generalized coordinates and velocities, as well as the Lagrange function of a mechanical system. Then, we state the least action principle, which is at the foot of physics. The Euler–Lagrange dynamic equations are finally deduced from the abovementioned tools.

1.2.1 Inertial frames and principle of relativity

Throughout this chapter, we will need the notion of frame of reference. This term refers to any system of axes which is liable to move in relation to another system and is fitted with a timer. Among the classical mechanics postulates is above all the invariance of the standards for measuring time and distances. Aside from that arbitrary change, we may state that time is universal, as are distances.[1] In the following, we will often consider the cases of two frames

[1] Every quantity derived from distances is also universal, just like volume.

of reference R and R' whose axes are aligned and whose origins coincide at $t = 0$, the second moving about at the *constant* velocity $\mathbf{V}$ in relation to the first. In accordance with the above, the laws of transformations of coordinates and time, when shifting from R to R', are written as

$$t = t'$$
$$\mathbf{r} = \mathbf{r}' + \mathbf{V}t \tag{1.1}$$

where $\mathbf{r}$ denotes the position vector, whereas the primes refer to the frame R'. The velocity of a body is defined as the derivative of its position vector with respect to time:

$$\mathbf{u} = \dot{\mathbf{r}} \doteqdot \frac{d\mathbf{r}}{dt} \tag{1.2}$$

Thus, deriving the coordinate transformation equation (1.1) with respect to time t (or t', which comes to the same thing), we get the Galilean transformation law for velocities:

$$\mathbf{u} = \mathbf{u}' + \mathbf{V} \tag{1.3}$$

From a mechanical standpoint, not all the frames of reference are equivalent, which means that the laws of mechanics do not have the same forms in them. Most of the frames are neither homogeneous nor isotropic in regard to these laws, and time does not play a regular role in them. If, for example, we consider the case of a revolving frame of reference (a carousel), then we can see that an object released on its periphery with a zero initial velocity will tend to drift away from the centre of rotation, whereas an object which is put on that very centre will remain motionless. There are some frames of reference, however, which prove to be homogeneous and isotropic, and for which time acts uniformly; they are known as Galilean or inertial frames. In such frames of reference, the laws of mechanics are *independent of time, position and direction of motion*. A significant postulate in classical mechanics is that a Galilean frame of reference is linked to every point in space. The first postulate of mechanics, as it is currently formulated, states that a body which is sufficiently distant from every other body (then the commonly used expression is an *isolated* system)[2] persists in its rectilinear and uniform motion in every Galilean frame of reference. The velocity composition law (1.3) then allows us to state that when R is Galilean, so R' is as well. The second postulate of mechanics, which is the most important one from a historical point of view, is *Galileo's principle of relativity*: the laws of mechanics have the same form in all the Galilean frames of reference. Or else, as Galileo himself stated: 'motion is like nothing'. That means a uniform motion has no effect on the formulation of the laws of mechanics. We can intuitively experience that principle, for example, when we are sitting in an aircraft flying at a constant and rectilinear velocity. That motion is then absolutely imperceptible, whatever its magnitude may be, and every mechanical experience[3] will take place in it just as at the ground level. If, on the other hand, the aircraft decelerates, accelerates or turns, then we feel that change of velocity. Hence, we can infer that a body moving at a constant velocity in a Galilean frame of reference will behave as though

[2] It is a priori important to make a difference with a *closed* system which inherently does not exchange matter with the outside. From a Lagrangian viewpoint, however, the closed property of the systems is implicit, since every system is once and for all composed of a given amount of matter.

[3] And even every physical experiment, according to the Einstein's principle of relativity.

it was at a standstill. If it is not subjected to any external influence, then it will indefinitely keep its velocity. This formulation is often called *principle of inertia* (or Newton's first law).

Since the laws of mechanics have the same form in all the Galilean frames of reference, the equations of motion are invariant with respect to transformation (1.1) (this is the Galilean invariance). The solutions of these equations, conversely, depend on the frame of reference as they depend on the boundary conditions (this question will be dealt with in the next section). The Galilean frames of reference play a major role in classical mechanics. Thus, the assumed existence of such a frame of reference is constraining, but only the theory of General Relativity will make it possible to construct invariant laws in any frame. On the contrary, the principle of relativity tells us that none of the Galilean frames of reference prevail over the others. In the following, unless stated otherwise, we will refer to such frames.

A final remark should be made. Invariance of the laws of mechanics in terms of time is disproved, even in a Galilean frame of reference, when friction occurs. It is well known that, in such conditions, due to macroscopic complications, the motions are not time-reversible. Lagrangian Mechanics, as it is set out in the initial paragraph of this chapter, does not take these facts into account; it can only be applied to non-dissipative processes (we will come back to that term when we speak about energy). The dissipative—that is non-reversible—systems shall be considered separately and will be discussed in Chapter 2, Section 2.4.

1.2.2 Generalized coordinates. Lagrangian of a system

From what we have written in the previous paragraph, we may infer that a body's velocity is not a quantity with an absolute physical significance, and that only acceleration (i.e. a changing velocity) is unambiguous. The laws of mechanics shall then refer to acceleration. Thus, in regard to those coordinates considered as time functions, differential equations of order 2 will describe the variation of a mechanical system. Consequently, mathematics tell us that two sets of boundary conditions are required for solving these equations. In other words, the future of a system can only be perfectly defined from an initial configuration provided that not only the positions of its constituents, but also their first derivatives with respect to time are known at any given point in time. Thus, a ball which is thrown vertically within an aircraft will describe a vertical trajectory as seen by the experimenter, whereas an observer standing on the ground would see it describing a parabola, because the initial ball velocity would have a non-zero horizontal component in this observer's frame of reference.

Generally speaking, we will consider a system whose *geometrical* description is given by N independent parameters known as the system's *generalized coordinates*, which we will denote $q_i(t)$ (however, we will often omit to explicitly write the dependence on time). The integer N is the system's number of geometrical degrees of freedom. For a particle, for example $N = 3$, and the q_i coincide with the x, y, z components of the particle's position vector $\mathbf{r}$ at a given point in time; for a rigid pendulum, only one coordinate is necessary, for example the angle θ it makes with the vertical. We can state from the above that the state of a system is fully determined, at any time, by the data of its generalized coordinates and their first derivatives with respect to time. The latter are called *generalized velocities* and are denoted $\dot{q}_i = dq_i/dt$. Then, for a particle, we have

$$\begin{pmatrix} q_1 & \dot{q}_1 \\ q_2, & \dot{q}_2 \\ q_3 & \dot{q}_3 \end{pmatrix} \doteqdot (\mathbf{r}, \mathbf{u}) \tag{1.4}$$

and for a pendulum:

$$(q, \dot{q}) \doteqdot (\theta, \dot{\theta}) \tag{1.5}$$

If there is a quantity describing the mechanical state of the system at a given point in time, then it is a function of the generalized coordinates and velocities. We will assume that such a function does exist, and we will refer to it as a *Lagrangian* (or *Lagrange function*) of the system:

$$L = L\left(\{q_i\}, \{\dot{q}_i\}\right) \tag{1.6}$$

it being understood that the subscript i covers the whole range of values from 1 to N, as symbolized by the brackets $\{\cdot\}$.

Now we will consider a system A which is subject to the action of such a big system B (in a sense which will be seen later on as linked to the notion of mass) that the latter is not affected by the presence of A (a 'standard'-sized body subject, for instance, to the Earth's gravitational influence). This is known as an *open* system,[4] and the coordinates and velocities characterizing the system B are known time functions as determined by the equations of mechanics (so far unknown in our explanations). The Lagrangian of the set (i.e. of $A \cup B$) will then be written as

[4] Our approach is quite analogous to that consisting in considering, in thermodynamics, that a non-isolated system should be taken as a part of a broader isolated system (see in Chapter 2).

$$\begin{aligned} L_{A\cup B} &= L\left(\{q_i^A\}, \{\dot{q}_i^A\}, \{q_i^B(t)\}, \{\dot{q}_i^B(t)\}\right) \\ &= L\left(\{q_i^A\}, \{\dot{q}_i^A\}, t\right) \end{aligned} \tag{1.7}$$

(here we have explicitly written the time in those parameters characterizing the system B, in order to emphasize that these quantities are known functions of time). Thus, we can see that a system which is subject to time-varying external influences exhibits an explicit dependency of its Lagrangian on time via the terms $q_i^B(t)$ and $\dot{q}_i^B(t)$, so that we will write from now on $L\left(\{q_i\}, \{\dot{q}_i\}, t\right)$. On the other hand, as regards an isolated system, there is no such dependency (in a Galilean frame of reference), since we have seen that the laws of mechanics are then uniform in time therein. Time, however, implicitly acts in L through the q_i and the $\dot{q}_i$, which are time-varying quantities.

The equations of mechanics shall a priori affect the Lagrangian, that is *in fine* the generalized coordinates. They should be associated with $2N$ boundary conditions to be solved (a couple of conditions for each coordinate). As previously explained, these conditions may be the initial values of coordinates and velocities. However, changes in the system can also be determined for other sets of boundary conditions, provided that they occur in a sufficient number.

1.2.3 Least action principle

In this paragraph, we temporarily refer to an N-dimensional abstract Cartesian space in which the system's generalized coordinates are transferred onto the coordinate axes. Such a space is known as a system *configuration space*. We will consider two system's geometrical states corresponding to a couple of sets of coordinates, denoted $q_i^{(1)}$ and $q_i^{(2)}$. Thus, these two states are related to two points M_1 and M_2 in the configuration space (refer to Fig. 1.1 for the case $N = 2$). The geometrical curves connecting these two points are called the *permissible trajectories* of the problem.

Now we will consider how the system should transition from state 1 at time t_1 to state 2 at time t_2. In other words, among the permissible trajectories, which is the single trajectory the system will follow according to the laws of mechanics? It is noteworthy that this problem is well formulated in terms of physics, since we have $2N$ boundary conditions, namely the initial and final values of the q_i. The answer to the question is given by the least action principle; to that purpose, we have to define the system's action $\widetilde{S}$:

$$\widetilde{S} = \int_{t_1}^{t_2} L\left(\{q_i\}, \{\dot{q}_i\}, t\right) dt \tag{1.8}$$

The principle is then expressed as follows: a minimum value of action is related to an actual motion. More precisely, this principle should be formulated in a slightly different way, because of the presently rather ambiguous definition of the Lagrangian: 'there is a function L of coordinates and velocities whose integral has a minimum value during a motion'. This principle may seem to be audacious; actually, it fairly well matches intuition. Nature, indeed, commonly minimizes[5] the integrals of some quantities, and the Lagrangian is the single presently available mechanical quantity, even though we have less information about the latter. The least action principle itself, however, will provide us with

[5] Strictly speaking, it is a minimum for sufficiently short paths, that is for sufficiently close times. Action can be maximum in some cases; that consideration, however, will not affect the following discussions, since the equations of movement to be introduced later on will in fact be based on a principle of *extremum*.

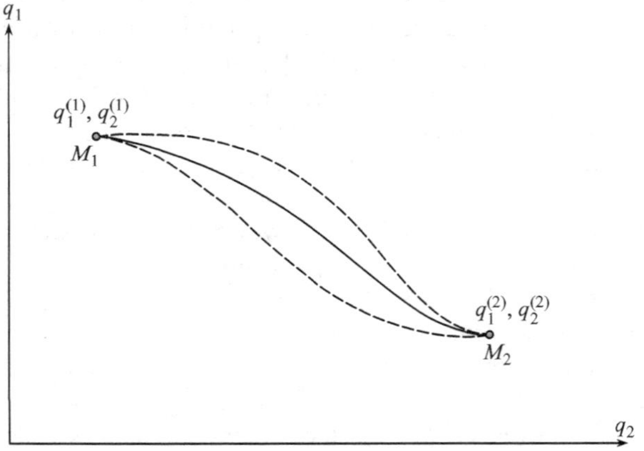

Fig. 1.1 Configuration space of a system with two geometrical parameters. Among all the permissible paths, only one corresponds to the physical reality, as shown in full line.

some information about L. First, it is necessarily a real function; otherwise the notion of minimum would be meaningless. Second, it is additive. To that purpose, we will consider a system consisting of two parts A and B. If we move them sufficiently apart from each other, their interaction will be negligible. We can then infer that the equations of motion of one part cannot include terms related to the other part. The whole system action should then take on the form of the sum of A and B respective actions.

Something else can be noticed: multiplying the Lagrangian by any positive constant does not affect the equations of motion, since the target being aimed at is only minimizing its integral. One may believe that this will result in an undetermination, because the Lagrangians of two systems could arbitrarily be multiplied by two different constants. Now, due to the additivity of the Lagrangian, we know that such a constant should be the same for all the systems. The resulting degree of freedom simply corresponds to the *arbitrary selection of units*. So far, indeed, the Lagrangian unit has not been specified.

Lastly, we can observe that, for the same reason as before, adding a constant to the Lagrangian does not alter the equations of motion. The Lagrangian is then only defined to within one additive constant. Further investigations are possible to make sure that adding L to the total derivative of a coordinate and time function does not affect the equations of motion. In that respect, let Φ be such a function, and let us define the amended Lagrangian

$$L' \doteqdot L + \frac{d\Phi}{dt} \tag{1.9}$$

The corresponding action is written as

$$\begin{aligned} \widetilde{S}' &= \int_{t_1}^{t_2} \left(L\left(\{q_i\}, \{\dot{q}_i\}, t\right) + \frac{d\Phi\left(\{q_i\}, t\right)}{dt} \right) dt \\ &= \int_{t_1}^{t_2} L\left(\{q_i\}, \{\dot{q}_i\}, t\right) dt + \left[\Phi\left(\{q_i\}, t\right)\right]_{t_1}^{t_2} \\ &= \widetilde{S} + \Phi\left(\left\{q_i^{(2)}\right\}, t_2\right) - \Phi\left(\left\{q_i^{(1)}\right\}, t_1\right) \end{aligned} \tag{1.10}$$

Thus, the minimum of $\widetilde{S}$ remains unchanged, since the last two terms of (1.10) are constants, the conditions $q_i^{(1)}$ and $q_i^{(2)}$ being prescribed. An important 'reverse' of that rule: one system, in two Galilean frames of reference, has two Lagrangians which differ from each other in the total derivative of a coordinate and time function. In fact, one should remember that the Lagrangian of a system depends on the selected frame of reference, since it is a function of its coordinates and velocities, which are defined in a given frame. Let us refer to the couple of presumably Galilean frames R and R' defined in Section 1.2.1. The principle of relativity tells us the equations of motion are the same, that is the actions have the same minimum in both R and R', which means they differ from a constant. This is only possible where the system's respective Lagrangians in either frame satisfy the relation (1.9), which is the action's most general additive undetermination. This result is the mathematical expression of the Galileo's principle of relativity as applied to the least action principle.

1.2.4 Lagrange equations

We are now going to seek a necessary condition on the Lagrangian L to get a least action $\widetilde{S}$. For the sake of simplicity, we will first consider the case of a single degree of freedom: $N = 1$. We then have a single coordinate q, which is associated with two boundary conditions $q^{(1)}$ and $q^{(2)}$ (initial and final points in time); the Lagrangian is written as $L(q, \dot{q}, t)$. First of all, let us consider the system's actual trajectory, denoted $q(t)$ in the configuration space, and let us consider a variation δq shifting the actual trajectory to another permissible trajectory $q(t) + \delta q(t)$. We can assess the variation $\delta\widetilde{S}$ of the action associated with δq:

$$\begin{aligned}
\delta\widetilde{S} &= \widetilde{S}(q+\delta q) - \widetilde{S}(q) \\
&= \int_{t_1}^{t_2} \delta L(q, \dot{q}, t)\, dt \\
&= \int_{t_1}^{t_2} \left(\frac{\partial L}{\partial q}\delta q + \frac{\partial L}{\partial \dot{q}}\delta \dot{q} \right) dt
\end{aligned} \tag{1.11}$$

(here, for providing clearer notation, we have introduced notations by a convenient abuse of language: it should be understood that $\partial L/\partial q$ and $\partial L/\partial \dot{q}$ are the derivatives of L with respect to its first and second variables, respectively). The variation of $\delta\dot{q}$ is, of course, equal to the time derivative of variation δq; thus, the second term of (1.11) can be integrated by parts to give

$$\begin{aligned}
\delta\widetilde{S} &= \int_{t_1}^{t_2} \left(\frac{\partial L}{\partial q}\delta q + \frac{\partial L}{\partial \dot{q}}\frac{d\delta q}{dt} \right) dt \\
&= \int_{t_1}^{t_2} \left(\frac{\partial L}{\partial q} - \frac{d}{dt}\frac{\partial L}{\partial \dot{q}} \right) \delta q dt + \left[\frac{\partial L}{\partial \dot{q}}\delta q \right]_{t_1}^{t_2}
\end{aligned} \tag{1.12}$$

However, the trajectory $q + \delta q$ should satisfy the boundary conditions $q(t_1) = q^{(1)}$ and $q(t_2) = q^{(2)}$ to be permissible, then $\delta q = 0$ for these two points in time; thus, the second term of (1.12) vanishes. The minimum of $\widetilde{S}$ occurs when a slight variation of q only affects $\widetilde{S}$ to order 2, which prescribes $\delta\widetilde{S} = 0$; that relation holding true for all the values of δq, the relation (1.12) immediately yields

$$\frac{d}{dt}\frac{\partial L}{\partial \dot{q}} - \frac{\partial L}{\partial q} = 0 \tag{1.13}$$

One could pertinently argue that the above demonstration is fairly inaccurate. The scientific literature abounds with rough (but wrong) demonstrations of the (right) result which we have reached. Although this way of reasoning is quite useful[6] to physicists, we have chosen to provide an accurate proof in order to reassure the mathematicians. To do this, let us now return to the terms of the problem and refer to space E of functions q with continuous second derivatives:

$$E = C^2([t_1, t_2], \mathbb{R}) \tag{1.14}$$

[6] We will use it, in particular, at the end of Chapter 3 as regards the continuous media.

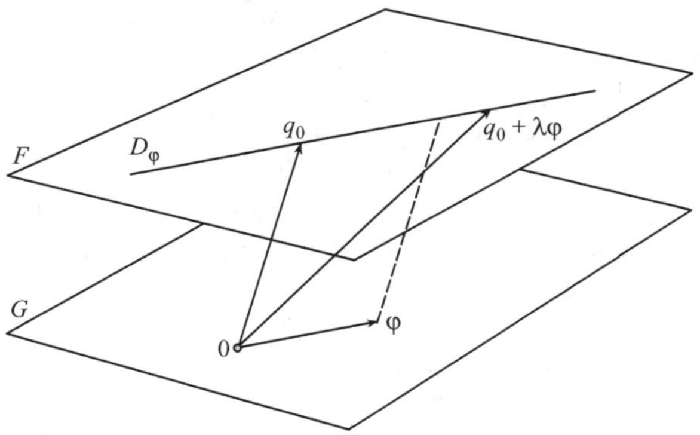

Fig. 1.2 Illustrated demonstration of the Lagrange equation.

E is a vector space which we can provide with the following scalar product:

$$\langle f, g \rangle = \int_{t_1}^{t_2} f(t)\, g(t)\, dt \tag{1.15}$$

We also define the set of permissible trajectories (i.e. satisfying the boundary conditions in the configuration space):

$$F = \left\{ q \in E, q(t_1) = q^{(1)}, q(t_2) = q^{(2)} \right\} \tag{1.16}$$

as well as the space of functions cancelling themselves out at the boundaries:

$$G = \{ q \in E, q(t_1) = 0, q(t_2) = 0 \} \tag{1.17}$$

Clearly enough, G is a Euclidean vector subspace of E, and F is an affine subspace of E directed by G. The problem consists in seeking the minima of action $\widetilde{S}$ over space F. The method is as follows: if $\widetilde{S}$ goes through a minimum over F for a function q_0, the latter is also a minimum for $\widetilde{S}$ over every affine subspace of dimension 1 passing through q_0 and contained in F (refer to Fig. 1.2). Then, let us consider the affine subspace of dimension 1 D_φ directed by any function φ of G . We have:

$$\forall \lambda \in \mathbb{R}, \ \widetilde{S}(q_0) \leq \widetilde{S}(q_0 + \lambda\varphi) \tag{1.18}$$

λ being a linear coordinate of the line D_φ. We can then derive the action with respect to λ in order to get a *necessary* condition for a minimum:

$$\left(\frac{d\widetilde{S}(q_0 + \lambda\varphi)}{d\lambda} \right)_{\lambda = 0} = 0 \tag{1.19}$$

Let us now consider the definition (1.8) of $\widetilde{S}$, in which $q_0 + \lambda\varphi$ are substituted for the q_i and $\dot{q}_0 + \lambda\dot{\varphi}$ for $\dot{q}_i$. A derivation[7] under the integral yields, letting $\lambda = 0$:

[7]The Lagrangian is assumed to be regular enough.

$$\int_{t_1}^{t_2} \left[\varphi \frac{\partial L}{\partial q}(q_0, \dot{q}_0, t) + \dot{\varphi} \frac{\partial L}{\partial \dot{q}}(q_0, \dot{q}_0, t) \right] dt = 0 \tag{1.20}$$

At present, let us integrate the second term by parts (from now on, for the sake of conciseness, we will omit the arguments q_0 and $\dot{q}_0$):

$$\int_{t_1}^{t_2} \varphi \frac{\partial L}{\partial q} dt + \left[\varphi \frac{\partial L}{\partial \dot{q}} \right]_{t_1}^{t_2} - \int_{t_1}^{t_2} \varphi \frac{d}{dt} \frac{\partial L}{\partial \dot{q}} dt = 0 \tag{1.21}$$

The second term cancels itself out in accordance with the boundary conditions, since φ is within G. By lumping the other two terms under the same integral, we finally obtain:

$$\left\langle \varphi, \frac{d}{dt} \frac{\partial L}{\partial \dot{q}} - \frac{\partial L}{\partial q} \right\rangle = 0 \tag{1.22}$$

This equation holds for any function φ of G, that is the right-hand side in the scalar product (1.22) is in the orthogonal subspace of space G. Now, one can easily realize that G is dense in E (in the sense of the norm derived from the scalar product (1.15)), and then that its orthogonal subspace is reduced to the zero vector. Thus, the necessary condition for an extremum is finally written as:

$$\frac{d}{dt} \frac{\partial L}{\partial \dot{q}} - \frac{\partial L}{\partial q} = 0 \tag{1.23}$$

which is definitely the equation (1.13). The method which we have applied comes within the calculation of variations and the equation (1.23) is the so-called *Euler equation*. In the basic register of Lagrangian mechanics, however, it is known as the *Lagrange equation*, and is the equation of motion being sought. Formally, it is, indeed, desirably a second order equation in q, since L—hence its derivatives—depends on the first derivatives of the generalized coordinate, and the first term is subjected to a time derivative. The latter derivative takes the possible implicit dependence of L on time into account through q and $\dot{q}$.

In the case of a system with several degrees of freedom, it can be shown likewise that each coordinate is governed by the same equation, which is intuitive from a physical point of view (since it varies independently from the others):

$$\forall i, \quad \frac{d}{dt} \frac{\partial L}{\partial \dot{q}_i} - \frac{\partial L}{\partial q_i} = 0 \tag{1.24}$$

Formally, (1.24) is a system of N differential equations of order 2, provided with $2N$ boundary conditions. For solving these equations, the common practice consists in taking the initial values of coordinates and generalized velocities into account.

Quantities $\partial L / \partial \dot{q}_i$ are usually called *generalized momenta*:

$$p_i \div \frac{\partial L}{\partial \dot{q}_i} \tag{1.25}$$

With this notation, the Lagrange equations are also written as:

$$\forall i, \quad \frac{dp_i}{dt} = \frac{\partial L}{\partial q_i} \tag{1.26}$$

Remarkably enough, the Lagrange equations (1.24) are written regardless of the selected system of coordinates. To make progress in our knowledge of the laws of mechanics (to provide a truly explicit equation on the q_i), however, we have to know the expression of L as a function of the q_i. The next paragraph aims to determine such a dependency in the case of a system of mutually interacting 'particles'.

In some instances, one should take into account a number of geometrical or kinematic constraints resulting from the kind of system being investigated. For a system of hinged bars or springs, for example, the degrees of freedom are not mutually independent. The problem will then consist of determining the minimum of the action while meeting the prescribed constraints. The problem of a minimization under constraints can be treated using the method of Lagrange multipliers. We initially will consider the case of a unique constraint to be respected, which results in a relation between the generalized positions and velocities:

$$\Psi\left(\{q_i\},\{\dot{q}_i\}\right)=0 \tag{1.27}$$

Let us now introduce an additional, possibly time-varing parameter A and let us consider the modified Lagrangian as yielded by

$$L'\left(\{q_i\},\{\dot{q}_i\},A,\dot{A},t\right) \doteqdot L\left(\{q_i\},\{\dot{q}_i\},t\right)+A\Psi\left(\{q_i\},\{\dot{q}_i\}\right) \tag{1.28}$$

The validity of that choice stems from the fact that L' is equal to L, since the value of $\Psi\left(\{q_i\},\{\dot{q}_i\}\right)$ is inherently zero (eqn (1.27)). Thus, the system can be described by the Lagrangian L' instead of L. Besides, the derivatives of L' with respect to A and $\dot{A}$ are written as

$$\begin{aligned}\frac{\partial L'}{\partial A}&=\Psi\left(\{q_i\},\{\dot{q}_i\}\right)\\ \frac{\partial L'}{\partial \dot{A}}&=0\end{aligned} \tag{1.29}$$

Thus, the Lagrange equation (1.23) concerning $q=A$ is written as $\Psi=0$, which is exactly tantamount to writing the condition being sought (1.27). The Lagrangian (1.28) is then suitable to treat the mechanical problem being considered. The equation governing the system's behaviour under constraints then remains unchanged (eqn (1.24)). In some cases, each degree of freedom i can be subjected to a specific constraint. The same number of Lagrange multipliers $\{A_i\}$ will then be defined to write

$$L'\left(\{q_i\},\{\dot{q}_i\},\{A_i\},\left\{\dot{A}_i\right\},t\right) \doteqdot L\left(\{q_i\},\{\dot{q}_i\},t\right)+\sum_{i=1}^{n} A_i\Psi_i\left(q_i,\dot{q}_i\right) \tag{1.30}$$

We will use this procedure on several occasions in this book, in Chapters 2, 3 and 6.

1.3 Mechanics of a system of particles

We now apply the previous tools to the case of a moving particle system, starting from a single, isolated particle to finish with a system of several interacting particles. After having deduced the Lagrangian of such a system, we apply the Lagrange equations to yield Newton's second law of motion. We introduce the concepts of mass, particle momentum and forces, as well as kinetic energy.

1.3.1 Lagrangian of an isolated particle

We are now going to determine the form of the Lagrangian L in the case of a system with N_p material points (or particles). First of all, it should be emphasized that when we speak of a 'particle', we do not necessarily mean an elementary particle, but any (possibly macroscopic) body which is liable to be described, on a gemetrical viewpoint, by a set of three independent generalized coordinates coinciding with the actual coordinates of its centre of inertia (to be precisely defined in Section 1.5.1). However, such a particle is not considered as point-sized, thus it possesses volume and mass. The Earth, for example, may be considered as such when one studies its revolution around the Sun, whereas it may not be seen as a material point when one intends to investigate its diurnal rotational motion. Thus, it is to be understood that, in this section, we ignore every effect inducing either a rotation of the body on itself[8] or an internal distortion. In such a case, the generalized coordinates associated with each particle a can be gathered in one vector $\mathbf{r}_a$, the generalized velocities taking the form of the N_p velocity vectors $\mathbf{u}_a$, and so $L = L(\{\mathbf{r}_a\}, \{\mathbf{u}_a\}, t)$, the subscript a covering all the values from 1 to N_p (the number of degrees of freedom is then $N = 3N_p$).[9] When they are put together in threes, the Lagrange equations (1.24) are then rearranged as (Landau and Lifshitz, 1976):

$$\forall a, \quad \frac{d\mathbf{p}_a}{dt} = \frac{d}{dt}\frac{\partial L}{\partial \mathbf{u}_a} = \frac{\partial L}{\partial \mathbf{r}_a} \tag{1.31}$$

where the derivative of L with respect to a vector is to be understood as the vector the components of which are the derivatives of L with respect to that vector's respective components (see Appendix A, eqn (A.54)). We have noted $\mathbf{p}_a = \partial L/\partial \mathbf{u}_a$, which we will simply call the *momentum vector* of particle a. It must immediately be stated that the right-hand side is the Lagrangian's gradient 'with respect to coordinates $\mathbf{r}_a$'. In compliance with these notations, we will seek the Lagrangian with respect to the velocity and position vectors, which advantageously ignores every particular system of coordinates. Remember that, throughout the following, we will refer to a *Galilean* frame of reference. In addition, the Galilean invariance of equation (1.31) is satisfied provided that the Lagrangian only depends on inter-particle distances.

Let us first consider an isolated particle (i.e. far from any external influence). In this case, we may omit the subscript a and we simply have $L = L(\mathbf{r}, \mathbf{u}, t)$, then exploit the properties of the inertial frames as specified in Section 1.2.1. Thus, we know that the Lagrangian should not explicitly depend on time, since

[8] This phenomenon will be dealt with in Section 1.5.1.

[9] In this book, the Latin subscripts a, b, c and so on will be kept as labels denoting particles, whereas i, j, k, and so on usually denote the generalized degrees of freedom or the Cartesian components.

the laws of mechanics do not depend on it in a Galilean frame of reference and the particle, being isolated, can in no way estimate its position in time.[10] Besides, during its travel, the particle occupies perfectly equivalent points in space, since the rest of the Universe is assumed to be located 'at infinity', and a Galilean frame of reference is homogeneous with respect to the laws of mechanics, so the particle cannot spot its position in space. Accordingly, the Lagrangian of an isolated particle should not depend on its position **r**. Lastly, the isotropic properties of space in a Galilean frame of reference show that the Lagrangian should not depend on the direction of particle velocity. L should then be sought in the form of a function either of the norm of velocity $u = |\mathbf{u}|$ or (without loss of generality) of the square of velocity:

$$L = L\left(u^2\right) \tag{1.32}$$

Let us return to the case of the two Galilean frames of reference R and R', the second moving in relation to the first one at the constant velocity **V**. The principle of relativity and the least action principle then tell us that the particle's Lagrangian, in both frames of reference, differs from a total derivative (refer to the end of Section 1.2.3). That means there is, according to equation (1.9), a function Φ of **r** and t which is such that:

$$L\left(u'^2\right) = L\left(u^2\right) + \frac{d\Phi\left(\mathbf{r}, t\right)}{dt} \tag{1.33}$$

We will briefly consider the case in which the norm V of **V** is small before u. Let us use the velocity composition law (1.3) to write u', and let us expand to the first order in V. We get:

$$L\left(u^2 - 2\mathbf{u}\cdot\mathbf{V} + V^2\right) = L\left(u^2\right) - 2L'\left(u^2\right)\mathbf{u}\cdot\mathbf{V} + O\left(V^2\right) \tag{1.34}$$

The condition being sought is then written as, to the first order in V:

$$-2L'\left(u^2\right)\mathbf{u}\cdot\mathbf{V} = \frac{d\Phi\left(\mathbf{r}, t\right)}{dt} \tag{1.35}$$

According to (1.2), we can see that the equality is verified if $L'\left(u^2\right)$ equals a constant C, with $\Phi\left(\mathbf{r}, t\right) = -2C\mathbf{r}\cdot\mathbf{V}$. It can easily be observed that such (sufficient) condition is also necessary, and so:

$$L = Cu^2 \tag{1.36}$$

the constant C being undetermined. It can readily be checked that such a form is suitable in all cases. For a velocity **V** with an arbitrary norm, (1.36) does yield

$$\begin{aligned} L\left(u'^2\right) &= Cu'^2 \\ &= Cu^2 - 2C\mathbf{u}\cdot\mathbf{V} + CV^2 \\ &= L\left(u^2\right) + \frac{d}{dt}\left(-2C\mathbf{r}\cdot\mathbf{V} + CV^2t\right) \end{aligned} \tag{1.37}$$

The Lagrangian of a free particle, in a Galilean frame of reference, is thus proportional to the square of its velocity. The quantity C is constant in that it

[10] Here we ignore every effect of body 'ageing', which would serve as an intrinsic measurement of time. These effects are inherently part of the irreversible processes which are dealt with in Chapter 2.

depends neither on time nor on the particle's position or velocity. However, it is not a universal constant in physics, since it is an intrinsic property of the particle being considered, which results, in particular, from the additivity of the Lagrangian (Section 1.2.3). The common practice consists in putting $C = m/2$, so that:

$$L = \frac{1}{2}mu^2 \tag{1.38}$$

By definition, m is called *mass*[11] of the particle. We can see it corresponds to the arbitrary choice of units, which was previously mentioned in Section 1.2.3. It should be noted that the law of conservation of mass is implicitly included in its definition, since m is defined as a constant with respect to time. One can also readily see that mass is necessarily positive. Otherwise, one could choose, indeed, within the configuration space,[12] a sufficiently long permissible trajectory linking the initial and final positions so that the material point, moving about at a constant velocity in it, has an arbitrarily large action in absolute value, while being negative. This value would then have no minimum, which is contrary to the least action principle. Ultimately, another important property of mass is its additivity, which stems from the additivity of the Lagrangian. The latter property is definitely relevant to particles with respective masses m_a and m_b,[13] which are sufficiently distant from each other and have the same velocity u, since we then have

$$L_{a\cup b} = \frac{1}{2}(m_a + m_b)\, u^2 \tag{1.39}$$

The conservative nature of mass, however, consequently ensures its additivity in any circumstances.

It should immediately be noted[14] that we then have $\partial L/\partial \mathbf{r} = 0$ and $\mathbf{p} = \partial L/\partial \mathbf{u} = m\mathbf{u}$, and so the Lagrange equation (1.31) gives us the equation of motion of the isolated particle as

$$\frac{d\mathbf{u}}{dt} = 0 \tag{1.40}$$

where the constant nature of mass has been used to extract it from the derivation term (we assume the mass is non-zero). Thus, the velocity of an isolated particle is constant, which is in accordance with the principle of inertia (refer to Section 1.2.1). We naturally come across this principle again, since we have used it implicitly in the construction of the Lagrangian of a free particle.

1.3.2 Lagrangian of a particle system

Let us now consider a system of N_p interacting particles, but which is isolated from the rest of the Universe and is once again an isolated system as defined above. If these particles are distant from each other to such an extent that the mutual interaction can still be ignored, then, as we have seen it, the Lagrangian is additive, and so:

[11] More precisely, it is the *inertial* mass, characterizing the difficulty to alter the motion of a body, as discussed later.

[12] In the case of the single particle, the configuration space obviously comes down to the ordinary space.

[13] As can already be felt, the value of an arbitrary physical quantity A linked to an indexed particle a will be denoted as A_a throughout the following.

[14] One should be careful with the derivatives of L: the derivation with respect to velocity is carried out with a fixed position and conversely, pursuant to the above introduced formalism.

$$L = \sum_{a=1}^{N_p} \frac{1}{2} m_a u_a^2 \tag{1.41}$$

It should be pointed out that only here does the notion of mass take on its true meaning. As stated in Section 1.2.3, the Lagrangian can still, indeed, be mutiplied by a positive constant without affecting the equations of motion. That multiplication, for function (1.41), is tantamount to changing the unit for measuring mass; the mass ratios of the various particles, which alone have a physical significance, then remain unchanged.

A very important remark should be made here. There are two kinds of mass in classical mechanics, namely the *gravitational* mass, which plays a part in the law of universal gravitation, and the *inertial* mass, which accounts for the fact that the motion of a given object can be more or less easily altered. We have just introduced the latter, and assuming it to be identical to the former would be groundless. Surprisingly enough, however, it can empirically be found that these two masses are proportional (not identical, as long as we have not set any unit and standard for the measurement of the inertial mass). Only the theory of general relativity enables one to set a theoretical framework in which this astonishing property occurs naturally.[15] Classical mechanics, on the contrary, assumes it is a chance, which is quite simplistic of course. In spite of the misleading principles which may result, that 'coincidence' is usually operated by giving the same unit to both inertial mass and gravitational mass and by selecting such a standard of measurement that they are likened to each other. The purpose of the arbitrary coefficient $\frac{1}{2}$ in equation (1.41) is then that of restoring the established equations of mechanics (as set out later) and the usual units. The units of measurement are then no longer arbitrary and the Lagrangian takes on the dimension of an energy, a quantity to be defined in Section 1.4.1.

[15] The appropriate phrase is then the *principle of equivalence*.

If the particles are interacting, a corrective term should be added to equation (1.41) to take these interactions into account; it is conventionally denoted $-E_p$, and is known as the system's *interaction potential*:

$$L(\{\mathbf{r}_a\}, \{\mathbf{u}_a\}) = \sum_{a=1}^{N_p} \frac{1}{2} m_a u_a^2 - E_p(\{\mathbf{r}_a\}) \tag{1.42}$$

(it being understood that all the particles play a part in E_p). The materialization of the interactions occurring through an additive function in the Lagrangian has to be considered as an empirical fact, and its form (which depends on the kind of interaction being investigated) can only be determined experimentally. One can easily realize, however, that it is only defined to within one additive constant, just like the Lagrangian. It is also obvious that E_p cannot depend on time, once again because the laws of mechanics do not depend on it in a Galilean frame of reference (for an isolated system). On the other hand, the dependence on the coordinates is caused by the sensitivity of interactions to the particle mutual positions. However, only the distances, and not the absolute positions $\mathbf{r}_a$, should have an influence (for reasons of isotropy and homogeneity, and in accordance with the principle of Galilean invariance). In particular, if the system is moved integrally, while still being assumed to be isolated, then the Lagrangian—and then the potential—should remain unchanged. For the moment, we will ignore this constraint which is prescribed on function E_p, and we will not discuss it until we deal with the conservation laws (Section 1.4.3), since it has no direct effect on the following discussions.

It is essential to immediately understand why E_p cannot depend on the particle velocities against the classical background being adopted. Let us assume, indeed, that these interactions propagate at a velocity V_I. Then, according to the principle of relativity, that velocity is a constant of mechanics, that is it is the same in both Galilean frames of reference R and R'. The law of composition of velocities then gives us, if the interaction propagates along the direction of displacement of R' in relation to R:

$$V_I = V_I + V \tag{1.43}$$

We can see that V_I should be infinite to satisfy this relation. Thus, the effect of the motion of one particle has immediate impacts upon the others. In such conditions, one can understand that the interactions cannot depend on the way the particles move about (i.e. on their velocities), but only on their immediate positions.[16]

We will provide two examples of empirical potentials. The first is associated with the gravity forces[17] and is written as

$$E_p^G(\{\mathbf{r}_a\}) = -G\sum_{a\neq b}\frac{m_a m_b}{r_{ab}} \tag{1.44}$$

where $G = 6.67 \times 10^{-11}\ kg^{-1}m^3s^{-2}$ is Newton's universal constant (we have omitted the summation bounds of the discrete subscripts a and b, it being understood that such a sum will from now on always apply to all the particles of a given system). The particle masses,[18] as well as their mutual distances, are once again involved in formula (1.44):

$$\begin{aligned} r_{ab} &\doteqdot |\mathbf{r}_{ab}| \\ \mathbf{r}_{ab} &\doteqdot \mathbf{r}_a - \mathbf{r}_b \end{aligned} \tag{1.45}$$

The second example provided is about the electromagnetically-induced forces which govern the interactions of molecules in a fluid. For one pair[19] of molecules a and b, that so-called *Lennard–Jones* potential is written as:

$$E_p^{LJ}(\mathbf{r}_a, \mathbf{r}_b) = E_0\left[\left(\frac{r_0}{r_{ab}}\right)^{P_1} - 2\left(\frac{r_0}{r_{ab}}\right)^{P_2}\right] \tag{1.46}$$

with $P_1 = 12$ and $P_2 = 6$. The reference potential E_0 and the characteristic distance r_0 depend on the fluid being considered. Both mentioned potentials (1.44) and (1.46) suitably illustrate the previously discussed idea of translational invariance, which results in a dependence on the mutual inter-particle distances.

Let us now return to more general considerations. In Section 1.2.3, we investigated the case of a system A being subject to the influence of a larger

[16] That result becomes wrong in relativistic mechanics, in which the standards for the measurement of time and distances are not universal. Thus, Galileo's law of composition of velocities has to be replaced by the Lorentz transformations, which allow a displacement at a *finite* universal velocity (the speed of light). It should also be said, however, that the notion of interactions cannot be dealt with through the notion of potential in relativistic mechanics.

[17] The notion of force will subsequently be defined from the present discussion.

[18] They are the *gravitational* masses, not the inertial mass which we have previously introduced.

[19] The interactions between larger sets of molecules (triplets, etc.) are fairly rare and can be ignored in a first approach. This question will be briefly dealt with in Chapter 2.

system B, the latter being unaffected by A. We can see that the mass ratio of B to A has become the criterion under which that hypothesis is valid. As mentioned before, it should then be considered that the coordinates and velocities of system B are supposedly known, so that the global system's Lagrangian can be written as:

$$L = \sum_{a=1}^{N_p} \frac{1}{2} m_a^A \left(u_a^A\right)^2 + \sum_{b=1}^{N'_p} \frac{1}{2} m_b^B \left[u_b^B(t)\right]^2 - E_p\left(\left\{\mathbf{r}_a^A\right\}, \left\{\mathbf{r}_b^B(t)\right\}\right) \quad (1.47)$$

where N_p and N'_p are the numbers of particles making up the two systems. Since the velocities u_b^B of subsystem B are known functions of time, the second term of (1.47) should be regarded as constant, and can then be omitted according to the considerations in Section 1.2.3. As regards the positions in subsystem B, which are known as well, they occur in the potential through an explicit time dependence. Thus, omitting the superscripts for the sake of simplicity:

$$L\left(\{\mathbf{r}_a\}, \{\mathbf{u}_a\}, t\right) = \sum_a \frac{1}{2} m_a u_a^2 - E_p\left(\{\mathbf{r}_a\}, t\right) \quad (1.48)$$

Thus, the Lagrangian's form remains unchanged, except that time now occurs explicitly in the potential. We can see that the Lagrangian of a system of interacting particles, in a Galilean frame of reference, happens to be the difference between two quantities, which we will respectively call *kinetic energy* (denoted here E_k) and *potential energy* E_p (refer to Section 1.4.1 for an explanation about these denominations). Velocities only occur in the first term (hence, so far, $\mathbf{p}_a = \partial E_k / \partial \mathbf{u}_a$), coordinates only in the second one, as well as time if the system is subjected to an external influence. We note that kinetic energy

$$E_k = \sum_a \frac{1}{2} m_a u_a^2 \quad (1.49)$$

is a positive definite quadratic form.

The previously introduced case of the gravity field becomes much simpler in the case where the very massive system B represents the Earth (or another very massive body). The very high number of particles b (with individual volumes V_b) making up system B (amounting to a sphere Ω) does make it possible to liken the discrete sum (1.44) to an integral.[20] If $\rho = m_b / V_b$ denotes the presumably constant Earth's density[21] for a single particle a ($N_p = 1$), then we get:

$$\begin{aligned} E_p^G(\mathbf{r}_a) &= -G m_a \sum_b \frac{m_b}{r_{ab}} \\ &\approx -m_a G \rho \int_\Omega \frac{d\mathbf{r}}{|\mathbf{r}_a - \mathbf{r}|} \quad (1.50) \\ &= -m_a \frac{G M_0}{z_a + R} \end{aligned}$$

[20] This procedure will be set up as a method in Chapter 6, to lay the foundations of the SPH method. The links between the discrete and continuous points of view, however, will appear as from Chapter 3, within the context of continuum mechanics.

[21] This notion will be discussed with further details in Chapter 1.5.4.

where M_0 and R denote the Earth's mass and radius, whereas z_a is the elevation of a above the Earth's surface. We have seen that the Lagrangian is only defined to within one constant, which means it is all the same for the potential. By mere convention, the practice consists in setting that constant so that the interactions tend towards zero when the inter-particle distances tend towards infinity; this is what we can do in (1.50). If $z_a \ll R$ (the relevant force is the force of gravity), an expansion to order 1 in z_a/R would, indeed, make appear a constant term which we will omit from now on; the Earth's gravitational potential will ultimately be written as:

$$E_p^g(z_a) = m_a g z_a \tag{1.51}$$

with $g = MG/R^2 \approx 9.807\ ms^{-2}$.

For a particle system exhibiting mutual forces of gravity which are insignificant when compared with the Earth's gravity, we get

$$E_p^g(\{z_a\}) = \sum_a m_a g z_a \tag{1.52}$$

in accordance with the Lagrangian additivity principle in the absence of interacting components in a system.

1.3.3 Newton's second law

From equation (1.48) providing the Lagrangian, we find, for the linear momentum of a particle belonging to a system:

$$\forall a,\ \mathbf{p}_a = \frac{\partial L}{\partial \mathbf{u}_a} = m_a \mathbf{u}_a \tag{1.53}$$

(as in the case of an isolated particle being discussed in Section 1.3.1). The Lagrange equations (1.31) then yield the equation of motion of each particle:

$$\forall a,\ m_a \frac{d\mathbf{u}_a}{dt} = -\frac{\partial E_p}{\partial \mathbf{r}_a} \tag{1.54}$$

We have used the invariance of mass to get it out of the derivation sign. The right-hand term, by definition, is called the *force* exerted on that particle, and denoted $\mathbf{F}_a$:

$$\mathbf{F}_a \doteq -\frac{\partial E_p}{\partial \mathbf{r}_a} = -\nabla_a E_p \tag{1.55}$$

The derivative of the potential with respect to the position vector then appears in the second term; the vector will be denoted in the form of a gradient ∇_a, that is taken 'with respect to the coordinates of particle a', which are involved in function $E_p(\{\mathbf{r}_b\}, t)$ when the dummy subscript b equates a. Equation (1.54) is finally written as

$$\forall a,\ m_a \boldsymbol{\gamma}_a = \frac{d\mathbf{p}_a}{dt} = \mathbf{F}_a(\{\mathbf{r}_b\}, t) \tag{1.56}$$

where $\boldsymbol{\gamma}_a \doteq d\mathbf{u}_a/dt$ is the particle acceleration. This formula expresses the so-called *Newton's second law*; once again, it is a second-order equation as regards the coordinates being taken as time functions.

The force, as the opposite of the potential gradient, is ultimately only determined experimentally. As with the Lagrange equation (1.31), the Galilean invariance requires forces which do not depend on the inter-particle distances. The gravitational forces derived from (1.44) and (1.51) are taken from (1.55) and are respectively given by

$$\begin{aligned} \mathbf{F}_a^G(\{\mathbf{r}_b\}) &= -G\sum_b \frac{m_a m_b}{r_{ab}^2}\mathbf{e}_{ab} \\ \mathbf{F}_a^g(z_a) &= m_a\mathbf{g} \end{aligned} \tag{1.57}$$

where $\mathbf{g}$ is a g-normed, downward vector and

$$\mathbf{e}_{ab} = \frac{\mathbf{r}_{ab}}{r_{ab}} \tag{1.58}$$

is the unit vector being directed from particle a to particle b. These forces satisfy the Galilean invariance principle. It should be noticed that where the forces are restricted to gravity, (1.56) is written as

$$\forall a,\ \gamma_a = \frac{d\mathbf{u}_a}{dt} = -G\sum_b \frac{m_b}{r_{ab}^2}\mathbf{e}_{ab} \tag{1.59}$$

Thus, the motion of a particle no longer depends on its mass m_a. This is an effect of the principle of equivalence as discussed in Section 1.3.2.

More generally, just like the potential, the force cannot depend on the particle velocity.[22] Thus, the frictional forces, in accordance with the remark at the end of Section 1.2.1, cannot be scrutinized through this treatment, since they depend on the system's component velocities (as regards this matter, refer to Section 2.4.3 in Chapter 2). Besides, according to (1.53), the kinetic energy (1.49) may now be written as

$$E_k = \frac{1}{2}\sum_a \mathbf{p}_a\cdot\mathbf{u}_a \tag{1.60}$$

A final remark has to be be made. If Cartesian points are not used when describing the system or, where it is not a system of material points the Lagrangian form is altered, then the form of Newton's second law (1.56) will change too. However, energy cannot have any form. Whichever system (system of jointed bars, springs...) is considered, indeed, it comprises molecules which, in a first approximation, may be likened to particles[23] that are globally subjected to a number of constraints necessarily making the degrees of freedom mutually dependent. Thus, the molecule positions $\mathbf{r}_a$ are well-known functions[24] of the system parameters q_i:

$$\forall a,\ \mathbf{r}_a = \mathbf{r}_a(\{q_k\}) \tag{1.61}$$

Kinetic energy is then expanded as follows

$$E_k = \sum_a \frac{1}{2}m_a\left(\frac{d\mathbf{r}_a}{dt}\right)^2$$

[22] Specifically, the Lorentz force, which expresses the effect of an electromagnetic field on an electrically charged body, cannot be constructed according to this formalism, since it depends on the body velocity. Then, the theory of electromagnetism cannot be fully established without the assistance of special relativity.

[23] Provided that their rotational and vibrational energies are included in the internal energy (refer to Chapter 2).

[24] We assume the so-called 'holonomic' constraints, that is which do not depend on time.

$$= \sum_a \frac{1}{2} m_a \left(\sum_i \frac{\partial \mathbf{r}_a}{\partial q_i} \frac{dq_i}{dt} \right)^2 \tag{1.62}$$

$$= \sum_a \frac{1}{2} m_a \sum_{i,j} \frac{\partial \mathbf{r}_a}{\partial q_i} \cdot \frac{\partial \mathbf{r}_a}{\partial q_j} \dot{q}_i \dot{q}_j$$

We can then define the quantities Λ_{ij} through

$$\Lambda_{ij} (\{q_k\}) \doteqdot \sum_a m_a \frac{\partial \mathbf{r}_a (\{q_k\})}{\partial q_i} \cdot \frac{\partial \mathbf{r}_a (\{q_k\})}{\partial q_j} \tag{1.63}$$

the subscript k covering the whole generally possible range of values. Definition (1.63) shows that the Λ_{ij} make up a symmetric matrix. Thus, the Lagrangian is still a quadratic function of the generalized velocities; the coordinates can now play a role in the expression of the kinetic energy. That can be generally written as:

$$\begin{aligned} E_k &= \frac{1}{2} \sum_{i,j} \Lambda_{ij} (\{q_k\}) \dot{q}_i \dot{q}_j \\ E_p &= E_p (\{q_k\}, t) \\ L &= E_k - E_p \end{aligned} \tag{1.64}$$

The E_k positiveness then tells us that the matrix $\mathbf{\Lambda}$ of coefficients Λ_{ij} is positive-definite. We get an equation of motion for each degree of freedom in the usual form

$$\forall i, \ \frac{dp_i}{dt} = F_i \tag{1.65}$$

with

$$\begin{aligned} \forall i, \ p_i &= \frac{\partial E_k}{\partial \dot{q}_i} = \sum_j \Lambda_{ij} (\{q_k\}) \dot{q}_j \\ F_i &= \frac{\partial L}{\partial q_i} = \frac{1}{2} \sum_{j,k} \frac{\partial \Lambda_{jk}}{\partial q_i} \dot{q}_j \dot{q}_k - \frac{\partial E_p}{\partial q_i} \end{aligned} \tag{1.66}$$

The quantities F_i are called *generalized forces*. As evidenced by the above written expression, they may, unlike the particle forces (1.55), depend on the generalized velocities. We will discuss major effects in Chapters 2 (Section 2.5.2) and 6 (Section 6.5.2).

1.4 Conservation laws

We consider here the conservation of fundamental quantities, through the symmetries (invariance) of the Lagrange function of an isolated system. We show how invariance according to time translation, space translations and rotations lead to the conservations laws of energy, linear and angular momenta, respectively. We also present the Hamilton equations. Finally, a generic example is briefly studied: the oscillating ordinary pendulum.

1.4.1 Energy and Hamiltonian

We return now to the general case of some system being defined by its generalized coordinates and velocities. From the very beginning, we have considered systems with N geometrical degrees of freedom, that is $2N$ physical degrees of freedom (the $2N$ initial conditions). Thus, there must be $2N - 1$ integrals of motion (time-invariant quantities). Only some of them, however, are interesting because of their additivity. As we will see, each of them results from a space-time regularity property with respect to the the laws of mechanics; this is a theorem which was formulated by Noether (1918). We will discuss the case of energy in this Section whereas the other conservation laws with be dealt with in Section 1.4.3.

Let us consider the case of an isolated system. Time has no explicit effect in the expression of its Lagrangian, which confirms that the laws of mechanics are time-invariant. We use this principle and write that the partial derivative of L with respect to time vanishes, that is its total derivative is written as:

$$\frac{dL}{dt} = \sum_i \left(\frac{\partial L}{\partial q_i} \dot{q}_i + \frac{\partial L}{\partial \dot{q}_i} \ddot{q}_i \right) \tag{1.67}$$

Let us use now the Lagrange equation (1.26) and the definition (1.25) of generalized momenta. The previous equation becomes:

$$\begin{aligned} \frac{dL}{dt} &= \sum_i \left(\frac{dp_i}{dt} \dot{q}_i + p_i \ddot{q}_i \right) \\ &= \frac{d}{dt} \sum_i p_i \dot{q}_i \end{aligned} \tag{1.68}$$

We see that the quantity E, as defined by:

$$E \doteq \sum_i p_i \dot{q}_i - L \tag{1.69}$$

is time-invariant for an isolated system. Hence

$$\frac{dE}{dt} = 0 \tag{1.70}$$

This quantity is referred to as the system *energy*; we find that its conservation is closely related to the invariance of the laws of mechanics with respect to time. In addition, when a system is not isolated, the dependence of the potential at t (eqn (1.48)) also results in a variation of E, the system exchanging energy with those external systems affecting it. In the case of a system which is subjected to the permanent influence of an external system (steady potential), however, energy is always conserved (such is the case, for instance, for a body being subjected to the Earth's gravitational field). For a non-isolated system, the energy variation over an infinitesimal lapse of time dt is given by

$$dE = \sum_i p_i dq_i - L dt \tag{1.71}$$

where dq_i is the position variation over the time dt.

In the case of a particle system, those formulas (1.48), (1.49) and (1.53) yielding the Lagrangian, the momenta and the kinetic energy result in:

$$\begin{aligned} E &= \sum_a m_a \mathbf{u}_a \cdot \mathbf{u}_a - \left(\sum_a \frac{1}{2} m_a u_a^2 - E_p \right) \\ &= \sum_a \frac{1}{2} m_a u_a^2 + E_p \\ &= E_k + E_p \end{aligned} \tag{1.72}$$

Energy happens to be the sum of kinetic energy and potential energy (for the sake of concision, we have omitted in it the explicit dependence on the particle positions). Besides, this formula is still valid in generalized coordinates, as can easily be checked from (1.64), (1.66) and (1.69). Using the fact that the potential only depends on position vectors whereas the kinetic energy does not, equation (1.72) also makes it possible to write the time derivative of energy as:

$$\begin{aligned} \frac{dE}{dt} &= \sum_a m_a \mathbf{u}_a \cdot \frac{d\mathbf{u}_a}{dt} + \frac{dE_p}{dt} \\ &= \sum_a \mathbf{u}_a \cdot \frac{d\mathbf{p}_a}{dt} + \sum_a \frac{dE_p}{d\mathbf{r}_a} \cdot \frac{d\mathbf{r}_a}{dt} \\ &= \sum_a \mathbf{u}_a \cdot \left(\frac{d\mathbf{p}_a}{dt} + \frac{\partial L}{\partial \mathbf{r}_a} \right) \\ &= \sum_a \mathbf{u}_a \cdot \left(\frac{d\mathbf{p}_a}{dt} - \mathbf{F}_a \right) \end{aligned} \tag{1.73}$$

In that form, once again that derivative vanishes, since each term of the last member vanishes in accordance with the equation of motion. The last equation also shows that

$$\frac{dE_k}{dt} = \sum_a \mathbf{F}_a \cdot \mathbf{u}_a \tag{1.74}$$

that is the kinetic energy is not constant. The right-hand side denotes its rate of variation over time, which is given as the sum of quantities $\mathbf{F}_a \cdot \mathbf{u}_a$. Such a term is known as the *power* of the force affecting the particle a, and denotes an energy flux between E_p and E_k. It is either positive or negative, as the case may be; importantly enough, the lost kinetic energy is fully recovered in the form of potential energy, and the converse is also true.

From the Lagragian's additivity immediately results that of energy, when several systems are sufficiently distant from each other. Otherwise, the interactions between these systems generate an additional energy which exactly corresponds to an interaction potential, according to (1.42).

Energy is expressed by a function of generalized positions and velocities; the former act *a minima* in the potential and the latter only in kinetic energy. The generalized positions and momenta may sometimes be advantageously considered as independent variables. The system state, indeed, can be described at any time by one pair of variables for each geometrical degree of freedom, but the option $(\{q_i\}, \{\dot{q}_i\})$ is not the only one.[25] Energy expressed in terms of variables $(\{p_i\}, \{q_i\})$ is referred to as the system's *Hamilton function* or *Hamiltonian* and denoted H:

$$H(\{p_i\}, \{q_i\}) \doteqdot E(\{q_i\}, \{\dot{q}_i\}) = \sum_i p_i \dot{q}_i - L \tag{1.75}$$

The general form (1.66) can be reversed as:

$$\dot{q}_i = \sum_j \Gamma_{ij}(\{q_k\})\, p_j \tag{1.76}$$

the Γ_{ij} being the coefficients of the inverse Λ^{-1} of matrix Λ defined in Section 1.3.3 (Λ^{-1} exists because the Λ is a symmetric positive definite matrix). By transferring (1.76) into the form (1.64) of energy, we find that the Hamilton function is tantamount to a quadratic form of the generalized momenta:

$$\begin{aligned} H &= \frac{1}{2} \sum_{i,j,l,m} \Lambda_{ij}(\{q_k\})\, \Gamma_{il}(\{q_k\})\, \Gamma_{jm}(\{q_k\})\, p_l p_m + E_p(\{q_k\}, t) \\ &= \frac{1}{2} \sum_{i,j} \Gamma_{ij}(\{q_k\})\, p_i p_j + E_p(\{q_k\}, t) \end{aligned} \tag{1.77}$$

Coefficients Γ_{ij} make up a symmetric positive definite matrix.

For an isolated particle, the Hamiltonian is written as

$$H(\mathbf{p}_a, \mathbf{r}_a) = \frac{|\mathbf{p}_a|^2}{2m_a} \tag{1.78}$$

Hence, in the case of a particle system, it reads

$$H(\{\mathbf{p}_a\}, \{\mathbf{r}_a\}, t) = \sum_a \frac{|\mathbf{p}_a|^2}{2m_a} + E_p(\{\mathbf{r}_a\}, t) \tag{1.79}$$

That relation may also be written as

$$H = \frac{1}{2} \sum_{a,b} \mathbf{p}_a^T \mathbf{\Gamma}_{ab} \mathbf{p}_b + E_p(\{\mathbf{r}_a\}, t) \tag{1.80}$$

where the $\mathbf{\Gamma}_{ab}$ are three-dimensional matrices[26] ascribed to each pair of particles. It should be pointed out[27] that the subscripts a and b do not represent the matrix subscripts, but particle labels; thus, a matrix $\mathbf{\Gamma}_{ab}$ of coefficients $(\mathbf{\Gamma}_{ab})_{ij}$ is attached to each pair (a, b) of particles. Likewise, matrices $\mathbf{\Lambda}_{ab} = \mathbf{\Gamma}_{ab}^{-1}$ can be introduced; we get

$$E_k = \frac{1}{2} \sum_{a,b} \mathbf{u}_a^T \mathbf{\Lambda}_{ab}(\{\mathbf{r}_c\})\, \mathbf{u}_b \tag{1.81}$$

[25] Formally, that change of variables is in every respect similar to the arbitrary selection of state variables in thermodynamics.

[26] For essential reasons in statistical physics (Chapter 2), we only consider, in the first part of this book, a three-dimensional space, even though some problems (e.g. refer to Section 4.4.5 in Chapter 4) are featured by a statistical invariance enabling work in two dimensions, particularly within the context of numerical simulation being dealt with in the second part of this book.

[27] Throughout the book, one should not forget the ambiguity in some notations. We will often use such quantities as $\mathbf{r}_{ab}$, which is defined as the difference $\mathbf{r}_a - \mathbf{r}_b$, whereas $\mathbf{\Gamma}_{ab}$ does not occur as a difference between two terms.

Comparing formula (1.79) with (1.77) reveals that these matrices are given by

$$\begin{aligned}\mathbf{\Gamma}_{ab} &= \frac{1}{m_a}\delta_{ab}\mathbf{I}_3 \\ \mathbf{\Lambda}_{ab} &= m_a\delta_{ab}\mathbf{I}_3\end{aligned} \tag{1.82}$$

where $\mathbf{I}_3$ is the identity matrix of dimension 3. Matrices $\mathbf{\Gamma}_{ab}$ and $\mathbf{\Lambda}_{ab}$ are symmetrical quantities with respect to the subscripts a and b, despite appearances. Actually, if $a \neq b$ then $\mathbf{\Gamma}_{ab} = 0$, whereas if $a = b$ then $\mathbf{\Lambda}_{ab} = m_a\mathbf{I}_3 = m_b\mathbf{I}_3$. Equation (1.66) is now written as

$$\begin{aligned}\forall a,\ \mathbf{p}_a &= \frac{\partial E_k}{\partial \mathbf{u}_a} = \sum_b \mathbf{\Lambda}_{ab}\left(\{\mathbf{r}_c\}\right)\mathbf{u}_b \\ \mathbf{F}_a &= \frac{\partial L}{\partial \mathbf{r}_a} = \frac{1}{2}\sum_{b,c}\mathbf{u}_b^T\frac{\partial \mathbf{\Lambda}_{bc}}{\partial \mathbf{r}_a}\mathbf{u}_c - \frac{\partial E_p}{\partial \mathbf{r}_a} = -\frac{\partial E_p}{\partial \mathbf{r}_a}\end{aligned} \tag{1.83}$$

We note that the matrices $\mathbf{\Lambda}_{bc}$ do not depend on the positions of particles, which means that the particle forces do not depend on the velocities, since the quantities $\partial\mathbf{\Lambda}_{bc}/\partial\mathbf{r}_a$ are then zero matrices. With the definition (1.82), we recover the equations (1.53) and (1.55). Thus, introducing this formalism may look unnecessary, but we will see in Section 2.5.2 of Chapter 2 how it can be generalized for describing the dissipative forces by introducing matrices $\mathbf{\Lambda}_{bc}$ which depend on the positions.

We will complete this Section by mentioning the virial theorem for a particle system, which will be useful in Chapter 2. We will temporarily consider the quantity[28]

[28] Which, of course, should not be mistaken for the previously mentioned universal gravitational constant.

$$G \doteq \sum_a \mathbf{p}_a \cdot \mathbf{r}_a \tag{1.84}$$

By differentiating that quantity with respect to time, we find, using (1.60):

$$\begin{aligned}\frac{dG}{dt} &= \sum_a \frac{d\mathbf{p}_a}{dt}\cdot\mathbf{r}_a + \sum_a \mathbf{p}_a\cdot\frac{d\mathbf{r}_a}{dt} \\ &= \sum_a \mathbf{F}_a\cdot\mathbf{r}_a + 2E_k\end{aligned} \tag{1.85}$$

Let us now consider a time-varying particle system while keeping *bounded* coordinates $\mathbf{r}_a$. The time average of (1.85) between point $t = 0$ and $t = T$ gives

$$\frac{G\,(t=T) - G\,(t=0)}{T} = \frac{1}{T}\int_0^T\left(\sum_a \mathbf{F}_a\cdot\mathbf{r}_a + 2E_k\right)dt \tag{1.86}$$

Contemplating a sufficiently long period of time T, with the quantity G being bounded, the left-hand side can arbitrarily be made small, so that the right-hand side should be zero:

$$\lim_{T\to+\infty}\frac{1}{T}\int_0^T E_k dt = -\frac{1}{2}\lim_{T\to+\infty}\frac{1}{T}\int_0^T\left(\sum_a \mathbf{F}_a\cdot\mathbf{r}_a\right)dt \qquad (1.87)$$

This argument can be used to reckon the time average of the kinetic energy in a space-bounded system, the prevailing forces $\mathbf{F}_a$ in it being known. We will provide an application of it in Chapter 2 (Section 2.3.3).

1.4.2 Hamilton equations

We return to the general case of an arbitrary system with generalized coordinates; let us write the differential of H in terms of dp_i and dq_i by differentiating (1.75):

$$\begin{aligned} dH &= \sum_i (\dot{q}_i dp_i + p_i d\dot{q}_i) - dL(q_i, \dot{q}_i) \\ &= \sum_i (\dot{q}_i dp_i + p_i d\dot{q}_i) - \frac{\partial L}{\partial t}dt - \sum_i\left(\frac{\partial L}{\partial q_i}dq_i + \frac{\partial L}{\partial \dot{q}_i}d\dot{q}_i\right) \\ &= -\frac{\partial L}{\partial t}dt + \sum_i(\dot{q}_i dp_i - \dot{p}_i dq_i) \\ &= \frac{\partial H}{\partial t}dt + \sum_i\left(\frac{\partial H}{\partial p_i}dp_i + \frac{\partial H}{\partial q_i}dq_i\right) \end{aligned} \qquad (1.88)$$

(we have used the definition of generalized momenta, as well as the equation of motion, whereas the last line is nothing but a definition of the partial derivatives of H). By equating the last two lines on a term-to-term basis, we get:

$$\begin{aligned} &\frac{\partial H}{\partial t} = -\frac{\partial L}{\partial t} \\ \forall i,\ &\frac{\partial H}{\partial p_i} = \dot{q}_i \\ &\frac{\partial H}{\partial q_i} = -\dot{p}_i \end{aligned} \qquad (1.89)$$

The first of these equations shows that, for an isolated system (i.e. whose Lagrangian is not time-dependent), energy is conserved, as is already known. The last two equations define a system comprising two series of N first-order linear differential equations (i.e. two equations for each geometrical degree of freedom) which are known as *Hamilton equations*. They are another form of the equations of motion and may be substituted for the Lagrange equations (1.26). Each of them should be provided with N initial conditions. Because of their symmetry and simplicity, they are often referred to as *canonical equations*. Both momenta and coordinates can then be seen as so-called *mechanically conjugate*[29] quantities. For a particle system, these equations may be gathered in a vector form, like the Lagrange equations (Section 1.3.1):

[29] This terminology will be clarified in Section 1.5.4.

$$\forall a, \; \frac{\partial H}{\partial \mathbf{p}_a} = \dot{\mathbf{r}}_a = \mathbf{u}_a$$
$$\frac{\partial H}{\partial \mathbf{r}_a} = -\dot{\mathbf{p}}_a = \mathbf{F}_a \tag{1.90}$$

The Hamilton equations can be written in a simple matrix form, provided that a two-dimensional mechanical state vector $\mathbf{X}_i$ be defined for each degree of freedom i:

$$\forall i, \; \mathbf{X}_i \doteqdot \begin{pmatrix} p_i \\ q_i \end{pmatrix} \tag{1.91}$$

The Hamilton's differential system (1.89) then becomes:

$$\forall i, \; \dot{\mathbf{X}}_i = \boldsymbol{\omega}_2 \frac{\partial H}{\partial \mathbf{X}_i} \tag{1.92}$$

where $\boldsymbol{\omega}_2$ is a two-dimensional matrix as given by:

$$\boldsymbol{\omega}_2 \doteqdot \begin{pmatrix} 0 & -1 \\ 1 & 0 \end{pmatrix} \tag{1.93}$$

and known as *unitary symplectic matrix*.[30] It verifies the following properties:

[30] Explanations about the word 'symplectic' will be provided in Section 2.2.2, Chapter 2.

$$\boldsymbol{\omega}_2 = -\boldsymbol{\omega}_2^T \tag{1.94}$$

and

$$\boldsymbol{\omega}_2^T = \boldsymbol{\omega}_2^{-1} \tag{1.95}$$

that is it is skew-symmetric and orthogonal. With such a formalism, energy conservation can be taken as resulting directly from the properties of $\boldsymbol{\omega}_2$:

$$\begin{aligned} \frac{dH}{dt} &= \sum_i \frac{\partial H}{\partial \mathbf{X}_i} \cdot \dot{\mathbf{X}}_i \\ &= \sum_i \left(\frac{\partial H}{\partial \mathbf{X}_i} \right)^T \boldsymbol{\omega}_2 \frac{\partial H}{\partial \mathbf{X}_i} \\ &= 0 \end{aligned} \tag{1.96}$$

(since $\boldsymbol{\omega}_2$ is a rotation matrix of angle $\pi/2$, and so any vector $\mathbf{A}$ is orthogonal to vector $\boldsymbol{\omega}_2\mathbf{A}$). Equation (1.96) is identical to (1.70). We notice that this argument is seemingly based on the property of orthogonality (1.95) of matrix $\boldsymbol{\omega}_2$, but skew-symmetry (1.94) is actually more important. For two arbitrary vectors $\mathbf{A}$ and $\mathbf{B}$, we may indeed define the usual scalar product as

$$\langle \mathbf{A}, \mathbf{B} \rangle = \sum_i A_i B_i = \mathbf{A} \cdot \mathbf{B} = \mathbf{A}^T \mathbf{B} = \langle \mathbf{B}, \mathbf{A} \rangle \tag{1.97}$$

We can then write

$$\begin{aligned}\langle \mathbf{A}, \boldsymbol{\omega}_2\mathbf{B}\rangle &= \mathbf{A}^T(\boldsymbol{\omega}_2\mathbf{B})\\ &= (\boldsymbol{\omega}_2\mathbf{B})^T\mathbf{A}\\ &= \mathbf{B}^T\boldsymbol{\omega}_2^T\mathbf{A}\\ &= -\mathbf{B}^T\boldsymbol{\omega}_2\mathbf{A}\\ &= -\langle \boldsymbol{\omega}_2\mathbf{A}, \mathbf{B}\rangle\end{aligned} \tag{1.98}$$

Matrix $\boldsymbol{\omega}_2$ is then said to be skew-self-adjoint. Applying that relation to the case $\mathbf{A} = \mathbf{B}$ yields $\langle \mathbf{A}, \boldsymbol{\omega}_2\mathbf{A}\rangle = 0$, which is a further evidence of (1.96).

We can do the same thing for a particle system, lumping the parameters related to each particle into a single six-component vector:

$$\forall a,\ \mathbf{X}_a \doteqdot \begin{pmatrix}\mathbf{p}_a\\ \mathbf{r}_a\end{pmatrix} \tag{1.99}$$

In such a case, the Hamiltonian's derivatives are written, starting from eqns (1.55) and (1.79) as:

$$\frac{\partial H}{\partial \mathbf{X}_a} = \begin{pmatrix}\mathbf{u}_a\\ -\mathbf{F}_a\end{pmatrix} \tag{1.100}$$

and the Hamilton equation (1.90) becomes

$$\forall a,\ \dot{\mathbf{X}}_a = \boldsymbol{\omega}_6\frac{\partial H}{\partial \mathbf{X}_a} \tag{1.101}$$

where $\boldsymbol{\omega}_6$ is a 6-dimensional symplectic matrix as given by

$$\boldsymbol{\omega}_6 \doteqdot \begin{pmatrix}\mathbf{0}_3 & -\mathbf{I}_3\\ \mathbf{I}_3 & \mathbf{0}_3\end{pmatrix} \tag{1.102}$$

$\mathbf{0}_3$ being the zero matrix 3 × 3. We then have

$$\frac{dH}{dt} = \sum_i \left(\frac{\partial H}{\partial \mathbf{X}_a}\right)^T \boldsymbol{\omega}_6 \frac{\partial H}{\partial \mathbf{X}_a} = 0 \tag{1.103}$$

since $\boldsymbol{\omega}_6$ satisfies the same properties as $\boldsymbol{\omega}_2$. Lastly, note that all the degrees of freedom of an arbitrary system might also be lumped into a single $2N$- sized vector:

$$\mathbf{X} \doteqdot \begin{pmatrix}\{p_i\}\\ \{q_i\}\end{pmatrix} \tag{1.104}$$

We may contemplate, for any given system, a $2N$-sized abstract space along the axes of which are put the state variables p_i and q_i $(i = 1, ..., N)$. We call it the system's *phase space*.[31] Within this space, a point $\mathbf{X}$ gathers the whole information about the system's *mechanical* state (unlike the configuration space introduced in Section 1.2.3, which only contains the geometrical information). The Hamilton equations for the whole system are then written in the condensed form:

[31] Strictly speaking, the coordinates and momenta would have to be made dimensionless (or at least be given identical physical dimensions) before being adjoined in one geometrical construction; this is because, as explained later on, some of the usual geometrical transformations related to the notion of measurement (rotations, calculation of volumes) are contemplated within that space. By dividing the p_i and q_i by a reference momentum and a reference length, respectively, energy would have, for instance, to be reckoned in s^{-1} in order for the Hamilton equations to keep the same form. The action would then be a dimensionless quantity, which is indeed neater. The results being discussed here would nevertheless not be affected.

Lots of remarks would still have to be made in this respect, particularly as regards the role of mass and Einstein's equivalence principle. Without intending to provide further insights, it is worth mentioning that resorting to the Planck units (1899) is the proper procedure for constructing the phase space.

$$\dot{\mathbf{X}} = \mathbf{\Omega}_{2N} \frac{\partial H}{\partial \mathbf{X}} \tag{1.105}$$

where a $2N \times 2N$-dimensional symplectic matrix has been defined:

$$\mathbf{\Omega}_{2N} = \begin{pmatrix} \mathbf{0}_N & -\mathbf{I}_N \\ \mathbf{I}_N & \mathbf{0}_N \end{pmatrix} \tag{1.106}$$

$\mathbf{I}_N$ and $\mathbf{0}_N$ respectively denoting the both N-dimensional identity matrix and zero matrix. Matrix $\mathbf{\Omega}_{2N}$ satisfies properties (1.94) and (1.95), and the energy conservation can also be written as

$$\begin{aligned} \frac{dH}{dt} &= \frac{\partial H}{\partial \mathbf{X}} \cdot \dot{\mathbf{X}} \\ &= \left(\frac{\partial H}{\partial \mathbf{X}}\right)^T \mathbf{\Omega}_{2N} \frac{\partial H}{\partial \mathbf{X}} = 0 \end{aligned} \tag{1.107}$$

The Hamiltonian plays a major role in Mechanics, since it makes it possible to set up a global formalism for writing down the equations of motion in an algebraically simple way. An account of its main features will be given to conlude this Section. To this purpose, we consider an arbitrary quantity f characterizing the system and consequently depending on the parameters $(\{p_i\}, \{q_i\})$. Its time-derivative is written, using (1.105) as:

$$\begin{aligned} \frac{df}{dt} &= \frac{\partial f}{\partial t} + \frac{\partial f}{\partial \mathbf{X}} \cdot \dot{\mathbf{X}} \\ &= \frac{\partial f}{\partial t} + \left(\frac{\partial f}{\partial \mathbf{X}}\right)^T \mathbf{\Omega}_{2N} \frac{\partial H}{\partial \mathbf{X}} \end{aligned} \tag{1.108}$$

The *Poisson bracket* of two arbitrary functions f and g are then defined by the following relation:

$$\begin{aligned} \{f, g\} &\doteq \left(\frac{\partial f}{\partial \mathbf{X}}\right)^T \mathbf{\Omega}_{2N} \frac{\partial g}{\partial \mathbf{X}} \\ &= \sum_i \left(\frac{\partial f}{\partial \mathbf{X}_i}\right)^T \boldsymbol{\omega}_2 \frac{\partial g}{\partial \mathbf{X}_i} \\ &= \sum_i \left(\frac{\partial f}{\partial q_i}\frac{\partial g}{\partial p_i} - \frac{\partial f}{\partial p_i}\frac{\partial g}{\partial q_i}\right) \\ &= -\{g, f\} \end{aligned} \tag{1.109}$$

Equation (1.108) is then rearranged as

$$\frac{df}{dt} = \frac{\partial f}{\partial t} + \{f, H\} \tag{1.110}$$

For a quantity f which does not explicitly depend on time ($\partial f/\partial t = 0$), we get

$$\dot{f} = \{f, H\} \tag{1.111}$$

The conservation of energy H for an isolated system once again stems from the general relation (1.111), since it is obvious that $\{H, H\} = 0$. Besides, considering the case in which f comes down[32] to the vector $\mathbf{X}$ as given by (1.104), we obtain Hamilton's equations of motion in the form

$$\dot{\mathbf{X}} = D_H \mathbf{X} \tag{1.112}$$

where D_H is a linear operator (*Poisson operator*) as defined by

$$D_H f \doteqdot \{f, H\} \tag{1.113}$$

More generally, the Poisson bracket of two quantities can be written in the form of the action exerted on the second quantity by a linear derivation operator as defined from the first quantity:

$$D_g f \doteqdot \{f, g\} = -\{g, f\} \tag{1.114}$$

The Poisson formalism could, of course, be applied either by separating each degree of freedom from the others or by considering a set of particles, provided that $\boldsymbol{\omega}_2$ and $\boldsymbol{\omega}_6$ be respectively substituted for $\boldsymbol{\Omega}_{2N}$ in (1.109). In the latter case, we would write, for example

$$\begin{aligned} \{f, g\} &= \sum_a \left(\frac{\partial f}{\partial \mathbf{X}_a}\right)^T \boldsymbol{\omega}_6 \frac{\partial g}{\partial \mathbf{X}_a} \\ &= \sum_a \left(\frac{\partial f}{\partial \mathbf{r}_a} \cdot \frac{\partial g}{\partial \mathbf{p}_a} - \frac{\partial f}{\partial \mathbf{p}_a} \cdot \frac{\partial g}{\partial \mathbf{r}_a}\right) \end{aligned} \tag{1.115}$$

and (1.112) becomes:

$$\forall a, \; \dot{\mathbf{X}}_a = D_H \mathbf{X}_a \tag{1.116}$$

The general system (1.112)[33] is advantageous in that it leads to the following formal solution:

$$\mathbf{X}(t_2) = \Psi_{t_1,t_2} \mathbf{X}(t_1) \tag{1.117}$$

where Ψ_{t_1,t_2} is a linear differential operator as defined by

$$\Psi_{t_1,t_2} \doteqdot \exp\left[\int_{t_1}^{t_2} D_H(t)\, dt\right] \tag{1.118}$$

(D_H is implicitly time-dependent via the system parameters). Here we have defined the exponential function $\exp(P) = e^P$ of an operator P through the following relation[34]:

$$\begin{aligned} e^P &\doteqdot \sum_{k=1}^{+\infty} \frac{1}{k!} P^k \\ &= Id + P + \frac{1}{2} P^2 + O\left(\|P\|^3\right) \end{aligned} \tag{1.119}$$

where Id is the identity operator. This formalism will be useful in Sections 5.4.1 and 5.4.2 in Chapter 5 to discuss the numerical solution of the Hamilton equations within the framework of the SPH method. We will then

[32] Formula (1.110) extends, of course, to the case of vectors.

[33] Note that this system is written in a seemingly linear form. That linearity, however, is only formal, since the operator D_H may depend on the positions of particles via the forces.

[34] The norm $\|P\|$ of an operator may be understood as an upper bound of its result's norm on an arbitrary quantity f, divided by the f norm.

resort to an important property of the Poisson bracket. To demonstrate it, we note that definition (1.114) gives:

$$\begin{aligned}\{g,\{h,f\}\}+\{h,\{f,g\}\} &= \{g,\{h,f\}\}-\{h,\{g,f\}\}\\ &= D_g\left(D_h f\right)-D_h\left(D_g f\right)\\ &= \left[D_g, D_h\right] f\end{aligned} \tag{1.120}$$

where the commutator of two operators has been defined by

$$[P, Q] \doteqdot PQ - QP \tag{1.121}$$

If the phase space parameters $(\{p_i\}, \{q_i\})$ are gathered under the global notation $\mathbf{X} = \{x_i\}$ $(i = 1, ..., 2N)$, the general form of a linear differential operator is a linear combination of the partial derivatives with respect to the x_i. We will then write, in a quite general way:

$$\begin{aligned} D_g &= \sum_i \alpha_i \frac{\partial}{\partial x_i}\\ D_h &= \sum_i \beta_i \frac{\partial}{\partial x_i}\end{aligned} \tag{1.122}$$

the $\{\alpha_i\}$ and $\{\beta_i\}$ being undetermined coefficients. Combining these expressions, we find

$$\begin{aligned} D_g D_h &= \sum_i \alpha_i \frac{\partial}{\partial x_i} \sum_j \beta_j \frac{\partial}{\partial x_j}\\ &= \sum_{i,j} \alpha_i \left(\frac{\partial \beta_j}{\partial x_i}\frac{\partial}{\partial x_j} + \beta_j \frac{\partial^2}{\partial x_i \partial x_j}\right)\end{aligned} \tag{1.123}$$

Likewise:

$$D_h D_g = \sum_{i,j} \beta_i \left(\frac{\partial \alpha_j}{\partial x_i}\frac{\partial}{\partial x_j} + \alpha_j \frac{\partial^2}{\partial x_i \partial x_j}\right) \tag{1.124}$$

Interverting the dummy subscripts i and j in (1.124), the difference between these two operators is written as

$$\begin{aligned} \left[D_g, D_h\right] f &= \sum_{i,j} \left(\alpha_i \frac{\partial \beta_j}{\partial x_i}\frac{\partial f}{\partial x_j} - \beta_j \frac{\partial \alpha_i}{\partial x_j}\frac{\partial f}{\partial x_i}\right)\\ &= \sum_{i,j} \left(\alpha_j \frac{\partial \beta_i}{\partial x_j} - \beta_j \frac{\partial \alpha_i}{\partial x_j}\right)\frac{\partial f}{\partial x_i}\end{aligned} \tag{1.125}$$

Note that the second derivatives are no longer in that formulation. Thus, the second derivatives of f cancel each other out in the left-hand side of (1.120). Nevertheless, that does not mean the expression (1.125) vanishes, since it includes bilinear combinations of the first derivatives of f, g and h. The operators D_g and D_h then a priori do not commute ($\left[D_g, D_h\right] \neq 0$). On the other hand, let us now consider the expression

$$\{f, \{g, h\}\} + \{g, \{h, f\}\} + \{h, \{f, g\}\} \qquad (1.126)$$

The Poisson bracket $\{g, h\}$ being a bilinear combination of the first derivatives of g and h, the double bracket $\{f, \{g, h\}\}$ is a bilinear combination of their second derivatives. The whole expression (1.126) is then a linear expression of the second derivatives of f, g and h. Now, the previous argument shows that the second derivatives of f disappear. When applied to f, g and h, the same argument shows that the quantity as given by (1.126) is identically zero:

$$\{f, \{g, h\}\} + \{g, \{h, f\}\} + \{h, \{f, g\}\} = 0 \qquad (1.127)$$

this is the *Jacobi's identity*. Starting from the definition (1.114), the antisymmetric relation it includes and the definition (1.121), that relation can be rearranged as

$$\left[D_g, D_h\right] = D_{\{h,g\}} \qquad (1.128)$$

In other words, the commutator of two derivation operators is also a derivation operator. This result confirms that two arbitrary operators do not commute, unless the Poisson bracket in the functions defining them vanishes.

1.4.3 Linear momentum and angular momentum

As discussed in Section 1.4.1, time independence of the Lagrangian in an isolated system leads to the law of conservation of energy. We will demonstrate that space homogeneity and isotropy also lead to conservation laws. For the sake of simplicity, we keep considering the case of an isolated particle system. We have mentioned a property of potential energy in Section 1.3.2: it only depends on mutual distances $r_{ab} = |\mathbf{r}_a - \mathbf{r}_b|$ between particles, and so the Lagrangian has the same, as yet unexploited, property. In particular, the Lagrangian of an isolated system should not be affected by any global translation of that system:

$$\forall \boldsymbol{\epsilon}, \quad L(\{\mathbf{r}_a\}, \{\mathbf{u}_a\}) = L(\{\mathbf{r}_a + \boldsymbol{\epsilon}\}, \{\mathbf{u}_a\}) \qquad (1.129)$$

Since the translations are additive, it suffices to satisfy the previous condition for an arbitrary infinitesimal translation; that is which is defined by a small vector $\boldsymbol{\epsilon}$ before each of the vectors $\mathbf{r}_a$. Expanding the right-hand side of (1.129) to the first order yields

$$\forall \boldsymbol{\epsilon}, \quad L(\{\mathbf{r}_a + \boldsymbol{\epsilon}\}, \{\mathbf{u}_a\}) = L(\{\mathbf{r}_a\}, \{\mathbf{u}_a\}) + \sum_a \frac{\partial L}{\partial \mathbf{r}_a} \cdot \boldsymbol{\epsilon} + o(|\boldsymbol{\epsilon}|) \qquad (1.130)$$

Subtracting (1.129) from (1.130), and observing that $\boldsymbol{\epsilon}$ is any vector, we achieve the Lagrangian invariance condition through a translation:

$$\sum_a \frac{\partial L}{\partial \mathbf{r}_a} = \mathbf{0} \qquad (1.131)$$

The Lagrange equations (1.31) then give:

$$\frac{d}{dt} \sum_a \mathbf{p}_a = \mathbf{0} \qquad (1.132)$$

Thus, vector **P**, as defined by

$$\mathbf{P} \doteqdot \sum_a \mathbf{p}_a = \sum_a m_a \mathbf{u}_a \tag{1.133}$$

is invariant over time for an isolated system:

$$\frac{d\mathbf{P}}{dt} = \mathbf{0} \tag{1.134}$$

It can then readily be found out that the Lagrangian is invariant for every translation, which is not necessarily infinitesimal.[35] Vector **P**, which is the sum of the particle linear momentum, is called the total linear momentum (or simply momentum) of the system. Its conservation results from the homogeneity of the laws of mechanics in a Galilean frame of reference. Additivity of the linear momentum is evident; however, unlike energy, that momentum is additive even in the case of interacting systems.

[35] Conversely, the origin of coordinates may as well be set in a sufficiently remote place from the system (if the latter is bounded), and then an arbitrary translation may be considered as infinitesimal as compared with the position vectors of the particles.

Equation (1.131) can be rearranged using the equation of motion (1.56):

$$\sum_a \mathbf{F}_a = \mathbf{0} \tag{1.135}$$

The sum of forces mutually exerted by the various parts of an isolated system is then zero. It is the law of action-reaction (or Newton's third law), another form of the law of conservation of linear momentum. In the formalism of the generalized coordinates and velocities (refer to Section 1.3.3), one could likewise demonstrate that the sum of generalized momenta is constant, which amounts to saying that the sum of generalized forces equals zero.

Let us now write that an infinitesimal rotation (about an arbitrary point) necessarily leaves the Lagrangian unchanged, because of the space isotropy. We may recall in that respect that a three-dimensional rotation of angle θ, about an axis which is oriented by the unit vector **n**, is a linear transformation associated with the following tensor:

$$\mathbf{R}_{\mathbf{n},\theta} = (\cos\theta)\,\mathbf{I}_3 + (1 - \cos\theta)\,\mathbf{n} \otimes \mathbf{n} + (\sin\theta)\,\hat{\mathbf{n}} \tag{1.136}$$

where $\hat{\mathbf{n}}$ is the skew-symmetric tensor as defined by

$$\hat{\mathbf{n}} \doteqdot \begin{pmatrix} 0 & -n_z & n_y \\ n_z & 0 & -n_x \\ -n_y & n_x & 0 \end{pmatrix} \tag{1.137}$$

For an infinitesimal angle, we will have, at first-order in θ, $\cos\theta \approx 1$ and $\sin\theta \approx \theta$, which yields

$$\begin{aligned} \mathbf{R}_{\mathbf{n},\theta} &= \mathbf{I}_3 + \theta\hat{\mathbf{n}} \\ \mathbf{R}_{\mathbf{n},\theta} \cdot \mathbf{r} &= \mathbf{r} + \theta\mathbf{n} \times \mathbf{r} \end{aligned} \tag{1.138}$$

Rotating a particle system through an infinitesimal angle θ then amounts to turning each position $\mathbf{r}_a$ into $\mathbf{r}_a + \boldsymbol{\epsilon} \times \boldsymbol{r}_a$, where $\boldsymbol{\epsilon} = \theta\mathbf{n}$. The same transformation, however, should be applied to the velocities, since

$$\begin{aligned}\frac{d\left(\mathbf{r}_a+\boldsymbol{\epsilon}\times\mathbf{r}_a\right)}{dt}&=\frac{d\mathbf{r}_a}{dt}+\boldsymbol{\epsilon}\times\frac{d\mathbf{r}_a}{dt}\\&=\mathbf{u}_a+\boldsymbol{\epsilon}\times\mathbf{u}_a\end{aligned}\tag{1.139}$$

Thus, the invariance of the Lagrangian under an infinitesimal rotation is written as

$$\forall\boldsymbol{\epsilon},\quad L\left(\{\mathbf{r}_a\},\{\mathbf{u}_a\}\right)=L\left(\{\mathbf{r}_a+\boldsymbol{\epsilon}\times\mathbf{r}_a\},\{\mathbf{u}_a+\boldsymbol{\epsilon}\times\mathbf{u}_a\}\right)\tag{1.140}$$

A first-order expansion yields:

$$\begin{aligned}&\forall\boldsymbol{\epsilon},\; L\left(\{\mathbf{r}_a+\boldsymbol{\epsilon}\times\mathbf{r}_a\},\{\mathbf{u}_a+\boldsymbol{\epsilon}\times\mathbf{u}_a\}\right)=L\left(\{\mathbf{r}_a\},\{\mathbf{u}_a\}\right)\\&+\sum_a\left[\frac{\partial L}{\partial\mathbf{r}_a}\cdot\left(\boldsymbol{\epsilon}\times\mathbf{r}_a\right)+\frac{\partial L}{\partial\mathbf{u}_a}\cdot\left(\boldsymbol{\epsilon}\times\mathbf{u}_a\right)\right]+o\left(|\boldsymbol{\epsilon}|\right)\end{aligned}\tag{1.141}$$

Using the permutation properties of the mixed product $\mathbf{A}\cdot(\mathbf{B}\times\mathbf{C})$, the combination of (1.140) and (1.141) yields:

$$\forall\boldsymbol{\epsilon},\quad\sum_a\left[\mathbf{r}_a\times\frac{\partial L}{\partial\mathbf{r}_a}+\mathbf{u}_a\times\frac{\partial L}{\partial\mathbf{u}_a}\right]\cdot\boldsymbol{\epsilon}=\mathbf{0}\tag{1.142}$$

Let us now resort to the Lagrange equations (1.31). Taking into account the definition of momenta as given by (1.53), and since that equation is valid for any $\boldsymbol{\epsilon}$, hence:

$$\sum_a\left(\mathbf{r}_a\times\frac{d\mathbf{p}_a}{dt}+\mathbf{u}_a\times\mathbf{p}_a\right)=\mathbf{0}\tag{1.143}$$

We notice that the second term of the left-hand side in (1.143) vanishes, in accordance with (1.53), which also reads:

$$\frac{d\left(\mathbf{r}_a\times\mathbf{p}_a\right)}{dt}=\mathbf{r}_a\times\frac{d\mathbf{p}_a}{dt}\tag{1.144}$$

We then have, for an isolated system:

$$\sum_a\mathbf{r}_a\times\frac{d\mathbf{p}_a}{dt}=\mathbf{0}\tag{1.145}$$

We introduce each particle's *angular momentum* $\mathbf{M}_a$ as defined by:

$$\mathbf{M}_a\doteqdot\mathbf{r}_a\times\mathbf{p}_a=m_a\mathbf{r}_a\times\mathbf{u}_a\tag{1.146}$$

We may also define the moment $\mathbf{M}$ of an isolated system as the sum of the individual moments of the particles which it comprises:

$$\mathbf{M}\doteqdot\sum_a\mathbf{M}_a=\sum_a m_a\mathbf{r}_a\times\mathbf{u}_a\tag{1.147}$$

Like the linear momentum, the angular momentum is, of course, additive in all cases. The angular momentum time derivative is written, thanks to (1.144) and (1.56):

$$\begin{aligned}\frac{d\mathbf{M}}{dt} &= \sum_a \mathbf{r}_a \times \frac{d\mathbf{p}_a}{dt} \\ &= \sum_a \mathbf{r}_a \times \mathbf{F}_a\end{aligned} \tag{1.148}$$

The term under the summation symbol of (1.148) is known as the *moment of force* (or torque) exerted on the particle a. The result (1.145) then shows that $\mathbf{M}$ is constant over time for an isolated system:

$$\frac{d\mathbf{M}}{dt} = \mathbf{0} \tag{1.149}$$

or else

$$\sum_a \mathbf{r}_a \times \mathbf{F}_a = \mathbf{0} \tag{1.150}$$

That conservation law results from the isotropy of the laws of mechanics in a Galilean frame of reference. Invariance under an infinitesimal rotation immediately results in the invariance by all the rotations.

We notice that an infinite number of moments would theoretically have to be defined, since we can choose the spatial origin. A change of origin, however, obviously only alters the moments through the addition of a constant vector. Actually, if a translation of vector $\mathbf{r}_0$ is made, the following is obtained, with (1.133):

$$\begin{aligned}\mathbf{M}' &= \sum_a (\mathbf{r}_a - \mathbf{r}_0) \times \mathbf{p}_a \\ &= \sum_a \mathbf{r}_a \times \mathbf{p}_a - \mathbf{r}_0 \times \sum_a \mathbf{p}_a \\ &= \mathbf{M} + \mathbf{P} \times \mathbf{r}_0\end{aligned} \tag{1.151}$$

Conserving the momenta $\mathbf{P}$ and $\mathbf{M}$ then implies conserving $\mathbf{M}'$.

We have highlighted two conservation laws for the total linear and angular momenta of every isolated system. Note that each of these conservation laws is written in two different but equivalent forms: one comes in the form of the time derivative of a quantity ($\mathbf{P}$ or $\mathbf{M}$, eqns (1.134) and (1.149)), the other one as a sum of individual contributions (forces or moments of forces, eqns (1.135) and (1.150)). The former forms of these conservation laws could be described as 'continuous' (since they involve a time derivative), the latter as 'discrete'.

Lastly, we have highlighted 7 invariants for one isolated system (energy as discussed in Section 1.4.1, as well as three components of linear momentum and those of angular momentum). Each conservation law stems from the invariance of the law of mechanics with respect to a transformation (time invariance, homogeneity, isotropy), which is valid in every inertial frame of reference. As a rule, in physics, each invariance property leads to a conservation law.[36]

[36] Some of these laws do not occur within the context of classical mechanics. In quantum mechanics, for instance, the invariance of the wave function under symmetry around a central point results in the conservation of a quantity which is known as *parity*.

1.4.4 Investigating a generic case: the simple pendulum

In order to illustrate these highlighted processes and the introduced notations, we will advantageously consider one of the simplest systems, the equivalent of the famous Drosophila fly in the dynamical systems, namely the pendulum consisting of a bob swinging back and forth at the end of a wire, as shown in Fig. 1.3. The bob position vector has an orthoradial component of $\ell\theta$. Given the expression (1.57) of the force of gravity, the projection of the equation of motion (1.56) onto that axis is then

$$m\ell^2 \frac{d^2\theta}{dt^2} = F_\theta = -mg\ell \sin\theta \tag{1.152}$$

which can be rearranged as

$$\ddot{\theta} = -\omega_0^2 \sin\theta \tag{1.153}$$

with

$$\omega_0 \doteq \sqrt{\frac{g}{\ell}} \tag{1.154}$$

The pendulum kinetic energic is determined from (1.49), its velocity being denoted $\ell\dot{\theta}$, whereas the value of its gravitational potential energy, according to (1.51), is mgz, with $z = 1 - \cos\theta$:

$$\begin{aligned} E_k &= \frac{m\ell^2\dot{\theta}^2}{2} \\ E_p &= mg\ell\,(1 - \cos\theta) \end{aligned} \tag{1.155}$$

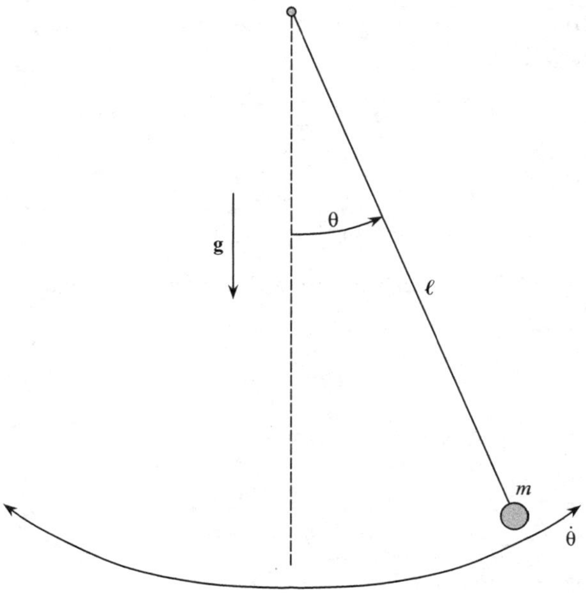

Fig. 1.3 Simple pendulum diagram.

As expected, the kinetic energy is a quadratic function of generalized velocity $\dot{\theta}$ (refer to Section 1.4.1). The system's total energy $E = E_k + E_p$ is then written as

$$E = mg\ell\left(\frac{\dot{\theta}^2}{2\omega_0^2} + 1 - \cos\theta\right) \tag{1.156}$$

and its Lagrangian

$$L\left(\theta, \dot{\theta}\right) = mg\ell\left(\frac{\dot{\theta}^2}{2\omega_0^2} - 1 + \cos\theta\right) \tag{1.157}$$

Here the Lagrange equation (1.23) is written as

$$\frac{d}{dt}\frac{\partial L}{\partial \dot{\theta}} = \frac{\partial L}{\partial \theta} \tag{1.158}$$

and, of course, would once again give (1.152). At present, a simple derivation of (1.156) yields

$$\begin{aligned}\frac{1}{mg\ell}\frac{dE}{dt} &= \dot{\theta}\left(\frac{\ddot{\theta}}{\omega_0^2} + \sin\theta\right)\\ &= 0\end{aligned} \tag{1.159}$$

according to (1.153). Thus, the system's energy is conserved, as expected from a system which is subject to the influence of a time-invariant external field. Since the system has only one geometrical degree of freedom, no other invariant can be constructed here.[37] In addition, with definition (1.25), we have

[37] This is because $2N - 1 = 1$. Besides, here we cannot apply the conservation of linear and angular momenta, since the system is subjected to the influence of an external body (the Earth).

$$p = \frac{\partial L}{\partial \dot{\theta}} = m\ell^2\dot{\theta} \tag{1.160}$$

With (1.160), and again using the notation $q \doteq \theta$, we also could write the Hamiltonian as

$$H(p, q) = mg\ell\left[\frac{1}{2}\left(\frac{p}{A}\right)^2 + 1 - \cos q\right] \tag{1.161}$$

with

$$A \doteq m\ell^2\omega_0 = m\sqrt{g\ell^3} \tag{1.162}$$

Equation (1.77) is applicable here with a matrix $\mathbf{\Gamma}$ comprising a single coefficient Γ as given by

$$\Gamma = \frac{1}{m\ell^2} \tag{1.163}$$

The Hamilton equation (1.89) becomes

$$\begin{aligned} p &= \dot{q}\\ mg\ell\sin q &= -\dot{p}\end{aligned} \tag{1.164}$$

and would once again normally yield (1.152).

If only the moderate amplitude motions are considered, the approximations $\sin\theta \approx \theta$ and $\cos\theta \approx 1 - \theta^2/2$ can be made and the second line of (1.155) becomes

$$E_p \approx \frac{mg\ell\theta^2}{2} \tag{1.165}$$

The system then becomes linear, the equation of motion (1.153) being reduced to

$$\ddot{\theta} = -\omega_0^2\theta \tag{1.166}$$

It is well known that the solutions of this differential equation are circular functions, and so the coordinates in the phase space (two-dimensional in this case) are given by

$$\begin{aligned} p &= -A\theta_0 \sin\left(\omega_0 t + \varphi\right) \\ q &= \theta_0 \cos\left(\omega_0 t + \varphi\right) \end{aligned} \tag{1.167}$$

Quantities θ_0 and φ are determined by the initial conditions (angle θ_0 and angular velocity $\dot{\theta}_0$). Developing the trigonometric functions occurring in (1.167), one can write as well

$$\begin{aligned} p &= -A\theta_0 \left[\sin\left(\omega_0 t\right)\cos\varphi + \cos\left(\omega_0 t\right)\sin\varphi\right] \\ q &= \theta_0 \left[\cos\left(\omega_0 t\right)\cos\varphi - \sin\left(\omega_0 t\right)\sin\varphi\right] \end{aligned} \tag{1.168}$$

At the inital point in time (at $t = 0$), we have in particular

$$\begin{aligned} p\left(t = 0\right) &= -A\theta_0 \sin\varphi \\ q\left(t = 0\right) &= \theta_0 \cos\varphi \end{aligned} \tag{1.169}$$

The system state at point t is then written as a linear function of its initial state. Using (1.91) as a model, we can introduce the system's state vector:

$$\mathbf{X} \doteqdot \begin{pmatrix} p \\ q \end{pmatrix} \tag{1.170}$$

Taking (1.169) into account, the system (1.168) is then rearranged as:

$$\mathbf{X}\left(t\right) = \mathbf{\Psi}\left(t\right)\mathbf{X}\left(t = 0\right) \tag{1.171}$$

with

$$\mathbf{\Psi}\left(t\right) \doteqdot \begin{pmatrix} \cos t^+ & -A\sin t^+ \\ \dfrac{1}{A}\sin t^+ & \cos t^+ \end{pmatrix} \tag{1.172}$$

and[38]

$$t^+ \doteqdot \omega_0 t \tag{1.173}$$

The system state at point t then linearly depends on its initial state, which illustrates the general relation (1.117). Matrix $\mathbf{\Psi}\left(t\right)$ has the same form as that of the operator $\Psi_{0,t}$ introduced by the latter formula. The system's equations may conveniently be rearranged non-dimensionally as:[39]

[38] The non-dimensionalized quantities will often be denoted by a superscript $^+$.

[39] We have previously mentioned the disadvantageous non-homogeneity of the phase space variables. We will come back to this matter in Chapter 2 (Section 2.2.1).

$$\mathbf{X}^+ \doteq \begin{pmatrix} p^+ = \dfrac{p}{A} = \dfrac{\dot{\theta}}{\omega_0} \\ q^+ = q = \theta \end{pmatrix} \tag{1.174}$$

which enables us to rearrange (1.172) as:

$$\begin{aligned} \mathbf{X}^+ \left(t^+\right) &= \boldsymbol{\Psi}^+ \left(t^+\right) \mathbf{X}^+ \left(t^+ = 0\right) \\ \boldsymbol{\Psi}^+ \left(t^+\right) &\doteq \begin{pmatrix} \cos t^+ & -\sin t^+ \\ \sin t^+ & \cos t^+ \end{pmatrix} \end{aligned} \tag{1.175}$$

With these notations, energy takes the following simple form:

$$\begin{aligned} H\left(p^+, q^+\right) &= \frac{mg\ell}{2}\left(p^{+2} + q^{+2}\right) \\ &= \frac{mg\ell}{2}\left|\mathbf{X}^+\right|^2 \end{aligned} \tag{1.176}$$

Hence, conservation of energy can be seen as the effect of the orthogonality of the matrix $\boldsymbol{\Psi}^+\left(t^+\right)$ as given by (1.175). This is because we have:

$$\left[\boldsymbol{\Psi}^+\left(t^+\right)\right]^{-1} = \left[\boldsymbol{\Psi}^+\left(t^+\right)\right]^T \tag{1.177}$$

and, consequently:

$$\begin{aligned} \left|\mathbf{X}^+\left(t^+\right)\right|^2 &= \left[\mathbf{X}^+\left(t^+\right)\right]^T \mathbf{X}^+\left(t^+\right) \\ &= \left[\boldsymbol{\Psi}^+\left(t^+\right)\mathbf{X}^+\left(t^+ = 0\right)\right]^T \boldsymbol{\Psi}^+\left(t^+\right)\mathbf{X}^+\left(t^+ = 0\right) \\ &= \left[\mathbf{X}^+\left(t^+ = 0\right)\right]^T \left[\boldsymbol{\Psi}^+\left(t^+\right)\right]^T \boldsymbol{\Psi}^+\left(t^+\right)\mathbf{X}^+\left(t^+ = 0\right) \\ &= \left[\mathbf{X}^+\left(t^+ = 0\right)\right]^T \mathbf{X}^+\left(t^+ = 0\right) \\ &= \left|\mathbf{X}^+(0)\right|^2 \end{aligned} \tag{1.178}$$

Thus, the conservation of energy is directly linked to the property of *orthogonality* (1.177), to be compared with (1.94) and (1.95).

Figure 1.4 illustrates the system's trajectory layout within the phase space. These curves, when taken together, make up a system's *phase portrait*. Those trajectories lying closer to the origin are circles[40] corresponding to the linear approximation (1.167). Each curve is distinguished from the others by the initial conditions. If a larger initial angle θ_0 (i.e. more potential energy) and/or a higher initial angular velocity $\dot{\theta}_0$ (i.e. more kinetic energy) is/are given to the pendulum, (1.156) shows that energy will be higher. Hence, each trajectory corresponds to a determined value of energy, being conserved along the course.

We will come back to the simple pendulum example on several occasions in the next chapters of this book, in order to illustrate other processes and other laws (Chapters 2 and 4), as well as to discuss numerical equation solution techniques (Chapter 5).

[40] Or rather ellipses, the scales of axes being arbitrary as long as the system has not been non-dimensionalized.

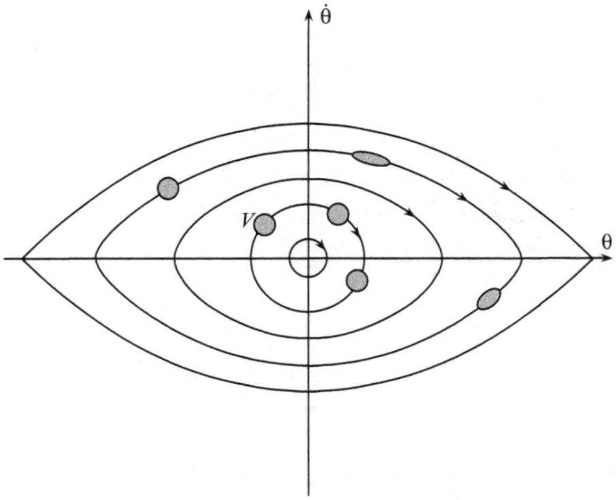

Fig. 1.4 Phase space of the simple pendulum and configuration of phase portrait.

1.5 Composite systems

We now consider composite systems, that is systems made of several subsystems. We first introduce the concepts of centre of mass and inertia tensor, in order to study the motion of rigid bodies. We then distinguish between internal and external forces, and define the internal energy. Finally, density and pressure are defined, and the Hamilton equations are updated to include their effects.

1.5.1 Centre of mass. Inertia tensor

In this Section, we will discuss the case of rigid bodies, which will be used, in particular, in Chapter 5 (Section 6.3.3). We will first set the quite general relations governing the total linear and angular momenta upon a change of frame of reference. In fact, the linear momentum of a global system takes different values in various Galilean frames of reference. Considering both previously mentioned frames of reference R and R' (R' moving about at the velocity $\mathbf{V}$ with respect to R), the velocity composition law (1.3) and the constant nature of the masses can be used to link the momenta $\mathbf{P}$ and $\mathbf{P}'$ of a system through:

$$\begin{aligned}\mathbf{P} &= \sum_b m_b \mathbf{u}_b \\ &= \sum_b m_b \left(\mathbf{u}'_b + \mathbf{V}\right) \\ &= \mathbf{P}' + M\mathbf{V}\end{aligned} \tag{1.179}$$

where M denotes the sum of the masses of particles b, that is the system total mass:

$$M \doteqdot \sum_b m_b \tag{1.180}$$

There is, in particular, a frame of reference R' in which the system momentum vanishes. Letting $\mathbf{P}' = \mathbf{0}$ in (1.179), we find that velocity (in R) in that frame of reference is given by

$$\begin{aligned}\mathbf{V} &= \frac{1}{M}\mathbf{P} \\ &= \frac{1}{M}\sum_b m_b \mathbf{u}_b\end{aligned} \tag{1.181}$$

that is the mass-weighted average of individual velocities. The system as a whole is said to be at rest with respect to that frame of reference; this is a generalization of the notion of isolated particle at rest. Velocity as given by (1.181) is then known as *overall system velocity*. This relation shows that the momentum of a system is related to its overall velocity by an equation analogous to (1.53), where the role of the system mass is played by the sum M of the masses of its constituents. It is another approach to the mass additivity. Lastly, we note that the right-hand term in (1.181) is the derivative with respect to time of the radius vector as defined by:

$$\begin{aligned}\mathbf{R} &\doteqdot \frac{1}{M}\sum_b m_b \mathbf{r}_b \\ \mathbf{V} &= \frac{d\mathbf{R}}{dt} = \dot{\mathbf{R}}\end{aligned} \tag{1.182}$$

The point as determined by vector $\mathbf{R}$ is referred to as the system's *centre of mass*, and the relevant Galilean frame of reference is its *inertial frame of reference*. We have

$$\begin{aligned}\forall b,\ \mathbf{r}_b &= \mathbf{r}'_b + \mathbf{R} \\ \mathbf{u}_b &= \mathbf{u}'_b + \mathbf{V}\end{aligned} \tag{1.183}$$

(all these quantities being implicitly time-dependent). The law of conservation of momentum of an isolated system may then be formulated by stating that its centre of mass undergoes a steady motion. This fact is a generalization of the law of inertia as set out in Section 1.3.1 for an isolated particle. Deriving (1.181) with respect to time, we get, taking into account the mass constancy and the equation of motion of a particle (1.56):

$$\begin{aligned}M\frac{d\mathbf{V}}{dt} &= \sum_b m_b \frac{d\mathbf{u}_b}{dt} \\ &= \sum_b \mathbf{F}_b\end{aligned} \tag{1.184}$$

This equation is identical to (1.54). This result can be expressed by stating that the forces are additive:[41]

$$\mathbf{F} = \sum_b \mathbf{F}_b \tag{1.185}$$

[41]From the definition of the force of gravity $\mathbf{F}_{g,a}$ as given by (1.57), this result matches the mass additivity, provided that the principle of equivalence is acknowledged.

Let us now consider what the angular momentum becomes upon such a change of frame of reference. Using the laws (1.183) and the definition (1.182) of the centre of mass, we find:

$$\begin{aligned}\mathbf{M} &= \sum_b m_b \mathbf{r}_b \times \mathbf{u}_b \\ &= \sum_b m_b \mathbf{r}_b \times \left(\mathbf{u}'_b + \mathbf{V}\right) \\ &= \sum_b m_b \left(\mathbf{r}'_b + \mathbf{R}\right) \times \mathbf{u}'_b + \left(\sum_b m_b \mathbf{r}_b\right) \times \mathbf{V} \\ &= \sum_b m_b \mathbf{r}'_b \times \mathbf{u}'_b + \mathbf{R} \times \sum_b m_b \mathbf{u}'_b + M\mathbf{R} \times \mathbf{V}\end{aligned} \tag{1.186}$$

Afterwards, taking the definition of the total momentum and the nullity of $\mathbf{P}'$ into account:

$$\begin{aligned}\mathbf{M} &= \mathbf{M}' + \mathbf{R} \times \mathbf{P}' + M\mathbf{R} \times \mathbf{V} \\ &= \mathbf{M}' + \mathbf{R} \times \mathbf{P}\end{aligned} \tag{1.187}$$

A rigid body is characterized in that the mutual distances between its constituents remain unchanged over time, which significantly reduces the number of independent degrees of freedom. This is because a solid may be likened to an object which is only just more sophisticated than a particle, comprising angular parameters in addition to its degrees of freedom for position. The geometrical parameters then come down to the data of the position vector $\mathbf{R}$ for the centre of mass, along with the three angles $\theta_1, \theta_2, \theta_3$ making it possible to characterize the orientation of the body with respect to a given system of coordinates. These angles may be gathered in one vector:

$$\boldsymbol{\theta}^T \doteqdot (\theta_1, \theta_2, \theta_3) \tag{1.188}$$

the *angular velocity* vector can then be defined by the relation

$$\boldsymbol{\Omega} \doteqdot \dot{\boldsymbol{\theta}} \tag{1.189}$$

The motion of a rigid body is composed of a translation being defined by its overall velocity $\mathbf{V}$ and a rotation. Between two very close points in time which are separated from the infinitesimal time dt, the latter makes the body revolve in the plane containing the vectors $\boldsymbol{\theta}(t)$ and $\boldsymbol{\theta}(t + dt)$, that is about the axis supported by $\dot{\boldsymbol{\theta}} = \boldsymbol{\Omega}$, in the inertial frame of reference. The value of infinitesimal angle of rotation is then $|\boldsymbol{\theta}(t + dt) - \boldsymbol{\theta}(t)| = |\boldsymbol{\Omega}|\, dt$. Using equation (1.138), we can infer that the position of each particle in the body happens to be incremented by

$$\mathbf{r}'_b\,(t+dt) = \mathbf{r}'_b\,(t) + \boldsymbol{\Omega} \times \mathbf{r}'_b\,(t)\,dt \tag{1.190}$$

Dividing by dt, we can see that the velocity of each of the particles making up the rigid body is governed at any time by the following law:

$$\begin{aligned} \mathbf{u}'_b &= \boldsymbol{\Omega} \times \mathbf{r}'_b \\ \mathbf{u}_b &= \mathbf{V} + \boldsymbol{\Omega} \times (\mathbf{r}_b - \mathbf{R}) \end{aligned} \tag{1.191}$$

(the two lines of (1.191) are equivalent owing to (1.183)). In particular, if we define

$$\mathbf{u}_{ab} \doteqdot \mathbf{u}_a - \mathbf{u}_b \tag{1.192}$$

then we have

$$\begin{aligned} \frac{d\mathbf{r}_{ab}}{dt} &= \mathbf{u}_{ab} \\ &= \boldsymbol{\Omega} \times \mathbf{r}_{ab} \end{aligned} \tag{1.193}$$

It is a rigid motion, as evidenced by the following argument. Let us consider two particles a and b from the body and let us observe how the distance between them, namely r_{ab}, as defined by (1.45), will change over time. We have, with (1.193):

$$\begin{aligned} \frac{dr_{ab}^2}{dt} &= 2\mathbf{u}_{ab} \cdot \mathbf{r}_{ab} \\ &= 2\left(\boldsymbol{\Omega} \times \mathbf{r}'_{ab}\right) \cdot \mathbf{r}'_{ab} \\ &= 0 \end{aligned} \tag{1.194}$$

(we have employed the vector product properties). The distribution (1.191) of velocities can be found again through the Lagrange multiplier technique which was set out in Section 1.2.4. To do this, we must endeavour to minimize the system action under the condition of rigidity of the body. That condition can be formulated by observing that the area of the triangle defined by vectors $\mathbf{u}'_b$ and $\mathbf{r}'_b$ remains unaltered during the motion; in other words, the determinant $\det\left(\mathbf{u}'_b, \mathbf{r}'_b, \boldsymbol{\Omega}\right) = \mathbf{u}'_b \cdot \left(\mathbf{r}'_b \times \boldsymbol{\Omega}\right)$ is constant. That must hold true for each particle, resulting in as many conditions to be observed, each being fitted with a Lagrange multiplier A_b, as in the general case provided by equation (1.30). Thus, the system's Lagrangian may be written as

$$L = E_k + L_{rigid} = \sum_b \left[\frac{1}{2} m_b u_b^2 + A_b\,(\mathbf{u}_b - \mathbf{V}) \cdot [(\mathbf{r}_b - \mathbf{R}) \times \boldsymbol{\Omega}]\right] \tag{1.195}$$

Here we have ignored the inter-particle forces, because they do not induce any relative motion. As to the total force exerted on the body, we omit it on purpose, since, for the moment, we do not try to write any complete equation for its motion (we will do it later on), but merely characterize the distribution of velocities from one particle to another. For the same reason, we will just derive the Lagrangian with respect to the velocities of the latter. Using definition (1.181), we find

$$\frac{\partial\left(\mathbf{u}_b-\mathbf{V}\right)}{\partial \mathbf{u}_a}=\left(1-\frac{m_a}{M}\right)\delta_{ab}\mathbf{I}_3 \tag{1.196}$$

Thus, the derivative of (1.195) is written as

$$\frac{\partial L}{\partial \mathbf{u}_a}=m_a\mathbf{u}_a+A_a\left(1-\frac{m_a}{M}\right)\left(\mathbf{r}_a-\mathbf{R}\right)\times\mathbf{\Omega} \tag{1.197}$$

Now, in the absence of forces, the principle of inertia (refer to Section 1.2.1) states that the momentum $\partial L/\partial \mathbf{u}_a$ is constant, which gives

$$m_a\mathbf{u}_a=\mathbf{A}+A_a\left(1-\frac{m_a}{M}\right)\mathbf{\Omega}\times\left(\mathbf{r}_a-\mathbf{R}\right) \tag{1.198}$$

where $\mathbf{A}$ is a constant vector. That equation is consistent with law (1.191), provided that $\mathbf{A}=m_a\mathbf{V}$ and $A_a=M/\left(M-m_a\right)$ are assumed.

The respective values of kinetic energy in an arbitrary frame of reference and in the inertial frame of reference can from now on be readily linked together through the properties of cross product:

$$\begin{aligned}E_k&=\sum_b\frac{1}{2}m_b\left(\mathbf{V}+\mathbf{\Omega}\times\mathbf{r}'_b\right)^2\\&=\frac{1}{2}\left(\sum_b m_b\right)V^2+\mathbf{V}\cdot\left(\mathbf{\Omega}\times\sum_b m_b\mathbf{r}'_b\right)+\sum_b\frac{1}{2}m_b\left(\mathbf{\Omega}\times\mathbf{r}'_b\right)^2\\&=\frac{1}{2}MV^2+\mathbf{V}\cdot\left(\mathbf{\Omega}\times\mathbf{P}'\right)+\frac{1}{2}\sum_b m_b\left(\mathbf{\Omega}\times\mathbf{r}'_b\right)^2\end{aligned} \tag{1.199}$$

Because of the nullity of $\mathbf{P}'$, that calculation is ultimately reduced to

$$E_k=E'_k+E_{rot} \tag{1.200}$$

where the energy of rotation E_{rot} is given by the last term of (1.199). To make progress in the calculation of that quantity, let us perform the following manipulations:

$$\begin{aligned}\left(\mathbf{\Omega}\times\mathbf{r}'_b\right)^2&=\Omega^2\left|\mathbf{r}'_b\right|^2-\left(\mathbf{\Omega}\cdot\mathbf{r}'_b\right)^2\\&=\Omega_i\Omega_i x'_{k,b}x'_{k,b}-\Omega_i\Omega_j x'_{i,b}x'_{j,b}\\&=\Omega_i\left(x'_{k,b}x'_{k,b}\delta_{ij}-x'_{i,b}x'_{j,b}\right)\Omega_j\end{aligned} \tag{1.201}$$

(we have used the Einstein summation convention for the implicit summation over every index appearing twice in one monomial; refer to Appendix 9). Through that result, (1.199) yields

$$E_{rot}=\frac{1}{2}\mathbf{\Omega}^T\mathbf{J}\mathbf{\Omega} \tag{1.202}$$

where the *inertia tensor* of the body has been defined by

$$\begin{aligned}\mathbf{J} &\doteqdot \sum_b m_b \left(x'_{k,b}x'_{k,b}\delta_{ij} - x'_{i,b}x'_{j,b}\right) \mathbf{e}_i \otimes \mathbf{e}_j \\ &= \sum_b m_b \left(\left|\mathbf{r}'_b\right|^2 \mathbf{I}_3 - \mathbf{r}'_b \otimes \mathbf{r}'_b\right)\end{aligned} \tag{1.203}$$

This tensor, which is constructed in a symmetric form, merely depends on the shape of the body, which is inherently invariable; thus, it is constant.[42] The form (1.202) of the rotational energy is similar to (1.81), and shows that $\mathbf{J}$ is a positive definite—that is invertible—matrix. Referring to the inertial frame of reference, the angular momentum of the body may also be related to the inertia tensor:

[42] We may recall that the position vectors $\{\mathbf{r}'_b\}$ of the particles are considered here in the inertial frame of reference of the solid. As a real symmetric tensor, $\mathbf{J}$ is orthogonal diagonalizable and is then ultimately determined by its eigenvalues, which are known as *principal moments of inertia*. Expressions of them for many standard bodies can be found in relevant literature.

$$\begin{aligned}\mathbf{M}' &= \sum_b m_b \mathbf{r}'_b \times \mathbf{u}'_b \\ &= \sum_b m_b \mathbf{r}'_b \times \left(\boldsymbol{\Omega} \times \mathbf{r}'_b\right) \\ &= \sum_b m_b \left[\left|\mathbf{r}'_b\right|^2 \boldsymbol{\Omega} - \left(\mathbf{r}'_b \cdot \boldsymbol{\Omega}\right) \mathbf{r}'_b\right]\end{aligned} \tag{1.204}$$

Besides, we may also write

$$\begin{aligned}\left|\mathbf{r}'_b\right|^2 \boldsymbol{\Omega} - \left(\mathbf{r}'_b \cdot \boldsymbol{\Omega}\right) \mathbf{r}'_b &= \left(x'_{k,b}x'_{k,b}\Omega_i - x'_{j,b}\Omega_j x'_{i,b}\right) \mathbf{e}_i \\ &= \left(x'_{k,b}x'_{k,b}\delta_{ij} - x'_{i,b}x'_{j,b}\right) \Omega_j \mathbf{e}_i\end{aligned} \tag{1.205}$$

Thus, definition (1.203) allows us to write (1.204) as:

$$\mathbf{M}' = \mathbf{J}\boldsymbol{\Omega} \tag{1.206}$$

Let us now consider what causes the motion of the rigid body. The energy of rotation E_{rot} adds to kinetic energy, hence to the Lagrangian. As to potential energy E_p, we already know it can only be a function of the positions of the constituents $\{\mathbf{r}_b\}$, and even of their mutual distances. As regards a rigid body, we have seen that these parameters come down to the pair $(\mathbf{R}, \boldsymbol{\theta})$. The body's Lagrangian then takes the following form:

$$L\left(\mathbf{R}, \mathbf{V}; \boldsymbol{\theta}, \boldsymbol{\Omega}\right) = \frac{1}{2}MV^2 + \frac{1}{2}\boldsymbol{\Omega}^T\mathbf{J}\boldsymbol{\Omega} - E_p\left(\mathbf{R}, \boldsymbol{\theta}, t\right) \tag{1.207}$$

As compared to the case of a particle, we can see that a rotation effect appears; it is formulated by the pair of state variables $(\boldsymbol{\theta}, \boldsymbol{\Omega})$. When writing (1.207), we have separated that pair from the first one (namely $(\mathbf{R}, \mathbf{V})$) by a semicolon in order to denote their mutual independence.[43] The generalized momentum corresponding to that parameter is written, according to the definition (1.25), by deriving the Lagrangian (1.207):

[43] The configuration space for the solid is then a six-dimensional space.

$$\begin{aligned}\mathbf{P}_{\boldsymbol{\theta}} &= \frac{\partial L}{\partial \boldsymbol{\Omega}} \\ &= \mathbf{J}\boldsymbol{\Omega} \\ &= \mathbf{M}'\end{aligned} \tag{1.208}$$

(we have used (1.206)). The body's Hamiltonian can then be written on the basis of (1.207) and (1.206), by analogy with (1.80):

$$H\left(\mathbf{P}, \mathbf{R}; \mathbf{M}', \boldsymbol{\theta}\right) = \frac{|\mathbf{P}|^2}{2M} + \frac{1}{2}\mathbf{M}'^T\mathbf{J}^{-1}\mathbf{M}' + E_p\left(\mathbf{R}, \boldsymbol{\theta}, t\right) \tag{1.209}$$

The state parameters of the body can be gathered under a vector looking like (1.99):

$$\mathbf{X}_S \doteqdot \begin{pmatrix} \mathbf{P} \\ \mathbf{R} \\ \mathbf{M}' \\ \boldsymbol{\theta} \end{pmatrix} \tag{1.210}$$

The equations of motion of a rigid body are obtained by applying the Lagrange equations (1.31) to each pair of parameters, which yields

$$\begin{aligned}\frac{d}{dt}\frac{\partial L}{\partial \mathbf{V}} &= \frac{\partial L}{\partial \mathbf{R}} \\ \frac{d}{dt}\frac{\partial L}{\partial \boldsymbol{\Omega}} &= \frac{\partial L}{\partial \boldsymbol{\theta}}\end{aligned} \tag{1.211}$$

With the Lagrangian (1.207), we get

$$\begin{aligned}M\frac{d\mathbf{V}}{dt} &= -\frac{\partial E_p}{\partial \mathbf{R}} \\ \mathbf{J}\frac{d\boldsymbol{\Omega}}{dt} &= -\frac{\partial E_p}{\partial \boldsymbol{\theta}}\end{aligned} \tag{1.212}$$

We are now going to calculate the right-hand sides of these equations. The derivative of potential energy with respect to the position of the centre of mass is written as

$$\frac{\partial E_p}{\partial \mathbf{R}} = \sum_b \frac{\partial \mathbf{r}_b}{\partial \mathbf{R}}\frac{\partial E_p}{\partial \mathbf{r}_b} \tag{1.213}$$

where $\partial \mathbf{r}_b / \partial \mathbf{R}$ is a matrix equalling the identity, since a rigid body is inherently undeformable. Thus, we get, with the definition of forces (1.55):

$$\begin{aligned}-\frac{\partial E_p}{\partial \mathbf{R}} &= \sum_b \mathbf{F}_b \\ &= \mathbf{F}\end{aligned} \tag{1.214}$$

In order to determine the derivative of potential energy with respect to the angular variables $\boldsymbol{\theta}$ occurring in (1.212), the simplest way consists of writing the variation of E_p according to the variation of $\boldsymbol{\theta}$. To do this, we may integrate the first line of (1.191) between two infinitesimally close time points:

$$d\mathbf{r}'_b = d\boldsymbol{\theta} \times \mathbf{r}'_b \tag{1.215}$$

We then perform the following calculation, which is based on definition (1.55), as well as (1.215) and the properties of the vector product:

$$\begin{aligned} dE_p &= \sum_b \frac{\partial E_p}{\partial \mathbf{r}'_b} \cdot d\mathbf{r}'_b \\ &= -\sum_b \mathbf{F}_b \cdot \left(d\boldsymbol{\theta} \times \mathbf{r}'_b\right) \\ &= -\left(\sum_b \mathbf{r}'_b \times \mathbf{F}_b\right) \cdot d\boldsymbol{\theta} \end{aligned} \tag{1.216}$$

Hence, we get

$$-\frac{\partial E_p}{\partial \boldsymbol{\theta}} = \sum_b \mathbf{r}'_b \times \mathbf{F}_b \doteqdot \mathbf{F}_\Omega \tag{1.217}$$

which defines the total moment $\mathbf{F}_\Omega$ with respect to the centre of mass. With (1.214) and (1.217), the equations of motion (1.212) take the ultimate form

$$\begin{aligned} M\frac{d\mathbf{V}}{dt} &= \mathbf{F} \\ \mathbf{J}\frac{d\boldsymbol{\Omega}}{dt} &= \mathbf{F}_\Omega \end{aligned} \tag{1.218}$$

Unsurprisingly, the first line of this system is identical to equation (1.184), provided with the total force exerted on the body (eqn (1.185)). That means the body is governed by a law of translation analogous to a simple particle. Its rotational motion is determined by the second line of equation (1.218), which could have been derived from (1.148), with (1.206), $\mathbf{J}$ being constant.[44] All things considered, we may state that the inertia tensor is related to $\boldsymbol{\Omega}$ just as mass is related to $\mathbf{V}$ and as linear momentum is related to angular momentum, as substantiated by (1.208). The moment of forces controls the rotation of the body just as the forces control its translation. Integrating equation (1.218) yields $\mathbf{V}$ and $\boldsymbol{\Omega}$ (remember that $\mathbf{J}$ is invertible) which then allows us to determine $\mathbf{R}$ and $\boldsymbol{\theta}$ owing to the second line of (1.182) and (1.189). The whole system of equations is also written as

$$\dot{\mathbf{X}}_S = \begin{pmatrix} \mathbf{F} \\ \mathbf{V} \\ \mathbf{F}_\Omega \\ \boldsymbol{\Omega} \end{pmatrix} \tag{1.219}$$

We will deal with a procedure for integrating (1.219) in Chapter 6 (Section 6.3.3).

[44] We may recall that the total angular momentum is zero just for an isolated system. Here we consider a solid which, on the contrary, is affected by its outer environment, for example a fluid in motion (for this case, refer to Section 6.3.3).

Note also that (1.148) is valid in the more general case of a body which is not necessarily rigid, but then $\mathbf{M}'$ is no longer governed by the simple law (1.206).

Lastly, note that, further on in this book, we will have opportunities to consider particular cases in which a fluid flows at least locally or temporarily in the same state of rotational motion of rigid body as (1.191). In this case, however, equation (1.218) cannot be applied, since the fluid motion will only adopt the state of a rigid body motion as a consequence of the particular processes taking place within the fluid (refer, in particular, to Chapters 2 and 3, Section 2.4.3 and 3.2.2).

1.5.2 Internal and external forces

We have seen that the particles that have been considered may be macroscopic bodies comprising a large number of atoms, and so each particle a can once again be divided into sub-particles b. In order to highlight the consequences of this fact upon the internal behaviour of such a body, we will now turn our attention to systems comprising sub-systems and discuss the effects of these sub-constituents on the overall system. Let us consider an isolated system comprising N_p particles a, which themselves are divided into small sub-systems (referred to as sub-particles), subscripted by letter b, and in a total amount of ℓ. Sub-particles may be the smallest constituents of the system, that is the molecules,[45] or else ensembles of molecules making up particles which are smaller than the original particles. We may generally assume (as experimentally confirmed) that the force experienced by a system component can be divided into different parts, each one representing the influence of another constituent on it.[46] In other words, we assume that each sub-particle b receives an individual force $\mathbf{F}_{b' \to b}$ from each other sub-particle b', the forces being additive. The equation of motion (1.56) then becomes, for the elementary constituents:

$$\forall b, \ \frac{d\mathbf{p}_b}{dt} = \sum_{b'=1}^{\ell} \mathbf{F}_{b' \to b} \tag{1.220}$$

This idea is clearly illustrated by the force resulting from the Lennard–Jones potential (1.46). In the case where the 'sub-particles' are the molecules, this force is exerted between two mutually interacting molecules α and β:

$$\mathbf{F}^{LJ}_{\beta \to \alpha} = \frac{E_0}{r_{ab}} \left[P_1 \left(\frac{r_0}{r_{ab}} \right)^{P_1} - 2P_2 \left(\frac{r_0}{r_{ab}} \right)^{P_2} \right] \mathbf{e}_{\alpha\beta} \tag{1.221}$$

where $\mathbf{e}_{\alpha\beta}$ is defined as in (1.58). In the case of particles being subject to gravity, formula (1.57) also highlights individual forces as given by

$$\mathbf{F}^{G}_{b \to a} = -G \frac{m_a m_b}{r_{ab}^2} \mathbf{e}_{ab} \tag{1.222}$$

This idea infers that the force potential too can be split into individual contributions involving a universal function of positions:

$$E_p\left(\{\mathbf{r}_a\}\right) = \sum_a E_p\left(\mathbf{r}_a\right) \tag{1.223}$$

[45] We already are on the margins of statistical mechanics, that is the behaviour of large systems, which is dealt with in Chapter 2. Nevertheless, we have decided to start on this subject as from Chapter 1, since the considerations we are going to develop now do not necessarily imply using the statistical tools of Chapter 2.

[46] This hypothesis is consistent with the additivity of forces, which was demonstrated in the previous section.

where E_p is some function. For each particle a, let us now add the equation (1.220) of motion of its constituents. The momentum is additive, so we get:

$$\forall a, \quad \frac{d\mathbf{p}_a}{dt} = \sum_{b'=1}^{\ell} \sum_{b\in a} \mathbf{F}_{b'\to b} \tag{1.224}$$

These forces may be divided into two groups: those exerted by the sub-particles b making up particle a on top of each other, and those exerted by the constituents of the other particles:

$$\begin{aligned} \sum_{b'=1}^{\ell} \mathbf{F}_{b'\to b} &= \sum_{b'\in a} \sum_{b\in a} \mathbf{F}_{b'\to b} + \sum_{b'\in a'\neq a} \sum_{b\in a} \mathbf{F}_{b'\to b} \\ &= \mathbf{F}_a^{int} + \mathbf{F}_a^{ext} \end{aligned} \tag{1.225}$$

where $\mathbf{F}_a^{int}$ and $\mathbf{F}_a^{ext}$ respectively denote the internal and external forces[47] with respect to particle a. Now, since the forces only depend on the distances between constituents, those exerted by the sub-particles making up a on top of each other would remain unchanged if that system was isolated. According to the law of action-reaction, their sum is then zero. Hence, the equation of motion of a sub-system becomes:

[47] In the case of a rigid body, as discussed in the previous section, the inner stresses are zero. Inner forces, however, nearly always occur within a seemingly solid body; we will calculate them in Chapter 3.

$$\forall a, \quad \frac{d\mathbf{p}_a}{dt} = \sum_{b'\in a'\neq a} \sum_{b\in a} \mathbf{F}_{b'\to b} = \sum_{a'\neq a} \mathbf{F}_{a'\to a} \tag{1.226}$$

where the force exerted by a' on a is defined by

$$\mathbf{F}_{a'\to a} = \sum_{b'\in a'\neq a}^{k} \sum_{b\in a} \mathbf{F}_{b'\to b} \tag{1.227}$$

We find that the law of motion remains unaltered with respect to the equation of a single particle, the *external* forces $\mathbf{F}_a^{ext}$ alone now exerting an influence on momentum. Apart from that detail, each sub-system is still governed by the Lagrange equations. This does not imply that the initially written Lagrange equations are now faulty, but that among the forces actually taking part in the motion of a system, the forces exerted among its constituents cancel each other out. The external forces, of course, may be written, starting from equation (1.55), as the gradient of a potential $E_{ext}(\{\mathbf{r}_a\}, t)$, depending on the coordinates of all the system particles (excluding the sub-particles), and possibly on time for a non-isolated system:

$$\mathbf{F}_a^{ext} = -\frac{\partial E_{ext}}{\partial \mathbf{r}_a} \tag{1.228}$$

The equation of motion is now

$$\forall a, \quad \frac{d\mathbf{p}_a}{dt} = \mathbf{F}_a^{ext} \tag{1.229}$$

As before, an isolated system ($\mathbf{F}^{ext} = \mathbf{0}$) is governed by the generalized inertia law as discussed in the previous paragraph. Similarly, it is obvious that only the external forces now occur in the Hamilton equation (1.100). It could likewise be shown that the derivative of the moment equals the sum of the moments of the external forces:

$$\forall a, \quad \frac{d\mathbf{M}_a}{dt} = \mathbf{r}_a \times \mathbf{F}_a^{ext} \tag{1.230}$$

once again, we get the conservation of angular momentum (1.149) for an isolated system. In the case where the external forces come down to gravity, the momentum for the whole system is written in a simple way, through definitions (1.57) and (1.182):

$$\begin{aligned}\mathbf{M}^g &= \sum_b \mathbf{r}_b \times \mathbf{F}_b^g \\ &= \sum_b \mathbf{r}_b \times m_b \mathbf{g} \\ &= M\mathbf{R} \times \mathbf{g}\end{aligned} \tag{1.231}$$

For a rigid body, the system (1.218) gives

$$\begin{aligned}M\frac{d\mathbf{V}}{dt} &= \sum_b \mathbf{F}_b^{ext} \\ \mathbf{J}\frac{d\boldsymbol{\Omega}}{dt} &= \sum_b \mathbf{r}_b \times \mathbf{F}_b^{ext}\end{aligned} \tag{1.232}$$

1.5.3 Internal energy

Just like linear and angular momenta, energy takes various forms in various frames of reference, as did the kinetic energy of a rigid body in Section 1.5.1. Referring to the same notations, we may deal with the total energy of a particle which is considered as being made up of sub-particles b. To do this, we use the division (1.225) to split the potential energy into two components, one—denoted $E_{p,int}$ (internal potential)—concerning the interactions among the system constituents, whereas the other—denoted E_{ext} (external potential)—represents the possible interactions with an external system; thus, potential energy is written as:

$$E_p\left(\{\mathbf{r}_a\}\right) = E_{p,int}\left(\{\mathbf{r}_a\}\right) + E_{ext}\left(\{\mathbf{r}_a\}, t\right) \tag{1.233}$$

somewhat as in (1.225). Only E_{ext} can explicitly depend on time if the system is not isolated. In the new frame of reference, using the notations of Section 1.5.1, we get:

$$
\begin{aligned}
E' &= \sum_b \frac{1}{2} m'_b u'^2_b + E_{p,int}\left(\left\{\mathbf{r}'_b\right\}\right) + E_{ext}\left(\{\mathbf{r}_a\}, t\right) \\
&= \sum_b \frac{1}{2} m_b \left(u_b^2 + 2\mathbf{u}_b \cdot \mathbf{V} + V^2\right) + E_{p,int}\left(\left\{\mathbf{r}'_b\right\}\right) + E_{ext}\left(\{\mathbf{r}_a\}, t\right) \quad (1.234) \\
&= E + \mathbf{V} \cdot \mathbf{P} + \frac{1}{2} M V^2 + E_{p,int}\left(\{\mathbf{r}_a\}, t\right)
\end{aligned}
$$

(we have used the fact that potential energy only depends on the distances between sub-particles, which are not altered by a change of frame). Let us now consider the case in which the sub-particles coincide with the molecules making up the system. The energy of the system in its inertial frame of reference is then called *internal energy* and denoted E_{int}. It lumps the potential energy $E_{p,int}$ of the system inner forces[48] and the relative motions of its constituents together in the whole system's inertial frame of reference:

$$E_{int}\left(\{\mathbf{r}_a\}\right) \doteqdot \sum_b \frac{1}{2} m_b u_b^2 + E_{p,int}\left(\{\mathbf{r}_a\}\right) \tag{1.235}$$

Velocities $\{\mathbf{u}_b\}$ here being defined in the inertial frame of reference, and the inner forces being unaltered by a change of frame, internal energy, unlike kinetic energy, has the form of a Galilean invariant. We find that total energy E may be related to E_{int} by:

$$E = \frac{1}{2} M V^2 + E_{int} + E_{ext} \tag{1.236}$$

Noting that the first term in the right-hand side of (1.236) is the kinetic energy of the whole system, we see that internal energy may be considered as a form of potential energy; hence it is an energy that should be taken into account in the equations of the overall motion of a system being made up of sub-systems. The potential energy of a particle can then be written as:

$$E_{p,a}\left(\{\mathbf{r}_b\}, t\right) = E_{int,a}\left(\{\mathbf{r}_b\}\right) + E_{ext}\left(\{\mathbf{r}_b\}, t\right) \tag{1.237}$$

Applying this argument to each particle a of a system, taking the division (1.223) into account and considering that energy is additive, the Lagrangian of a system of particles which themselves are made up of molecules in motion is ultimately given by

$$L\left(\{\mathbf{r}_a\}, \{\mathbf{u}_a\}, t\right) = \sum_a \frac{1}{2} m_a u_a^2 - \sum_a E_{int,a}\left(\{\mathbf{r}_b\}\right) - \sum_a E_{ext}\left(\mathbf{r}_a, t\right) \tag{1.238}$$

This is the most general form of the Lagrange function of a system of discrete interacting particles within the framework of classical mechanics.

Like the potential of outer forces E_{ext}, the form of internal energies $E_{int,a}$ depends on the positions of its constituents. Since $E_{int,a}$ characterizes a whole particle, however, one may consider it may come down to a function of its experimentally accessible physical properties which results in specific state variables for each particle a; the variables are as yet unknown and will be

[48] In the particular case where the system sub-constituents are the molecules, $E_{p,int}$ represents the Lennard–Jones potential of forces (1.46) and the potential of mutual gravity forces (1.44) which, however, are much lower than them. Both potentials depend on the intermolecular distances, then on their positions $\{\mathbf{r}_a\}$. If these molecules are polyatomic, their rotational and vibrational energies would still have to be added to them.

If the system sub-constituents are macroscopic particles, possible motions at intermediate scales between those of atoms and particles a (turbulent fluctuations within a fluid, for instance) would result in an additional kinetic energy which could also be treated as a form of internal energy, but which is usually preferably kept as an independent variable (refer to Chapter 4 about this matter).

denoted $Z_{a,i}$. We will provide details about these internal variables in the next section and in Chapter 2 (Section 2.3.2):

$$E_{int,a} = E_{int}\left(\left\{Z_{a,i}\right\}_i\right) \tag{1.239}$$

where E_{int} is a so far undetermined function (notation $\left\{Z_{a,i}\right\}_i$ means that the curly braces enclose all the parameters $Z_{a,i}$ for each value of i, the particle label a remaining fixed). The Lagrangian (1.238) yields the equation of motion as

$$\begin{aligned}\forall a, \quad \frac{d\mathbf{p}_a}{dt} &= -\frac{\partial E_{int,a}}{\partial \mathbf{r}_a} - \frac{\partial E_{ext}}{\partial \mathbf{r}_a} \\ &= \mathbf{F}_a^{int} + \mathbf{F}_a^{ext} = \mathbf{F}_a\end{aligned} \tag{1.240}$$

Within this context, we must consider that $E_{int,a}$ depends on position $\mathbf{r}_a$ through the internal variables $\left\{Z_{a,i}\right\}_i$ (the relevant process will be discussed in Chapters 2 and 3). In this form, it appears that the internal energy gives rise to the presence of a new inner force. In accordance with the discussion in the previous section, this force vanishes for an isolated system, whereas it is usually (for a system interacting with the outside) non-zero, since it happens to be an effect of the motions of external particles, which influence the position of a. In that respect, and although if is fitted with the superscript ^{int} meaning 'internal', one should be aware it is some kind of external force, since it results from processes taking place outside the particle that is involved in them.

Total energy (1.72) henceforth reads as

$$E\left(\{\mathbf{r}_a\}, \{\mathbf{u}_a\}, t\right) = \sum_a \frac{1}{2} m_a u_a^2 + \sum_a E_{int}\left(\left\{Z_{a,i}\right\}_i\right) + \sum_a E_{ext}\left(\mathbf{r}_a, t\right) \tag{1.241}$$

and the Hamiltonian (1.79) becomes

$$H\left(\{\mathbf{p}_a\}, \{\mathbf{r}_a\}, t\right) = \sum_a \frac{|\mathbf{p}_a|^2}{2m_a} + \sum_a E_{int}\left(\left\{Z_{a,i}\right\}_i\right) + \sum_a E_{ext}\left(\mathbf{r}_a, t\right) \tag{1.242}$$

For an isolated system, the law of evolution of the total kinetic energy (1.74) remains unchanged, the power being from now on exerted by the total force $\mathbf{F}_a^{int} + \mathbf{F}_a^{ext}$:

$$\begin{aligned}\frac{dE_k}{dt} &= \sum_a \mathbf{F}_a^{int} \cdot \mathbf{u}_a + \sum_a \mathbf{F}_a^{ext} \cdot \mathbf{u}_a \\ &= P^{int} + P^{ext}\end{aligned} \tag{1.243}$$

As regards the total potential energy, it complies with the law

$$\begin{aligned}\frac{dE_{ext}}{dt} &= \sum_a \frac{\partial E_{ext}}{\partial \mathbf{r}_a} \cdot \frac{d\mathbf{r}_a}{dt} \\ &= -\sum_a \mathbf{F}_a^{ext} \cdot \mathbf{u}_a \\ &= -P^{ext}\end{aligned} \tag{1.244}$$

If both kinetic and potential energies change over time, the conservation of total energy shows that the total internal energy should change in the same proportion, with an opposite sign. In the absence of outer forces, we then get:

$$\begin{aligned}\frac{dE_{int}}{dt} &= -\frac{dE_k}{dt} - \frac{dE_{ext}}{dt} \\ &= -P_{int} \\ &= -\sum_a \mathbf{F}_a^{int} \cdot \mathbf{u}_a\end{aligned} \tag{1.245}$$

All these formulas mean that, within the context of composite systems, there are exchanges between the kinetic, potential and internal energies. The signs of the relevant flows, that is the powers of forces, are undetermined, but their values offset each other.

We observe, from (1.241), that each particle has its own energy:

$$\begin{aligned}E_a &= \frac{1}{2} m_a u_a^2 + E_{int}\left(\left\{Z_{a,i}\right\}_i\right) + E_{ext}\left(\mathbf{r}_a, t\right) \\ &= E_{k,a} + E_{int,a} + E_{ext,a}\end{aligned} \tag{1.246}$$

The energy attached to each particle changes in accordance with laws that are much like those of a whole system. If the particles move about, during the period of time dt, along a set of infinitesimal vectors $\{d\mathbf{r}_a = \mathbf{u}_a dt\}$, calculations analogous to (1.243) and (1.244) show that the corresponding variations of energy are written as

$$\begin{aligned}dE_{k,a} &= \left(\mathbf{F}_a^{int} + \mathbf{F}_a^{ext}\right) \cdot d\mathbf{r}_a \\ dE_{ext,a} &= -\mathbf{F}_a^{ext} \cdot d\mathbf{r}_a\end{aligned} \tag{1.247}$$

These quantites are referred to as *works* of the internal and external forces. By adding them, we find

$$dE_{k,a} + dE_{ext,a} = \mathbf{F}_a^{int} \cdot d\mathbf{r}_a \tag{1.248}$$

Relations (1.247) are analogous to those of an isolated system, although an arbitrary particle is a priori an open system. A particle exchanging forces with its surroundings is not governed, however, by the law of conservation of energy, since it is not an isolated system. Actually, we may write

$$\begin{aligned}dE_a &= dE_{k,a} + dE_{int,a} + dE_{ext,a} \\ &= dE_{int,a} + \mathbf{F}_a^{int} \cdot d\mathbf{r}_a\end{aligned} \tag{1.249}$$

Now, it is noteworthy that formula (1.245) does not allow us to infer $dE_{int,a} = -\mathbf{F}_a^{int} \cdot d\mathbf{r}_a$ for each particle of a system, since the internal energy depends on the so far unknown state variables $\left\{Z_{a,i}\right\}_i$. If such was the case, then the energy of each particle would remain constant, according to (1.249), which only holds true for an isolated particle ($dE_a = 0$). Thus, the state variables account for the interparticle energy transfer. In the next paragraph, we will see how the variation $dE_{int,a}$ can be written within a simplified framework, since we will discuss the general case in Chapter 2 (Section 2.3.2). On the other hand, we may write, owing to (1.245):

$$\sum_a dE_{int,a} = -\sum_a \mathbf{F}_a^{int} \cdot d\mathbf{r}_a \tag{1.250}$$

1.5.4 Pressure and density

From the arguments in Section 1.5.3, we can infer the form of the force which a particle should receive per unit area due to its internal energy $E_{int,a}$. In the absence of other (external) forces, this energy plays the role of a potential from which the force being sought arises (refer to eqn (1.240)):

$$\mathbf{F}_a^{int} = -\frac{\partial E_{int,a}}{\partial \mathbf{r}_a} \tag{1.251}$$

This force should be understood as resulting from a possible variation of its internal energy under the effect of a gentle virtual variation of its position. Now, we have seen that the evolution of internal energy takes place through the motions of the various particles, which *in fine* affect their mutual positions, that is their respective volumes. If the particle volume is denoted V_a, we get:

$$\mathbf{F}_a^{int} = -\frac{\partial E_{int,a}}{\partial V_a}\frac{dV_a}{d\mathbf{r}_a} \tag{1.252}$$

The derivative of volume with respect to a virtual displacement $d\mathbf{r}_a$ is obviously the surface Σ_a of the particle as seen from the virtual point towards which $d\mathbf{r}_a$, or more precisely $\Sigma_a \mathbf{n}_a$, where $\mathbf{n}_a$ is the unit vector as directed by $d\mathbf{r}_a$. Hence, the received force is normal to the surface and proportional to its area, the factor of proportionality being denoted p_a and referred to as the particle *pressure*:[49]

$$\frac{\left|\mathbf{F}_a^{int}\right|}{\Sigma_a} = p_a \doteqdot -\frac{\partial E_{int,a}}{\partial V_a} \tag{1.253}$$

Pressure is defined as a force per unit area, exerted due to the internal energy, that is the motion of molecules making up the particles. Given its definition, the internal energy and the volume being Galilean invariants, pressure also verifies that property.

[49] We hope that since pressure p is a scalar quantity, it will not be mistaken for the momentum $\mathbf{p}$ of a particle. Besides, the generalized momenta will still be fitted with subscripts i, j, etc.

Let us now introduce the internal energy per unit mass $e_{int,a}$ and density ρ_a by

$$\begin{aligned} e_{int,a} &\doteqdot \frac{E_{int,a}}{m_a} \\ \rho_a &\doteqdot \frac{m_a}{V_a} \end{aligned} \tag{1.254}$$

By definition, density is positive. Mass conservativity immediately implies

$$\frac{dV_a}{V_a} = -\frac{d\rho_a}{\rho_a} \tag{1.255}$$

Equation (1.253) is then transformed to yield

$$p_a = \rho_a^2 \frac{\partial e_{int,a}}{\partial \rho_a} \tag{1.256}$$

Hence, the specific internal energy $e_{int,a}$ should be considered with respect to both position and density ρ_a, which is then one of the state variables $\left\{Z_{a,i}\right\}_i$ mentioned in the previous paragraph:

$$e_{int,a} = e_{int}\left(\rho_a\right) \tag{1.257}$$

The volume of a particle, however, depends on the positions of the surrounding particles, and so internal energy happens to be an implicit function of the positions of all the particles, in accordance with what was written in the previous section:

$$e_{int,a} = e_{int,a}\left(\{\mathbf{r}_b\}\right) \tag{1.258}$$

The way the positions affect the densities will be discussed in Chapter 3 within the context of continua, then in Chapter 5 in SPH method formalism. As in Section 1.5.2, the external force exerted on a particle is then written in the form of a sum of contributions from the various particles:

$$\mathbf{F}_a^{int} = \sum_b \frac{\partial E_{int,a}}{\partial \mathbf{r}_b} = \sum_b \mathbf{F}_{b\to a}^{int} \tag{1.259}$$

Relation (1.253) shows that the action of a pressure changes the internal energy to the extent of $-p_a dV_a$. Developing the internal energy according to its variable starting from (1.257), we then get

$$\begin{aligned} dE_{int,a} &= \frac{\partial E_{int,a}}{\partial \rho_a} d\rho_a \\ &= -p_a dV_a \end{aligned} \tag{1.260}$$

that is with (1.249):

$$dE_a = -p_a dV_a - \mathbf{F}_a^{int} \cdot d\mathbf{r}_a \tag{1.261}$$

This equation shows that the relation of pressure to volume is the same as that of force to position. This close relation is formulated by stating that pressure is thermodynamically conjugate to volume. As a rule, by analogy with

(1.253), the conjugate variable[50] X_a of every variable x_a attached to a particle a is defined by

$$X_a \doteq \frac{\partial E_a}{\partial x_a} \qquad (1.262)$$

[50]By analogy with the forces F_i introduced in Section 1.3.3, the X_a are sometimes referred to as 'generalized forces'.

With that formalism, the variation of energy (1.261) is written in the general form

$$\begin{aligned} dE_a &= \sum_i \frac{\partial E_a}{\partial x_{a,i}} dx_{a,i} \\ &= \sum_i X_{a,i} dx_{a,i} \end{aligned} \qquad (1.263)$$

The quantity $X_a dx_a$ is known as *generalized work* associated with the variation of x_a. Relation (1.71), which is analogous to (1.262), shows, for instance, that momenta p_i and positions q_i are reciprocally conjugate pairs.

Let us return to relation (1.261). For an isolated particle ($dE_a = 0$), $-p_a dV_a$ is then nothing but the work of internal forces, that is $\mathbf{F}_a^{int} \cdot d\mathbf{r}_a$. For an ensemble of isolated particles ($dE = 0$), we will then have

$$\sum_a \mathbf{F}_a^{int} \cdot d\mathbf{r}_a = -\sum_a p_a dV_a \qquad (1.264)$$

The forces at work within such a system presently depend on the density, since internal energy $e_{int,a}$ depends on ρ_a. With (1.258), the Hamiltonian (1.242) should then be arranged as

$$H(\{\mathbf{p}_a\}, \{\mathbf{r}_a\}, \{\rho_a\}, t) = \sum_a \frac{|\mathbf{p}_a|^2}{2m_a} + \sum_a m_a e_{int}(\rho_a) + \sum_a E_{ext}(\mathbf{r}_a, t) \qquad (1.265)$$

This is positively a system whose each particle a from now on will have an additional parameter (or state variable), namely ρ_a. We are going to see what the effects are within the context of the Hamilton equations. Each particle is now characterized by the state vector

$$\mathbf{Y}_a \doteq \begin{pmatrix} \mathbf{p}_a \\ \mathbf{r}_a \\ \rho_a \end{pmatrix} \qquad (1.266)$$

which supersedes the vector $\mathbf{X}_a$ as defined by (1.99). Hence, in all such motion-related equations as the Hamilton equations (1.90), the partial derivative of the Hamiltonian with respect to the position should be replaced by

$$\frac{\partial H}{\partial \mathbf{r}_a} + \sum_b \frac{\partial H}{\partial \rho_b} \frac{\partial \rho_b}{\partial \mathbf{r}_a} \qquad (1.267)$$

Besides, the density of a particle changes over time. Since ρ_a is a function of the $\{\mathbf{r}_b\}$, its time derivative may be written as

$$\forall a,\ \dot{\rho}_a \doteqdot \frac{d\rho_a}{dt} = \sum_b \frac{\partial \rho_a}{\partial \mathbf{r}_b} \cdot \frac{d\mathbf{r}_b}{dt} = \sum_b \frac{\partial \rho_a}{\partial \mathbf{r}_b} \cdot \mathbf{u}_b \tag{1.268}$$

Thus, the Hamilton equations are now written as

$$\forall a,\ \dot{\mathbf{p}}_a = -\frac{\partial H}{\partial \mathbf{r}_a} - \sum_b \frac{\partial \rho_b}{\partial \mathbf{r}_a}\frac{\partial H}{\partial \rho_b} \qquad \dot{\mathbf{r}}_a = \frac{\partial H}{\partial \mathbf{p}_a} \qquad \dot{\rho}_a = \sum_b \frac{\partial \rho_a}{\partial \mathbf{r}_b} \cdot \frac{\partial H}{\partial \mathbf{p}_b} \tag{1.269}$$

Using the relations (1.53), (1.240), (1.254) and (1.256), the definition (1.265) of the Hamiltonian makes it possible to write its derivative with respect to the state vector $\mathbf{Y}_a$:

$$\frac{\partial H}{\partial \mathbf{Y}_a} = \begin{pmatrix} \mathbf{u}_a \\ -\mathbf{F}_a \\ \dfrac{V_a p_a}{\rho_a} \end{pmatrix} \tag{1.270}$$

Equation (1.269) then take a matrix form which is analogous to (1.101) (refer to Section 1.4.2):

$$\forall a,\ \dot{\mathbf{Y}}_a = \sum_b (\boldsymbol{\omega}_7)_{ab} \frac{\partial H}{\partial \mathbf{Y}_b} \tag{1.271}$$

where a 7×7 matrix $(\boldsymbol{\omega}_7)_{ab}$ has been introduced for each pair of particles (a, b) as given by

$$(\boldsymbol{\omega}_7)_{ab} \doteqdot \begin{pmatrix} [\mathbf{0}_3] & -\delta_{ab}[\mathbf{I}_3] & -\dfrac{\partial \rho_b}{\partial \mathbf{r}_a} \\ \delta_{ab}[\mathbf{I}_3] & [\mathbf{0}_3] & \mathbf{0} \\ \left(\dfrac{\partial \rho_a}{\partial \mathbf{r}_b}\right)^T & \mathbf{0}^T & 0 \end{pmatrix} \tag{1.272}$$

(here we have highlighted the matricial nature of the zero and identity matrices $\mathbf{0}_3$ and $\mathbf{I}_3$ by inserting them between square brackets in order to distinguish them from the zero vector $\mathbf{0}$, for example). Unlike the matrix $\boldsymbol{\omega}_6$ used in Section 1.4.2 (eqn (1.102)), $(\boldsymbol{\omega}_7)_{ab}$ contains terms which depend on the particles a and *b*. The Hamiltonian's derivative with respect to time is now written, by analogy with (1.103), as:

$$\begin{aligned}\frac{dH}{dt} &= \sum_a \frac{\partial H}{\partial \mathbf{Y}_a} \cdot \dot{\mathbf{Y}}_a \\ &= \sum_{a,b} \left(\frac{\partial H}{\partial \mathbf{Y}_a}\right)^T (\boldsymbol{\omega}_7)_{ab} \frac{\partial H}{\partial \mathbf{Y}_b}\end{aligned} \tag{1.273}$$

The conservation of energy, as in Section 1.4.1, is then ensured by the property of skew-self-adjunction which immediately appears in the definition (1.272), that is

$$(\boldsymbol{\omega}_7)_{ba} = -(\boldsymbol{\omega}_7)_{ab}^T \tag{1.274}$$

As compared with formula (1.94), it is worth noting that the transposition involves not only the matrix subscripts, but also the particle labels.

As in Section 1.4.2, we could lump the state variables of the whole system together by defining, by analogy with (1.104):

$$\mathbf{Y} \doteqdot \begin{pmatrix} \{\mathbf{p}_a\} \\ \{\mathbf{r}_a\} \\ \{\rho_a\} \end{pmatrix} \tag{1.275}$$

A condensed form of the Hamilton equations then appears:

$$\dot{\mathbf{Y}} = \boldsymbol{\Omega}_{7N_p} \frac{\partial H}{\partial \mathbf{Y}} \tag{1.276}$$

with a $7N_p \times 7N_p$ matrix (N_p being, as previously explained, the number of particles) as given by

$$\boldsymbol{\Omega}_{7N_p} \doteqdot \begin{pmatrix} [\mathbf{0}] & -\left[\mathbf{I}_{3N_p}\right] & -\left[\dfrac{\partial \rho_b}{\partial \mathbf{r}_a}\right]_{a,b} \\ \left[\mathbf{I}_{3N_p}\right] & [\mathbf{0}] & [\mathbf{0}] \\ -\left[\dfrac{\partial \rho_a}{\partial \mathbf{r}_b}\right]_{a,b}^T & [\mathbf{0}] & [\mathbf{0}] \end{pmatrix} \tag{1.277}$$

and the immediate properties of skew-self-adjunction, and hence of conservation of energy. Note that this proposed formalism has nothing more than that in Section 1.4.2, where the forces are exclusively written as a function of the positions, since the density itself is a function of the latter via (1.258). However, it makes it possible to emphasize the role of pressure as a potential of the internal forces. We will see a continuous form of it in Chapter 3 (Section 3.6.3), which will clarify the role of pressure loads linked to mass conservation.

2 Statistical mechanics

2.1 Introduction

Sections 1.5.3 and 1.5.4 in Chapter 1 demonstrate that large systems need to be dealt with separately. We will dedicate this chapter to this and consider the usual kinematic quantities from a macroscopic point of view. The effects of the microscopic matter constituents (the molecules) will then result in additional variables (thermodynamical state variables), density being an example. We will see that, because of this approach, extra forces expressing the microscopic effects on a macroscopic scale should be introduced.

Since Boltzmann's pioneering works (for example, one may read the 1964 US issue), the study of large system behaviour, which is part of statistical physics (or thermodynamics), has been plentifully documented and has such a broad scope that it cannot be dealt with in one chapter. In order to provide a transition from the notions of Chapter 1 (about discrete particle systems) to the reflections in Chapters 3 and 4 (about continuous media, particularly fluids), however, one has, even incompletely, to consider thermodynamics. Since we cannot review all the 'tricks' of statistical mechanics, we will merely outline its main principles, laying stress, however, on the most important items, namely the thermodynamical quantities governing fluid dynamics. The notion of energy dissipation will also be a core topic throughout this chapter. We will intentionally mainly restrict our discussions to the case of gases, for which simple statistical laws can be formulated, assuming that most of the notions and laws being dealt with may, at least qualitatively, be extended to the case of liquids.

In this chapter, we will pay attention to a particle a comprising a very high number of sub-particles in the form of molecules or atoms (taken as classical objects). The values of the mechanical state parameters of each of these sub-systems are then barely predictable, and even practically unpredictible because of the very high number of degrees of freedom. A statistical approach is then the only feasible solution to assess the mechanical future of such bodies. Such an approach is made possible by the high number of involved molecules, the statistical fluctuations of the average quantities decreasing with that number N_m inversely as its square root, in accordance with the central limit theorem.

According to this approach, a particle is henceforth necessarily macroscopic. From now on, the position of a particle is then likened to that of its centre of mass, whereas its velocity is the overall velocity of its constituents (refer to Section 1.5.1). In order to prevent confusion between particles (macroscopic systems) and molecules (microscopic objects), we reserve the notations $\mathbf{r}_a$ and $\mathbf{u}_a$ for the former, the latter being fitted with Greek subscripts and tilded quantities (for instance $\tilde{\mathbf{r}}_\alpha$, $\tilde{\mathbf{u}}_\alpha$). We will sometimes have to return to the more general concepts of Lagrangian or Hamiltonian mechanics, in which case we will keep the notations p_i and q_i to describe the parameters of the relevant system. Sometimes as well, when we consider a given particle, we will merely use subscriptless notations for a clearer formulation.

In this chapter, we will employ the notion of probability densities for a molecule and for a whole system, which will make it possible to set up Liouville and Boltzmann's equations from the elements of Chapter 1. The notion of entropy will be clearly defined, then we will explain the second law of thermodynamics through Boltzmann's equation for a gas. Our discussion will then lead to a definition of the Bolzmann statistical distributions in equilibrium and out of equilibrium, then to the concept of heat and work. These elements will advantageously be used to set the principle of state equation. The case of out-of-equilibrium systems will then be thoroughly explored, introducing the notion of kinetic coefficients, then of dissipative forces. We will apply these ideas to the modelling of a damped pendulum, then of particle friction. This will enable us to complete the equations of Chapter 1, through the dissipative Liouville and Hamilton equations (for the macroscopic systems). Lastly, two specific techniques for irreversible process modelling will be outlined, namely the concept of mirror systems, then the Langevin and Fokker–Planck equations.

2.2 Statistical behaviour of large systems

This section aims at introducing the fundamentals of the mechanics of large systems, or composite systems with many elements. The probability density functions are introduced in the so-called phase space, then the Liouville theorem is demonstrated. We then introduce the concept of entropy and prove that it always increases for an isolated system of particles forming a gaseous fluid.

2.2.1 Probability densities

Since we investigate systems with a very high number of parameters in this chapter, we shall use the notion of probability density for various quantities. Although the equations of mechanics (more specifically the Hamilton equations) are perfectly deterministic, if the system has a very high number of degrees of freedom, we cannot get a thorough knowledge of its initial physical state though measurements. In addition, if, for example, we consider a collection of a very high number of molecules making up a fluid, the mutual collisions they experience gradually alter their respective momenta, and so a small initial error increases indefinitely over time.[1] Accordingly, the accurate system state (i.e. the accurate knowledge of both coordinates and momenta of all its individual constituents) has no desirable usefulness or meaning. Lastly,

[1] For further details about that kind of process, refer to Section 4.2.2 in Chapter 4.

even though we could get a perfect knowledge of such a complex system as a fluid as regards the detailed motion of all its molecules, simultaneously treating such plentiful information would unquestionably be beyond our reach and would probably be of little interest. We must then contemplate describing its state in a statistical way,[2] by considering the probability density which is associated with the presence of a system at a given point in the phase space. In this chapter, we will then investigate the way the probability that the system constituents have induced p_i and q_i values over time; practically, however, only the mean value of those quantities, as well as all the mean quantities they will enable us to construct, will be attractive for us from a physical point of view. This is what should be understood as *statistical mechanics*.

[2]We may recall that the famous, much debated classical Laplacian derminism, was set out for the first time in 1814 in Laplace's book dealing with the theory of probability. Laplace only formulated his idea for better emphasizing that it was practically useless: 'The curve described by a simple molecule of air or vapour is regulated in a manner just as certain as the planetary orbits; the only difference between them is that which comes from our ignorance. Probability is relative, in part to this ignorance, in part to our knowledge' (translated from French by T. Pain).

We will denote the aforementioned probability density function as $\Phi\left(\{p_i\},\{q_i\},t\right)$. Accordingly,

$$\Phi\left(\{p_i\},\{q_i\},t\right)\prod_i dp_i dq_i \tag{2.1}$$

accounts, at time t, for the probability for the point representing the system in the phase space to occur within the volume of that $\{p_i\}$, $\{q_i\}$-centred point and be measured as equal to the product of the $dp_i dq_i$.

When the system consists of N_m molecules in relative motion, the probability density $\Phi\left(\{\tilde{\mathbf{p}}_\alpha\},\{\tilde{\mathbf{r}}_\alpha\},t\right)$ may, of course, be defined as well. It is, however, more convenient to assume the probability $\varphi\left(\tilde{\mathbf{p}},\tilde{\mathbf{r}},t\right)$ that an individual molecule occurring in $\tilde{\mathbf{r}}$ at time t has a momentum $\tilde{\mathbf{p}}$. More precisely,

$$\varphi\left(\tilde{\mathbf{p}},\tilde{\mathbf{r}},t\right)d\tilde{\mathbf{p}} \tag{2.2}$$

accounts for the average number of molecules, in place $\tilde{\mathbf{r}}$ and at time t being considered,[3] having a momentum within the range $\left[\tilde{\mathbf{p}},\tilde{\mathbf{p}}+d\tilde{\mathbf{p}}\right]$. This function should be unsderstood as being defined over the set $\Omega_p \times \Omega_r = \mathbb{R}^6$ of possible momenta and positions. The average number of molecules in one point (per unit volume) is obtained by integrating that probability density over all the possible values of the momentum vector:

[3]More precisely, it should thereby be understood that the number of molecules occurring within the $\tilde{\mathbf{r}}$-centred volume $d\tilde{\mathbf{r}}$ between times t and $t+dt$ equals $\varphi\left(\tilde{\mathbf{p}},\tilde{\mathbf{r}},t\right)d\tilde{\mathbf{p}}d\tilde{\mathbf{r}}dt$. $d\tilde{\mathbf{p}}$ should be understood as an element of integration 'volume' within the momentum space, which should strictly be denoted $d^3\tilde{\mathbf{p}}$. Likewise, $d\tilde{\mathbf{r}}$ is sized like a three-dimensional volume.

$$n\left(\tilde{\mathbf{r}},t\right)=\int_{\Omega_p}\varphi\left(\tilde{\mathbf{p}},\tilde{\mathbf{r}},t\right)d\tilde{\mathbf{p}} \tag{2.3}$$

where Ω_p denotes the space of all the possible momenta, that is theoretically $\mathbb{R}^3$ or a subset of that space. More generally, the statistical average of a quantity A, at a given point in space and a given time point is written as

$$\langle A\rangle\left(\tilde{\mathbf{r}},t\right)=\frac{1}{n\left(\tilde{\mathbf{r}},t\right)}\int_{\Omega_p}A\varphi\left(\tilde{\mathbf{p}},\tilde{\mathbf{r}},t\right)d\tilde{\mathbf{p}} \tag{2.4}$$

The distribution functions should, of course, be normalized:

$$\begin{gathered}\int_{\Omega_p\times\Omega_r}\Phi\left(\{p_i\},\{q_i\},t\right)\prod_i dp_i dq_i = 1\\ \int_{\Omega_p\times\Omega_r}\varphi\left(\tilde{\mathbf{p}},\tilde{\mathbf{r}},t\right)d\tilde{\mathbf{p}}d\tilde{\mathbf{r}} = N_m\end{gathered} \tag{2.5}$$

If one considers a fluid particle from a macroscopic point of view, it can be said that its velocity $\mathbf{u}_a(t)$ is defined by a mean velocity $\tilde{\mathbf{p}}/m$ of the molecules it contains, m being the mass of molecules:[4]

$$\begin{aligned}\mathbf{u}_a(t) &= \frac{1}{m}\int_{\Omega_p\times\Omega_a} \tilde{\mathbf{p}}\varphi\left(\tilde{\mathbf{p}},\tilde{\mathbf{r}},t\right)d\tilde{\mathbf{p}}d\tilde{\mathbf{r}} \\ &= \int_{\Omega_a} n\left(\tilde{\mathbf{r}},t\right)\langle\tilde{\mathbf{u}}\rangle\left(\tilde{\mathbf{r}},t\right)d\tilde{\mathbf{r}}\end{aligned} \tag{2.6}$$

[4]It will be assumed hereinafter that the molecules are identical. In particular, they all have the same mass, though a more general approach could be set up without further difficulties.

where Ω_a denotes the part of space which is occupied by the particle. That holds true as well for every quantity; for instance, the energy of a particle may be defined by

$$E_a(t) = \int_{\Omega_p\times\Omega_a} \tilde{E}\left(\tilde{\mathbf{p}},\tilde{\mathbf{r}}\right)\varphi\left(\tilde{\mathbf{p}},\tilde{\mathbf{r}},t\right)d\tilde{\mathbf{p}}d\tilde{\mathbf{r}} \tag{2.7}$$

where $\tilde{E}$ denotes the energy of a molecule. We may mention that since the particle has a well determined identity, the relevant quantities, in particular velocity, now only depend on time, in accordance with Lagrangian formalism (the spatial dependence is implicitly included in the particle label a). Its position $\mathbf{r}_a(t)$ may be defined as the centre of the mass of the molecules it contains, in accordance with Section 1.5.1. This approach, however, is ambiguous, so a remark needs be made regarding the choice of the 'macroscopic scale'. The choice of the characteristic size of those entities which we refer to as 'particles' is not the only one and, strictly speaking, is defined in a rather vague manner. One may think that this choice is appropriate if it does not affect the statistical treatment, that is to say if the average quantities are fairly insensitive to the choice of particle scale.[5] As a last resort, however, one had better stick to the definition (2.4) of the statistical quantities, based on function φ, which shows how important it is for macroscopically describing a fluid. We will resume these arguments in Chapter 3, particularly in Section 3.5.

[5]This 'macroscopic' point of view is sometimes known as *coarse-graining* in scientific literature.

In any case, two scales, and hence two categories of parameters describing the fluid, are to be distinguished: on the one hand those molecules described by the $\left(\{\tilde{\mathbf{p}}_\alpha\},\{\tilde{\mathbf{r}}_\alpha\}\right)$, on the other hand those particles described by the $\left(\{\mathbf{p}_a\},\{\mathbf{r}_a\}\right)$ (or, as well, the $\{\mathbf{X}_a\}$, using again the notation (1.99)). Using once again the generalized notations $(\{p_i\},\{q_i\})$, their averages could be defined $(\{P_i\},\{Q_i\})$ in the same way.

2.2.2 Liouville's theorem. Boltzmann's equation

In the first place, we will consider a quite general Hamiltonian system as defined by its generalized coordinates (refer to Section 1.4.2). In the course of time, the point representing the system state evolves in the phase space along with the Hamilton equations which represent the respective rates of change of the p_i and q_i. According to the selected initial condition, we then obtain a set of trajectories representing a phase portrait of the system (Fig. 1.4, in Section 1.4.4 of Chapter 1, illustrates an example of it for the simple pendulum). We may think that the vector having components $(\dot{p}_i,\dot{q}_i)$ represents the velocity vector of an 'abstract continuous medium' in every point along

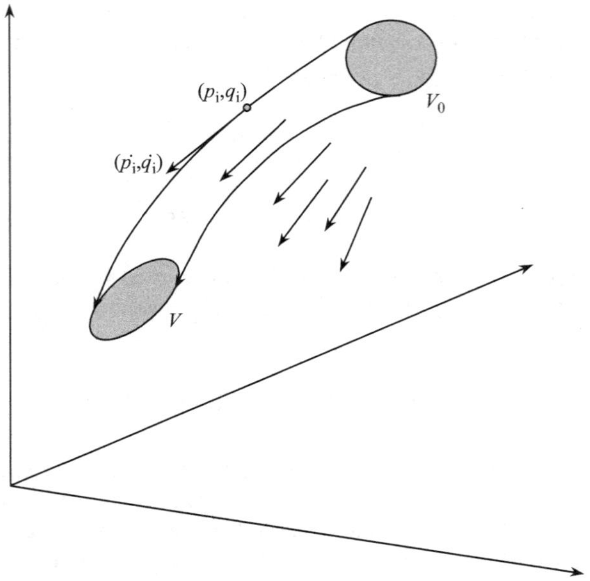

Fig. 2.1 Flow in the phase space, with conservation of volume (Liouville's theorem).

a trajectory. This is what is known as the phase space *flow*. If we mentally follow the motion of the points included in a small given volume V of the phase space, (Fig. 2.1), then we can see that the total 'mass' included in that volume corresponds to the probability that the system has a state corresponding to a point in that volume. Thus, the probability Φ for the presence in the phase space formally plays the part of a density. If we follow the volume V as it evolves dynamically, the total probability for the system to lie in V is, of course, kept over time, and so we can observe an abstract 'fluid' whose mass is conserved, just as for a genuine fluid.[6] Going slightly further into Chapter 3 (Section 3.2.3), we may apply the continuity equation (3.43) to that velocity and density field and write

[6]Aside from the fact that, as must be repeated, we are in a space of very high dimension; nevertheless, this has no effect on the next mathematical developments.

$$\frac{1}{\Phi}\frac{d\Phi}{dt} + \sum_i \left(\frac{\partial \dot{p}_i}{\partial p_i} + \frac{\partial \dot{q}_i}{\partial q_i} \right) = 0 \tag{2.8}$$

The term under the summation symbol denotes the ($2N$-dimensional) 'divergence' of the abstract 'velocity' field ($\{\dot{p}_i\}, \{\dot{q}_i\}$). Through the Hamilton equation (1.89), we quickly get:

$$\begin{aligned} \frac{1}{\Phi}\frac{d\Phi}{dt} &= \sum_i \left(\frac{\partial^2 H}{\partial p_i \partial q_i} - \frac{\partial^2 H}{\partial q_i \partial p_i} \right) \\ &= 0 \end{aligned} \tag{2.9}$$

This equation demonstrates that the probability density of the whole system in the phase space remains unchanged over the course of time; this is Liouville's theorem:

$$\Phi(t) = cst \tag{2.10}$$

It should be emphasized that this property merely results from the equations of motion and that it is caused by the symplectic nature of the Hamilton equations, which was only half seen in Section 1.4.2. Using the matrix formulation (1.92), indeed, the calculations (2.8) and (2.9) can be rearranged by advantageously applying the tensor calculation rules (refer to eqn (A.68) in Appendix A). We then use the fact that the symplectic matrix $\boldsymbol{\omega}_2$ is constant and find

$$\begin{aligned} \frac{1}{\Phi}\frac{d\Phi}{dt} &= \sum_i \frac{\partial}{\partial \mathbf{X}_i} \cdot \dot{\mathbf{X}}_i \\ &= \sum_i \frac{\partial}{\partial \mathbf{X}_i} \cdot \left(\boldsymbol{\omega}_2 \frac{\partial H}{\partial \mathbf{X}_i} \right) \\ &= \boldsymbol{\omega}_2 : \sum_i \frac{\partial^2 H}{\partial \mathbf{X}_i \partial \mathbf{X}_i} \\ &= 0 \end{aligned} \tag{2.11}$$

The latter result is due to the fact that the twofold contracted product of two matrices equals zero when one matrix is skew-symmetric whereas the other one is symmetric. Thus, the property of skew-symmetry (1.94) of the matrix $\boldsymbol{\omega}_2$ is at the origin of that result. Liouville's theorem is then closely related to the conservation of energy for an isolated system.

If the probability (i.e. the 'density') is conserved, then the conservation of the 'mass' ensures that of the 'volume'. In other words, the abstract 'fluid' which we have imagined is incompressible. We can then conclude that a given volume V of the phase space moves about as per the dynamics of the Hamilton equations, while keeping its measurement (Fig. 2.1). Actually, this property does not suffice to set up the symplectic nature of the geometry resulting from the Hamilton equations. The calculation (2.9), indeed, exhibits a result which is stronger than the conservation of volume[7] in the phase space, that is the conservation of the area[8] projected by that volume onto every plane (p_i, q_i), which, by definition, determines the symplectic nature of the equations. We may recall that the calculation (2.9) is valid for every system based on the Hamiltonian formalism and not merely for a collection of molecules. In order to illustrate this result, let us resume the case of the simple pendulum as introduced in Section 1.4.4. As stated, that pendulum is defined by parameters $(q, \dot{q}) = (\theta, \dot{\theta})$, whose evolution over time is given by (1.167) when taking the approximation of the small motions into account. The system evolves as per equation (1.175), and so the evolution of the volume in the phase space is given by the determinant of the matrix[9] $\boldsymbol{\Psi}^+ (t^+)$. Hence we will have[10]

$$\begin{aligned} \frac{V}{V_0} &= \det \boldsymbol{\Psi}^+ \left(t^+\right) \\ &= 1 \end{aligned} \tag{2.12}$$

From a geometrical point of view, this result may be interpreted as follows. Let us recall that, in the approximation of the pendulum's small oscillations,

[7] We have used the word 'volume' to denote a hypervolume in a space with an arbitrary number of dimensions. Since this is a two-dimensional phase space for the pendulum case, it is obviously a surface.

[8] This result is a theorem which was formulated by Poincaré.

[9] Since this system is linear, this matrix is the Jacobian of the transformation yielding the state of the system as a function of its initial state.

[10] This is another early insight into Chapter 3 (eqn (3.22)).

the phase trajectories are circles as shown in Fig. 1.4. The fact that the volumes remain constant in them is evidenced by nearly parallel trajectories in the vicinity of the phase origins. For larger oscillations, however, the linear approximation (1.167) is no longer valid. Nevertheless, the volumes in the phase space are preserved owing to the Hamilton equations, but the phase trajectories are no longer parallel.

Within the scope of an ensemble of molecules, we will now try to apply the principle as given by equation (2.12), by analogy with Rief (1965). To that purpose, we will consider an ensemble of N_m molecules whose coordinates and momenta at time point $t' = t + dt$ are denoted $\tilde{\mathbf{r}}'_\alpha$ and $\tilde{\mathbf{p}}'_\alpha$ and related to their values at time point t by relations derived from (1.2) and (1.56):

$$\begin{aligned} \tilde{\mathbf{p}}'_\alpha &= \tilde{\mathbf{p}}_\alpha + \tilde{\mathbf{F}}_\alpha\left(\{\tilde{\mathbf{r}}_\beta\}, t\right) dt \\ \tilde{\mathbf{r}}'_\alpha &= \tilde{\mathbf{r}}_\alpha + \frac{\tilde{\mathbf{p}}_\alpha}{m} dt \end{aligned} \tag{2.13}$$

The volume in the phase space is then the product of quantities $d\tilde{\mathbf{r}}_\alpha d\tilde{\mathbf{p}}_\alpha$ Each of these quantites evolves according to the determinant of the Jacobian function $\mathbf{J}_{\alpha\beta}$ of the transformation $\left(\tilde{\mathbf{p}}_\beta, \tilde{\mathbf{r}}_\beta\right) \to \left(\tilde{\mathbf{p}}'_\alpha, \tilde{\mathbf{r}}'_\alpha\right)$ (i.e. $\tilde{\mathbf{X}}_\beta \to \tilde{\mathbf{X}}_\alpha$, introducing a notion which is analogous to (1.99)). Through (2.13), this matrix is rearranged as:

$$\mathbf{J}_{\alpha\beta} \doteq \begin{pmatrix} \dfrac{\partial \tilde{\mathbf{p}}'_\alpha}{\partial \tilde{\mathbf{p}}_\beta} = \mathbf{I}_3 \delta_{\alpha\beta} & \dfrac{\partial \tilde{\mathbf{p}}'_\alpha}{\partial \tilde{\mathbf{r}}_\beta} = \dfrac{\partial \tilde{\mathbf{F}}_\alpha}{\partial \tilde{\mathbf{r}}_\beta} dt \\ \dfrac{\partial \tilde{\mathbf{r}}'_\alpha}{\partial \tilde{\mathbf{p}}_\beta} = \dfrac{dt}{m} \mathbf{I}_3 \delta_{\alpha\beta} & \dfrac{\partial \tilde{\mathbf{r}}'_\alpha}{\partial \tilde{\mathbf{r}}_\beta} = \mathbf{I}_3 \delta_{\alpha\beta} \end{pmatrix} \tag{2.14}$$

The full Jacobian matrix of the transformation of the molecular system $\left(\{\tilde{\mathbf{p}}_\beta\}, \{\tilde{\mathbf{r}}_\beta\}\right) \to \left(\{\tilde{\mathbf{p}}'_\alpha\}, \{\tilde{\mathbf{r}}'_\alpha\}\right)$ then has a ($2N_m \times 2N_m$-sized) determinant which is written as

$$J \doteq \frac{\partial\left(\{\tilde{\mathbf{p}}'_\alpha\}, \{\tilde{\mathbf{r}}'_\alpha\}\right)}{\partial\left(\{\tilde{\mathbf{p}}_\beta\}, \{\tilde{\mathbf{r}}_\beta\}\right)} = \begin{vmatrix} \mathbf{I}_3 & \mathbf{0} & \cdots & \dfrac{\partial \tilde{\mathbf{F}}_\alpha}{\partial \tilde{\mathbf{r}}_\alpha} dt & \dfrac{\partial \tilde{\mathbf{F}}_\alpha}{\partial \tilde{\mathbf{r}}_\beta} dt & \cdots \\ \mathbf{0} & \mathbf{I}_3 & & \dfrac{\partial \tilde{\mathbf{F}}_b}{\partial \tilde{\mathbf{r}}_\alpha} dt & \dfrac{\partial \tilde{\mathbf{F}}_b}{\partial \tilde{\mathbf{r}}_\beta} dt & \\ \vdots & & \ddots & \vdots & & \ddots \\ \frac{dt}{m}\mathbf{I}_3 & \mathbf{0} & \cdots & \mathbf{I}_3 & \mathbf{0} & \cdots \\ \mathbf{0} & \frac{dt}{m}\mathbf{I}_3 & & \mathbf{0} & \mathbf{I}_3 & \\ \vdots & & \ddots & \vdots & & \ddots \end{vmatrix} \tag{2.15}$$

The $N_m \times N_m$ matrices of size 3×3 occurring in the bottom left-hand part of J are proportional to dt, and so the Jacobian only differs from one by a $6N_m$-

order term in dt. Since the time increment dt is arbitrarily small, the invariance of volume is then exact. More precisely, we can see that the determinant in the Jacobian $\mathbf{J}_{\alpha\beta}$ is governed by the same property:

$$\begin{aligned} J_{\alpha\beta} &\doteqdot \det \mathbf{J}_{\alpha\beta} \\ &= \frac{\partial\left(\tilde{\mathbf{p}}'_{\alpha}, \tilde{\mathbf{r}}'_{\alpha}\right)}{\partial\left(\tilde{\mathbf{p}}_{\beta}, \tilde{\mathbf{r}}_{\beta}\right)} \\ &= \frac{\partial \tilde{\mathbf{X}}_{\alpha}}{\partial \tilde{\mathbf{X}}_{\beta}} \\ &= \begin{vmatrix} \mathbf{I}_3 \delta_{\alpha\beta} & \dfrac{\partial \tilde{\mathbf{F}}_{\alpha}}{\partial \tilde{\mathbf{r}}_{\beta}} dt \\ \dfrac{dt}{m} \mathbf{I}_3 \delta_{\alpha\beta} & \mathbf{I}_3 \delta_{\alpha\beta} \end{vmatrix} = \delta_{\alpha\beta} \end{aligned} \tag{2.16}$$

$J_{\alpha\beta}$ then equals 1 when $\beta = \alpha$. Thus, the volumes associated with the parameters of each molecule remain constant within the phase space in addition to the volume associated with the set of degrees of freedom. This result is consistent with the symplectic nature of the equations.

We will now consider an individual molecule and try to describe the microscopic evolution of the unknown function φ given by eqn (2.2). A particular case occurs: it is a gas which is rarefied to such an extent that the molecules do not interact nor collide (it is known as a *perfect (or ideal) gas*). One may consider that the molecular interactions are very weak within a gas, except when one or several of them come very close to each other. In the absence of collisions, the interaction potential may then be considered as being zero. Each of the molecules is then an isolated system satisfying Liouville's theorem (2.9) as applied to the individual probability density $\varphi\left(\tilde{\mathbf{p}}, \tilde{\mathbf{r}}, t\right)$, that is $d\varphi/dt = 0$ or, in an expanded form:

$$\frac{\partial \varphi}{\partial t} + \frac{\partial \varphi}{\partial \tilde{\mathbf{r}}} \cdot \frac{d\tilde{\mathbf{r}}}{dt} + \frac{\partial \varphi}{\partial \tilde{\mathbf{p}}} \cdot \frac{d\tilde{\mathbf{p}}}{dt} = 0 \tag{2.17}$$

With the definition (1.2) of velocity and the equation of motion (1.56), this equation also gives

$$\frac{\partial \varphi}{\partial t} + \frac{\partial \varphi}{\partial \tilde{\mathbf{r}}} \cdot \tilde{\mathbf{u}} + \frac{\partial \varphi}{\partial \tilde{\mathbf{p}}} \cdot \tilde{\mathbf{F}} = 0 \tag{2.18}$$

where $\tilde{\mathbf{F}}$ is the force received from a possible external system (e.g. the Earth's gravity) by a molecule lying in $\tilde{\mathbf{r}}$. This equation is the simplest form of the Boltzmann equation,[11] which is applicable to a perfect gas (i.e. without any collision).

In a true gas, as we said earlier, the collisions between particles affect their respective momenta, hence function φ. Equation (2.18) should then be modified; we are now going to discuss the way it is modified. Each collision between two molecules[12] β and γ may be characterized by a set of four momenta, namely those, denoted $\tilde{\mathbf{p}}_{\beta}$ and $\tilde{\mathbf{p}}_{\gamma}$, of both molecules before the collision, as well as those characterizing them after the collision, which are

[11] The left-hand side of the equation is sometimes referred to as the *Liouville operator* applied to distribution φ.

[12] We will ignore those collisions which involve more than two molecules.

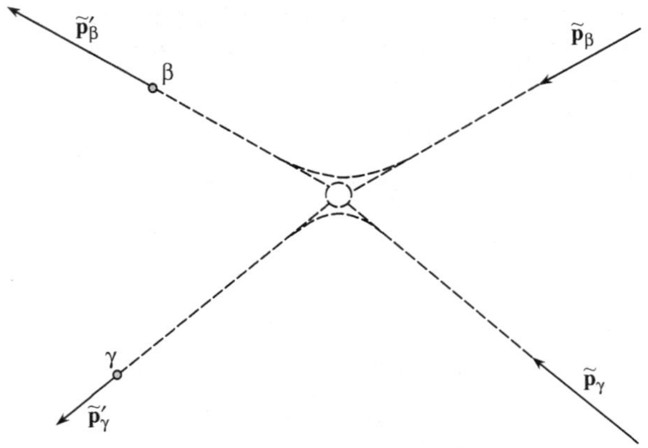

Fig. 2.2 Collision of two molecules.

denoted $\tilde{\mathbf{p}}'_\beta$ and $\tilde{\mathbf{p}}'_\gamma$ (refer to Fig. 2.2). If the fluid is a gas, it may also be thought that the interactions are restricted to pairs of comparatively adjacent molecules. Immediately prior to their collision, that is when their mutual interaction can no longer be ignored, they make up an isolated system, then verifying the law of conservation of total momentum (Section 1.4.3):

$$\tilde{\mathbf{p}}_\beta + \tilde{\mathbf{p}}_\gamma = \tilde{\mathbf{p}}'_\beta + \tilde{\mathbf{p}}'_\gamma \tag{2.19}$$

The average number of such collisions is proportional to the number of pairs of molecules with initial momenta $\tilde{\mathbf{p}}_\beta$ and $\tilde{\mathbf{p}}_\gamma$, as yielded by the product $\varphi\left(\tilde{\mathbf{p}}_\beta, \tilde{\mathbf{r}}, t\right)\varphi\left(\tilde{\mathbf{p}}_\gamma, \tilde{\mathbf{r}}, t\right) d\tilde{\mathbf{p}}_\beta d\tilde{\mathbf{p}}_\gamma$. More precisely, there is an unknown (positive) function $w\left(\tilde{\mathbf{p}}_\beta, \tilde{\mathbf{p}}_\gamma; \tilde{\mathbf{p}}'_\beta, \tilde{\mathbf{p}}'_\gamma\right)$ such that this number equals

$$w\left(\tilde{\mathbf{p}}_\beta, \tilde{\mathbf{p}}_\gamma; \tilde{\mathbf{p}}'_\beta, \tilde{\mathbf{p}}'_\gamma\right)\varphi_\beta\varphi_\gamma d\tilde{\mathbf{p}}_\beta d\tilde{\mathbf{p}}_\gamma d\tilde{\mathbf{p}}'_\beta d\tilde{\mathbf{p}}'_\gamma \tag{2.20}$$

where $\varphi_\beta \doteqdot \varphi\left(\tilde{\mathbf{p}}_\beta, \tilde{\mathbf{r}}, t\right)$ and $\varphi_\gamma \doteqdot \varphi\left(\tilde{\mathbf{p}}_\gamma, \tilde{\mathbf{r}}, t\right)$ have been noted. We already notice that, because the laws of mechanics are reversible over time, the 'reverse' collisions in the course of time, turning $\left(-\tilde{\mathbf{p}}'_\beta, -\tilde{\mathbf{p}}'_\gamma\right)$ into $\left(-\tilde{\mathbf{p}}_\beta, -\tilde{\mathbf{p}}_\gamma\right)$, occcur in equal numbers, which implies

$$w\left(\tilde{\mathbf{p}}_\beta, \tilde{\mathbf{p}}_\gamma; \tilde{\mathbf{p}}'_\beta, \tilde{\mathbf{p}}'_\gamma\right) = w\left(-\tilde{\mathbf{p}}'_\beta, -\tilde{\mathbf{p}}'_\gamma; -\tilde{\mathbf{p}}_\beta, -\tilde{\mathbf{p}}_\gamma\right) \tag{2.21}$$

Here we must use a space symmetry which has not been employed yet in our discussions. It is the invariance through central symmetry about an arbitrary point that transforms every vector into its opposite. If we accept that the laws of mechanics are invariant through this kind of transformation, then the probability of a collision is not changed by central symmetry. Thus, we may write

$$w\left(-\tilde{\mathbf{p}}'_\beta, -\tilde{\mathbf{p}}'_\gamma; -\tilde{\mathbf{p}}_\beta, -\tilde{\mathbf{p}}_\gamma\right) = w\left(\tilde{\mathbf{p}}'_\beta, \tilde{\mathbf{p}}'_\gamma; \tilde{\mathbf{p}}_\beta, \tilde{\mathbf{p}}_\gamma\right) \tag{2.22}$$

that is combining the last two equalities:

$$w\left(\tilde{\mathbf{p}}_\beta, \tilde{\mathbf{p}}_\gamma; \tilde{\mathbf{p}}'_\beta, \tilde{\mathbf{p}}'_\gamma\right) = w\left(\tilde{\mathbf{p}}'_\beta, \tilde{\mathbf{p}}'_\gamma; \tilde{\mathbf{p}}_\beta, \tilde{\mathbf{p}}_\gamma\right) \tag{2.23}$$

We also need to mention another property of w, which will be of use later on:

$$w\left(\tilde{\mathbf{p}}_\beta, \tilde{\mathbf{p}}_\gamma; \tilde{\mathbf{p}}'_\beta, \tilde{\mathbf{p}}'_\gamma\right) = w\left(\tilde{\mathbf{p}}_\gamma, \tilde{\mathbf{p}}_\beta; \tilde{\mathbf{p}}'_\gamma, \tilde{\mathbf{p}}'_\beta\right) \tag{2.24}$$

This relation only holds true because the molecules are identical, and so if their respective characteristics are exchanged, the probabilities remain unaffected.

We are now going to count the collisions increasing or reducing the number of molecules lying[13] in $\tilde{\mathbf{r}}$ at time t and having a momentum within the range $\left[\tilde{\mathbf{p}}, \tilde{\mathbf{p}} + d\tilde{\mathbf{p}}\right]$. This is because a collision is liable to modify the momentum of β so that it leaves that range. The number of such 'losses' per unit volume is yielded by the integral of quantity (2.20) over all the possible momenta of the impacting molecule γ and over all the pairs of possible momenta resulting from the collision, that is

[13] Once gain, we mean those molecules occurring within the range $\left[\tilde{\mathbf{r}}, \tilde{\mathbf{r}} + d\tilde{\mathbf{r}}\right]$.

$$\int w\left(\tilde{\mathbf{p}}, \tilde{\mathbf{p}}_\gamma; \tilde{\mathbf{p}}'_\beta, \tilde{\mathbf{p}}'_\gamma\right) \varphi\varphi_\gamma d\tilde{\mathbf{p}}_\gamma d\tilde{\mathbf{p}}'_\beta d\tilde{\mathbf{p}}'_\gamma \tag{2.25}$$

with $\varphi \doteqdot \varphi\left(\tilde{\mathbf{p}}, \tilde{\mathbf{r}}, t\right)$. Integration is performed over all the values taken by $\tilde{\mathbf{p}}_\gamma$, $\tilde{\mathbf{p}}'_\beta$ and $\tilde{\mathbf{p}}'_\gamma$, that is a priori over Ω_p^3, a nine-dimensional space. Actually, integration sould be performed over a six-dimensional sub-variety $\bar{\Omega}$ of that space, since the vector relation (2.19) prescribes three relations among $\tilde{\mathbf{p}}_\gamma$, $\tilde{\mathbf{p}}'_\beta$ and $\tilde{\mathbf{p}}'_\gamma$ for each value of vector $\tilde{\mathbf{p}}$. Contrary to this process, some collisions change the momentum of a molecule so that it comes within the range $\left[\tilde{\mathbf{p}}, \tilde{\mathbf{p}} + d\tilde{\mathbf{p}}\right]$. These collisions are 'gains' reckoned through the quantity

$$\int w\left(\tilde{\mathbf{p}}'_\beta, \tilde{\mathbf{p}}'_\gamma; \tilde{\mathbf{p}}, \tilde{\mathbf{p}}_\gamma\right) \varphi'_\beta\varphi'_\gamma d\tilde{\mathbf{p}}_\gamma d\tilde{\mathbf{p}}'_\beta d\tilde{\mathbf{p}}'_\gamma \tag{2.26}$$

with $\varphi'_\beta \doteqdot \varphi\left(\tilde{\mathbf{r}}, \tilde{\mathbf{p}}'_\beta, t\right)$ and $\varphi'_\gamma \doteqdot \varphi\left(\tilde{\mathbf{r}}, \tilde{\mathbf{p}}'_\gamma, t\right)$. Thus, the evolution $d\varphi/dt$ of φ caused by the collisional processes is yielded by the difference of integrals (2.26) and (2.25), which then should be added into the right-hand side of (2.18). Hence, using the relation (2.23), the full Boltzmann's equation for a gas is given by

$$\frac{\partial\varphi}{\partial t} + \frac{\partial\varphi}{\partial\tilde{\mathbf{r}}} \cdot \tilde{\mathbf{u}} + \frac{\partial\varphi}{\partial\tilde{\mathbf{p}}} \cdot \tilde{\mathbf{F}} = C\left(\varphi\right) \tag{2.27}$$

a *collision integral* being given by

$$C\left(\varphi\right) \doteqdot \int_{\bar{\Omega}} w\left(\tilde{\mathbf{p}}, \tilde{\mathbf{p}}_\gamma; \tilde{\mathbf{p}}'_\beta, \tilde{\mathbf{p}}'_\gamma\right)\left(\varphi'_\beta\varphi'_\gamma - \varphi\varphi_\gamma\right) d\tilde{\mathbf{p}}_\gamma d\tilde{\mathbf{p}}'_\beta d\tilde{\mathbf{p}}'_\gamma \tag{2.28}$$

Equation (2.27), which is valid at every point in space and time, theoretically yields the evolution of φ. It is an integro-differential equation, because it involves an integral depending on the unknown function φ on the right-hand side. For a non-gaseous fluid, the problem should be addressed in a different way; since the particle interactions are not restricted to the collision phases, their proximity induces a markedly different behaviour. We should then consider Boltzmann's equation, in this written form, as a convenient model which will allow us to highlight a number of important collective phenomena.

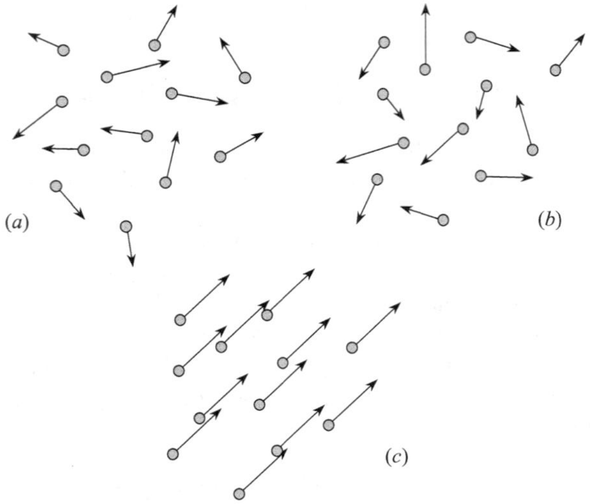

Fig. 2.3 Three possible configurations for a collective motion of molecules. Configurations (a) and (b) are statistically equivalent, whereas (c) is highly unlikely.

2.2.3 Entropy

Having in mind the statistical behaviour of systems with a very high number of degrees of freedom, we now have to consider sizes which are more macroscopic than the molecular-scaled ones; accordingly, we must change our point of view. Statistically, indeed, one system may exhibit different microscopic states which, however, correspond to one macroscopic state. In Fig. 2.3, for example, the configurations (a) and (b) may be considered as statistically identical. Thus, we are led to view a macrocopic system as an entity which a probability density which is denoted $\overline{\Phi}$. The states (a) and (b) in 2.3 correspond to the same value of $\overline{\Phi}$, whereas (c), which is intuitively much rarer from a statistical point of view, corresponds to a much lower value of $\overline{\Phi}$. The number of microscopic configurations corresponding to a macroscopic state with a probability $\overline{\Phi}$ is denoted $N\left(\overline{\Phi}\right)$. Since the number of molecules making up the system is very high, the quantity $\overline{\Phi}$ varies a lot from one macroscopic state to another; thus, one can more conveniently work with a quantity which is known as *entropy* and defined by

$$S \doteqdot k_B \ln N\left(\overline{\Phi}\right) \tag{2.29}$$

where k_B is referred to as *Boltzmann constant.*[14] It can be readily understood that entropy[15] is additive for non-interacting systems, since the probabilities are multiplicative for two statistically independent systems. $\overline{\Phi}$ is obviously equal to the $N\left(\overline{\Phi}\right)/N_{tot}$ ratio, where N_{tot} is the total number of possible microscopic states; thus, (2.29) can be rearranged as

$$\overline{\Phi} = \frac{1}{N_{tot}} \exp\left(\frac{S}{k_B}\right) \tag{2.30}$$

We are going to try to relate entropy to the function φ characterizing a given molecule, still in the case of a relatively undense gas. To do this, we have to

[14]The latter is taken as equal to $1.3806503 \times 10^{-23}$ kg.m^{2}s^{-2}K^{-1} in the standard system of units, but some authors take it to equal unity, which comes down to ascribing a unit of energy to the temperature, as can be seen in Section 2.3.1.

[15]Strictly speaking, and as noted in Section 2.2.1, here the probability density for the system presence would have to be non-dimensional, which requires us to reduce it to a factor which, as evidenced by quantum mechanics, equals h^n, where n is the number of degrees of freedom and h is Planck's constant.

characterize the overall state of a large system as a function of its constituents' individual states' data.

Let us consider an isolated system which is made up of N_m identical molecules, each of them being able to occur in a number of possible states $(\tilde{\mathbf{p}}, \tilde{\mathbf{r}})$. For an easier approach, we will assume the possible states of the individual molecules are not distributed over a continuum, but are liable to occupy values which are close to discrete points[16] $\left(\tilde{\mathbf{p}}^{(i)}, \tilde{\mathbf{r}}^{(i)}\right)$ in the phase space. This is tantamount to dividing the phase space of an individual molecule into $d\tilde{\mathbf{p}}d\tilde{\mathbf{r}}$-sized elementary volumes centred on each of the points $\left(\tilde{\mathbf{p}}^{(i)}, \tilde{\mathbf{r}}^{(i)}\right)$. Each of these small volumes is once again split into k small, infinitesimal sub-volumes which, by convention, are identically equal to unity, so that $d\tilde{\mathbf{p}}d\tilde{\mathbf{r}} = k$. Each sub-volume defines a configuration corresponding to the i-th state. We now will use that schematic formalism to reckon the microscopic states (i.e. all the individual states of the molecules) corresponding to a system's given macroscopic state. For each possible state i of a molecule, corresponding to one of the phase space small elementary volumes lying about $\left(\tilde{\mathbf{p}}^{(i)}, \tilde{\mathbf{r}}^{(i)}\right)$, the number of molecules being in this state is denoted N_i. A macroscopic state of the system clearly corresponds to the data $\{N_i\}$ of the set of numbers N_i (the sum equalling N_m). We must first note that

[16] This approach is only necessary in quantum mechanics, but makes our present approach easier.

$$\varphi\left(\tilde{\mathbf{p}}^{(i)}, \tilde{\mathbf{r}}^{(i)}, t\right) d\tilde{\mathbf{p}}d\tilde{\mathbf{r}} = N_i \tag{2.31}$$

Besides, it is obvious that

$$\varphi\left(\tilde{\mathbf{p}}^{(i)}, \tilde{\mathbf{r}}^{(i)}, t\right) = \frac{N_i}{k} \tag{2.32}$$

The number of possible microscopic configurations corresponding to the presence of N_i molecules in state i equals k^{N_i} as divided by $N_i!$, because all the (identical) molecules being in a given state i may be arbitrarily permuted. Thus, the probability for the system to be in the macroscopic state corresponding to the datum of the $\{N_i\}$ equals

$$N\left(\overline{\Phi}\right) = \prod_i \frac{k^{N_i}}{N_i!} \tag{2.33}$$

Through definition (2.29), we can then formulate the entropy as

$$\begin{aligned} S &= k_B \sum_i \left(N_i \ln k - \ln N_i!\right) \\ &\approx k_B \sum_i N_i \left(\ln \frac{k}{N_i} + 1\right) \end{aligned} \tag{2.34}$$

(we have used the fact that the 'occupation numbers' N_i are very high, which allows us to come nearer to $\ln N_i!$ through $N_i\left(\ln N_i - 1\right)$). In this form, it already happens that the contributions from energy levels for which N_i is (comparatively) small contribute towards reducing entropy. Now, relatively moderate values of some N_i are only possible if other N_i, on the contrary, are very high, since the total number of molecules is set. Thus, entropy may be viewed as a measurement of the system's macroscopic homogeneity.[17] By making the state number tend towards infinity[18] in (2.34), the discrete sum can be approximated with an integral. With (2.31) and (2.32), we get:

$$S = k_B \int_{\Omega_p \times \Omega_r} \varphi\,(1 - \ln \varphi)\, d\tilde{\mathbf{p}} d\tilde{\mathbf{r}} \tag{2.35}$$

Let us now calculate the temporal derivative of entropy. Note that, for every arbitrary function ψ and every variable θ, we have:

$$\frac{\partial}{\partial \theta}\left[\psi\,(1 - \ln \psi)\right] = -\frac{\partial \psi}{\partial \theta} \ln \psi \tag{2.36}$$

Hence, with $\psi \equiv \varphi$ and $\theta \equiv t$, we get

$$\frac{dS}{dt} = -k_B \int_{\Omega_p \times \Omega_r} \frac{\partial \varphi}{\partial t} \ln \varphi d\tilde{\mathbf{p}} d\tilde{\mathbf{r}} \tag{2.37}$$

The derivative of φ is given to us by Boltzmann's equation (2.27). One can easily understand that the last two terms on the left-hand side do not play any part in (2.37). Actually, resuming (2.35) with $\psi \equiv \varphi$ and $\theta \equiv \tilde{\mathbf{r}}$:

$$\begin{aligned} &\int_{\Omega_p \times \Omega_r} \left(\frac{\partial \varphi}{\partial \tilde{\mathbf{r}}} \cdot \tilde{\mathbf{u}}\right) \ln \varphi d\tilde{\mathbf{p}} d\tilde{\mathbf{r}} \\ &= \int_{\Omega_p \times \Omega_r} \left(\frac{\partial}{\partial \tilde{\mathbf{r}}}\left[\varphi\,(1 - \ln \varphi)\right] \cdot \tilde{\mathbf{u}}\right) d\tilde{\mathbf{p}} d\tilde{\mathbf{r}} \\ &= \int_{\Omega_p} \tilde{\mathbf{u}} \cdot \left(\int_{\Omega_r} \frac{\partial}{\partial \tilde{\mathbf{r}}}\left[\varphi\,(1 - \ln \varphi)\right] d\tilde{\mathbf{r}}\right) d\tilde{\mathbf{p}} \\ &= 0 \end{aligned} \tag{2.38}$$

The integral in the centre of the second line of (2.38) is, indeed, turned into an area integral at the 'boundary' of the domain of integration, according to the Gauss theorem (refer to Appendix A, eqn (A.73)). Now, if it happens that such a boundary is driven away towards infinity and, clearly enough, no particle can lie at infinity nor have an infinite velocity; then, φ vanishes there.

For the same reasons, with $\theta \equiv \tilde{\mathbf{p}}$:

$$\begin{aligned} &\int_{\Omega_p \times \Omega_r} \left(\frac{\partial \varphi}{\partial \tilde{\mathbf{p}}} \cdot \tilde{\mathbf{F}}\right) \ln \varphi d\tilde{\mathbf{p}} d\tilde{\mathbf{r}} \\ &= \int_{\Omega_p \times \Omega_r} \left(\frac{\partial}{\partial \tilde{\mathbf{p}}}\left[\varphi\,(1 - \ln \varphi)\right] \cdot \tilde{\mathbf{F}}\right) d\tilde{\mathbf{p}} d\tilde{\mathbf{r}} \\ &= \int_{\Omega_r} \tilde{\mathbf{F}} \cdot \left(\int_{\Omega_p} \frac{\partial}{\partial \tilde{\mathbf{p}}}\left[\varphi\,(1 - \ln \varphi)\right] d\tilde{\mathbf{p}}\right) d\tilde{\mathbf{r}} \\ &= 0 \end{aligned} \tag{2.39}$$

[17] Many authors, particularly in the popular scientific works, speak about 'disorder'. We believe the word 'homogeneity' is more relevant here. This is because the case of a gas in thermodynamic equilibrium (refer to Section 2.3.1) corresponds to a macroscopically highly homogeneous state, whereas the disorder is intuitively maximal at the microscopic scale.

[18] This is theoretically impossible if high values of occupation numbers are kept. We will just imagine that the state number is high enough to make this approximation valid, which is consistent with quantum reality.

Thus, the single contribution to variation of entropy of the system is given by the collision integral $C(\varphi)$ of (2.27):

$$\frac{dS}{dt} = -k_B \int_{\Omega_r} \int_{\bar{\Omega}_p} w\left(\tilde{\mathbf{p}}, \tilde{\mathbf{p}}_\gamma; \tilde{\mathbf{p}}'_\beta, \tilde{\mathbf{p}}'_\gamma\right)\left(\varphi'_\beta \varphi'_\gamma - \varphi\varphi_\gamma\right) \times \ln\varphi d\tilde{\mathbf{p}}_\gamma d\tilde{\mathbf{p}}'_\beta d\tilde{\mathbf{p}}'_\gamma d\tilde{\mathbf{p}} d\tilde{\mathbf{r}} \tag{2.40}$$

where $\bar{\Omega}_p = \Omega_p \times \bar{\Omega}$ denotes the set of quadruplets $\left(\tilde{\mathbf{p}}, \tilde{\mathbf{p}}_\gamma, \tilde{\mathbf{p}}'_\beta, \tilde{\mathbf{p}}'_\gamma\right)$ satisfying the relation (2.19) for $\tilde{\mathbf{p}}_\beta \equiv \tilde{\mathbf{p}}$, that is $\tilde{\mathbf{p}} + \tilde{\mathbf{p}}_\gamma = \tilde{\mathbf{p}}'_\beta + \tilde{\mathbf{p}}'_\gamma$. We now have to make some changes in the integral of the right-hand side of (2.40). First of all, we notice that the change of variable $\tilde{\mathbf{p}}'_\beta \leftrightarrows \tilde{\mathbf{p}}$ and $\tilde{\mathbf{p}}'_\gamma \leftrightarrows \tilde{\mathbf{p}}_\gamma$ can be performed because relation (2.19) is then still satisfied. Relation (2.23) then yields:

$$\begin{aligned}
&\int_{\bar{\Omega}_p} w\left(\tilde{\mathbf{p}}, \tilde{\mathbf{p}}_\gamma; \tilde{\mathbf{p}}'_\beta, \tilde{\mathbf{p}}'_\gamma\right)\left(\varphi'_\beta \varphi'_\gamma - \varphi\varphi_\gamma\right) \\
&\quad \times \ln\varphi d\tilde{\mathbf{p}}_\gamma d\tilde{\mathbf{p}}'_\beta d\tilde{\mathbf{p}}'_\gamma d\tilde{\mathbf{p}} \\
&= \int_{\bar{\Omega}_p} w\left(\tilde{\mathbf{p}}'_\beta, \tilde{\mathbf{p}}'_\gamma; \tilde{\mathbf{p}}, \tilde{\mathbf{p}}_\gamma\right)\left(\varphi'_\beta \varphi'_\gamma - \varphi\varphi_\gamma\right) \ln\varphi d\tilde{\mathbf{p}}_\gamma d\tilde{\mathbf{p}} d\tilde{\mathbf{p}}'_\gamma d\tilde{\mathbf{p}}'_\beta \\
&= \int_{\bar{\Omega}_p} w\left(\tilde{\mathbf{p}}, \tilde{\mathbf{p}}_\gamma; \tilde{\mathbf{p}}'_\beta, \tilde{\mathbf{p}}'_\gamma\right)\left(\varphi\varphi_\gamma - \varphi'_\beta \varphi'_\gamma\right) \ln\varphi'_\beta d\tilde{\mathbf{p}}_\gamma d\tilde{\mathbf{p}}'_\beta d\tilde{\mathbf{p}}'_\gamma d\tilde{\mathbf{p}} \\
&= \frac{1}{2}\int_{\bar{\Omega}_p} w\left(\tilde{\mathbf{p}}, \tilde{\mathbf{p}}_\gamma; \tilde{\mathbf{p}}'_\beta, \tilde{\mathbf{p}}'_\gamma\right)\left(\varphi'_\beta \varphi'_\gamma - \varphi\varphi_\gamma\right) \\
&\quad \times \left(\ln\varphi - \ln\varphi'_\beta\right) d\tilde{\mathbf{p}}_\gamma d\tilde{\mathbf{p}}'_\beta d\tilde{\mathbf{p}}'_\gamma d\tilde{\mathbf{p}}
\end{aligned} \tag{2.41}$$

(the latter equality can be obtained by averaging the expression achieved in the third line with the initial integral). Besides, we may rearrange (2.41) by replacing the variables by $\tilde{\mathbf{p}} \leftrightarrows \tilde{\mathbf{p}}_\gamma$ and $\tilde{\mathbf{p}}'_\beta \leftrightarrows \tilde{\mathbf{p}}'_\gamma$ (which is still consistent with (2.19)) to get

$$\begin{aligned}
&\int_{\bar{\Omega}_p} w\left(\tilde{\mathbf{p}}, \tilde{\mathbf{p}}_\gamma; \tilde{\mathbf{p}}'_\beta, \tilde{\mathbf{p}}'_\gamma\right)\left(\varphi'_\beta \varphi'_\gamma - \varphi\varphi_\gamma\right) \\
&\quad \times \ln\varphi d\tilde{\mathbf{p}}_\gamma d\tilde{\mathbf{p}}'_\beta d\tilde{\mathbf{p}}'_\gamma d\tilde{\mathbf{p}} \\
&= \frac{1}{2}\int_{\bar{\Omega}_p} w\left(\mathbf{p}_\gamma, \mathbf{p}; \mathbf{p}'_\gamma, \mathbf{p}'_\beta\right)\left(\varphi'_\gamma \varphi'_\beta - \varphi_\gamma\varphi\right) \\
&\quad \times \left(\ln\varphi_\gamma - \ln\varphi'_\gamma\right) d\tilde{\mathbf{p}}_\gamma d\tilde{\mathbf{p}}'_\beta d\tilde{\mathbf{p}}'_\gamma d\tilde{\mathbf{p}} \\
&= \frac{1}{2}\int_{\bar{\Omega}_p} w\left(\mathbf{p}, \mathbf{p}_\gamma; \mathbf{p}'_\beta, \mathbf{p}'_\gamma\right)\left(\varphi'_\beta \varphi'_\gamma - \varphi\varphi_\gamma\right) \\
&\quad \times \left(\ln\varphi_\gamma - \ln\varphi'_\gamma\right) d\tilde{\mathbf{p}}_\gamma d\tilde{\mathbf{p}}'_\beta d\tilde{\mathbf{p}}'_\gamma d\tilde{\mathbf{p}}
\end{aligned} \tag{2.42}$$

(we have used (2.24)). Adding (2.41) and (2.42), then returning to (2.40), we find

$$\frac{dS}{dt} = -\frac{k_B}{2}\int_{\Omega_r}\int_{\bar{\Omega}_p} w\left(\tilde{\mathbf{p}}, \tilde{\mathbf{p}}_\gamma; \tilde{\mathbf{p}}'_\beta, \tilde{\mathbf{p}}'_\gamma\right)\left(\varphi'_\beta\varphi'_\gamma - \varphi\varphi_\gamma\right) \times \ln\frac{\varphi\varphi_\gamma}{\varphi'_\beta\varphi'_\gamma} d\tilde{\mathbf{p}}_\gamma d\tilde{\mathbf{p}}'_\beta d\tilde{\mathbf{p}}'_\gamma d\tilde{\mathbf{p}} d\tilde{\mathbf{r}} \tag{2.43}$$

Now, the logarithm is an increasing function, and so $(x - y)(\ln y - \ln x)$ is always negative. Thus, the equation which we have just obtained teaches us that

$$\frac{dS}{dt} \geqslant 0 \tag{2.44}$$

This result, which is known as the[19] *H-theorem*, shows that the entropy of an isolated system should increase over the course of time. For a non-gaseous fluid, the previous arguments do not apply as strictly as for a gas, but the main result persists: entropy increases, giving the evolution of the system an irreversible nature. One could predict the existence of an irreversibly changing quantity by returning to the literal expression of Boltzmann's equation. Mentally changing the time sign, the molecule velocities and momenta do change sign, which, of course, changes the sign of the left-hand side in (2.27), whereas the collision integral (2.28) remains unchanged. Thus, Boltzmann's equation belongs to the class of *kinetic equations*.

[19] That is what L. Boltzmann called it, H being the notation he used to denote entropy.

It must be stated that the property of increasing entropy holds true for every system (specifically every particle) that can be viewed as isolated. However, the entropy of an arbitrary particle being part of an ensemble can be defined, but nothing will then ensure it is increasing; only the entropy of the whole system certainly increases. Entropy of a part of a more global system may temporarily decrease provided that some entropy is transferred to other parts, so that the total entropy keeps rising. Besides, entropy cannot increase indefinitely and, for a system being left to itself, should then tend towards a maximal value corresponding to a so-called state of *thermodynamical equilibrium*.

The purely statistical result which we have just discussed has major effects. Definition (2.29), indeed, shows that, unlike the microscopic probability Φ, which is constant according to Liouville's theorem (2.10), the macoscopic probability $\overline{\Phi}$ varies over time. We will review the reasons for the process, as well as some of its consequences, in the following Section.

Some mechanisms evolve sufficiently slowly to allow us to consider each part of the system isolated, since its entropy remains constant over the course of time:

$$\frac{dS}{dt} = 0 \tag{2.45}$$

In this case, it is referred to as an *isentropic* mechanism. We will study the conditions for isentropicity in Chapter 3 (Section 3.3.1). Taking into account our last remark about Boltzmann's equation' irreversibility, all those processes in which the molecular collisions play a negligible part may be considered as isentropic. Relation (2.45), of course, is also valid at thermodynamical equilibrium.

2.3 Thermodynamical quantities

We start here by looking for the probability density function of a system near equilibrium, leading to the Boltzmann probability distribution. The latter is used to compute the pressure and internal energy of a perfect fluid. We also define the concepts of heat and work, and prove the second law of thermodynamics. We then deduce a state equation for perfect fluids. The notions of sound speed and mean free path are defined.

2.3.1 Boltzmann distribution

What was said above shows that, in addition to the conservation laws of classical mechanics as reviewed in Chapter 1 (Section 1.4), there is a quantity which irremediably increases over the time for an isolated system, namely entropy.

First of all, we must recall that this phenomenon is caused by collisions. In the absence of collisions, entropy would be a conservative quantity. The mechanism bringing about the increase of entropy can be briefly described as follows. The intermolecular impacts account for intermolecular transfers of momentum (and then of kinetic energy) without any change in total momentum and energy. In the course of time, the changes of velocity as experienced by the molecules because of the impacts induce great disorder in the velocity distribution, distributing the velocity vectors as disparately as possible among all the molecules (refer to Fig. 2.3). Entropy measures that disorder.[20] Definition (2.29), indeed, explains that entropy is higher when the system is in a more probable macroscopic state. Now, as we have seen it, configurations (*a*) and (*b*) in Fig. 2.3 are much more frequent than configuration (*c*). To be convinced, it suffices to consider the simple model of a system being determined by N parameters which can take N determined values. There are $N!$ configurations in which all the parameters take different values, whereas there are only N configurations in which all the parameters take the same value. In a fluid, consisting of a very high number of molecules, the ratio of the two kinds of configuration is huge. In other words, as time elapses, a system tends to adopt a microscopic configuration coresponding to the most probable macroscopic state. Simultaneously, the energy of the system constituents is homogeneously distributed (on an average) among its various molecular degrees of freedom.[21]

[20] According to the margin note in Section 2.2.3, it is a microscopic disorder corresponding to an extensive macroscopic homogeneity.

[21] This is known as energy *equipartition*.

Let us consider an isolated system, which may consist of only one particle or be extended to a whole fluid, gathering a collection of macroscopic particles. The system is assumed to be in its state of thermodynamical equilibrium; then its entropy does not vary. Such a state occurs when the right-hand side of equation (2.43) vanishes, that is when the product of the individual probabilities is left unchanged by the collisions, that is $\varphi'_\beta \varphi'_\gamma = \varphi \varphi_\gamma$, or else:

$$\varphi\left(\tilde{\mathbf{p}}'_\beta, \tilde{\mathbf{r}}'_\beta\right) \varphi\left(\tilde{\mathbf{p}}'_\gamma, \tilde{\mathbf{r}}'_\gamma\right) = \varphi\left(\tilde{\mathbf{p}}, \tilde{\mathbf{r}}\right) \varphi\left(\tilde{\mathbf{p}}_\gamma, \tilde{\mathbf{r}}_\gamma\right) \tag{2.46}$$

(the dependence on t has been omitted here, since a state of equilibrium is assumed). It should now be recalled that all four involved momenta satisfy the conservation laws to be reviewed. The first, which has already been discussed, is the conservation of momentum $\tilde{\mathbf{p}}'_\beta + \tilde{\mathbf{p}}'_\gamma = \tilde{\mathbf{p}} + \tilde{\mathbf{p}}_\gamma$. We must also make sure, however, that the impact also conserves the energy of the couple of molecules being considered as an isolated sub-system. The energy of the pair of molecules $\tilde{E}(\tilde{\mathbf{p}}, \tilde{\mathbf{r}})$, according to Chapter 1, is the sum of kinetic energy $\tilde{E}_k(\tilde{\mathbf{p}})$, potential energy $E_{ext}(\tilde{\mathbf{r}})$ of external forces (e.g. gravity) and their internal energies[22] $\tilde{E}_{int}$. It is considered that the potential interaction energy is only non-zero within short distances beween the two molecules, and that the possible internal rotational and vibrational energy is unchanged upon collision.[23] The mass m of a molecule is naturally conserved. As regards the angular momentum, its conservation is automatically provided if the force $\tilde{\mathbf{F}}$ derived from that potential is central, since the terms in the form $\tilde{\mathbf{r}}_\beta \times \tilde{\mathbf{F}}_\beta$ are then identically zero. At equilibrium, we then have the following relations:

[22]Which include, as explained in Section 1.5.3, the potential of their mutual interactions, as well as the vibrational and rotational energy of each molecule.

[23]This is referred to as an elastic collision.

$$\varphi\left(\tilde{\mathbf{p}}'_\beta, \tilde{\mathbf{r}}'_\beta\right)\varphi\left(\tilde{\mathbf{p}}'_\gamma, \tilde{\mathbf{r}}'_\gamma\right) = \varphi(\tilde{\mathbf{p}}, \tilde{\mathbf{r}})\,\varphi(\tilde{\mathbf{p}}_\gamma, \tilde{\mathbf{r}}_\gamma)$$

$$\tilde{\mathbf{p}}'_\beta + \tilde{\mathbf{p}}'_\gamma = \tilde{\mathbf{p}} + \tilde{\mathbf{p}}_\gamma \tag{2.47}$$

$$\tilde{E}\left(\tilde{\mathbf{p}}'_\beta, \tilde{\mathbf{r}}'_\beta\right) + \tilde{E}\left(\tilde{\mathbf{p}}'_\gamma, \tilde{\mathbf{r}}'_\gamma\right) = \tilde{E}(\tilde{\mathbf{p}}, \tilde{\mathbf{r}}) + \tilde{E}(\tilde{\mathbf{p}}_\gamma, \tilde{\mathbf{r}}_\gamma)$$

Thus, function φ should transform sums of momenta and energy into products of these very quantities. This is only possible if it occurs as an exponential law of the two variables $\tilde{\mathbf{p}}$ and $\tilde{E}$, whose more general form is yielded by

$$\begin{aligned}\varphi(\tilde{\mathbf{p}}, \tilde{\mathbf{r}}) = \varphi_0(\tilde{\mathbf{p}}, \tilde{\mathbf{r}}) &\doteqdot \Theta \exp\left(-\frac{\mathbf{a}\cdot\tilde{\mathbf{p}} + \tilde{E}}{k_B T}\right) \\ &= \Theta \exp\left[-\frac{\mathbf{a}\cdot\tilde{\mathbf{p}} + \tilde{E}_k(\tilde{\mathbf{p}}) + \tilde{E}_p(\tilde{\mathbf{r}})}{k_B T}\right]\end{aligned} \tag{2.48}$$

where Θ, T and $\mathbf{a}$ are two unknown scalars and one unknown vector. The potential $\tilde{E}_p$ includes the internal energy of the molecules (rotation, vibration), their possible interaction energies (Lennard–Jones potential) and the potential of external forces (refer to Chapter 1, Section 1.5.3). It can be observed that, at equilibrium, the fluid is necessarily globally at rest,[24] that is perfectly isotropic, and so the vector $\mathbf{a}$ should be zero; indeed, no direction of molecule momentum is given prevalence. In such a case, the probability density is reduced to a mere function of energy:

[24]In some Galilean frame of reference.

$$\varphi_0(\tilde{\mathbf{p}}, \tilde{\mathbf{r}}) = \Theta \exp\left[-\frac{\tilde{E}(\tilde{\mathbf{p}}, \tilde{\mathbf{r}})}{k_B T}\right] \tag{2.49}$$

This law can be found again by expressing the fact that the thermodynamical equilibrium corresponds to a maximal value of entropy, according to (2.45). To that purpose, we try to maximize the integral (2.35), which can be done through the Lagrange equation (1.23), by substituting $\varphi(1-\ln\varphi)$, which is taken as a function of the variable φ, for the Lagrangian L. One should remember, however, that this extremum has to be sought under constraints. One the one hand, the probability density should meet the normalization condition as yielded by the second line of (2.5). Moreover, the sum of energies of the molecules being included in the relevant particle should equal the whole energy of the particle, that is according to (2.7). We then have to seek the extremum of an integral by satisfying two constraints which themselves are given by integrals. To do this, we use the Lagrange multiplier technique as set out in Section 1.2.4, which here consists in seeking an extremum of the following quantity:

$$S + A\left(\int_{\Omega_p\times\Omega_r} \varphi d\tilde{\mathbf{p}}d\tilde{\mathbf{r}} - N_m\right) + B\left(\int_{\Omega_p\times\Omega_r} \tilde{E}\varphi d\tilde{\mathbf{p}}d\tilde{\mathbf{r}} - E\right) \qquad (2.50)$$

where A and B are the multipliers associated with the constraints (2.5) and (2.7). Since the number of molecules and the total energy are set, we then seek an extremum of the following integral:

$$\begin{aligned} &\int_{\Omega_p\times\Omega_r} \Lambda(\varphi)\, d\tilde{\mathbf{p}}d\tilde{\mathbf{r}} \\ &\Lambda(\varphi) \doteqdot \varphi\left[k_B(1-\ln\varphi) + A + B\tilde{E}\right] \end{aligned} \qquad (2.51)$$

where Λ acts as a 'Lagrangian'. It does not depend on the derivatives of φ, and the associated Lagrange equation is then reduced to $(\partial\Lambda/\partial\varphi)(\varphi=\varphi_0)=0$, that is

$$-k_B\ln\varphi_0 + A + B\tilde{E} = 0 \qquad (2.52)$$

We ultimately get

$$\begin{aligned} \varphi_0(\tilde{\mathbf{p}},\tilde{\mathbf{r}}) &= \exp\left(\frac{A+B\tilde{E}}{k_B}\right) \\ &= \Theta\exp\left(-\frac{\tilde{E}}{k_BT}\right) \end{aligned} \qquad (2.53)$$

which coincides with (2.48) if we let[25] $A \doteqdot k_B\ln\Theta$ and $B \doteqdot -1/T$. This is the so-called *Maxwell–Boltzmann distribution*. One should remember it is the distribution of a system in a state of thermodynamical equilibrium, that is a stationary solution of the Boltzmann's equation for a gas, then satisfying $d\varphi_0/dt=0$ and $dS/dt=0$.

[25] A is identified as the opposite of entropy at equilibrium, as discussed in Section 2.4.2.

To make progress in our arguments, we will from now on consider the case without an external field nor an internal molecular internal energy and without molecular interactions, that is without collisions (perfect gas). Energy is then reduced to its kinetic component, that is $\tilde{E}=\tilde{E}_k=|\tilde{\mathbf{p}}|^2/2m$ (according to (1.78)). The normalization factor Θ is then obtained by writing that the total probability is equal to one, according to (2.5):

$$
\begin{aligned}
N_m &= \int_{\Omega_p \times \Omega_r} \varphi_0 \left(\tilde{\mathbf{p}}, \tilde{\mathbf{r}}\right) d\tilde{\mathbf{p}} d\tilde{\mathbf{r}} \\
&= \Theta V \int_{\Omega_p} \exp\left(-\frac{\left|\tilde{\mathbf{p}}\right|^2}{2mk_B T}\right) d\tilde{\mathbf{p}}
\end{aligned} \tag{2.54}
$$

To calculate the integral, we observe that the integrand is isotropic, that is it only depends on the norm $\left|\tilde{\mathbf{p}}\right|$. Sphere-integrating the three-dimensional volume element $d\tilde{\mathbf{p}}$ gives $4\pi \left|\tilde{\mathbf{p}}\right|^2 d\left|\tilde{\mathbf{p}}\right|$. Consequently, we have the following general relation, which is valid for every integer k and every real a:

$$
\begin{aligned}
\int_{\mathbb{R}^3} |\mathbf{x}|^{2k} e^{-a|\mathbf{x}|^2} d\mathbf{x} &= 4\pi \int_0^{+\infty} |\mathbf{x}|^{2k+2} e^{-a|\mathbf{x}|^2} d\,|\mathbf{x}| \\
&= \frac{4\pi}{a^{k+3/2}} \int_0^{+\infty} y^{2k+2} e^{-y^2} dy
\end{aligned} \tag{2.55}
$$

A mathematical induction then readily yields

$$
\int_0^{+\infty} y^{2k} e^{-y^2} dy = \frac{(2k)!\sqrt{\pi}}{2^{2k+1} k!} \tag{2.56}
$$

Thus, (2.55) yields

$$
\int_{\mathbb{R}^3} |\mathbf{x}|^{2k} e^{-a|\mathbf{x}|^2} d\mathbf{x} = \left(\frac{\pi}{a}\right)^{3/2} \frac{(2k+2)!}{2\,(k+1)!\,(4a)^k} \tag{2.57}
$$

For $k = 0$ and $a = 1/\left(2mk_B T\right)$, the formula (2.54) becomes

$$
N_m = \Theta V \left(2\pi m k_B T\right)^{3/2} \tag{2.58}
$$

($V = N_m/n$ is the total volume of the supposedly homogeneous fluid, and so $n\left(\tilde{\mathbf{r}}, t\right) = n = cte$). Hence:

$$
\begin{aligned}
\Theta &= \frac{n}{\left(2\pi m k_B T\right)^{3/2}} \\
&= \frac{\rho}{\sqrt{\left(2\pi k_B T\right)^3 m^5}}
\end{aligned} \tag{2.59}
$$

We can estimate the pressure by considering a wall which would be immersed in a fluid and would experience impacts from molecules running against it. According to Newton's equation (1.56), a molecule with a momentum $\tilde{\mathbf{p}}$ will transmit to the wall with a normal component $\mathbf{n}$ a force $\tilde{\mathbf{f}}$ equalling its variation of momentum during the required impact time dt. When bouncing off the wall, the molecule will fly away with a momentum whose component normal to the wall will have changed sign; that makes it possible to write the variation of normal momentum as $\tilde{\mathbf{p}}\,(t + dt) \cdot \mathbf{n} - \tilde{\mathbf{p}}\,(t) \cdot \mathbf{n} = 2\tilde{\mathbf{p}}\,(t) \cdot \mathbf{n}$. The elementary normal force exerted by that molecule is then written as

$$
\tilde{\mathbf{f}} \cdot \mathbf{n} = \frac{2\tilde{\mathbf{p}} \cdot \mathbf{n}}{dt} \tag{2.60}
$$

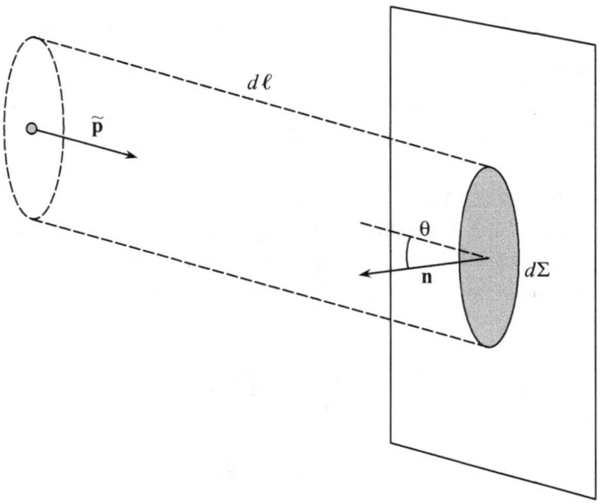

Fig. 2.4 Schematic diagram for statistically calculating the pressure.

The time taken by the molecule to follow that trajectory is given by its distance $d\ell$ from the wall:

$$\begin{aligned} dt &= \frac{d\ell}{|\tilde{\mathbf{u}}|\cos\theta} \\ &= \frac{m d\ell}{\tilde{\mathbf{p}}\cdot\mathbf{n}} \end{aligned} \tag{2.61}$$

where θ is the angle its trajectory makes with the wall (refer to Fig. 2.4). Hence, we have

$$\tilde{\mathbf{f}}\cdot\mathbf{n} = \frac{2\left(\tilde{\mathbf{p}}\cdot\mathbf{n}\right)^2}{m d\ell} \tag{2.62}$$

To find the total force as experienced by the wall, the integration should be made over all the possible momenta of those molecules occurring in the volume $d\ell d\Sigma$, where $d\Sigma$ is the elementary area of the wall portion being considered, multiplying by the probability density:

$$\mathbf{f}\cdot\mathbf{n} = \frac{2d\Sigma}{m}\int_{\Omega_p^+}\left(\tilde{\mathbf{p}}\cdot\mathbf{n}\right)^2\varphi\left(\tilde{\mathbf{p}},\tilde{\mathbf{r}}\right)d\tilde{\mathbf{p}} \tag{2.63}$$

Ω_p^+ denotes the set of momenta of the molecules running towards the wall, that is such that $\tilde{\mathbf{p}}\cdot\mathbf{n}\geqslant 0$. Pressure is derived therefrom as the ratio of that integral to the area $d\Sigma$. The integration may be taken over Ω_p by symmetry, provided that the factor of 2 is deleted. Lastly, since $\tilde{\mathbf{p}}\cdot\mathbf{n}$ equals the component $\tilde{p}_x$ of $\tilde{\mathbf{p}}$ normal to the wall, an argument of isotropy allows us to write

$$\begin{aligned} p &= \frac{1}{m}\int_{\Omega_p}\left(\tilde{\mathbf{p}}\cdot\mathbf{n}\right)^2\varphi\left(\tilde{\mathbf{p}},\tilde{\mathbf{r}}\right)d\tilde{\mathbf{p}} \\ &= \frac{1}{3m}\int_{\Omega_p}\left|\tilde{\mathbf{p}}\right|^2\varphi\left(\tilde{\mathbf{p}},\tilde{\mathbf{r}}\right)d\tilde{\mathbf{p}} \end{aligned} \tag{2.64}$$

because $|\tilde{\mathbf{p}}|^2 = \tilde{p}_x^2 + \tilde{p}_y^2 + \tilde{p}_z^2$. This is a rather general argument which amounts to stating that the molecule energy variation $|\tilde{\mathbf{p}}|^2/2m$ equals the quantity $-pdV$, in accordance with Section 1.5.4 in Chapter 1. Formula (2.64) shows that the pressure is positive and will be advantageously utilized in Section 2.3.3.

Let us now consider the internal energy and try to assess it in the same manner. Since this fluid is assumed to be globally at rest, the average kinetic energy of the molecules is likened to E_{int} (eqn (1.235) in the absence of external forces). Now, according to (2.4), that energy is obtained by calculating the integral

$$\begin{aligned} E_{int} &= \left\langle \tilde{E}_k \right\rangle = \frac{1}{n}\int_{\Omega_p} \tilde{E}_k\left(\tilde{\mathbf{p}}\right)\varphi_0\left(\tilde{\mathbf{p}},\tilde{\mathbf{r}}\right)d\tilde{\mathbf{p}} \\ &= \frac{1}{2nm}\int_{\Omega_p} |\tilde{\mathbf{p}}|^2\varphi_0\left(\tilde{\mathbf{p}},\tilde{\mathbf{r}}\right)d\tilde{\mathbf{p}} \\ &= \frac{\Theta}{2nm}\int_{\Omega_p} |\tilde{\mathbf{p}}|^2\exp\left(-\frac{|\tilde{\mathbf{p}}|^2}{2mk_BT}\right)d\tilde{\mathbf{p}} \end{aligned} \tag{2.65}$$

With (2.55) for $k = 1$, we ultimately find

$$E_{int} = \frac{3k_BT}{2} \tag{2.66}$$

Quantity T then happens to be proportional to the average kinetic energy of a molecule, and is then positive. That quantity is called the medium temperature and is measured in Kelvins.

The rms (root mean square) molecular velocity is a statistically interesting quantity as given by

$$\begin{aligned} \mathring{u} &\doteqdot \left\langle|\tilde{\mathbf{u}}|^2\right\rangle^{1/2} \\ &= \sqrt{\frac{2\left\langle\tilde{E}_k\right\rangle}{m}} \\ &= \sqrt{\frac{3k_BT}{m}} \end{aligned} \tag{2.67}$$

One should remember that our last considerations are only valid for a perfect gas at equilibrium. If the gas is out of equilibrium, for example if it is subject to a velocity which varies in space due to external loads (external fields, domain boundary conditions, etc.), the above results can be matched through the following crucial finding. The collision processes returning the system to its state of equilibrium have a characteristic time scale on the order of

$$\tau_m \doteqdot \frac{\ell}{\mathring{u}} \tag{2.68}$$

where ℓ is the *mean free path*, that is the average distance between collisions. Given the usual values of these quantities in the ordinary fluids, this characteristic time is much shorter than the time during which the macroscopic quantities happen to be modified. Likewise, the spatial scale characterizing the variations of the macroscopic quantities is very large with respect to the typical intermolecular distance. It can then be assumed, in many cases, that each fluid particle a is permanently close to the state of local equilibrium, which, apart from a few details, allows us to keep the distribution of probability of equilibrium. It should be specified, indeed, that out of the global equilibrium, the statistical properties of the microscopic motion of molecules no longer have any reason for complying with a law of isotropy, and so the factor $\mathbf{a} \cdot \tilde{\mathbf{p}}$ of (2.48) should be kept. This is because if the fluid particle locally moves at a velocity $\mathbf{u}_a$, then the molecules in its frame of reference move about at the velocity $\tilde{\mathbf{p}}/m - \mathbf{u}_a$; hence, their individual kinetic energies now equal

$$\frac{1}{2}m\left(\frac{\tilde{\mathbf{p}}}{m} - \mathbf{u}_a\right)^2 = \frac{1}{2}mu_a^2 - \mathbf{u}_a \cdot \tilde{\mathbf{p}} + \frac{|\tilde{\mathbf{p}}|^2}{2m} \tag{2.69}$$

instead of $|\tilde{\mathbf{p}}|^2/2m$. The particle velocity $\mathbf{u}_a$ is likened to the vector $\mathbf{a}$, whereas the parameter $mu_a^2/2$ occurs as extra energy which can logically be absorbed into the normalization coefficient Θ. For reasons to be explained in Chapter 3 (Section 3.5.3), it is more convenient, however, to keep this term. The distribution function within a particle a is then reduced to:

$$\varphi\left(\tilde{\mathbf{p}}, \tilde{\mathbf{r}}, t\right) = \Theta \exp\left[-\frac{\frac{1}{2}m\left(\tilde{\mathbf{p}}/m - \mathbf{u}_a\right)^2 + \tilde{E}_p\left(\tilde{\mathbf{r}}\right)}{k_B T_a}\right] \tag{2.70}$$

This is what is called the *non-equilibrium Maxwell–Boltzmann distribution*, generalizing (2.49). The temperature of particle a has been denoted T_a; this quantity can now vary in both time and space, just like $\mathbf{u}_a$, since the fluid is out of equilibrium. The distribution (2.70) then does not verify $d\varphi/dt = 0$, as explained in Section 3.5.3. For this reason, applying the formula (2.35) is useless for explicitly calculating the entropy, since it involves an integral on space (here being reduced to the extent of a particle). Theoretically, the distribution (2.70) would make it possible to calculate all the desired average quantities. However, it is only strictly applicable to gases, as explained earlier (for further details, refer to Rief, 1965). As to relation (2.67), it is still valid for an out-of-equilibrium gas, but, of course, it is concerned with the root mean square value of the molecule velocity in the particle's inertial frame of reference:

$$\begin{aligned} \mathring{u} &\doteqdot \left\langle \left|\tilde{\mathbf{u}} - \mathbf{u}_a\right|^2 \right\rangle^{1/2} \\ &= \sqrt{\frac{2E_{int,a}}{m}} \\ &= \sqrt{\frac{3k_B T_a}{m}} \end{aligned} \tag{2.71}$$

That quantity occurs, as expected, as a Galilean invariant.

Although not all the above results apply to the non-gaseous fluids, they will subsequently be used for determining other essential results.

2.3.2 Heat and work

We will temporarily return to the case of a particle a. It is noteworthy that its entropy S_a depends on other characterizing thermodynamical quantities. Actually, if a particle is at local equilibrium, its entropy is kept. Since it is also additive, S_a may then be nothing but an additive quantity of well-known conservative quantities. We have identified seven quantities in Section 1.4: E_a, $\mathbf{p}_a$, $\mathbf{M}_a$. Entropy, however, is obviously a Galilean invariant, since it results from the application of mechanics equations to an ensemble of molecules, regardless of the work frame of reference. Thus, only internal energy (refer to Section 1.5.3) may act among the conservative quantities which entropy depends on:

$$S_a = S_a\left(E_{int,a}\right) \tag{2.72}$$

For an isolated particle at equilibrium, the conservation of internal energy then leads to the conservation of entropy, which is evidence of a thermodynamical equilibrium, as expected. Entropy per unit mass σ_a, which is also known as *specific entropy*, can be defined as:

$$\sigma_a \doteq \frac{S_a}{m_a} \tag{2.73}$$

Using the definition (1.254), (2.72) can then be rearranged as:

$$\sigma_a = \sigma_a\left(e_{int,a}\right) \tag{2.74}$$

Internal energy can be expressed with respect to σ_a by inverting this relation. However, one should consider the relation (1.257), which states that internal energy per unit mass $e_{int,a}$ also depends on density. We ultimately get the general relation

$$e_{int,a} = e_{int}\left(\rho_a, \sigma_a\right) \tag{2.75}$$

Specific entropy σ_a is the second (and last) of the state variables $\left\{Z_{a,i}\right\}_i$ mentioned in Section 1.5.3, that is $\left\{Z_{a,i}\right\}_i = (\rho_a, \sigma_a)$. For an isentropic process, since entropy σ_a is constant, $e_{int,a}$ only depends on ρ_a, and equation (1.258) of Section 1.5.4 is recovered. By writing down the differential of (2.75), we get the variation of a particle internal energy as[26]

$$de_{int,a} = \left(\frac{\partial e_{int,a}}{\partial \rho_a}\right)_\sigma d\rho_a + \left(\frac{\partial e_{int}}{\partial \sigma}\right)_\rho d\sigma_a \tag{2.76}$$

(the subscripts after the parentheses mean that the partial derivatives are considered while keeping the other variable constant). Returning to the true internal energy $E_{int,a}$ of the particle, we may also write, with (1.254), (1.255) and (2.73):

$$dE_{int,a} = -\left(\frac{\partial e_{int,a}}{\partial \rho_a}\right)_\sigma dV_a + \left(\frac{\partial e_{int}}{\partial \sigma}\right)_\rho dS_a \tag{2.77}$$

As the disorder becomes established, the increase of entropy should alter the internal energy. Actually, since the molecules move about in a more and more disparate manner, the variations of their velocities grow and internal energy

[26] We consider, here and after in the present chapter, that the fluid is made of one single chemical substance. In other circumstances, one should also take account of the mass fractions of the various substances.

should a priori rise. We will investigate the reasons for this; we will first scrutinize the possible causes of the changes in the total energy of a particle interacting with an external medium. To do this, we consider a given particle and from now on omit its label a for the sake of clarity. The discussions about entropy, as developed in Section 2.2.3, show that the energy of an open thermodynamical system varies according to the distribution of energy of its constituents, the energy being the sum of a (global) kinetic energy and an internal energy corresponding to the kinetic and potential energies of the constituents (molecules) in the inertial frame of reference. In order to get a deeper insight into these notions, let us resume the formalism of Section 2.2.3, where the state parameters are discretized, but within a slightly modified context as regards energy alone. To that purpose, we divide all the possible values taken by energy into small segments which are centred on values E_i. Since the Maxwell–Boltzman probability density φ is a function of energy alone (eqn (2.49)), the number N_i of molecules being in that state of energy is given by

$$N_i = \exp\left(-\frac{E_i}{k_B T}\right) \tag{2.78}$$

the total number $N_m = \sum_i N_i$ of molecules being kept. The total energy of the particle can be written using a formula which is analogous to (2.65), that is

$$\begin{aligned} E &= \int_{\Omega_E} \tilde{E}\varphi\left(\tilde{E}\right) d\tilde{E} \\ &= \sum_i N_i E_i \end{aligned} \tag{2.79}$$

The variation of energy is then given by

$$dE = \sum_i N_i dE_i + \sum_i E_i dN_i \tag{2.80}$$

The first term of (2.80) corresponds to a variation of the individual energies making up the system. Hence it can be nothing but a variation caused by the internal force power, as per Section 1.5.3. Now, Section 1.5.4 explained that this work results from the action of pressure and is written as $-pdV$, according to (1.261). This result is confirmed by the calculation having led to equation (2.64). This work will be denoted δW. The second term comes in addition to (1.261) and represents a possible redistribution of energies among all the molecules. Then it characterizes the collective disorder and is referred to as *amount of received heat*, denoted δQ:

$$\begin{aligned} \delta W &\doteq \sum_i N_i dE_i = -pdV \\ \delta Q &\doteq \sum_i E_i dN_i \end{aligned} \tag{2.81}$$

Resuming the total variation of energy as calculated in (2.80), we ultimately get:

$$dE = -pdV + \delta Q \tag{2.82}$$

This relation is the *first law of thermodynamics*. As expected, the energy of a non-isolated particle is then not constant, but changes provided that some work and/or heat is exchanged with its immediate surroundings.[27] From what was said before, heat happens to be a variation of the number of molecules exhibiting each energy state, the energy levels remaining constant. In order to relate that magnitude to those quantities introduced in the previous paragraph, let us write the variation of entropy as a function of the dN_i. To that purpose, we note that the total number of molecules is constant, which gives

[27] We may recall that the discussion is focused on identical molecules. Otherwise, the reactions liable to occur between different kinds of molecules would give rise to another term in the previous calculations, that term being related to the notion of chemical potential (refer, for instance, to Landau and Lifshitz, 1980).

$$\sum_i dN_i = dN_m = 0 \tag{2.83}$$

Afterwards, the property (2.34) enables us to write

$$\begin{aligned} dS &= k_B \sum_i \left[dN_i \left(\ln \frac{k}{N_i} + 1 \right) + N_i d \ln \frac{k}{N_i} \right] \\ &= k_B \sum_i (\ln k dN_i - \ln N_i dN_i + N_i d \ln k) \\ &\approx -k_B \sum_i \ln N_i dN_i \end{aligned} \tag{2.84}$$

In the case of a perfect gas at equilibrium, using again the Maxwell–Boltzmann distribution (2.78), we have $\ln N_i = -E_i/k_B T$, and (2.84) yields

$$\begin{aligned} dS &= \frac{1}{T} \sum_i E_i dN_i \\ &= \frac{\delta Q}{T} \end{aligned} \tag{2.85}$$

In the more general case of any fluid being non necessarily at equilibrium, relation (2.85) does not hold true any longer, since the intermolecular interaction processes (particularly the collisions) contribute to an increase in entropy. Besides, if the collisions are non-elastic (refer to Section 2.3.1), a part of the molecule energy can be converted into internal molecular energy (vibration, rotation) during collisions. These phenomena are then irreversible and inevitably increase the entropy in relation to equation (2.85). Thus, we may write the more general inequality

$$dS \geqslant \frac{\delta Q}{T} \tag{2.86}$$

That property constitutes the *second law of Thermodynamics*. Thus, the variation of entropy involves heat exchanges and is associated with a redistribution of the molecules among the various available energy states, which is a disorder appearance process, as discussed in Section 2.3.1. The redistribution of energy which occurs along with the increase in entropy is then closely linked to temperature, that is to the kinetic energy of the molecular motion, according to (2.67).

Relations (2.82) and (2.85) show that, in the case of a reversible process, the variation of energy caused by the variation of entropy is given by $\delta Q = TdS$, whereas that relation becomes an inequality in the irreversible case. Hence we write:

$$\begin{aligned} \delta Q_{rev} &= TdS \\ \delta Q_{irr} &\leqslant TdS \end{aligned} \tag{2.87}$$

Among the various components of energy, only internal energy (2.77) explicitly depends on entropy. We can then infer that, in equation (2.77), the term which is proportional to dS should be likened to T. With equation (1.256), we can then write

$$\begin{aligned} p &= \rho^2 \left(\frac{\partial e_{int}}{\partial \rho} \right)_\sigma \\ T &= \left(\frac{\partial e_{int}}{\partial \sigma} \right)_\rho \end{aligned} \tag{2.88}$$

which generalizes (1.256), or else:

$$\begin{aligned} p &= -\left(\frac{\partial E_{int}}{\partial V} \right)_S \\ T &= \left(\frac{\partial E_{int}}{\partial S} \right)_V \end{aligned} \tag{2.89}$$

which generalizes (1.253). Equation (2.77) now reads as

$$dE_{int} = -pdV + TdS \tag{2.90}$$

Combining that relation with (2.82) and utilizing (2.86), we get:

$$\begin{aligned} dE_k + dE_{ext} &= dE - dE_{int} \\ &= \delta Q - TdS \\ &\leqslant 0 \end{aligned} \tag{2.91}$$

If the system is isolated or subject to a constant external potential ($dE_{ext} = 0$), then kinetic energy changes by an amount $\delta Q - TdS$, which equals zero for a reversible process, according to (2.85). In the case of an irreversible process, kinetic energy decreases as internal energy increases. The energy of an isolated system is then conserved while distributing itself in a disordered way among the molecules β making it up. Thus, the internal energy should increase over the course of time, and so the overall kinetic energy of the system should decrease to conserve the total energy. The increase of entropy then happens to be a mechanism converting the overall kinetic energy of a particle into internal energy during the return to a state of equilibrium. Accordingly, there is a flux of energy from E_k to E_{int}, possessing a determined sign, unlike the general mechanism of energetic equilibrium being contemplated in Section 1.5.4, which is only caused by the pressure loads. The two contributions from dE

in equation (2.82) are then of quite different natures. We will resume this discussion in Section 2.4. In the case of a reversible process, $dS = 0$ for an isolated system, and the amount of heat vanishes as well.

We will now use again the notation A_a to refer to every quantity A with respect to particle a. If a particle is not isolated, its entropy varies by an undetermined sign. Thus, the particles making up a system may experience energy variations $T_a dS_a$ of arbirary sign; only the entropy of the (supposedly isolated) total system certainly increases. Let us consider two particles a and b making up an isolated system at thermodynamical equilibrium; thus, their total mutual entropy is maximal. It is then constant, and so the internal energy is maximal too according to (2.72). Hence we have:

$$\begin{aligned} 0 &= \frac{d\left(S_a + S_b\right)}{dt} \\ &= \frac{dS_a\left(E_{int,a}\right)}{dt} + \frac{dS_b\left(E_{int,b}\right)}{dt} \end{aligned} \tag{2.92}$$

The total internal energy $E_{int} = E_{int,a} + E_{int,b}$ being conserved, however, we have $dE_{int,a} = -dE_{int,b}$. Then, according to (2.92) and (2.89), we can write

$$\begin{aligned} 0 &= \frac{dS_a}{dE_{int,a}} \frac{dE_{int,a}}{dt} + \frac{dS_b}{dE_{int,b}} \frac{dE_{int,b}}{dt} \\ &= \left(\frac{1}{T_a} - \frac{1}{T_b}\right) \frac{dE_{int,a}}{dt} \end{aligned} \tag{2.93}$$

which leads to $T_a = T_b$. More generally, temperature is then constant within a medium at equilibrium. When, on the contrary, two particles have different temperatures, they do not make up a system at equilibrium; their state variables evolve over time and their total entropy increases. The preceding calculation then shows that

$$\left(\frac{1}{T_a} - \frac{1}{T_b}\right) \frac{dE_{int,a}}{dt} > 0 \tag{2.94}$$

If, for example, $T_a > T_b$, $dE_{int,a}/dt < 0$: there is a flux of internal energy from a to b which, according to (2.89) is written as $dE_{int,a} = T_a dS_a$. Both energy and entropy of particle a increase, whereas it is the opposite for b. In the absence of any other external influence, this process lasts until the two particles have the same temperature. The system comprising the two particles is then at thermodynamical equilibrium and both its internal energy and its entropy have reached their maximal values. This argument first shows that irreversible phenomena tend to make the quantities homogeneous in space over the time; this is an excessively irreversible mechanism which we will discuss again in Chapters 3 and 4.

In the following, we will contemplate constant temperature systems, that is without heat exchanges ($\delta Q = 0$), and (2.91) becomes

$$dE_{k,a} + dE_{ext,a} = -T_a dS_a \leqslant 0 \tag{2.95}$$

A particle at thermodynamical equilibrium ($dS_a = 0$) will then have a constant energy. For a system of particles, it is noteworthy that the work of the internal forces still obeys the relation (1.248), which also yields

$$\mathbf{F}_a^{int} \cdot d\mathbf{r}_a = -T_a dS_a \tag{2.96}$$

With (1.250), we also get

$$dE_{int} = \sum_a dE_{int,a} = \sum_a T_a dS_a \tag{2.97}$$

For an isolated particle, we then have $dE_{int,a} = T_a dS_a$. Since the entropy naturally tends to increase, dS_a should be positive and tends to make the internal energy increase to the detriment of the macroscopic kinetic energy, as we have already explained it. At the thermodynamical equilibrium, $dS_a = 0$ means that the internal energy remains unchanged.

2.3.3 State equation

We will keep discussing the thermodynamical properties of a large system; once again the particle labels a, b, and so on will be ignored. Pressure, as introduced in Section 1.5.4, probably depends on the root mean square velocity of the molecules as given by (2.71), that is on temperature; we will investigate the process. Internal energy e_{int}, as we have seen, depends on entropy σ and density ρ. The variables could be changed and e_{int} could be expressed with respect to p and ρ, or we could even, by inverting that relationship, seek the pressure in the form

$$p = p\,(e_{int}, \rho) \tag{2.98}$$

In the case of an isentropic mechanism, we have seen that the internal energy only depends on ρ, which allows us to write

$$\begin{aligned} e_{int} &= e_{int}\,(\rho) \\ p &= p\,(\rho) \end{aligned} \tag{2.99}$$

Such a relation as (2.98) or (2.99) is known as the *state equation* of the fluid being studied.[28] We will look for such a law under some hypotheses. For an isolated system being subject to an irreversible transformation, energy is constant, and so the relations (2.82) and (2.87) give us $pdV = TdS$, that is

[28] Equation (2.99) is of the 'barotropic' type.

$$p = T\frac{dS}{dV} \tag{2.100}$$

In order to determine the dependence of entropy on volume, let us discuss the macroscopic probability density $\overline{\Phi}$ of the particle making up the system being investigated. We saw in Section 2.3.1 that the molecule distribution function is linked to energy E. For a gas without interactions, the potential energy being zero, this function may only depend on the molecule positions (since the kinetic energy and the internal energy do not depend on it). Through a simple probabilistic reasoning, we then may state that $\overline{\Phi}$ is proportional to V^{-N_m}. Using definition (2.29), we find:

$$\begin{aligned}\frac{dS}{dV} &= -k_B \frac{d \ln \left(\Phi_0 V^{-N_m}\right)}{dV} \\ &= \frac{N_m k_B}{V}\end{aligned} \tag{2.101}$$

where Φ_0 is a constant. Ultimately, (2.100) leads to

$$p = nk_B T = \frac{\rho k_B T}{m} \tag{2.102}$$

(with, as explained earlier, $n = N_m/V$ and $\rho = mN_m/V$). This state equation yields a positive pressure, as expected. We need to emphasize that this form is only valid for a perfect gas, that is without interactions between its molecules. We find that the root mean square molecular velocity, as given by (2.71), then becomes

$$\mathring{u} = \sqrt{\frac{3p}{\rho}} \tag{2.103}$$

Thus, pressure is a possible measurement of the molecular energy, as suggested by the discussions in the previous two sections, for instance by the equations (2.64) and (2.82). Formula (2.103) also confirms the invariance of pressure by Galilean transformation.

Formula (2.102) can also be derived from equations (2.64), (2.65) and (2.66). This result may also be found from the virial theorem (1.87) applied to a fairly large ensemble of molecules β making up a particle a, as we are about to see. Assuming that the system will not significantly vary over a given period of time, the statistical average (2.4) may, indeed, be likened to a temporal average[29] analogous to the left-hand side of (1.87) (divided by the number N_m of molecules). Besides, it may be thought that the discrete sum occurring in the right-hand side of (1.87) is equivalent, in this case, to a volume integral:

[29] The steady turbulent flows will be similarly dealt with (on an average) in Chapter 4.

$$\left\langle \tilde{E}_k \right\rangle = -\frac{1}{2N_m} \int_\Omega \left\langle \tilde{\mathbf{F}}_\beta \right\rangle \cdot d\mathbf{r}_b \tag{2.104}$$

Ω denoting the domain constituting the particle.[30] In the case of a perfect gas (i.e. in the absence of collisions), the integral is reduced to a surface integral along the boundary of the particle, and the average force $\left\langle \tilde{\mathbf{F}}_\beta \right\rangle$ comes down (per unit area) to the pressure loads $-p\mathbf{n}$ exerted along the domain boundaries. If the contemplated particle is small enough, it may be considered that its pressure is uniform. From definition (1.253), we can then write:

[30] It has become obvious that we are gradually coming closer to the formalism of continuous media, to be discussed in Chapter 3.

$$\left\langle \tilde{E}_k \right\rangle = \frac{p}{2N_m} \oint_{\partial\Omega} \mathbf{r} \cdot \mathbf{n} \, d\Gamma \tag{2.105}$$

Through Gauss' theorem, the surface integral is reduced to

$$\begin{aligned}\oint_{\partial\Omega} \mathbf{r}\cdot\mathbf{n}\,d\Gamma &= \int_{\Omega} \mathrm{div}\mathbf{r}\,d\Omega \\ &= \int_{\Omega} 3d\Omega \\ &= 3V\end{aligned} \tag{2.106}$$

Putting together (2.65), (2.66), (2.105) and (2.106), we find again the relation (2.102).

Many equations of state are more suitable than the perfect gas simplistic law (2.102). Van der Wals (1873) suggests, for example, an alternative which is based on the idea that each molecule occupies a non-zero volume. More generally, one may intuitively feel that pressure should be an increasing function of density. For some fluids, the following procedure can be applied. Let us consider the quantity

$$K = -\rho\frac{\partial p}{\partial \rho} \tag{2.107}$$

known as the fluid *incompressibility modulus*. That modulus, indeed, measures the effect of density on pressure, that is it measures a change in volume. For a fluid with a rather stable density,[31] this coefficient is then slowly variable too, and so it can be expanded into a first-order Taylor series in p:

$$K = K_0 + \gamma p + O\left(p^2\right) \tag{2.108}$$

[31] The conditions for incompressibility of a fluid will be discussed in Section 3.4.3 of Chapter 3. The next arguments are valid as long as the relative volume deviations do not exceed approx. 10%.

with

$$\begin{aligned}K_0 &\doteqdot \left(\rho\frac{\partial p}{\partial \rho}\right)_{p=0} \\ \gamma &\doteqdot \left(\frac{\partial K}{\partial p}\right)_{p=0}\end{aligned} \tag{2.109}$$

Only keeping the below first-order terms, the relation (2.108) is also written as

$$\rho\frac{\partial p}{\partial \rho} - \gamma p - K_0 = 0 \tag{2.110}$$

It is a differential equation with respect to $p\,(\rho)$, whose solution is written as

$$p\,(\rho) = C\rho^{\gamma} - \frac{K_0}{K_0'} \tag{2.111}$$

where C is a constant. It can be noticed that pressure vanishes for a reference density ρ_0 as given by

$$\rho_0 \doteqdot \left(\frac{K_0}{CK_0'}\right)^{\frac{1}{\gamma}} \tag{2.112}$$

Density is then necessarily higher than ρ_0. A common practice consists of putting $c = (K_0/\rho_0)^{1/2}$. Inserting that relation and (2.112) into (2.111), we ultimately get the Tait (1888) state equation (see also Murnaghan, 1944):

$$p(\rho) = \frac{\rho_0 c^2}{\gamma}\left[\left(\frac{\rho}{\rho_0}\right)^{\gamma} - 1\right] \tag{2.113}$$

which yields, as expected, a positive pressure. Note that this equation yields

$$\frac{\partial p}{\partial \rho} = c^2\left(\frac{\rho}{\rho_0}\right)^{\gamma-1} \tag{2.114}$$

Still within the frame of an isotropic transformation, the relation (2.88) allows us to derive the internal energy of pressure (2.113), according to the density:

$$\begin{aligned} e_{int} &= cst + \int \frac{p}{\rho^2} d\rho \\ &= e_0 + \frac{c^2}{\gamma(\gamma-1)}\left[\left(\frac{\rho}{\rho_0}\right)^{\gamma-1} + (\gamma-1)\frac{\rho_0}{\rho} - \gamma\right] \end{aligned} \tag{2.115}$$

where e_0 denotes the internal energy for $\rho = \rho_0$. The latter equation also allows us to express the internal energy as a function of pressure, through (2.113):

$$e_{int}(p) = e_0 + \frac{c^2}{\gamma(\gamma-1)}\left[\begin{array}{c}\left(\dfrac{\gamma p}{\rho_0 c^2} + 1\right)^{\frac{\gamma-1}{\gamma}} \\ + (\gamma-1)\left(\dfrac{\gamma p}{\rho_0 c^2} + 1\right)^{-\frac{1}{\gamma}} - \gamma\end{array}\right] \tag{2.116}$$

The physical significance of the presently introduced quantity c will appear in Chapter 3 (Section 3.4.3), and we will provide several values of it for common fluids (Table 3.2). We can, however, already achieve an interesting result in this respect. Relation (2.103) suggests, indeed, that $(p/\rho)^{1/2}$ is a magnitude of velocity $\mathring{u}$, characterizing the motion of molecules within the inertial frame of reference of the particle being studied. In such conditions, (2.113) shows that c accounts for another assessment of that quantity and that the order of magnitude of pressure is

$$p \sim \rho c^2 \tag{2.117}$$

For most fluids, γ is of the order of several units. It is considered that $\gamma = 7$ is a satisfactory value for water. From a physical point of view, the action of the state equation is as follows: when molecules come closer to each other to such an extent that density is locally increased, the growth of law (2.113) also results in a higher pressure. Pressure loads are then generated and drive the molecules away, which returns the fluid to a state where density remains close to its reference value ρ_0. This phenomenon is all the more strong as c is higher, and is further increased by the comparatively high value of the superscript γ. The fluid is then hardly compressible, according to the hypotheses.

2.4 Dissipative systems

In order to treat the case of macroscopic systems, we introduce the idea of kinetic coefficients, satisfying Onsager's symmetry conditions. This concept leads to dissipative forces, since macroscopic energy is no longer conserved, as is proved. We apply these developments to the pendulum case, then to particle systems. The symmetry properties of space are used to write the more general mathematical form of particle friction.

2.4.1 Kinetic coefficients

The irreversible nature of the dynamics of large systems, as evidenced above, now incites us to contemplate the modelling of those processes accompanying that irreversibility. The remark made at the end of Section 2.2.3 allows us to see that the collective molecular motion affects the macroscopic motion (i.e. the average motion after Boltzmann). In other words, the macroscopic irreversibility of large systems should lead to the occurrence of processes which can be watched on a macroscopic scale, though they are caused by the microscopic motions. For additional features about the concepts developed in this section, the reader may refer to Español (2003).

We have highlighted the irreversibility of large system physics, which prompts us to conduct, as a first step, a deeper investigation into the recovery of the thermodynamical equilibrium. We first of all will examine some of the notions prevailing in the behaviour of a system tending to recover that equilibrium (e.g. a particle, but without using the label a).

As in Section 1.4.2, we will initially gather all the parameters $(\{p_i\}, \{q_i\})$ characterizing the system under one notation $\mathbf{X} = \{x_i\}$, where the subscript i ranges from 1 to $2N = 6N_m$). At thermodynamical equilibrium, these parameters[32] take values being denoted x_i^0. If the system is slightly out of equilibrium, entropy tends to increase in order to return it to equilibrium, and so the quantities $\xi_i \doteqdot x_i - x_i^0$ change over time. Since the system is fully characterized by the x_i (and hence by the ξ_i), the variation rates $\dot{\xi}_i$ of these quantites depend upon the ξ_i themselves, through the Hamilton equations. However, we consider the more general situation in which those equations determining the $\dot{\xi}_i$ are unknown. The latter, of course, are zero at equilibrium. Besides, since the system is generally slowly evolving towards equilibrium, the $\dot{\xi}_i$ are low and can then be linearized with a good approximation:

$$\begin{aligned} \dot{\xi}_i &= f(\{\xi_k\}) \\ &= -\sum_j A_{ij}\xi_j \end{aligned} \tag{2.118}$$

[32] The following discussion could be applied to any quantity, provided that it is liable to change over time, that is by preventing the conservative quantities.

This relation defines a matrix $\mathbf{A}$ (the 'minus' sign is used just as a convention). In a matrix form, if $\boldsymbol{\xi} = \mathbf{X} - \mathbf{X}_0$ denotes the ($2N$-dimensioned) vector consisting of all the $\{\xi_k\}$, one will write

$$\dot{\boldsymbol{\xi}} = -\mathbf{A}\boldsymbol{\xi} \tag{2.119}$$

Besides, since entropy increased during that process, its variations according to the ξ_i are undoubtedly important for the description of the return to equilibrium. Still under the assumption of slow variations, we will expand entropy to the second order about its equilibrium value S_0 and write

$$\begin{aligned} S &= S_0 + \sum_j b_j \xi_j - \frac{1}{2} \sum_{i,j} B_{ij} \xi_i \xi_j \\ &= S_0 + \mathbf{b}\boldsymbol{\xi} - \frac{1}{2} \boldsymbol{\xi}^T \mathbf{B} \boldsymbol{\xi} \end{aligned} \tag{2.120}$$

which defines a vector **b** and another matrix **B**, which can still be chosen as symmetrical. Then we find

$$\frac{\partial S}{\partial \boldsymbol{\xi}} = \mathbf{b} - \mathbf{B}\boldsymbol{\xi} \tag{2.121}$$

Since entropy is constant at equilibrium ($\boldsymbol{\xi} = \mathbf{0}$), the vector **b** should be zero. We then introduce the quantities Ξ_i making up a vector as given by

$$\begin{aligned} \boldsymbol{\Xi} &\doteq -\frac{\partial S}{\partial \boldsymbol{\xi}} = -\frac{\partial S}{\partial \mathbf{X}} \\ &= \mathbf{B}\boldsymbol{\xi} \end{aligned} \tag{2.122}$$

(deriving with respect to either $\boldsymbol{\xi}$ or **X** will give the same result). It can be noticed that, apart from one factor, these quantities are thermodynamically conjugate to the ξ_i, as understood from Section 1.5.4 in Chapter 1. This is because (2.95) makes it possible to write

$$\boldsymbol{\Xi} = \frac{1}{T} \frac{\partial \left(E_k + E_{ext}\right)}{\partial \boldsymbol{\xi}} \tag{2.123}$$

to be compared with (1.262). Since equation (2.122) is linear, that conjugation relation is reciprocal. Ultimately, entropy is written as:

$$\begin{aligned} S &= S_0 - \frac{1}{2} \boldsymbol{\xi}^T \mathbf{B} \boldsymbol{\xi} \\ &= S_0 - \frac{1}{2} \boldsymbol{\xi} \cdot \boldsymbol{\Xi} \end{aligned} \tag{2.124}$$

Matrix **B** can then be seen as substantially positive-definite, and hence invertible, since entropy increases towards its equilibrium value S_0. The parameter rate of variation can then be expressed in a linear way as a function of the conjugate parameters, using (2.119) and (2.122) :

$$\begin{aligned} \dot{\boldsymbol{\xi}} &= -\mathbf{A}\mathbf{B}^{-1} \boldsymbol{\Xi} \\ &= -\mathbf{C} \boldsymbol{\Xi} \\ &= \mathbf{C} \frac{\partial S}{\partial \boldsymbol{\xi}} \end{aligned} \tag{2.125}$$

The matrix coefficients C_{ij} given by

$$\mathbf{C} \doteq \mathbf{A}\mathbf{B}^{-1} \tag{2.126}$$

are known as *kinetic coefficients* with respect to the process being considered and do not depend on parameters ξ_i. The rates of variation of the conjugate quantities obviously obey the same kind of law:

$$\begin{aligned}\dot{\boldsymbol{\Xi}} &= \mathbf{B}\dot{\boldsymbol{\xi}} \\ &= -\mathbf{D}\boldsymbol{\xi} \\ &= -\mathbf{B}\mathbf{C}\boldsymbol{\Xi}\end{aligned} \tag{2.127}$$

with

$$\mathbf{D} \doteqdot \mathbf{B}\mathbf{A} \tag{2.128}$$

From the elements above, we will now turn our attention to such statistical correlations as

$$\varphi_{ij}\left(t', t''\right) \doteqdot \left\langle \xi_i\left(t'\right) \xi_j\left(t''\right)\right\rangle \tag{2.129}$$

where t' and t'' are two arbitrary time points. The statistical average, which is denoted here by brackets, is defined over the whole system, that is by a formula analogous to (2.4); however, φ is substituted for by the distribution of macroscopic probability $\overline{\Phi}$ of the whole system, since we are only interested in macroscopic states; thus, in this section for every quantity A, we will have:

$$\langle A\rangle = \int_{\Omega_\xi} A\left(\boldsymbol{\xi}\right) \overline{\Phi}\left(\boldsymbol{\xi}\right) d\boldsymbol{\xi} \tag{2.130}$$

where Ω_ξ denotes the domain containing the values which the vector $\boldsymbol{\xi}$ may take. In a matrix form, (2.129) is written as

$$\boldsymbol{\varphi}\left(t', t''\right) = \left\langle \boldsymbol{\xi}\left(t'\right) \otimes \boldsymbol{\xi}\left(t''\right)\right\rangle \tag{2.131}$$

The distribution function is given by (2.30). The relation (2.124) then yields

$$\begin{aligned}\overline{\Phi}\left(\boldsymbol{\xi}\right) &= \frac{1}{N_{tot}} \exp\left(\frac{S}{k_B}\right) \\ &= \Theta \exp\left(-\frac{1}{2k_B}\boldsymbol{\xi}^T \mathbf{B}\boldsymbol{\xi}\right) \\ &= \Theta \exp\left(-\frac{1}{2k_B}\boldsymbol{\xi}\cdot\boldsymbol{\Xi}\right)\end{aligned} \tag{2.132}$$

with $\Theta \doteqdot \exp\left(S/k_B\right)/N_{tot}$ (note it is closely related to the Maxwell–Boltzmann distribution (2.70) which characterizes a molecule; we will further discuss that point in the next section). Thus, coming back to the definition (2.130), the average value of a quantity A is written as

$$\langle A\rangle = \Theta \int_{\Omega_\xi} A\left(\boldsymbol{\xi}\right) \exp\left(-\frac{1}{2k_B}\boldsymbol{\xi}^T \mathbf{B}\boldsymbol{\xi}\right) d\boldsymbol{\xi} \tag{2.133}$$

Let us now work on the correlation tensor $\boldsymbol{\varphi}$ from what has been written above. Since the laws of mechanics are time-invariant, these quantities will obviously only depend on $t = t' - t''$, and then we have, through a translation of $-t$ over the time:

$$\begin{aligned}\boldsymbol{\varphi}(t) &= \langle \boldsymbol{\xi}(t) \otimes \boldsymbol{\xi}(0) \rangle \\ &= \langle \boldsymbol{\xi}(0) \otimes \boldsymbol{\xi}(t) \rangle^T \\ &= \langle \boldsymbol{\xi}(-t) \otimes \boldsymbol{\xi}(0) \rangle^T \\ &= \boldsymbol{\varphi}^T(-t)\end{aligned} \tag{2.134}$$

In addition, according to an important law, the equations governing the motion of molecules do not vary through a time direction change. It does not mean, however, that parameters ξ_i remain unaffected. Two cases occur, indeed, when swapping the time trajectory sign: the quantities ξ_i may remain unchanged or their sign may change. The latter case happens particularly if ξ_i denotes a momentum component, the former if ξ_i denotes a position component. Thus, we will generally write

$$\xi_i(-t) = \epsilon_i \xi_i(t) \tag{2.135}$$

with $\epsilon_i = \pm 1$. Hence we get $\varphi_{ji}(-t) = \epsilon_i \epsilon_j \varphi_{ji}(t)$, that is

$$\langle \boldsymbol{\xi}(t) \otimes \boldsymbol{\xi}(0) \rangle = \left\langle \bar{\boldsymbol{\xi}}(t) \otimes \bar{\boldsymbol{\xi}}(0) \right\rangle^T \tag{2.136}$$

where $\bar{\boldsymbol{\xi}}$ is the vector of components[33] $\epsilon_i \xi_i$. Deriving with respect to time and utilizing (2.125), we get:

[33] There is no sum over subscript i in the following formula.

$$\langle [\mathbf{C}\boldsymbol{\Xi}(t)] \otimes \boldsymbol{\xi}(0) \rangle = \left\langle \left[\mathbf{C}\bar{\boldsymbol{\Xi}}(t)\right] \otimes \bar{\boldsymbol{\xi}}(0) \right\rangle^T \tag{2.137}$$

with $\bar{\Xi} \doteqdot \epsilon_i \Xi_i$. For $t = 0$, that yields

$$\mathbf{C} \langle \boldsymbol{\Xi} \otimes \boldsymbol{\xi} \rangle = \left\langle \bar{\boldsymbol{\Xi}} \otimes \bar{\boldsymbol{\xi}} \right\rangle^T \mathbf{C}^T \tag{2.138}$$

where vectors $\boldsymbol{\xi}$ and $\boldsymbol{\Xi}$ are considered at $t = 0$ (we have used the fact that matrix $\mathbf{C}$ is statistically constant, since it does not depend upon $\boldsymbol{\xi}$). Since the time point $t = 0$ is arbitrarily chosen, relation (2.138) is valid at every point in time. We will now calculate the averages appearing in it through the distribution of probability of state of the system; with (2.122) and (2.133), we get:

$$\langle \boldsymbol{\Xi} \otimes \boldsymbol{\xi} \rangle = \Theta \mathbf{B} \int_{\Omega_\xi} \boldsymbol{\xi} \otimes \boldsymbol{\xi} \exp\left(-\frac{1}{2k_B} \boldsymbol{\xi}^T \mathbf{B} \boldsymbol{\xi}\right) d\boldsymbol{\xi} \tag{2.139}$$

For calculating the second side of that equality, we will use a procedure as proposed by Landau and Lifshitz (1980). Let us introduce the following vector function:

$$\mathbf{f}(\langle \boldsymbol{\xi} \rangle) \doteqdot \Theta \int_{\Omega_\xi} \boldsymbol{\xi} \exp\left[-\frac{1}{2k_B} (\boldsymbol{\xi} - \langle \boldsymbol{\xi} \rangle)^T \mathbf{B} (\boldsymbol{\xi} - \langle \boldsymbol{\xi} \rangle)\right] d\boldsymbol{\xi} \tag{2.140}$$

Integrating over the $\xi_i - \langle \xi_i \rangle$ rather than the ξ_i is of no importance, since it is just a translation over an infinite domain of integration. We then simply get:

$$\begin{aligned} \mathbf{f}(\langle \boldsymbol{\xi} \rangle) &= \Theta \int_{\Omega_\xi} \boldsymbol{\xi} \exp\left(-\frac{1}{2k_B} \boldsymbol{\xi}^T \mathbf{B} \boldsymbol{\xi}\right) d\boldsymbol{\xi} \\ &= \langle \boldsymbol{\xi} \rangle \end{aligned} \tag{2.141}$$

Let us now differentiate (2.140) with respect to $\langle \boldsymbol{\xi} \rangle$:

$$\begin{aligned} \frac{\partial \mathbf{f}}{\partial \langle \boldsymbol{\xi} \rangle} &= \frac{\Theta}{k_B} \int_{\Omega_\xi} \boldsymbol{\xi} \otimes [\mathbf{B}(\boldsymbol{\xi} - \langle \boldsymbol{\xi} \rangle)] \\ &\quad \times \exp\left[-\frac{1}{2k_B} (\boldsymbol{\xi} - \langle \boldsymbol{\xi} \rangle)^T \mathbf{B} (\boldsymbol{\xi} - \langle \boldsymbol{\xi} \rangle)\right] d\boldsymbol{\xi} \end{aligned} \tag{2.142}$$

For $\langle \boldsymbol{\xi} \rangle = \mathbf{0}$, we get

$$\begin{aligned} \frac{\partial \mathbf{f}}{\partial \langle \boldsymbol{\xi} \rangle} (\boldsymbol{\xi} = \mathbf{0}) &= \frac{\Theta}{k_B} \int_{\Omega_\xi} \boldsymbol{\xi} \otimes (\mathbf{B}\boldsymbol{\xi}) \exp\left(-\frac{1}{2k_B} \boldsymbol{\xi}^T \mathbf{B} \boldsymbol{\xi}\right) d\boldsymbol{\xi} \\ &= \frac{1}{k_B} \langle \boldsymbol{\xi} \otimes \boldsymbol{\Xi} \rangle \end{aligned} \tag{2.143}$$

(using (2.122)). However, (2.141) shows that $\partial \mathbf{f} / \partial \langle \boldsymbol{\xi} \rangle$ equals the identity matrix. Hence:

$$\langle \boldsymbol{\xi} \otimes \boldsymbol{\Xi} \rangle = k_B \mathbf{I}_{2N} \tag{2.144}$$

Lastly, returning to (2.138), we find

$$\epsilon_i C_{ij} = \epsilon_j C_{ji} \tag{2.145}$$

Matrix $\mathbf{C}$ is then either symmetric or antisymmetric in blocks, depending on whether the ϵ_i has the same sign or not. This result of symmetry of the kinetic coefficients C_{ij} is known as *Onsager's theorem*. Lastly, note that the relation (2.144), fitted with (2.122) and (2.126), also yields

$$\mathbf{A} \langle \boldsymbol{\xi} \otimes \boldsymbol{\xi} \rangle = k_B \mathbf{C} \tag{2.146}$$

We put henceforth

$$\alpha_{ij} \doteqdot -\frac{1}{T} \epsilon_i C_{ij} \tag{2.147}$$

Matrix $\boldsymbol{\alpha}$ is then symmetric. By misuse of language, it is sometimes called *matrix of kinetic coefficients* instead of $\mathbf{C}$.

We are going to further discuss the case where all the ϵ_i are negative, as happens with the momentum components. We then have $\mathbf{C} = T\boldsymbol{\alpha}$, and matrix $\mathbf{C}$ is symmetric. Since energy only depends of momenta via kinetic energy, (2.123) is also written as

$$\begin{aligned} \boldsymbol{\Xi} &= \frac{1}{T} \frac{\partial E_k}{\partial \boldsymbol{\xi}} \\ &= \frac{1}{T} \frac{\partial H}{\partial \mathbf{X}} \end{aligned} \tag{2.148}$$

With this equation, relation (2.125) is written as

$$\begin{aligned}\dot{\boldsymbol{\xi}} &= -T\boldsymbol{\alpha}\boldsymbol{\Xi} \\ &= -\boldsymbol{\alpha}\frac{\partial H}{\partial \mathbf{X}}\end{aligned} \tag{2.149}$$

where $H = E$ is energy. This equation gives the evolution of the isolated system (i.e. in the absence of external forces) upon its return to the thermodynamical equilibrium. The right-hand side then represents new forces accompanying the return to equibrium, that is inherently irreversible forces. In the general case of an ordinary system, these forces are exerted in addition to the internal and external forces, as set out in the next sections.

A major consequence of Onsager's theorem (2.145) is that relation (2.125) may also be written as[34]

[34]Hereinafter, the factor T is used by mere convention.

$$\dot{\boldsymbol{\xi}} = -\frac{1}{T}\frac{\partial F}{\partial \boldsymbol{\Xi}} \tag{2.150}$$

with

$$\begin{aligned}F &\doteq \frac{T}{2}\boldsymbol{\Xi}^T\mathbf{C}\boldsymbol{\Xi} \\ &= \frac{T}{2}\boldsymbol{\xi}^T\mathbf{D}\boldsymbol{\xi} \\ &= \frac{1}{2}(T\boldsymbol{\Xi})^T\boldsymbol{\alpha}(T\boldsymbol{\Xi})\end{aligned} \tag{2.151}$$

(we have successively used used the relations (2.122), (2.126) and (2.128), as well as the properties of symmetry of matrices **C** and **B** and the definition (2.147)). In the relation consisting of the first line of (2.151), the antisymmetric components of **C** and **D** play no part, whereas the symmetric components make up a positive definite sub-matrix. To demonstrate it, we can observe that function F immediately gives the variation of entropy of the system, owing to (2.122) and (2.125):

$$\begin{aligned}T\frac{dS}{dt} &= T\frac{\partial S}{\partial \boldsymbol{\xi}}\cdot\dot{\boldsymbol{\xi}} \\ &= -T\boldsymbol{\Xi}^T\dot{\boldsymbol{\xi}} \\ &= T\boldsymbol{\Xi}^T\mathbf{C}\boldsymbol{\Xi} \\ &= 2F\end{aligned} \tag{2.152}$$

Matrix $\boldsymbol{\alpha}$ is then positive definite. Relation (2.95) specifies that $2F$ also corresponds to the dissipation of the macroscopic energy $E_k + E_{ext}$:

$$\begin{aligned}\frac{d(E_k + E_{ext})}{dt} &= -T\frac{dS}{dt} \\ &= -2F\end{aligned} \tag{2.153}$$

From a macroscopic point of view, the forces occurring in the right-hand side of (2.149) are then *dissipative*, unlike the conservative forces as introduced in Chapter 1. In addition, it should be pointed out that, in Chapter 1, it appeared that the property of conservation of energy directly stemmed from the invariant nature of the laws of mechanics with respect to time. Thus, the inherently irreversible process of increase of entropy logically involves to some extent a non-compliance with the conservation of energy.[35] We will provide further details about this mechanism in the next Sections.

[35] Once and for all, we must say that, on a microscopic scale, energy is actually conserved.

The calculations we have made in this section regard, as we have said, the microscopic quantities $\mathbf{X} \equiv (\{p_i\}, \{q_i\})$. Identical developments could be made with the macroscopic quantities $\langle \mathbf{X} \rangle \equiv (\{P_i\}, \{Q_i\})$, that is averaged quantities as per Boltzmann (refer to the end of Section 2.2.1). All the results which we achieved in this paragraph would then remain unchanged, except for one detail: since mean parameters are considered, it happens that the kinetic coefficients depend on the mean generalized positions, as in the calculation made in Section 1.3.3 to determine the kinetic energy (1.62) of a compound object:

$$\alpha_{ij} = \alpha_{ij} (\{Q_k\}) \tag{2.154}$$

2.4.2 Dissipative forces

In order to go deeper into the conclusions of the previous section, let us now resume the Hamilton equations (1.89). We will show that they fall within the general scope we have just described, in the case of a system without internal forces. In this case, the Hamiltonian H equals the macroscopic energy $E_k + E_{ext}$. Using relation (2.95), the Hamilton equations are also written as:

$$\begin{aligned} \dot{p}_i &= T \frac{\partial S}{\partial q_i} \\ \dot{q}_i &= -T \frac{\partial S}{\partial p_i} \end{aligned} \tag{2.155}$$

(here we substitute an indicial formalism for the tensor form of equations, for better highlighting the separation of the parameters into two groups: momenta and coordinates, i.e. $\boldsymbol{\xi} = (\{p_i\}, \{q_i\}) \equiv \mathbf{X}$). These equations are then written like (2.125), with two series of ξ_i:

$$\begin{aligned} \xi_{p_i} &\div p_i, & \xi_{q_i} &\div q_i \\ \Xi_{p_i} &= -\frac{\partial S}{\partial p_i}, & \Xi_{q_i} &= -\frac{\partial S}{\partial q_i} \end{aligned} \tag{2.156}$$

We have the case where the sign of the first series of parameters changes through a time reversal (momenta, $\epsilon_{p_i} = -1$), whereas the second series is unchanged (positions, $\epsilon_{q_i} = 1$). We notice that where the parameters characterize a molecule (i.e. $(\{p_i\}, \{q_i\}) \equiv (\tilde{\mathbf{p}}_\alpha, \tilde{\mathbf{r}}_\alpha)$), their thermodynamicaly conjugate quantities are given by

$$\begin{aligned}\Xi_{\tilde{\mathbf{p}}_\alpha} &= -\frac{\partial S}{\partial \tilde{\mathbf{p}}_\alpha} = \frac{1}{T}\frac{\partial \tilde{E}_k}{\partial \tilde{\mathbf{p}}_\alpha} = \frac{1}{mT}\tilde{\mathbf{p}}_\alpha \\ \Xi_{\tilde{\mathbf{r}}_\alpha} &= -\frac{\partial S}{\partial \tilde{\mathbf{r}}_\alpha} = \frac{1}{T}\frac{\partial \tilde{E}_{ext}}{\partial \tilde{\mathbf{r}}_\alpha} = -\frac{1}{T}\tilde{\mathbf{F}}_\alpha^{ext}\end{aligned} \tag{2.157}$$

The momenta then occur as the conjugate quantities of the coordinates, in compliance with what has already been stated in Sections 1.4.1 and 1.5.4 in Chapter 1. The probability density (2.132) is then written as

$$\begin{aligned}\overline{\Phi}(\boldsymbol{\xi}) &= \Theta \exp\left[-\frac{1}{2k_B}\sum_\alpha \left(\tilde{\mathbf{p}}_\alpha \cdot \Xi_{\tilde{\mathbf{p}}_\alpha} + \tilde{\mathbf{r}}_\alpha \cdot \Xi_{\tilde{\mathbf{r}}_\alpha}\right)\right] \\ &= \Theta \exp\left[-\frac{1}{2k_B T}\sum_\alpha \left(\frac{|\tilde{\mathbf{p}}_\alpha|^2}{m} - \tilde{\mathbf{r}}_\alpha \cdot \tilde{\mathbf{F}}_\alpha^{ext}\right)\right]\end{aligned} \tag{2.158}$$

In the case of a field of external forces with such a central harmonic potential as (1.44), the forces take a form analogous to (1.57) and we find once again the Maxwell–Boltzmann distribution (2.49) for the whole system.

Let us resume the general case. Equations (2.155) are then rearranged as

$$\begin{aligned}\dot{p}_i &= -\sum_j C_{p_i q_j}\Xi_{q_j} \\ \dot{q}_i &= -\sum_j C_{q_i p_j}\Xi_{p_j}\end{aligned} \tag{2.159}$$

with

$$\begin{aligned}C_{p_i p_j} &= C_{q_i q_j} = 0 \\ C_{p_i q_j} &= -C_{q_i p_j} = T\delta_{ij}\end{aligned} \tag{2.160}$$

The increase of entropy is then given by (2.152) :

$$\begin{aligned}\frac{dS}{dt} &= -\sum_i \left(\Xi_{p_i}\dot{\xi}_{p_i} + \Xi_{q_i}\dot{\xi}_{q_i}\right) \\ &= \sum_i \left(\frac{\dot{q}_i}{C_{q_i p_i}}\dot{p}_i + \frac{\dot{p}_i}{C_{p_i q_i}}\dot{q}_i\right) \\ &= 0\end{aligned} \tag{2.161}$$

(we have advantageously utilized (2.156), (2.159) and (2.160)). This is due to the fact that matrix $\mathbf{C}$ is not symmetric positive definite here, because of the negative ϵ. Within a microscopic context, where the parameters represent the molecules, entropy is then a conserved quantity, which is consistent with the fact that it represents a purely macroscopic quantity, as defined from average probability densities, in accordance with (2.29). As demonstrated by that calculation, the increase of entropy cannot be accurately modelled from the Hamilton equations, which is quite normal. We must remember, indeed, that the latter lead to Liouville's theorem, then to the Hamilton equation without collisions (2.18), which conserves entropy. As regards the collisions, they alter the molecule velocities; thus, on a macroscopic scale, they generate dissipative forces which consequently cannot fall within the scope of the previously discussed Hamilton equations.

Let us now consider the case of a particle (i.e. $(\{p_i\}, \{q_i\}) \equiv (\mathbf{p}_a, \mathbf{r}_a)$) for which the irreversible effects would be ignored. Equations (2.157) and (2.159) are then identically applied and yield

$$\begin{aligned} \dot{\mathbf{p}}_a &= \frac{1}{T}\sum_b \mathbf{C}_{\mathbf{p}_a\mathbf{r}_b}\mathbf{F}_a^{ext} \\ \dot{\mathbf{r}}_a &= -\frac{1}{T}\sum_b \mathbf{C}_{\mathbf{r}_a\mathbf{p}_b}\mathbf{u}_a \end{aligned} \tag{2.162}$$

where the $\mathbf{C}_{\mathbf{r}_a\mathbf{p}_b}$ and $\mathbf{C}_{\mathbf{p}_a\mathbf{r}_b}$ are 3×3 matrices. Taking the definitions (1.99) and (1.100) in Section 1.4.2 into account, these equations are also written in the form of a matrix:

$$\dot{\mathbf{X}}_a = \frac{1}{T}\sum_b \begin{pmatrix} \mathbf{0}_3 & \mathbf{C}_{\mathbf{p}_a\mathbf{r}_b} \\ \mathbf{C}_{\mathbf{r}_a\mathbf{p}_b} & \mathbf{0}_3 \end{pmatrix} \frac{\partial H}{\partial \mathbf{X}_a} \tag{2.163}$$

This corresponds to equation (1.101), if we put

$$\mathbf{C}_{\mathbf{p}_a\mathbf{r}_b} = -\mathbf{C}_{\mathbf{r}_a\mathbf{p}_b} = T\delta_{ab}\mathbf{I}_3 \tag{2.164}$$

Taking (2.147) into account, the matrix $\boldsymbol{\alpha}$ of those kinetic coefficients occurring in (2.163) is then likened to $-\boldsymbol{\omega}_6$. The conservation of entropy (2.161) is then still valid. Since the matrix $\boldsymbol{\alpha} = -\boldsymbol{\omega}_6$ is asymmetric, indeed, it cannot be positive definite.

In order to build a model reflecting the second law of thermodynamics, we must adopt another approach. As is well known, we may consider that the transition from the microscopic level (molecules) to the macroscopic level (thermodynamical system) takes place by looking at the mean values. As suggested in Section 2.4.1, the kinetic energy is then dissipated via new forces. Besides, the discussions in Chapter 1 have illustrated the conservation of energy as an effect of the fact that the elementary forces originate from a potential function. We then have to acknowledge that the dissipative forces are not yielded by a potential function. The calculation (1.253) asserts that the pressure loads, being conservative, are orthogonal to the surface onto which they are exerted. The presently introduced forces cannot be studied via a potential because they exert friction forces which are non-orthogonal to the surface on which they act.[36]

[36] We will provided further detail on this subject in Chapter 3.

Taking the previous observations into account, the friction forces can be investigated within the frame of the formalism we have just introduced, by reasoning as follows. In the very broad case of a system being defined by its generalized parameters, we will henceforth consider the average (i.e. macroscopic parameters $(\{P_i\}, \{Q_i\})$ (refer to the end of Section 2.2.1), with relations analogous to (2.156) and (2.157). In particular, we have

$$\begin{aligned} \xi_{P_i} &= P_i \\ \Xi_{P_i} &= \frac{1}{T}\dot{Q}_i \end{aligned} \tag{2.165}$$

because, in the case of molecules, (2.157) yields $\Xi_{\tilde{\mathbf{p}}_\alpha} = \frac{1}{T}\tilde{\mathbf{u}}_\alpha$. We must from now on add additional forces to the right-hand side in the Hamilton equation (2.155). However, the second set of equations making up the definition of the generalized momenta should not be modified. Only the first set of equations (equations of motion) will be changed. The calculation (2.161) shows that the conservation of entropy results from the absence of diagonal terms in the matrix of the kinetic coefficients (2.159), that is of non-zero terms $C_{P_iP_j}$.[37] As in the previous section, the dissipative forces will then be seen as linear terms[38] with respect to the quantities Ξ_{P_i}:

$$\begin{aligned} \dot{P}_i &= -\sum_j C_{P_iQ_j}\Xi_{Q_j} - \sum_j C_{P_iP_j}\Xi_{P_j} \\ &= T\frac{\partial S}{\partial Q_i} - \frac{1}{T}\sum_j C_{P_iP_j}\dot{Q}_j \end{aligned} \tag{2.166}$$

now with non-zero coefficients $C_{P_iP_j}$, which modifies the Hamilton equations. This is because the generalized friction forces, which will hereinafter be denoted F_i^{diss} (the superscript ^{diss} highlights their dissipative nature), appear on the right-hand side:

$$F_i^{diss} \doteqdot -\frac{1}{T}\sum_j C_{P_iP_j}\dot{Q}_j \tag{2.167}$$

We will note $\alpha_{ij} = C_{P_iP_j}/T$, these parameters serving as kinetic coefficients for the friction phenomenon, in accordance with the general relation (2.147). The matrix $\boldsymbol{\alpha}$ of coefficients α_{ij} is then $N \times N$-dimensional. In the equation governing $\dot{Q}_i$, no new term appears with respect to the non-dissipative case, which means that the $C_{Q_iQ_j}$ is zero. Lastly, the matrix C will from now on be written as

$$\mathbf{C} = T\begin{pmatrix} \boldsymbol{\alpha} & \mathbf{0}_N \\ \mathbf{0}_N & \mathbf{0}_N \end{pmatrix} \tag{2.168}$$

As shown by formula (2.154), we must consider that the kinetic coefficients depend on the average generalized coordinates $\{Q_k\}$. With (2.165) and (2.168), we can write (2.167) as follows:

$$F_i^{diss} = -\sum_j \alpha_{ij}\left(\{Q_k\}\right)\dot{Q}_j \tag{2.169}$$

consistently with (2.149). The α_{ij} verifies Onsager's theorem (2.145):

$$\alpha_{ji} = \alpha_{ij} \tag{2.170}$$

because the ϵ_{P_i} have the same sign. The dissipative forces depend on the generalized momenta, as expected. They can be arranged in a form analogous to (2.150):

$$F_i^{diss} = -\frac{\partial F}{\partial \dot{Q}_i} \tag{2.171}$$

[37] Here can be noticed one of the unpleasant consequences of not having made the units of both momenta and coordinates homogeneous: the $C_{P_iP_j}$ are not homogeneous with the $C_{P_iQ_j}$.

[38] As a first approximation, the linearity of these terms is also suggested by the examination of the Boltzmann equation with the collision terms (refer, for example, to Landau and Lifshitz, 1980). We will discuss that matter again in Chapter 3.

where F is a quadratic form as given by

$$F = \frac{1}{2}\sum_{i,j} \alpha_{ij}\left(\{Q_k\}\right)\dot{Q}_i\dot{Q}_j \tag{2.172}$$

by analogy with (2.151). As regards Lagrange's equation of motion (1.65), it becomes

$$\begin{aligned}\forall i,\ \frac{dP_i}{dt} &= F_i^{cons} + F_i^{diss} \\ &= \frac{\partial L}{\partial Q_i} - \frac{\partial F}{\partial \dot{Q}_i}\end{aligned} \tag{2.173}$$

the (conservative) forces F_i^{cons} playing the part of the forces derived from a potential, which were introduced in Chapter 1.

Let us now resume the calculation (2.161) of the evolution of entropy with the dissipative forces. Only the contribution $-\Xi_{P_i}\dot{\xi}_{P_i}$ now persists and yields:

$$\begin{aligned}\frac{dS}{dt} &= -\frac{1}{T}\sum_i \dot{Q}_i\dot{P}_i \\ &= -\frac{1}{T}\sum_i F_i^{diss}\dot{Q}_i \\ &= \frac{1}{T}\sum_i \frac{\partial F}{\partial \dot{Q}_i}\dot{Q}_i \\ &= \frac{2F}{T}\end{aligned} \tag{2.174}$$

owing to the quadratic nature of F, and in accordance with (2.152). This formula indicates that $\boldsymbol{\alpha}$ is positive definite, and so F is strictly positive in the presence of dissipative forces, making entropy increase as expected. That increase is accompanied, of course, by a decrease of the macroscopic energy, as mentioned in Section 2.4.1. The calculation as given at the beginning of Section 1.4.1 is, indeed, modified, which leads to

$$\begin{aligned}\frac{d\left(E_k + E_{ext}\right)}{dt} &= \sum_i \left(\dot{P}_i\dot{Q}_i + P_i\ddot{Q}_i\right) - \frac{dL}{dt} \\ &= -\sum_i \frac{\partial F}{\partial \dot{Q}_i}\dot{Q}_i \\ &= \sum_i F_i^{diss}\dot{Q}_i\end{aligned} \tag{2.175}$$

which shows that the power of the generalized friction forces accounts for the loss of energy. Because of the quadratic nature of F, we may infer

$$\frac{d\left(E_k + E_{ext}\right)}{dt} = -2F \tag{2.176}$$

and the macroscopic energy decreases over time in accordance with (2.153). The friction forces being introduced here are then seemingly non-compliant with the law of conservation of energy. This is because the formalism we use in this section does not involve any internal energy, since it theoretically recovers the dissipated macroscopic energy. Formula (2.174) shows, indeed, that the right-hand side of (2.176) corresponds, in accordance with (2.152), to $-TdS/dt$, that is to the flux of internal energy coming from the macroscopic energy under the action of the irreversible processes. We will discuss, in the next section, the case of a particle system in which the evolution of the internal energy will make it possible to formally recover the conservation of total energy.

An elementary example of the application of these principles can be given in the case of the pendulum as set out in Section 1.4.4. The friction which ambient air[39] necessarily imparts to it accounts for a resistance force taking, by analogy with (2.169), the form $F^{diss} = -\alpha\dot{\theta}$. Thus, the equation of motion (1.153) now takes the following form:

[39]The calculation of the friction force exerted by a fluid on a rigid body will be examined in Chapter 3, Section 3.4.3.

$$m\ell^2\ddot{\theta} = -mg\ell\sin\theta - \alpha\dot{\theta} \tag{2.177}$$

Since we have here only one degree of freedom, the positive definite matrix α_{ij} is reduced to a scalar $\alpha > 0$ serving as a single kinetic coefficient. The amount of energy dissipated in the course of time can easily be derived from (2.177) by means of a calculation generalizing (1.159), but it suffices to call upon the equations (2.172) and (2.176) to immediately get:

$$\frac{dE}{dt} = -\alpha\dot{\theta}^2 \tag{2.178}$$

This is a damped or dissipative pendulum, which is more realistic than the simple model of Section 1.4.4. We will discuss some of its consequences in Chapter 4 (Section 4.2.1).

2.4.3 Particle friction

In order to utilize the arguments of the previous section in the case of a particle system, we apply the achieved results to the case in which the average parameters represent the particle positions and momenta, that is $(\{P_i\}, \{Q_i\}) \equiv (\{\mathbf{p}_a\}, \{\mathbf{r}_a\})$. We can observe, however, that the fricion forces do not depend on the absolute velocities, but upon velocity discrepancies among particles, since no friction occurs in an ensemble of particles moving about with a uniform velocity.[40] Thus, the friction force received by particle a is written as:

[40]The Galilean invariance (Section 1.2.1) is at the core of this argument.

$$\mathbf{F}_a^{diss}\left(\{\mathbf{u}_b\}\right) = -\sum_{b\neq a} \boldsymbol{\alpha}_{ab}\left(\{\mathbf{r}_c\}\right)\mathbf{u}_{ab} \tag{2.179}$$

with

$$\mathbf{u}_{ab} \doteqdot \mathbf{u}_a - \mathbf{u}_b \tag{2.180}$$

Quantities $\boldsymbol{\alpha}_{ab}$ are now three-dimensional positive definite matrices serving as the kinetic coefficients. The subscripts a and b do not represent their matrix subscripts, but labels attached to the particles, as in Section 1.4.1, equation (1.80). Onsager's symmetry here imposes two relations :

$$\begin{aligned} \boldsymbol{\alpha}_{ab}^{T} &= \boldsymbol{\alpha}_{ab} \\ \boldsymbol{\alpha}_{ab} &= \boldsymbol{\alpha}_{ba} \end{aligned} \tag{2.181}$$

The first line of (2.181) strictly results from the discussion in Section 2.4.1, whereas the second line has a simple interpretation. Since the friction forces result from macroscopic complications of some microscopic phenomena, they originate from mechanisms which obey the law of conservation of momentum. Now, unlike energy, the latter is always additive, and so the macroscopic resultant $\mathbf{F}_a^{diss}$ of these phenomena should verify that principle too. Thus, the kinetic matrices $\boldsymbol{\alpha}_{ab}$ verify the relation $\boldsymbol{\alpha}_{ab}\mathbf{u}_{ab} = -\boldsymbol{\alpha}_{ba}\mathbf{u}_{ba}$, ie., since $\mathbf{u}_{ba} = -\mathbf{u}_{ab}$, $\boldsymbol{\alpha}_{ab} = \boldsymbol{\alpha}_{ba}$, in accordance with (2.181). It can then be seen that the dissipation force is written as a sum of individual contributions, like (1.259) for the conservative forces:[41]

$$\mathbf{F}_a^{diss} = \sum_b \mathbf{F}_{b\to a}^{diss} \tag{2.182}$$

[41] The summation may be formally extended to all the particles b, since $\mathbf{u}_{aa} = \mathbf{0}$.

with a local law of action and reaction:

$$\begin{aligned} \mathbf{F}_{b\to a}^{diss} &= -\boldsymbol{\alpha}_{ab}\mathbf{u}_{ab} \\ &= -\mathbf{F}_{a\to b}^{diss} \end{aligned} \tag{2.183}$$

In particular, $\mathbf{F}_{a\to a}^{diss} = 0$, which is consistent with the fact that the sum (2.179) only extends over those particles b which differ from a. The dependence of kinetic matrices $\boldsymbol{\alpha}_{ab}$ upon positions $\{\mathbf{r}_c\}$ can only imply particles a and b involved in the friction process being studied, which means that $\boldsymbol{\alpha}_{ab} = \boldsymbol{\alpha}_{ab}(\mathbf{r}_a, \mathbf{r}_b)$. Besides, it is obvious, for reasons of translational spatial invariance, that these two position vectors can only act through the inter-particle vector $\mathbf{r}_{ab} = \mathbf{r}_a - \mathbf{r}_b$. Thus, the friction force can be formulated in a form which is similar to (2.171):

$$\mathbf{F}_a^{diss} = -\boldsymbol{\alpha}_{ab}(\mathbf{r}_{ab})\,\mathbf{u}_{ab} = -\frac{\partial F}{\partial \mathbf{u}_a} \tag{2.184}$$

where

$$F \doteqdot \frac{1}{2}\sum_{a\neq b} \mathbf{u}_{ab}^{T}\boldsymbol{\alpha}_{ab}(\mathbf{r}_{ab})\,\mathbf{u}_{ab} \tag{2.185}$$

has been defined.[42]

[42] The symmetry of quantities $\boldsymbol{\alpha}_{ab}$ in relation to subscripts a and b plays a major role in this arrangement.

The definite positiveness of the kinetic matrices then provides for the positiveness of F. The equation of motion (1.240) becomes henceforth

$$\begin{aligned} \forall a,\ \frac{d\mathbf{p}_a}{dt} &= \mathbf{F}_a^{int} + \mathbf{F}_a^{diss} + \mathbf{F}_a^{ext} \\ &= \sum_b \left(\mathbf{F}_{b\to a}^{int} + \mathbf{F}_{b\to a}^{diss}\right) + \mathbf{F}_a^{ext} \\ &= -\frac{\partial E_{int,a}}{\partial \mathbf{r}_a} - \frac{\partial F}{\partial \mathbf{u}_a} - \frac{\partial E_{ext}}{\partial \mathbf{r}_a} \end{aligned} \tag{2.186}$$

The dissipative force power is calculated from (2.184) and (2.185):

$$\begin{aligned}\sum_a \mathbf{F}_a^{diss} \cdot \mathbf{u}_a &= \sum_{a \neq b} \boldsymbol{\alpha}_{ab} (\mathbf{r}_{ab}) \mathbf{u}_{ab} \cdot \mathbf{u}_a \\ &= -\frac{1}{2} \sum_{a \neq b} \mathbf{u}_{ab}^T \boldsymbol{\alpha}_{ab} (\mathbf{r}_{ab}) \mathbf{u}_{ab} \\ &= -F\end{aligned} \tag{2.187}$$

(we have split the third line term into a couple of identical contributions, then changed the dummy subscripts in one of these resulting two terms; lastly, we have taken advantage of the properties of symmetry of the various quantities). Relation (2.187) differs from (2.176) by a factor of 2, which is due to the fact that (2.179) is based on velocity differences, not on such absolute velocities as in (2.169). An analogous calculation will obviously change (2.174) for

$$\frac{dS}{dt} = \frac{F}{T} \tag{2.188}$$

We also may introduce the macroscopic dissipation of particle a, verifying

$$\begin{aligned}T_a \frac{dS_a}{dt} &= \frac{1}{2} \sum_{b \neq a} \mathbf{u}_{ab}^T \boldsymbol{\alpha}_{ab} (\mathbf{r}_{ab}) \mathbf{u}_{ab} \\ &= -\mathbf{F}_a^{diss} \cdot \mathbf{u}_a \\ &\doteqdot F_a\end{aligned} \tag{2.189}$$

With (2.189), the internal energy balance (2.97) will, of course, become:

$$\frac{dE_{int}}{dt} = -\sum_a \mathbf{F}_a^{diss} \cdot \mathbf{u}_a = F \tag{2.190}$$

Equation (2.190) shows that the total internal energy increases, under the effect of the irreversible processes, by an amount $F > 0$. We may also write the variation of internal energy of a given particle as:

$$\frac{dE_{int,a}}{dt} = \frac{1}{2} \sum_{b \neq a} \mathbf{u}_{ab}^T \boldsymbol{\alpha}_{ab} (\mathbf{r}_{ab}) \mathbf{u}_{ab} = F_a \tag{2.191}$$

The law of evolution of total energy (for an isolated system) may be applied as in (2.175), except that the internal energy evolves as per formula (2.90). Using (1.247), (2.186), (2.90) and (2.188), we get:

$$\begin{aligned}\frac{dE}{dt} &= \frac{d}{dt} \sum_a \left(\frac{1}{2} m_a u_a^2 + E_{int,a} + E_{ext,a} \right) \\ &= \sum_a \left(m_a \frac{d\mathbf{u}_a}{dt} \cdot \mathbf{u}_a + \frac{dE_{int,a}}{dt} + \frac{dE_{ext,a}}{dt} \right)\end{aligned} \tag{2.192}$$

$$
\begin{aligned}
&= \sum_a \left[\left(\mathbf{F}_a^{int} + \mathbf{F}_a^{diss} + \mathbf{F}_a^{ext} \right) \cdot \mathbf{u}_a - p_a \frac{dV_a}{dt} + T_a \frac{dS_a}{dt} - \mathbf{F}_a^{ext} \cdot \mathbf{u}_a \right] \\
&= \sum_a \left(\mathbf{F}_a^{int} \cdot \mathbf{u}_a - p_a \frac{dV_a}{dt} + T_a \frac{dS_a}{dt} - F_a \right) \\
&= 0
\end{aligned}
$$

(the last terms cancel each other owing to (1.264) and (2.189)). The macroscopic friction forces then represent a model reflecting the flux of macroscopic energy towards internal energy, under the action of processes accounting for the increase of entropy.

Relation (1.243) is obviously still valid here, all the forces $\mathbf{F}_a^{int} + \mathbf{F}_a^{ext} + \mathbf{F}_a^{diss}$ taking part in the variation of kinetic energy through their respective powers:

$$
\begin{aligned}
\frac{dE_k}{dt} &= \sum_a \mathbf{F}_a^{int} \cdot \mathbf{u}_a + \sum_a \mathbf{F}_a^{ext} \cdot \mathbf{u}_a + \sum_a \mathbf{F}_a^{diss} \cdot \mathbf{u}_a \\
&= P^{int} + P^{ext} - F
\end{aligned}
\tag{2.193}
$$

Thus, the kinetic energy loses the amount F which is recovered by the internal energy, according to what was said above.

Let us go deeper into the form which the kinetic matrices should take. We have seen that the friction forces verify the action and reaction principle which is closely related to the momentum additivity. Since the angular momentum has the same property, it should immediately be inferred that the friction forces should verify the law of conservation of total motion, that is that the moment of the forces should change its sign when the particle labels are swapped:

$$
\sum_{a \neq b} \mathbf{r}_a \times (\boldsymbol{\alpha}_{ab} \mathbf{u}_{ab}) = - \sum_{a \neq b} \mathbf{r}_b \times (\boldsymbol{\alpha}_{ba} \mathbf{u}_{ba}) \tag{2.194}
$$

or else, using the previously established properties of symmetry:

$$
\sum_{a \neq b} \mathbf{r}_{ab} \times (\boldsymbol{\alpha}_{ab} \mathbf{u}_{ab}) = \mathbf{0} \tag{2.195}
$$

This is only possible if $\boldsymbol{\alpha}_{ab}\mathbf{u}_{ab}$ is aligned with $\mathbf{r}_{ab}$, hence if there is a scalar quantity β_{ab} which is symmetric with respect to subscripts a and b such that $\boldsymbol{\alpha}_{ab}\mathbf{u}_{ab} = \beta_{ab}\mathbf{r}_{ab}$. Moreover, β_{ab} should be a scalar linearly depending on $\mathbf{u}_{ab}$, that is a value in the form $\beta_{ab} = \boldsymbol{\gamma}_{ab} \cdot \mathbf{u}_{ab}$, where $\boldsymbol{\gamma}_{ab}$ is an asymmetric vector with respect to the subscripts a and b, depending on $\mathbf{r}_{ab}$ like $\boldsymbol{\alpha}_{ab}$. The latter argument corresponds to

$$
\begin{aligned}
\boldsymbol{\alpha}_{ab}(\mathbf{r}_{ab}) &= \mathbf{r}_{ab} \otimes \boldsymbol{\gamma}_{ab}(\mathbf{r}_{ab}) \\
\boldsymbol{\gamma}_{ba}(\mathbf{r}_{ab}) &= -\boldsymbol{\gamma}_{ab}(\mathbf{r}_{ab})
\end{aligned}
\tag{2.196}
$$

and yields

$$
\mathbf{F}_a^{diss} = - \sum_{b \neq a} \left(\boldsymbol{\gamma}_{ab} \cdot \mathbf{u}_{ab} \right) \mathbf{r}_{ab} \tag{2.197}
$$

Formula (2.197) does not yet possess all the required properties for a physically sound particle friction law. In particular, the matrices $\boldsymbol{\alpha}_{ab}$ as given by (2.196) do not satisfy the first property of (2.181). For a deeper insight into this question, let us now temporarily consider a set of particles making up an isolated system, that is at thermodynamical equilibrium. Both its total linear and angular momentum are conserved, whereas its entropy is maximal. Let us refer to the inertial frame of reference, so that energy only comes down to internal energy (Section 1.5.3). We should then maximize the entropy of the whole system as

$$\begin{aligned} S &= S(E_{int}) \\ &= S(E - E_k) \\ &= S\left(E - \sum_a \frac{|\mathbf{p}_a|^2}{2m_a}\right) \end{aligned} \tag{2.198}$$

The calculation of minimum, however, should be carried out under the constraints of conservation of momenta (Section 1.4.3):

$$\begin{aligned} \sum_a \mathbf{p}_a &= \mathbf{P} \\ \sum_a \mathbf{r}_a \times \mathbf{p}_a &= \mathbf{M} \end{aligned} \tag{2.199}$$

To that purpose, we will use the Lagrange multiplier technique as introduced in Section 1.2.4, which here consists of minimizing not $S(E_{int})$ but the value

$$\begin{aligned} S(\{\mathbf{p}_a\}, \mathbf{A}, \mathbf{B}) =& S\left(E - \sum_a \frac{|\mathbf{p}_a|^2}{2m_a}\right) \\ &+ \mathbf{A} \cdot \left(\sum_a \mathbf{p}_a - \mathbf{P}\right) - \mathbf{B} \cdot \left(\sum_a \mathbf{r}_a \times \mathbf{p}_a - \mathbf{M}\right) \end{aligned} \tag{2.200}$$

where $\mathbf{A}$ and $\mathbf{B}$ are unknown auxiliary vectors representing the multipliers associated with the constraints (2.199). In accordance with what was said at the end of Section 1.2.4, we may recall that the value occurring in the right-hand side duly equals S, since the additive terms are zero according to (2.199). Besides, deriving (2.200) with respect to $\mathbf{A}$ and $\mathbf{B}$ will restore both desired constraints (2.199). In that form, S is a function of the particle momenta; deriving with respect to $\mathbf{p}_a$ gives

$$\frac{\partial S}{\partial \mathbf{p}_a} = -\frac{\partial S}{\partial E}\frac{\mathbf{p}_a}{m_a} + \mathbf{A} + \mathbf{B} \times \mathbf{r}_a \tag{2.201}$$

Since the system is isolated, there is no heat exchange with the outside, and then $dE = -pdV + TdS$, which yields $\partial S/\partial E = 1/T$. By cancelling the expression (2.201), we find that the velocity of each particle is written as

$$\mathbf{u}_a = \frac{\mathbf{p}_a}{m_a} = T(\mathbf{A} + \mathbf{B} \times \mathbf{r}_a) \tag{2.202}$$

This is a motion of a rigid body, according to equation (1.191). Now, energy is constant at equilibrium, then dissipation vanishes. We can conclude that no friction should occur within the velocity field (2.202), which is consistent with intuition (friction has completed its dissipative work by homogenizing the velocities to achieve the rigid rotation state).[43] Thus, another condition should be prescribed on the friction forces, namely force vanishing for a global rotation of the particle system, that is for a velocity distribution in the form $\mathbf{u}_a = \mathbf{V} + \boldsymbol{\Omega} \times \mathbf{r}_a$, where $\mathbf{V}$ ($\equiv T\mathbf{A}$) and $\boldsymbol{\Omega}$ ($\equiv T\mathbf{B}$) are arbitrary vectors, which gives $\mathbf{u}_{ab} = \boldsymbol{\Omega} \times \mathbf{r}_{ab}$ (eqn (1.193)). That condition will be satisfied it, and only if

$$\forall \boldsymbol{\Omega}, \ \sum_{b \neq a} \left[\boldsymbol{\gamma}_{ab} \cdot (\boldsymbol{\Omega} \times \mathbf{r}_{ab}) \right] \mathbf{r}_{ab} = \mathbf{0} \tag{2.203}$$

Using the mathematical properties of the mixed product and the tensors, the term under the summation symbol is rearranged in the form $\left[\mathbf{r}_{ab} \otimes \left(\mathbf{r}_{ab} \times \boldsymbol{\gamma}_{ab}\right)\right] \boldsymbol{\Omega}$. Condition (2.203) can then only be used again if the $\boldsymbol{\gamma}_{ab}$ are aligned with $\mathbf{r}_{ab}$, so that there are coefficients $\gamma_{ab} = \gamma_{ba}$ each of which depends on $\mathbf{r}_{ab}$, and such that $\boldsymbol{\gamma}_{ab} = \gamma_{ab} \mathbf{r}_{ab}$. Besides, for reasons of isotropy, γ_{ab} can only depend on the distance r_{ab}. Thus, the kinetic matrices are ultimately written as

$$\boldsymbol{\alpha}_{ab} \left(\mathbf{r}_{ab}\right) = \gamma_{ab} \left(r_{ab}\right) \mathbf{r}_{ab} \otimes \mathbf{r}_{ab} \tag{2.204}$$

which yields

$$\mathbf{F}_a^{diss} = -\sum_{b \neq a} \gamma_{ab} \left(r_{ab}\right) \left(\mathbf{u}_{ab} \cdot \mathbf{r}_{ab}\right) \mathbf{r}_{ab} \tag{2.205}$$

and the first symmetry of (2.181) is satisfied. Henceforth, the dissipation function is written as

$$F = \frac{1}{2} \sum_{a \neq b} \gamma_{ab} \left(r_{ab}\right) \left(\mathbf{u}_{ab} \cdot \mathbf{r}_{ab}\right)^2 \tag{2.206}$$

A zero dissipation in the case of a rigid rotation ($\mathbf{u}_{ab} \cdot \mathbf{r}_{ab} = 0$, after (1.194)) is found again. It can be seen, however, that the friction force $\mathbf{F}_{b \to a}^{diss}$ exerted by b on a is aligned with $\mathbf{r}_{ab}$, contrary to the intuition that it would have to be aligned with $\mathbf{u}_{ab}$ (as will be explained when we deal with continuous media in Chapter 3). The latter case would correspond to kinetic matrices which would be proportional to identity, but it would not satisfy the law of conservation of angular momentum. It should be pointed out that dissipation occurs, through the equations (2.185) and (2.206), as a quadratic function of the velocity differences, as in the previous paragraph. Coefficients γ_{ab} should be strictly positive in order to provide the definite positiveness of the kinetic matrices $\boldsymbol{\alpha}_{ab}$. They may depend on such local thermodynamical properties of the particles as their temperatures.[44]

[43] This is only possible, of course, if there is no external influence, that is especially in the absence of walls imposing a friction condition (refer to Chapter 3).

[44] That dependence will appear through the notion of viscosity, a quantity to be defined in Chapter 3.

2.5 Further considerations on dissipation

We further develop the ideas of the last section to deduce a more general form of Hamilton equations, for the case of non-conservative systems, on the basis of entropy. This is illustrated with the dissipative pendulum case. The idea of mirror systems is then introduced to build an elegant framework for irreversible processes. Eventually, we investigate the effects of thermodynamical fluctuations through the Langevin and Fokker–Planck equations, in a stochastic formalism.

2.5.1 Dissipative Hamilton equations

We want to apply the concepts of Section 1.4.2 in Chapter 1 to the more general case of a system which is subject to conservative and dissipative forces. The number of degrees of freedom being equal to $2N$, we will denote those kinetic matrices defined in Section 2.4.1 as $\mathbf{C}_{2N}$ and $\boldsymbol{\alpha}_{2N}$, echoing in this way the matrix $\boldsymbol{\Omega}_{2N}$ introduced in Section 1.4.2. (1.105) can then be generalized by summing the dissipative forces (2.149), remembering that $\boldsymbol{\xi} = \mathbf{X} - \mathbf{X}_0$ represents the deviation of the system parameters at equilibrium, which gives $\dot{\boldsymbol{\xi}} = \dot{\mathbf{X}}$. We get

$$\begin{aligned} \dot{\mathbf{X}} &= \boldsymbol{\Omega}_{2N} \frac{\partial H}{\partial \mathbf{X}} + \mathbf{C}_{2N} \frac{\partial S}{\partial \mathbf{X}} \\ &= \left(\boldsymbol{\Omega}_{2N} - \boldsymbol{\alpha}_{2N} \right) \frac{\partial H}{\partial \mathbf{X}} \end{aligned} \tag{2.207}$$

where $\mathbf{X}$ is defined as in (1.104), but from averages $(\{P_i\}, \{Q_i\})$ (Español, 2007). As set out in Chapter 1, the first term in the right-hand side corresponds to forces satisfying the law of conservation of energy. As stated above, the second term can be seen as forces verifying the entropy growth principle. Ottinger and Grmela (1997) show it is the most general formulation allowing us to satisfy these two principles, $\boldsymbol{\Omega}_{2N}$being asymmetric whereas $\mathbf{C}_{2N}$ and $\boldsymbol{\alpha}_{2N}$ are symmetric positive-definite. The meanings of these matrices are quite different, the former reflecting a reversible dynamics whereas the latter represent a mechanism which is irreversible on a macroscopic scale.

We may recall that the initially contemplated molecule system is governed by the Hamilton equations (Section 1.4.2) and Liouville's theorem (Section 2.2.2), which express the conservation of energy and volume in the phase space (or of the probability density), on a microscopic scale. Nevertheless, if we contemplate particle systems which no longer obey the conservation of macroscopic energy, and whose macroscopic probability densities vary, then these two theorems become wrong on that scale. A 'macroscopic' phase space may be associated with the parameters $(\{P_i\}, \{Q_i\})$, and the system may be described in it using amended Hamilton equations, in order to take into account the dissipative effect induced by the statistical viewpoint on a macroscopic scale. Thus, we return to the Hamilton equations, providing them with the above calculated friction forces, in order to update the arguments given in Section 1.4.1. The calculation (1.88) becomes:

$$dH = -\frac{\partial L}{\partial t}dt + \sum_i \left[\dot{Q}_i dP_i - \left(\dot{P}_i + \frac{\partial F}{\partial \dot{Q}_i} \right) dQ_i \right] \tag{2.208}$$

(note that, by dividing that equation by dt and considering the case of an isolated system ($\partial L/\partial t = 0$), (2.175) would immediately be recovered with $H = E_k + E_{ext}$). The Hamilton equations (1.89) are presently rearranged in the following macroscopic form:

$$\begin{aligned} \frac{\partial H}{\partial t} &= -\frac{\partial L}{\partial t} \\ \frac{\partial H}{\partial P_i} &= \dot{Q}_i \\ \frac{\partial H}{\partial Q_i} &= -\dot{P}_i - \frac{\partial F}{\partial \dot{Q}_i} \end{aligned} \tag{2.209}$$

They are then unchanged, except for the equations of motion (the third set of equations), according to the dissipative Lagrange equation (2.173). We may notice that the dissipative forces (2.169) (or (2.171)) are also written as

$$-\frac{\partial F}{\partial \dot{Q}_i} = -\sum_j \alpha_{ij} (\{Q_k\}) \frac{\partial H}{\partial P_i} \tag{2.210}$$

By analogy with (1.105), the system (2.209) may also be written in a matrix form:

$$\dot{\mathbf{X}} = \overline{\boldsymbol{\Omega}}_{2N} \frac{\partial H}{\partial \mathbf{X}} \tag{2.211}$$

where $\overline{\boldsymbol{\Omega}}_{2N}$ is a $2N \times 2N$ matrix as given by:

$$\overline{\boldsymbol{\Omega}}_{2N} = \begin{pmatrix} -\boldsymbol{\alpha}(\{Q_k\}) & -\mathbf{I}_N \\ \mathbf{I}_N & \mathbf{0}_N \end{pmatrix} = \boldsymbol{\Omega}_{2N} - \boldsymbol{\alpha}_{2N} \tag{2.212}$$

where $\boldsymbol{\alpha}$ is the $N \times N$ matrix whose coefficients are the α_{ij}, $\boldsymbol{\alpha}_{2N} = \mathbf{C}_{2N}/T$ being defined as in (2.168). As expected, equations (2.211) and (2.212) are identical to (2.207). The non-conservation of the macroscopic energy explicitly appears in this formulation. Matrix $\overline{\boldsymbol{\Omega}}_{2N}$, indeed, does not obey the property of skew-symmetry (1.94), since that would come down to the skew-symmetry of $\boldsymbol{\alpha}$. With the relation (2.170), the matrix $\boldsymbol{\alpha}$ would then be zero and could not be positive definite (we made a similar observation in Section 2.4.2 in the case of a particle system). Thus, the calculation (1.107) is altered to give

$$\begin{aligned} \frac{dH}{dt} &= \left(\frac{\partial H}{\partial \mathbf{X}}\right)^T \overline{\boldsymbol{\Omega}}_{2N} \frac{\partial H}{\partial \mathbf{X}} \\ &= -\sum_{i,j} \frac{\partial H}{\partial P_i} \alpha_{ij} (\{Q_k\}) \frac{\partial H}{\partial P_j} \end{aligned} \tag{2.213}$$

$$= -\sum_{i,j} \alpha_{ij}\left(\{Q_k\}\right) \dot{Q}_i \dot{Q}_j$$

$$= -2F$$

(we have used the second equation of the system (2.209)). The energy dissipation rate (2.176) is, indeed found again. Here, however, dissipation seems to be applied to the total energy, and not to the mere macroscopic energy. This is because we have omitted the internal energy and its dependence on entropy. We will then more accurately reformulate the preceding principles, referring to the notations of Section 1.5.4. From now on, the Hamiltonian is written as

$$\begin{aligned} H\left(\{\mathbf{p}_a\}, \{\mathbf{r}_a\}, \{\rho_a\}, \{\sigma_a\}, t\right) &= \sum_a \frac{|\mathbf{p}_a|^2}{2m_a} \\ &+ \sum_a m_a e_{int,a}(\rho_a, \sigma_a) \\ &+ \sum_a E_{ext}\left(\mathbf{r}_a, t\right) \end{aligned} \tag{2.214}$$

The state vector (1.266) now possesses an additional state variable, namely the specific entropy, and we will differentiate it using the notation $\mathbf{Z}_a$:

$$\mathbf{Z}_a \doteqdot \begin{pmatrix} \mathbf{p}_a \\ \mathbf{r}_a \\ \rho_a \\ \sigma_a \end{pmatrix} \tag{2.215}$$

Relations (2.73) and (2.88) show that the derivative of the Hamiltonian (2.214) with respect to entropy equals

$$\frac{\partial H}{\partial \sigma_a} = m_a T_a \tag{2.216}$$

Thus, the energy with respect to the state vector, that is (1.270), becomes

$$\frac{\partial H}{\partial \mathbf{Z}_a} = \begin{pmatrix} \mathbf{u}_a \\ -\mathbf{F}_a^{int} - \mathbf{F}_a^{ext} \\ \dfrac{V_a p_a}{\rho_a} \\ m_a T_a \end{pmatrix} \tag{2.217}$$

and the Hamilton equations (1.269) should be altered. To that purpose, we note that the spatial derivatives of entropy should be added to the term (1.267). Afterwards, we must add the dissipative forces to the equation of motion (with respect to $\dot{\mathbf{p}}_a$). Lastly, we use (2.189) to write an equation for the temporal derivative $\dot{\sigma}_a$ of entropy, by adding a term caused by the motion of particles, as we did for density in (1.268). Taking (2.216) and (2.217) into account, we find

$$\begin{aligned}
\forall a,\ \dot{\mathbf{p}}_a &= -\frac{\partial H}{\partial \mathbf{r}_a} - \sum_b \left(\frac{\partial \rho_b}{\partial \mathbf{r}_a}\frac{\partial H}{\partial \rho_b} + \frac{\partial \sigma_b}{\partial \mathbf{r}_a}\frac{\partial H}{\partial \sigma_b} \right) \\
&\quad - \frac{1}{m_a T_a}\mathbf{F}_a^{diss}\frac{\partial H}{\partial \sigma_a} \\
\dot{\mathbf{r}}_a &= \frac{\partial H}{\partial \mathbf{p}_a} \\
\dot{\rho}_a &= \sum_b \frac{\partial \rho_a}{\partial \mathbf{r}_b}\cdot\frac{\partial H}{\partial \mathbf{p}_b} \\
\dot{\sigma}_a &= \sum_b \frac{\partial \sigma_a}{\partial \mathbf{r}_b}\cdot\frac{\partial H}{\partial \mathbf{p}_b} - \frac{1}{m_a T_a}\mathbf{F}_a^{diss}\cdot\frac{\partial H}{\partial \mathbf{p}_a}
\end{aligned} \tag{2.218}$$

This system takes the following matrix form, which is analogous to (1.271), but includes the dissipative forces like (2.207):

$$\forall a,\ \dot{\mathbf{Z}}_a = \sum_b (\boldsymbol{\omega}_8)_{ab}\frac{\partial H}{\partial \mathbf{Z}_b} \tag{2.219}$$

where a 8 × 8 matrix has been defined by

$$(\boldsymbol{\omega}_8)_{ab} \doteqdot \begin{pmatrix} [\mathbf{0}_3] & -\delta_{ab}[\mathbf{I}_3] & -\dfrac{\partial \rho_b}{\partial \mathbf{r}_a} & -\mathbf{B}_a \\ \delta_{ab}[\mathbf{I}_3] & [\mathbf{0}_3] & \mathbf{0} & \mathbf{0} \\ \left(\dfrac{\partial \rho_a}{\partial \mathbf{r}_b}\right)^T & \mathbf{0}^T & 0 & 0 \\ \mathbf{B}_a^T & \mathbf{0}^T & 0 & 0 \end{pmatrix} \tag{2.220}$$

with

$$\mathbf{B}_a \doteqdot \frac{\partial \sigma_a}{\partial \mathbf{r}_b} - \frac{\delta_{ab}}{m_a T_a}\mathbf{F}_a^{diss} \tag{2.221}$$

This matrix may be compared to $(\boldsymbol{\omega}_7)_{ab}$, as defined by (1.272) for the case without any variation of entropy. Owing to the fact that δ_{ab} vanishes if $a \neq b$, the term $\mathbf{F}_b^{diss}$ may be substituted for $\mathbf{F}_a^{diss}$, and so that matrix verifies the property $(\boldsymbol{\omega}_8)_{ba} = -(\boldsymbol{\omega}_8)_{ab}^T$, which implies the conservation of total energy, the lost kinetic energy being recovered in the form of internal energy.

Let us now consider what Liouville's theorem becomes in that formalism, first returning to the generalized notations. As explained at the end of Section 2.2.2, the macroscopic probability $\overline{\Phi}$ changes over time along with entropy. The calculation (2.9), applied to the macroscopic system's probability density, now gives

$$\frac{1}{\overline{\Phi}}\frac{d\overline{\Phi}}{dt} = \sum_i \left(\frac{\partial^2 H}{\partial P_i \partial Q_i} + \frac{\partial^2 F}{\partial P_i \partial \dot{Q}_i} - \frac{\partial^2 H}{\partial Q_i \partial P_i} \right) \tag{2.222}$$

$$= \sum_i \frac{\partial^2 F}{\partial P_i \partial \dot{Q}_i} \tag{2.223}$$

that is taking the form (2.172) of F into account:

$$\begin{aligned} \frac{1}{\overline{\Phi}} \frac{d\overline{\Phi}}{dt} &= \sum_{i,j} \frac{\partial \alpha_{ij} \left(\{Q_k\}\right) \dot{Q}_j}{\partial P_i} \\ &= \sum_{i,j} \alpha_{ij} \left(\{Q_k\}\right) \frac{\partial^2 H}{\partial P_i \partial P_j} \end{aligned} \tag{2.224}$$

(we may recall that the derivatives with respect to P_i are made with a constant Q_i in the phase space). Through formula (1.77) we can conclude:

$$\frac{\partial^2 H}{\partial P_i \partial P_j} = \Gamma_{ij} \tag{2.225}$$

which yields

$$\frac{1}{\overline{\Phi}} \frac{d\overline{\Phi}}{dt} = \sum_{i,j} \alpha_{ij} \Gamma_{ij} = \mathrm{tr}\left(\boldsymbol{\alpha} \boldsymbol{\Gamma}\right) \tag{2.226}$$

where the matrices are here of the N-th order, N being the number of particles. Now, the trace of the product of two positive definite matrices is strictly positive. Thus, $\overline{\Phi}$ increases over time. The variation of volume $\overline{V}$ in the macroscopic phase space (refer to eqn (1.255), as well as eqn (3.20) of Chapter 3, for the analogy with the real continua), is given by

$$\frac{1}{\overline{V}} \frac{d\overline{V}}{dt} = -\frac{1}{\overline{\Phi}} \frac{d\overline{\Phi}}{dt} < 0 \tag{2.227}$$

We may recall that, in the absence of dissipative terms, equation (2.9) shows that the probability density Φ in the phase space, acting as a density, remains constant, and so the 'volumes' V are constant in it, as illustrated by Fig. 2.1. In contrast, the dissipative term reduces the volumes $\overline{V}$ in the macroscopic quantity phase space so much faster as the dissipation of energy is higher, which is the dissipative Liouville theorem. This is because as the system takes more and more disordered—that is more and more probable—configurations, the number $N\left(\overline{\Phi}\right)$ of microscopic configurations corresponding to the current macroscopic configuration increases, which increases entropy (eqn (2.29)) and probability $\overline{\Phi}$ (eqn (2.30)).

In order to illustrate this result, let us resume the case of the simple dissipative pendulum as set out at the end of Section 2.4.2. In the case of the small oscillations ($\sin\theta \approx \theta$, voir § 1.4.4), the equation of motion (2.177) can be solved analytically, since it becomes linear:

$$m\ell^2 \ddot{\theta} = -mg\ell\theta - \alpha\dot{\theta} \tag{2.228}$$

We can observe that the system's equations are then written, in a non-dimensional form (refer to definition (1.174)), as:

$$\dot{\mathbf{X}}^+ = \omega_0 \begin{pmatrix} -\alpha/A & -1 \\ 1 & 0 \end{pmatrix} \mathbf{X}^+ \tag{2.229}$$

This arrangement may be compared with (2.212). A solution in the following form is sought

$$\theta = \theta_0 \exp(i\omega t + \varphi) \tag{2.230}$$

By injecting (2.230) into (2.228), we get the following characteristic equation:

$$\omega^2 - i\frac{\alpha}{m\ell^2}\omega - \frac{g}{\ell} = 0 \tag{2.231}$$

We will consider the case where the friction is comparatively moderate, i.e. verifies

$$\alpha \leqslant 2m\sqrt{g\ell^3} = 2A \tag{2.232}$$

where A is defined by (1.162). In this case, the solution[45] of (2.231) is written as

$$\begin{aligned} \omega &= \sqrt{\frac{g}{\ell} - \left(\frac{\alpha}{2m\ell^2}\right)^2} + i\frac{\alpha}{2m\ell^2} \\ &= \omega_1 + i\omega_2 \end{aligned} \tag{2.233}$$

which defines ω_1 and ω_2, both of them being positive. In the frictionless case ($\alpha = 0$), we find $\omega_1 = \omega_0$, in accordance with (1.154). Referring to the notations in Section 1.4.4 (eqns (1.160) and (1.162)), the system's components in the phase space are the following:

$$\begin{aligned} p &= -A\theta_0 \exp(-\omega_2 t) \sin(i\omega_1 t + \varphi) \\ q &= \theta_0 \exp(-\omega_2 t) \cos(i\omega_1 t + \varphi) \end{aligned} \tag{2.234}$$

which gives again (1.168) in the frictionless case ($\omega_2 = 0, \omega_1 = \omega_0$). Thus, we find a dampened oscillatory motion.[46] The first line of (1.175) is still valid, with a transformation matrix which is now given by

$$\mathbf{\Psi}^+\left(t^+\right) = \exp\left(-\frac{\omega_2}{\omega_1}t^+\right)\begin{pmatrix} \cos t^+ & -\sin t^+ \\ \sin t^+ & \cos t^+ \end{pmatrix} \tag{2.235}$$

provided that $t^+ = \omega_1 t$ is put, which gives again (1.173) in the frictionless case.

The fact that the volumes of the macroscopic phase space decrease over time is illustrated by the calculation of the determinant in (2.235), from the equation (2.12) and the definition (2.233) of ω_2:

[45] There are, of course, two solutions, which only differ from one another by the sign before the radical, determinating the pendulum's direction of rotation. Here we will arbitrarily adopt the positive sign.

[46] If the relation (2.232) was not satisfied, it would be easily demonstrated that the motion is so dampened that it cannot exhibit oscillations.

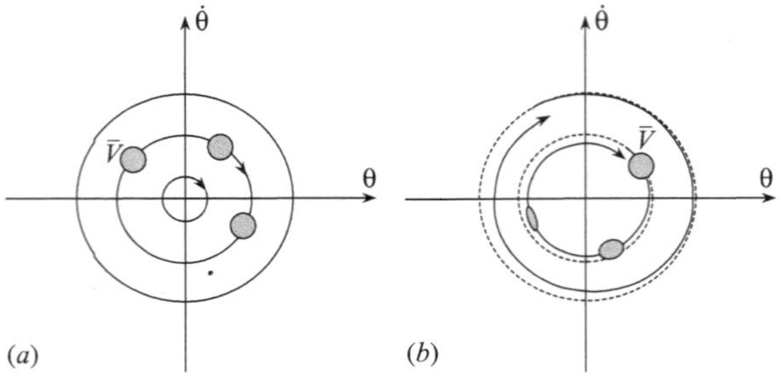

Fig. 2.5 Shapes of the phase portrait of the non-dissipative (*a*) and dissipative (*b*) linear pendulum.

$$\begin{aligned}\frac{\overline{V}}{\overline{V}_0} &= \det \boldsymbol{\Psi}^+\left(t^+\right) \\ &= \exp\left(-2\omega_2 t\right) \\ &= \exp\left(-\frac{\alpha t}{m\ell^2}\right)\end{aligned} \tag{2.236}$$

which gives again (2.12) in the non-dissipative case being studied in Section 2.2.2. That result can be made more explicit using the equations (1.163), (2.226) and (2.227). The matrices $\boldsymbol{\alpha}$ and $\boldsymbol{\Gamma}$ here include one coefficient each, which gives the decrease of the volumes in the phase space as

$$\begin{aligned}\frac{1}{\overline{V}}\frac{d\overline{V}}{dt} &= -\alpha\Gamma \\ &= -\frac{\alpha}{m\ell^2} < 0\end{aligned} \tag{2.237}$$

Integrating that equation with respect to time, we find

$$\ln\frac{\overline{V}}{\overline{V}_0} = -\frac{\alpha t}{m\ell^2} \tag{2.238}$$

which is equivalent to (2.236). Figure 2.5 shows the shape of the phase portrait near the origin without (*a*) and with (*b*) dissipation, highlighting the mechanism of phase volume reduction in the dissipative case. The non-dissipative case is a close-up view of Fig. 1.4 near the origin. The irreversible nature of the dissipative motion is illustrated by the phase trajectory tendency to inexorably head for the origin, corresponding to a 'final' situation of rest.

As a rule, one could wonder why the macroscopic approach modifies the evolution of the volumes within the phase space of a mechanical system. Figure 2.6 illustrates this process: as time elapses, since the molecule energy is distributed among a higher and higher number of values, the system additionally becomes homogeneous on a macroscopic scale. The parameters characterizing the particles are then distributed in a increasingly homogeneous way, and so the macroscopic phase trajectories come closer to each other. Although the actual volume of the phases remains constant, its shape is increasingly spread into outgrowths, and its average after Boltzmann decreases. That figure may be compared with 2.1. We must conclude from that analysis that the dissipative processes homogenize the parameters on a macroscopic scale; we will provide

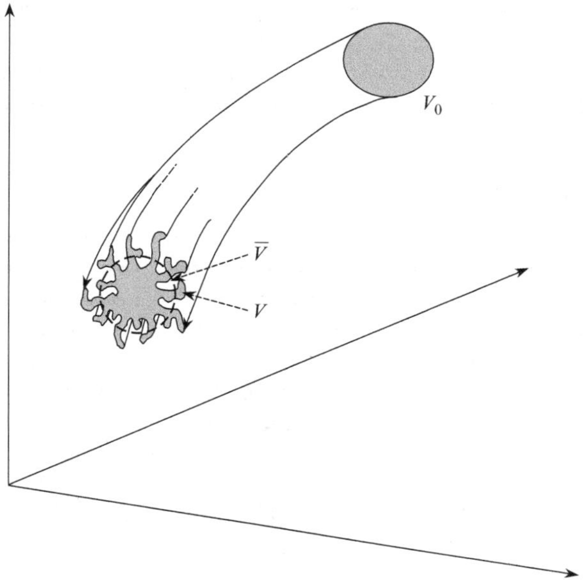

Fig. 2.6 Evolution in the phase space: while the volume corresponding to the microscopic (molecular) motion remains constant, in constrast the volume associated with the (macroscopic) average motion (here a thick dotted line) decreases over time (volume shrinkage).

further details on that mechanism[47] in Chapters 3 and 4. It should be pointed out that this homogenizing nature is intuitively consistent with the principle of an irreversible phenomenon over time.

[47] The homogenization process taking place along with the increase of entropy was mentioned at the end of Section 2.3.2 as regards temperature.

In the case of a particle system, formula (2.226) should be rearranged. The dependence of the particle friction forces on the differences among velocities $\mathbf{u}_{ab}$ (instead of $\dot{Q}_i$) amounts to substituting $\frac{1}{m_a}\mathbf{I}_3 - \mathbf{\Gamma}_{ab}$ for $\mathbf{\Gamma}_{ab}$. With (1.82), we get

$$\begin{aligned} \frac{1}{\overline{\Phi}}\frac{d\overline{\Phi}}{dt} &= \sum_{a\neq b} \mathrm{tr}\left[\boldsymbol{\alpha}_{ab}\left(\frac{1}{m_a}\mathbf{I}_3 - \mathbf{\Gamma}_{ab}\right)\right] \\ &= \sum_{a\neq b} \mathrm{tr}\left[\frac{1}{m_a}\left(1-\delta_{ab}\right)\boldsymbol{\alpha}_{ab}\right] \\ &= \sum_{a\neq b} \frac{\mathrm{tr}\,\boldsymbol{\alpha}_{ab}}{m_a} > 0 \end{aligned} \tag{2.239}$$

This form proves more clearly that the dissipative nature of the system (positiveness of the kinetic matrices $\boldsymbol{\alpha}_{ab}$) does induce a reduction of the volumes in the macroscopic phase space. By choosing (2.204), we find:

$$\frac{1}{\overline{V}}\frac{d\overline{V}}{dt} = -\sum_{a\neq b} \frac{\gamma_{ab}\left(r_{ab}\right) r_{ab}^2}{m_a} < 0 \tag{2.240}$$

since the masses are always positive (refer to Section 1.3.1).

As can be seen, whereas the macroscopic point of view makes it possible to become unhindered by the complex individual motions of a large number of molecules, it adds an element into the equations of motion—namely the dissipative forces—and infringes the law of conservation of energy, as well as the law of conservation of volumes within the phase space, on a macroscopic scale. Let us remember, however, that these values are still conserved on a microscopic scale, and that their loss results from the applied statistical approach, revealing a flux of energy from the macroscopic scale to the molecule scale.

2.5.2 Mirror systems

Due to the discussions in the previous paragraphs, we may consider that a dissipative system consists of two sub-systems, one of which, representing the macroscopic point of view, would give energy to the other one, representing the microscopic point of view. Such an approach can be formalized using the notion of mirror systems.[48] To that purpose, we consider an isolated system of average generalized coordinates $\{Q_k\}$ (it is called a *basic system* and represents the average—i.e. macroscopic—quantities), and a *mirror coordinate* Q_i^* is associated to each Q_i. We consider the following Lagrangian:

[48] Here we borrow ideas from Basdevant, (2005), while keeping an eye on the discrete turbulence model of Section 6.5.2, which is built in a particle formalism.

$$L = \frac{1}{2}\sum_i m_i \dot{Q}_i \dot{Q}_i^* - E_p(\{Q_k\}) \tag{2.241}$$

where the $\{m_i\}$ are positive constant values. If the mirror coordinates $\{Q_k^*\}$ represent the macroscopic system, we may generally consider that there is a linear relation which turns each basic velocity $\dot{Q}_k$ into its mirror velocity $\dot{Q}_k^*$. That relation, for instance, may be an overall average whose coefficients may depend on the positions, as when we established the formula (1.83) in Section 1.3.3:

$$\dot{Q}_i^* = \sum_j m_j A_{ij}(\{Q_k\})\dot{Q}_j \tag{2.242}$$

This relation may be considered a definition of the A_{ij}, about which we will also assume they make a symmetric positive definite tensor $\mathbf{A}$. We can then rearrange the Lagrangian in a form which only involves the basic system parameters:

$$L = \frac{1}{2}\sum_{i,j} m_i m_j A_{ij}(\{Q_k\})\dot{Q}_i\dot{Q}_j - E_p(\{Q_k\}) \tag{2.243}$$

The generalized momentum of that system can easily be derived therefrom:

$$\begin{aligned} P_i = \frac{\partial L}{\partial \dot{Q}_i} &= \sum_j m_i m_j A_{ij}(\{Q_k\})\dot{Q}_j \\ &= m_i \dot{Q}_i^* \end{aligned} \tag{2.244}$$

Therefore, the generalized velocities of the mirror system do provide the generalized momenta of the basic system. Note that the relation (2.242) between velocities defines an identical relation between the variations of coordinates dQ_i^* and dQ_j; thus, we can write

$$\frac{\partial Q_i^*}{\partial Q_j} = m_i m_j A_{ij}(\{Q_k\}) \tag{2.245}$$

Thus, the mirror coordinates can be expressed as a function of the basic coordinates. It will then be possible for every function $\{Q_k\}$ to be expressed in accordance with the $\{Q_k^*\}$. In particular, the derivatives of an arbitrary quantity C can be expressed in relation to the mirror coordinates:

$$\begin{aligned}\frac{\partial C}{\partial Q_i^*} &= \sum_j \frac{\partial C}{\partial Q_j}\frac{\partial Q_j^*}{\partial Q_i} \\ &= \sum_j m_i m_j A_{ij}\left(\{Q_k\}\right)\frac{\partial C}{\partial Q_j}\end{aligned} \tag{2.246}$$

We may also invert the relation (2.244) by putting $\mathbf{B} = \mathbf{A}^{-1}$. The symmetric positive definite matrix $\mathbf{B}$ will define a relation making it possible to switch from the mirror system to the basic system; its components will then preferably be expressed as a function of the $\{Q_k^*\}$:

$$\dot{Q}_i = \sum_j \frac{B_{ij}\left(\{Q_k^*\}\right)}{m_i m_j} P_j = \frac{1}{m_i}\sum_j B_{ij}\left(\{Q_k^*\}\right)\dot{Q}_j^* \tag{2.247}$$

Returning to the definition (2.243), we then find a third formulation of the Lagrangian, now regarding the mirror system velocities, by also expressing the potential in accordance with the $\{Q_k^*\}$:

$$L = \frac{1}{2}\sum_{i,j} B_{ij}\left(\{Q_k^*\}\right)\dot{Q}_i^*\dot{Q}_j^* - E_p\left(\{Q_k^*\}\right) \tag{2.248}$$

It should be kept in mind that the system does not consist of the two sets of data (basic parameters, mirror parameters), but can be seen as consisting of *either* basic parameters *or* mirror parameters, it being possible to express the Lagrangian as a function of either set of parameters. Formula (2.248) now gives, by analogy with (2.244), the generalized mirror momenta:

$$P_i^* = \frac{\partial L}{\partial \dot{Q}_i^*} = m_i \dot{Q}_i \tag{2.249}$$

Equations (2.244) and (2.249) result in some kind of duality between the two systems. Returning to (2.242) and (2.247), one may also write:

$$\begin{aligned}\dot{Q}_i^* &= \sum_j A_{ij}\left(\{Q_k\}\right)P_j^* \\ P_i^* &= \sum_j B_{ij}\left(\{Q_k^*\}\right)\dot{Q}_j^*\end{aligned} \tag{2.250}$$

The Lagrange equations characterizing the system motions may equally be written in both component systems:

$$\begin{aligned}\forall i,\ \frac{dP_i}{dt} &= F_i = \frac{\partial L}{\partial Q_i} \\ \frac{dP_i^*}{dt} &= F_i^* = \frac{\partial L}{\partial Q_i^*}\end{aligned} \tag{2.251}$$

Based on formulas (2.243), (2.244), (2.248) and (2.249), we find:

$$\begin{aligned}\forall i,\ m_i\frac{d\dot{Q}_i^*}{dt} &= \frac{1}{2}\sum_{j,k} m_j m_k \frac{\partial A_{jk}}{\partial Q_i}\dot{Q}_j\dot{Q}_k - \frac{\partial E_p}{\partial Q_i}\\ m_i\frac{d\dot{Q}_i}{dt} &= \frac{1}{2}\sum_{j,k}\frac{\partial B_{jk}}{\partial Q_i^*}\dot{Q}_j^*\dot{Q}_k^* - \frac{\partial E_p}{\partial Q_i^*}\end{aligned} \tag{2.252}$$

We note that in addition to the conservative forces, which we are familiar with, appear forces depending on the generalized velocities; they then occur as friction forces, in agreement with the idea of an energy flux from a system towards its mirror system. These forces, however, oddly occur as quadratic forms of the velocities, whereas they were linear forms in Section 2.4.3. It will be explained in Chapter 3 (Section 3.6.4), however, that such a formalism is useful for a representation of scattering through a variational principle, whereas Chapter 6 (Section 6.5.2) will show it provides a neat approach to particle turbulence within the context of the SPH method.

Let us turn to the system energy. As usual, it is defined by

$$\begin{aligned}E &= \sum_i P_i\dot{Q}_i - L\\ &= \sum_i P_i^*\dot{Q}_i^* - L\\ &= E_k + E_p\end{aligned} \tag{2.253}$$

with, on completion of calculations and taking all the forms we have introduced for the Lagrangian into account:

$$\begin{aligned}E_k &= \frac{1}{2}\sum_i m_i\dot{Q}_i\dot{Q}_i^*\\ &= \frac{1}{2}\sum_{i,j} m_i m_j A_{ij}\left(\{Q_k\}\right)\dot{Q}_i\dot{Q}_j\\ &= \frac{1}{2}\sum_{i,j} B_{ij}\left(\{Q_k^*\}\right)\dot{Q}_i^*\dot{Q}_j^*\end{aligned} \tag{2.254}$$

The positiveness of matrices **A** and **B** ensures that of E, about which we also can state that it is constant, without having to demonstrate it once again, because this contemplated system fulfils all the conditions implied by the demonstration of the conservation of energy (1.73). Moreover, we can define an altered 'energy' as follows:

$$E^* \doteqdot \frac{1}{2}\sum_i m_i\dot{Q}_i^{*2} + E_p\left(\{Q_k^*\}\right) \tag{2.255}$$

This quantity, which could represent the mirror system energy, is not conserved, as evidenced by the following calculation, which is based on (2.244) and (2.252):

$$
\begin{aligned}
\frac{dE^*}{dt} &= \sum_i m_i \dot{Q}_i^* \frac{d\dot{Q}_i^*}{dt} + \sum_i \frac{\partial E_p}{\partial Q_i^*}\frac{dQ_i^*}{dt} \\
&= \sum_i \dot{Q}_i^* \left(\frac{dP_i}{dt} + \frac{\partial E_p}{\partial Q_i^*}\right) \\
&= \frac{1}{2}\sum_{i,j,k} m_j m_k \frac{\partial A_{jk}}{\partial Q_i}\dot{Q}_i^* \dot{Q}_j \dot{Q}_k + \sum_i \left(\frac{\partial E_p}{\partial Q_i^*} - \frac{\partial E_p}{\partial Q_i}\right)\dot{Q}_i^*
\end{aligned}
\tag{2.256}
$$

Thus, the mirror system either gains or loses energy, according to the sign of dE^*/dt. This loss (gain) is made to the advantage (to the detriment) of the basic system. In this respect, the basic system and the mirror system can be seen as two parts of one set. There is then a transfer of energy between the two systems which, for instance, may represent the loss of macroscopic energy (dissipation) to the advantage of thermal agitation in a continuous medium, while ensuring the conservation of total energy. The very formal nature of that approach, however, should be pointed out. We note that the quadratic nature of the friction forces leads, in that formalism, to a cubic dependence of dissipation with respect to the generalized velocities, as shown by (2.256).

The above formalism can easily be adapted to the case of a particle system, with notations which speak for themselves. The relations between basic system and mirror system are written as:[49]

$$
\begin{aligned}
\mathbf{p}_a &= \sum_b m_a m_b A_{ab}\left(\{\mathbf{r}_c\}\right)\mathbf{u}_b \\
\mathbf{u}_a &= \sum_b \frac{B_{ab}\left(\{\mathbf{r}_c^*\}\right)}{m_a m_b}\mathbf{p}_b \\
\mathbf{u}_a^* &= \sum_b A_{ab}\left(\{\mathbf{r}_c\}\right)\mathbf{p}_b^* \\
\mathbf{p}_a^* &= \sum_b B_{ab}\left(\{\mathbf{r}_c^*\}\right)\mathbf{u}_b^*
\end{aligned}
\tag{2.257}
$$

[49]The general nature of the following approach could be extended to the case where matrices $\mathbf{A}_{ab}$ and $\mathbf{B}_{ab}$ are substituted for the coefficients A_{ab} and B_{ab}. This complexity, however, will be useless for what we intend to do with it.

where the momenta are given by

$$
\begin{aligned}
\mathbf{p}_a &= m_a \mathbf{u}_a^* \\
\mathbf{p}_a^* &= m_a \mathbf{u}_a
\end{aligned}
\tag{2.258}
$$

In the formulation (2.257), the quantities A_{ab} and B_{ab} are elements of two $N_p \times N_p$-dimensional mutually inverse tensors (N_p being the number of particles). An important remark should be made here: unlike the notations in Sections 1.4.1 and 2.4.3, the sub-scripts a and b do represent here the (scalar) components of these two tensors. The Lagrangian is written as:

$$
L = \frac{1}{2}\sum_a m_a \mathbf{u}_a \cdot \mathbf{u}_a^* - E_p
$$

$$
\begin{aligned}
&= \frac{1}{2}\sum_{a,b} m_a m_b A_{ab}\left(\{\mathbf{r}_c\}\right)\mathbf{u}_a\cdot\mathbf{u}_b - E_p \\
&= \frac{1}{2}\sum_{a,b} B_{ab}\left(\{\mathbf{r}_c^*\}\right)\mathbf{u}_a^*\cdot\mathbf{u}_b^* - E_p
\end{aligned}
\tag{2.259}
$$

Note that the first line of (2.257) and the second line of (2.259) correspond to the general form (1.81), with

$$
\mathbf{\Lambda}_{ab}\left(\{\mathbf{r}_c\}\right) = m_a m_b A_{ab}\left(\{\mathbf{r}_c\}\right)\mathbf{I}_3 \tag{2.260}
$$

We may recall that, according to Section 1.4.1, the ordinary case is featured by $\mathbf{\Lambda}_{ab} = m_a\delta_{ab}\mathbf{I}_3$, which corresponds to the particular case $A_{ab} = \delta_{ab}/m_a$. The equations of motion occur in an analogous form to (1.83), but now with forces depending upon the velocities, as in (2.252):

$$
\begin{aligned}
\forall a,\ m_a\frac{d\mathbf{u}_a^*}{dt} &= \frac{1}{2}\sum_{b,c} m_b m_c \frac{\partial A_{bc}}{\partial \mathbf{r}_a}\left(\mathbf{u}_b\cdot\mathbf{u}_c\right) - \frac{\partial E_p}{\partial \mathbf{r}_a} \\
m_a\frac{d\mathbf{u}_a}{dt} &= \frac{1}{2}\sum_{b,c} \frac{\partial B_{bc}}{\partial \mathbf{r}_a^*}\left(\mathbf{u}_b^*\cdot\mathbf{u}_c^*\right) - \frac{\partial E_p}{\partial \mathbf{r}_a^*}
\end{aligned}
\tag{2.261}
$$

Once again, the forces occurring in (2.261) exhibit a quadratic dependence on velocities, whereas the particle friction forces (2.183) as defined in Section 2.4.3 have been built on the basis of a linear dependence. The model we disclose here is thus only valid when the linear terms of the friction forces are either zero or negligible. This model actually amounts to contemplating kinetic coefficients $\boldsymbol{\alpha}_{ab}$ linearly depending on the velocities. The suitability of that quadratic formulation will appear within the context of Section 6.5.2.

As regards energy, it is defined by

$$
\begin{aligned}
E &= \sum_a \mathbf{p}_a\cdot\mathbf{u}_a - L \\
&= \sum_a \mathbf{p}_a^*\cdot\mathbf{u}_a^* - L
\end{aligned}
\tag{2.262}
$$

and is reduced to analogous to (2.259), '+' signs being substituted for the '−' signs. Lastly, the energy flux between the two systems is written as

$$
\begin{aligned}
&\frac{d}{dt}\left(\frac{1}{2}\sum_a m_a u_a^{*2} + E_p\right) \\
&\quad = \frac{1}{2}\sum_{a,b,c} m_b m_c\left(\frac{\partial A_{bc}}{\partial \mathbf{r}_a}\cdot\mathbf{u}_a^*\right)\left(\mathbf{u}_b\cdot\mathbf{u}_c\right) \\
&\quad + \sum_a\left(\frac{\partial E_p}{\partial \mathbf{u}_a^*} - \frac{\partial E_p}{\partial \mathbf{u}_a}\right)\cdot\mathbf{u}_a^*
\end{aligned}
\tag{2.263}
$$

Thus, it exhibits a cubic dependence on the velocities, as in (2.256).

2.5.3 Thermodynamical fluctuations

In this section, we will turn to the effect of the statistical fluctuations on the motion of a fluid or solid particle. It may happen, indeed, that some fluctuations have an individual effect on top of the average pressure (Section 1.5.4) and friction (Section 2.4.3) forces. This happens, for example, when a small-sized body is subject to the molecular agitation in a fluid; this is known as a Brownian motion (e.g. refer to[50] Einstein, 1956). We also contemplate arguments which will be useful for modelling the effect of the turbulent fluctuations in a fluid, as dealt with in Chapter 4, though the notations are then slightly different (refer to Section 4.6.3). For more information about the following developments, the reader may refer to Español and Öttinger (1993), or Ellero et al. (2003).

[50] The first satisfactory explanation of the Brownian motion was given, indeed, by A. Einstein, and was the subject matter of one of his famous papers in 1905.

In order to get clear ideas, let us consider the general case of a particle a which is subject to collisions caused by the erratic motion of the molecules. By analogy with Section 2.2.1, we denote by $\varphi_a\left(\mathbf{p}_a, \mathbf{r}_a, t\right)$ (or, more simply, $\varphi_a\left(\mathbf{p}_a, \mathbf{r}_a\right)$) the probability for the particle to have parameters $\left(\mathbf{p}_a, \mathbf{r}_a\right)$ at time point t. Between time points t and $t + dt$, there is some probability for its momentum $\mathbf{p}_a$ and its position $\mathbf{r}_a$ to change by an amount $\delta\mathbf{p}_a$ and an amount $\delta\mathbf{r}_a$ respectively; this probability will be denoted $w\left(\mathbf{p}_a, \mathbf{r}_a; \delta\mathbf{p}_a, \delta\mathbf{r}_a\right)$. Probability φ_a is then reduced by an amount $w\left(\mathbf{p}_a, \mathbf{r}_a; \delta\mathbf{p}_a, \delta\mathbf{r}_a\right)\varphi_a\left(\mathbf{p}_a, \mathbf{r}_a\right)$ during the short lapse of time dt; that holds true for all the possible values of $\delta\mathbf{p}_a$ and $\delta\mathbf{r}_a$. Conversely, during that lapse of time, φ_a may increase whenever these quantities change by $\delta\mathbf{p}_a$ and $\delta\mathbf{r}_a$ and take the values $\left(\mathbf{p}_a, \mathbf{r}_a\right)$ for the particle having parameters $\left(\mathbf{p}_a - \delta\mathbf{p}_a, \mathbf{r}_a - \delta\mathbf{r}_a\right)$. The total variation of φ_a over dt is then given by

$$d\varphi_a = \int_{\Omega_r}\int_{\Omega_p} \delta w\left(\mathbf{p}_a, \mathbf{r}_a; \delta\mathbf{p}_a, \delta\mathbf{r}_a\right) d\delta\mathbf{p}_a d\delta\mathbf{r}_a \tag{2.264}$$

where

$$\begin{aligned} \delta w \doteq & w\left(\mathbf{p}_a - \delta\mathbf{p}_a, \mathbf{r}_a - \delta\mathbf{r}_a; \delta\mathbf{p}_a, \delta\mathbf{r}_a\right) \\ & \times \varphi_a\left(\mathbf{p}_a - \delta\mathbf{p}_a, \mathbf{r}_a - \delta\mathbf{r}_a\right) \\ & - w\left(\mathbf{p}_a, \mathbf{r}_a; \delta\mathbf{p}_a, \delta\mathbf{r}_a\right)\varphi_a\left(\mathbf{p}_a, \mathbf{r}_a\right) \end{aligned} \tag{2.265}$$

has been defined.

If $\delta\mathbf{p}_a$ and $\delta\mathbf{r}_a$ are high, then, δw is necessarily very low, since the particle can only experience slight variations of its parameters because it has a much greater mass than the surrounding molecules. Only the contributions from the small values of $\delta\mathbf{p}_a$ and $\delta\mathbf{r}_a$ can then be kept in the integral (2.264). A first-order expansion yields[51]

[51] There is no summation symbol above the subscripts a in the following, though they are repeated in many terms (they are not dummy sub-scripts).

$$\begin{aligned} \delta w \approx & -\delta\mathbf{r}_a \cdot \frac{\partial w\left(\mathbf{p}_a, \mathbf{r}_a; \delta\mathbf{p}_a, \delta\mathbf{r}_a\right)\varphi_a\left(\mathbf{p}_a, \mathbf{r}_a\right)}{\partial\mathbf{r}_a} \\ & - \delta\mathbf{p}_a \cdot \frac{\partial w\left(\mathbf{p}_a, \mathbf{r}_a; \delta\mathbf{p}_a, \delta\mathbf{r}_a\right)\varphi_a\left(\mathbf{p}_a, \mathbf{r}_a\right)}{\partial\mathbf{p}_a} \end{aligned} \tag{2.266}$$

Thus, equation (2.264) gives (omitting the dependence of w and φ_a on $\mathbf{p}_a$ and $\mathbf{r}_a$ for brevity):

$$\begin{aligned}
d\varphi_a = &- \int_{\Omega_r}\int_{\Omega_p} \delta\mathbf{r}_a \cdot \frac{\partial w\left(\delta\mathbf{p}_a, \delta\mathbf{r}_a\right)\varphi_a}{\partial \mathbf{r}_a} \\
&+ \delta\mathbf{p}_a \cdot \frac{\partial w\left(\delta\mathbf{p}_a, \delta\mathbf{r}_a\right)\varphi_a}{\partial \mathbf{p}_a} d\delta\mathbf{p}_a d\delta\mathbf{r}_a \\
= &- \frac{\partial}{\partial \mathbf{r}_a} \cdot \left(\varphi_a \int_{\Omega_r}\int_{\Omega_p} w\left(\delta\mathbf{p}_a, \delta\mathbf{r}_a\right)\delta\mathbf{r}_a d\delta\mathbf{p}_a d\delta\mathbf{r}_a\right) \\
&- \frac{\partial}{\partial \mathbf{p}_a} \cdot \left(\varphi_a \int_{\Omega_r}\int_{\Omega_p} w\left(\delta\mathbf{p}_a, \delta\mathbf{r}_a\right)\delta\mathbf{p}_a d\delta\mathbf{p}_a d\delta\mathbf{r}_a\right)
\end{aligned} \tag{2.267}$$

The two dual integrals between brackets define the average quantities in a statistical averaging

$$\begin{aligned}
&\int_{\Omega_r}\int_{\Omega_p} w\left(\delta\mathbf{p}_a, \delta\mathbf{r}_a\right)\delta\mathbf{r}_a d\delta\mathbf{p}_a d\delta\mathbf{r}_a = \langle\delta\mathbf{r}_a\rangle = \frac{\mathbf{p}_a}{m_a}dt \\
&\int_{\Omega_r}\int_{\Omega_p} w\left(\delta\mathbf{p}_a, \delta\mathbf{r}_a\right)\delta\mathbf{p}_a d\delta\mathbf{p}_a d\delta\mathbf{r}_a = \left\langle\delta\mathbf{p}_a\right\rangle = \mathbf{F}_a dt
\end{aligned} \tag{2.268}$$

This way, we find the evolution of the probability density φ_a over time in the following form

$$\frac{\partial\varphi_a}{\partial t} + \frac{\partial}{\partial \mathbf{r}_a} \cdot (\varphi_a \mathbf{u}_a) + \frac{\partial}{\partial \mathbf{p}_a} \cdot (\varphi_a \mathbf{F}_a) = 0 \tag{2.269}$$

The previous paragraphs have shown that the force experienced by particle a can be split up into a conservative force (internal force, or pressure load, depending on all the positions), a dissipative force (friction force, depending on all the velocities, i.e. on momenta) and an external force (only depending on the position of the particle being studied):

$$\mathbf{F}_a = \mathbf{F}_a^{int}\left(\{\mathbf{r}_b\}\right) + \mathbf{F}_a^{diss}\left(\left\{\mathbf{p}_b\right\}\right) + \mathbf{F}_a^{ext}\left(\mathbf{r}_a\right) \tag{2.270}$$

Since the first one does not depend on $\mathbf{u}_a$, and the derivation with respect to $\mathbf{r}_a$ is made under a fixed $\mathbf{u}_a$, we get the transport equation φ_a as

$$\frac{\partial\varphi_a}{\partial t} + \frac{\partial\varphi_a}{\partial \mathbf{r}_a} \cdot \mathbf{u}_a + \frac{\partial\varphi_a}{\partial \mathbf{p}_a} \cdot \left(\mathbf{F}_a^{int} + \mathbf{F}_a^{ext}\right) + \frac{\partial}{\partial \mathbf{p}_a} \cdot \left(\varphi_a \mathbf{F}_a^{diss}\right) = 0 \tag{2.271}$$

We can see that the equation is analogous to a Boltzmann equation (2.27), accompanied by a dissipative force. The latter term reflects, on a macroscopic scale, the integral of collisions (2.28) which is the cause of the dissipative processes. In that form, equation (2.271), which could be referred to as a *dissipative Boltzmann equation*, only models average processes, and hence cannot reflect the fluctuations being discussed. Now Section 2.4.1 has shown that the effect of such fluctuations involves a statistical correlation matrix which is linked to the dissipative forces by equation (2.146). Thus, the statistical correlations (the second-order moments) of the velocities do determine the efforts caused by the erratic motion of the molecules. It seems that the Taylor expansion (2.266) should then be extended further than the first-order. The second-order term is written as

$$\frac{1}{2}\delta\mathbf{p}_a^T \frac{\partial^2 w\left(\mathbf{p}_a, \mathbf{r}_a; \delta\mathbf{p}_a, \delta\mathbf{r}_a\right)\varphi_a\left(\mathbf{p}_a, \mathbf{r}_a\right)}{\partial\mathbf{p}_a^2}\delta\mathbf{p}_a \tag{2.272}$$

The momenta correlation tensor $\left\langle\delta\mathbf{p}_a \otimes \delta\mathbf{p}_a\right\rangle$ should then be of the first-order in dt to exert an influence over the particle average motion, which suggests that the statistical velocity fluctuations $\delta\mathbf{u}_a$ as caused by the impacts exerted by the molecules are proportional to $dt^{1/2}$:

$$\delta\mathbf{u}_a = \sqrt{Z_a dt}\boldsymbol{\xi}_a(t) \tag{2.273}$$

where Z_a is unknown and $\boldsymbol{\xi}_a(t)$ represents a 'random' signal which greatly fluctuates over time and reflects the erratic nature of the motion as determined by the molecular impacts. This *stochastic* approach implies that the infinitesimal time dt be judiciously chosen. It should, indeed, be substantially longer than the characteristic time τ_m of the molecular motion so that the latter can be considered as a random signal. It should also be sufficiently short as compared to the characteristic time of the average particle evolution, as denoted T_R here; its definition will be specified later on:

$$\tau_m \ll dt \ll T_R \tag{2.274}$$

The signal $\boldsymbol{\xi}_a(t)$ then has a zero average, since the average effect of the molecular impacts vanishes for reasons of statistical isotropy. Besides, Z_a may also be chosen so that the autocorrelation tensor of the stochastic signal $\boldsymbol{\xi}_a(t)$ be equal to one:

$$\begin{aligned} \left\langle\boldsymbol{\xi}_a\right\rangle &= \mathbf{0} \\ \left\langle\boldsymbol{\xi}_a \otimes \boldsymbol{\xi}_a\right\rangle &= \mathbf{I}_3 \end{aligned} \tag{2.275}$$

For example, independent standard normal distributions may be adopted for the components of $\boldsymbol{\xi}_a$.

The molecular motion keeps the memory of its past over times on the order of τ_m. Subsequently, because of the collisions, the molecules will completely forget their past trajectories. We can therefore consider, if we take two time points t' and t'' which are sufficiently far from each other, that the signal $\boldsymbol{\xi}_a(t')$ is fully decorrelated from $\boldsymbol{\xi}_a(t'')$, which can be written as[52]

[52] It is then called *white noise*.

$$\tau_m \ll t' - t'', \quad \left\langle\boldsymbol{\xi}_a(t') \otimes \boldsymbol{\xi}_a(t'')\right\rangle = \delta\left(t' - t''\right) dt\mathbf{I}_3 \tag{2.276}$$

where the Dirac distribution δ is defined in Appendix B. In the particular case $t' - t'' = 0$, (2.276) gives once again the second line of (2.275).

Being provided with that model, let us return to the equation about φ_a. With the second-order term as given by (2.272), the right-hand side in equation (2.267) is increased by the amount

$$-\frac{\partial}{\partial\mathbf{p}_a}\cdot\left(\mathbf{Z}_a\varphi_a\right)$$

with a matrix $\mathbf{Z}_a$ as defined by

$$\mathbf{Z}_a \doteqdot \int_{\Omega_r}\int_{\Omega_p} w\left(\mathbf{p}_a, \mathbf{r}_a; \delta\mathbf{p}_a, \delta\mathbf{r}_a\right)\delta\mathbf{p}_a \otimes \delta\mathbf{p}_a d\delta\mathbf{p}_a d\delta\mathbf{r}_a \\ = \left\langle \delta\mathbf{p}_a \otimes \delta\mathbf{p}_a \right\rangle \tag{2.277}$$

With the relations (2.273) and (2.275), we find

$$\mathbf{Z}_a = m_a^2 Z_a \left\langle \boldsymbol{\xi}_a \otimes \boldsymbol{\xi}_a \right\rangle dt \\ = m_a^2 Z_a dt \mathbf{I}_3 \tag{2.278}$$

Equation (2.271) then takes the final form[53]

$$\frac{\partial \varphi_a}{\partial t} + \frac{\partial \varphi_a}{\partial \mathbf{r}_a} \cdot \mathbf{u}_a + \frac{\partial \varphi_a}{\partial \mathbf{p}_a} \cdot \left(\mathbf{F}_a^{int} + \mathbf{F}_a^{ext}\right) \\ + \frac{\partial}{\partial \mathbf{p}_a} \cdot \left(\varphi_a \mathbf{F}_a^{diss} - \frac{m_a^2}{2}\frac{\partial Z_a \varphi_a}{\partial \mathbf{p}_a}\right) = 0 \tag{2.279}$$

It is called the *Fokker–Planck equation* (Pitaevskii and Lifshitz, 1981; Rief, 1965). In addition to the dissipative forces resulting from the mean particle velocity, then comes a new force reflecting the effect of the multiple impacts exerted on the relevant particle by the molecules. Equation (2.279) shows that these forces are similar to dissipative forces. We will show that these two types of forces are actually related. The equation of motion of a macroscopic particle is now written as[54]

$$\forall a,\ m_a \frac{d\mathbf{u}_a}{dt} = \mathbf{F}_a^{int} + \mathbf{F}_a^{diss} + \mathbf{F}_a^{stoc} + \mathbf{F}_a^{ext} \tag{2.280}$$

The stochastic force determining a variation of velocity as given by (2.273) is written as $\mathbf{F}_a^{stoc} = \delta\mathbf{p}_a/dt$, i.e.[55]

$$\mathbf{F}_a^{stoc} = m_a \sqrt{\frac{Z_a}{dt}}\boldsymbol{\xi}_a(t) \tag{2.281}$$

Owing to both that relation and (2.276), the stochastic force autocorrelation tensor may be written as

$$\left\langle \mathbf{F}_a^{stoc}(t') \otimes \mathbf{F}_a^{stoc}(t'')\right\rangle = \frac{m_a^2 Z_a}{dt}\left\langle \boldsymbol{\xi}_a(t') \otimes \boldsymbol{\xi}_a(t'')\right\rangle \\ = m_a^2 Z_a \delta\left(t' - t''\right)\mathbf{I}_3 \tag{2.282}$$

We now refer to the inertial frame of reference of the ambient fluid; so, the velocity of the latter is locally zero. The dissipative force exerted onto the particle can be formed from the model of the expression (2.179) as seen in Section 2.4.3. However, here we consider the effect of the average friction caused by the action of molecules, not of the other particles; thus, the contributions from all those particles acting upon the particle a can be lumped together into one term:

$$\mathbf{F}_a^{diss} = -m_a \mathbf{G}_a \mathbf{u}_a \tag{2.283}$$

[53] We will subsequently explain that those terms strictly above the second-order in the Taylor expansion (2.267) are negligible.

[54] The dynamic equations might be written in the form of a matrix as in the expansions of Section 2.5.1 (refer, for example, to Español, 2007; Vasquez-Quesada et al., 2009).

[55] The following formula reveals that this force is not properly defined for too small time increments. This 'force' shall consequently be treated with the utmost care in the next calculations. As throughout this book, it is a physicist's viewpoint. The proper mathematical foundations of the Brownian motion theory have been laid, in particular by Ito (e.g. refer to Gardiner, 1985).

where $m_a\mathbf{G}_a$ serves as a kinetic matrix $\boldsymbol{\alpha}$. In an arbitrary frame of reference, $\mathbf{u}_a$ would have to be substituted for by $\mathbf{u}_a - \langle \mathbf{u}_\alpha \rangle$ in (2.283), $\langle \mathbf{u}_\alpha \rangle$ being the mean velocity of the molecules acting upon particle a. Equation (2.283) is then formally similar to (2.169) rather than (2.179). Anyway, the Onsager's symmetry theorem (2.170) results in

$$\mathbf{G}_a = \mathbf{G}_a^T \tag{2.284}$$

Moreover, $\mathbf{G}_a$ is necessarily symmetric positive-definite, that is invertible. With the model (2.283), the equation of motion (2.280) is rearranged as

$$\forall a,\ m_a \frac{d\mathbf{u}_a}{dt} = \mathbf{F}_a^{int} - m_a\mathbf{G}_a\mathbf{u}_a + \mathbf{F}_a^{stoc}(t) + \mathbf{F}_a^{ext} \tag{2.285}$$

where we have highlighted the fast time-dependence of the stochastic force. The other terms are liable to vary, but much more slowly, with a characteristic time T_R. With the condition (2.274), we may then assume within an excellent approximation that the parameters $\mathbf{F}_a^{int}$, $\mathbf{F}_a^{ext}$ and $\mathbf{G}_a$ in equation (2.285) are constant over the time interval dt, and seek a solution by introducing the auxiliary vector variable

$$\mathbf{v}_a \doteqdot \mathbf{u}_a^T \exp(\mathbf{G}_a t) \tag{2.286}$$

Differentiating (2.286) then introducing into (2.285), we find

$$\begin{aligned} m_a \frac{d\mathbf{v}_a}{dt} &= m_a\left(\frac{d\mathbf{u}_a}{dt} + \mathbf{G}_a\mathbf{u}_a\right)^T \exp(\mathbf{G}_a t) \\ &= \left[\mathbf{F}_a^{int} + \mathbf{F}_a^{stoc}(t) + \mathbf{F}_a^{ext}\right]^T \exp(\mathbf{G}_a t) \end{aligned} \tag{2.287}$$

The solution of that equation is written as

$$\mathbf{v}_a(t) = \mathbf{v}_a^0 + \frac{1}{m_a}\int_0^t \left[\mathbf{F}_a^{int} + \mathbf{F}_a^{stoc}(t') + \mathbf{F}_a^{ext}\right]^T \exp(\mathbf{G}_a t')\, dt' \tag{2.288}$$

where $\mathbf{v}_a^0 \doteqdot \mathbf{v}_a(t=0)$. Whence

$$\begin{aligned} \mathbf{u}_a(t) &= \mathbf{u}_a^0 \exp(-\mathbf{G}_a t) \\ &\quad + \frac{1}{m_a}\int_0^t \exp\left[\mathbf{G}_a(t'-t)\right]\cdot\left[\mathbf{F}_a^{int} + \mathbf{F}_a^{stoc}(t') + \mathbf{F}_a^{ext}\right] dt \\ &= \mathbf{u}_a^0 \exp(-\mathbf{G}_a t) + \frac{1}{m_a}\mathbf{G}_a^{-1}\left[\mathbf{I}_3 - \exp(-\mathbf{G}_a t)\right]\cdot\left(\mathbf{F}_a^{int} + \mathbf{F}_a^{ext}\right) \\ &\quad + \frac{1}{m_a}\int_0^t \exp\left[\mathbf{G}_a(t'-t)\right]\cdot\mathbf{F}_a^{stoc}(t')\, dt' \end{aligned} \tag{2.289}$$

with $\mathbf{u}_a^0 \doteqdot \mathbf{u}_a(t=0)$, the particle's initial velocity. The stochastic force varies so quickly over the time that we may consider that the average of the last integral of that solution vanishes. Thus, the mean velocity obeys the law

$$\langle \mathbf{u}_a \rangle(t) = \mathbf{u}_a^0 \exp(-\mathbf{G}_a t) + \frac{1}{m_a}\mathbf{G}_a^{-1}\left[\mathbf{I}_3 - \exp(-\mathbf{G}_a t)\right]\cdot\left(\mathbf{F}_a^{int} + \mathbf{F}_a^{ext}\right) \tag{2.290}$$

That result manifestly shows that the characteristic time of the average particle evolution can be assessed through

$$T_R \sim \frac{1}{|\mathbf{G}_a|} \tag{2.291}$$

Before making further calculations, it is important to recall that this solution is only valid if $dt \ll T_R$, so that the parameters in the equation of motion can be regarded as constant. Then, by expanding (2.290) to the first-order in dt/T_R, we get:

$$\begin{aligned} dt \ll T_R, \quad \langle \mathbf{u}_a \rangle (dt) &= \mathbf{u}_a^0 \left(\mathbf{I}_3 - \mathbf{G}_a dt \right) + \frac{1}{m_a} \left(\mathbf{F}_a^{int} + \mathbf{F}_a^{ext} \right) dt \\ &= \mathbf{u}_a^0 + \frac{1}{m_a} \left(\mathbf{F}_a^{int} + \mathbf{F}_a^{diss} + \mathbf{F}_a^{ext} \right) dt \end{aligned} \tag{2.292}$$

Unsurprisingly, we can find that the particle's mean velocity is determined by the internal, dissipative and external forces, from its initial velocity. The solution (2.290), however, remains exact at every point of time in the particular case where parameters $\mathbf{F}_a^{int}$, $\mathbf{F}_a^{ext}$ and $\mathbf{G}_a$ are given as constant at every point of time. Examining the particular case where $T_R \ll t$ is then interesting:

$$T_R \ll t, \quad \langle \mathbf{u}_a \rangle = \frac{1}{m_a} \mathbf{G}_a^{-1} \cdot \left(\mathbf{F}_a^{int} + \mathbf{F}_a^{ext} \right) \tag{2.293}$$

This velocity is constant over time. Thus, the particle asymptotically tends towards a state of equilibrium with the relaxation time T_R, by the end of which the initial condition $\mathbf{u}_a^0$ is forgotten. The dissipative forces are then zero, since the particle is moving about at the ambient velocity of the fluid consisting of molecules.

Subtracting the mean velocity (2.290) from the solution (2.289), we get the fluctuation of velocity as caused by the molecular impacts:

$$\begin{aligned} \mathbf{u}'_a (t) &\doteqdot \mathbf{u}_a (t) - \langle \mathbf{u}_a \rangle (t) \\ &= \frac{1}{m_a} \int_0^t \exp\left[\mathbf{G}_a \left(t' - t \right) \right] \cdot \mathbf{F}_a^{stoc} \left(t' \right) dt' \end{aligned} \tag{2.294}$$

Since the statistical correlations of molecular velocities play a major role in this process, calculating the correlation tensor of the particle velocity fluctuation might be attractive:

$$\begin{aligned} \left(\mathbf{u}'_a \otimes \mathbf{u}'_a \right) (t) = \frac{1}{m_a^2} \int_0^t \int_0^t &\left\{ \exp\left[\mathbf{G}_a \left(t' - t \right) \right] \otimes \exp\left[\mathbf{G}_a \left(t'' - t \right) \right] \right\} \\ &: \left\{ \mathbf{F}_a^{stoc} \left(t' \right) \otimes \mathbf{F}_a^{stoc} \left(t'' \right) \right\} dt' dt'' \end{aligned} \tag{2.295}$$

(the meaning of the : symbol can be found in the Appendix A). With the model (2.281) and the property (2.282), we get

$$
\begin{aligned}
&\left\langle \mathbf{u}'_a \otimes \mathbf{u}'_a \right\rangle (t) \\
&= \int_0^t \int_0^t \left\{ \exp\left[\mathbf{G}_a\left(t'-t\right)\right] \otimes \exp\left[\mathbf{G}_a\left(t''-t\right)\right]\right\} \\
&: Z_a \delta\left(t''-t'\right) \mathbf{I}_3 dt' dt'' \\
&= Z_a \int_0^t \left\{ \exp\left[\mathbf{G}_a\left(t'-t\right)\right] \otimes \exp\left[\mathbf{G}_a\left(t'-t\right)\right]\right\} : \mathbf{I}_3 dt' \\
&= Z_a \int_0^t \exp\left[2\mathbf{G}_a\left(t'-t\right)\right] dt' \\
&= \frac{Z_a}{2} \mathbf{G}_a^{-1} \left[\mathbf{I}_3 - \exp\left(-2\mathbf{G}_a t\right)\right]
\end{aligned}
\tag{2.296}
$$

In the case where the parameters are constant over time, this tensor asymptotically tends towards the following limiting value:

$$
T_R \ll t, \quad \left\langle \mathbf{u}'_a \otimes \mathbf{u}'_a \right\rangle \approx \frac{Z_a}{2} \mathbf{G}_a^{-1}
\tag{2.297}
$$

(a link may be established between this formula and (2.146).) Now, we have seen that the particle is at thermodynamical equilibrium for high values of t, and the arguments in Section 2.3.2 bias us towards stating that its kinetic energy equals the internal energy of the surrounding fluid, that is $E_{int} = E_{k,a}$. Besides, at thermodynamical equilibrium, the fluctuations of the particle velocity are necessarily isotropic, and the relation (2.297) leads to assert that $\mathbf{G}_a$ is proportional to the identity tensor, that is

$$
\mathbf{G}_a = \frac{1}{T_R} \mathbf{I}_3
\tag{2.298}
$$

according to (2.291). With (2.297) and (2.298), as well as the relation (A.35) in Appendix A, we then have

$$
\begin{aligned}
E_{int} &= E_{k,a} \\
&= \frac{1}{2} m_a \left\langle \mathbf{u}'_a \cdot \mathbf{u}'_a \right\rangle \\
&= \frac{1}{2} m_a \mathrm{tr} \left\langle \mathbf{u}'_a \otimes \mathbf{u}'_a \right\rangle \\
&= \frac{m_a Z_a}{4} \mathrm{tr}\left(\mathbf{G}_a^{-1}\right) \\
&= \frac{3 T_R m_a Z_a}{4}
\end{aligned}
\tag{2.299}
$$

Energy E_{int} is given by the formula (2.66), the fluid temperature being equal to that of the particle at thermodynamical equilibrium ($T = T_a$). By making it identical to (2.299), we get a formula giving the relaxation time T_R:

$$
T_R = \frac{2 k_B T_a}{m_a Z_a}
\tag{2.300}
$$

It is time to pause for a comment about this significant result, which is known as the *first fluctuation-dissipation theorem* (refer to Pottier, 2007 or Rief, 1965). This result reveals that both friction force and stochastic force are the consequences of one process, namely the molecular motion. It should be kept in mind, however, that this theorem results from the application of the law (2.296) for sufficiently long times. This is theoretically only possible if the internal and external forces, as well as time T_R, are constant, which only happens with a fluid which is globally at thermodynamical equilibrium. It is noteworthy that the fluctuation-dissipation theorem relates the asymptotic energy (at equilibrium, i.e. for $T_R \ll t$) to the friction force exerted upon the return of the system to equilibrium, that is during the first instants ($dt \ll T_R$).

The terms above the second-order in the Taylor expansion (2.266) still have to be investigated. The ratio of the $(n+1)$-th order term to the n-th order term is of the order of magnitude of $\left|\delta \mathbf{p}_a\right| / \left|\mathbf{p}_a\right|$. $\left|\delta \mathbf{p}_a\right| \sim m_a \sqrt{Z_a dt}$ can be assessed through the assessments (2.277) and (2.278), whereas (2.300) gives $\left|\mathbf{p}_a\right| \sim m_a \sqrt{Z_a T_R}$. Thus:

$$\frac{\left|\delta \mathbf{p}_a\right|}{\left|\mathbf{p}_a\right|} \sim \sqrt{\frac{dt}{T_R}} \tag{2.301}$$

These terms are then negligible as compared with the previously considered first and second order terms and very quickly decrease with n, according to condition (2.274) (Pottier, 2007; Gardiner, 1985). With (2.283) and (2.298), the Fokker–Planck equation (2.279) ultimately takes the folowing shape:

$$\begin{aligned} &\frac{\partial \varphi_a}{\partial t} + \frac{\partial \varphi_a}{\partial \mathbf{r}_a} \cdot \mathbf{u}_a + \frac{\partial \varphi_a}{\partial \mathbf{p}_a} \cdot \left(\mathbf{F}_a^{int} + \mathbf{F}_a^{ext}\right) \\ &- \frac{\partial}{\partial \mathbf{u}_a} \cdot \left(\frac{1}{T_R} \varphi_a \mathbf{u}_a + \frac{1}{2} \frac{\partial Z_a \varphi_a}{\partial \mathbf{u}_a}\right) = 0 \end{aligned} \tag{2.302}$$

The last term, which occurs as a second derivative, can be formally likened to a scattering of information $Z_a \varphi_a$ within the phase space (refer to Chapter 3, Section 3.4.2). The stochastic forces then induce a scattering of the volumes in the phase space, as in Fig. 2.6.

Let us now return to the velocity correlations. With equation (2.300), we can henceforth write (2.296) as

$$\left\langle \mathbf{u}'_a \otimes \mathbf{u}'_a \right\rangle (t) = \frac{T_R Z_a}{2} \left[1 - \exp\left(-\frac{2t}{T_R}\right)\right] \mathbf{I}_3 \tag{2.303}$$

In the case $t \equiv dt \ll T_R$, we find again (2.278). With (2.303), the fluctuation (2.294) can be written as:

$$\mathbf{u}'_a(t) = \sqrt{\frac{T_R Z_a}{2} \left[1 - \exp\left(-\frac{2t}{T_R}\right)\right]} \boldsymbol{\xi}_a(t) \tag{2.304}$$

It can be observed that the random fluctuation of velocity equals zero at the initial time, because we have decided to solve the problem with a determined initial velocity $\mathbf{u}_a^0$. The solution (2.289) can ultimately be arranged as follows:

$$\begin{aligned}\mathbf{u}_a(t) = &\exp\left(-\frac{t}{T_R}\right)\mathbf{u}_a^0\\ &+\frac{T_R}{m_a}\left[1-\exp\left(-\frac{t}{T_R}\right)\right]\cdot\left(\mathbf{F}_a^{int}+\mathbf{F}_a^{ext}\right)\\ &+\sqrt{\frac{T_R Z_a}{2}\left[1-\exp\left(-\frac{2t}{T_R}\right)\right]}\boldsymbol{\xi}_a(t)\end{aligned} \tag{2.305}$$

At the short points of times, this law becomes

$$\begin{aligned}dt \ll T_R,\quad \mathbf{u}_a(dt) = &\,\mathbf{u}_a^0+\frac{1}{m_a}\left(\mathbf{F}_a^{int}+\mathbf{F}_a^{ext}\right)dt\\ &-\frac{1}{T_R}\mathbf{u}_a^0 dt+\sqrt{Z_a dt}\,\boldsymbol{\xi}_a(t)\end{aligned} \tag{2.306}$$

Thus, the velocity variation $d\mathbf{u}_a = \mathbf{u}_a(dt) - \mathbf{u}_a^0$ can be written, by making $\mathbf{u}_a^0$ identical to $\mathbf{u}_a$, as:

$$d\mathbf{u}_a = \left[\frac{1}{m_a}\left(\mathbf{F}_a^{int}+\mathbf{F}_a^{ext}\right)-\frac{1}{T_R}\mathbf{u}_a\right]dt+\sqrt{Z_a dt}\,\boldsymbol{\xi}_a(t) \tag{2.307}$$

which is quite exactly (2.285), with the formulas (2.281) and (2.300). This is then called the *Langevin equation*.

Lastly, we will turn to the correlations of velocities at different points of time. A calculation analogous to (2.296) yields, taking (2.300) into account:

$$\begin{aligned}\left\langle \mathbf{u}'_a(t')\otimes\mathbf{u}'_a(t'')\right\rangle = &\int_0^{t'}\int_0^{t''}\exp\frac{t^{(1)}-t'}{T_R}\\ &\times\exp\frac{t^{(2)}-t''}{T_R}Z_a\delta\left(t^{(1)}-t^{(2)}\right)\mathbf{I}_3 dt^{(1)}dt^{(2)}\end{aligned} \tag{2.308}$$

The Dirac distribution, as in (2.296), deletes one of the integrals, the other one ranging from 0 to $\min\left(t',t''\right)$:

$$\begin{aligned}\left\langle \mathbf{u}'_a(t')\otimes\mathbf{u}'_a(t'')\right\rangle &= Z_a\int_0^{\min(t',t'')}\exp\frac{2t^{(1)}-t'-t''}{T_R}dt^{(1)}\mathbf{I}_3\\ &=\frac{T_R Z_a}{2}\exp\left(-\frac{t'+t''}{T_R}\right)\left[\exp\frac{2\min\left(t',t''\right)}{T_R}-1\right]\mathbf{I}_3\end{aligned} \tag{2.309}$$

Note that, in the particular case where $t' = t'' \equiv t$, we get

$$\left\langle \mathbf{u}'_a\otimes\mathbf{u}'_a\right\rangle(t) = \frac{T_R Z_a}{2}\left[1-\exp\left(-\frac{2t}{T_R}\right)\right]\mathbf{I}_3 \tag{2.310}$$

in accordance with (2.296) and (2.300). The expression of the tensor, denoted $\mathbf{D}_u\left(t',t''\right)$, of correlation of the differences of velocities at two points of time can be derived from the equality (2.309):

$$\begin{aligned}\mathbf{D}_u\left(t',t''\right) &\doteq \left\langle\left[\mathbf{u}'_a\left(t''\right)-\mathbf{u}'_a\left(t'\right)\right]\otimes\left[\mathbf{u}'_a\left(t''\right)-\mathbf{u}'_a\left(t'\right)\right]\right\rangle \\ &= \left\langle\mathbf{u}'_a\left(t''\right)\otimes\mathbf{u}'_a\left(t''\right)\right\rangle+\left\langle\mathbf{u}'_a\left(t'\right)\otimes\mathbf{u}'_a\left(t'\right)\right\rangle \\ &\quad -\left\langle\mathbf{u}'_a\left(t''\right)\otimes\mathbf{u}'_a\left(t'\right)\right\rangle-\left\langle\mathbf{u}'_a\left(t'\right)\otimes\mathbf{u}'_a\left(t''\right)\right\rangle\end{aligned} \tag{2.311}$$

Using the relation $2\min\left(t',t''\right)-t'-t''=-\left|t''-t'\right|$, the equations (2.309) and (2.311) give

$$\mathbf{D}_u\left(t',t''\right)=T_R Z_a\left\{\begin{array}{c}1-\frac{1}{2}\left[\exp\left(-\frac{t'}{T_R}\right)-\exp\left(-\frac{t''}{T_R}\right)\right]^2 \\ -\exp\left(-\frac{\left|t''-t'\right|}{T_R}\right)\end{array}\right\}\mathbf{I}_3 \tag{2.312}$$

With the condition (2.274), we get the following approximation:

$$t',t''\ll T_R,\quad \mathbf{D}_u\left(t',t''\right)\approx Z_a\left|t''-t'\right|\mathbf{I}_3 \tag{2.313}$$

The fact that these correlations only depend on the time deviation $\left|t''-t'\right|$ is a consequence of the stochastic process invariance through a translation over time. The fact that all the correlation tensors are proportional to the identity tensor results from the isotropy of that process.

For completing this chapter, we will establish some results regarding the energy of a particle being subject to a stochastic bias. Multiplying the equation of motion (2.307) by $\mathbf{u}_a$, then taking the average of the result, we get an equation giving the average variation of kinetic energy of a particle:

$$\left\langle dE_k\right\rangle=\left[\left(\mathbf{F}_a^{int}+\mathbf{F}_a^{ext}\right)\cdot\left\langle\mathbf{u}_a\right\rangle-\frac{2\left\langle E_k\right\rangle}{T_R}\right]dt+m_a\sqrt{Z_a dt}\left\langle\boldsymbol{\xi}_a(t)\cdot\mathbf{u}'_a(t)\right\rangle \tag{2.314}$$

(we have used (2.281) and the fact that only the fluctuation of velocity is statistically correlated to the stochastic signal). For calculating the last term of (2.314), we establish a calculation analogous to (2.296) from the relations (2.276) and (2.281):

$$\begin{aligned}&\left\langle\boldsymbol{\xi}_a(t)\otimes\mathbf{u}'_a\left(t'\right)\right\rangle \\ &=\frac{1}{m_a}\int_0^{t'}\exp\frac{t''-t'}{T_R}\left\langle\boldsymbol{\xi}_a(t)\otimes\mathbf{F}_a^{stoc}\left(t''\right)\right\rangle dt'' \\ &=\sqrt{\frac{Z_a}{dt}}\int_0^{t'}\exp\frac{t''-t'}{T_R}\left\langle\boldsymbol{\xi}_a(t)\otimes\boldsymbol{\xi}_a\left(t''\right)\right\rangle dt'' \\ &=\sqrt{Z_a dt}\int_0^{t'}\exp\frac{t''-t'}{T_R}\delta\left(t-t''\right)dt''\mathbf{I}_3\end{aligned} \tag{2.315}$$

That integral is only non-zero if the time point t occurs within the segment $\left[0,t'\right]$. For $t'=t$, we therefore get

$$\left\langle\boldsymbol{\xi}_a(t)\otimes\mathbf{u}'_a(t)\right\rangle=\mathbf{0} \tag{2.316}$$

This result means that the fluctuating velocity is statistically independent of the stochastic force. It is well-founded, because $\mathbf{u}'_a(t)$ can only depend on the values of $\mathbf{F}_a^{stoc}(t')$ at the preceding time points ($t' < t$), which values are decorrelated by $\boldsymbol{\xi}_a(t)$ according to (2.282). The trace of (2.316) then gives $\left\langle \boldsymbol{\xi}_a(t) \cdot \mathbf{u}'_a(t) \right\rangle = 0$, and equation (2.314) becomes

$$\frac{\langle dE_k \rangle}{dt} = \left(\mathbf{F}_a^{int} + \mathbf{F}_a^{ext} \right) \cdot \langle \mathbf{u}_a \rangle - \frac{2\langle E_k \rangle}{T_R} \tag{2.317}$$

A parallel should be established between this equation and (2.193). It can be found that the stochastic forces do not dissipate the energy. It should be recalled, however, that they originate from the same process as the friction forces, which here dissipate the energy with a relaxation time T_R, as evidenced by the equation which we have just established. It should be emphasized, however, that (2.317) does not yield the Lagrangian derivative of $\langle E_k \rangle$, since the average does not commute with the derivative. We will establish a number of significant results for that subject, against the background of a continuous model of turbulent flows, in Chapter 4, particularly in Section 4.6.3. In anticipation of this forthcoming chapter, the temporal derivative of the fluctuating energy of a particle is usefully written from the trace of (2.310):

$$\begin{aligned} \frac{d}{dt}\left(\frac{1}{2} m_a \left\langle \left| \mathbf{u}'_a \right|^2 \right\rangle \right) &= \frac{3 m_a Z_a}{2} \exp\left(-\frac{2t}{T_R} \right) \\ &\approx -\frac{3 m_a Z_a}{2} \\ &= -\frac{3 k_B T_a}{T_R} \end{aligned} \tag{2.318}$$

(the second line only holds true for $t \ll T_R$, and the last one results from the fluctuation-dissipation theorem (2.300)). It is noteworthy that the amplitude Z_a of the fluctuations directly determines the decrease of energy of the fluctuating velocity.

3 Continuous media and viscous fluids

3.1 Introduction

The first two chapters have gradually provided us with elements allowing us to apprehend the fluids through a discrete approach to their spatial discretization. We first considered an array of macroscopic particles (Chapter 1), before regarding a fluid as a set of particles behaving like very large systems which, in turn, obey statistical laws; that made it possible to introduce the notion of dissipation of energy (Chapter 2). The logical continuation of our investigations is the concept of continuous medium, for which it is possible to consider the particles which are sufficiently small to be likened to material points continuously adjacent in space, yet still large enough to be governed by the thermodynamical laws.

The continua have been extensively studied since the works by Augustin-Louis Cauchy in the 1830s. Although the study of fluids can be derived from them through Cauchy's equations together with an appropriate behaviour law, Leonhard Euler could already formulate the equation of motion of a fluid in the absence of viscosity as far back as 1755. In fact, the Navier–Stokes equations were given a compound name as a tribute to both Louis Marie Henri Navier, who discovered them from heuristic considerations in 1823, and Sir Georges Gabriel Stokes, who produced a definite evidence thereof from the theory of elasticity over 20 years later.[1]

[1]The name of Adhémar Barré de Saint-Venant should in principle accompany that of Navier, for he derived a rigorous proof of these equations two years before Stokes; his name, however, was given recognition under different circumstances (in particular in the context of the shallow water equations, see Section 8.3.2).

The continuum theory will be disclosed here with the reservations we had already formulated in the introduction to Chapter 2, that is that a fairly thorough account is impossible within one chapter. This is why we will go straight to the essential points, laying stress on such mathematical notions as the balances in continuous formalism and the flux of physical quantities. Links with both preceding chapters will be regularly set up in view of globally logical explanations without forgetting, before the second part of this book, the discrete modelling of the fluids using particle-oriented notions connected to the discussions in the first two chapters.

At the beginning of this chapter, we will prepare suitable tools for investigating the continuum kinematics (Lagrangian derivative, strain rate tensor, etc.), which will lead to the continuity equation. The notion of flux, which is underlying throughout that account, will be set forth within the frame of the scalar transfer-diffusion equation, then of Cauchy's equation, from the definition of the Cauchy stress tensor. Through an analysis of the energy balances to be made in connection with Chapters 1 and 2, the law governing the behaviour of a viscous fluid will subsequently be formulated; this will lead to the Navier–Stokes equations. Afterwards, the similarity laws and the role of pressure in the notion of incompressibility will be reviewed, then we will briefly set out surface tension theory. We will subsequently temporarily return to the statistical notions of Chapter 2 for explaining how the equations of the fluid can be recovered from the Boltzmann equation. Last, we will discuss some elements regarding the variational principles for the fluids from the arguments of Chapter 1 extended to the continuous formalism.

Readers unfamiliar with the tensorial notations are invited to refer to Appendix A.

3.2 Continuum kinematics

We start our study of continuous media with the definition of fields, then of advection. The strain tensor is introduced as the basic concept for studying the media deformation, leading to dilatation rate, scalar rate-of-strain and vorticity. The method of integral balance is stated, then applied to the density, leading to the continuity equation. It is finally applied to scalar fields.

3.2.1 Mesoscopic particles and continuous fields

Owing to the notations in Chapter 2, we have learnt to go without particle subscripts, which provides an easy transition to the continuum formalism through the notion of field. In order to work out a relevant continuum formalism, we must first note that this very notion is nothing but a model, since a genuine medium is made up of distant molecules with quite variable and sometimes very high velocities, even within a medium globally at rest (in a given frame of reference). To achieve a suitable formalism for getting quantities slowly and steadily varying in space, we may proceed as in the beginning of Chapter 2 (Section 2.2.1), and define all the necessary quantities[2] using the probabilistic averaging operator (2.4). There is, however, another less rigorous but more intuitive point of view which we will briefly describe.

To that purpose, we consider elementary spatial entities which are large enough with respect to the intermolecular distance to allow a statistical averaging of all the useful quantities (velocity, mass, etc.). Statistical physics (Landau and Lifshitz, 1980) then shows that when such an element includes a sufficiently large number of molecules, the mean velocity varies little if its size is enlarged even further.[3] It should be kept in mind, however, that these elements should remain small enough[4] to allow us to take them as material points on the observer's scale; so, they may be interpreted as particles, in the meaning defined in Section 1.3.1 and all the results obtained so far can be applied to them. Thus, we may consider a velocity field $\mathbf{u}(\mathbf{r})$ which can be considered as being defined at every point $\mathbf{r}$ (coinciding with the centre of mass of each particle at the point of time being considered). In addition, it appears from what has just been said that the field being defined this way is continuous and derivable (in most cases[5]). Likewise, by adding up the masses of molecules

[2] We will adopt an analogous approach for the theory of turbulence (Chapter 4).

[3] We may recall that, throughout this book, we refer to a classical description of the molecules. When we desire to describe the molecules by means of quantum mechanics, the contemplated approach to the definition of the continuous media quantities is required for properly describing the velocities. Actually, considering too small elements of volume would come down to locating the contemplated molecules in space; thus, according to the Heisenberg's uncertainty principle, too little information would be available about their velocities to calculate an average. These arguments are based on a rigorous theory in statistical mechanics (Landau and Lifshitz, 1980).

[4] The typical size of these small elements is on the order of 10^{-4} m. The efficiency of this method implies that this characteristic distance is still larger than the particles' mean free path. This condition may be invalid in such cases as, for example, the theory of rarefied gases or in microfluidics. It is then necessary to return to a probabilistically average approach, as mentioned earlier.

[5] Except, for instance, when a shock wave occurs in a compressible fluid. The regularity properties of the solutions of viscous fluid equations have been extensively investigated. Refer to the fundamental paper by Leray (1934).

included within a particle, then dividing by its volume, a regular density field $\rho(\mathbf{r})$ is defined. It is also possible to define this way all kinds of extensive or intensive fields, as well as such purely statistical quantities as pressure or internal energy $e_{int}(\mathbf{r})$ (refer to Sections 1.5.3 and 1.5.4), subsequently counted per unit mass ($e_{int} = E_{int}/m$, m being the mass of a mesoscopic particle).

The resulting fields may be integrated over a domain Ω to get integral quantities. For example, the total mass, kinetic energy, linear momentum and angular momentum of the domain Ω will be defined by means of the following integrals:

$$\begin{aligned} M &= \int_\Omega \rho d\Omega \\ E_k &= \int_\Omega \rho e_k d\Omega \\ \mathbf{P} &= \int_\Omega \rho \mathbf{u} d\Omega \\ \mathbf{M} &= \int_\Omega \rho \mathbf{r} \times \mathbf{u} d\Omega \end{aligned} \tag{3.1}$$

by analogy with (1.180), (1.49), (1.133) and (1.147), respectively. Furthermore, we have defined the kinetic energy per unit mass from the model of the specific internal energy as defined in Section 1.5.4:

$$e_k \doteq \frac{1}{2}u^2 \tag{3.2}$$

In this chapter (and the next), because of the continuous nature of the fields being discussed, we will never refer to the label of a particle a, the latter being only identified by its position $\mathbf{r}$, which is a continuous function of time. A distinction should be made, however, between the coordinates of a particle at a given time point and the components of a stationary point in the space. These are Lagrange's and Eulers' respective points of view: we will state that a particle being at a reference time point[6] in the point $\mathbf{x}$ of space with components $\{x_i\} = (x, y, z)$ has a location $\mathbf{r}(\mathbf{x}, t)$ at time point t. The velocity of a particle is then, in accordance with (1.2):

[6] This often called the 'initial configuration'.

$$\mathbf{u} = \frac{d\mathbf{r}}{dt} = \frac{\partial \mathbf{r}(\mathbf{x}, t)}{\partial t} \tag{3.3}$$

and represents the momentum per unit mass. The Jacobian tensor of the space transformation allowing us to switch from the initial configuration to an arbitrary configuration is denoted as

$$\mathbf{J} \doteq \frac{\partial \mathbf{r}}{\partial \mathbf{x}} \tag{3.4}$$

Let us now consider a scalar field A. Each particle being located in $\mathbf{r}(\mathbf{x}, t)$ at a given time point, every field can be considered as a function of time $A(\mathbf{x}, t) = A(\mathbf{r}(\mathbf{x}, t))$. When a particle is moving about, the value of A characterizing that particle then changes for two reasons: first, because the field A may explicitly vary in the course of time under the action of various processes;

second, because the particle has moved from $\mathbf{r}(t)$ to $\mathbf{r}(t+dt)$ in the space and A explicitly depends on $\mathbf{r}$. We write the total differential of A as

$$dA = \frac{\partial A}{\partial t}dt + \frac{\partial A}{\partial x}dx + \frac{\partial A}{\partial y}dy + \frac{\partial A}{\partial z}dz \tag{3.5}$$

where the dx_i are the components of $d\mathbf{r} \doteqdot \mathbf{r}(t+dt) - \mathbf{r}(t)$, the partial derivatives being taken at the point which is occupied at time point t by the particle. Dividing that formula by dt, we get

$$\frac{dA}{dt} = \frac{\partial A}{\partial t} + \frac{\partial A}{\partial x_i}\frac{dx_i}{dt} \tag{3.6}$$

(here we have applied the Einstein convention on repeated indices, as we will fairly often do in this chapter and the next). Since dx_i/dt is the i-th component u_i of the particle velocity, we can ultimately write, in a vector form:

$$\frac{dA}{dt} = \frac{\partial A}{\partial t} + \mathbf{grad}\, A \cdot \mathbf{u} \tag{3.7}$$

Thus, the Lagrangian derivative (i.e. following the motion of a particle) of a quantity is the sum of a Eulerian derivative (i.e. at a fixed point in space) and of a transfer term resulting from the particle motion, which term is known as *advection*. The latter is interpreted in a simple way: each particle carries a piece of information about A during its motion.

A similar calculation would give an analogous result for the Lagrangian derivative of a vector; in particular, the velocity vector derivative gives the acceleration of a particle as:

$$\frac{d\mathbf{u}}{dt} = \frac{\partial \mathbf{u}}{\partial t} + \mathbf{grad}\,\mathbf{u} \cdot \mathbf{u} \tag{3.8}$$

The last term in that formula is a matrix-vector product; we may recall that the gradient of a vector is a second-order tensor. This advection term is called *inertia*, since it represents the fact that a particle carries a piece of information about its own velocity, that is its tendency to conserve its motion (refer to Chapter 1). Note that another formulation of that term is given by

$$\mathbf{grad}\,\mathbf{u} \cdot \mathbf{u} = \mathbf{div}\,(\mathbf{u} \otimes \mathbf{u}) - (\mathrm{div}\mathbf{u})\,\mathbf{u} \tag{3.9}$$

the first 'divergence' symbol denoting a vectorial operator acting upon a tensor. It can be observed that inertia exhibits quadratic terms with respect to velocities, since the components of the tensor $\mathbf{u} \otimes \mathbf{u}$ are the quantities $u_i u_j$. The continuum equations are therefore inherently nonlinear.

Lastly, it should be pointed out that the temporal derivative of the Jacobian tensor (3.4) is nothing but the velocity gradient, since the material derivative commutes with the partial derivatives in x_i, according to (3.5):

$$\frac{d\mathbf{J}}{dt} = \frac{\partial \mathbf{u}}{\partial \mathbf{x}} = \mathbf{grad}\,\mathbf{u} \tag{3.10}$$

Taking the trace of the preceding equation, we get

$$\frac{d\mathrm{tr}\,\mathbf{J}}{dt} = \mathrm{div}\mathbf{u} \tag{3.11}$$

3.2.2 Strain of a medium

We will now prepare a suitable tool for describing the features of the deformation of a continuous medium during its evolution as time elapses. In the next pages, we will frequently employ variously dimensioned objects (segments, surfaces, volumes). The main trend in our arguments will consist in observing *material* entities, that is those which are made up of a preset array of particles to be followed over time (Lagrangian point of view).

Let us first consider an infinitesimal material vector $d\mathbf{r}$. After a short lapse of time dt, this small vector remains infinitesimal and rectilinear (to the first-order in dt). The tail of this vector has been displaced by a vector $\mathbf{u}(\mathbf{r})dt$, its head by $\mathbf{u}(\mathbf{r}+d\mathbf{r})dt$. Thus, the vector $d\mathbf{r}\,(t+dt) - d\mathbf{r}\,(t)$ is written as

$$\begin{aligned} d\mathbf{r}\,(t+dt) - d\mathbf{r}\,(t) &= [\mathbf{u}(\mathbf{r}+\mathbf{dr}) - \mathbf{u}(\mathbf{r})]\,dt \\ &= (\mathbf{gradu}\cdot d\mathbf{r})\,dt \end{aligned} \tag{3.12}$$

then we get

$$\frac{d\,(d\mathbf{r})}{dt} = \mathbf{grad\,u}\cdot d\mathbf{r} \tag{3.13}$$

The derivative of the scalar product of two infinitesimal vectors $d\mathbf{r}$ and $d\mathbf{r}'$ is immediately found from it, using the relation (A.42) of the Appendix A, as

$$\begin{aligned} \frac{d\left(d\mathbf{r}\cdot d\mathbf{r}'\right)}{dt} &= \frac{d\,(d\mathbf{r})}{dt}\cdot d\mathbf{r}' + d\mathbf{r}\cdot\frac{d\left(d\mathbf{r}'\right)}{dt} \\ &= (\mathbf{grad\,u}\cdot d\mathbf{r})\cdot d\mathbf{r}' + d\mathbf{r}\cdot\left(\mathbf{grad\,u}\cdot d\mathbf{r}'\right) \\ &= 2d\mathbf{r}^T\cdot\mathbf{s}\cdot d\mathbf{r}' \end{aligned} \tag{3.14}$$

where $\mathbf{s}$ is the tensor as defined by

$$\begin{aligned} \mathbf{s} &\doteq \frac{1}{2}\left[\mathbf{grad\,u} + (\mathbf{grad\,u})^T\right] \\ &= (\mathbf{grad\,u})^S \\ &= \frac{1}{2}\left(\frac{\partial u_i}{\partial x_j} + \frac{\partial u_j}{\partial x_i}\right)\mathbf{e}_i\otimes\mathbf{e}_j \end{aligned} \tag{3.15}$$

In accordance with the Appendix A, we have denoted by $\mathbf{A}^S$ the symmetric part of a tensor $\mathbf{A}$, that is the average of that tensor and its transpose (eqn (A.21)). The tensor $\mathbf{s}(\mathbf{r}, t)$, which is symmetric by construction, is called *strain rate tensor* and include all the information about the way the medium is deformed over time at a given point and at a given instant (its components have the dimension of a frequency). Thus, putting $d\mathbf{r}' = d\mathbf{r}$ in (3.14), we can see that the expansion rate per unit length along the direction of the unit vector $\mathbf{e}$ carrying a vector $d\mathbf{r}$ is given by

$$\begin{aligned} \frac{1}{|d\mathbf{r}|}\frac{d\,(|d\mathbf{r}|)}{dt} &= \frac{1}{2\,|d\mathbf{r}|^2}\frac{d\left(|d\mathbf{r}|^2\right)}{dt} \\ &= \mathbf{e}^T\mathbf{s}\mathbf{e} \end{aligned} \tag{3.16}$$

A similar calculation would show that the rate of variation of the angle θ between two initially orthogonal infinitesimal vectors $d\mathbf{r}$ and $d\mathbf{r}'$ which are respectively carried by the unit vectors $\mathbf{e}$ and $\mathbf{e}'$ is given by

$$\frac{d\theta}{dt} = 2\mathbf{e}^T\mathbf{s}\mathbf{e}' \tag{3.17}$$

This is the rate of distorsion of the medium in a plane formed by these two vectors[7].

Lastly, let us turn to the deformation of an infinitesimal cubic volume $V = dxdydz$. Since the tensor $\mathbf{s}$ is symmetrical and real, it is diagonalizable in an orthogonal basis whose directions represent the *principal directions of strain* of the medium at the contemplated location and at the time point being considered. We can see, in that base, that the temporal variation dV of the volume V is only caused by the length variation of the dx_i:

$$\frac{1}{V}\frac{dV}{dt} = \frac{1}{dx_i}\frac{d\,(dx_i)}{dt} \tag{3.18}$$

(with a sum on i). Using the relation (3.16), we get:

$$\begin{aligned}\frac{1}{V}\frac{dV}{dt} &= \mathbf{e}_i^T\mathbf{s}\mathbf{e}_i \\ &= s_{ii} \\ &= \text{tr}\,\mathbf{s} \\ &= \text{div}\mathbf{u}\end{aligned} \tag{3.19}$$

This result remains true in every base, since the trace is a tensor invariant. Thus, the strain rate tensor trace yields the medium's volume expansion rate in each point; the rate may be either positive (expansion) or negative (contraction). The next to last equality in (3.19) results from the definition (3.15) of that tensor. Invoking (1.255) and (3.11), we may also write

$$\begin{aligned}\frac{1}{\rho}\frac{d\rho}{dt} &= -\frac{1}{V}\frac{dV}{dt} \\ &= -\frac{d\text{tr}\,\mathbf{J}}{dt}\end{aligned} \tag{3.20}$$

Integrating that relation with respect to time, we get

$$\text{tr}\,\mathbf{J} = 3 - \ln\frac{\rho}{\rho_0} \tag{3.21}$$

where ρ_0 denotes the value of the density being considered in those conditions corresponding to the initial configuration, for which $\mathbf{J} = \mathbf{I}_3$. In some cases (which will subsequently often be referred to), the material behaves incompressibly (i.e. it verifies $\rho = \rho_0$). Equation (3.19) then yields tr $\mathbf{s} = 0$ everywhere and at every time point. Relation (3.21) then shows that tr $\mathbf{J}$ is constant, which corresponds to $V = V_0$.

Yet another tool can measure the evolution of an infinitesimal volume, namely the determinant in the Jacobian matrix $\mathbf{J}$. The volume at time point

[7] At the points being considered in both space and time, which is implicit in all the discussions in this chapter, since we consider fields.

t is then related to the initial volume by

$$V(t) = (\det \mathbf{J})\, V_0 \tag{3.22}$$

Thus, for an incompressible medium, we have $\det \mathbf{J} = 1$. Let us now consider a small deformation from the initial configuration, resulting from a motion exerted during the infinitesimal instant dt. The preceding formula gives

$$\frac{V(dt) - V_0}{dt} = \frac{\det \mathbf{J} - 1}{dt} \tag{3.23}$$

Since the deformation is small, the Jacobian tensor is close to identity:

$$\mathbf{J} = \mathbf{I}_3 + d\mathbf{J} \tag{3.24}$$

Thus, its determinant is close to 1:

$$\det \mathbf{J} \approx 1 + \mathrm{tr}\,(d\mathbf{J}) \tag{3.25}$$

Injecting that identity into (3.23) and making dt tend towards zero, we find again (3.20).

Let us return to the strain rate tensor $\mathbf{s}$. Like every symmetric three-dimensional tensor, $\mathbf{s}$ has three invariants, that is three quantities which do not depend on the chosen basis. They have a strong physical meaning, since the isotropic space properties (Section 1.2.1) teach us that the choice of the basis should not affect the equations of mechanics. These three quantities may be, for instance, the eigenvalues of $\mathbf{s}$. It is more convenient, however, to use invariants which can be calculated from the literal expression of the tensor matrix in any basis. This is the case with such quantities as $\mathrm{tr}\left(\mathbf{s}^k\right)$, k being an arbitrary integer. This is because the trace of a second-order tensor is a zero-order tensor (see Appendix A, eqn (A.47)). We will show that the first three values of k $(1, 2, 3)$ alone characterize[8] the tensor $\mathbf{s}$:

[8] It is important to write, for example, $s_{ij}s_{ij}$ instead of s_{ij}^2, in order to apply the sub-script summation convention.

$$\begin{aligned} I_{\mathbf{s}} &\doteq \mathrm{tr}\,\mathbf{s} = s_{ii} \\ II_{\mathbf{s}} &\doteq \mathrm{tr}\left(\mathbf{s}^2\right) = s_{ij}s_{ij} \\ III_{\mathbf{s}} &\doteq \mathrm{tr}\left(\mathbf{s}^3\right) = s_{ij}s_{jk}s_{ki} \end{aligned} \tag{3.26}$$

To that purpose, we invoke the Cayley–Hamilton's theorem which stipulates that a second-order tensor is the solution of its characteristic polynomial. For a three-dimensional second-order tensor, this is written as:

$$\begin{aligned} &\mathbf{s}^3 - (\mathrm{tr}\,\mathbf{s})\,\mathbf{s}^2 + \frac{1}{2}\left[\mathrm{tr}\left(\mathbf{s}^2\right) - (\mathrm{tr}\,\mathbf{s})^2\right]\mathbf{s} \\ &\quad + \frac{1}{3}\left[\mathrm{tr}\left(\mathbf{s}^3\right) - \frac{2}{3}(\mathrm{tr}\,\mathbf{s})\,\mathrm{tr}\left(\mathbf{s}^2\right) + \frac{1}{2}(\mathrm{tr}\,\mathbf{s})^3\right]\mathbf{I}_3 = \mathbf{0} \end{aligned} \tag{3.27}$$

the coefficient before the identity tensor $\mathbf{I}_3$ being nothing but $\det \mathbf{s}$. Multiplying (3.27) by $\mathbf{s}$ and taking the trace of the result being obtained, we observe that $\mathrm{tr}\left(\mathbf{s}^4\right)$ is expressed as a function of $I_{\mathbf{s}}$, $II_{\mathbf{s}}$ and $III_{\mathbf{s}}$. By mathematical induction, one can easily see that every invariant of order k (i.e. in the form $\mathrm{tr}\,\mathbf{s}^k$) is related to all three quantities as given by (3.26). It is little wonder that there are three invariants:[9] they are recombinations of the three eigenvalues λ_i of $\mathbf{s}$:

[9] It may be argued against that the tensor $\mathbf{s}$ originally had six independent components. The other three degrees of freedom are the angles defining the axes of the diagonalization base which also depends both on the location being considered and the selected instant.

$$\mathrm{tr}\left(\mathbf{s}^k\right) = \sum_{i=1}^{3} \lambda_i^k \tag{3.28}$$

We have just seen that the first invariant in (3.26) represents the volume expansion rate. For an incompressible medium, this invariant vanishes and consequently provides no information about the strain of the medium. The second conversely, occurs as the square of a norm; then, it is only zero for a zero deformation. Thus, it is a simplified measurement of strain; we will subsequently often use this invariant in the following form:

$$s \doteqdot \sqrt{2 s_{ij} s_{ij}} = \sqrt{2 II_{\mathbf{s}}} = \sqrt{2\mathbf{s} : \mathbf{s}} \tag{3.29}$$

which will be referred to as *scalar rate-of-strain* (once again, it should be pointed out that it is a field, which then varies in both time and space). Its inverse gives an order of magnitude of the characteristic time for the local strain of the medium.[10]

We may notice, to complete this section, that the symmetric part $\boldsymbol{\omega}$ of the tensor **grad u** comprises three independent components which are the components of the vector **curl u** as well:

$$\boldsymbol{\omega} \doteqdot \frac{1}{2}\left[\mathbf{grad\ u} - (\mathbf{grad\ u})^T\right] = \frac{1}{2}\begin{pmatrix} 0 & \dfrac{\partial u}{\partial y} - \dfrac{\partial v}{\partial x} & \dfrac{\partial u}{\partial z} - \dfrac{\partial w}{\partial x} \\ \dfrac{\partial v}{\partial x} - \dfrac{\partial u}{\partial y} & 0 & \dfrac{\partial v}{\partial z} - \dfrac{\partial w}{\partial y} \\ \dfrac{\partial w}{\partial x} - \dfrac{\partial u}{\partial z} & \dfrac{\partial w}{\partial y} - \dfrac{\partial v}{\partial z} & 0 \end{pmatrix} \tag{3.30}$$

where the components of **u** have been denoted as (u, v, w). The vorticity tensor $\boldsymbol{\omega}$, which is skew-symmetric, has invariants too. The second-order invariant alone, analogous to (3.384), will subsequently be useful:

$$\omega \doteqdot \sqrt{2\boldsymbol{\omega} : \boldsymbol{\omega}} = \sqrt{2\omega_{ij}\omega_{ij}} \tag{3.31}$$

This tensor accounts for the locally rotating nature of a velocity field. In particular, for a purely rotational field, which corresponds to the rigid body rotational motion (1.191), we have $\mathbf{u} = \mathbf{V} + \boldsymbol{\Omega} \times \mathbf{r}$, where $\mathbf{V}$ and $\boldsymbol{\Omega}$ are two constant vectors, and the vorticity tensor is written as:

$$\boldsymbol{\omega} = \begin{pmatrix} 0 & -\Omega_z & \Omega_y \\ \Omega_z & 0 & -\Omega_x \\ -\Omega_y & \Omega_x & 0 \end{pmatrix} \tag{3.32}$$

(i.e. $\boldsymbol{\omega} = \hat{\boldsymbol{\Omega}}$, according to the definition (1.137)). We then have the following identities:

[10] In cosmology, the Universe may be likened to a continuous medium which is expanding since the Big Bang. Observation reveals a great homogeneity in the fastness of that expansion, with a rate-of-strain which is known as the *Hubble's constant*, $s = H_0$. It value is close to 2.5 $10^{-18}\ s^{-1}$, which corresponds to a characteristic time of 13 billion years, consistently with the estimated Universe age.

$$
\begin{aligned}
\mathbf{u} &= \mathbf{V} + \boldsymbol{\omega}\mathbf{r} \\
\mathbf{grad}\,\mathbf{u} &= \boldsymbol{\omega} \\
\mathbf{s} &= \mathbf{0} \\
\mathbf{curl}\,\mathbf{u} &= 2\boldsymbol{\Omega} \\
\omega &= 2\,|\boldsymbol{\Omega}|
\end{aligned}
\tag{3.33}
$$

Every velocity gradient is written as the sum of a pure expansion (or contraction), a pure deformation and a pure rotation:

$$
\begin{aligned}
\mathbf{grad}\,\mathbf{u} &= \mathbf{s} + \boldsymbol{\omega} \\
&= \frac{1}{3}\,(\mathrm{tr}\,\mathbf{s})\,\mathbf{I}_3 + \mathbf{s}^D + \boldsymbol{\omega}
\end{aligned}
\tag{3.34}
$$

We may recall that the deviator $\mathbf{A}^D$ of a tensor $\mathbf{A}$ is defined by the relation (A.23) in Appendix A (here with $n = 3$). In particular, we have, with (3.19):

$$
\mathbf{s}^D \doteqdot \mathbf{s} - \frac{1}{3}\,(\mathrm{div}\mathbf{u})\,\mathbf{I}_3 \tag{3.35}
$$

3.2.3 Continuity equation

We will frequently employ volume balances in the next pages. To that purpose, we consider a regular and orientable volume Ω having a boundary $\partial\Omega$, with an external normal unit vector $\mathbf{n}$, moving about along a velocity field denoted as $\mathbf{v}\,(\mathbf{r})$, a priori unrelated to the velocity of those particles which it consists of. Contemplating a physical quantity A, we then calculate the variation rate of its volume integral over time in the following form

$$
\begin{aligned}
&\int_{\Omega(t+dt)} A\,(\mathbf{r}, t+dt)\,d\Omega - \int_{\Omega(t)} A\,(\mathbf{r}, t)\,d\Omega = \\
&\int_{\Omega(t+dt)} A\,(\mathbf{r}, t+dt)\,d\Omega - \int_{\Omega(t)} A\,(\mathbf{r}, t+dt)\,d\Omega \\
&+ \int_{\Omega(t)} A\,(\mathbf{r}, t+dt)\,d\Omega - \int_{\Omega(t)} A\,(\mathbf{r}, t)\,d\Omega
\end{aligned}
\tag{3.36}
$$

At the first-order in dt, the difference of the last two integrals equals the integral of the partial derivative of A as multiplied by dt. As regards the first two integrals in the right-hand side of (3.36), t can be substituted for $t + dt$ in its terms $A\,(\mathbf{r}, t+dt)$ (still at the first-order). Figure 3.1 then shows that the difference between these integrals is reduced to a surface integral as calculated on the boundary $\partial\Omega$. The integrand value is A as multiplied by the volume based on the small element with a surface $d\Gamma$ and a height $\mathbf{v}dt \cdot \mathbf{n}$. Dividing (3.36) by dt and making dt tend towards zero, we ultimately get

$$
\frac{d}{dt}\int_{\Omega} A\,(\mathbf{r}, t)\,d\Omega = \int_{\Omega} \frac{\partial A}{\partial t} d\Omega + \oint_{\partial\Omega} A\mathbf{v} \cdot \mathbf{n}\,d\Gamma \tag{3.37}
$$

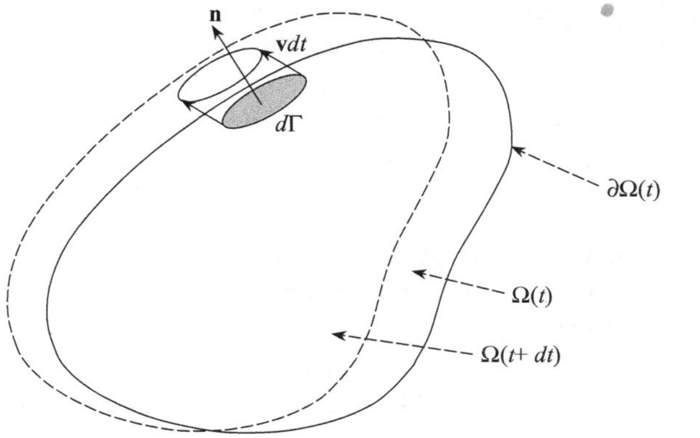

Fig. 3.1 Schematic diagram for the demonstration of the Leibniz's theorem (3.37).

That formula is known as *Leibniz's theorem*. If a material volume is chosen, that is a volume enclosing a given collection of continuous medium particles which is followed up over time, velocity $\mathbf{v}$ merges with the field $\mathbf{u}$, and we get

$$\frac{d}{dt}\int_{\Omega} A\,(\mathbf{r},t)\,d\Omega = \int_{\Omega}\frac{\partial A}{\partial t}d\Omega + \oint_{\partial\Omega} A\mathbf{u}\cdot\mathbf{n}\,d\Gamma \tag{3.38}$$

As for the Lagrangian variation of a quantity linked to a particle (Section 3.2.1), the integral of a field then varies both because A explicitly depends on time and the volume moves along with the motion $\mathbf{u}$ of the particles which compose it. The scalar product $\mathbf{u}\cdot\mathbf{n}$ represents that effect of displacement: if the motion of those particles located on the boundary is tangential to the latter, no transfer across the edge occurs in that place. Applying Gauss' theorem (refer to Appendix A) to (3.38), we find:

$$\frac{d}{dt}\int_{\Omega} A d\Omega = \int_{\Omega}\left[\frac{\partial A}{\partial t} + \operatorname{div}\,(A\mathbf{u})\right]d\Omega \tag{3.39}$$

The same equation will, of course, describe the balance of a vector quantity:

$$\frac{d}{dt}\int_{\Omega} \mathbf{A} d\Omega = \int_{\Omega}\left[\frac{\partial \mathbf{A}}{\partial t} + \mathbf{div}\,(\mathbf{A}\otimes\mathbf{u})\right]d\Omega \tag{3.40}$$

with the adequate operators.

We will subsequently apply the above results to formulate in the continuum formalism those laws which were established in Chapter 1. Let us first contemplate the case of mass. The sum of elementary masses $dm = \rho d\Omega$ of its constituents has an integral which is the total mass M of Ω, as given by (3.1). Now, as discussed in Section 1.3.1, mass is inherently conserved. Since the volume Ω is material, its mass is then constant ($dM/dt = 0$), for any chosen volume Ω. Consequently, applying (3.39) to $A = \rho$, we find:

$$\forall\Omega,\quad \int_{\Omega}\left[\frac{\partial \rho}{\partial t} + \operatorname{div}\,(\rho\mathbf{u})\right]d\Omega = 0 \tag{3.41}$$

which immediately yields:

$$\frac{\partial \rho}{\partial t} + \operatorname{div}\,(\rho\mathbf{u}) = 0 \tag{3.42}$$

This equation represents the local conservation of mass in a continuous medium and is called the *continuity equation*. Using (3.7), we also may write

$$\frac{1}{\rho}\frac{d\rho}{dt} = -\mathrm{div}\mathbf{u} \tag{3.43}$$

Unsurprisingly, this equation corresponds to (3.20) with (3.11). As expected, we find that a locally diverging velocity field (i.e. div $\mathbf{u} > 0$) reduces the density in a volume with a given mass, and conversely.

By the way, note that the acceleration (3.8) of a particle may now be formulated otherwise:

$$\begin{aligned}\rho\left(\frac{\partial \mathbf{u}}{\partial t} + \mathbf{grad}\,\mathbf{u}\cdot\mathbf{u}\right) &= \frac{\partial \rho\mathbf{u}}{\partial t} - \frac{\partial \rho}{\partial t}\mathbf{u} \\ &\quad + \mathbf{div}\,(\rho\mathbf{u}\otimes\mathbf{u}) - \mathrm{div}\,(\rho\mathbf{u})\,\mathbf{u}\end{aligned} \tag{3.44}$$

that is with equation (3.42):

$$\rho\frac{d\mathbf{u}}{dt} = \frac{\partial \rho\mathbf{u}}{\partial t} + \mathbf{div}\,(\rho\mathbf{u}\otimes\mathbf{u}) \tag{3.45}$$

(that relation may be compared with 3.9). In the case of an incompressible medium, eqns (3.45) and (3.42) are respectively reduced to

$$\frac{d\mathbf{u}}{dt} = \frac{\partial \mathbf{u}}{\partial t} + \mathbf{div}\,(\mathbf{u}\otimes\mathbf{u}) \tag{3.46}$$

and

$$\mathrm{div}\mathbf{u} = 0 \tag{3.47}$$

The latter equation may also be written in the following form using subscripts:

$$\frac{\partial u_i}{\partial x_i} = 0 \tag{3.48}$$

Let us put $A = \rho B$, which denotes the mass density of a quantity. Through the continuity equation (3.42) the right-hand side of (3.39) can be rearranged as:

$$\begin{aligned}\frac{\partial A}{\partial t} + \mathrm{div}\,(A\mathbf{u}) &= \rho\frac{\partial B}{\partial t} + \frac{\partial \rho}{\partial t}B + B\mathrm{div}\,(\rho\mathbf{u}) \\ &\quad + \mathbf{grad}B\cdot\rho\mathbf{u} \\ &= \rho\frac{\partial B}{\partial t} + \mathbf{grad}B\cdot\rho\mathbf{u} \\ &= \rho\frac{dB}{dt}\end{aligned} \tag{3.49}$$

(the two central terms on the right-hand side of the first line cancel each other out, because of eqn (3.42)). Thus, the form (3.39) of generalized balance is arranged as:

$$\frac{d}{dt}\int_\Omega \rho B d\Omega = \int_\Omega \rho\frac{dB}{dt}d\Omega \tag{3.50}$$

Thus, the integration over a material volume commutes with the Lagrangian derivative, provided that it is integrated as a function of the element of mass $dm = \rho d\Omega$ contained in the element of volume $d\Omega$. Besides, this result becomes obvious when considered from that angle, since the mass is conserved; furthermore, the demonstration (3.49) indirectly uses that conservation law. It could, of course, be easily evidenced that the relation (3.50) is also valid for a vector field $\mathbf{B}$.

3.3 Balances and fluxes

The idea of flux is defined here, and we build the local advection-diffusion equation of a scalar field. The method of integral balance is then applied to the linear momentum, on the basis of the Cauchy stress tensor. We prove that the latter is symmetric, thanks to the conservation of angular momentum. The balance method is then applied to energy. We thus introduce the concept of the continuous media dissipation rate.

3.3.1 Flux of a physical quantity

We have described, in the previous section, the suitable mathematical tools for expressing the temporal variation of various quantities and their advective motions. This is just a matter of kinematics, and we must now turn to the description of the physical processes which induce these motions. In other words, as regards the evolution of the momentum, we have only written the left-hand side of the equation of motion (1.240) as applied to the continua, whereas the right-hand side is still missing.

The physical evolution of a scalar (temperature, concentration, etc.) or vector (momentum) is governed, as explained in Section 1.5.3 of Chapter 1 and throughout Chapter 2, on the one hand by microscopic processes which are represented by state variables and, on the other hand, by external processes (external fields). The internal processes, in turn, may be split into two species, namely the transfer terms, which merely redistribute the relevant quantity in the course of time in space, and the source terms, which locally foster or delete it. As regards temperature, for instance, heat conduction belongs to the first class, whereas the exothermal chemical reactions fall within the second. We will state that the transfer processes (molecular diffusion, heat conduction, etc.) may be modelled through the local notion of fluxes, whereas the second species processes are featured by a local mass source term S_B.

Let us briefly dwell on the notion of flux. Considering a volume Ω with a boundary $\partial\Omega$ and an infinitesimal material surface $d\Gamma$, the amount of a substance B flowing across $d\Gamma$ per unit time and per unit mass, under the action of the molecular transfer processes, is equal to $-\mathbf{q}^B \cdot \mathbf{n}$, $\mathbf{q}^B$ denoting the outflow of B across the boundary Ω (the 'minus' sign is in accordance with the convention of an outlet flux). Thus, for a scalar quantity, we will have:

$$\frac{d}{dt}\int_{\Omega} \rho B d\Omega = -\oint_{\partial\Omega} \mathbf{q}^B \cdot \mathbf{n}\, d\Gamma + \int_{\Omega} \rho S_B d\Omega \qquad (3.51)$$

For an isolated medium, that is when the flux along the boundary vanishes and in the absence of any source term, we get the following simple global conservation law:

$$\frac{d}{dt}\int_{\Omega} B dm = 0 \tag{3.52}$$

Let us return to the overall balance (3.51), then invoke Gauss' theorem, the law (3.50) and the arbitrary nature of volume Ω. Through the definition (3.7), we find:

$$\frac{\partial B}{\partial t} + \mathbf{grad} B \cdot \mathbf{u} = -\frac{1}{\rho}\mathrm{div}\mathbf{q}^B + S_B \tag{3.53}$$

This equation accounts for the *advection-diffusion* of quantity B within the continuous medium. The term $\mathbf{grad} B \cdot \mathbf{u}$ accounts for advection (as in (3.7)), whereas the term div $\mathbf{q}^B$ is a diffusion, that is to say a 'contagion' of quantity B through molecular processes. The evolution of a quantity in one point in a continuous medium may then be caused not merely by its transfer by the medium local velocity, but also by a spatial redistribution which is modelled by the diffusion term.

Note that the case $B = 1$ corresponds to the conservation of mass, both flux and source terms then being inherently zero. We may also emphasize that the flux is defined within a Lagrangian context, which means it is an exchange of information between particles, not between two spatially fixed places.

Adding to (3.53) the continuity equation (3.42) as multiplied by B, the balance can be written as

$$\frac{\partial \rho B}{\partial t} = -\mathrm{div}\left(\mathbf{q}^B + \rho B \mathbf{u}\right) + S_B \tag{3.54}$$

This just shows that the flux of B is $\mathbf{q}^B + \rho B \mathbf{u}$ in a Eulerian frame of reference, the first term denoting a displacement (the advection). When $B = 1$, it is obviously the single mass flux, since there is no diffusive mass transfer (as evidenced by eqn (3.42)); by definition, that flux vanishes in a Lagrangian frame of reference. The balance (3.51) can then be rearranged as

$$\int_{\Omega} \frac{\partial \rho B}{\partial t} d\Omega = -\oint_{\partial\Omega} \left(\mathbf{q}^B + \rho B \mathbf{u}\right) \cdot \mathbf{n}\, d\Gamma + \int_{\Omega} \rho S_B d\Omega \tag{3.55}$$

Applying the same arguments to a vector quantity would give such an equation as:

$$\frac{\partial \mathbf{B}}{\partial t} + \mathbf{grad\ B} \cdot \mathbf{u} = -\frac{1}{\rho}\mathbf{div\ q^B} + \mathbf{S}_B \tag{3.56}$$

or else:

$$\frac{\partial \rho \mathbf{B}}{\partial t} = -\mathbf{div}\left(\mathbf{q^B} + \rho \mathbf{B} \otimes \mathbf{u}\right) + \mathbf{S}_B \tag{3.57}$$

where $\mathbf{q^B}$ is a second-order tensor representing the vector $\mathbf{B}$ flux (each of its components, as a scalar, being diffused by a vector composing one the tensor columns).

Let us return to the case of the scalar field. We may notice that equation (3.53) is not closed, in the sense that the velocity data would not suffice to solve it without knowing the dependence of the diffusive flux $\mathbf{q}^B$, which we are about to attempt to do. To get a clear idea, let us contemplate the case where B denotes the measure of a chemical substance (mass concentration). The flux then undoubtedly depends primarily on the distribution of concentration (particularly its derivatives, since it can be felt that there is more diffusion when the variations of concentrations are large). Writing a first-order Taylor series expansion,[11] we get:

$$\mathbf{q}^B(B) \approx \mathbf{q}_0^B - \mathbf{J}_B \cdot \mathbf{grad} B \tag{3.58}$$

where $\mathbf{J}_B$ is a second-order tensor (the 'minus' sign is used by convention). The constant term $\mathbf{q}_0^B$ is identically zero, because a constant concentration in space will not give rise to any diffusion (refer to Section 2.3.2). As to the tensor $\mathbf{J}_B$, one can easily understand that in the case of diffusion within a material with an isotropic molecular structure (for example a fluid), it should itself be isotropic[12] (i.e. proportional to the identity tensor). Moreover, because of the extreme smallness of the spatial scale characterizing the molecular motion, it can escape every local influence of the boundary conditions which is liable to break such isotropy (particulary the shape of the solid edges, if any). Last, the effects of the subsequently discussed external fields (gravity), although they are not isotropic, do not affect the molecular motions, in which inertia does prevail. We may consequently write (3.58) as

$$\mathbf{q}^B = -J_B \, \mathbf{grad} B \tag{3.59}$$

The scalar quantity J_B is called the *dynamical molecular diffusion coefficient* of the quantity B. Since the diffusion is exerted, of course, from the highest concentrations to lowest, the coefficient is necessarily positive. The practice shows it is usually independent of B but variable as a function of temperature (refer to Section 3.4.1). Practically, it is generally very low (refer to Section 3.5.3); thus, the terms above the second order have *a posteriori* reasonably been ignored in the development (3.58).

It should be kept in mind that the latest arguments are a model of reality, which is quite reliable indeed in the present case. Being provided with (3.59), equation (3.53) now yields

$$\frac{\partial B}{\partial t} + \mathbf{grad} B \cdot \mathbf{u} = \frac{1}{\rho} \mathrm{div} \left(J_B \mathbf{grad} B \right) + S_B \tag{3.60}$$

In the case where J_B can be taken as constant, (3.60) is reduced to

$$\frac{\partial B}{\partial t} + \mathbf{grad} B \cdot \mathbf{u} = K_B \nabla^2 B + S_B \tag{3.61}$$

with

$$K_B \doteqdot \frac{J_B}{\rho} \tag{3.62}$$

K_B is called the *kinematic* molecular diffusion coefficient. In this form, the diffusion term tends to homogenize the spatial distribution of B, and then

[11] This procedure looks like that of the calculation beginning with eqn (2.118).

[12] This statement is applicable throughout this chapter. In Chapter 4, however, we will see that the turbulent effects do not follow the forthcoming discussion, because of their macroscopic nature. These arguments are no longer also valid in a material with an anisotropic structure (e.g. a crystal).

It might be objected that we are not in a Galilean frame of reference, since we follow up the Lagrangian motion of a particle; we should not forget, however, that statistics of the molecular motions do not depend on the selected frame of reference.

happens to be a basically *irreversible* term, since the gradients of B generate processes which reduce them.

Specific entropy σ (refer to Chapter 2, Section 2.3.2) is an example of a useful scalar field. In the case of a continuous medium governed by an isentropic process, we may state that the entropy of each mesoscopic particle is constant over time, which is expressed as

$$\frac{d\sigma}{dt} = \frac{\partial \sigma}{\partial t} + \mathbf{grad}\sigma \cdot \mathbf{u} = 0 \tag{3.63}$$

In a more general case, it may be stated that there is a flux which redistributes entropy among the various particles over the course of time,[13] or that the entropy spontaneously increases under irreversible processes. Through the arguments in Section 2.3.2, we can assume that this flux depends on the temperature differences, which allows to write a specific entropy transport equation by analogy with (3.60):

[13] In the absence of such a flux, the process is said *adiabatic*.

$$\frac{\partial \sigma}{\partial t} + \mathbf{grad}\sigma \cdot \mathbf{u} = \frac{1}{\rho T}\text{div}\,(J_T\mathbf{grad}T) \tag{3.64}$$

where K_T is called the *heat conduction coefficient* of the relevant medium (it should be considered that the factor T in the denominator results from the definition of J_T). We may now turn to a first condition of isentropicity in a mechanism being related to the evolution of a continuous medium. Equation (3.64), indeed, shows that entropy varies little when temperature is comparatively homogeneous; we henceforth will assume that this condition is fulfilled. We will see in Section 3.3.3 a more complete form of the entropy governing equation.

3.3.2 Cauchy stress tensor

The arguments set out in the previous section should now be extended to the case of momentum, that is a vector quantity. Since it is additive (refer to Section 1.4.3), we may introduce the force $\mathbf{T}$ exerted, per unit area, by the environment of Ω on a boundary element $d\Gamma$. There are then two kinds of external forces being exerted on Ω: the integral of $\mathbf{T}$ (contact forces involved by the environment) over $\partial\Omega$ and the integral of the Earth's gravitational force per unit mass $\mathbf{F}_g$ (external field) over Ω. The equation of dynamics (1.229) will then be written as:

$$\frac{d}{dt}\int_\Omega d\mathbf{p} = \oint_{\partial\Omega} \mathbf{T}d\Gamma + \int_\Omega \mathbf{F}_g dm \tag{3.65}$$

$d\mathbf{p}$ being the momentum of a particle. $\mathbf{F}_g$ is written in the conservative form as given, pursuant to Section 1.3.2, by (1.57):

$$\mathbf{F}_g = -\nabla e_g = \mathbf{g} \tag{3.66}$$

where the potential per unit mass e_g is given by equation (1.51):

$$e_g = gz = -\mathbf{g}\cdot\mathbf{r} \tag{3.67}$$

(z being the vertical component). We may notice, as compared with the notations in Chapter 1, that ∇ (gradient) and ∇_a (gradient with respect to the components of a particle) become one and a same vector because the potential is stationary. The last integral of (3.65) might come down to a surface integral, as required for an external force. The sum of the two right-hand integrals represents the total external force. For an isolated continuous medium, which does not receive any external contribution (no contact stress, a negligible effect of gravity), the right-hand side of (3.65) vanishes to achieve the conservation of total momentum:

$$\frac{d}{dt}\int_{\Omega} d\mathbf{p} = \mathbf{0} \tag{3.68}$$

which is the continuous analogous of (1.132).

As explained in Section 1.5.3, it should be possible to express the contact forces in the form of the spatial variation of a given quantity. To that purpose, let us consider the set of particles making up the material volume Ω, and let us review the forces they exchange. The sum of these forces can be written in the form of a volume integral. It should be pointed out, however, that these forces act substantially over a very short range,[14] and that only considering the forces exchanged by 'neighbouring' particles is then enough. Besides, as set out in Section 1.5.2, according to the action-reaction law (1.135), these forces cancel each other out, except for those received by the particles occurring along the domain boundary, which are subject to forces exerted by the particles making up the environment. As a matter of fact, the sum of these resultant forces does yield the contact stress integral in (3.65). Thus, we are faced with a volume integral which can be turned into a surface integral, which means that the volume-integrated term occurs in the form of a divergence, because of Gauss' theorem.[15] Hence, there is such a field of tensors $\boldsymbol{\sigma}$ as:

$$\int_{\Omega} \mathbf{div}\, \boldsymbol{\sigma}\, d\Omega = \oint_{\partial\Omega} \mathbf{T} d\Gamma \tag{3.69}$$

Then, we necessarily have $\mathbf{T} = \boldsymbol{\sigma} \cdot \mathbf{n}$. This relation shows that the amount $\sigma_{ij} n_j$ (with a sum over j) is the i-th component of the force being exerted on an infinitesimal surface oriented by $\mathbf{n}$, per unit area, or even that σ_{ij} denotes the i-th component of the force (per unit area) being exerted on an infinitesimal surface oriented by the vector $\mathbf{e}_j$. The tensor $\boldsymbol{\sigma}$ is known as the *Cauchy stress tensor*. With $d\mathbf{p} = \mathbf{u} dm$ and $dm = \rho d\Omega$ (eqn (3.1)), writing all the integrals in a volume form, we get:

$$\frac{d}{dt}\int_{\Omega} \rho \mathbf{u} d\Omega = \int_{\Omega} (\mathbf{div}\, \boldsymbol{\sigma} + \rho \mathbf{g})\, d\Omega \tag{3.70}$$

Through both rule (3.50) and formula (3.8), we can ultimately achieve a local form of the equation of dynamics, which is referred to as *Cauchy's equation*:

$$\frac{\partial \mathbf{u}}{\partial t} + \mathbf{grad}\, \mathbf{u} \cdot \mathbf{u} = \frac{1}{\rho}\mathbf{div}\, \boldsymbol{\sigma} + \mathbf{g} \tag{3.71}$$

This equation is written in the form given by (3.56) (if the density was constant, the term $\mathbf{g}$ could be integrated into the 'divergence' sign by putting

[14] In particular, the mutual gravitational forces exerted over each other by the particles (refer to Section 1.3.1) are quite negligible compared to the contact stresses.

[15] A similar demonstration could be used to justify the fact that the flux of a scalar B as contemplated in Section 3.3.1 is written as $\mathbf{q}^B \cdot \mathbf{n}$.

$\boldsymbol{\sigma}^* \doteqdot \boldsymbol{\sigma} - gz\mathbf{I}_3$, as discussed concerning the fluids in Section 3.4.2). Thus, the stress tensor denotes (to within the sign) the momentum flux, which means that the amount $-\mathbf{T} = -\boldsymbol{\sigma} \cdot \mathbf{n}$ is the momentum transferred per unit time and unit mass at a given point, across a unit area which is centred on that point and is oriented by $\mathbf{n}$ (with eqn (1.56), it was already obvious that the forces account for momentum transfers). As with the diffusion of a scalar discussed in Section 3.3.1, it should, however, be emphasized that the latter assertion is valid in a Lagrangian frame of reference. In a Eulerian frame, equation (3.45) shows that the momentum flux would be written as $\rho\mathbf{u} \otimes \mathbf{u} - \boldsymbol{\sigma}$, and the balance (3.70) becomes

$$\int_\Omega \frac{\partial \rho \mathbf{u}}{\partial t} d\Omega = \oint_{\partial\Omega} (\boldsymbol{\sigma} - \rho\mathbf{u} \otimes \mathbf{u})\, \mathbf{n}\, d\Gamma + \int_\Omega \rho \mathbf{g}\, d\Omega \tag{3.72}$$

(to be compared with (3.55)). Cauchy's equation may advantageously be written in a form involving the tensor subscripts:

$$\forall i, \; \frac{\partial u_i}{\partial t} + \frac{\partial u_i}{\partial x_j} u_j = \frac{1}{\rho} \frac{\partial \sigma_{ij}}{\partial x_j} + g_i \tag{3.73}$$

The evolution of the continuous medium total angular momentum (as defined by (3.1)) is obtained by taking the vector product of $\rho\mathbf{r}$ with (3.71), then through integration:

$$\int_\Omega \rho \mathbf{r} \times \frac{d\mathbf{u}}{dt} d\Omega = \int_\Omega (\mathbf{r} \times \mathbf{div}\, \boldsymbol{\sigma} + \rho \mathbf{r} \times \mathbf{g})\, d\Omega \tag{3.74}$$

Invoking (1.144) and (3.50), we get

$$\frac{d}{dt} \int_\Omega \rho\, (\mathbf{r} \times \mathbf{u})\, d\Omega = \int_\Omega (\mathbf{r} \times \mathbf{div}\, \boldsymbol{\sigma} + \rho \mathbf{r} \times \mathbf{g})\, d\Omega \tag{3.75}$$

which is analogous to (1.148). If the medium is fully isolated from external influences, we get a continuous equivalent for the conservation law (1.149):

$$\frac{d}{dt} \int_\Omega d\mathbf{M} = \mathbf{0} \tag{3.76}$$

where $d\mathbf{M} = \mathbf{r} \times d\mathbf{p}$ is the angular momentum of a particle,[16] that is of the elementary volume $d\Omega$.

[16] We are establishing, from the beginning of this chapter, continuous equations which are analogous to the discrete equations of Chapters 2 and 3 for a continuous medium, before applying them to the case of a fluid. Needless to say the case of rigid bodies, which is studied within a discrete context in Section 1.5.1, could be treated in the same way.

We have previously provided an argument leading to the definition of $\boldsymbol{\sigma}$, based on the required equality between a volume integral and a surface integral; this property should still exist, of course, for the total angular momentum, due to both its conservation and additivity. Let us now write in two different ways the angular momentum received by the volume Ω via the contact forces:

$$\int_\Omega \mathbf{r} \times \mathbf{div}\, \boldsymbol{\sigma}\, d\Omega = \oint_{\partial\Omega} \mathbf{r} \times \mathbf{T} d\Gamma \tag{3.77}$$

Let us integrate the left-hand side piece-wise:

$$\int_\Omega \mathbf{r} \times \mathbf{div}\boldsymbol{\sigma}\, d\Omega = \oint_{\partial\Omega} \mathbf{r} \times \mathbf{T} d\Gamma - \int_\Omega \left(\mathbf{grad}\, \mathbf{r} \cdot \boldsymbol{\sigma} - \boldsymbol{\sigma}^T \cdot \mathbf{grad}\, \mathbf{r}\right) d\Omega \tag{3.78}$$

The property (3.77) is then satisfied provided that the last integral in (3.78) vanishes. Since **grad r** $= \mathbf{I}_3$, we quickly get:

$$\boldsymbol{\sigma} = \boldsymbol{\sigma}^T \tag{3.79}$$

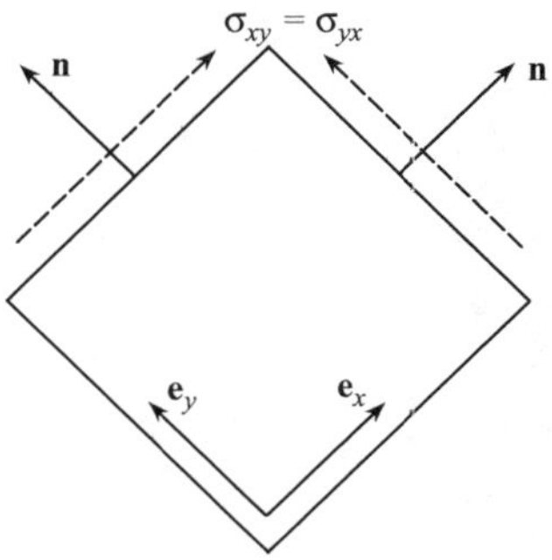

Fig. 3.2 Illustrated symmetry of the Cauchy stress tensor.

Hence, the stress tensor is symmetrical. Mechanically interpreting that result is easy: if the continuous medium sample mentioned in Fig. 3.2 received stresses σ_{xy} and σ_{yx} of different magnitudes, it would spontaneously begin revolving, which would only be possible provided that an external influence exerts a moment on the sample by means of the surface by which it is bounded.[17]

Lastly, note that the stress tensor (like every tensor) can be written in one way as the sum of an isotropic tensor (i.e. proportional to the identity tensor) and a tensor having a zero trace:

$$\boldsymbol{\sigma} = -p\mathbf{I}_3 + \boldsymbol{\tau} \tag{3.80}$$

with

$$\begin{aligned} p &\doteqdot -\frac{1}{3}\mathrm{tr}\,\boldsymbol{\sigma} \\ \boldsymbol{\tau} &\doteqdot \boldsymbol{\sigma}^D = \boldsymbol{\sigma} + p\mathbf{I}_3 \end{aligned} \tag{3.81}$$

where $\boldsymbol{\sigma}^D$ is the deviator of $\boldsymbol{\sigma}$, which represents the friction stresses (shear), whereas the isotropic part of stresses represents pressure/tensile loads (depending on whether p is positive or negative, respectively)[18]. Like pressure (refer to Section 2.3.3), the Cauchy stress tensor is a Galilean invariant, because the contact forces do not depend on the frame of reference. This result fully agrees with the fact that they derive from the internal energy via (3.100).

Like the flux of a scalar quantity, the stresses occur here in an undetermined form which gives equation (3.71) a non-closed nature. This is because that equation of motion is quite general and is indiscriminately applicable to all the media. Now, different media do not identically react to a given stress; thus, a particular medium is featured by a specific dependence of $\boldsymbol{\sigma}$ on its kinematic parameters, which provides a *behaviour law*. We will determine, in Section 3.4.1, a suitable law for the viscous fluids. Such work, like the search for the flux of a scalar quantity (Section 3.3.1) is based on a reflection conducted on the statistical properties of the material behaviour on a microscopic scale, which regards the energetic behaviour of the continua. The next section will then deal with the energy-related matters.

[17]However, the piezoelectric materials have crystal properties generating magnetic fields under the action of an internal stress; thus, moments spontaneously occur within them; the property (3.79) is then invalidated.

[18]This is a matter of convention linked to the sign of $\boldsymbol{\sigma}$ being chosen. This convention looks like that for the flux sign of a scalar (Section 3.3.1).

3.3.3 Energy balances

Let us turn to the continuous forms of the energy balances[19] as set forth in Chapters 1 and 2. First of all, multiplying Cauchy's equation (3.71) by $\rho\mathbf{u}$, we get:

$$\begin{aligned} \rho\frac{d}{dt}\frac{u^2}{2} &= (\mathbf{div}\,\boldsymbol{\sigma})\cdot\mathbf{u} + \rho\mathbf{g}\cdot\mathbf{u} \\ &= \mathrm{div}\,(\boldsymbol{\sigma}\cdot\mathbf{u}) - \boldsymbol{\sigma}:\mathbf{s} + \rho\mathbf{g}\cdot\mathbf{u} \end{aligned} \tag{3.82}$$

[19]We neglect, here and after, all sources of heat (radiation, nuclear reactions) except chemical sources.

(the first two terms of the right-hand side are obtained by deriving the product $\boldsymbol{\sigma} \cdot \mathbf{u}$, taking the symmetry of $\boldsymbol{\sigma}$ into account). Let us integrate that relation over a volume, using Gauss' theorem and the law (3.50):

$$\frac{d}{dt}\int_{\Omega} \rho \frac{u^2}{2} d\Omega = \oint_{\partial\Omega} (\boldsymbol{\sigma} \cdot \mathbf{u}) \cdot \mathbf{n}\, d\Gamma - \int_{\Omega} \boldsymbol{\sigma} : \mathbf{s}\, d\Omega + \int_{\Omega} \rho \mathbf{g} \cdot \mathbf{u}\, d\Omega \quad (3.83)$$

Now, the symmetry of $\boldsymbol{\sigma}$ yields $(\boldsymbol{\sigma} \cdot \mathbf{u}) \cdot \mathbf{n} = (\boldsymbol{\sigma} \cdot \mathbf{n}) \cdot \mathbf{u}$, and the left-hand integral, of course, is the total kinetic energy included in the volume Ω (as given by (3.1)), and so, ultimately:

$$\frac{dE_k}{dt} = \oint_{\partial\Omega} \mathbf{T} \cdot \mathbf{u}\, d\Gamma + \int_{\Omega} \mathbf{F}_g \cdot \mathbf{u}\, dm - \int_{\Omega} \boldsymbol{\sigma} : \mathbf{s}\, d\Omega \quad (3.84)$$

The first two integrals represent the power of the external forces (which is quite normal because this procedure is analogous to the calculation (1.73)). As $\mathbf{F}_g \cdot \mathbf{u} = -\nabla e_g \cdot d\mathbf{r}/dt = -de_g/dt$, the gravity force power integral equals the time derivative of the total gravitational potential energy, as expected. As regards the last integral of (3.84), it represents a dissipation of energy, since it cannot be turned into a surface integral. Thus, the field

$$f \doteq \frac{1}{\rho} \boldsymbol{\sigma} : \mathbf{s} \quad (3.85)$$

has a simple physical meaning, namely the energy dissipation rate per unit mass at a given point in the medium at a given time point. As explained in Chapter 2, the energy which is lost by the macroscopic system is recovered in the form of internal energy (Section 2.3.2). Returning to the stress tensor form (3.81) and owing to (3.19) and (3.43), (3.85) is rearranged as

$$\begin{aligned} f &= -\frac{p}{\rho} \operatorname{tr} \mathbf{s} + \frac{1}{\rho} \boldsymbol{\tau} : \mathbf{s} \\ &= \frac{p}{\rho^2} \frac{d\rho}{dt} + \widetilde{f} \end{aligned} \quad (3.86)$$

putting

$$\widetilde{f} \doteq \frac{1}{\rho} \boldsymbol{\tau} : \mathbf{s} \quad (3.87)$$

In the absence of shear ($\boldsymbol{\tau} = \mathbf{0}$) we find again, in (3.86), a term of energy variation in p/ρ^2, analogous to (1.256), which is then expressed by (3.87) in the continuous formalism. Thus, p is likened to the thermodynamic pressure as defined in Section 1.5.4, as previously suggested by (3.81). As to $\widetilde{f}$, it corresponds to the dissipated energy per unit mass and unit time. Its integral is similar to the quantity being denoted as F in Chapter 2. (3.84) can then be rearranged in a form analogous to equation (2.193):

$$\begin{aligned} \frac{dE_k}{dt} &= \int_{\Omega} \frac{p}{\rho} \operatorname{div} \mathbf{u}\, d\Omega + \oint_{\partial\Omega} \mathbf{T} \cdot \mathbf{u}\, d\Gamma \\ &\quad + \int_{\Omega} \mathbf{F}_g \cdot \mathbf{u} dm - \int_{\Omega} \rho \widetilde{f}\, d\Omega \\ &= P^{int} + P^{ext} - F \end{aligned} \quad (3.88)$$

Equation (3.82) ultimately yields a local equation for the kinetic equation per unit mass (defined by (3.2)) as

$$\frac{de_k}{dt} = \frac{1}{\rho}\text{div}\,(\boldsymbol{\sigma}\cdot\mathbf{u}) + \mathbf{g}\cdot\mathbf{u} - f \tag{3.89}$$

With (3.64) and (3.85), the relation (2.90) giving the flux received by the internal energy now becomes

$$\begin{aligned}\frac{de_{int}}{dt} &= f + T\frac{d\sigma}{dt}\\ &= \frac{1}{\rho}\text{div}\,(J_T\,\mathbf{grad}T) + \frac{1}{\rho}\boldsymbol{\sigma}:\mathbf{s}\end{aligned} \tag{3.90}$$

where σ is, as previously explained, the entropy per unit mass[20]. Note that with (3.86), the first line of (3.90) is also written as

[20]Once again, we hope there is no likelihood of confusion between the scalar σ and the tensor $\boldsymbol{\sigma}$.

$$\frac{de_{int}}{dt} = \tilde{f} + \frac{p}{\rho^2}\frac{d\rho}{dt} + T\frac{d\sigma}{dt} \tag{3.91}$$

Based on (3.89), the equation (3.82) can also be arranged so as to determine the Lagrangian variation of the total energy mass density, taking the definition (3.67) into account:

$$\begin{aligned}\frac{de}{dt} &\doteq \frac{d}{dt}\left(e_k + e_{int} + e_g\right)\\ &= \frac{de_k}{dt} + \frac{de_{int}}{dt} - \mathbf{g}\cdot\frac{d\mathbf{r}}{dt}\\ &= \frac{1}{\rho}\text{div}\,(\boldsymbol{\sigma}\cdot\mathbf{u}) - f + \frac{de_{int}}{dt}\end{aligned} \tag{3.92}$$

The first term in this equation represents the contact stress power, whereas the variation of internal energy is given by (3.86). We ultimately get

$$\frac{de}{dt} = \frac{1}{\rho}\text{div}\,(\boldsymbol{\sigma}\cdot\mathbf{u} + J_T\,\mathbf{grad}T) \tag{3.93}$$

As for equation (3.64), it now takes the following more general form,[21] based on (3.90):

[21]One could also extend the second law of thermodynamics (2.86) to continuous media.

$$\frac{\partial\sigma}{\partial t} + \mathbf{grad}\sigma\cdot\mathbf{u} = f + \frac{1}{\rho T}\text{div}\,(J_T\,\mathbf{grad}T) \tag{3.94}$$

The time has come to take stock of the continuous equations we have. Equations (3.61), (3.71), (3.89), (3.90) and (3.93), describing the evolution of a scalar, the momentum as well as the kinetic, internal and total energies, are written in similar forms. We may actually put them together through the general definition (3.7) of a Lagrangian derivative:

$$\begin{aligned}
\frac{\partial B}{\partial t} + \mathbf{grad} B \cdot \mathbf{u} &= \frac{1}{\rho}\mathrm{div}\,(J_B \mathbf{grad} B) + S_B \\
\frac{\partial \mathbf{u}}{\partial t} + \mathbf{grad}\,\mathbf{u} \cdot \mathbf{u} &= \frac{1}{\rho}\mathbf{div}\,\boldsymbol{\sigma} + \mathbf{g} \\
\frac{\partial e_k}{\partial t} + \mathbf{grad}\,\mathbf{u} \cdot e_k &= \frac{1}{\rho}\mathrm{div}\,(\boldsymbol{\sigma} \cdot \mathbf{u}) + \mathbf{g} \cdot \mathbf{u} - f \\
\frac{\partial e_{int}}{\partial t} + \mathbf{grad}\,\mathbf{u} \cdot e_{int} &= \frac{1}{\rho}\mathrm{div}\,(J_T \mathbf{grad} T) + f \\
\frac{\partial e}{\partial t} + \mathbf{grad}\,\mathbf{u} \cdot e &= \frac{1}{\rho}\mathrm{div}\,(\boldsymbol{\sigma} \cdot \mathbf{u} + J_T \mathbf{grad} T)
\end{aligned} \tag{3.95}$$

Disregarding gravity,[22] all the equations are formally identical to (3.53) (or (3.56) for the momentum), with fluxes (in a Lagrangian frame of reference) and source terms displayed on Table 3.1. Once again, the stress tensor represents the momentum flux, whereas the vector $\boldsymbol{\sigma} \cdot \mathbf{u}$ happens to be the energy flux, that is a work, as expressed in Section 1.5.4 of Chapter 1 and Section 2.3.2 of Chapter 2. We may notice that the source terms of e_k and e_{int} are provided with determined signs.[23] This is consistent with the second law of thermodynamics, which states that the kinetic energy of an isolated system is irreversibly dissipated into internal energy. It should be pointed out that when exothermal or endothermal chemical reactions take place, a further source term should be added to the internal energy; it would represent the gain or the loss of corresponding heat.

[22] As previously mentioned, it may be formally included in the stress tensor.

[23] Since the dissipation f is positive, the source term of e_k is negative and may then be referred to as a 'sink term'.

Table 3.1 Fluxes and source terms of the set of variables treated through the balance method.

Variable	Flux	Source
B	$J_B \mathbf{grad} B$	S_B
$\mathbf{u}$	$\boldsymbol{\sigma}$	0
e_c	$\boldsymbol{\sigma} \cdot \mathbf{u}$	$-f$
e_{int}	$J_T \mathbf{grad} T$	f
e	$\boldsymbol{\sigma} \cdot \mathbf{u} + J_T \mathbf{grad} T$	0

As explained in Section 3.3.1, a flux represents a spatial redistribution of the quantity being considered, which does not contribute to the total amount of the quantity occurring in the medium. On that account, the conserved quantities (momentum and total energy) have zero source terms.

For a continuous medium obeying an isentropic mechanism. Equations (3.90) and (3.93) become, through (3.63):

$$\begin{aligned}
\frac{de_{int}}{dt} &= f \\
\frac{de}{dt} &= \frac{1}{\rho}\mathrm{div}\,(\boldsymbol{\sigma} \cdot \mathbf{u})
\end{aligned} \tag{3.96}$$

To complete this paragraph, we introduce a quantity which is denoted as h and is referred to as *enthalpy*. It is defined by

$$\begin{aligned}
dh &\doteq \frac{1}{\rho}dp \\
&= d\frac{p}{\rho} + \frac{p}{\rho^2}d\rho
\end{aligned} \tag{3.97}$$

With (1.256), we get, for an isentropic transformation:[24]

$$h = e_{int} + \frac{p}{\rho} \tag{3.98}$$

[24] In a more general framework, one could build an equation governing h, similar to (3.95).

3.4 Viscous fluids

We now introduce the behaviour law of a viscous fluid, from relatively simple principles. The notion of viscosity is thus introduced. The Navier–Stokes equations are deduced, and their boundary conditions stated. The concept of similarity is presented through the definition of dimensionless numbers. We study more closely the approximations of perfect and incompressible flows. The role of pressure in the incompressibility property is clarified, leading to an algorithm for solving the equations numerically. Finally, surface tension is investigated.

3.4.1 Behaviour law of a viscous fluid

In order to determine the behaviour law of a fluid, let us first turn to the temporal variation $\widetilde{f}$ of internal energy per unit volume, as given by (3.87). We have seen it is a continuous form of the particle dissipation F_a as given in Section 2.5. Now, equations (2.184) and (2.189) state that the particle friction force is expressed as the derivative of F_a with respect to velocity. By analogy, we may then state that the friction stress should be written as the derivative of $\widetilde{f}$ with respect to the strain tensor. Unlike the case of the particle formalism, we have to take a factor 2 into account, as evidenced by the general calculation (2.153). This formal analogy then takes the following form:

$$
\begin{aligned}
F_a &= -\mathbf{F}_a^{diss} \cdot \mathbf{u}_a & \mathbf{F}_a^{diss} &= -\frac{\partial F_a}{\partial \mathbf{u}_a} \\
\widetilde{f} &= \frac{1}{\rho}\boldsymbol{\tau} : \mathbf{s} & \frac{1}{\rho}\boldsymbol{\tau} &= \frac{1}{2}\frac{\partial \widetilde{f}}{\partial \mathbf{s}}
\end{aligned}
\tag{3.99}
$$

The discussion in Chapter 2, however, shows that such a derivation should be made under a constant temperature. Hence, we ultimately get:

$$
\boldsymbol{\tau} = \frac{\rho}{2}\left(\frac{\partial \widetilde{f}}{\partial \mathbf{s}}\right)_T \tag{3.100}
$$

which means that such a derivation occurs for each of the respective components in both tensors. Formula (3.100) is formally analogous to equations (2.89) of Chapter 2. We will use this general relation to seek $\boldsymbol{\tau}$ as a function of $\mathbf{s}$. To that purpose, we must try to express $\widetilde{f}$ as a function of $\mathbf{s}$.

The shears[25] within a continuous medium are directly related, of course, to the velocity differences between those particles making it up. Thus, $\boldsymbol{\tau}$ should depend on the tensors $\mathbf{s}$ and $\boldsymbol{\omega}$. Now, a fluid is primarily featured by its isotropy, that is by its rotation invariant behaviour. Thus, the stress tensor should not depend on $\boldsymbol{\omega}$. Another effect is that those scalars characterizing a fluid should not depend on the orientation of the selected coordinate system, and so the function $\widetilde{f}$ should only be sought as a function of those *invariants* of the strain rate tensor as given in Section 3.2.2:

$$
\widetilde{f} = \widetilde{f}\left(\operatorname{tr}\mathbf{s}, \operatorname{tr}\mathbf{s}^2, \operatorname{tr}\mathbf{s}^3\right) \tag{3.101}
$$

[25] We take this argument from Landau et al. (1986), with a slightly different point of view.

As for the treatment of a scalar flux discussed in Section 3.3.1, we are going to perform an expansion of the stress tensor to the first-order as regards the components of **s**, which implies a second-order expansion of (3.101):

$$\rho\widetilde{f} = 2\kappa \text{tr}\,\mathbf{s} + 2\mu \text{tr}\,\mathbf{s}^2 + \lambda\,(\text{tr}\,\mathbf{s})^2 + O\left(|\mathbf{s}|^3\right) \tag{3.102}$$

where κ, μ and λ are scalar coefficients. Equation (3.100) then gives

$$\begin{aligned}
\boldsymbol{\tau} &\simeq \kappa\frac{\partial\,(\text{tr}\,\mathbf{s})}{\partial\mathbf{s}} + \mu\frac{\partial\left(\text{tr}\,\mathbf{s}^2\right)}{\partial\mathbf{s}} + \frac{\lambda}{2}\frac{\partial\left[(\text{tr}\,\mathbf{s})^2\right]}{\partial\mathbf{s}} \\
&= \left(\kappa\frac{\partial s_{kk}}{\partial s_{ij}} + 2\mu s_{kl}\frac{\partial s_{kl}}{\partial s_{ij}} + \lambda s_{kk}\frac{\partial s_{ll}}{\partial s_{ij}}\right)\mathbf{e}_i\otimes\mathbf{e}_j \\
&= \left(\kappa\delta_{ik}\delta_{jk} + 2\mu s_{kl}\delta_{ik}\delta_{jl} + \lambda s_{kk}\delta_{il}\delta_{jl}\right)\mathbf{e}_i\otimes\mathbf{e}_j \\
&= \left(\kappa\delta_{ij} + 2\mu s_{ij} + \lambda s_{kk}\delta_{ij}\right)\mathbf{e}_i\otimes\mathbf{e}_j \\
&= (\kappa + \lambda\text{tr}\,\mathbf{s})\,\mathbf{I}_3 + 2\mu\mathbf{s}
\end{aligned} \tag{3.103}$$

When the strain equals zero, the shear should be zero too, and so the coefficient κ must be zero. Ultimately, (3.103) yields

$$\begin{aligned}
\boldsymbol{\tau} &= \lambda\,(\text{div}\mathbf{u})\,\mathbf{I}_3 + 2\mu\mathbf{s} \\
&= \zeta\,(\text{div}\mathbf{u})\,\mathbf{I}_3 + 2\mu\mathbf{s}^D
\end{aligned} \tag{3.104}$$

with

$$\zeta \doteqdot \lambda + \frac{2}{3}\mu \tag{3.105}$$

Returning to definition (3.81), we may write the stress tensor of a viscous fluid as follows:

$$\begin{aligned}
\boldsymbol{\sigma} &= (-p + \lambda\text{tr}\,\mathbf{s})\,\mathbf{I}_3 + 2\mu\mathbf{s} \\
&= (-p + \zeta\text{tr}\,\mathbf{s})\,\mathbf{I}_3 + 2\mu\mathbf{s}^D
\end{aligned} \tag{3.106}$$

If the fluid is incompressible[26] ($\text{tr}\,\mathbf{s} = \text{div}\mathbf{u} = 0$), we find:

$$\begin{aligned}
\boldsymbol{\sigma} &= -p\mathbf{I}_3 + 2\mu\mathbf{s} \\
\boldsymbol{\tau} &= 2\mu\mathbf{s}
\end{aligned} \tag{3.107}$$

[26] We may notice that only in this case will $\text{tr}\,\boldsymbol{\sigma} = -3p$, according to (3.81). In the general case, the definition of pressure would have to be amended for keeping this equality.

This is sometimes referred to as *Stokes' behaviour law*. The normal stresses in a fluid necessarily correspond to pressure (not tensile) loads. Thus, p is positive, according to what was seen in Section 2.3.3. The coefficient μ is called fluid's *dynamical molecular viscosity*; ζ is sometimes referred to as *bulk viscosity*. Since the derivation as written in (3.100) is made under a constant temperature, we infer that p, μ and ζ vary according to the temperature T. The dependence of p with respect to T has already been mentioned in Chapter 2.

It should be pointed out that the first line in calculation (3.103) is an approximation. Actually, we could have directly derived (3.101) and written

$$\frac{1}{\rho}\boldsymbol{\tau} = \frac{\partial\widetilde{f}}{\partial\,(\text{tr}\,\mathbf{s})}\frac{\partial\,(\text{tr}\,\mathbf{s})}{\partial\mathbf{s}} + \frac{\partial\widetilde{f}}{\partial\left(\text{tr}\,\mathbf{s}^2\right)}\frac{\partial\left(\text{tr}\,\mathbf{s}^2\right)}{\partial\mathbf{s}} + \frac{\partial\widetilde{f}}{\partial\left(\text{tr}\,\mathbf{s}^3\right)}\frac{\partial\left(\text{tr}\,\mathbf{s}^3\right)}{\partial\mathbf{s}} \tag{3.108}$$

so that the coefficients κ, λ, μ in (3.103) represent the first terms of Taylor series expansions of quantities $\partial \widetilde{f}/\partial \left(\mathrm{tr}\,\mathbf{s}^k\right)$ with respect to the s_{ij}. In this form, it turns out that the coefficients of viscosity should depend on the invariants via the function $\widetilde{f}$. There are, however, fluids for which that dependence is very slow, and their coefficients of viscosity may then be regarded as constant with respect to $\mathbf{s}$, which will always be assumed in the following.[27] This is what is known as a Newtonian fluid. For most common fluids, the small values of λ and μ then justify the limited expansion (3.103), since the viscous forces are moderate as compared with the pressure loads.

We find that using the law of thermodynamics (3.100) leads us to a linear form of the shear tensor $\boldsymbol{\tau}$ with respect to the strain rate tensor $\mathbf{s}$ (eqn (3.107)). This dependence[28] is analogous to that giving the flux of a scalar quantity (3.59). We may recall in that respect that the stress tensor represents a momentum flux; viscosity then serves as a momentum diffusion coefficient whereas the strain rate tensor plays the role of the diffused quantity gradients. It should be pointed out that, as in Section 3.3.1, isotropic considerations result in a single coefficient before the linear term in $\mathbf{s}$. In the absence of that assumption, the most general form relating $\boldsymbol{\tau}$ to $\mathbf{s}$ would involve a second-order tensor (in that respect, refer to Section 4.5.1 in Chapter 4). The analogy between the relations linking flux and gradient for a scalar and for momentum teach that viscosity brings about a homogenization of the velocity field; thus, it is an inherently ireversible term.

Let us return to the kinetic energy dissipation rate. With the above discussion, (3.102) yields

$$\begin{aligned} \widetilde{f} &= \frac{\lambda}{\rho}\left(\mathrm{div}\mathbf{u}\right)^2 + \nu s^2 \\ f &= -\frac{p}{\rho}\mathrm{div}\mathbf{u} + \frac{\lambda}{\rho}\left(\mathrm{div}\mathbf{u}\right)^2 + \nu s^2 \end{aligned} \tag{3.109}$$

where

$$\nu \doteqdot \frac{\mu}{\rho} \tag{3.110}$$

is the fluid's *molecular kinematic viscosity*, which depends on T too. Since we do not take the thermal variations into account, however, we consider, for a given problem, that the kinematic viscosity is constant. Its respective values for water and air at 20°C are given by Table 3.2. The relations (3.109) show that, for an incompressible flow, f and $\widetilde{f}$ coincide:

$$f = \widetilde{f} = \nu s^2 \tag{3.111}$$

In addition, let us establish the following calculation, based on the definition (3.35), as well as on the relations (3.19):

$$\begin{aligned} &\zeta\left(\mathrm{div}\mathbf{u}\right)^2 + 2\mu\mathbf{s}^D : \mathbf{s}^D \\ &= \zeta\left(\mathrm{div}\mathbf{u}\right)^2 + 2\mu\left[\mathbf{s} - \frac{1}{3}\left(\mathrm{div}\mathbf{u}\right)\mathbf{I}_3\right]^2 \end{aligned} \tag{3.112}$$

[27] Once again, that remark is not applicable to the case of turbulence modelling, which is dealt with in Chapter 4.

[28] This stems directly from the first-order expansion made upon the calculation (3.102).

$$= \zeta\,(\mathrm{div}\mathbf{u})^2 + \mu\left[s^2 - \frac{4}{3}\,(\mathrm{div}\mathbf{u})\,\mathbf{I}_3 : \mathbf{s} + \frac{2}{3}\,(\mathrm{div}\mathbf{u})^2\right]$$

$$= \mu s^2 + \left(\zeta - \frac{2}{3}\mu\right)(\mathrm{div}\mathbf{u})^2$$

We can then rearrange (3.109) as follows, with the definition (3.105):

$$\widetilde{f} = \frac{\zeta}{\rho}\,(\mathrm{div}\mathbf{u})^2 + 2\nu\mathbf{s}^D : \mathbf{s}^D \tag{3.113}$$

Equation (3.88) can then be written as

$$\frac{dE_k}{dt} = P_{int} + P_{ext} - \int_{\Omega}\left[\zeta\,(\mathrm{div}\mathbf{u})^2 + 2\mu\mathbf{s}^D : \mathbf{s}^D\right]d\Omega \tag{3.114}$$

Since the kinetic energy dissipation is always positive for an isolated medium, this relation and (3.109) show that the viscosities μ and ζ are positive, as well as λ (which additionally yields $\zeta \geqslant 2\mu/3$).

It is thus found that only viscous forces dissipate energy in this case, which is evidence of their irreversible nature. It is noteworthy that the scalar strain rate (as defined in (3.29)) accounts for this dissipation. Besides, dissipation varies as the square of the velocity, since the components of tensor $\mathbf{s}$ are linear in $\mathbf{u}$ (eqn (3.15)), which is consistent with the law (2.185) as determined in Chapter 2. Lastly, for a purely rotational velocity field at a given point, we have explained in Section 3.2.2 that $\mathbf{s} = \mathbf{0}$, which demonstrates that the shear tensor $\boldsymbol{\tau}$ equals zero, as well as the dissipation f. Once again, a purely rotational field does not generate any friction nor energy loss, as discussed within a discrete context in Section 2.4.3. This conclusion would remain valid in the presence of a uniform overall translational motion without any effect of the velocity gradients and coming down to changing the inertial frame of reference.

Let us now establish some results as regards the energy balances from the general equations briefly mentioned in Section 3.3.3. Taking what was previously explained into account, the balances (3.96) become

$$\begin{aligned} \frac{de_{int}}{dt} &= -\frac{p}{\rho}\mathrm{tr}\,\mathbf{s} + \frac{\lambda}{\rho}\,(\mathrm{tr}\,\mathbf{s})^2 + \nu s^2 \\ \frac{de}{dt} &= \frac{1}{\rho}\mathrm{div}\left[(-p + \lambda\mathrm{tr}\,\mathbf{s})\,\mathbf{u} + 2\mu\mathbf{s}\mathbf{u}\right] \end{aligned} \tag{3.115}$$

A final important remark should be made about pressure. If we had applied the procedure adopted for calculating $\boldsymbol{\tau}$ to the tensor $\boldsymbol{\sigma}$, formula (3.100) would have been applied substantially without any change, by substituting f for $\widetilde{f}$. Pressure would then have occurred in the behaviour law through the factor κ of the calculation (3.103). From that point of view, it appears that pressure formally originates from the derivative of tr $\mathbf{s}$. Now, that quantity subsequently vanishes in the incompressible case. Pressure, however, subsists in the law (3.107), whereas it would not occur in it if we had put tr $\mathbf{s} = 0$ from the first line of the calculation onwards (3.103). This argument shows that pressure is linked to the (weakly) compressible nature of a fluid (as previously suggested by relation (1.256) in Chapter 1). When particles come closer to each other, the field of their velocities is locally convergent, and (3.43) teaches that density increases. The discussion in Section 2.3.3 then allows us to state that pressure increases too and that its variations are all the larger and that the quantity c occurring in the state equation (2.113) is greater.

An 'incompressible' fluid[29] should then be considered as actually hardly compressible and corresponds to a very high value of the parameter c, whose meaning will be understood in Section 3.4.3. Besides, the utmost sensitivity of pressure to slight variations of density shows that even in this case, though the density may be regarded as constant, pressure is still fairly variable. Thus, p will in all cases be a field varying in both space and time. The respective densities of clear water and air at 20 °C are provided by the Table 3.2. Such a fluid is fully featured by its kinematic viscosity and its density, which are fixed values for a given temperature.

[29]Speaking of an incompressible *flow* would be more appropriate, as explained in the next section.

Table 3.2 Water and air features at 20 °C.

Fluid	Water	Air
ρ (kg.m^{-3})	998	1.204
ν (m^{2}s^{-1})	10^{-6}	1.56×10^{-5}
c (ms^{-1})	1480	343

Last, there is a class of flows for which the viscous forces may be ignored (at least in some area of the flow, as explained in Section 3.4.3). In such a case, the stress tensor (3.107) reduces to

$$\boldsymbol{\sigma} = -p\mathbf{I}_3 \tag{3.116}$$

that is it is isotropic. The dissipation rate is then zero, as expected. It is referred to as a *perfect fluid model*. The energy balances (3.115) then take the following forms:

$$\begin{aligned} \frac{de_{int}}{dt} &= -\frac{p}{\rho}\mathrm{div}\mathbf{u} \\ \frac{de}{dt} &= -\frac{1}{\rho}\mathrm{div}\,(p\mathbf{u}) \end{aligned} \tag{3.117}$$

When the flow is both incompressible and perfect, this is also written as

$$\begin{aligned} \frac{de_{int}}{dt} &= 0 \\ \frac{de}{dt} &= -\frac{1}{\rho}\mathbf{u}\cdot\mathbf{grad}\,p \end{aligned} \tag{3.118}$$

Thus, a perfect, incompressible fluid flow corresponds to a constant internal energy.

In many cases, the flow compressibility is too high to ignore the density variations, but low enough to disregard the second-order term in tr $\mathbf{s}$ in equation (3.115), which amounts to putting $\lambda = 0$. Hence we get the following formulas:

$$\begin{aligned} \frac{de_{int}}{dt} &= -\frac{p}{\rho}\mathrm{div}\mathbf{u} + \nu s^2 \\ \frac{de}{dt} &= \frac{1}{\rho}\mathrm{div}\,(-p\mathbf{u} + 2\mu\mathbf{s}\mathbf{u}) \end{aligned} \tag{3.119}$$

3.4.2 Navier–Stokes equations

The equation of motion of a viscous fluid is achieved by inserting the behaviour law (3.106) into Cauchy's equation (3.71):

$$\begin{aligned}\frac{\partial \mathbf{u}}{\partial t} + \mathbf{grad}\,\mathbf{u}\cdot\mathbf{u} &= \frac{1}{\rho}\mathbf{div}\left[(-p+\lambda \mathrm{tr}\,\mathbf{s})\,\mathbf{I}_3\right] \\ &\quad + \frac{2}{\rho}\mathbf{div}\,(\mu \mathbf{s}) + \mathbf{g} \\ &= \frac{1}{\rho}\mathbf{grad}\,(-p+\lambda \mathrm{div}\mathbf{u}) + \frac{1}{\rho}\mathbf{div}\,(\mu\,\mathbf{grad}\,\mathbf{u}) \\ &\quad + \frac{1}{\rho}\mathbf{div}\left[\mu\,(\mathbf{grad}\,\mathbf{u})^T\right] + \mathbf{g}\end{aligned} \tag{3.120}$$

The last term in (3.120) may also be written as

$$\begin{aligned}\mathbf{div}\left[\mu\,(\mathbf{grad}\,\mathbf{u})^T\right] &= \frac{\partial}{\partial x_j}\left(\mu\frac{\partial u_j}{\partial x_i}\right)\mathbf{e}_i \\ &= \left(\frac{\partial \mu}{\partial x_j}\frac{\partial u_j}{\partial x_i} + \mu\frac{\partial^2 u_j}{\partial x_i \partial x_j}\right)\mathbf{e}_i \\ &= (\mathbf{grad}\,\mathbf{u})^T\,\mathbf{grad}\,\mu + \mathbf{grad}\,(\mu \mathrm{div}\mathbf{u})\end{aligned} \tag{3.121}$$

If we ignore the variations of temperature, and if ever we contemplate the motion of a single fluid, the viscosities may be assumed to be constant in both space and time. This being the case, (3.120) becomes, with (3.121):

$$\frac{\partial \mathbf{u}}{\partial t} + \mathbf{grad}\,\mathbf{u}\cdot\mathbf{u} = -\frac{1}{\rho}\mathbf{grad}\,p + \nu\nabla^2\mathbf{u} + \frac{\lambda+\mu}{\rho}\mathbf{grad}\,\mathrm{div}\,\mathbf{u} + \mathbf{g} \tag{3.122}$$

Equation (3.122), together with the continuity equation (3.42), makes up the Navier–Stokes equation system for a compressible Newtonian viscous fluid (Landau and Lifshitz, 1987). These equations should also be accompanied either with a state equation (for instance, (2.113)) for calculating pressure or with equation (3.115) about internal energy, together with such a law as (2.116). All of these equations, particularly the equation of motion (3.122), obey the principle of Galilean invariance,[30] like Cauchy's equation (3.71). This result is due, in particular, to the fact that pressure, and more generally the Cauchy stress tensor, are Galilean invariants (refer to Sections 2.3.3 and 3.5.2), just like density. The property of Galilean invariance also results from the fact that the velocity derivatives do not depend on the frame of reference, in accordance with the discrete modelling of the friction forces as set out in Section 2.4.3 in Chapter 2.

[30] For a more general insight into the invariance properties of the Navier–Stokes equations, please refer to Frisch (1995) and Kambe (2008).

When the flow can be regarded as incompressible, we can easily treat the case of several fluids, and more generally of a fluid with a space variable viscosity. With the continuity equation (3.47), indeed, the equation of motion (3.120) is reduced to

$$\frac{d\mathbf{u}}{dt} = \frac{\partial \mathbf{u}}{\partial t} + \mathbf{grad}\,\mathbf{u}\cdot\mathbf{u} = -\frac{1}{\rho}\mathbf{grad}\,p + \frac{2}{\rho}\mathbf{div}\,(\mu \mathbf{s}) + \mathbf{g} \tag{3.123}$$

If the temperature variations are negligible and if the fluid is uniform, then μ may once again be regarded as constant in space, which ultimately yields, once again using (3.47):

$$\frac{\partial \mathbf{u}}{\partial t} + \mathbf{grad}\,\mathbf{u}\cdot\mathbf{u} = -\frac{1}{\rho}\mathbf{grad}\;p + \nu\nabla^2\mathbf{u} + \mathbf{g} \tag{3.124}$$

that is in an indicial form:

$$\forall i,\quad \frac{\partial u_i}{\partial t} + \frac{\partial u_i}{\partial x_j}u_j = -\frac{1}{\rho}\frac{\partial p}{\partial x_j} + \nu\frac{\partial^2 u_i}{\partial x_j \partial x_j} + g_i \tag{3.125}$$

A couple of remarks should be made, however, about the use of the incompressible Navier–Stokes equations. When the flow is featured by a slightly variable density, keeping the continuity equation in its compressible form (3.42) may be useful, even though the compression terms are ignored in the equation of motion; then we may refer to a *weakly compressible* flow, which we will consider throughout the following. Besides, it may be useful to keep the viscous term in its form (3.123) rather than as (3.124), in order to take into account a possible mixing of several fluids resulting in a space variable viscosity. For a single fluid, with (3.42) and (2.113), the system of equations governing the motion of such a fluid is written as

$$\begin{aligned}
&\frac{\partial \mathbf{u}}{\partial t} + \mathbf{grad}\,\mathbf{u}\cdot\mathbf{u} = -\frac{1}{\rho}\mathbf{grad}\;p + \nu\nabla^2\mathbf{u} + \mathbf{g}\\
&\frac{\partial \rho}{\partial t} + \operatorname{div}(\rho\mathbf{u}) = 0\\
&p = \frac{\rho_0 c^2}{\gamma}\left[\left(\frac{\rho}{\rho_0}\right)^{\gamma} - 1\right]
\end{aligned} \tag{3.126}$$

When numerically solving this system (discussed, in particular, in Chapters 5 and 6), we are faced with five unknowns (the three components of $\mathbf{u}$, as well as p and ρ) for five equations, and the procedure for solving them may be as follows: the equation of motion makes it possible to calculate $\mathbf{u}$, the continuity equation then gives ρ, lastly p is obtained via the state equation. In the strictly incompressible case (with all the restrictions involved by that denomination), the system is different and will be dealt with in Section 3.4.4.

Let us now more closely consider the equation of motion. The pressure gradient term has a simple physical meaning: it tends to direct the flow towards the decreasing pressures, according to intuition. Note that pursuant to (3.97), we have

$$\rho\,\mathbf{grad}\;h = \mathbf{grad}\;p \tag{3.127}$$

The pressure term in the Navier–Stokes equation of motion could then also be written as $-\mathbf{grad}\;h$, and so enthalpy represents the pressure force potential (regarless of the compressible or incompressible nature).

The viscosity term occurs, as expected, in the form of a diffusion which is analogous to the right-hand side of the scalar equation (3.61), which substantiates its irreversible nature, as already mentioned in the preceding paragraph with the dissipation rate (3.111). Thus, the fluid layers exchange momentum via this term, the faster ones speeding up the slower ones and conversely, in a velocity homogenizing process. There is another way to demonstrate the irreversibility of this term: if we change the time sign, the velocity sign changes too, and so $\mathbf{div}\,(\mu\mathbf{s})$ is the single term which changes sign in (3.123). Remember that velocity diffusion and energy dissipation are two jointly specific phenomena in a viscous process. We have previously observed that these forces exhibit a linear shape with respect to the velocities, as in Section 2.4.3 in Chapter 2. Viscosity then plays the role of a kinetic coefficient, after the fashion of quantities $\boldsymbol{\alpha}_{ab}$ as defined in this paragraph or, instead, this role is assumed by the quantities

$$\alpha \equiv \frac{\nu}{L^2} \tag{3.128}$$

where L is a distance between two mesoscopic particles.[31]

[31] This relation will become clearer with the SPH method, in Chapter 5 (Section 6.2).

As mentioned in the previous paragraph, the viscous forces may be ignored for some flows. The system (3.126) then becomes

$$\begin{aligned} \frac{\partial \mathbf{u}}{\partial t} + \mathbf{grad}\,\mathbf{u}\cdot\mathbf{u} &= -\frac{1}{\rho}\mathbf{grad}\,p + \mathbf{g} \\ \frac{\partial \rho}{\partial t} + \operatorname{div}(\rho\mathbf{u}) &= 0 \\ p &= \frac{\rho_0 c^2}{\gamma}\left[\left(\frac{\rho}{\rho_0}\right)^{\gamma} - 1\right] \end{aligned} \tag{3.129}$$

The equation of motion is then referred to as the *Euler equation*. A criterion for determining the relevance of the perfect fluid model will be provided in the next section.

We have pointed out, at the very beginning of Chapter 1, that if the frame of reference then chosen does not affect the equations of motion, then on the contrary it affects their solutions through the initial conditions. Within the context of the continuous medium equations, which can be seen as partial derivative equations, this holds true for the boundary conditions too. To complete this section, we will briefly review them. The action-reaction law shows that the stress $\mathbf{T} = \boldsymbol{\sigma}\cdot\mathbf{n}$ should be continuous at every point of an interface with a normal $\mathbf{n}$ between two inherently different media. Thus, at an interface between two non-miscible fluids, we will write

$$\begin{aligned} \mathbf{0} &= [(-p\mathbf{I}_3 + \boldsymbol{\tau})\,\mathbf{n}] \\ &= [-p\mathbf{n} + \boldsymbol{\tau}\cdot\boldsymbol{n}] \end{aligned} \tag{3.130}$$

where the bracket $[A]$ represents the discontinuity in quantity A, that is the difference between its values immediately on either side of the interface at the point being considered. Projecting onto the normal and the plane P locally tangential to the interface, we get, in the case of an incompressible fluid

$$\begin{aligned} &[p + 2\mu\mathbf{sn}] = 0 \\ \forall\mathbf{t}\in P,\ &\left[\mu\mathbf{t}^T\mathbf{sn}\right] = 0 \end{aligned} \tag{3.131}$$

where **t** is any unit vector of P (let us specify that the two fluids do not necessarily have the same viscosity). If either fluid has a much smaller viscosity than the other one,[32] the shear stresses $2\mu\mathbf{s}$ will possibly be ignored in the first medium, and so the second condition in (3.131) shows that the shear equals zero in a position closely adjacent to the interface. At the interface, the stress tensor is then diagonal (as in (3.116)), and the first condition in (3.131) is then reduced to the pressure continuity, whereas the second one expresses the tangential stress continuity:

$$\begin{aligned} [p] &= 0 \\ [\boldsymbol{\tau} \cdot \boldsymbol{n}] &= 0 \end{aligned} \tag{3.132}$$

In particular, on a free surface, provided that it is assumed that the atmospheric pressure p_{atm} prevailing just above is constant in space, we will get a pressure p_{surf} as given by

$$p_{surf} = p_{atm} \tag{3.133}$$

Since, however, pressure only plays a role in the equations of motion through its gradient, p can be substituted for $p - p_{atm}$,[33] which yields the equivalent condition:

$$p_{surf} = 0 \tag{3.134}$$

As regards a fluid at rest due to gravity, pressure and gravity are the only two remaining terms of the Navier–Stokes equation (3.124). The boundary condition (3.133) can then be used for writing down the distribution of *hydrostatic pressure*:

$$p_{hydr} = p_{atm} + \rho g \left(z_{surf} - z \right) \tag{3.135}$$

where z_{surf} denotes the free surface elevation. This pressure distribution, which varies vertically, accounts for a force $\mathbf{F}^A$ exerted on a body immersed in the fluid, as given by the pressure integral:

$$\begin{aligned} \mathbf{F}^A &= -\oint_{\partial\Omega} p\mathbf{n}d\Gamma \\ &= -\int_{\Omega} \mathbf{grad}\ p\ d\Omega \end{aligned} \tag{3.136}$$

(we have used the formula (A.74) of Appendix (1)). Since the fluid is at rest, the equation of motion (3.129) is reduced to the equilibrium of forces **grad** $p = \rho\mathbf{g}$, which yields

$$\begin{aligned} \mathbf{F}^A &= -\int_{\Omega} \rho\mathbf{g}d\Omega \\ &= -\rho V\mathbf{g} \end{aligned} \tag{3.137}$$

where V is the volume of the body. This force, which is known as the *buoyancy*, opposes gravity.

[32] Such is the case of air with respect to water, taking the above values into account. The discussion below is then applicable, to a first approximation, to a free water surface due to gravity.

[33] This amounts to stating that pressure can only be defined to within one constant.

The pressure and gravity terms of (3.124) can be unified by introducing the *dynamical pressure* p^*:

$$\begin{aligned} p^* &\doteqdot p - \rho \mathbf{g} \cdot \mathbf{r} \\ &= p - p_{hydr} + \rho g z_{surf} + p_{atm} \end{aligned} \tag{3.138}$$

Thus, within a trivial constant, it is the difference between the pressure and the hydrostatic pressure. This justifies its designation, since p^* represents the portion of pressure which is directly linked to the occurrence of a fluid motion.[34] Hence, the Navier–Stokes equation of motion takes a simpler form:

[34] More exactly, a motion with a non-zero vertical component.

$$\frac{\partial \mathbf{u}}{\partial t} + \mathbf{grad}\,\mathbf{u} \cdot \mathbf{u} = -\frac{1}{\rho}\mathbf{grad}\,p^* + \nu \nabla^2 \mathbf{u} \tag{3.139}$$

Another free-surface condition may also be written, because the free surface may be characterized as the locus of points verifying

$$\Phi_s(\mathbf{r}, t) \doteqdot z_{surf} - z = 0 \tag{3.140}$$

which defines Φ_s. Since the relation constantly holds true, we may say that the derivative of Φ_s following the local motion of the surface vanishes, which, by analogy with (3.7), is written as:

$$\frac{\partial \Phi_s}{\partial t} + \mathbf{grad}\Phi_s \cdot \mathbf{u}_{surf} = 0 \tag{3.141}$$

where $\mathbf{u}_{surf}$ is the surface velocity at the point being considered. After definition (3.140), the partial derivative of Φ_s with respect to time still equals $\partial z_{surf}/\partial t$. Besides, the gradient of Φ_s is, of course, a vector which is locally normal (but not unitary) to the free-surface and is directed towards the inside of the fluid, which will be denoted by $\mathbf{n}_{surf}$. The above relation can then be formulated as:

$$\frac{\partial z_{surf}}{\partial t} + \mathbf{u}_{surf} \cdot \mathbf{n}_{surf} = 0 \tag{3.142}$$

Unlike the so-called *dynamic* condition (3.133), the condition (3.142) is said to be *kinematic*, since it is only based on the local motion of the surface. If the fluid velocity is locally aligned with the surface, we have $\mathbf{u}_{surf} \cdot \mathbf{n}_{surf} = 0$, and this condition shows that z_{surf} is unchanged, as expected in such a case. If, on the contrary, the surface velocity is normal to it and is directed towards the outside of the fluid, we get $\mathbf{u}_{surf} \cdot \mathbf{n}_{surf} < 0$, and (3.142) shows that z_{surf} intuitively increases.

Let us now turn to the wall boundary conditions. The friction exerted by a viscous fluid on a rigid wall results in the continuity of velocity in each point $\mathbf{r}_{wall}$ of that wall, and so the velocity of the fluid on the wall equals the wall velocity, which is denoted by $\mathbf{u}_{wall}$:

$$\mathbf{u}(\mathbf{r} = \mathbf{r}_{wall}) = \mathbf{u}_{wall}(\mathbf{r}_{wall}) \tag{3.143}$$

(adherence or wall *no-slip* condition). Here we have written $\mathbf{u}_{wall}(\mathbf{r}_{wall})$, in order to point out that the wall may have an arbitrary rigid motion (refer to Chapter 1, 1.5.1), so that the velocity may differ from one point to another at its surface. In the particular case of a motionless wall,[35] (3.143) yields

[35] It is noteworthy that the condition (3.143) is incompatible with the motion of an interface between two fluids slipping along a wall. The processes involved by this phenomenon, which is simple indeed, are as yet poorly known.

$\mathbf{u}\left(\mathbf{r}=\mathbf{r}_{wall}\right)=\mathbf{0}$. We may recall, however, that such a condition is only valid upon a friction. With a perfect fluid model, the absence of friction implies a mere condition of impermeability (*free-slip* condition):

$$\left[\mathbf{u}\left(\mathbf{r}=\mathbf{r}_{wall}\right)-\mathbf{u}_{wall}\left(\mathbf{r}_{wall}\right)\right]\cdot\mathbf{n}_{wall}\left(\mathbf{r}_{wall}\right)=0 \tag{3.144}$$

where $\mathbf{n}_{wall}$ is the unit vector normal to the wall in the point being considered. This assumption amounts to ignoring the boundary layer developing near a wall because of the friction. This is, however a theoretically improper approach, since the velocity gradients do arise in the vicinity of the solid boundaries and give rise to significant viscous effects. We will discuss the consequences in Chapter 4, when establishing the boundary conditions for the turbulent flows (Section 4.4.6). For a motionless wall, the condition (3.144) yields

$$(\mathbf{u}\cdot\mathbf{n})\left(\mathbf{r}_{wall}\right)=0 \tag{3.145}$$

The normal to the wall may depend on time, if the wall is mobile. With a rigid motion as given by (1.191), considering that the unit normal is a 'rigid' fictitious vector moving along with the solid, we can apply the law (1.193) to get the following dynamical equation:

$$\frac{d\mathbf{n}_{wall}}{dt}=\boldsymbol{\Omega}\times\mathbf{n}_{wall} \tag{3.146}$$

The condition on the velocity at the wall makes it possible to write a condition on pressure. To do this, let us choose a point M on the wall; we can observe the following relation, which is valid for every vector field $\mathbf{u}$:

$$\mathbf{grad}\,\mathbf{u}\cdot\mathbf{u}=\mathbf{grad}\frac{|\mathbf{u}|^2}{2}+(\mathbf{curl}\,\mathbf{u})\times\mathbf{u} \tag{3.147}$$

Being provided with that equality, the scalar product of the equation of motion (3.139) times the vector $\mathbf{n}_{wall}$ in point M gives

$$\frac{\partial\mathbf{u}}{\partial t}\cdot\mathbf{n}_{wall}+\mathbf{grad}\left(\frac{u^2}{2}+\frac{p^*}{\rho}\right)\cdot\mathbf{n}_{wall}=\nu\nabla^2\mathbf{u}\cdot\mathbf{n}_{wall}+\det\left[\mathbf{u},\mathbf{curl}\,\mathbf{u},\mathbf{n}_{wall}\right] \tag{3.148}$$

Let us now refer to the frame of reference related to M in its motion.[36] In this frame of reference, the fluid velocity in point M vanishes pursuant to (3.143), whereas the vector $\mathbf{n}_{wall}$ in point M is constant. Hence we get

$$\frac{\partial}{\partial\mathbf{n}_{wall}}\left(\frac{u^2}{2}+\frac{p^*}{\rho}\right)=\nu\left(\nabla^2\mathbf{u}\right)_{\mathbf{r}=\mathbf{r}_{wall}}\cdot\mathbf{n}_{wall} \tag{3.149}$$

where $\partial A/\partial\mathbf{n}\doteqdot\mathbf{grad}\,A\cdot\mathbf{n}$. One may often think, however, that the wall is locally quite planar; thus, we may ignore the velocity components normal to the wall, as well as the partial derivatives regarding the spatial components located within its plane.[37] Among the terms in (3.149), only pressure may then be kept as a first approximation. The pressure condition at the wall is then written as

[36] The fact that this frame of reference may be non-Galilean does not matter here, since we consider the equation at a given point in time, the wall motion being frozen during a short lapse of time. More strictly, we must refer to a Galilean frame of reference the motion of which coincides with the wall motion at that very moment.

[37] Such are the planar boundary layer hypothèses (refer to Landau and Lifshitz, 1987).

$$\frac{\partial p^*}{\partial \mathbf{n}_{wall}} \approx \mathbf{0} \qquad (3.150)$$

Remember it is only an approximation, which does not hold true in the case of a very curved wall, *a fortiori* in those places where the wall has corners.

3.4.3 Similarity. Orders of magnitude

The universality of the laws of mechanics, that is their invariance over a number of changes as perceived by the observer, was evidenced on several occasions in Chapter 1. We have just recalled a major feature of them, namely the Galilean invariance, but the equations must also remain universal in spite of the chosen measurement standards. In order to clarify this issue against the background of the Navier–Stokes equations, they may usefully be written in a dimensionless form. To that purpose, we contemplate a given flow and we choose characteristic scales for length (L), time (T), velocity[38] (U) and pressure (p_0). Strictly speaking, these choices may initially be made quite arbitrarily. The space and time components, as well as the velocities and pressure fields, can then be non-dimensionalized as follows, providing the dimensionless quantities with a $^+$ as a superscript:

[38]Referring to Chapter 1, we must mention the non-Galilean invariant feature of that definition. The scale U is meaningless unless it is defined as a difference in velocities rather than as an absolute velocity.

$$\begin{aligned} x_i^+ &\doteqdot \frac{x_i}{L} \\ t^+ &\doteqdot \frac{t}{T} \\ \mathbf{u}^+ &\doteqdot \frac{\mathbf{u}}{U} \\ p^+ &\doteqdot \frac{p}{p_0} \end{aligned} \qquad (3.151)$$

The derivation operators should be non-dimensionalized too; we may write, for example:

$$\begin{aligned} \nabla^+ = \mathbf{grad}^+ &\doteqdot \frac{\partial}{\partial x_i^+}\mathbf{e}_i \\ &= L\mathbf{grad} \end{aligned} \qquad (3.152)$$

Using these notations, the Navier–Stokes equation of motion (3.124) can easily be rearranged in the following dimensionless form:

$$\begin{aligned} St\frac{\partial \mathbf{u}^+}{\partial t^+} + \mathbf{grad}^+\mathbf{u}^+ \cdot \mathbf{u}^+ &= -\frac{1}{Ma^2}\mathbf{grad}^+ p^+ \\ &+ \frac{1}{Re}\left(\nabla^+\right)^2 \mathbf{u}^+ - \frac{1}{Fr^2}\mathbf{e}_z \end{aligned} \qquad (3.153)$$

where four dimensionless numbers have been defined:

$$\begin{aligned} St &\doteqdot \frac{L}{UT} \\ Re &\doteqdot \frac{UL}{\nu} \end{aligned} \qquad (3.154)$$

$$Fr \doteqdot \frac{U}{\sqrt{gL}}$$

$$Ma \doteqdot \sqrt{\frac{\rho U^2}{p_0}}$$

They are commonly referred to as the *Strouhal number* (St), the *Reynolds number* (Re), the *Froude number* (Fr) and the *Mach number* (Ma) of the relevant flow. It can be noticed that the dimensionless equation of motion (3.153) takes the same form as the original dimensional equation. Within that context, the factor $1/Re$ now plays the role of viscosity.

In order to profitably employ equation (3.153), the L, T, U and p_0 scales should be suitably chosen in such a way that they represent the flow: L may represent the diameter of a solid body, T the period of a wave system, U the flow rate in a pipe and so on. Hence, it can easily be found that each of these numbers provides us with an order of magnitude of the ratio between a term in the equation and the inertia term **grad u** · **u**. Suitably chosen scales makes it possible, indeed, to state that the non-dimensionalized quantities as given by (3.151) are of the order of magnitude of one. For instance, we have

$$\begin{aligned}\frac{|\partial \mathbf{u}/\partial t|}{|\mathbf{grad\,u}\cdot\mathbf{u}|} &= \frac{U\left|\partial \mathbf{u}^+/\partial t^+\right|/T}{U^2\left|\mathbf{grad}^+\mathbf{u}^+\cdot\mathbf{u}^+\right|/L} \\ &\sim \frac{L}{UT} \\ &= St\end{aligned} \tag{3.155}$$

Likewise:

$$\begin{aligned}Re &\sim \frac{|\mathbf{grad\,u}\cdot\mathbf{u}|}{\left|\nu\nabla^2\mathbf{u}\right|} \\ Fr^2 &\sim \frac{|\mathbf{grad\,u}\cdot\mathbf{u}|}{|\mathbf{g}|}\end{aligned} \tag{3.156}$$

The usefulness of these dimensionless numbers is obvious: for example, if the Froude number is high enough before 1, inertia will prevail over gravity and **g** will possibly be ignored in the equation of motion. As a rule, it can be stated that when Re and Fr are very high, their numerical values no longer affect the flow, because they occur at the denominator of (3.153). For the same reason, the value of St is unimportant when it is very low, which corresponds to a very large time scale T , that is to a nearly steady flow.[39]

Apparently, equation (3.148) confirms that the reference pressure p_0 can be assessed by ρU^2, which is tantamount to stating that p^+ is of the order of magnitude of one. More exactly, it is the pressure deviation δp between two points due to the effects of inertia:

$$\delta p \sim \rho U^2 \tag{3.157}$$

Comparing (3.157) and (2.117) will show that the pressure deviations are much smaller than pressure itself.[40] Besides, equation (3.135) suggests that, due to gravity, the pressure deviations have an order of magnitude of

[39]Nevertheless, one should be careful with the notion of permanent flow, since that notion should be redefined within the context of turbulence (refer to Chapter 4).

[40]As previously explained, these pressure deviations do have a physical meaning for the Navier–Stokes equations, since pressure plays a role in it through its gradient. Although an absolute definition was achieved in Chapter 2 (refer to eqn (2.64)), we should remember, indeed, that it serves as a potential (Section 3.4.2) and can therefore only be defined to within an additive constant.

$$\delta p \sim \rho g H \tag{3.158}$$

where H is a characteristic vertical scale of flow (water head in a tank, etc.). Where inertia and gravity produce similar effects, both assessments (3.157) and (3.158) coincide if H is likened to L, because Fr is of the order of magnitude of one according to (3.156). Whatever, (2.117) suggests the choice $p_0 = \rho c^2$, giving

$$Ma \doteqdot \frac{U}{c} \tag{3.159}$$

A couple of useful approximations are mentioned in Sections 3.4.1 and 3.4.2 for (weakly) compressible flows and a perfect fluid. We are going to try to take advantage of the notion of dimensionless number as to provide suitable criteria for characterizing the flows liable to enter either category.

A criterion for judging the validity of the perfect fluid model is based on an assessment of the viscous term $\nu \nabla^2 \mathbf{u}$ in relation to the other terms of (3.124). The Reynolds number Re is generally very high, in view of the very low values of ν for the usual fluids (refer to Table 3.2). We may infer that the molecular viscous forces may generally be ignored. However, two remarks are to be made against this argument. Through the relation (3.156) the spatial scale L_ν beyond which the viscous effects become significant can be estimated:

$$L_\nu \doteqdot \frac{\nu}{U} \tag{3.160}$$

That scale is typically of the order of magnitude of one micrometre or several tens of micrometres. Thus, the viscous forces may generally not be ignored in close proximity to a wall, where the velocity gradients are very high (an application will be discussed in Chapter 4, Section 4.4.5). With this definition, the Reynolds number may also be written as

$$Re = \frac{L}{L_\nu} \tag{3.161}$$

Besides, when Re is high enough, a flow becomes highly unsteady and gives rise to turbulent effects: the velocity field is then so quickly variable in space that the viscous effects may not always be ignored, though the Reynolds number value is unimportant. This paradox will be specified and cleared up in Chapter 4. Anyway, the perfect fluid model is, indeed, a property related to a flow, not to a fluid.

Lastly, an essential remark should be made about this model: even where the viscous forces may be ignored, the viscosity cannot vanish and even exerts a significant action. This is because the viscous forces may only be ignored when the velocity field is so regular that the second derivatives of velocity (i.e. $\nabla^2 \mathbf{u}$) are quite small. Now, we have learnt in Section 3.4.2 that, as a matter of fact, the viscous forces exert a homogenizing action on the velocity field. An interesting process then occurs: viscosity tends to make the system enter a state in which the viscous effects are as small as possible, to the extent allowed by the other constraints of the system (e.g. the presence of walls). This process is analogous to the thermodynamic equilibrium resulting from the increased entropy and making a particle adopt a motion of a solid at equilibrium (refer to Section 2.4.3). Thus, a fluid which is free of any external influence will be subject to the effect of the viscous forces until it reaches an equilibrium corresponding to a motion of a solid (Section 1.5.1), where dissipation (unlike the viscosity) vanishes. Putting $\nu = 0$ is then nothing but a mathematical trick.[41]

Let us now examine the incompressibility hypothesis. As shown in Chapter 2 (Section 2.3.3), it is valid when the speed of sound is very large, which suggests a necessary condition like $Ma^2 << 1$. To precise this assumption, let us now contemplate the particular case of a flow in the form of a slight variation of the parameters $(\mathbf{u}, \rho, p)$ about a state of equilibrium corresponding to the rest, accompanied by a density which everywhere equals its reference value ρ_0. The system state is then featured by the fields $(\mathbf{u}, \rho_0 + \delta\rho, p = 0)$ with $\delta\rho \ll \rho_0$, which allows us to linearize the system (3.126).[42] We find:

$$\begin{aligned}
\frac{\partial \delta\rho}{\partial t} &= -\rho_0 \mathrm{div}\mathbf{u} \\
\frac{\partial \mathbf{u}}{\partial t} &= -\frac{1}{\rho_0}\frac{\partial p}{\partial \rho}\mathbf{grad}\ \delta\rho + \nu\nabla^2\mathbf{u} \\
\frac{\partial p}{\partial \rho} &= c^2
\end{aligned} \tag{3.162}$$

Combining these three equations, and using the fact that the Laplace operator commutes with divergence (refer to Appendix A), we get an equation with partial derivatives with respect to ρ:

$$\frac{\partial^2 \delta\rho}{\partial t^2} = c^2\nabla^2\delta\rho + \nu\frac{\partial \nabla^2 \delta\rho}{\partial t} \tag{3.163}$$

It is a damped wave equation accompanied by the propagation velocity c. The fluid compressibility, which acts through the state equation, then accounts for density (and then pressure) waves which are also known as acoustic waves.[43] c is therefore referred to as the *sound velocity* of the fluid being considered.[44]

Solutions of (3.163) can be sought in the form of planar waves:

$$\delta\rho = R_0 \exp\left[i\left(\omega t - \mathbf{K}\cdot\mathbf{r}\right)\right] \tag{3.164}$$

where R_0 is a very small constant before ρ_0, in order to satisfy the hypothesis of slightly varying parameters. By injecting (3.164) into (3.163), we can find a simple formula relating the frequency ω to the norm K of the wave vector $\mathbf{K}$ via viscosity:

$$\omega^2 - i\omega K^2\nu - c^2K^2 = 0 \tag{3.165}$$

The solutions provide the following dispersion relation:

$$\omega = \frac{\nu K^2}{2}\left(\pm\sqrt{\left(\frac{2Re}{Ma.KL}\right)^2 - 1} + i\right) \tag{3.166}$$

[41] No wonder the seeming paradox. The following analogy could be put forward: a freely falling body is subject to a driving work exerted by gravity. If it ultimately lands on a horizontal plane, gravity will no longer affect its motion, and so we may put $g = 0$ in the equations ; for all that, g is not zero in reality. On the contrary, in the absence of gravity, the body could not lie on the horizontal plane. In both cases, we observe a process which makes the system enter a state in which its effects are minimized.

[42] The effect of gravity is ignored here. Besides, it was shown in Section 3.4.2 that it can be formally included within the pressure term. Besides, let us observe that the non-linear terms will be readily accounted for in Chapter 4 on turbulence.

[43] They characterize the sound propagation within the fluid.

[44] Strictly speaking, it is the sound velocity at rest and for $\rho = \rho_0$.

In the particular case of the perfect fluid model, (3.166) yields the simple relation

$$\omega = cK \tag{3.167}$$

In relation to the latter equality, the viscous case has an imaginary part of ω, which results in an actual exponential in (3.164):

$$\delta\rho \propto \exp(i\omega t) \propto \exp\left(-\frac{K^2 \nu t}{2}\right) \tag{3.168}$$

The viscous forces then tend to damp the density waves, as expected, all the more since viscosity is high. This behaviour is consistent with the damped pendulum model (2.233) and (2.234), which substantiates the role of viscosity as a kinetic coefficient.

We can now establish a criterion allowing us to judge the validity of the incompressible approach in the general case. The third equation in the system (3.162) shows that the relative variations of density can be assessed by

$$\begin{aligned} \frac{\delta\rho}{\rho_0} &= \frac{\partial\rho}{\partial p}\frac{\delta p}{\rho_0} \\ &\sim \frac{L}{c^2}\left|\frac{1}{\rho}\mathbf{grad}\ p\right| \end{aligned} \tag{3.169}$$

Returning to equation (3.124), which is still without the viscous terms, we can see that the pressure term is balanced by the time variation term $\partial \mathbf{u}/\partial t$ to which inertia and gravity are added. We get the following estimate:

$$\begin{aligned} \frac{\delta\rho}{\rho_0} &\sim \frac{L}{c^2}\max\left(\left|\frac{\partial \mathbf{u}}{\partial t}\right|, |\mathbf{grad}\ \mathbf{u}\cdot\mathbf{u}|, |\mathbf{g}|\right) \\ &\sim \max\left(\frac{LU}{c^2 T}, \frac{U^2}{c^2}, \frac{gL}{c^2}\right) \\ &= Ma^2 \max\left(St, 1, \frac{1}{Fr^2}\right) \end{aligned} \tag{3.170}$$

The incompressibility criterion (actually, the near-incompressibility criterion, as explained in the previous paragraph), then comprises three conditions:

$$\begin{aligned} Ma^2 &\ll 1 \\ St\,Ma^2 &\ll 1 \\ \frac{Ma^2}{Fr^2} &\ll 1 \end{aligned} \tag{3.171}$$

In the case where the gravity forces are negligible ($1 \ll Fr$) and the flow is steady ($T \to +\infty$), the first of these conditions subsists alone:[45] the flow velocity should remain small enough before the sound velocity. A Mach number of 0.1 will yield relative variations of density on the order of 1%, which is reasonable to speak of near-incompressibility. Because of the very high values of c for the ordinary fluids (refer to Table 3.2, Section 3.4.1), this condition is not so constraining.[46] The theoretical limit of an incompressible fluid formally

[45] In the case of very rapidly varying flows, conversely the second condition is dominant. This is the case of some tsunamis, for which a compressible approach is necessary to understand the process of wave generation (Levin and Nosov, 2009).

[46] We may recall that we do not contemplate the temperature variations, which would add a further criterion into the list as given by (3.171).

corresponds to an infinite sound velocity, which shows again that it is an ideal concept.

The dimensionless numbers get a major theoretical significance in fluid mechanics. They are suitable, indeed, for greatly simplifying a number of problems. The non-linearity of the Navier–Stokes equations (i.e. of the advection term) generally makes their analytical solution impossible even in geometrically simple cases, since the solutions may be extremely complex.[47] For illustration purposes, let us consider the issue of the calculation of the force exerted on a solid body moving about at a steady velocity with an 'infinite' fluid, that is having very large dimensions before that of the body[48]. If the motion may be considered weakly compressible, an exact calculation would imply the solution of the equations of motion (3.126), from which the stress tensor would be derived using the formula (3.107). Integrating this tensor in the form of an integral analogous to (3.69), extended over the body's surface, would then provide the force value being sought. Such a work is usually impossible[49]; we can, however, simplify the terms of the problem through the following argument. Let us stick to the component of the force opposite to the motion of the body, which is known as *drag*[50] and is denoted here by F^D. It can obviously only depend on the fluid and body characteristics. The first are ρ and ν, according to Section 3.4.1, whereas the body is fully featured by its spatial dimensions[51] $\{L_i\}$ $(i = 1, \ldots, k)$ and its velocity U. Thus, we can write the following general relation:

$$F^D = \phi_{dim}\left(\rho, \nu, U, L_1, \ldots, L_k\right) \tag{3.172}$$

where ϕ_{dim} is an unknown dimensional function[52]. The important thing is that this relation is valid whichever unit system may be chosen. Thus, if L_1 is taken as a length unit, L_1/U as a time unit and ρL_1^3 as a mass unit, the parameters in equations (3.172) are amended and become

$$\begin{aligned} F^D &\longleftarrow \frac{F^D}{\rho L_1^2 U^2} \\ \rho &\longleftarrow 1 \\ \nu &\longleftarrow \frac{\nu}{U L_1} = \frac{1}{Re} \\ U &\longleftarrow 1 \\ \forall i,\ L_i &\longleftarrow \frac{L_i}{L_1} \end{aligned} \tag{3.173}$$

Equation (3.172), in this new system of units, is written as:

$$\begin{aligned} F^D &= \rho L_1^2 U^2 \phi_{dim}\left(1, \frac{1}{Re}, 1, 1, \frac{L_2}{L_1}, \ldots, \frac{L_k}{L_1}\right) \\ &= \frac{1}{2}\rho S U^2 \frac{2L_1^2}{S} \phi_{adim}\left(Re, \frac{L_2}{L_1}, \ldots, \frac{L_k}{L_1}\right) \end{aligned} \tag{3.174}$$

into which we have introduced the solid *reference area*, that is the surface S it offers to an observer towards whom it is moving. The dimensionless

[47]This aspect will be addressed in Chapter 4. Analytical solutions, however, do exist in some cases, particularly when the Reynolds number is very low (see below). An example of approximate analytical solution with a high Reynolds number will be provided, however, in Section 4.4.5, but with the assistance of a turbulence model.

[48]The hypotheses which are raised here (constant velocity, fluid 'infiniteness') are essential. In all cases, the following discussion would not suffice to characterize the whole force which the solid is subject to.

[49]Stokes has provided an exact solution in the case of a sphere with a very low Reynolds number (for example, refer to Landau and Lifshitz, 1987).

[50]The same view could be supported for the other force component, which is known as *lift*.

[51]The solid is assumed to be undeformable. We will also raise the simplifying hypothesis that its shape data fully comes down to a countable set of lengths.

[52]It is to be understood that it depends on the physical units.

function ϕ_{adim} is derived from ϕ_{dim} and then remains unknown. We may notice, however, that the product $2L_1^2\phi_{adim}/S$ only depends on the distance ratios L_i/L_1 and therefore occurs in the form of a coefficient representing the solid's shape and henceforth only depending on Re. It is denoted as C_D, and it is called the *drag coefficient*. Since the force vector, whose norm is F^D, impedes the motion of the body, it is ultimately written as:

$$\mathbf{F}^D = -\frac{1}{2}\rho S C_D\,(Re, shape)\,U\mathbf{U} \tag{3.175}$$

where $\mathbf{U}$ is the body velocity vector, with a norm U. This very general relation yields the friction (drag) of a body with a given shape as the product of the fluid density, its reference area, the square of velocity and a coefficient only depending on the Reynolds number. Thus, the number of pertinent variables could be significantly reduced through simple dimensional reasoning, which is practically very attractive (examples of such laws will be discussed in Chapter 4, then in Sections 7.3.1, 7.4.1, 7.4.3, 8.2.1, 8.4.1 and 8.5.1). As a rule, reducing a relation between various parameters to its dimensionless form amounts to giving a universal form of it, which does not depend on the chosen measurement standards, and highlights the dependence of the mechanisms on the scales.[53]

[53] This principle is at the basis of the theory of scale models in Hydraulics (refer to Section 8.5.1).

It is not surprising that a motion countering force depends on the Reynolds number, that is ultimately on the fluid viscosity. This force, indeed, exerts a negative power $\mathbf{F}^D \cdot \mathbf{U}$ which tends to reduce the body's kinetic energy (and then its velocity), according to equation (1.74). Thus, to keep the solid velocity constant, a driving power must be supplied to the solid which will transfer it to the fluid, pursuant to the law of action-reaction (Chapter 1, Section 1.4.3). Thus, the fluid kinetic energy tends to increase, according to law (3.89). Hence, there is an energy flux from the solid towards the fluid; it is directly related to the stress $\boldsymbol{\sigma}$ occurring at their interface. In the frame of reference for the solid,[54] however, the fluid has a steady flow regime,[55] and so its energy is constant. Thus, the fluid should get rid of the energy it receives from the solid by dissipating it. Such is the role of the quantity f in equation (3.89), which accounts for the importance of viscosity because of the existing relation between f and ν (eqn (3.111)).

[54] This view is valid in this frame of reference because it is Galilean, since U is assumed to be constant.

[55] With the above restriction, which does not change our counclusions.

The dependence of the drag coefficient on the Reynolds number may be further discussed. In the particular case where Re is very low, viscous forces prevail in the flow and equation (3.153) shows that the advection term may be ignored. If the gravity forces are ignored (e.g. if Fr is high), the equation of motion (3.124) in steady flow conditions becomes

$$\mathbf{grad}\ p = \mu\nabla^2\mathbf{u} \tag{3.176}$$

(the definition (3.110) has been used). Equation (3.176)[56] only involves dynamical viscosity to characterize the fluid; thus, the earlier analysis can be simplified. From now on, an analogous argument makes it possible to write

[56] It can be noted that it is linear, which explains why, under such conditions, exact solutions of it can sometimes be determined.

$$F^D = \mu L_1 U f_{adim}\,(shape) \tag{3.177}$$

Since the list of parameters which the force may depend on has been reduced by one unit, the number of dimensionless quantities occurring in the final formula can be reduced to the same extent, which accounts for the disappearance of Re. It should be pointed out that the formula (3.175) is still valid, which allows us to write

$$\begin{aligned} C_D\,(Re, shape) &= \frac{2FD}{\rho S U^2} \\ &= \frac{\nu}{UL_1}\frac{2L_1^2}{S}\phi_{adim}\,(shape) \\ &= \frac{1}{Re}\psi_{adim}\,(shape) \end{aligned} \tag{3.178}$$

We thus have shown that C_D varies inversely to Re when the latter is very low. Conversely, if Re is very high,[57] we have previously seen that its value no longer has any importance. Hence, we may state that C_D asymptotically tends towards a constant which depends on the solid's shape. One may wonder why the drag force, which we have just shown is closely related to viscosity, does not depend on its value in such conditions. This paradox will be cleared up in Chapter 4 (Section 4.4.3).

[57] More exactly, if it is high enough for the flow to be fully turbulent (refer to Chapter 4).

A final remark should be made as regards the force exerted on a solid. The latest discussion, together with the law (3.175), shows that, at a high Reynolds number, the force is a quadratic function of the body's velocity. Conversely, (3.177) proves that the force varies linearly as a function of the velocity with a low Reynolds number.[58] One may logically wonder why initially linear forces ($\nu\nabla^2\mathbf{u}$) result, in some cases, in a quadratic form for the force exerted on a solid. Since this force is related to the energy flux between the body and the fluid, that is $\mathbf{F}^D\cdot\mathbf{U}$, ultimately dissipated by the latter, this means the energy dissipation rate varies as U^2 at a low Reynolds number and as U^3 at a high Reynolds number. If the first assessment is consistent both with the law (3.111) providing f and with the law (3.128), the second seems mysterious and reminds us of the quadratic dissipative forces which were dealt with in Section 2.5.2 in Chapter 2. This finding is far from being minor and will be amply discussed again in Chapter 4.

[58] This result is consistent with the linearity of the friction forces as introduced in Chapter 2, Section 2.4. In particular, we applied that principle at the end of Section 2.4.2 .

We already can generalize the procedure being used for assessing the force received by a body. Let us consider a problem[59] featured by n parameters, out of which m parameters are dimensionally independent. Using a procedure analogous to the calculation (3.173) and (3.174), it can readily be shown that $n-m$ independent diimensionless numbers can be built from the n initial independent dimensional parameters. If the latter are linked by an undetermined function, it will be the same for the dimensionless numbers. Besides, if $n-m=1$, it can be inferred that the single dimensionless number is featured by a constant value. This is the Vaschy–Buckingham theorem (Vaschy, 1892); we will immediately provide an example taken from celestial mechanics. The revolution period T of a planet around the Sun depends on the semi-major axis a of its ellipse of revolution, the Sun mass M the Newton's gravitational constant G occurring in equation (1.44) (the mass of the planet itself cannot exert any action, as explained by eqn (1.59)). We then have $n=4$ parameters, out of which $m=3$ are dimensionally independent, because here we have three fundamental units (namely mass, length, time). Thus, we are in a case where $n-m=1$, from which we derive

[59] It may be any physics problem.

$$T^2 \propto \frac{a^3}{GM} \tag{3.179}$$

This is Kepler's Third Law. The single advantage of completely solving the equation of motion (1.59) would consist in determining the multiplicative constant (here being equal to $4\pi^2$, refer, for example, to Meriam, 1966). In fluid mechanics, such an approach is very valuable because the equations can not be solved rigorously, as we will see on many occasions in Chapter 4.

The above considerations raise the more general issue of the relationship between time and distances in mechanics. In other words, we have to ask ourselves the following question: how do the time scales vary as a function of the spatial scales for a given problem? The answer to that question primarily lies is the discussions conducted in Chapter 1: insofar as the evolution of a mechanical system is determined by its Lagrangian (or its Hamiltonian), and since the spatial scale cannot affect the equations of motion, it may be stated that the ratio between the various kinds of energy remains unaltered by a change of (time and space) scales. Now, we have seen in Section 1.3.2 that the potential energy only depends on the position parameters q_i. In the following, we will assume such a dependence as $E_{ext} \propto q_i^p$, p being an arbitrary power. The law (1.44) corresponds, for example, to $p = -1$. We then have $E_{ext} \propto L^p$ whereas E_k varies in L^2/T^2; the universality of the (kinetic energy)/(potential energy) ratio then yields

$$T \propto L^{1-p/2} \tag{3.180}$$

Let us provide three examples of this very general law. In the case of general gravity (1.44), with $p = -1$ we get $T \propto L^{3/2}$, in accordance with (3.179). For a body which is subject to Earth's gravity (1.52), we have $p = 1$, resulting in $T \propto L^{1/2}$. We come across the well-known result of the law of falling bodies. In the particular case of a fluid subject to gravity, we can also find this result from the definition (3.154) of the Froude number Fr. Lastly, the law (1.165) yields $p = 2$ for a pendulum subjected to small oscillations, that is $T \propto 1$. Once again, we find the law according to which the period of the small oscillations of a pendulum is independent of its amplitude.

In the case of a dissipative process, the same discussion is to be conducted with the (time variation of kinetic energy)/(energy dissipation rate) ratio, according to either (2.193) or (3.89). Denoting by α an order of magnitude of kinetic coefficients which is associated with the dissipative forces (refer to Section 2.4), the former term varies in L^2/T^3, whereas the latter varies like $\alpha U^2 \propto \alpha L^2/T^2$, as set out in Chapter 2 (eqn (2.185)). Thus, we get the general relation

$$T \propto \frac{1}{\alpha} \tag{3.181}$$

For the Navier–Stokes equations, equation (3.128) then makes it impossible to write

$$T \propto \frac{L^2}{\nu} \tag{3.182}$$

Thus, the time required for the diffusion of momentum over a given characteristic distance varies as the square of the latter. According to that formula, a higher viscosity will reduce time T, enhancing the homogeneity of the velocity field.

3.4.4 Pressure and incompressibility

Although it is only an approximated reality, the model of a strictly incompressible fluid can be used for advantageously treating the hardly compressible flows, that is with a quite small Mach number (refer to Section 3.4.3). The system (3.126) is rearranged, with the continuity equation (3.47), as:

$$\begin{aligned} &\frac{\partial \mathbf{u}}{\partial t} + \mathbf{grad}\, \mathbf{u} \cdot \mathbf{u} = -\frac{1}{\rho}\mathbf{grad}\; p + \nu \nabla^2 \mathbf{u} + \mathbf{g} \\ &\mathrm{div}\mathbf{u} = 0 \end{aligned} \tag{3.183}$$

A difficulty then arises, since the continuity equation now only includes the velocity $\mathbf{u}$, which makes the solution techniques as described in Section 3.4.2 non-applicable. Truly speaking, ρ is then no longer an unknown, but the state equation therefore becomes null and void and pressure should be calculated otherwise. Let us assume, for illustration purposes, that we want to solve the equation of motion consisting of the first line of (3.183) by means of an explicit first order temporal numerical scheme.[60] For example, for each fluid particle, we will get an approximate acceleration (left-hand term, according to (3.8)) at time m through

[60] This issue will be more thoroughly dealt with in Chapter 5.

$$\left(\frac{d\mathbf{u}}{dt}\right)^m \approx \frac{\mathbf{u}^{m+1} - \mathbf{u}^m}{\delta t} \tag{3.184}$$

where $\mathbf{u}^{m+1}$ denotes velocity at the next time point and δt a sufficiently small temporal increment (the time step). Hence, (3.183) allows us to assess the velocity at time $m+1$ as a function of the various quantities at the preceding time point through

$$\mathbf{u}^{m+1} = \mathbf{u}^m + \left[\left(-\frac{1}{\rho}\mathbf{grad}\; p + \nu \nabla^2 \mathbf{u}\right)^m + \mathbf{g}\right]\delta t \tag{3.185}$$

Unfortunately, p cannot be calculated by means of that scheme, the continuity equation $\mathrm{div}\mathbf{u} = 0$ being useless within that context.

We will briefly set down the theoretical basis of a method which is often used for remedying this problem. To that purpose, let us apply the operator div to the equation of motion, in which acceleration is written in the form (3.46); $\mathbf{g}$ being constant, we find

$$\frac{\partial \mathrm{div}\mathbf{u}}{\partial t} + \mathrm{div}\left[\mathbf{div}\left(\mathbf{u} \otimes \mathbf{u}\right)\right] = -\frac{1}{\rho}\nabla^2 p + \nu \nabla^2 \mathrm{div}\mathbf{u} \tag{3.186}$$

(let us recall that the Laplace operator commutes with the divergence). The continuity equation then shows that two terms vanish, which yields[61]

[61] Equation (3.187) makes it possible to rediscover the estimation (3.157).

The presently employed procedure is analogous to what was done in Section 3.4.3 to set up the wave equation (3.163), but with a constant density and without linearization.

$$\begin{aligned}\nabla^2 p &= -\rho \mathrm{div}\left[\mathbf{div}\left(\mathbf{u}\otimes\mathbf{u}\right)\right] \\ &= -\rho\frac{\partial^2 u_i u_j}{\partial x_i \partial x_j}\end{aligned} \tag{3.187}$$

(3.187) is a pressure *Poisson's equation.* Its right-hand side can be written in a more expressive form by developing the derivative, then exploiting the continuity equation:

$$\begin{aligned}\frac{\partial^2 u_i u_j}{\partial x_i \partial x_j} &= \frac{\partial}{\partial x_i}\left(\frac{\partial u_i}{\partial x_j}u_j + u_i\frac{\partial u_j}{\partial x_j}\right) \\ &= \frac{\partial}{\partial x_i}\left(\frac{\partial u_i}{\partial x_j}u_j\right) \\ &= \frac{\partial^2 u_i}{\partial x_i \partial x_j}u_j + \frac{\partial u_i}{\partial x_j}\frac{\partial u_j}{\partial x_i} \\ &= \frac{\partial u_i}{\partial x_j}\frac{\partial u_j}{\partial x_i}\end{aligned} \tag{3.188}$$

Let us now calculate the difference between the respective squares of the scalar deformation rate and vorticity, from their definitions (3.29) and (3.31), acting upon the dummy sub-scripts:

$$\begin{aligned}s^2 - \omega^2 &= 2s_{ij}s_{ij} - 2\omega_{ij}\omega_{ij} \\ &= \frac{1}{2}\left(\frac{\partial u_i}{\partial x_j} + \frac{\partial u_j}{\partial x_i}\right)\left(\frac{\partial u_i}{\partial x_j} + \frac{\partial u_j}{\partial x_i}\right) \\ &\quad - \frac{1}{2}\left(\frac{\partial u_i}{\partial x_j} - \frac{\partial u_j}{\partial x_i}\right)\left(\frac{\partial u_i}{\partial x_j} - \frac{\partial u_j}{\partial x_i}\right) \\ &= \left(\frac{\partial u_i}{\partial x_j}\frac{\partial u_i}{\partial x_j} + \frac{\partial u_j}{\partial x_i}\frac{\partial u_i}{\partial x_j}\right) - \left(\frac{\partial u_i}{\partial x_j}\frac{\partial u_i}{\partial x_j} - \frac{\partial u_j}{\partial x_i}\frac{\partial u_i}{\partial x_j}\right) \\ &= 2\frac{\partial u_i}{\partial x_j}\frac{\partial u_j}{\partial x_i}\end{aligned} \tag{3.189}$$

With (3.188) and (3.189), the Poisson's equation (3.187) is ultimately rearranged as:

$$\nabla^2 p = \rho\frac{\omega^2 - s^2}{2} \tag{3.190}$$

We know that the solution of this equation is written as (e.g. refer to Roach, 1982):

$$p\left(\mathbf{r}\right) = -\frac{\rho}{8\pi}\int_{\Omega}\frac{\left(\omega^2 - s^2\right)\left(\mathbf{r}'\right)}{\left|\mathbf{r} - \mathbf{r}'\right|}d^3\mathbf{r}' \tag{3.191}$$

We may notice that pressure therefore depends on the velocity field data throughout the flow. With the incompressible fluid model, the information about pressure then instantly propagates.[62]. Considering those calculations made in Section 3.4.3 on the role of the sound velocity c, the incompressible

[62] This is a characteristic property of the elliptical differential equations.

case formally corresponds, indeed, to an infinite sound velocity, that is to a zero Mach number. That result is consistent with the estimate (3.170) of density variations, which are zero here. The formula (3.191) can be interpreted by deriving it with respect to **r**, which gives the pressure gradient, that is the pressure force as received by the fluid particle located in point **r** (Minier and Peirano, 2001):

$$\left[-\frac{1}{\rho}\mathbf{grad}\, p\right](\mathbf{r}) = -\frac{1}{8\pi}\int_{\Omega}\frac{\left(\omega^2 - s^2\right)\left(\mathbf{r}'\right)}{|\mathbf{r}-\mathbf{r}'|^2}d^3\mathbf{r}' \tag{3.192}$$

We may notice the formal analogy of that equation with the theory of electrostatics.[63] Within that context, the quantity $\omega^2 - s^2$ plays a role analogous to the electric charge. Now, we have seen in Section 3.2.2 that ω represents the locally revolving nature of the flow. Hence we can see that the high vorticity area (which is often referred to as a *vortex*) corresponds to a positive 'charge'. A particle occurring in it will be subject to an attractive force keeping it within that area. We may then state that a vortex is stable. Equation (3.190) teaches us that such an area corresponds to $\nabla^2 p > 0$, that is to a low pressure area.

[63] For example, refer to Landau and Lifshitz, 1975).

In order to solve the system (3.183), pressure could be calculated by solving the Poisson equation (3.190), that is directly applying (3.191). We will see that using a slightly different approach is attractive. Through this method, combined with a numerical scheme, an attractive solution of the problem can be found while clarifying the mathematical role of pressure in the Navier–Stokes equations (Chorin, 1968; Temam, 1968).

To that purpose, let us consider a fluid volume Ω. Let us then consider the set[64] of vector fields **u** and scalar fields p, which are sufficiently regular (in other words, continuously derivable):

[64] In this case, in particular, we must carefully avoid any confusion between the velocity vector at a point and at a given instant with the corresponding vector field, although both of them are denoted by **u**. It is the same with p.

$$\begin{aligned} E &= \left\{\mathbf{u} \in C^1\left(\Omega, \mathbb{R}^3\right)\right\} \\ F &= \left\{p \in C^1\left(\Omega, \mathbb{R}\right)\right\} \end{aligned} \tag{3.193}$$

We note that the differential operator div is defined from E towards F, and the operator **grad** from F towards E. These two vector spaces can be provided with scalar products and their related norms:

$$\begin{aligned} \langle \mathbf{u}, \mathbf{v}\rangle &\doteq \int_{\Omega}\mathbf{u}\cdot\mathbf{v}\, d\Omega \qquad \|\mathbf{u}\|_E^2 = \int_{\Omega}|\mathbf{u}|^2\, d\Omega \\ (p, q) &\doteq \int_{\Omega} pq\, d\Omega \qquad \|p\|_F^2 = \int_{\Omega} p^2 d\Omega \end{aligned} \tag{3.194}$$

where $u = |\mathbf{u}|$ is the modulus of vector **u**. Let us assume that the boundary $\partial\Omega$ of Ω is split up into two subsets $\partial\Omega_u$ and $\partial\Omega_p$ on which we have respective conditions of friction or sliding (wall) and prescribed pressure (free surface), according to equations (3.145)[65] and (3.134). We then consider the following subspaces of E and F:

[65] We now consider the case of a free sliding condition, although a viscous term is used in the following, in order to encompass the case of 'perfect fluid'-type wall conditions. The strongest condition $\mathbf{u} = \mathbf{0}$ would not change our argument.

$$\begin{aligned} E' &= \left\{\mathbf{u} \in E, (\mathbf{u}\cdot\mathbf{n}_{wall})_{\partial\Omega_u} = 0\right\} \\ F' &= \left\{p \in F, p_{\partial\Omega_p} = 0\right\} \end{aligned} \tag{3.195}$$

For a pair ($\mathbf{u}$, p) belonging to $E' \times F'$, let us now contemplate the following calculation:

$$\int_\Omega \text{div}\,(p\mathbf{u})\,d\Omega = \int_\Omega p\text{div}\mathbf{u}\,d\Omega + \int_\Omega (\mathbf{grad}\ p)\cdot\mathbf{u}\,d\Omega \\ = (p, \text{div}\mathbf{u}) + \langle \mathbf{grad}\ p, \mathbf{u}\rangle \tag{3.196}$$

Besides, the Gauss formula and the boundary conditions (3.195) yield

$$\int_\Omega \text{div}\,(p\mathbf{u})\,d\Omega = \oint_{\partial\Omega} p\mathbf{u}\cdot\mathbf{n}_{wall}d\Gamma = 0 \tag{3.197}$$

(3.196) then gives the following simple expression:

$$(p, \text{div}\mathbf{u}) = -\,\langle \mathbf{grad}\ p, \mathbf{u}\rangle \tag{3.198}$$

which means that the operators div and $-$**grad** are *adjoint* in the scalar product sense (3.194) (the div and **grad** operators are also said to be *skew-adjoint*). As a significant consequence, the kernel of the former is orthogonal to the image of the latter, and so they are in direct sum. Thus, every vector $\mathbf{u}$ field can be uniquely decomposed as follows:

$$\forall\mathbf{u}\in E',\ \exists!\left(\mathbf{u}^d, \mathbf{u}^g\right)\in \ker(\text{div})\times \text{Im}\,(\mathbf{grad}),\ \ \mathbf{u} = \mathbf{u}^d + \mathbf{u}^g \tag{3.199}$$

which may also be written as

$$\forall\mathbf{u}\in E',\ \exists!\left(\mathbf{u}^d, q\right)\in \ker(\text{div})\times F,\ \begin{cases}\mathbf{u} = \mathbf{u}^d + \mathbf{grad}\ q\\ \text{div}\mathbf{u}^d = 0\end{cases} \tag{3.200}$$

Thus, every velocity field is written as the unique sum of an incompressible field and the gradient of a scalar field (*Helmholtz–Hodge decomposition*). Applying the div operator to (3.200), we get

$$\text{div}\mathbf{u} = \nabla^2 q \tag{3.201}$$

or else

$$\mathbf{u}^d = \mathbf{u} - \mathbf{grad}\left[\left(\nabla^2\right)^{-1}\text{div}\mathbf{u}\right] \tag{3.202}$$

where $\left(\nabla^2\right)^{-1}$ is the inverse of the Laplace operator. The preceding relation shows that $\mathbf{I}_3-\mathbf{grad}\left[\left(\nabla^2\right)^{-1}\text{div}\right]$ is the operator of projection onto the ker (div) subspace of E, and can be used to get $\mathbf{u}^d$. Lastly, the orthogonality of ker (div) and Im (**grad**) allows us to invoke the Pythagorean theorem and write

$$\|\mathbf{u}\|_E^2 = \left\|\mathbf{u}^d\right\|_E^2 + \|\mathbf{grad}\ q\|_E^2 \tag{3.203}$$

Furthermore, we know that $\mathbf{u}^d$ is the vector[66] which minimizes the norm $\|\mathbf{grad}\ q\|_E$, that is (with (3.194)) which yields a minimum value of the difference

$$\int_\Omega |\mathbf{u}|^2\,d\Omega - \int_\Omega \left|\mathbf{u}^d\right|^2 d\Omega = \frac{2}{\rho}\left[E_k(\mathbf{u}) - E_k\left(\mathbf{u}^d\right)\right] \tag{3.204}$$

[66] In the 'vector field, element of E' sense, in accordance with the above remark.

Thus, solving the Poisson equation (3.201) and applying (3.202) makes it possible to determine the incompressible velocity field $\mathbf{u}^d$ as close as possible to $\mathbf{u}$ in the kinetic energy sense.

Let us now assume that we want to solve the equation of motion (3.183) using a more sophisticated numerical scheme than (3.185), comprising two steps. The first step consists in only deriving an estimate $\mathbf{u}^{m*}$ of the velocity[67] from the viscosity and gravity forces:

$$\begin{aligned} \mathbf{u}^{m*} &= \mathbf{u}^m + \left[\nu\left(\nabla^2\mathbf{u}\right)^m + \mathbf{g}\right]\delta t \\ \mathbf{u}^{m+1} &= \mathbf{u}^{m*} - \frac{1}{\rho}\left(\mathbf{grad}\ p\right)^{m+1}\delta t \end{aligned} \tag{3.205}$$

(we have used the fact that ρ is constant). The second step is implicit, since pressure is considered at the final time point $m+1$ and is then still to be calculated. Now, during the process (3.205), we want, starting from an incompressible initial field $\mathbf{u}^m$, to come to a final field $\mathbf{u}^{m+1}$ satisfying the same property. Applying the div operator to the second equation of (3.205), we then find:

$$\left(\nabla^2 p\right)^{m+1} = \frac{\rho}{\delta t}\mathrm{div}\mathbf{u}^{m*} \tag{3.206}$$

Thus, pressure p^{m+1} still obeys a Poisson equation, whose right-hand side is now known via the first step.[68] The second equation of (3.205) actually occurs in the same form as (3.202), $\mathbf{u}^{m+1}$ playing the role of $\mathbf{u}^d$ being projected onto the space of incompressible fields from the estimated velocity field $\mathbf{u}^{m*}$, which plays the role of $\mathbf{u}$. As such, the analogy may be pursued by likening q to $p^{m+1}\delta t/\rho$. It can be seen that pressure is the quantity which makes the velocity field incompressible. Once again, we find the link between pressure and the notion of incompressibility, as already briefly explained in Section 2.3.3 and 3.4.1.

Since the ultimate velocity vector should satisfy the condition (3.195), that is $\mathbf{u}^{m+1}\cdot\mathbf{n}_{wall} = 0$, taking the scalar product of the second equation in (3.205) with the vector normal to wall $\mathbf{n}$, we get

$$\left(\frac{\partial p}{\partial \mathbf{n}_{wall}}\right)^{m+1} = \frac{\rho}{\delta t}\mathbf{u}^{m*}\cdot\mathbf{n}_{wall} \tag{3.207}$$

which may be compared to (3.150). This relation is the wall boundary condition for pressure with this suggested scheme. We will profitably use the considerations of this section in Section 6.2.5, within the context of the SPH numerical method.

[67] Here we avoid the common notation $\mathbf{u}^{m+\frac{1}{2}}$, for reasons which will become clear in Section 6.2.5.

[68] Applying the divergence operator to the first equation of (3.205) would be useless since, within the context of a numerical method, the discrete operators div and ∇^2 do not necessarily commute any longer (refer to Chapters 5 and 6).

3.4.5 Surface tension

When two immiscible fluids (such as water and air) are involved in one flow, one may wonder how their interfaces will evolve. If pressure varies by a deviation Δp on either side of a boundary between two fluids, the latter is not in a thermodynamical equilibrium; it will therefore be the place where so-called *surface tension* forces will occur, as determined by the arrangement of

Table 3.3 Value of the surface tension coefficient between water and air for several temperature values.

T (°C)	0	20	37
β (Pa.m)	7.6×10^{-2}	7.3×10^{-2}	7.0×10^{-2}

molecules adjacent to that boundary. Just as the work of the pressure forces is given by the relation (1.260), that is $dE_{int} = -\Delta p dV$, the surface force work will obviously be proportional to the variation of the interface area. If one point of the contact surface Σ is moved by an infinitesimal vector $d\mathbf{r}$, the surface area A of the surface Σ varies by an amount dA. For obvious isotropy reasons, we then may think that the potential energy (per unit area) varies by an amount βdA, where β is referred to as the *surface tension coefficient* relevant to both fluids, and is measured in Pa.m. Whereas energy decreases as the volume increases, it seems obvious that it increases when the surface is stretched, as for a spring; thus, β is positive.[69] At rest, the above discussion allows us to write the following thermodynamical equilibrium:

[69] Through a Lagrangian formulation of the surface tension forces, based on Chapter 1 developments, this result could be achieved from the least action principle, using a procedure analogous to the way we could state that mass is positive in Section 1.3.1.

$$-\Delta p dV + \beta dA = 0 \tag{3.208}$$

Definition (3.208) shows that β is thermodynamical conjugate to surface area A, according to the definition given in Section 1.5.4. For thermodynamical reasons, it depends, like viscosity, on temperature (refer to Section 3.4.1). For an interface between water and air, Table 3.3 provides several values of that coefficient.

We henceforth will assume that β is constant, which is justified if the temperature variations are ignored. Integrating all over the contact surface, we will then write the potential of those forces being sought:[70]

[70] Thus, they are conservative. The superscript ST refers, of course, to surface tension.

$$E_p^{ST} = \beta \int_{\Sigma} dA \tag{3.209}$$

Let us now assume that the surface Σ is small enough for the various geometrical and thermodynamical quantities making it up to be taken as constant. According to the discussion in Section 1.3.3, the force corresponding to the potential (3.209) is then obtained by deriving (3.209) with respect to the position of the point being considered, that is $\mathbf{F}^{ST} = \partial E_p^{ST} / \partial \mathbf{r}$. If the force for unit area is denoted as $\mathbf{f}^{ST} = \mathbf{F}^{ST} / A$, then we have

$$\mathbf{f}^{ST} = -\frac{\beta}{A} \frac{\partial A}{\partial \mathbf{r}} \tag{3.210}$$

In order to calculate the derivative of the surface area A with respect to the position $\mathbf{r}$ of one of its points, we employ differential geometry. We consider the vector basis which is locally defined by the main curvature directions of Σ, which are oriented by the unit tangent vectors $\mathbf{t}_1$ and $\mathbf{t}_2$, the unit vector normal to the interface being denoted as $\mathbf{n}^I$, as in Fig. 3.3. Two main local curvatures $\kappa_1$et κ_2 can then be defined by the following differential geometrical relations:

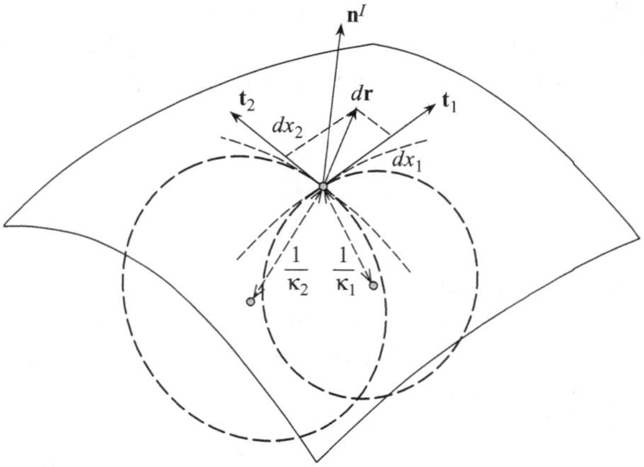

Fig. 3.3 Local coordinate system for calculating the surface tension forces.

$$
\begin{aligned}
d\mathbf{t}_1 &= \kappa_1 \mathbf{n}^I dx_1 \\
d\mathbf{t}_2 &= \kappa_2 \mathbf{n}^I dx_2
\end{aligned}
\tag{3.211}
$$

In Fig. 3.3, the circle of radius $1/\kappa_1$ (respectively $1/\kappa_2$) is located within the plane determined by vectors $\mathbf{n}^I$ and $\mathbf{t}_1$ (respectively $\mathbf{t}_2$). The curvatures, which are reckoned in m^{-1}, also verify the relation

$$
d\mathbf{n}^I = -\kappa_1 \mathbf{t}_1 dx_1 - \kappa_2 \mathbf{t}_2 dx_2 \tag{3.212}
$$

$dx_1 \doteqdot d\mathbf{r} \cdot \mathbf{t}_1$ and $dx_2 \doteqdot d\mathbf{r} \cdot \mathbf{t}_2$ being the components of $d\mathbf{r}$ in the tangential plane. When the tangent vectors vary by $d\mathbf{t}_1$ and $d\mathbf{t}_2$, the corresponding relative variation of surface area, that is dA/A, is proportional to their product. With (3.211), we obtain $dA/A = \kappa \mathbf{n}^I \cdot d\mathbf{r}$, where $\kappa \doteqdot \kappa_1 + \kappa_2$ is the local curvature of the contact surface. Thus, we get $\partial A/\partial \mathbf{r} = A\kappa \mathbf{n}^I$. Besides, owing to the relation (3.212), we may write $\operatorname{div}\mathbf{n}^I = -\kappa_1 - \kappa_2 = -\kappa$, which ultimately yields, with (3.210), two formulations of the surface tension forces per unit area:

$$
\begin{aligned}
\mathbf{f}^{ST} &= \beta \kappa \mathbf{n}^I \\
&= -\beta \left(\operatorname{div}\mathbf{n}^I\right) \mathbf{n}^I
\end{aligned}
\tag{3.213}
$$

The equilibrium (3.208) then gives the pressure difference on either side of the interface when at rest:

$$
\Delta p = \beta \kappa \tag{3.214}
$$

a formula which is usually credited to Laplace. The surface tension forces are then locally normal to the interface. A curvature of the interface results in a pressure difference which is higher when the curvature is larger.

It may prove useful to seek the stress tensor $\boldsymbol{\sigma}^{ST}$ associated with these forces. To do this, we may note that the molecular processes which induce them are statistically isotropic in the tangential plane, and so only the normal vector $\mathbf{n}^I$ can be involved in the formulation of $\boldsymbol{\sigma}^{ST}$. Thus, the only second-order tensors which are invariant under rotation about $\mathbf{n}^I$ and can be constructed from the problem data are $\mathbf{I}_3$ and $\mathbf{n}^I \otimes \mathbf{n}^I$, and $\boldsymbol{\sigma}^{ST}$ should then be sought as a linear combination of these two tensors. It can quickly be noticed, by means of the above geometrical considerations, that the proper combination in the chosen basis is:

$$\mathbf{I}_3 - \mathbf{n}^I \otimes \mathbf{n}^I = \begin{pmatrix} 1 & 0 & 0 \\ 0 & 1 & 0 \\ 0 & 0 & 0 \end{pmatrix} \tag{3.215}$$

However, there still remains one issue: the forces being sought occur on the surface Σ, and so the expression being found would have to be multiplied by a Dirac distribution $\delta_\Sigma(\mathbf{r})$ equalling 1 on Σ and 0 elsewhere (the general definition of this distribution is provided in Appendix B). We would then be faced, however, with another problem, since this found tensor would have the same physical dimension as β, that is a length would be missing to give it the dimension of a stress. Now, no length characterizes the relevant problem, because the interface should a priori be considered as having an infinitely small thickness. In order to address these two difficulties, we introduce a (dimensionless) function C identifying the two fluids: $C = 0$ for one of them, $C = 1$ for the other. Mathematically, the gradient of this function is, of course, the distribution $\delta_\Sigma(\mathbf{r})$. It should then be considered that the interface is 'fuzzy', so that C can still be approximated through a derivable function having a very steep gradient near Σ. We then notice that the gradient of C provides the orientation of the local normal to the interface:

$$\mathbf{n}^I = \frac{\mathbf{grad}\, C}{|\mathbf{grad}\, C|} \tag{3.216}$$

The norm of the gradient of C then takes on the form of the inverse of a local characteristic length which represents an order of magnitude of the interface thickness and can be likened to the distribution δ_Σ:

$$\delta_\Sigma(\mathbf{r}) = |\mathbf{grad}\, C| \tag{3.217}$$

We then will write, in accordance with (3.215):

$$\boldsymbol{\sigma}^{ST} = \frac{\beta}{|\mathbf{grad}\, C|}\left(|\mathbf{grad}\, C|^2\, \mathbf{I}_3 - \mathbf{grad}\, C \otimes \mathbf{grad}\, C\right) \tag{3.218}$$

or else, in an indicial form:

$$\sigma_{ij}^{ST} = \frac{\beta}{\sqrt{\dfrac{\partial C}{\partial x_k}\dfrac{\partial C}{\partial x_k}}}\left(\frac{\partial C}{\partial x_k}\frac{\partial C}{\partial x_k}\delta_{ij} - \frac{\partial C}{\partial x_i}\frac{\partial C}{\partial x_j}\right) \tag{3.219}$$

As such, the tensor $\boldsymbol{\sigma}^{ST}$ vanishes outside of the interface, since the characteristic function C is constant (piecewise) in it. As expected, $\boldsymbol{\sigma}^{ST}$ is symmetric too (refer to Section 3.3.2).

The above formalism is advantageous in that the surface tension forces are expressed per unit volume, as explained henceforth. We will first note that every function B may be written in two integral forms per unit area and volume, owing to the distribution (3.217):

$$\int_{\Sigma} B dA = \int_{\Omega} B\delta_{\Sigma}(\mathbf{r})\, d^3\mathbf{r} = \int_{\Omega} B\,|\mathbf{grad}\ C|\, d^3\mathbf{r} \tag{3.220}$$

For $B = 1$, we get the surface area of Σ:

$$\int_{\Sigma} dA = \int_{\Omega} |\mathbf{grad}\ C|\, d^3\mathbf{r} \tag{3.221}$$

Thus, the potential (3.209) is rearranged in the per unit volume form

$$E_p^{ST} = \beta \int_{\Omega} |\mathbf{grad}\ C|\, d^3\mathbf{r} \tag{3.222}$$

The total surface tension force may also be considered by integrating the force per unit area $\mathbf{f}^{ST}$ and utilizing (3.220):

$$\int_{\Sigma} \mathbf{f}^{ST} dA = \int_{\Omega} \mathbf{f}^{ST}\, |\mathbf{grad}\ C|\, d^3\mathbf{r} = \int_{\Omega} \mathbf{F}^{ST} \rho d^3\mathbf{r} \tag{3.223}$$

where $\mathbf{F}^{ST}$ is a surface tension force per unit mass. Likening the quantities under the last two integrals and utilizing (3.213) and (3.216), we find:

$$\mathbf{F}^{ST} = \frac{\beta\kappa}{\rho}\mathbf{grad}\ C = -\frac{\beta}{\rho}\left(\mathrm{div}\mathbf{n}^I\right)\mathbf{grad}\ C \tag{3.224}$$

Mathematically, this force is first-order infinitely great, since it is reckoned per unit volume rather than per unit area; this is consistent with the fact that the interface, in reality, is three-dimensional but has a very small thickness.

The surface tension forces, because they appear on thin membranes and result from molecular processes, have moderate amplitudes. As for the other forces, the order of magnitude of their ratio to the inertia forces can be assessed (refer to Section 3.4.3). We then may write, in an analogous manner to (3.156):

$$\frac{|\rho\mathbf{grad}\ \mathbf{u}\cdot\mathbf{u}|}{|\beta\kappa\mathbf{grad}\ C|} \sim \frac{\rho U^2 L}{\beta} \doteqdot We \tag{3.225}$$

Because of the low values of β for the usual fluids, the Weber number We takes very high values, which often allows us to ignore the surface tension forces. As we did for the viscous forces, it is possible to define, from the relation (3.225), a length scale beyond which the latter become higher:

$$L_\beta \doteqdot \frac{\beta}{\rho U^2} \tag{3.226}$$

it is usually one millimetre or so. The Weber number is then also written as

$$We = \frac{L}{L_\beta} \tag{3.227}$$

with can be compared with (3.161).

All the above formulas may easily be taken over an arbitrary number of fluids by introducing as many characteristic functions as the number of fluids, for instance.[71]

[71]We assume that the various phases are not subject to any chemical reaction, in accordance with a remark made in Section 2.3.2.

3.5 From Boltzmann to the continua

We show here how to go from the Boltzmann equation to the Navier–Stokes equations. First, we investigate the time evolution of Boltzmann-averaged quantities, in order to build a general form of balance integral equations. By this method, we recover the continuity and momentum equations of a continuous medium, and prove the symmetry of the Cauchy stress tensor. We also find again the state law of perfect fluids and the behaviour law of viscous fluids.

3.5.1 Evolution of statistical averages

At the beginning of this chapter, we said that the quantities regarding a point in a continuous medium can be obtained from an average in the sense of the Boltzmann definition, in accordance with the formalism applied in Chapter 2. Based on this idea, we may think that such defined quantities (velocity, density, etc.) vary in accordance with equations which could be derived from the Boltzmann equation as discussed in Section 2.2.2. So far, in this chapter, we have preferentially not applied that viewpoint, but instead followed a 'balances' type approach to material volumes of continuous medium. We are going to explain the way that Boltzmann's approach can lead to the same equations, which will make it possible to provide further details for some issues.

As mentioned in Section 2.2.1, the mean quantities for a mesoscopic particle in a fluid are strictly defined, at every point in space and time, from the probability density associated with an individual molecule. We may recall the definition for an arbitrary quantity A (now explicitly writing all the time, space and momentum dependencies, even though we will subsequently sometimes drop them):[72]

[72]In this and the next two section, the position of a molecule $\tilde{\mathbf{r}}$ and the space variable $\mathbf{r}$ for the discussion of continuous media are one and the same.

$$\langle A \rangle (\mathbf{r}, t) = \frac{1}{n(\mathbf{r}, t)} \int_{\Omega_p} A\left(\tilde{\mathbf{p}}, \mathbf{r}, t\right) \varphi\left(\tilde{\mathbf{p}}, \mathbf{r}, t\right) d\tilde{\mathbf{p}} \tag{3.228}$$

where $n(\mathbf{r}, t)$ represents the mean number of molecules occurring within a $\mathbf{r}$-centred infinitesimal volume, as reckoned per unit volume:

$$n(\mathbf{r}, t) = \int_{\Omega_p} \varphi\left(\tilde{\mathbf{p}}, \mathbf{r}, t\right) d\tilde{\mathbf{p}} \tag{3.229}$$

whereas Ω_p denotes the space of possible values of a molecule's momentum $\tilde{\mathbf{p}}$, which is likened to $\mathbb{R}^3$. When multiplying $n(\mathbf{r}, t)$ by the mass of a molecule, the fluid density is obtained in the form:[73]

[73]If the fluid consists of several molecule species, then the mass m (variable) should remain within the integral.

$$\rho(\mathbf{r}, t) = mn(\mathbf{r}, t) = m \int_{\Omega_p} \varphi\left(\tilde{\mathbf{p}}, \mathbf{r}, t\right) d\tilde{\mathbf{p}} \tag{3.230}$$

Likewise, we will write the velocity field as:

$$\mathbf{u}(\mathbf{r},t) = \langle \tilde{\mathbf{u}} \rangle (\mathbf{r},t) = \frac{1}{n(\mathbf{r},t)} \int_{\Omega_p} \tilde{\mathbf{u}} \varphi (\tilde{\mathbf{p}}, \mathbf{r}, t)\, d\tilde{\mathbf{p}} \tag{3.231}$$

Lastly, the fluid's kinetic energy per unit mass e_k is given by

$$\begin{aligned} e_k(\mathbf{r},t) &= \frac{\langle \tilde{E}_k \rangle (\mathbf{r},t)\, n(\mathbf{r},t)}{\rho(\mathbf{r},t)} \\ &= \frac{1}{\rho(\mathbf{r},t)} \int_{\Omega_p} \frac{1}{2} m \left|\tilde{\mathbf{u}}\right|^2 \varphi(\tilde{\mathbf{p}},\mathbf{r},t)\, d\tilde{\mathbf{p}} \end{aligned} \tag{3.232}$$

The statistical averaging operator as defined this way has several properties which we are going to set out. First of all, the average is obviously additive. Besides, since the integral only regards the momentum, we find, for every quantity independent of momentum $\tilde{\mathbf{p}}$

$$\langle A(\mathbf{r},t) \rangle = A(\mathbf{r},t) \tag{3.233}$$

In particular, it is left unchanged by the average of an already averaged quantity:

$$\begin{aligned} \langle \langle A \rangle \rangle &= \langle A \rangle \\ \langle \langle A \rangle B \rangle &= \langle A \rangle \langle B \rangle \end{aligned} \tag{3.234}$$

We will sometimes note $A' \doteqdot A - \langle A \rangle$, with the obvious property:

$$\langle A' \rangle = 0 \tag{3.235}$$

For the same reason, the partial derivatives with respect to both time and coordinates $\mathbf{r}$ commute with the integration symbol:

$$\begin{aligned} \int_{\Omega_p} \frac{\partial A}{\partial t} d\tilde{\mathbf{p}} &= \frac{\partial}{\partial t} \int_{\Omega_p} A d\tilde{\mathbf{p}} \\ \int_{\Omega_p} \frac{\partial A}{\partial \mathbf{r}} d\tilde{\mathbf{p}} &= \frac{\partial}{\partial \mathbf{r}} \int_{\Omega_p} A d\tilde{\mathbf{p}} \end{aligned} \tag{3.236}$$

We will now utilize the Boltzmann equation (2.27) to determine the variation of the mean quantities. Let us first write this equation as

$$\frac{\partial \varphi}{\partial t} + \frac{\partial \varphi \tilde{u}_i}{\partial x_i} + \frac{\partial \varphi \tilde{F}_i}{\partial \tilde{p}_i} = C(\varphi) \tag{3.237}$$

(here we use the repeated-index summation convention, the vector $\mathbf{r}$ having the x_i as a component). The above formulation is justified by the fact that the partial derivative with respect to the position $\mathbf{r}$ is achieved with a fixed $\tilde{\mathbf{u}}$ in (2.27). We may also recall that the force $\tilde{\mathbf{F}}$ refers to the external fields, which do not depend on the velocities (i.e. on the momenta). Multiplying by any field A, then applying the statistical average, we find

$$\int_{\Omega_p} A \frac{\partial \varphi}{\partial t} d\tilde{\mathbf{p}} + \int_{\Omega_p} A \frac{\partial \varphi \tilde{u}_i}{\partial \tilde{x}_i} d\tilde{\mathbf{p}} + \int_{\Omega_p} A \frac{\partial \varphi \tilde{F}_i}{\partial \tilde{p}_i} d\tilde{\mathbf{p}} = \int_{\Omega_p} A C(\varphi)\, d\tilde{\mathbf{p}} \tag{3.238}$$

An important remark should be made. The conservative quantities of mechanics (mass, momentum, energy, for example) are not affected by the collision integral $C(\varphi)$ (as yielded by eqn (2.28)), because the collision processes satisfy the total conservation of these quantities,[74] as explained in Section 2.3.1. It is the same with the number of involved molecules. Hence, for $A = m$, $\tilde{\mathbf{p}}$ or $\tilde{E}_k$, the right-hand side of (3.238) will be identically zero:

[74] Except for energy if the collisions are not elastic.

$$\int_{\Omega_p} A\frac{\partial \varphi}{\partial t} d\tilde{\mathbf{p}} + \int_{\Omega_p} A\frac{\partial \varphi \tilde{u}_i}{\partial \tilde{x}_i} d\tilde{\mathbf{p}} + \int_{\Omega_p} A\frac{\partial \varphi \tilde{F}_i}{\partial \tilde{p}_i} d\tilde{\mathbf{p}} = 0 \tag{3.239}$$

For $A \equiv m$, the equation (3.239) readily yields, using the properties (3.236) and the definitions (3.230) and (3.231):

$$\frac{\partial \rho}{\partial t} + \frac{\partial \rho u_i}{\partial x_i} + m\int_{\Omega_p} \frac{\partial \varphi \tilde{F}_i}{\partial \tilde{p}_i} d\tilde{\mathbf{p}} = 0 \tag{3.240}$$

Since the integrand is a divergence, the integral is reduced to the 'boundary' of Ω_p, where φ vanishes (refer to Chapter 2). Thus, we find once again

$$\frac{\partial \rho}{\partial t} + \operatorname{div}(\rho \mathbf{u}) = 0 \tag{3.241}$$

that is the continuity equation (3.42) of a variable density continuous medium. In the following sections of this chapter, we will apply the preceding ideas to the main equations being dealt with in this chapter.

3.5.2 Balances and fluxes through Boltzmann

Putting $A \equiv \tilde{\mathbf{p}} = m\tilde{\mathbf{u}}$ in (3.239), we get

$$m\int_{\Omega_p} \frac{\partial \varphi}{\partial t}\tilde{\mathbf{u}} d\tilde{\mathbf{p}} + m\int_{\Omega_p} \frac{\partial \varphi \tilde{u}_i}{\partial \tilde{x}_i}\tilde{\mathbf{u}} d\tilde{\mathbf{p}} + m\int_{\Omega_p} \frac{\partial \varphi \tilde{F}_i}{\partial \tilde{p}_i}\tilde{\mathbf{u}} d\tilde{\mathbf{p}} = 0 \tag{3.242}$$

Since the partial derivatives with respect to time and space are achieved with a fixed velocity, we may also write, using (3.231), (3.236) and (3.230):

$$\frac{\partial \rho \mathbf{u}}{\partial t} + m\frac{\partial}{\partial x_i}\int_{\Omega_p} \tilde{u}_i \tilde{\mathbf{u}} \varphi d\tilde{\mathbf{p}} + m\int_{\Omega_p} \frac{\partial \varphi \tilde{F}_i}{\partial \tilde{p}_i}\tilde{\mathbf{u}} d\tilde{\mathbf{p}} = 0 \tag{3.243}$$

In order to treat the central integral of that equation, we have to introduce a new notation. The molecules have velocities which differ from the average local velocity $\mathbf{u}$; we then will put $\tilde{\mathbf{u}}' \doteqdot \tilde{\mathbf{u}} - \langle \tilde{\mathbf{u}} \rangle = \tilde{\mathbf{u}} - \mathbf{u}$ ($\tilde{\mathbf{u}}'$ stands for the 'random' part of the molecule velocity). We get

$$\begin{aligned}\int_{\Omega_p} \tilde{u}_i \tilde{\mathbf{u}} \varphi d\tilde{\mathbf{p}} &= n\left\langle u_i \mathbf{u} + \tilde{u}_i' \mathbf{u} + u_i \tilde{\mathbf{u}}' + \tilde{u}_i' \tilde{\mathbf{u}}' \right\rangle \\ &= n\left(u_i \mathbf{u} + \left\langle \tilde{u}_i' \right\rangle \mathbf{u} + u_i \left\langle \tilde{\mathbf{u}}' \right\rangle + \left\langle u_i' \tilde{\mathbf{u}}' \right\rangle\right) \\ &= n\left(u_i \mathbf{u} + \left\langle u_i' \tilde{\mathbf{u}}' \right\rangle\right)\end{aligned} \tag{3.244}$$

(we have used (3.234) and (3.235)). Then we have

$$m\frac{\partial}{\partial x_i}\int_{\Omega_p}\tilde{u}_i\tilde{\mathbf{u}}\varphi d\tilde{\mathbf{p}} = \frac{\partial \rho u_i \mathbf{u}}{\partial x_i} + \frac{\partial \rho\left\langle \tilde{u}'_i\tilde{\mathbf{u}}'\right\rangle}{\partial x_i}$$
$$= \mathbf{div}\left(\rho\mathbf{u}\otimes\mathbf{u}\right) + \mathbf{div}\left(\rho\left\langle\tilde{\mathbf{u}}'\otimes\tilde{\mathbf{u}}'\right\rangle\right) \tag{3.245}$$

Lastly, let us calculate the last integral of (3.243) using the rule of derivation of a product:

$$m\int_{\Omega_p}\frac{\partial\varphi\tilde{F}_i}{\partial\tilde{p}_i}\tilde{\mathbf{u}}d\tilde{\mathbf{p}} = m\int_{\Omega_p}\frac{\partial\varphi\tilde{F}_i\tilde{\mathbf{u}}}{\partial\tilde{p}_i}d\tilde{\mathbf{p}} - \int_{\Omega_p}\varphi\tilde{F}_i\frac{\partial\tilde{\mathbf{p}}}{\partial\tilde{p}_i}d\tilde{\mathbf{p}} \tag{3.246}$$

The first integral of the right hand side vanishes pursuant to Gauss' theorem, whereas $\partial\tilde{\mathbf{p}}/\partial\tilde{p}_i = \mathbf{e}_i$, the i-th unit vector of the work base. We also can turn into a good account the fact that the external force does not depend on the momentum in order to take it out of the integral. From now on, we will write it as $\mathbf{F}$, since it depends on the position $\mathbf{r}\equiv\tilde{\mathbf{r}}$. Thus, using (3.229) and (3.230):

$$m\int_{\Omega_p}\frac{\partial\varphi\tilde{F}_i}{\partial\tilde{p}_i}\tilde{\mathbf{u}}d\tilde{\mathbf{p}} = -\left(\int_{\Omega_p}\varphi d\tilde{\mathbf{p}}\right)\mathbf{F}$$
$$= -\rho\frac{\mathbf{F}}{m} \tag{3.247}$$

Lastly, combining (3.243), (3.245) and (3.247), we get

$$\frac{\partial\rho\mathbf{u}}{\partial t} + \mathbf{div}\left(\rho\mathbf{u}\otimes\mathbf{u}\right) = -\mathbf{div}\left(\rho\left\langle\tilde{\mathbf{u}}'\otimes\tilde{\mathbf{u}}'\right\rangle\right) + \rho\frac{\mathbf{F}}{m} \tag{3.248}$$

Using (3.45) and taking the gravity $\mathbf{F} = m\mathbf{g}$ as external forces, we ultimately find:

$$\frac{d\mathbf{u}}{dt} = -\frac{1}{\rho}\mathbf{div}\left(\rho\left\langle\tilde{\mathbf{u}}'\otimes\tilde{\mathbf{u}}'\right\rangle\right) + \mathbf{g} \tag{3.249}$$

This is the Cauchy equation (3.71), provided that we put

$$\boldsymbol{\sigma} = -\rho\left\langle\tilde{\mathbf{u}}'\otimes\tilde{\mathbf{u}}'\right\rangle \tag{3.250}$$

We notice that the stress tensor is written as the statistical autocorrelation tensor of the velocity fluctuations as exhibited by the molecules with respect to the continuous velocity field. This is unsurprising; if all the molecules moved at the same velocity within a fluid particle, no internal force would be found in it. We may, however, surprisingly and interestingly note that the possible intermolecular impacts do not take any explicit part in the stresses. Their influence, however, is implicit, since they account for the velocity fluctuations $\tilde{\mathbf{u}}'$. The components of this tensor equal $\sigma_{ij} = -\rho\left\langle\tilde{u}'_i\tilde{u}'_j\right\rangle$ and, as was well established, are a priori unknown. The definition (3.250) shows that the tensor $\boldsymbol{\sigma}$ is inherently symmetrical (with the presently adopted construction), in accordance with what we explained in Section 3.3.2 based on the conservation of the angular momentum (eqns (3.77) to (3.79)). In addition, this definition highlights the invariant nature of the stresses through a Galilean transformation, as explained in Section 3.3.2.

Since the case of a fluid is being considered, we can explicitly calculate the diagonal values of $\boldsymbol{\sigma}$. The isotropic nature of the laws of mechanics provides the fluctuating motion of molecules, under the effect of the collisions, with a feature which is independent of the direction being considered. Thus, the σ_{ii} may be related to the molecule root mean square velocity $\mathring{u}$ as defined by (2.71):[75]

[75]There is no summation sign above i in the next expression.

$$\begin{aligned}\forall i,\ \sigma_{ii} &= -\rho\left\langle \tilde{u}_i'^2 \right\rangle \\ &= -\frac{1}{3}\rho \mathring{u}^2\end{aligned} \tag{3.251}$$

In the case of the perfect gas, the equation (2.103) yields the expected relation:

$$\forall i,\ \sigma_{ii} = -p \tag{3.252}$$

according to (3.116). The preceding expression may be formulated otherwise. If we consider that applying the principles of Section 2.4.3 to the case of an isolated molecule is pertinent, we can model the effect of the collisions to which it is subject through a 'molecular friction' force, by analogy with (2.179) or (2.283):

$$\tilde{\mathbf{F}}^{diss} = -\alpha \tilde{\mathbf{u}}' \tag{3.253}$$

This force determines the derivative of momentum $d\tilde{\mathbf{p}}/dt$. We then are definitely in the same circumstances as in Section 2.4.1, with a matrix $\mathbf{A}$ (as defined by the general relation (2.119)) being given by

$$\mathbf{A} = \frac{1}{m}\boldsymbol{\alpha} \tag{3.254}$$

Formula (2.146), together with (2.147), then yields

$$\left\langle \tilde{\mathbf{u}}' \otimes \tilde{\mathbf{u}}' \right\rangle = \frac{k_B T}{m}\mathbf{I}_3 \tag{3.255}$$

This result is consistent with (3.250) and (3.252), provided that we put

$$p = \frac{\rho k_B T}{m} \tag{3.256}$$

Thus, we get back to the state equation (2.102) of a perfect gas. The other components of the stress tensor will be estimated in the next section, in the general case of any fluid.

Let us now turn to the kinetic energy equation,[76] which is tantamount to putting $A \equiv \tilde{E}_k = \frac{1}{2}m\left|\tilde{\mathbf{u}}\right|^2$ in (3.239). After some manipulations such as those we used for establishing the Cauchy equation, we get:

[76]Treating the internal energy, for instance, would not raise further difficulties.

$$\begin{aligned}&\frac{\partial}{\partial t}\int_{\Omega_p}\frac{1}{2}m\left|\tilde{\mathbf{u}}\right|^2\varphi d\tilde{\mathbf{p}} + \frac{\partial}{\partial x_i}\int_{\Omega_p}\frac{1}{2}m\tilde{u}_i\left|\tilde{\mathbf{u}}\right|^2\varphi d\tilde{\mathbf{p}} \\ &\qquad -\frac{1}{2}mF_i\int_{\Omega_p}\frac{\partial\left|\tilde{\mathbf{u}}\right|^2}{\partial\tilde{p}_i}\varphi d\tilde{\mathbf{p}} = 0\end{aligned} \tag{3.257}$$

The first integral simply equals ρe_k, as per (3.232). The second is treated by breaking down the molecule velocity into its average and fluctuation, as previously:

$$\int_{\Omega_p} \tilde{u}_i \left|\tilde{\mathbf{u}}\right|^2 \varphi d\tilde{\mathbf{p}} = n\left(u_i\left\langle\left|\tilde{\mathbf{u}}\right|^2\right\rangle + \left\langle\tilde{u}'_i\left|\tilde{\mathbf{u}}\right|^2\right\rangle\right) \tag{3.258}$$

The first of these terms equals $2ne_k u_i$. Afterwards, writing down $\left|\tilde{\mathbf{u}}\right|^2 = |\mathbf{u}|^2 + 2\mathbf{u}\cdot\tilde{\mathbf{u}}' + \left|\tilde{\mathbf{u}}'\right|^2$, we find

$$\left\langle\tilde{u}'_i\left|\tilde{\mathbf{u}}\right|^2\right\rangle = 2\mathbf{u}\cdot\left\langle\tilde{u}'_i\tilde{\mathbf{u}}'\right\rangle + \left\langle\tilde{u}'_i\left|\tilde{\mathbf{u}}'\right|^2\right\rangle \tag{3.259}$$

Thus, we find

$$\int_{\Omega_p} \frac{1}{2}m\tilde{u}_i\left|\tilde{\mathbf{u}}\right|^2\varphi d\tilde{\mathbf{p}} = \rho e_k u_i + \rho\mathbf{u}\cdot\left\langle\tilde{u}'_i\tilde{\mathbf{u}}'\right\rangle + \frac{1}{2}\rho\left\langle\tilde{u}'_i\left|\tilde{\mathbf{u}}'\right|^2\right\rangle \tag{3.260}$$

The third integral of (3.257) yields

$$\begin{aligned} F_i\int_{\Omega_p}\frac{\partial\left|\tilde{\mathbf{u}}\right|^2}{\partial\tilde{p}_i}\varphi d\tilde{\mathbf{p}} &= \frac{F_i}{m^2}\int_{\Omega_p}\frac{\partial\tilde{p}^2}{\partial\tilde{p}_i}\varphi d\tilde{\mathbf{p}} \\ &= \frac{2}{m^2}\mathbf{F}\cdot\int_{\Omega_p}\tilde{\mathbf{p}}\varphi d\tilde{\mathbf{p}} \\ &= \frac{2}{m}n\mathbf{F}\cdot\mathbf{u} \end{aligned} \tag{3.261}$$

Combining these relations, equation (3.257) becomes

$$\begin{aligned} \frac{\partial\rho e_k}{\partial t} + \operatorname{div}\left(\rho e_k\mathbf{u}\right) &= -\operatorname{div}\left(\mathbf{u}\cdot\rho\left\langle\tilde{\mathbf{u}}'\otimes\tilde{\mathbf{u}}'\right\rangle\right) \\ &\quad - \frac{\partial}{\partial x_i}\left(\frac{1}{2}\rho\left\langle\tilde{u}'_i\left|\tilde{\mathbf{u}}'\right|^2\right\rangle\right) + n\mathbf{F}\cdot\mathbf{u} \end{aligned} \tag{3.262}$$

Then, using the equality (3.49) and $\mathbf{F} = m\mathbf{g}$, we come to

$$\frac{de_k}{dt} = \frac{1}{\rho}\operatorname{div}\left(\boldsymbol{\sigma}\cdot\mathbf{u}\right) + \mathbf{g}\cdot\mathbf{u} - \frac{1}{\rho}\frac{\partial}{\partial x_i}\left(\frac{1}{2}\rho\left\langle\tilde{u}'_i\left|\tilde{\mathbf{u}}'\right|^2\right\rangle\right) \tag{3.263}$$

which is identical to (3.89), the kinetic energy dissipation rate being given by

$$f = \frac{\partial}{\partial x_i}\left(\frac{1}{2}\rho\left\langle\tilde{u}'_i\left|\tilde{\mathbf{u}}'\right|^2\right\rangle\right) \tag{3.264}$$

The consistency of this relation with (3.85) can be ascertained through an exact (fairly lengthy) calculation which, however, is of no relevance since the kinetic energy equation stems from the Cauchy equation, as explained in Section 3.3.3.

We may lastly note that the scalar transport equation (3.60) could be determined through the same procedure. The scalar concentration may be likened to the integral, over all the possible velocities, of the probability of presence of a molecule of the substance being sought, by analogy with (3.229). If the scalar accounts for a population of small bodies carried by the fluid, the scalar diffusion term can be found again from the Fokker–Planck's equation (2.302) (refer to Gardiner, 1985).

3.5.3 Viscous stresses through Boltzmann

We know from Chapter 2 (Section 2.4.1) that the friction forces are generated by an out-of-equilibrium state. The viscous stresses can therefore not be assessed through the Boltzmann distribution of thermodynamical equilibrium (2.49). On the contrary, an approximate solution of the Boltzmann equation (2.27) should be sought in an out-of-equilibrium state, that is taking the collision integral (2.28) into account. We learnt in the previous section that the term in $\partial\varphi/\partial\tilde{\mathbf{p}}$ leads to the external forces of the Navier–Stokes equation; that is why we will drop it in the following. We then are seeking a solution of the following equation:

$$\frac{d\varphi}{dt} \doteqdot \frac{\partial\varphi}{\partial t} + \frac{\partial\varphi}{\partial\mathbf{r}}\cdot\tilde{\mathbf{u}} = C(\varphi) \tag{3.265}$$

with, as previously explained, a collisional integral as given by equation (2.28), which we provide again for the sake of convenience:

$$C(\varphi) \doteqdot \int_{\bar{\Omega}} w\left(\tilde{\mathbf{p}}, \tilde{\mathbf{p}}_\gamma; \tilde{\mathbf{p}}'_\beta, \tilde{\mathbf{p}}'_\gamma\right)\left(\varphi'_\beta\varphi'_\gamma - \varphi\varphi_\gamma\right) d\tilde{\mathbf{p}}_\gamma d\tilde{\mathbf{p}}'_\beta d\tilde{\mathbf{p}}'_\gamma \tag{3.266}$$

We will seek a solution, assuming that it is only slightly different from the equilibrium, according to those principles as set out in Section 2.4.1. Thus, we will put

$$\varphi\left(\tilde{\mathbf{p}}, \mathbf{r}, t\right) = \varphi_0\left(\tilde{\mathbf{p}}, \mathbf{r}\right) + \delta\varphi\left(\tilde{\mathbf{p}}, \mathbf{r}, t\right) \tag{3.267}$$

where φ_0 is the equilibrium distribution, as given by (2.49), but without any potential, since we ignore the external forces,that is:

$$\varphi_0\left(\tilde{\mathbf{p}}\right) \doteqdot \Theta\exp\left[-\frac{\tilde{E}_k\left(\tilde{\mathbf{p}}\right)}{k_B T}\right] \tag{3.268}$$

and with the condition

$$\delta\varphi \ll \varphi_0 \tag{3.269}$$

First of all, we notice that the distribution φ_0 no longer depends on $\mathbf{r}$ and that it verifies

$$\frac{\partial\varphi_0}{\partial\tilde{E}} = -\frac{\varphi_0\left(\tilde{\mathbf{p}}\right)}{k_B T} \tag{3.270}$$

Afterwards, we define the auxiliary distribution χ by

$$\begin{aligned}\chi\left(\tilde{\mathbf{p}}, \mathbf{r}, t\right) &\doteqdot -\frac{\delta\varphi\left(\tilde{\mathbf{p}}, \mathbf{r}, t\right)}{\partial\varphi_0/\partial\tilde{E}}\\ &= k_B T\frac{\delta\varphi\left(\tilde{\mathbf{p}}, \mathbf{r}, t\right)}{\varphi_0\left(\tilde{\mathbf{p}}\right)}\end{aligned} \tag{3.271}$$

Thus, (3.267) is also written as

$$\varphi\left(\tilde{\mathbf{p}}, \mathbf{r}, t\right) = \varphi_0\left(\tilde{\mathbf{p}}\right)\left[1 + \frac{\chi\left(\tilde{\mathbf{p}}, \mathbf{r}, t\right)}{k_B T}\right] \tag{3.272}$$

Let us initially determine the influence of $\delta\varphi$ upon the collision integral (3.266). We may first recall that, in this integral, φ_γ denotes $\varphi\left(\tilde{\mathbf{p}}_\gamma, \mathbf{r}_\gamma\right)$, φ'_γ denotes $\varphi\left(\tilde{\mathbf{p}}'_\gamma, \mathbf{r}'_\gamma\right)$, and so on, the primes referring to the data about molecules after collision.[77] Each of these terms will be affected by the out-of-equilibrium state. We will then write the variation of $\varphi'_\beta\varphi'_\gamma - \varphi\varphi_\gamma$ caused by $\delta\varphi$ in the form

$$\begin{aligned}\delta\left(\varphi'_\beta\varphi'_\gamma - \varphi\varphi_\gamma\right) &= \left(\varphi'_{0,\beta} + \delta\varphi'_\beta\right)\left(\varphi'_{0,\gamma} + \delta\varphi'_\gamma\right) \\ &\quad - \left(\varphi_0 + \delta\varphi\right)\left(\varphi_{0,\gamma} + \delta\varphi_\gamma\right) - \varphi'_{0,\beta}\varphi'_{0,\gamma} + \varphi_0\varphi_{0,\gamma} \\ &\approx \varphi'_{0,\beta}\delta\varphi'_\gamma + \varphi'_{0,\gamma}\delta\varphi'_\beta - \varphi_0\delta\varphi_\gamma - \varphi_{0,\gamma}\delta\varphi\end{aligned} \tag{3.273}$$

[77] We use the same notations as in Sections 2.2.2 and 2.2.3. One must not mistake these primes for those which, in the preceding section, were used to denote the differences between molecule velocities and average velocity as meant by Boltzmann.

Afterwards, we introduce new notations by putting $\chi_\gamma \doteqdot \chi\left(\tilde{\mathbf{p}}_\gamma, \mathbf{r}_\gamma\right)$, and so on. With the definition (3.271), and using the relation (2.46), which is now written as $\varphi'_{0,\beta}\varphi'_{0,\gamma} = \varphi_0\varphi_{0,\gamma}$, we get

$$\begin{aligned}\delta\left(\varphi'_\beta\varphi'_\gamma - \varphi\varphi_\gamma\right) &\approx \frac{1}{k_B T}(\varphi'_{0,\beta}\varphi'_{0,\gamma}\chi'_\gamma + \varphi'_{0,\gamma}\varphi'_{0,\beta}\chi'_\beta \\ &\quad - \varphi_0\varphi_{0,\gamma}\chi_\gamma - \varphi_{0,\gamma}\varphi_0\chi) \\ &= \frac{\varphi_0\varphi_{0,\gamma}}{k_B T}\left(\chi'_\gamma + \chi'_\beta - \chi_\gamma - \chi\right)\end{aligned} \tag{3.274}$$

The variation δC of the collision integral is then given by

$$\begin{aligned}\delta C &= \int_{\bar{\Omega}} w\left(\tilde{\mathbf{p}}, \tilde{\mathbf{p}}_\gamma ; \tilde{\mathbf{p}}'_\beta, \tilde{\mathbf{p}}'_\gamma\right) \\ &\quad \times \delta\left(\varphi'_\beta\varphi'_\gamma - \varphi\varphi_\gamma\right) d\tilde{\mathbf{p}}_\gamma d\tilde{\mathbf{p}}'_\beta d\tilde{\mathbf{p}}'_\gamma \\ &= \frac{\varphi_0}{k_B T} I\left(\chi\right)\end{aligned} \tag{3.275}$$

with

$$\begin{aligned}I\left(\chi\right) \doteqdot &\int_{\bar{\Omega}} w\left(\tilde{\mathbf{p}}, \tilde{\mathbf{p}}_\gamma ; \tilde{\mathbf{p}}'_\beta, \tilde{\mathbf{p}}'_\gamma\right) \\ &\times \varphi_{0,\gamma}\left(\chi'_\gamma + \chi'_\beta - \chi_\gamma - \chi\right) d\tilde{\mathbf{p}}_\gamma d\tilde{\mathbf{p}}'_\beta d\tilde{\mathbf{p}}'_\gamma\end{aligned} \tag{3.276}$$

(the factor φ_0 could be got out of the integral because the latter has no integration over $\tilde{\mathbf{p}}$). The equilibrium distribution φ_0 once again verifies $d\varphi_0/dt = 0$, and (3.265) then yields $C\left(\varphi_0\right) = 0$. Hence:

$$\begin{aligned}C\left(\varphi\right) &\approx C\left(\varphi_0\right) + \delta C \\ &= \frac{\varphi_0}{k_B T} I\left(\chi\right)\end{aligned} \tag{3.277}$$

In order to make progress in our calculations, we now will endeavour to assess the left-hand side in equation (3.265). As stated in Section 2.3.1, we are seeking an out-of-equilibrium Boltzmann distribution, as given by formula (2.70). In order to keep true to our hypotheses, we drop the external force potential as well as the molecular interactions and the internal molecular energy. Within the context of the continua, however, we must remove the particular labels $_a$ from that definition and explicitly emphasize the dependence of both velocity and temperature on position and time:

$$\varphi\left(\tilde{\mathbf{p}}, \mathbf{r}, t\right) = \Theta \exp\left\{-\frac{m}{2k_B T\left(\mathbf{r}, t\right)}\left[\frac{\tilde{\mathbf{p}}}{m} - \mathbf{u}\left(\mathbf{r}, t\right)\right]^2\right\} \tag{3.278}$$

The derivatives of φ with respect to time and position are easily calculated:

$$\begin{aligned} \frac{1}{\varphi}\frac{\partial \varphi}{\partial t} &= \frac{m}{k_B T}\left[\frac{1}{2T}\frac{\partial T}{\partial t}\left(\tilde{\mathbf{u}} - \mathbf{u}\right)^2 + \frac{\partial \mathbf{u}}{\partial t}\cdot\left(\tilde{\mathbf{u}} - \mathbf{u}\right)\right] \\ \frac{1}{\varphi}\frac{\partial \varphi}{\partial \mathbf{r}} &= \frac{m}{k_B T}\left[\frac{1}{2T}\left(\tilde{\mathbf{u}} - \mathbf{u}\right)^2\left(\mathbf{grad}\ T\right) + \left(\tilde{\mathbf{u}} - \mathbf{u}\right)^T\left(\mathbf{grad}\ \mathbf{u}\right)\right] \end{aligned} \tag{3.279}$$

A significant remark now has to be made. We are seeking local quantities, which allows to get into the inertial frame of reference of the relevant mesoscopic particle. We may then put $\mathbf{u} = \mathbf{0}$, but, nevertheless, the velocity gradient persists. The left-hand side in (3.265) is then written, taking (3.279) into account, as:

$$\frac{d\varphi}{dt} = \frac{m\varphi_0}{k_B T}\left[\frac{|\tilde{u}|^2}{2T}\left(\frac{\partial T}{\partial t} + \mathbf{grad}\ T\cdot\tilde{\mathbf{u}}\right) + \frac{\partial \mathbf{u}}{\partial t}\cdot\tilde{\mathbf{u}} + \tilde{\mathbf{u}}^T\left(\mathbf{grad}\ \mathbf{u}\right)\cdot\tilde{\mathbf{u}}\right] \tag{3.280}$$

Still considering phenomena without any temperature variation, we will only keep, from that equation, those quantities depending on $\mathbf{u}$. The term $\partial\mathbf{u}/\partial t$ in (3.280) is balanced by the other terms of the Navier–Stokes equation of motion which, among others, contains the quantity being sought, that is the viscous stress tensor $\mathbf{div}\ \boldsymbol{\tau}$. In the approximation (3.269), however, we may state that this term would only provide our calculations with a correction of the immediately lower order of magnitude, and so we may safely ignore it. It is the same with the Navier–Stokes inertial term. As regards the pressure gradient, it would yield again the isotropic part of the stress tensor which we have calculated in the previous paragraph (eqn (3.251)). We may notice, however, that when we ignore this term, we deliberately forget those processes related to the fluid compressibility about which we have said at least twice that they are closely related to pressure (Section 3.4.1 and 3.4.4). This is an approximation of an incompressible fluid. We may thus eventually do without $\partial\mathbf{u}/\partial t$ in (3.280). Lastly, we can make the following transformation:

$$\tilde{\mathbf{u}}^T\left(\mathbf{grad}\ \mathbf{u}\right)\cdot\tilde{\mathbf{u}} = \tilde{u}_i\frac{\partial u_i}{\partial x_j}\tilde{u}_j \tag{3.281}$$

$$= \frac{1}{2}\left(\frac{\partial u_i}{\partial x_j} + \frac{\partial u_j}{\partial x_i}\right)\tilde{u}_i\tilde{u}_j$$
$$= \left(\tilde{\mathbf{u}} \otimes \tilde{\mathbf{u}}\right) : \mathbf{s}$$

Ultimately, (3.280) takes quite a simple form:

$$\frac{d\varphi}{dt} = \frac{m\varphi_0}{k_B T}\left(\tilde{\mathbf{u}} \otimes \tilde{\mathbf{u}}\right) : \mathbf{s} \tag{3.282}$$

Taking the above discussion into account, and with the help of (3.277), equation (3.265) yields

$$m\left(\tilde{\mathbf{u}} \otimes \tilde{\mathbf{u}}\right) : \mathbf{s} = I\left(\chi\right) \tag{3.283}$$

The left-hand side in that equality is linear in $\mathbf{s}$, whereas $I(\chi)$ is linear in χ, as evidenced by its definition (3.276). We then have to seek a linear dependence of χ on $\mathbf{s}$, that is

$$\chi\left(\tilde{\mathbf{p}}, \mathbf{r}, t\right) = \mathbf{A}\left(\tilde{\mathbf{p}}\right) : \mathbf{s}\left(\mathbf{r}, t\right) \tag{3.284}$$

where $\mathbf{A}$ is an unknown second-order tensor (we assume in that expression that the dependence of χ on time and space is fully taken on by $\mathbf{s}$). Like the statistical distributions, $\mathbf{A}$ most certainly depends on the energy of the molecules, that is to say, since we are in the particle's inertial frame of reference, on its temperature T.

Owing to the definition of $I(\chi)$ and the relation (3.284), we may write

$$I\left(\chi\right) = \mathbf{I}\left(\mathbf{A}\right) : \mathbf{s} \tag{3.285}$$

where $\mathbf{I}(\mathbf{A})$ is a second-order tensor as defined by analogy with (3.276):

$$\begin{aligned}\mathbf{I}\left(\mathbf{A}\right) \doteqdot \int_{\bar{\Omega}} & w\left(\tilde{\mathbf{p}}, \tilde{\mathbf{p}}_\gamma; \tilde{\mathbf{p}}'_\beta, \tilde{\mathbf{p}}'_\gamma\right) \\ & \times \varphi_{0,\gamma}\left(\mathbf{A}'_\gamma + \mathbf{A}'_\beta - \mathbf{A}_\gamma - \mathbf{A}\right) d\tilde{\mathbf{p}}_\gamma d\tilde{\mathbf{p}}'_\beta d\tilde{\mathbf{p}}'_\gamma\end{aligned} \tag{3.286}$$

with $\mathbf{A}'_\gamma \doteqdot \mathbf{A}\left(\tilde{\mathbf{p}}'_\gamma\right)$, and so on Considering that the relation (3.283) is valid whatever the local values of the components of $\mathbf{s}$ may be, we ultimately find

$$m\tilde{\mathbf{u}} \otimes \tilde{\mathbf{u}} = \mathbf{I}\left(\mathbf{A}\right) \tag{3.287}$$

We now have all the required elements for calculating the viscous stress tensor. Since we are in the local inertial frame of reference of the relevant particle, the random part of the molecular velocity equals $\mathbf{u}' = \tilde{\mathbf{u}} - \mathbf{u} = \tilde{\mathbf{u}}$. The definitions of both stress tensor (3.250) and mean quantities (3.228) induce

$$\begin{aligned}\boldsymbol{\tau}\left(\mathbf{r}, t\right) &= -\rho\left\langle\tilde{\mathbf{u}} \otimes \tilde{\mathbf{u}}\right\rangle \\ &= -\frac{1}{mk_B T}\int_{\Omega_p}\left(\tilde{\mathbf{p}} \otimes \tilde{\mathbf{p}}\right)\varphi_0\left(\tilde{\mathbf{p}}\right)\chi\left(\tilde{\mathbf{p}}, \mathbf{r}, t\right) d\tilde{\mathbf{p}} \\ &= \mathbf{B} : \mathbf{s}\left(\mathbf{r}, t\right)\end{aligned} \tag{3.288}$$

where **B** is a fourth-order tensor as defined by

$$\mathbf{B} \doteqdot -\frac{1}{mk_BT}\int_{\Omega_p}\left[\tilde{\mathbf{p}}\otimes\tilde{\mathbf{p}}\otimes A\left(\tilde{\mathbf{p}}\right)\right]\varphi_0\left(\tilde{\mathbf{p}}\right)d\tilde{\mathbf{p}} \tag{3.289}$$

B is then symmetric with respect to its first two indices and depends on T via **A**, but does not explicitly depend on space and time.

From now on, we can only make progress in our calculations through further hypotheses. Let us take a closer look at a monatomic gas, the rotational invariance then indicates that the tensor **B** is isotropic. The most general form satisfying these conditions is as follows:

$$\mathbf{B} = \mu\left(\delta_{ik}\delta_{jl} + \delta_{il}\delta_{jk}\right)\mathbf{e}_i\otimes\mathbf{e}_j\otimes\mathbf{e}_k\otimes\mathbf{e}_l \tag{3.290}$$

where μ is a scalar quantity depending on temperature. Equation (3.288) ultimately yields, using the symmetry of **s**:

$$\begin{aligned}\boldsymbol{\tau} &= \mu\left(\delta_{ik}\delta_{jl} + \delta_{il}\delta_{jk}\right)s_{lk}\mathbf{e}_i\otimes\mathbf{e}_j\\ &= \mu\left(s_{ji} + s_{ij}\right)\mathbf{e}_i\otimes\mathbf{e}_j\\ &= 2\mu\mathbf{s}\end{aligned} \tag{3.291}$$

This is once again the incompressible viscous behaviour law (3.107), and μ is nothing but the molecular dynamical viscosity, which occurs as a mere function of temperature, as discussed in Section 3.4.2. The other, compressibility-related, viscous terms in (3.104), could be derived through a similar procedure (refer to Landau and Lifshitz, 1981, and to Rief, 1965).

Viscosity can be assessed by twice contracting the tensor **B** (3.289). By comparison with (3.290), we find

$$\mu = -\frac{1}{12mk_BT}\int_{\Omega_p}\left[\tilde{\mathbf{p}}^T\mathbf{A}\left(\tilde{\mathbf{p}}\right)\tilde{\mathbf{p}}\right]\varphi_0\left(\tilde{\mathbf{p}}\right)d\tilde{\mathbf{p}} \tag{3.292}$$

An order of magnitude of the tensor **A** can be estimated from the relation (3.287). We may initially note that the integrals of such a kind as $C\left(\varphi\right)$, $I\left(\chi\right)$, and so on account for processes of return to equilibrium through the collisions, which are then exerted with a time scale corresponding to the collision characteristic time τ_m. We then can provide the following estimate:

$$\mathbf{I}\left(\mathbf{A}\right) \sim \frac{|\mathbf{A}|}{\tau_m} \tag{3.293}$$

As explained in Section 2.3.1, time τ_m is linked to the rms molecular velocity $\mathring{u}$ and the mean free trajectory ℓ by the relation (2.68). With equation (3.287), (3.293) then yields

$$|\mathbf{A}| \sim \tau_m m\mathring{u}^2 \sim \frac{k_BT\ell}{\mathring{u}} \tag{3.294}$$

(the latter estimate stems from the fact that the kinetic energy of molecules is on the order of k_BT, as evidenced by (2.67)). Through the definition (3.292), the following order of magnitude can then be achieved:

$$\mu \sim m\ell\mathring{u}\int_{\Omega_p}\varphi_0\left(\tilde{\mathbf{p}},\mathbf{r}\right)d\tilde{\mathbf{p}} \tag{3.295}$$

Lastly, the formula (3.230) giving the density provides, with (3.295), an estimate of the kinematic viscosity:

$$\nu \sim \ell \mathring{u} \tag{3.296}$$

The molecular viscosity then happens to be the product of a length times a velocity, which would directly result from a dimensional analysis (refer to Section 3.4.3). Although the rms velocity is very high (we may recall it is close to the sound velocity, according to Section 2.3.3), the quite small value of the mean free trajectory accounts for the very small values[78] of ν as given in Table 3.2, Section 3.4.1. As regards a perfect gas, the relation (2.71) lastly yields

$$\nu \sim \ell \sqrt{\frac{k_B T}{m}} \tag{3.297}$$

Viscosity then happens to be an increasing function of temperature. For a liquid, this result is no longer valid, in the first place because the relation (2.71) no longer occurs. The close proximity of molecules within a liquid plays a major part in its viscous nature. As temperature increases, molecular agitation drives them apart, which induces a lower viscosity (Landau et al., 1986).

[78] Chapter 4 will introduce the notion of eddy viscosity, which obeys the same dimensional principles but takes substantially higher values.

3.6 Variational principles for fluids

The least action principle is here applied to continuous systems. After introducing a four-dimensional formalism, we derive the continuous Lagrange equations, as well as the energy-momentum four-tensor of a medium. We first apply these ideas to a rigid medium, then to a strained medium, eventually to an incompressible perfect fluid. We show how it is possible to find the role of pressure in the incompressibility phenomenon, and how we can state its implication in wall boundary conditions. The case of irreversible processes (viscosity, diffusion) is finally briefly treated.

3.6.1 Generals. Rigid media

We are now going to derive the equations of continua and fluids from the least action principle as discussed in Chapter 1. This approach will provide clear answers for several issues and will sometimes prove useful upon the numerical modelling of incompressible flows (Chapters 5 and 6). The mathematical content of the following paragraphs, however, is somewhat tedious, and time-pressed readers may temporarily skip this section without compromising their understanding of the next chapters. The ideas have been developed in a rather uncommon way, from presently acknowledged considerations (for instance, refer to Swaters, 2000 and Kambe, 2008).

A continuous medium should be characterized by an infinite number of parameters q_i representing, for instance, the coordinates of those particles making it up.[79] More exactly, the medium is described by three continuous fields of parameters q_i representing the components of the position vectors $\mathbf{r}$ of its particles at every point in time, as a function of their initial positions $\mathbf{x}$, in accordance with the beginning of Chapter 1:

[79] Within the context of relativistic field theory, these parameters may also be the components of the electromagnetic field quadrivector.

$$q_i = q_i\left(\mathbf{x}, t\right), \quad i = 1, 2, 3 \tag{3.298}$$

Since each particle has its own characteristics (velocity, density, etc.), the medium's Lagrangian will be written as the integral, over its volume, of a Lagrangian density $\mathscr{L}$. This function, as with a discrete set of particles, should be a function of the q_i and $\dot{q}_i$, as well as of the spatial derivatives of the q_i, in order to take the continuous nature of the medium into account. We will denote the spatial derivatives as:

$$q_{i,j} \doteqdot \frac{\partial q_i}{\partial x_j} \tag{3.299}$$

Thus, if Ω denotes the volume as occupied by the medium, we will write

$$L = \int_\Omega \mathscr{L}\left(\{q_i\}, \{q_{i,j}\}, \{\dot{q}_i\}, t\right) d\Omega \tag{3.300}$$

the curly braces are provided for recalling that i and j range over all three values which are assigned to them. Time explicitly plays a part if the medium is an open system, but we will hereafter drop it for the sake of conciseness. The vector $\{q_i\} = (q_1, q_2, q_3)^T$ will henceforth be denoted as $\mathbf{q}$.

The action is written as the time integral of the Lagrangian, which can be written in a condensed form by putting $q_{i,0} \doteqdot \dot{q}_i$ as a convention, which is tantamount to working in a four-dimensional space[80] $(i = 0, 1, 2, 3)$. A *four-vector* will then comprise four components and may be written as $\mathbf{A} = (A_0, \mathbf{a}^T)^T$, A_0 being its 'temporal component' and $\mathbf{a}$ an ordinary spatial vector. Likewise, a tensor 3×4 will be the data $\mathbf{B} = (\mathbf{B}_0, [\mathbf{b}])$ of an ordinary tensor[81] $\mathbf{b}$, with a vector $\mathbf{B}_0$ on its left-hand side:

[80] This is the same approach as that adopted in special relativity for the treatment of the electromagnetic field (Landau and Lifshitz, 1975).

[81] We write $\mathbf{b}$ in brackets in order to remember its tensorial feature 3×3, unlike the vector $\mathbf{B}_0$. We will use this notation throughout the remaining part of this section.

$$\begin{aligned} \mathbf{A} &= \begin{pmatrix} A_0 \\ \mathbf{a} \end{pmatrix} \doteqdot \begin{pmatrix} A_0 \\ a_1 \\ a_2 \\ a_3 \end{pmatrix} \\ \mathbf{B} &= (\mathbf{B}_0, [\mathbf{b}]) \doteqdot \begin{pmatrix} B_{0,1} & b_{11} & b_{12} & b_{13} \\ B_{0,2} & b_{21} & b_{22} & b_{23} \\ B_{0,3} & b_{31} & b_{32} & b_{33} \end{pmatrix} \end{aligned} \tag{3.301}$$

Such conventions being defined, the action of the continuous medium is written as an integral over a four-dimensional hypervolume:

$$\begin{aligned} \tilde{S} &= \int_{t_1}^{t_2} \int_\Omega \mathscr{L}\left(\{q_i\}, \{q_{i,j}\}, \{\dot{q}_i\}\right) d\Omega dt \\ &= \int_\Theta \mathscr{L}\left(\mathbf{q}, \mathbf{grad}^*\mathbf{q}\right) d^4\mathbf{x}^* \end{aligned} \tag{3.302}$$

We have adopted a notation convention according to which a star in superscript refers to the four-dimensional formalism[82] which we have just introduced, and so $\mathbf{x}^* \doteqdot (t, \mathbf{x})^T$ and

[82] It should be kept in mind that the vector $\mathbf{q}$ is three-dimensional, unlike the quadrivector $\mathbf{x}^*$.

$$\mathbf{grad}^*\mathbf{q} \doteqdot \left(\dot{q}\ [\mathbf{grad}\ \mathbf{q}]\right)$$

$$= \begin{pmatrix} \frac{\partial q_1}{\partial t} & \frac{\partial q_1}{\partial x} & \frac{\partial q_1}{\partial y} & \frac{\partial q_1}{\partial z} \\ \frac{\partial q_2}{\partial t} & \frac{\partial q_2}{\partial x} & \frac{\partial q_2}{\partial y} & \frac{\partial q_2}{\partial z} \\ \frac{\partial q_3}{\partial t} & \frac{\partial q_3}{\partial x} & \frac{\partial q_3}{\partial y} & \frac{\partial q_3}{\partial z} \end{pmatrix} = \begin{pmatrix} \dot{q}_1\ q_{1,1}\ q_{1,2}\ q_{1,3} \\ \dot{q}_2\ q_{2,1}\ q_{2,2}\ q_{2,3} \\ \dot{q}_2\ q_{3,1}\ q_{3,2}\ q_{3,3} \end{pmatrix} \tag{3.303}$$

is a 3×4 tensor, with $\dot{q} = \partial\mathbf{q}/\partial t$, $\mathbf{grad}\ \mathbf{q}$ being the 3×3 tensor whose coefficients are the $\{q_{i,j}\}$ $(j = 1, 2, 3)$. The notation $\Theta \doteqdot [t_1, t_2] \times \Omega$ in equation (3.302) denotes the integration quadrivolume.

In the following, we will use the 3×4 *momentum tensor* as defined by

$$\mathbf{p} \doteqdot \frac{\partial\mathscr{L}}{\partial\mathbf{grad}^*\mathbf{q}} = \left(\frac{\partial\mathscr{L}}{\partial\dot{q}}\left[\frac{\partial\mathscr{L}}{\partial\mathbf{grad}\ \mathbf{q}}\right]\right)$$

$$= \begin{pmatrix} \frac{\partial\mathscr{L}}{\partial\dot{q}_1} & \frac{\partial\mathscr{L}}{\partial q_{1,1}} & \frac{\partial\mathscr{L}}{\partial q_{1,2}} & \frac{\partial\mathscr{L}}{\partial q_{1,3}} \\ \frac{\partial\mathscr{L}}{\partial\dot{q}_2} & \frac{\partial\mathscr{L}}{\partial q_{2,1}} & \frac{\partial\mathscr{L}}{\partial q_{2,2}} & \frac{\partial\mathscr{L}}{\partial q_{2,3}} \\ \frac{\partial\mathscr{L}}{\partial\dot{q}_3} & \frac{\partial\mathscr{L}}{\partial q_{3,1}} & \frac{\partial\mathscr{L}}{\partial q_{3,2}} & \frac{\partial\mathscr{L}}{\partial q_{3,3}} \end{pmatrix} \tag{3.304}$$

We will launch our investigations with the case of an undeformable medium, adopting the 'simple but inexact' approach as set out at the beginning of Section 1.2.4; though this simplistic case is still unsuitable for fluids, it will be suitable for becoming familiar with those notations and concepts which we will develop in the subsequent sections. For a rigid medium, the $d^4\mathbf{x}^*$ do not change over time, so that the variation of action merely results from the Lagrangian variation:

$$\begin{aligned} \delta\tilde{S} &= \int_\Theta \left(\frac{\partial\mathscr{L}}{\partial q_i}\delta q_i + \frac{\partial\mathscr{L}}{\partial q_{i,j}}\delta q_{i,j}\right) d^4\mathbf{x}^* \\ &= \int_\Theta \left[\frac{\partial\mathscr{L}}{\partial\mathbf{q}} \cdot \delta\mathbf{q} + \mathbf{p} : \left(\mathbf{grad}^*\delta\mathbf{q}\right)^T\right] d^4\mathbf{x}^* \end{aligned} \tag{3.305}$$

(the first line implies the repeated-index summation convention, as well as throughout the following discussion). The second term in the integrand may be rearranged as

$$\int_\Theta \mathbf{p} : \left(\mathbf{grad}^*\delta\mathbf{q}\right)^T d^4\mathbf{x}^* = \int_\Theta \left[\mathrm{div}^*\left(\mathbf{p}^{*T}\delta\mathbf{q}\right) - \mathbf{div}^*\mathbf{p}\cdot\delta\mathbf{q}\right] d^4\mathbf{x}^* \tag{3.306}$$

The 3×4-dimensional divergence operators, for a field of four-vectors $\mathbf{A} = (A_0, \mathbf{a}^T)^T$ and a field of 3×4 tensors denoted as $\mathbf{B} = (\mathbf{B}_0, [\mathbf{b}])$, are defined by[83]

[83] Note that, with the following notations, the compressible continuity equation (3.42) simply reads $\mathrm{div}^*(\rho, \rho\mathbf{u}^T)^T = 0$.

$$\text{div}^*\mathbf{A} \doteqdot \frac{\partial A_0}{\partial t} + \text{div}\mathbf{a}$$
$$\mathbf{div}^*\mathbf{B} \doteqdot \frac{\partial \mathbf{B}_0}{\partial t} + \mathbf{div}\,\mathbf{b} \tag{3.307}$$

The former is a scalar, the latter is an ordinary vector. The divergence integral in (3.306) can be reduced to an integral over the hypersurface enclosing the quadrivolume Θ, so that, if an external normal unit four-vector $\mathbf{n}^*$ is defined over that surface $\partial\Theta$, we have

$$\int_\Theta \mathbf{p} : \left(\mathbf{grad}^*\delta\mathbf{q}\right)^T d^4\mathbf{x}^* = -\int_\Theta \mathbf{div}^*\mathbf{p}\cdot\delta\mathbf{q}\, d^4\mathbf{x}^* + \oint_{\partial\Theta}\left(\mathbf{p}^T\delta\mathbf{q}\right)\cdot\mathbf{n}^* d^3\Gamma^* \tag{3.308}$$

Thus, the variation of action (3.305) is written as

$$\delta\tilde{S} = \int_\Theta \left(\frac{\partial \mathscr{L}}{\partial \mathbf{q}} - \mathbf{div}^*\mathbf{p}\right)\cdot\delta\mathbf{q}\, d^4\mathbf{x}^* + \oint_{\partial\Theta}\left(\mathbf{p}^T\delta\mathbf{q}\right)\cdot\mathbf{n}^* d^3\Gamma^* \tag{3.309}$$

The least action principle consists in investigating minimal solutions for the action, with preset boundary conditions (here, the boundaries are both temporal and spatial); thus, $\delta\mathbf{q}$ vanishes at the quadriboundary $\partial\Theta$. The preceding relation then yields, since it is valid for any variation $\delta\mathbf{q}$:

$$\frac{\partial \mathscr{L}}{\partial \mathbf{q}} - \mathbf{div}^*\mathbf{p} = \mathbf{0} \tag{3.310}$$

This four-dimensional equation is the Lagrange equation being sought. In an indicial form, it is written as

$$\forall i,\ \frac{\partial \mathscr{L}}{\partial q_i} - \frac{\partial}{\partial x_j}\frac{\partial \mathscr{L}}{\partial q_{i,j}} = 0 \tag{3.311}$$

or else, returning to the separation of time and space:

$$\forall i,\ \frac{\partial}{\partial t}\frac{\partial \mathscr{L}}{\partial \dot{q}_i} = \frac{\partial \mathscr{L}}{\partial q_i} - \frac{\partial}{\partial x_j}\frac{\partial \mathscr{L}}{\partial q_{i,j}} \tag{3.312}$$

The equations (3.312) may once again be lumped together, now in threes, in order to get a standard vector form:

$$\frac{\partial}{\partial t}\frac{\partial \mathscr{L}}{\partial \dot{\mathbf{q}}} = \frac{\partial \mathscr{L}}{\partial \mathbf{q}} - \mathbf{div}\frac{\partial \mathscr{L}}{\partial\, \mathbf{grad}\,\mathbf{q}} \tag{3.313}$$

We recognize the ordinary form of the Lagrange equations (1.24) in the first two terms, except that it is a *partial* time derivative which is involved, since we have got into a rigid medium. The third term in (3.313) stands for the spatial variation of a tensor which is directly linked to the stress tensor, as evidenced by the following calculation. Let us first write the following identity:

$$\begin{aligned}\forall j,\ \frac{\partial \mathscr{L}}{\partial x_j} &= \frac{\partial \mathscr{L}}{\partial q_i}\frac{\partial q_i}{\partial x_j} + \frac{\partial \mathscr{L}}{\partial q_{i,k}}\frac{\partial q_{i,k}}{\partial x_j} \\ &= \frac{\partial \mathscr{L}}{\partial q_i} q_{i,j} + \frac{\partial \mathscr{L}}{\partial q_{i,k}}\frac{\partial q_{i,k}}{\partial x_j}\end{aligned} \tag{3.314}$$

The first term in this equation can be altered taking the Lagrange equation (3.311) into account to yield

$$\forall j,\ \frac{\partial \mathscr{L}}{\partial x_j} = \frac{\partial}{\partial x_k}\frac{\partial \mathscr{L}}{\partial q_{i,k}} q_{i,j} + \frac{\partial \mathscr{L}}{\partial q_{i,k}}\frac{\partial q_{i,j}}{\partial x_k} = \frac{\partial}{\partial x_k}\left(\frac{\partial \mathscr{L}}{\partial q_{i,k}} q_{i,j}\right) \tag{3.315}$$

We come to the following conclusion:

$$\forall j,\ \frac{\partial}{\partial x_k}\left(\frac{\partial \mathscr{L}}{\partial q_{i,k}} q_{i,j} - \mathscr{L}\delta_{jk}\right) = 0 \tag{3.316}$$

This relation may also be written, in a four-dimensional form:

$$\mathbf{Div}^* \boldsymbol{\Sigma} = \mathbf{0} \tag{3.317}$$

with

$$\boldsymbol{\Sigma} \doteqdot (\mathbf{grad}^* \mathbf{q})^T \cdot \mathbf{p} - \mathscr{L}\mathbf{I}_4^* \tag{3.318}$$

$\mathbf{I}_4^*$ being the *identity four-tensor*. A four-tensor is a 4×4 tensor; hence, for example

$$\mathbf{I}_4^* \doteqdot \begin{pmatrix} 1 & \mathbf{0}^T \\ \mathbf{0} & [\mathbf{I}_3] \end{pmatrix} = \begin{pmatrix} 1 & 0 & 0 & 0 \\ 0 & 1 & 0 & 0 \\ 0 & 0 & 1 & 0 \\ 0 & 0 & 0 & 1 \end{pmatrix} \tag{3.319}$$

where $\mathbf{0}$ is the zero vector. In other words, a four-tensor consists of a four-vector standing above a 3×4 tensor. The *quadridivergence* operator[84] occurring in equation (3.317) is defined, for a four-tensor, from the definitions (3.307) by

[84] With a capital **D**, in order to distinguish it from the operator as defined in (3.307), which acts upon the 3×4 tensors.

$$\mathbf{Div}^* \begin{pmatrix} A & \mathbf{C}^T \\ \mathbf{B} & [\mathbf{D}] \end{pmatrix} \doteqdot \begin{pmatrix} \operatorname{div}^*\left(A,\ \mathbf{C}^T\right)^T \\ \mathbf{div}^*\left(\mathbf{B},\ [\mathbf{D}]\right) \end{pmatrix} = \begin{pmatrix} \dfrac{\partial A}{\partial t} + \operatorname{div}\mathbf{C} \\ \dfrac{\partial \mathbf{B}}{\partial t} + \mathbf{div}\,\mathbf{D} \end{pmatrix} \tag{3.320}$$

Thus, it is a four-vector.

The calculation we have just made is in every respect analogous to the calculation (1.68) having governed the establishment of the energy conservation law (1.69) for a system of material points. Equation (3.317) is then a four-vectorial conservation law and $\boldsymbol{\Sigma}$ is referred to as the medium's *energy-momentum four-tensor*.[85] In a developed form, it is written as

[85] Once again referring to the theory of relativity.

$$\boldsymbol{\Sigma} = \begin{pmatrix} \dfrac{\partial \mathscr{L}}{\partial \dot{\mathbf{q}}} \cdot \dot{\mathbf{q}} - \mathscr{L} & \dot{\mathbf{q}}^T \dfrac{\partial \mathscr{L}}{\partial \mathbf{grad}\,\mathbf{q}} \\ (\mathbf{grad}\,\mathbf{q})^T \dfrac{\partial \mathscr{L}}{\partial \dot{\mathbf{q}}} & \left[(\mathbf{grad}\,\mathbf{q})^T \dfrac{\partial \mathscr{L}}{\partial \mathbf{grad}\,\mathbf{q}} - \mathscr{L}\mathbf{I}_3\right] \end{pmatrix} \tag{3.321}$$

We may note that its doubly temporal component Σ_{00} is analogous to the energy of a discrete system (1.69). The conservation equations (3.317), in a developed four-vectorial form, reads as

$$\frac{\partial}{\partial t}\left(\frac{\partial \mathscr{L}}{\partial \dot{\mathbf{q}}}\cdot\dot{\mathbf{q}}-\mathscr{L}\right)=\mathbf{div}\left(\dot{\mathbf{q}}^T\frac{\partial \mathscr{L}}{\partial \mathbf{grad}\,\mathbf{q}}\right)$$
$$\frac{\partial}{\partial t}\left[(\mathbf{grad}\,\mathbf{q})^T\frac{\partial \mathscr{L}}{\partial \dot{\mathbf{q}}}\right]=\mathbf{div}\left[(\mathbf{grad}\,\mathbf{q})^T\frac{\partial \mathscr{L}}{\partial \mathbf{grad}\,\mathbf{q}}-\mathscr{L}\mathbf{I}_3\right] \tag{3.322}$$

The first stands for the medium's law of conservation of energy, and the vector under the divergence symbol then represents an energy flux, that is the force power, pursuant to Section 3.3.3. The second is an equation of motion including the same physics as the Lagrange equation (3.313), and the term under the divergence stands for a momentum flux, that is a stress tensor.[86]

Interestingly enough, in that formalism, both temporal and spatial translations take on the same form and lead to a condensed formulation of the laws of conservation of energy and momentum as discussed in Chapter 1 in the discrete case.

[86] It is the Cauchy tensor in the case of the fluid to be discussed later on; in the case of an electromagnetic field, it is referred to as the Maxwell tensor. As in relativity, after more complex calculations, the quadritensor $\mathbf{\Sigma}$ can be given a manifestly symmetric form, unlike in the present case.

3.6.2 Strained media

The case of such a strained medium (with an constant density) is built after the previous model, with some modifications. The vector $\mathbf{q}$ henceforth happens to be the position $\mathbf{r}$ of a particle, whereas according to (3.3) the vector $\dot{\mathbf{q}}$ is nothing but the Lagrangian derivative[87] $\mathbf{u}$ of $\mathbf{r}$. Besides, pursuant to (3.4) the deformation of the medium leads to:

[87] Some authors make the Lagrangian derivative appear from the notion of local invariance. We think, however, that our present approach is more convenient.

$$\mathbf{grad}\,\mathbf{q}=\frac{\partial \mathbf{r}}{\partial \mathbf{x}}=\mathbf{J} \tag{3.323}$$

The definition (3.303) then yields

$$\mathbf{grad}^*\mathbf{q}=\left(\mathbf{u},\ [\mathbf{J}]\right) \tag{3.324}$$

We will make our calculations, however, keeping the general notation $\mathbf{q}$, for a reason which will become clear later on. We also must take another consideration into account. Since the mass (not the volume) of each particle is conserved, a Lagrangian *mass* density has to be introduced;[88] then, instead of (3.302) we will write

[88] A volume density could also be conserved by varying the volume too, through the determinand in the Jacobian matrix $\mathbf{J}$, as it is done in general relativity through the determinant of the so-called metric tensor. The calculations, however, are far from simplified.

$$\tilde{S}=\int_{t_1}^{t_2}\int_{\Omega(t)}\mathscr{L}\left(\{q_i\},\{q_{i,j}\},\{\dot{q}_i\}\right)\rho d\Omega dt$$
$$=\int_{\Theta}\mathscr{L}\left(\mathbf{q},\mathbf{grad}^*\mathbf{q}\right)\rho d^4\mathbf{r}^* \tag{3.325}$$

We may note that the integration volume $\Omega(t)$, as a material volume, now explicitly depends on time,[89] as shown in Fig. 3.4. The integration in space is therefore presently made according to $\mathbf{r}^*=\left(t,\mathbf{r}^T\right)^T$ and not $\mathbf{x}^*=\left(t,\mathbf{x}^T\right)^T$. Since the mass $\rho d\Omega$ is constant, the variation of action is now given by

[89] We will keep this explicit dependence in our notations throughout this paragraph, in order to emphasize the Lagrangian feature of the argument.

$$\delta\tilde{S}=\int_{\Theta}\left(\frac{\partial \mathscr{L}}{\partial \mathbf{q}}\cdot\delta\mathbf{q}+\mathbf{p}:\mathbf{grad}^*\delta\mathbf{q}\right)\rho d^4\mathbf{r}^* \tag{3.326}$$

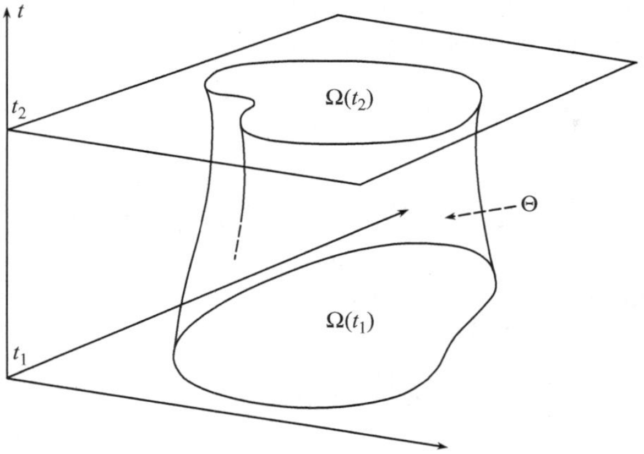

Fig. 3.4 Integration domain for the calculation of a continuous medium Lagrangian. The physical space is reduced to two dimensions for the sake of drawing convenience.

Here, because of the Lagrangian aspect of the motion, we must apply a different treatment to spatial and temporal variations in order to treat the second term:

$$\begin{aligned}&\int_{\Theta} \mathbf{p} : \mathbf{grad}^* \delta\mathbf{q} \, \rho d^4\mathbf{r}^* \\ &= \int_{t_1}^{t_2} \int_{\Omega(t)} \frac{\partial \mathcal{L}}{\partial \dot{\mathbf{q}}} \cdot \frac{d\delta\mathbf{q}}{dt} \rho d\Omega dt + \int_{\Theta} \frac{\partial \mathcal{L}}{\partial \mathbf{grad}\, \mathbf{q}} : \mathbf{grad}\, \delta\mathbf{q} \, \rho d^4\mathbf{r}^*\end{aligned} \tag{3.327}$$

The first term in (3.327) is rearranged as

$$\begin{aligned}&\int_{t_1}^{t_2} \int_{\Omega(t)} \frac{\partial \mathcal{L}}{\partial \dot{\mathbf{q}}} \cdot \frac{d\delta\mathbf{q}}{dt} \rho d\Omega dt \\ &= \int_{t_1}^{t_2} \int_{\Omega(t)} \left[\frac{d}{dt} \left(\frac{\partial \mathcal{L}}{\partial \dot{\mathbf{q}}} \cdot \delta\mathbf{q} \right) - \left(\frac{d}{dt} \frac{\partial \mathcal{L}}{\partial \dot{\mathbf{q}}} \right) \cdot \delta\mathbf{q} \right] \rho d\Omega dt\end{aligned} \tag{3.328}$$

The first term in the right-hand side of this equality is the integral over a material volume of a total derivative whose integration differential is the mass element, and then equals (pursuant to (3.50)) the total derivative of that integral:

$$\begin{aligned}&\int_{t_1}^{t_2} \int_{\Omega(t)} \frac{d}{dt} \left(\frac{\partial \mathcal{L}}{\partial \dot{\mathbf{q}}} \cdot \delta\mathbf{q} \right) \rho d\Omega dt \\ &= \int_{t_1}^{t_2} \frac{d}{dt} \int_{\Omega(t)} \left(\frac{\partial \mathcal{L}}{\partial \dot{\mathbf{q}}} \cdot \delta\mathbf{q} \right) \rho d\Omega dt \\ &= \left[\int_{\Omega(t)} \frac{\partial \mathcal{L}}{\partial \dot{\mathbf{q}}} \cdot \delta\mathbf{q} \, \rho d\Omega \right]_{t_t}^{t_2} \\ &= 0\end{aligned} \tag{3.329}$$

(the latter equation results from the fact that the variation $\delta\mathbf{q}$ vanishes at both initial and final points in time, as in the discrete case discussed in Chapter 1).

The second term in (3.327) is transformed through a combination of differential operators, then is integrated by parts:

$$\begin{aligned}
&\int_{\Theta} \frac{\partial \mathscr{L}}{\partial \mathbf{grad\,q}} : \mathbf{grad}\,\delta\mathbf{q}\,\rho d^4\mathbf{r}^* \\
&= \int_{\Theta} \left\{ \mathrm{div}\left[\rho \left(\frac{\partial \mathscr{L}}{\partial \mathbf{grad\,q}}\right)^T \delta\mathbf{q}\right] - \mathbf{div}\left(\rho \frac{\partial \mathscr{L}}{\partial \mathbf{grad\,q}}\right)\cdot \delta\mathbf{q}\right\} d^4\mathbf{r}^* \\
&= -\int_{\Theta} \mathbf{div}\left(\rho \frac{\partial \mathscr{L}}{\partial \mathbf{grad\,q}}\right)\cdot \delta\mathbf{q}\, d^4\mathbf{r}^* \\
&\quad + \oint_{\partial\Theta} \rho \left[\left(\frac{\partial \mathscr{L}}{\partial \mathbf{grad\,q}}\right)^T \delta\mathbf{q}\right]\cdot \mathbf{n}^* d^3\Gamma^*
\end{aligned} \tag{3.330}$$

The boundary integral, of course, vanishes pursuant to the spatial boundary conditions. Recombining (3.326), (3.328), (3.329) and (3.330), we get

$$\delta\tilde{S} = \int_{\Theta} \left[\rho\left(\frac{\partial \mathscr{L}}{\partial \mathbf{q}} - \frac{d}{dt}\frac{\partial \mathscr{L}}{\partial \dot{\mathbf{q}}}\right) - \mathbf{div}\left(\rho \frac{\partial \mathscr{L}}{\partial \mathbf{grad\,q}}\right)\right]\cdot \delta\mathbf{q}\, d^4\mathbf{r}^* \tag{3.331}$$

Once again, relying on the indifferent nature of the variation of parameters, we come to the equation of motion:

$$\rho\frac{\partial \mathscr{L}}{\partial \mathbf{q}} - \mathbf{D}^*\mathbf{p} = \mathbf{0} \tag{3.332}$$

to be compared with (3.310). Here we have introduced a new operator $\mathbf{D}^*$, playing the same role as the quadridivergence $\mathbf{Div}^*$ as defined by (3.320) in the case of a moving medium. It is defined, for a four-tensor, by:[90]

[90] This operator is analogous to the covariant derivative in general relativity.

$$\mathbf{D}^*\begin{pmatrix} A & \mathbf{C}^T \\ \mathbf{B} & [\mathbf{D}] \end{pmatrix} \doteqdot \begin{pmatrix} \rho\dfrac{dA}{dt} + \mathrm{div}\,(\rho\mathbf{C}) \\ \rho\dfrac{d\mathbf{B}}{dt} + \mathbf{div}\,(\rho\mathbf{D}) \end{pmatrix} \tag{3.333}$$

It then differs from the operator $\mathbf{Div}^*$ by the presence of density. Being provided with that notation, the calculation we have just made may be generalized for a field of 3×4 tensors $\mathbf{A} = (A_0, [\mathbf{a}])$ and a field of four-vectors $\mathbf{B}$ in the form of a Lagrangian four-dimensional version of Gauss' theorem (refex to Appendix A):

$$\begin{aligned}
&\int_{\Theta} \mathbf{A} : \mathbf{grad}^*\mathbf{B}\,\rho d^4\mathbf{r}^* = -\int_{\Theta} \mathbf{D}^*\mathbf{A}\cdot\mathbf{B}\,d^4\mathbf{r}^* \\
&+ \left[\int_{\Omega(t)} \mathbf{A}_0\cdot\mathbf{B}\,\rho d\Omega\right]_{t_t}^{t_2} + \oint_{\partial\Theta} \rho\left(\mathbf{a}^T\mathbf{B}\right)\cdot\mathbf{n}^* d^3\Gamma^*
\end{aligned} \tag{3.334}$$

Let us return to equation (3.332), which is the Lagrange equation for a compressible moving medium. When rearranged in its three-dimensional vector form, it reads as follows:

$$\rho\frac{d}{dt}\frac{\partial \mathscr{L}}{\partial \dot{\mathbf{q}}} = \rho\frac{\partial \mathscr{L}}{\partial \mathbf{q}} - \mathbf{div}\left(\rho\frac{\partial \mathscr{L}}{\partial \mathbf{grad\,q}}\right) \tag{3.335}$$

which may be compared with (3.313). Using the subscripts, we get

$$\forall i,\ \rho \frac{d}{dt}\frac{\partial \mathscr{L}}{\partial \dot{q}_i} = \rho \frac{\partial \mathscr{L}}{\partial q_i} - \frac{\partial}{\partial x_\alpha}\left(\rho \frac{\partial \mathscr{L}}{\partial q_{i,\alpha}}\right) \tag{3.336}$$

In the latter equation, the subscripts α (and, according to a convention to be applied throughout the following, all the dummy Greek sub-scripts) only range over the spatial values 1, 2 and 3.

Returning to the definition of parameters $(\mathbf{q}, \dot{\mathbf{q}}, \mathbf{grad\ q}) \equiv (\mathbf{r}, \mathbf{u}, \mathbf{J})$, equation (3.335) is also written as

$$\rho \frac{d}{dt}\frac{\partial \mathscr{L}}{\partial \mathbf{u}} = \rho \frac{\partial \mathscr{L}}{\partial \mathbf{r}} - \mathbf{div}\left(\rho \frac{\partial \mathscr{L}}{\partial \mathbf{J}}\right) \tag{3.337}$$

The work done in the previous paragraph for establishing the conservation laws may be generalized to get

$$\mathbf{D}^*\boldsymbol{\Sigma} = \mathbf{0} \tag{3.338}$$

where $\boldsymbol{\Sigma}$ is defined, like a rigid medium, by formulas which are analogous to (3.318) or (3.321). After being developed according to the sub-scripts, the tensorial equation reads as

$$\forall j,\ \rho \frac{d}{dt}\left(\frac{\partial \mathscr{L}}{\partial \dot{q}_i} q_{i,j} - \mathscr{L}\delta_{0j}\right) + \frac{\partial}{\partial x_\alpha}\left[\rho\left(\frac{\partial \mathscr{L}}{\partial q_{i,\alpha}} q_{i,j} - \mathscr{L}\delta_{\alpha j}\right)\right] = 0 \tag{3.339}$$

which here plays the role of (3.316). Instead of equation (3.322), we find the four-vectorial form

$$\begin{aligned} &\rho \frac{d}{dt}\left(\frac{\partial \mathscr{L}}{\partial \dot{\mathbf{q}}}\cdot \dot{\mathbf{q}} - \mathscr{L}\right) = \mathbf{div}\left(\rho \dot{\mathbf{q}}^T \frac{\partial \mathscr{L}}{\partial \mathbf{grad\ q}}\right) \\ &\rho \frac{d}{dt}\left[(\mathbf{grad\ q})^T \frac{\partial \mathscr{L}}{\partial \dot{\mathbf{q}}}\right] = \mathbf{div}\left[\rho\,(\mathbf{grad\ q})^T \frac{\partial \mathscr{L}}{\partial \mathbf{grad\ q}} - \rho \mathscr{L}\mathbf{I}_3\right] \end{aligned} \tag{3.340}$$

A developed formulation would once again give the main conservation laws for the continua occurring in the system (3.95) (except for the scalar transport equation, to be dealt with in Section 3.6.4).

3.6.3 Incompressible flows

Let us now turn to the case of an incompressible fluid. In accordance with what was said in the first two chapters, the Lagrangian per unit mass is written as

$$\mathscr{L} = \mathscr{L}(\mathbf{r}, \mathbf{u}, \rho, \sigma) = \frac{1}{2}u^2 - e_{int}(\rho, \sigma) + \mathbf{g}\cdot\mathbf{r} \tag{3.341}$$

with $u = |\mathbf{u}|$, σ denoting the specific entropy. Here, however, the internal energy does not have to be taken into account, since it is well-known it is constant in the isentropic compressible case (refer to Section 3.4.1). We then will ultimately write

$$\mathscr{L} = \mathscr{L}(\mathbf{r}, \mathbf{u}) = \frac{1}{2}u^2 + \mathbf{g}\cdot\mathbf{r} \tag{3.342}$$

The forces caused by the contribution of e_{int} (pressure forces) will naturally appear though the following procedure. Among the difficulties to be settled when treating a fluid through a variational approach is the fact that the spatial components of the particles are not independent, since they move about satisfying the continuity equation. One will then attempt to minimize the action, provided that the continuity equation is complied with. In the case of an incompressible fluid,[91] due to equation (3.11), we will observe the following condition for each particle:

[91] The general case of the incompressible fluid is treated in a similar way, although the calculations are somewhat more complex.

$$\mathrm{div}\mathbf{u} = \frac{d\,(\mathrm{tr}\,\mathbf{J})}{dt} = 0 \tag{3.343}$$

More exactly, tr $\mathbf{J}$ is a constant, but we must ultimately specify that condition so that $\mathbf{J}$ can be varied in our subsequent calculations.[92] We will solve this problem through the Lagrange multiplier method as introduced in Section 1.2.4. Since the condition (3.343) should be satisfied at every point in space and at each point in time, we will not have one multiplier, but a field of multipliers[93] $A\,(\mathbf{r}, t)$, and the quantity to be minimized will be written as

[92] This argument is based on findings which are similar to the remark immediately after eqn (3.115). We may note that Holm (1999) replaces the condition (3.23) with $\det \mathbf{J} - 1 = 0$, according to (3.22). We think that the method set out here is clearer.

[93] It is a generalization of eqn (1.30) to the case of a continuous medium.

$$\begin{aligned} S' = & \int_{t_1}^{t_2} \int_{\Omega(t)} \left(\frac{1}{2}u^2 + \mathbf{g}\cdot\mathbf{r}\right) \rho d\Omega dt \\ & - \int_{t_1}^{t_2} \int_{\Omega(t)} A \frac{d\,(\mathrm{tr}\,\mathbf{J})}{dt} \rho d\Omega dt \end{aligned} \tag{3.344}$$

The second integral can be modified through (3.50) to yield

$$\begin{aligned} & \int_{t_1}^{t_2} \int_{\Omega(t)} A \frac{d\,(\mathrm{tr}\,\mathbf{J})}{dt} \rho d\Omega dt \\ & = \int_{t_1}^{t_2} \int_{\Omega(t)} \frac{d A\,(\mathrm{tr}\,\mathbf{J})}{dt} \rho d\Omega dt - \int_{t_1}^{t_2} \int_{\Omega(t)} \dot{A}\,(\mathrm{tr}\,\mathbf{J})\,\rho d\Omega dt \\ & = \left[\int_{\Omega(t)} A\,(\mathrm{tr}\,\mathbf{J})\,\rho d\Omega\right]_{t_1}^{t_2} - \int_{t_1}^{t_2} \int_{\Omega(t)} \dot{A}\,(\mathrm{tr}\,\mathbf{J})\,\rho d\Omega dt \end{aligned} \tag{3.345}$$

Both initial and final conditions being determined, the integral in the first term is set for $t = t_1$ and $t = t_2$, and so this term is constant and does not play any part in the minimization process (refer to Section 1.2.3). The Lagrangian can therefore ultimately be written as

$$\mathscr{L} = \mathscr{L}\left(\mathbf{r}, \mathbf{u}, \mathbf{J}; A, \dot{A}, \mathbf{grad}\,A\right) = \frac{1}{2}u^2 + \dot{A}\,\mathrm{tr}\,\mathbf{J} + \mathbf{g}\cdot\mathbf{r} \tag{3.346}$$

It then happens that the parameter vector should be provided with a further degree of freedom corresponding to the Lagrange multiplier:

$$\begin{aligned} \mathbf{q} &= \begin{pmatrix} \mathbf{r}\,(\mathbf{x}, t) \\ A\,(\mathbf{x}, t) \end{pmatrix} \\ \dot{\mathbf{q}} &= \begin{pmatrix} \mathbf{u}\,(\mathbf{x}, t) \\ \dot{A}\,(\mathbf{x}, t) \end{pmatrix} \end{aligned} \tag{3.347}$$

We then have, instead of (3.324):

$$\mathbf{grad}^*\mathbf{q} = \begin{pmatrix} \mathbf{u} & [\mathbf{J}] \\ \dot{A} & (\mathbf{grad}\ A)^T \end{pmatrix} \tag{3.348}$$

(the subscript i now ranges over the values $i = 0, 1, 2, 3, 4$, the value 4 being ascribed to the parameter A). Knowing the Lagrangian will also provide the momentum tensor,[94] from its definition (3.304):

$$\mathbf{p} = \begin{pmatrix} \mathbf{u} & \left[\dot{A}\mathbf{I}_3\right] \\ \text{tr}\ \mathbf{J} & \mathbf{0}^T \end{pmatrix} \tag{3.349}$$

[94]Like $\mathbf{grad}^*\mathbf{q}$, it is a quadritensor, not a 3×4 tensor as in the preceding Section This results from the introduction of the new parameter A.

As regards the Lagrange equations (3.335), they are divided into two parts, namely a vector equation for momentum, provided with a scalar equation for the behaviour of the multiplier:

$$\begin{aligned} \rho\frac{d}{dt}\frac{\partial\mathscr{L}}{\partial\mathbf{u}} &= \rho\frac{\partial\mathscr{L}}{\partial\mathbf{r}} - \mathbf{div}\left(\rho\frac{\partial\mathscr{L}}{\partial\mathbf{J}}\right) \\ \rho\frac{d}{dt}\frac{\partial\mathscr{L}}{\partial\dot{A}} &= \rho\frac{\partial\mathscr{L}}{\partial A} - \text{div}\left(\rho\frac{\partial\mathscr{L}}{\partial\mathbf{grad}\ A}\right) \end{aligned} \tag{3.350}$$

The Lagrangian's form (3.346) can be used for quickly determining the following identities:

$$\begin{aligned} \frac{\partial\mathscr{L}}{\partial\mathbf{u}} &= \mathbf{u} & \frac{\partial\mathscr{L}}{\partial\dot{A}} &= \text{tr}\ \mathbf{J} \\ \frac{\partial\mathscr{L}}{\partial\mathbf{r}} &= \mathbf{g} & \frac{\partial\mathscr{L}}{\partial A} &= 0 \\ \frac{\partial\mathscr{L}}{\partial\mathbf{J}} &= \dot{A}\mathbf{I}_3 & \frac{\partial\mathscr{L}}{\partial\mathbf{grad}\ A} &= \mathbf{0} \end{aligned} \tag{3.351}$$

(we have used the equality $\partial\left(\text{tr}\ \mathbf{J}\right)/\partial\mathbf{J} = \mathbf{I}_3$). The Lagrange equations (3.350) then yield

$$\begin{aligned} \frac{d\mathbf{u}}{dt} &= -\frac{1}{\rho}\mathbf{grad}\left(\rho\dot{A}\right) + \mathbf{g} \\ \frac{d\text{tr}\ \mathbf{J}}{dt} &= 0 \end{aligned} \tag{3.352}$$

As expected, the second of these equations gives the continuity equation in its form (3.343), whereas the first is the Euler equation of motion as given by the first line of the system (3.129), provided that we put

$$p = \rho\dot{A} \tag{3.353}$$

We naturally come across the equation of a perfect fluid, since the discussion in Chapter 1 shows that the least action principle may only lead to conservative processes (in the next section, we will explain how the Lagrange equations can be rewritten in the viscous case).

The considerations above show that the incompressible fluid Lagrangian (3.346) becomes[95]

[95]With the condition (3.343) which we have chosen for expressing the incompressibility.

$$\mathscr{L} = \frac{1}{2}u^2 + \frac{p}{\rho}\,\mathrm{tr}\,\mathbf{J} + \mathbf{g}\cdot\mathbf{r} \tag{3.354}$$

Thus, it happens that pressure is the Lagrange multiplier associated with the condition of incompressibility; we will have an opportunity to discuss this result again in Section 6.2.5, within the frame of the SPH method. As previously mentioned, particularly in Section 3.4.4, pressure is a quantity which is closely related to the fluid incompressibility. We may recall that this feature already occurred in the equation of state (Section 2.3.3), as well as in the arguments prevailing in the preparation of the fluid behaviour law (Section 3.4.1). From now on, the contribution of pressure as the 'mainspring' of incompressibility in the total Lagrangian of the fluid system can be assessed. Both calculation (3.345) and relation (3.21) show it is established as[96]

$$\begin{aligned} L_{incomp} &= \int_{\Omega} p\,(\mathrm{tr}\,\mathbf{J})\,d\Omega \\ &= -\int_{\Omega} p \ln\frac{\rho}{\rho_0}\,d\Omega \end{aligned} \tag{3.355}$$

(the constant 3 of (3.21) was removed, since it can be included in the reference density). In every calculation based on (3.355), ρ should be kept virtually variable, and its constant value ρ_0 is then only specified by the end, just as when we have kept the quantity tr $\mathbf{J}$ unknown.

We also have to explain how the influence of the boundary condition at the wall (Section 3.4.2) can be assessed using the present variational formalism. Since we presently ignore the viscous effects, we have to consider the condition of impermeability (3.144), which should be satisfied at every point of the wall. We then, once again, have to solve a problem under constraints, which is done by introducing a field of Lagrange multipliers $B(\mathbf{r}, t)$ as defined at wall[97] $\partial\Omega$. The following surface integral should then be added to the system's Lagrangian:

$$L_{wall} = -\oint_{\partial\Omega} B\,(\mathbf{u} - \mathbf{u}_{wall})\cdot\mathbf{n}\,\rho d\Gamma \tag{3.356}$$

$\mathbf{u}_{wall}$ being, as we may recall it, the wall velocity at the point being considered. Applying Gauss' theorem, we may split it up into two volume integrals:

$$\begin{aligned} L_{wall} &= -\int_{\Omega} \mathrm{div}\,[B\,(\mathbf{u} - \mathbf{u}_{wall})]\,\rho d\Omega \\ &= -\int_{\Omega} B\,(\mathrm{div}\mathbf{u})\,\rho d\Omega - \int_{\Omega} (\mathbf{u} - \mathbf{u}_{wall})\cdot(\mathbf{grad}B)\,\rho d\Omega \end{aligned} \tag{3.357}$$

It should be kept in mind that the field B is only defined along the boundary of the domain, and so its gradient is normal to the wall, according to a relation which is identical to equation (3.216) built for treating the interfaces in Section 3.4.5. Thus, the second integral in (3.357) vanishes owing to the condition of impermeability (3.144). As regards the first one, it occurs in the same form as the condition of impermeability (3.343) and, because of the calculations we have just made, we can state that the Lagrange multiplier is likened to pressure. Since B is only defined at the wall, we may write, by analogy with (3.353):

[96] Based on (3.20), the following expression defines a Lagrangian density which exactly equals pdV, as expected (refer to Chapter 2).

[97] In relation to Section 3.4.4, we are considering the case where the fluid is surrounded by walls, i.e. in the absence of a free surface.

$$p\left(\mathbf{r}_{wall}\right) = \rho\dot{B} \tag{3.358}$$

at each point $\mathbf{r}_{wall}$ of the wall. The same calculation as in (3.345) then makes it possible to reduce the wall's Lagrangian (3.356) to the following form:

$$L_{wall} = \oint_{\partial\Omega} p\left(\mathbf{r}_{wall}\right)\left(\mathbf{r} - \mathbf{r}_{wall}\right)\cdot\mathbf{n}\left(\mathbf{r}_{wall}\right)d\Gamma \tag{3.359}$$

It is possible to provide another form of this, which will be useful in Chapter 6 (Section 6.3.2), within the frame of the SPH method. To that purpose, let us integrate (3.359) over the whole fluid domain, which implies multiplying times the Dirac distribution $\delta\left(\left|\mathbf{r}-\mathbf{r}'\right|\right)$, where $\mathbf{r}'$ stands for $\mathbf{r}_{wall}$. It is then possible to substitute $p\left(\mathbf{r}\right)$ for $p\left(\mathbf{r}_{wall}\right)$ to get

$$L_{wall} = \int_{\Omega}\oint_{\partial\Omega}\delta\left(\left|\mathbf{r}-\mathbf{r}'\right|\right)p\left(\mathbf{r}\right)\left(\mathbf{r}-\mathbf{r}'\right)\cdot\mathbf{n}\left(\mathbf{r}'\right)d\Gamma\left(\mathbf{r}'\right)d\Omega\left(\mathbf{r}\right) \tag{3.360}$$

To complete this section, let us briefly turn to the question of energy. Let us first calculate the energy-momentum four-tensor of a perfect incompressible fluid from the definition (3.318) and the formulas (3.348) and (3.349):

$$\begin{aligned}\mathbf{\Sigma} &= \begin{pmatrix}\mathbf{u}^T & \dot{A}\\ \left[\mathbf{J}^T\right] & \mathbf{grad}\,A\end{pmatrix}\begin{pmatrix}\mathbf{u} & \left[\dot{A}\mathbf{I}_3\right]\\ \mathrm{tr}\,\mathbf{J} & \mathbf{0}^T\end{pmatrix}\\ &\quad -\left(\frac{1}{2}u^2 + \dot{A}\,\mathrm{tr}\,\mathbf{J} + \mathbf{g}\cdot\mathbf{r}\right)\begin{pmatrix}1 & \mathbf{0}\\ \mathbf{0} & \left[\mathbf{I}_3\right]\end{pmatrix}\\ &= \begin{pmatrix}\frac{1}{2}u^2 - \mathbf{g}\cdot\mathbf{r} & \dot{A}\mathbf{u}^T\\ \mathbf{J}^T\mathbf{u} + \left(\mathrm{tr}\,\mathbf{J}\right)\mathbf{grad}\,A & \left[\dot{A}\mathbf{J}^T - \left(\frac{1}{2}u^2 + \dot{A}\,\mathrm{tr}\,\mathbf{J} + \mathbf{g}\cdot\mathbf{r}\right)\mathbf{I}_3\right]\end{pmatrix}\end{aligned} \tag{3.361}$$

The laws of conservation (3.338) then read as

$$\begin{cases}\rho\dfrac{d}{dt}\left(\dfrac{1}{2}u^2 - \mathbf{g}\cdot\mathbf{r}\right) = -\mathrm{div}\left(\rho\dot{A}\mathbf{u}\right)\\ \rho\dfrac{d}{dt}\left[\mathbf{J}^T\mathbf{u} + \left(\mathrm{tr}\,\mathbf{J}\right)\mathbf{grad}\,A\right] = -\mathbf{div}\left[\rho\dot{A}\mathbf{J}^T - \rho\begin{pmatrix}\frac{1}{2}u^2 + \dot{A}\,\mathrm{tr}\,\mathbf{J}\\ +\mathbf{g}\cdot\mathbf{r}\end{pmatrix}\mathbf{I}_3\right]\end{cases} \tag{3.362}$$

The first one, taking the definition (3.353) of p into account, is rearranged as:

$$\frac{de_k}{dt} = -\frac{1}{\rho}\mathrm{div}\left(p\mathbf{u}\right) + \mathbf{g}\cdot\mathbf{u} \tag{3.363}$$

which is identical to (3.89) in the absence of friction forces.[98] As regards the second law of conservation of (3.362), we know it represents the equation of motion. Rather lengthy (and unattractive) calculations allow us to suitably rearrange the latter equation as (3.352).

We are going to explain how the law of conservation of the total energy of a perfect fluid can be related to the (conservative) formalism stemming from the Hamilton equations and discussed in Section 1.5.4. To do this, we

[98] We may note, however, that energy does not include the internal energy term in that formalism, which is consistent with the fact that it is constant for an incompressible flow.

temporarily adopt the compressible formalism (the incompressible case is a limit case), which allows us to take the pressure forces into account through the internal energy in order to write the total energy mass density in accordance with (3.92):

$$e = \frac{1}{2}u^2 + e_{int} - \mathbf{g} \cdot \mathbf{r} \tag{3.364}$$

Thus, the total energy (Hamiltonian) of the system is given by the following integral:

$$H = \int_{\Omega(t)} e\rho d\Omega \tag{3.365}$$

As in Section 1.5.4 (eqn (1.266)), we now introduce the state vector $\mathbf{y}$ for each point in the fluid, as defined by

$$\mathbf{y}(\mathbf{r}, t) \doteqdot \begin{pmatrix} \mathbf{u} \\ \mathbf{r} \\ \rho \end{pmatrix} \tag{3.366}$$

(unlike the formalism of Section 1.5.4, velocity is substituted for momentum, which is equivalent to the momentum per unit mass). Using (1.256), we get

$$\frac{\partial e}{\partial \mathbf{y}} = \begin{pmatrix} \mathbf{u} \\ -\mathbf{g} \\ \dfrac{p}{\rho^2} \end{pmatrix} \tag{3.367}$$

by analogy with (1.270). With the theorem (3.50), the calculation (1.273) providing the evolution of the Hamiltonian becomes, within the continuous context:

$$\begin{aligned} \frac{dH}{dt} &= \int_{\Omega(t)} \frac{de}{dt} \rho d\Omega \\ &= \int_{\Omega(t)} \frac{\partial e}{\partial \mathbf{y}} \cdot \dot{\mathbf{y}} \rho d\Omega \\ &= \int_{\Omega(t)} \left(\mathbf{u} \cdot \rho \frac{d\mathbf{u}}{dt} - \rho \mathbf{g} \cdot \frac{d\mathbf{r}}{dt} + \frac{p}{\rho} \frac{d\rho}{dt} \right) d\Omega \end{aligned} \tag{3.368}$$

Taking into account the equation of motion without dissipative forces and the continuity equation as given by (3.129), and considering that $\mathbf{u} = d\mathbf{r}/dt$, we get:

$$\begin{aligned} \frac{dH}{dt} &= \int_{\Omega(t)} (-\mathbf{u} \cdot \mathbf{grad}\, p - p\mathrm{div}\mathbf{u})\, d\Omega \\ &= -\langle \mathbf{grad}\, p, \mathbf{u} \rangle - (p, \mathrm{div}\mathbf{u}) \\ &= 0 \end{aligned} \tag{3.369}$$

This calculation was successfully completed by means of the relation (3.198) of adjunction of the differential operators div and $-$**grad**. That property then accounts for the conservation of energy and is the effect, in the continuous formalism, of the property of asymmetry (1.274) of the matrix $(\omega_7)_{ab}$. Section 1.4.2 had already evidenced the skew-self-adjoint nature of an asymmetric matrix (calculation (1.98)). We will have an opportunity to come back to these properties in Chapter 5, in order to understand how the laws of conservation can be expressed in the language of a Lagrangian numerical method.

3.6.4 Irreversible processes

To complete this chapter, let us qualitatively explain how the irreversible processes can be addressed according to the just explained principles; the dissipative forces will be investigated first. Equation (2.186) in Chapter 2 teaches us that the latter, in the discrete case, should occur in the Lagrange equation in the form of the derivative of the dissipation function with respect to velocities, which are the temporal derivatives $\dot{q}_i$ of the system parameters. In the continuous case, this principle should be generalized by deriving the dissipation rate with respect to the temporal derivatives of those parameters being used for determining the forces. The Lagrange equation of motion shows that, for a fluid, these parameters are $\mathbf{r}$ and $\mathbf{J}$; the dissipation rate of kinetic energy f with respect to $\dot{\mathbf{r}} = \mathbf{u}$ and $\dot{\mathbf{J}} = \mathbf{grad}\ \mathbf{u}$ (eqn (3.10)) should then be derived. Now, we have seen in Section 3.4.1 that the dissipation rate of the mass unit of an incompressible fluid is given by

$$f = \nu s^2 \tag{3.370}$$

It therefore only depends on the velocity gradients, that is on the quantities $\dot{q}_{i,\alpha}$ $(i = 1, 2, 3)$; f then has to be derived with respect to these quantities. This principle is identical to equation (3.99), provided that a further $1/2$ factor be added before the derivative symbol. The dissipative forces (with a minus sign) add to the ordinary forces occurring under the divergence symbol in the Lagrange equations. Applying this principle to the Lagrange equations of a continuous medium[99] (3.336), we get

$$\forall i, \quad \frac{d}{dt}\frac{\partial \mathscr{L}}{\partial \dot{q}_i} = \frac{\partial \mathscr{L}}{\partial q_i} - \frac{\partial}{\partial x_\alpha}\left(\frac{\partial \mathscr{L}}{\partial q_{i,\alpha}} - \frac{1}{2}\frac{\partial f}{\partial \dot{q}_{i,\alpha}}\right) \tag{3.371}$$

[99] For the sake of simplicity in the following formulas we have deleted the density, since this is an incompressible case.

The arguments which prevailed when developing the law of conservation of the energy-momentum four-tensor in Section 3.6.2 should be amended as well. The equation of conservation (3.339) becomes, indeed

$$\begin{aligned}\forall j, \quad &\frac{d}{dt}\left(\frac{\partial \mathscr{L}}{\partial \dot{q}_i} q_{i,j} - \mathscr{L}\delta_{0j}\right) \\ &+ \frac{\partial}{\partial x_\alpha}\left(\frac{\partial \mathscr{L}}{\partial q_{i,\alpha}} q_{i,j} - \mathscr{L}\delta_{\alpha j}\right) = \frac{1}{2}\frac{\partial}{\partial x_\alpha}\frac{\partial f}{\partial \dot{q}_{i,\alpha}} q_{i,j}\end{aligned} \tag{3.372}$$

For $j = 0$, corresponding to energy, we derive

$$\frac{d}{dt}\left(\frac{\partial \mathscr{L}}{\partial \dot{q}_i}\dot{q}_i - \mathscr{L}\right) + \frac{\partial}{\partial x_\alpha}\left(\frac{\partial \mathscr{L}}{\partial q_{i,\alpha}}\dot{q}_i\right) = \frac{1}{2}\frac{\partial}{\partial x_\alpha}\frac{\partial f}{\partial \dot{q}_{i,\alpha}}\dot{q}_i \tag{3.373}$$

which can be rearranged as

$$\frac{d}{dt}\left(\frac{\partial \mathscr{L}}{\partial \dot{q}_i}\dot{q}_i - \mathscr{L}\right) = -\frac{\partial}{\partial x_\alpha}\left[\left(\frac{\partial \mathscr{L}}{\partial q_{i,\alpha}} - \frac{\partial f}{\partial \dot{q}_{i,\alpha}}\right)\dot{q}_i\right] - \frac{1}{2}\frac{\partial f}{\partial \dot{q}_{i,\alpha}}\dot{q}_{i,\alpha} \quad (3.374)$$

In the case of a fluid, the quadratic form of f makes is possible to write

$$\frac{\partial f}{\partial \dot{q}_{i,\alpha}}\mathbf{e}_i \otimes \mathbf{e}_\alpha = \frac{\partial f}{\partial\, \mathbf{grad}\, \mathbf{u}} = \frac{\partial f}{\partial \mathbf{s}} \quad (3.375)$$

The Lagrange vector equations (3.350) will then take on the following shape:

$$\frac{d}{dt}\frac{\partial \mathscr{L}}{\partial \mathbf{u}} = \frac{\partial \mathscr{L}}{\partial \mathbf{r}} - \mathbf{div}\left(\frac{\partial \mathscr{L}}{\partial \mathbf{J}} - \frac{1}{2}\frac{\partial f}{\partial \mathbf{s}}\right) \quad (3.376)$$

(3.376) is then the equation of motion of a viscous fluid in that formalism, fitted with the Lagrange and dissipation functions (3.354) and (3.370). It is a continuous form which is analogous to the discrete equation (2.186). The term under the divergence symbol (to within the sign and multipled times the density) obviously stands for the Cauchy stress tensor, as already mentioned in Section 3.6.1 (The Lagrangian's gradient stands for the external forces, namely gravity in this case). With the quadratic form of f, we find

$$\frac{\partial s^2}{\partial \mathbf{s}} = 4\mathbf{s} \quad (3.377)$$

(refer to the calculation (3.103)). As to the term $\partial\mathscr{L}/\partial\mathbf{J}$, it is given by (3.351). With (3.353), the stress tensor is then written as

$$-\frac{1}{\rho}\boldsymbol{\sigma} = \frac{p}{\rho}\mathbf{I}_3 - 2\nu\mathbf{s} \quad (3.378)$$

according to the law of behaviour of an incompressible viscous fluid (3.107). Lastly, the energy balance (3.340) will take corrective terms analogous to those in (3.374) and will be written as

$$\frac{d}{dt}\left(\frac{\partial \mathscr{L}}{\partial \mathbf{u}}\cdot\mathbf{u} - \mathscr{L}\right) = -\mathbf{div}\left[\mathbf{u}^T\left(\frac{\partial \mathscr{L}}{\partial \mathbf{J}} - \frac{1}{2}\frac{\partial f}{\partial \mathbf{s}}\right)\right] - \frac{1}{2}\frac{\partial f}{\partial \mathbf{s}} : \mathbf{s} \quad (3.379)$$

The first term in the right-hand side stands for an energy diffusion associated to the flux

$$\mathbf{u}^T\left(\frac{\partial \mathscr{L}}{\partial \mathbf{J}} - \frac{1}{2}\frac{\partial f}{\partial \mathbf{s}}\right) = -\frac{1}{\rho}\boldsymbol{\sigma}\cdot\mathbf{u} \quad (3.380)$$

This term is identical to the flux occurring in the energy transport equation (3.92), and yields again the flux $p\mathbf{u}$ of (3.363) in the absence of viscous forces. Lastly, the last term in (3.379), standing for dissipation, is once again obviously equal to f because of its quadratic form. As explained in Section 2.5.1, the dissipative processes are accompanied by volume shrinkage in the phase space. It is all the same within the continuous context.

Let us disclose another way of treating the irreversible processes from a variational principle. We have seen in Section 2.5.2 that the viscous forces can be derived from such a principle using the mirror system formalism. We may recall that the principle consists of considering that a Hamiltonian (i.e.

conservative) system comprises two subsystems one of which is determined by the parameters Q_i, and provides energy to the other one, being represented by the 'mirror' parameters Q_i^*. Through this trick, a loss of energy can be represented for the first subsystem, whereas the energy of the second one increases. Holm (1999) and other authors[100] utilize this procedure for building a variational formalism of the turbulent flow equations (refer to Chapter 4, Section 4.6.2). In this section, we will merely explain how this approach can be used to represent the diffusion[101] of a scalar $B(\mathbf{r}, t)$. To that purpose, we consider the Lagrangian[102]

$$\mathscr{L}\left(B, \dot{B}, B^*, \dot{B}^*\right) \doteqdot \frac{1}{2}\left(B^*\dot{B} - B\dot{B}^*\right) + K_B\,(\mathbf{grad}\;B)\cdot\left(\mathbf{grad}\;B^*\right) \quad (3.381)$$

where B^* denotes the concentration of a 'mirror scalar'. Here we have a single Lagrange equation for each system, which is written after the model of (3.335):

$$\begin{aligned} \rho\frac{d}{dt}\frac{\partial\mathscr{L}}{\partial\dot{B}} &= \rho\frac{\partial\mathscr{L}}{\partial B} - \mathbf{div}\left(\rho\frac{\partial\mathscr{L}}{\partial\mathbf{grad}\;B}\right) \\ \rho\frac{d}{dt}\frac{\partial\mathscr{L}}{\partial\dot{B}^*} &= \rho\frac{\partial\mathscr{L}}{\partial B^*} - \mathbf{div}\left(\rho\frac{\partial\mathscr{L}}{\partial\mathbf{grad}\;B^*}\right) \end{aligned} \quad (3.382)$$

The definition (3.381) then gives

$$\begin{aligned} \frac{\partial\mathscr{L}}{\partial B} &= -\frac{1}{2}\dot{B}^* \\ \frac{\partial\mathscr{L}}{\partial\dot{B}} &= \frac{1}{2}B^* \\ \frac{\partial\mathscr{L}}{\partial\mathbf{grad}\;B} &= K_B\mathbf{grad}\;B^* \end{aligned} \quad (3.383)$$

and similar formulas for the derivatives related to B^*, the first two relations changing sign. When provided with these relations, the Lagrange equations (3.382) yield

$$\begin{aligned} \frac{dB^*}{dt} &= -\frac{1}{\rho}\mathrm{div}\left(\rho K_B\mathbf{grad}\;B^*\right) \\ \frac{dB}{dt} &= \frac{1}{\rho}\mathrm{div}\left(\rho K_B\mathbf{grad}\;B\right) \end{aligned} \quad (3.384)$$

The second equation is identical to (3.60) without any source term; this equation was built in Section 3.3.1 and represents the transfer and the diffusion of the scalar B. It can be observed that the mirror scalar experiences an inverse diffusion, associated with a negative diffusion coefficient, which highlights the purely formal nature of that construction while providing for the conservativity of the total system. We must say that the laws of conservation (3.338) could easily be established here, but without any benefit.

[100] Readers may refer to Morse and Feshbach (1953), as well as Foias et al. (2001).

[101] Here we are inspired by Basdevant, (2005).

[102] As mentioned in Section 2.5.2, the considerations about the mirror systems come within some speculative formalism. No justification of the origin of the Lagrangian is then given.

4 Turbulent flows

4.1 Introduction

Fluid mechanics is often—and wrongly—regarded by the general public as a scientific field which does not contain mystery any longer, mostly because it basically deals with processes in line with our observation scale. Those having interest in science then turned their attention either to the infinitely small (quantum mechanics and its nearly metaphysical mysteries, but the industrial applications of which have drastically changed our everyday lives) or to the infinitely large (cosmology, against the background of the amazing construction of general relativity). Some of the deepest mysteries of contemporary physics now persist on the daily scale of fluids. Actually, fluid turbulence has long been a fascinating topic that's complexity, however, never allowed any adequate mathematical approach before the late nineteenth century, particularly with the research works conducted by Reynolds and Boussinesq. Throughout the twentieth century, it was at the core of physicists' concerns and was specifically focused on when numerical calculation,[1] then chaos theory emerged.

[1]Numerical simulation even owes a lot to the scientists' eagerness to predict the turbulent motion of fluids, since one of the first serious relevant attempts was conducted by Richardson in order to try to make atmospheric forecasts (1922).

Chaos, indeed, is the side from which we will introduce turbulence, considering in the first place the behaviour of disordered nonlinear dynamical systems. This analysis will provide mechanisms underlying the fluid behaviour in the turbulent regime; afterwards, we will turn to the specific equations allowing them to be mathematically treated. We do not intend, however, to provide a catalogue of turbulence patterns which, by the way, would be an awkward (or even unfeasible) task. As in the preceding chapters, we have no ambition to compile an exhaustive textbook dealing with this difficult and much documented topic. Our purpose is to identify the mechanisms and establish the fundamental equations which govern the flows in the turbulent regime, in order to work out the most relevant models in view of their Lagrangian numerical application in Part II of this book.

Before tackling the theory of turbulence, we must ask why this discipline is nearly ignored by many fluid mechanics theoricists and practitioners, particularly in the field of free surface hydraulics. More generally, it may be stated that the integration of turbulence developments into software has experienced two quite different rates in the fields of industrial and environmental simulation, a great deal of effort having been directed towards industrial simulation in that field, whereas environmental modelling is still looking forsaken in that respect. It seems that among the reasons for this, backwardness is a major one.

We will see, indeed, in this chapter that turbulence induces a mixing which can be modelled by diffusion terms looking like further forces in the equations of motion. Depending on the kind of flow being investigated, the order of magnitude of these forces may be compared with those of the other forces in the Navier–Stokes equations, starting from the dimensional analysis as set out in Chapter 3 (Section 3.4.3). It then happens that turbulence plays a prevailing part in almost all the often steady or nearly steady industrial flows, which are therefore much influenced by the diffusion processes. Moreover, lots of environmental flows have a free surface,[2] generating phenomena in which gravity, pressure and inertia forces are prevailing; thus, at first glance, the turbulent effects may be ignored. It should be kept in mind, however, that turbulence remains important in a number of cases of free surface flows, particularly in a slowly varying regime. Although we cannot provide any thorough review of these ideas, we will briefly illustrate them in Chapter 7 (in particular, refer to Section 7.3.2).

[2]Except for the atmospheric flows, whose scientific bases played a major part in the development of the theory of turbulence, as well as in the above mentioned development of numerical simulation.

Among the free surface flows which are closely related to the turbulent mixing processes, the case of flows with a variable density (estuaries, lakes, etc.), which is not discussed in this chapter, is also worth mentioning. From now on, indeed, we consider nearly incompressible flows, that is having such characteristics that very moderate relative variations of density can be provided, on the basis of the elements described in Chapter 3 (Sections 3.4.2 and 3.4.3).

This chapter will initially define the notion of chaos through the example of a damped and sustained pendulum, by reference to the first two chapters. We will then turn to the notion of strange attractor for a dynamical system. Using the Reynolds average concept, we will establish the Reynolds equations for the point-averaged models from the equations given in Chapter 3. A first closure model indroducing the notion of eddy viscosity will subsequently be explained. A brief overview of Kolmogorov's ideas will then lead us to propose several conventional models (zero, one and two equation models) for the calculation of that quantity. Studying the case of a stable flow in an infinite channel will provide us with some information about the turbulent boundary conditions. After reviewing the shortcomings of the models disclosed so far, we will set out some second order models, particularly an explicit algebraic model. We will complete this discussion with the disclosure of several non-averaged models: the large eddy simulation (LES), the α model, and lastly the stochastic models (the latter two being based on the notions as introduced at the end of Chapter 2).

4.2 Chaos dynamics

In order to introduce turbulence as a mechanical process, we start by investigating a modified pendulum including energy production (Van der Pol oscillator). We then consider more general dynamical systems, and introduce the concept of attractor in the phase space. From the experimental results of Reynolds, we conclude that a turbulent flow must have a strange attractor, the dimension of which is computed from the so-called Lyapunov exponents. We finally present the Fourier modes of turbulence and the process of energy cascade.

4.2.1 Example: the Van der Pol pendulum

Before going into further detail as regards the turbulent flow dynamics, we may usefully temporarily ignore the fluids and focus on the possible range of complexity of the dynamical systems in general. Complex oscillations, indeed, occur within the turbulent flows and can be more easily understood where a purely mechanical reference case is investigated. It is then advisable to initially come back to the simple pendulum as discussed in Section 1.4.4, provided with the friction force as described in Section 2.4.2 (eqn (2.177)). Since energy decreases according to the law (2.178), we have explained in Section 2.5.1 that the volumes then decrease in the phase space (eqn (2.227)), and so the trajectories come closer to each other as time elapses, as in Fig. 2.5. By the end of an infinite time (practically, long enough), all the phase trajectories converge to the origin point corresponding to immobility ($\dot{\theta} = 0$) in a vertical position ($\theta = 0$). This point towards which the trajectories inevitably (and irreversibly) converge is known as the system *attractor*.

This is a zero-dimensional, that is rather unattractive, object. Do systems having a non-zero-dimensional attractor exist? To go more deeply into that question, let us resume the damped pendulum case, but provide it with the energy it lacks by imparting it a small momentum at regular time intervals. This amounts to imparting it a driving force (as opposed to friction) when the oscillation amplitude becomes too small, which can be modelled through a variable friction coefficient α (we explained in Section 2.4.2 that the macroscopic kinetic coefficients α may depend on the generalized coordinates). We then must prescribe a coefficient taking negative values when θ is below a given threshold, for example:

$$m\ell^2\ddot{\theta} = -mg\ell\sin\theta - \alpha\left[\left(\frac{\theta}{\theta_1}\right)^2 - 1\right]\dot{\theta} \tag{4.1}$$

This is known as the *Van der Pol oscillator* (1920). The energy in this system can vary over time according to the law

$$\frac{dE}{dt} = \alpha\dot{\theta}^2 - \alpha\left(\frac{\theta\dot{\theta}}{\theta_1}\right)^2 \tag{4.2}$$

Since the system is sustained, its average energy over a sufficiently long time span, however, should be constant. If the time-averaged quantities are bar-topped, then we get

$$\overline{\frac{dE}{dt}} = \underbrace{\alpha\overline{\dot{\theta}^2}}_{P} - \underbrace{\frac{\alpha}{\theta_1^2}\overline{\left(\theta\dot{\theta}\right)^2}}_{\varepsilon} = 0 \tag{4.3}$$

Two terms can be found in that equality, namely P, which denotes a production of energy, and ε, which denotes a dissipation. These two terms are obviously positive and in equilibrium, so that the system can get a periodic regime, at least after some relaxation time. Equation (4.2) cannot be solved analytically, but a numerical solution would actually give a phase portrait representing trajectories all of which are converging towards a one-dimensional

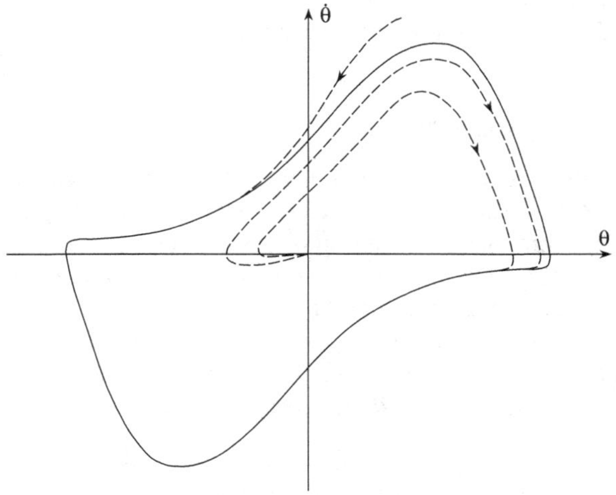

Fig. 4.1 Configuration of the phase portrait of the Van der Pol oscillator attractor (damped and sustained pendulum).

attractor as in Fig. 4.1. This is a limit cycle over which the system indefinitely evolves in the long run, *whatever initial conditions may be*. It is then important to distinguish it from the periodic cycles of Fig. 1.4, each of which corresponds to a level of energy which depends on the initial conditions being chosen. Here, there is no set of possible periodic trajectories, but a single limit cycle, making up a one-dimensional attractor towards which all the trajectories asymptotically tend.

Several comments are to be made about this brief analysis:

- A periodic dynamical system features closed trajectories (but they are not always attractors).
- A system can only have an attractor if it is subject to an energy dissipation, and the phase trajectories all the more quickly converge to the said attractor since the dissipation is greater (pursuant to (2.178)).
- The geometrical dimension of an attractor is (strictly) less than that of the phase space, because the continuous reduction of the volumes gives the attractor a zero volume.[3] This phenomenon makes the phase paths come closer to each other.
- An attractor reveals *forgotten initial conditions*, because the system moves towards it whichever place it comes from. Another way to see this consists of observing that, in accordance with the preceding remark, the number of degrees of freedom is reduced in the long run by the presence of the attractor. Thus, the Van der Pol pendulum initially has two degrees of freedom (θ and $\dot{\theta}$), and keeps only one in the long run (namely the curvilinear abscissa along the limit cycle).
- An attractor may not be reduced to one point (i.e. have a non-zero size), provided that the system enjoys an input of energy from the outside, exactly compensating for the friction-induced losses.

[3] A theorem formulated by Poincaré illustrates this outcome.

We additionally could mention that the main properties of a system can be found again in its attractor, or, more precisely, that the characteristics of the physical space can be seen in the phase space—for example the cyclic nature of the attractor in a periodic system. Thus, it can be felt that the more complex is a system, the more complex is its possible attractor.

4.2.2 Strange attractors

Let us come back to the case of fluids. In the late nineteenth century, O. Reynolds (1883) conducted the experiment which is presently known by his name; we will summarize its main results,[4] illustrating them with the teachings from the previous section. A flow corresponding to a given fluid (with a molecular viscosity ν), a given geometry (characterized by a length scale L) and a velocity scale[5] U is featured by a single dimensionless number, namely the Reynolds number Re (refer to Section 3.4.3). It may behave in three possible ways, according to the value of Re (as defined by (3.154)):

- When Re is low enough (below a threshold $Re_{c,1}$ ranging from several tens to several thousands, depending on the geometry), the fluid particle trajectories remain well-ordered and parallel; this is referred to as a *laminar* flow. In the fluid phase space, this regime corresponds to a punctual attractor. The flow, indeed, is stable, which means that a sufficiently gentle disturbance would quickly disappear owing to the dissipation exerted by the viscous forces. This case is similar to that of the damped pendulum.
- When Re reaches the threshold $Re_{c,1}$ while remaining below a second threshold $Re_{c,2}$ (which also depends on the geometry[6]), an unsteadiness arises, as evidenced by regular oscillations which are generally periodic in both time and space. In the phase space, the attractor then becomes a closed curve (periodicity). This behaviour is similar to that of the Van der Pol pendulum.[7]
- When Re reaches the threshold $Re_{c,2}$ (ranging from several hundreds to several thousands), the flow becomes extremely complex and irregular in both time and space; this is called a *turbulent* flow. It is featured by great variations of the velocity and pressure fields, generating complex and mobile three-dimensional structures with various sizes and frequencies. We are for the first time faced with a behaviour beyond the scope of the investigations of Section 4.2.1. The attractor is now very complex (an example of it will be disclosed later on), exhibiting fluctuations over numerous frequencies; it is the so-called *chaotic* kind of regime in mechanics.

The above results should be considered as stemming from experience, but may as well be subjected to theoretical analyses from the Euler and Navier–Stokes equations. The mathematical discussions of the instabilities in fluids and the processes giving rise to turbulence partly come within the scope of the theoretical study of change in the dynamical system regime, and go far beyond the scope of this book (for instance, refer to Goldstein et al., 2002). Among the many works conducted in this field, we will merely mention the general methodology to investigate the behaviour of a fluid system in such conditions that an unsteadiness arises[8] (Swaters, 2000). The idea consists of linearizing the equation of motion around a state which represents a solution,

[4] Strictly speaking, the Reynolds experiment concerns a flow in a circular pipe. We will generalize its results from various, more general observations.

[5] As mentioned in Section 3.4.3, U would be more suitably defined as an order of magnitude of the *spatial variations* of velocity within the flow. The flows separated into several distinct areas, however, should be treated carefully. This is the case of a fast-moving body in a fluid at rest at infinity, this flow being non-viscous out of the wake and the boundary layer and hightly turbulent in them.

[6] In the case of the Reynolds experiment proper (circular pipe), $Re_{c,1}$ is estimated at 2100, whereas $Re_{c,2}$ is on the order of 3000. More recently, however, further experiments demonstrate that these values depend on such hardly controllable parameters as the wall roughness.

In the case of a flow downstream of a finite-sized solid body, the critical values range from several tens to several hundreds, depending on the body's geometry (an example is disclosed in Section 7.4.1).

[7] For some types of flows, and depending on the type of prescribed disturbance, the transition does not obey any periodic regime, but takes place at intervals (intermittency).

[8] Examples of theoretical studies of the stability of a discrete fluid system will be set out in the second part of this book, within the context of the SPH numerical method (in particular, refer to Section 5.4.4).

the disturbance supposedly having a low amplitude. This method is then identical to that we applied for treating the pressure waves in Section 3.4.3, but uses a linearization around a dynamical situation, not at rest. The linear nature of those equations governing the behaviour of that disturbance makes it possible to seek the latter in the form of a harmonic function, which amounts to only keeping one Fourier component. Using the boundary conditions then leads to a closure relation of frequency versus wave number, that is a dispersion relation, like equation (3.166). Many results were achieved through that procedure in the early twentieth century, particularly by Orr (1907a and 1907b) and Sommerfeld (1908). One of the most impressive achievements in that respect was made by Orszag (1971), who explained, on a theoretical basis, the results of the Reynolds experiment; the value of $Re_{c,1}$, however, was higher than the experimental value.[9]

Some remarks can be made right now about turbulence. Because of the very low viscosity values of the usual fluids, the crushing majority of natural and industrial flows are turbulent, because the Reynolds number is usually very high.[10] Besides, the fact that a flow becomes more unsteady as the Reynolds number grows is consistent with intuition. The viscous term in the Navier–Stokes equation (3.123) is actually stabilizing as a dissipative term returning the system towards its attractor. We will subsequently explain that the advection term, which occurs in the numerator of the definition (3.154) of Re, on the contrary, is destabilizing, and so the Reynolds number happens to be a measure of the system unsteadiness. It is noteworthy, however, that the very eddy structures characterizing a turbulent flow are stable, as explained at the beginning of Section 3.4.4. We may conclude that the turbulent regime itself, although it results from an unsteadiness, is steady since a disturbance does not return the flow to another regime; this is what justifies the existence of an attractor for a turbulent flow.

A turbulent flow is very sensitive to the initial conditions, as can be seen for instance, when lighting up two 'identical' cigarettes: the microscopic but inevitable mutual differences will undoubtedly yield two significantly different smoke plumes. We will get further insight into the latter issue by temporarily coming back to a simpler dynamical system. Figure 4.2 (*a*) depicts a rectangular billiard provided with a ball which follows a trajectory according to Descartes' laws. If a second ball is launched after the first one with a slightly different initial position, it will follow a trajectory which will be parallel to the first one, and so the final mutual distance will be unchanged. Let us now consider a billiard fitted with a circular obstacle (Fig. 4.2 (*b*)); the rotundity of the latter 'diffuses' (this word is not used by chance) the balls at completely different angles, which results in a considerable increase of any initial uncertainty, small as it may be. This system is sensitive to the initial conditions because it exhibits a *non-linearity*. This non-linear feature is then destabilizing, which is a property of the advection term in the Navier–Stokes equations. In the phase space of the billiard with a central obstacle, rather quickly diverging trajectories can be seen. If several initial conditions are set this way on this billiard, then they soon will be dispersed all over the available area, in a mechanism looking like the diffusion of a gas in an enclosure. The non-linearity then has introduced some disorder and the results in Chapter 2 show, from a statistical point of view, that the (mean) macroscopic energy has

[9] Orszag predicts $Re_{c,1} = 5772$, which can be compared with the above mentioned value of 2100.

[10] Some major exceptions, however, can be mentioned, such as the flows in some blood vessels or in the field of industrial microfluidics.

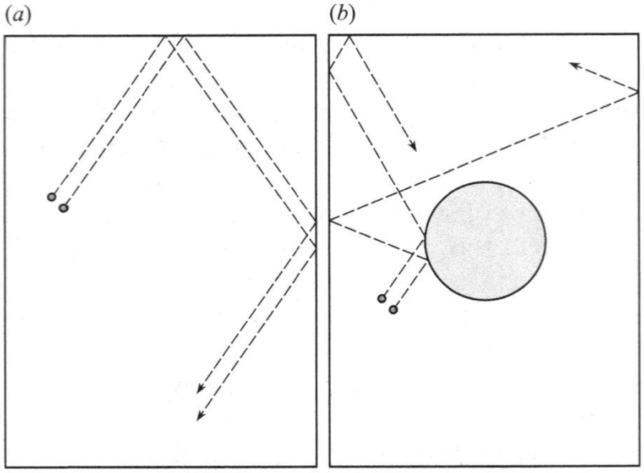

Fig. 4.2 Two billiards, a linear one (a) and a non-linear one (b). The second is sensitive to the initial conditions, unlike the first.

necessarily decreased whereas entropy has increased. This is *on an average* an irreversible process, though the detailed mechanism is quite reversible.

Thus, we come to the conclusion that the sensitivity to the initial conditions (trajectories diverging in the phase space) involves an energy dissipation. Now, the discussion in Section 4.2.1 states, on the contrary, that the energy losses go hand in hand with an *oversight* of the initial conditions, that is with closer trajectories in the phase space, which seems conflicting. We will try to explain this seeming paradox, which is peculiar to the chaotic systems. To that purpose, we will first ask how many degrees of freedom are necessary to get a chaotic (turbulent) regime. We have seen that the switching from the laminar regime to the oscillatory regime, when Re reaches the threshold $Re_{c,1}$, corresponds to an initially punctual attractor which would become a closed curve. This is the onset of a first oscillation frequency. If a second frequency, incommensurable with the first one, occurred, a possible attractor would have to take the shape of a torus, which can only be deployed in an at least three-dimensional space. More generally, according to the remarks made at the end of Section 4.2.1, a number n of frequencies *a minima* implies a number $n + 1$ of degrees of freedom. It seems this is allowed for a fluid, whose number of degrees of freedom is huge (three coordinates and three velocity components for each particle, the number of which is ambiguous but enormous). Thus, in the 1940s, Landau (1944) endeavoured to explain the onset of turbulence as a sequence of frequency splittings, giving rise to the above mentioned eddies. This hypothesis, however, blatantly contradicts the Reynolds' experiment, in which the turbulence occurs abruptly without implying any second frequency.

In the 1970s, the publications by Newhouse, Ruelle and Takens (1978) supplied[11] an answer element by demonstrating that a chaotic regime may arise as soon as three degrees of freedom are present ('period three implies chaos'). We should then acknowledge that the attractor of such a system (particularly the attractor of a turbulent flow) may be strictly less than three-dimensional, and we will adopt this case. We then immediately think of the torus of Fig. 4.3 (a), on which two trajectories could diverge while staying on it. Within that

[11] Based on a work initiated by Ruelle and Takens as early as in 1971.

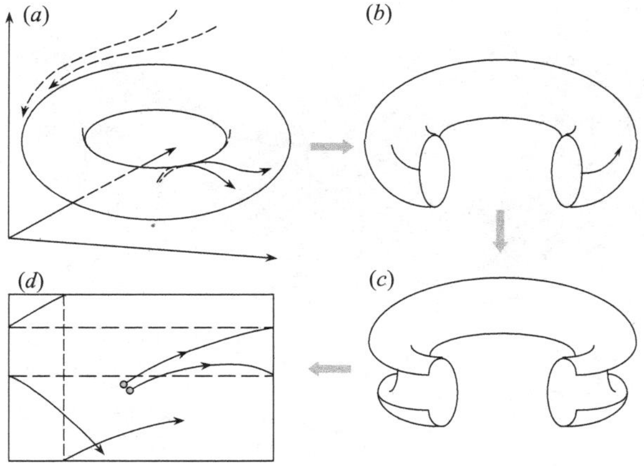

Fig. 4.3 Deployment of a toric attractor over a square. The sensitivity to the initial conditions would here infringe the mechanical determinism.

context, the oversight of the initial conditions shows itself in the trajectories inevitably getting closer to each other, from the outside of the attractor to the latter, whereas the sensitivity to the initial conditions is allowed by the fact that the trajectories on the attractor diverge. It then seems that the paradox is settled through a two-dimensional attractor. We must, however, drop this hypothesis for two reasons: on the one hand we know from experience that there are many more than two oscillatory frequencies in a turbulent flow, and on the other hand such a system could not exhibit any sensitivity to the initial conditions, as evidenced just by unfolding the torus to turn it into a square (Fig. 4.3 (*b*), (*c*) and (*d*)). It immediately appears that two arbitrarily diverging trajectories will ultimately meet each other, by periodicity. Now, according to a major property of the phase trajectories, the mechanical determinism prevents the two of them crossing each other; otherwise, the system would have two possible futures at the meeting point.

We should then acknowledge that the attractor of a fluid in a turbulent regime has a geometric dimension d which is strictly higher than 2 and strictly lower than 3, that is not integer:

$$2 < d < 3 \tag{4.4}$$

It then necessarily is a fractal object (for example, refer to Mandelbrot, 1982), which is usually known as a *strange attractor*. E. Lorentz (1963) first established such an attractor from a system of differential equations governing the evolution of velocity, pressure and temperature within a much simplified model of the atmosphere; this famous attractor is shown in Fig. 4.4. Its fractal structure is evidenced by an infinite number of 'plies' giving it a mille-feuille-like aspect. Oversight of the initial conditions and sensitivity to initial conditions coexist in it, and the laminated nature of this attractor makes the sensitivity possible without intersecting trajectories (they successively run across different plies).

Let us also observe the twin 'eyelet'[12] which takes part in the divergence of the trajectories before they converge again. The combination of successive temporary convergence and divergence of the trajectory looks like the

[12]Historically, Lorentz was prompted by that shape to speak of the *butterfly effect*. It seems that he had initially thought of the flapping wings of a seagull.

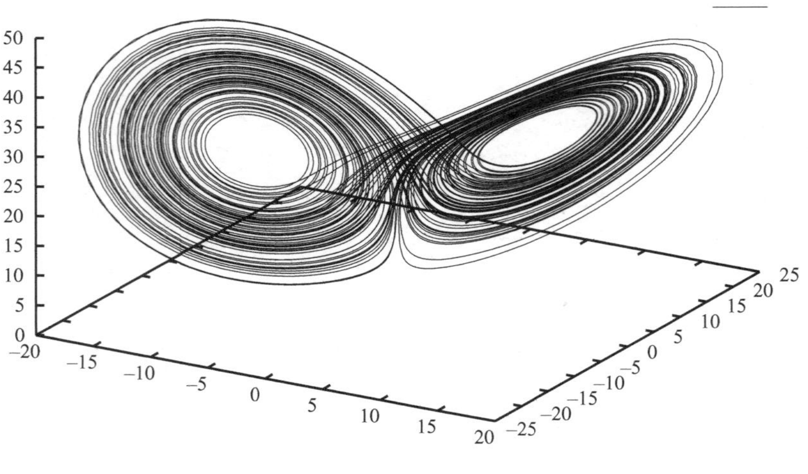

Fig. 4.4 The Lorentz attractor (image by R. Mizuno, www.mizuno.org).

mechanism of *stretching–folding over* in a flaky pastry recipe. This mechanism provides the attractor with its fractal aspect. We know from experience that this process is an outstanding procedure for *mixing* (flour and butter, for instance), and we can see that the structure of the Lorentz attractor has something in common with a diffusion (a mixing). This phenomenon is at the core of the chaotic processes (as in the case of the previously mentioned non-linear billiard) and the turbulent flow dynamics, and we will come back to it many a time. The stretching and folding over phenomenon which has just been described can be physically found again in the fluctuating motion of the fluids, and this is the way the mixing takes place in it, through the significant deformation of the medium. The attractor of a turbulent flow should have, if not the shape, at least the properties of the Lorentz attractor, and the actual physical properties of such a flow occur in the geometric structure of its attractor.

In order to go deeper into these notions, let us more quantitatively return to the notion of energy dissipation, which increasingly appears to be essential in the presently involved mechanisms. The volume reduction in the phase space is given, as earlier explained, by equation (2.226), which is just a variant of the dissipative Liouville's theorem and which we now rearrange as:

$$\frac{1}{\overline{V}}\frac{d\overline{V}}{dt} = -tr\,(\boldsymbol{\alpha}\boldsymbol{\Gamma}) \doteqdot -\frac{1}{T} < 0 \tag{4.5}$$

which is a definition of the time scale T. Assuming it is constant,[13] we get a locally exponential decrease of the volumes:

$$\overline{V} = \overline{V}_0 \exp\left(-\frac{t}{T}\right) \tag{4.6}$$

(to be compared with eqn (2.236)). For further insights into this behaviour, let us return to those quantities describing the system, and let us contemplate the particular case of a system having three degrees of freedom denoted as X_i and making up a three component-vector denoted as **X**. Like the Hamilton equations (1.92) or the Navier–Stokes equations (3.126), the equations

[13]This hypothesis is generally wrong, because there is no reason for the matrices $\boldsymbol{\alpha}$ and $\boldsymbol{\Gamma}$ to be constant, except in such particular cases as the damped pendulum, which was dealt with at the end of Section 2.5.1. It may be assumed that this hypothesis is valid to the first approximation in the vicinity of a particular point on the attractor, as discussed hereinafter. The result will then be qualitatively generalizable.

governing these quantities must be differential and of the first order, which can[14] be reduced to the following condensed form:

$$\dot{\mathbf{X}} = \mathbf{f}(\mathbf{X}) \tag{4.7}$$

where **f** is a vector function.[15] Since we are considering the system behaviour in the short run, it is advisable to expand this function into a first-order Taylor series around an arbitrary point $\mathbf{X}_0$:

$$\frac{d(\mathbf{X}-\mathbf{X}_0)}{dt} = \mathbf{G}\cdot(\mathbf{X}-\mathbf{X}_0) + O\left(|\mathbf{X}-\mathbf{X}_0|^2\right) \tag{4.8}$$

where **G** is the gradient matrix of vector **f** at point $\mathbf{X}_0$. The approximation we have just made amounts to introducing a matrix which has the dimension of a frequency T^{-1} and is assumed to be constant, as we did in Section 1.4.4 with the pendulum frequency being denoted as $\omega_0 \equiv 2\pi/T$, or else in Section 2.5.3 with the matrix $\mathbf{G}_a = T_R^{-1}\mathbf{I}_3$ (eqns (2.285) and (2.298)). Actually, the following discussion is locally valid, in the vicinity of an arbitrary point $\mathbf{X}_0$. The local solution of the linearized system (4.8) is written as[16]

$$\mathbf{X}-\mathbf{X}_0 = \mathbf{Y}^T \exp(\mathbf{G}t) \tag{4.9}$$

where **Y** is an unknown vector. Let us assume that the matrix **G** is diagonalizable and has real eigenvalues λ_i which we will arrange so that $\lambda_1 \geqslant \lambda_2 \geqslant \lambda_3$. These quantities are known as a system's *Lyapunov exponents*.[17] We get, out of the **G** eigenbasis:

$$\mathbf{X} = \mathbf{X}_0 + \mathbf{Y}^T \begin{pmatrix} e^{\lambda_1 t} & 0 & 0 \\ 0 & e^{\lambda_2 t} & 0 \\ 0 & 0 & e^{\lambda_3 t} \end{pmatrix} \tag{4.10}$$

Moreover, the vector field **f** represents the field of evolution 'velocities' of the dynamical system in the phase space. If we take up again the idea of the abstract model as introduced in Section 2.2.2 in which this field is likened to an abstract fluid, matrix **G** represents the velocity gradient of this 'fluid' at point $\mathbf{X}_0$, by formal analogy with (3.13). Its trace, equal to the trace of the 'abstract deformation rate'[18], represents the evolution rate of volumes $\overline{V}$ (still in the abstract phase space), pursuant to (3.19). Hence, we have

$$\frac{1}{\overline{V}}\frac{d\overline{V}}{dt} = \operatorname{tr}\mathbf{G} = \lambda_1 + \lambda_2 + \lambda_3 < 0 \tag{4.11}$$

the sign being determined through the comparison with equation (4.5). Hence we get $\lambda_1 + \lambda_2 + \lambda_3 = -1/T$, and the Jacobian determinant of the transformation (4.10) makes it possible to find (4.6) again.[19] Consequently, at least one of the Lyapunov exponents must be negative, namely λ_3. In the local eigenbasis (at point $\mathbf{X}_0$) of **G**, it means that those trajectories which are locally separated by a small eigenvector of λ_3 come closer to each other (refer to Fig. 4.5 (*a*)). Since these trajectories are 1-dimensional objects, no curve is expanded nor contracted along its own local direction. It means that one of the factors $e^{\lambda_i t}$ must equal 1. One of the exponents (i.e. λ_2) must then equal 0. The exponent λ_1 is then strictly positive so that a sensitivity to the initial conditions, that is the mutual distance between the trajectories in a given direction, illustrated in

[14] If the differential system is of the second order, then it can still be expressed in a different form to be reduced to a first-order system. This is what we did when reducing the Lagrange equations to the Hamilton equations in Chapter 1.

The presently used formalism is advantageous in that it is fairly general, which allows us to formally tackle all kinds of dynamic systems, including, to some extent, biological, chemical or economic systems.

[15] In the case of a system governed by the Hamilton equations, **f** is nothing but the operator D_H given by (1.112).

[16] We may notice that the following formula is closely related to the formal solution of the Hamilton equations (1.118), as well as the solution (2.290) of the Fokker–Planck equation.

[17] Let us specify: at point $\mathbf{X}_0$.

[18] It is the divergence of $\dot{\mathbf{X}}$ as well, by analogy with (3.19).

[19] This may be compared with the case (2.12) of a conservative system.

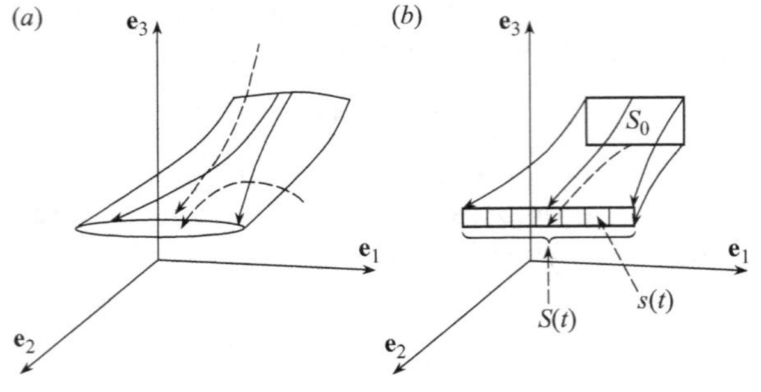

Fig. 4.5 Calculation of the fractal dimension of a strange attractor through the Lyapunov exponents.

Fig. 4.5 (*a*), can exist. The higher $|\lambda_3|$ is, the more 'attractive' the attractor is; the higher λ_1 is, the more dissipative the system is. However, the following restriction persists: λ_1 is necessarily strictly lower than the absolute value of λ_3 for the sign of (4.11) to be actually negative. Hence:

$$\lambda_3 < \lambda_2 = 0 < \lambda_1 < |\lambda_3| \tag{4.12}$$

This clearly explains why at least three degrees of freedom are required for the attractor and the sensitivity to initial conditions to coexist.[20]

The Lyapunov exponents determine the sensitivity of a process to small uncertainties as regards its initial conditions. Equation (4.10) shows that the exponents have the dimension of a time inverse. Quantity $\tau_1 \doteqdot 1/\lambda_1$ stands, for example, for the characteristic time scale for the distance between trajectories to increase in the phase space. It is sometimes referred to as the *Lyapunov time* or the *Lyapunov horizon*, since it provides a measure of the time from which the system is hardly predictable.[21] For a sufficiently long time before that horizon, the system may be considered a random system, because no memory of the earlier time points is retained.

By means of the concepts we have just set out, the fractal dimension of the system's attractor can be calculated[22] through the following procedure. Let us first consider a d-dimensioned hypercube (i.e. a square for $d = 2$, an ordinary cube for $d = 3$, etc.). If all the distances are cut into k smaller segments, then we get $N = k^d$ hypercubes having the same dimension d as the original one, but which are smaller. Generalizing that approach to any object, its dimension can be written as

$$d = \log_k N \tag{4.13}$$

Let us now consider the attractor, an enlarged schematic view of which is shown in Fig. 4.5 (*b*), and let us estimate its fractal dimension. To this purpose let us first turn to a *Poincaré section*, that is the intersection of the attractor with an arbitrary transverse plane. Each point in that section evolves in accordance with the law (4.7), and its area $S(t)$ varies according to the law

$$S(t) = S_0 e^{(\lambda_1+\lambda_3)t} < S_0 \doteqdot S(t=0) \tag{4.14}$$

This decrease of areas within the phase space directly results from the decrease of volumes, provided that $\lambda_2 = 0$. The section comprises N smaller elements, all of which are identical to the initial section S_0 and each one has an area

[20] It should be pointed out that such an analysis cannot be conducted in dimension 2. With the Van der Pol pendulum as set out in Section 4.2.1, for example, matrix **G** would be yielded by eqn (2.229). Since the sign of α may be positive in this case (refer to (2.178)), either a decrease or an increase of the volumes in the phase space can be observed, depending on the point in time being considered.

[21] In meteorology, for instance, the Lyapunov time is on the order of 2–3 days. In astronomy, where the systems are more slowly variable, it can be as long as several thousand years.

[22] There are several definitions of it. The procedure we are disclosing ends up in the so-called Hausdorff–Besicovitch dimension (Besicovitch and Ursell, 1937).

$$s(t) = S_0 e^{2\lambda_3 t} \tag{4.15}$$

Hence, we get

$$\begin{aligned} N &= \frac{S(t)}{s(t)} \\ &= e^{(\lambda_1 - \lambda_3)t} > 1 \end{aligned} \tag{4.16}$$

(we may recall, indeed, that λ_3 is negative). Between the initial time point $t = 0$ and the time point t, the distance shortening factor k is given by the shortening along axis x_3, that is

$$\begin{aligned} k &= \frac{\ell_3(t=0)}{\ell_3(t)} \\ &= e^{-\lambda_3 t} > 1 \end{aligned} \tag{4.17}$$

Formula (4.13) can be used to derive therefrom the dimension of the Poincaré section. To get the attractor dimension, one also must add 1, corresponding to its dimension along direction x_3. With (4.16) and (4.17), we get

$$\begin{aligned} d &= 1 + \frac{\ln N}{\ln k} \\ &= 2 + \frac{\lambda_1}{|\lambda_3|} \end{aligned} \tag{4.18}$$

The property (4.12) then states that the relation (4.4) is duly satisfied.

Like the other attractor's properties, the fractal nature can theoretically be found again in the geometry describing the motion of the system being investigated; we will provide an insight into it in the case of a turbulent regime flow in Section 4.4.3.

4.2.3 Scales and spectra

We have explained, in the previous section, that a system in a chaotic regime features a host of frequencies. As soon as that word is used, it is echoed by the name of J. Fourier. The tools named after him will be useful to go more deeply into the dynamics of these systems, particularly of the fluids in a turbulent regime. We may recall that since every quantity A depends on time t, its temporal Fourier transform (refer to Appendix B is a function of frequency space ω as yielded by

$$\widehat{f}(\omega) \doteqdot \int_{-\infty}^{+\infty} f(t)\, e^{-i\omega t} dt \tag{4.19}$$

Thus, a system obeying an oscillatory regime in the form of a circular function with a frequency ω_0 (like the non-damped pendulum in Section 1.4.4) is featured by a signal whose Fourier transform is a localized Dirac distribution at $\omega = \omega_0$ (refer to eqn (B.60)). A system having two distinct periods has a transform exhibiting two peaks, and so on. Conversely, a chaotic system is featured by a continuous band of frequencies, which means that its Fourier transform has a continuous shape on a given support. We do not intend to review all the causes and effects of such a behaviour within the general frame of the dynamics of non-linear systems, but we are about to explain why a fluid in a turbulent regime can give rise to such a great complexity.

We have previously said that the advection term in the Navier–Stokes equations, because of its non-linearity, is destabilizing, and it is therefore the primary cause of a chaos within a fluid. Let us define how it may act upon an an elementary oscillation of the velocity field. To that purpose, let us consider the ideal case of a one-dimensional and monochromatic velocity field with a wavelength λ:

$$u(x) = U \sin \frac{2\pi x}{\lambda} \tag{4.20}$$

The corresponding advection term is written as[23]

$$\begin{aligned} \frac{\partial u}{\partial x} u &= \frac{2\pi U}{\lambda} \cos \frac{2\pi x}{\lambda} \times U \sin \frac{2\pi x}{\lambda} \\ &= \frac{\pi U^2}{\lambda} \sin \frac{4\pi x}{\lambda} \end{aligned} \tag{4.21}$$

[23] One could also apply the Fourier transform and deduce the forthcoming result from its properties.

Thus, this term reveals a wavelength which is half the first one: $\lambda/2$. The same process obviously persists in three dimensions, although it is orchestrated in a more complex way. Thus, an oscillation (one eddy or a series of eddies) with a characteristic size λ is split up into smaller eddies due to the action of advection. The eddies generated by that dislocation, of course, are subjected to the same process, and we can observe a sequence of smaller and smaller eddies. This mechanism induces the spectral complexity of the turbulence, as well as its fractal nature. We will come back to this question in Section 4.4.3.

The process we have just described cannot, of course, indefinitely continue, since the eddy sizes are inevitably larger than the molecular mean free path. Actually, the process is interrupted on a substantially larger scale,[24] as we are going to see. We know, indeed, that the Reynolds number, which is a key parameter in the dynamical behaviour of a fluid, represents a measure of the conflict existing between the (destabilizing) advection and the (stabilizing) viscous forces. Let us try to quantify these two forces on the scale of an eddy with a size λ and a velocity u_λ, by analogy with (3.156):

[24] Except in the excessively rarefied gases, as happens in the upper layers of the Earth's atmosphere.

$$\left(\frac{|\mathbf{grad\ u} \cdot \mathbf{u}|}{|\nu \nabla^2 \mathbf{u}|} \right)_{\lambda,u} \sim \frac{u_\lambda \lambda}{\nu} \doteqdot Re_\lambda \tag{4.22}$$

Although we have no information so far about the way u_λ depends on λ, we can intuitively guess that Re_λ decreases along with λ (Section 4.4.3 will substantiate this result). Thus, as the eddies become smaller, the viscous forces play an increasingly major part in it versus the inertia forces. As the eddies are dislocated, such a small scale occurs that the former forces prevail over the latter, and so the velocity field becomes smooth in them, viscosity resulting in a more homogeneous velocity field (refer to Section 3.4.2). On that scale, we can then observe the dying out of the eddies through viscous dissipation. That scale λ_0 is then the characteristic size of the smallest turbulent eddies; it is known as the *Kolmogorov scale* and corresponds to a Re_λ on the order of the unit. Denoting as u_0 the velocity scale of these small eddies, we then get

$$\lambda_0 \sim \frac{\nu}{u_0} \tag{4.23}$$

Let us now briefly introduce the formalism for describing the spectral nature of turbulence at a given point and time of an arbitrary flow. To each λ-sized eddy corresponds a wave number

$$K \doteqdot \frac{2\pi}{\lambda} \tag{4.24}$$

More precisely, one can build a wave vector $\mathbf{K}$ having K as a norm and aligned with the eddy's rotational axis. If the 'turbulent' velocity field representing the eddies is denoted as $\mathbf{u}'$,[25] its spatial Fourier transform can then be built (refer to Appendix B), as defined by

[25] The exact definition of $\mathbf{u}'$ will be given in the next section.

$$\hat{\mathbf{u}}'(\mathbf{K}, t) = \int_{\mathbb{R}^3} \mathbf{u}'(\mathbf{r}, t)\, e^{-i\mathbf{K}\cdot\mathbf{r}} d^3\mathbf{r} \tag{4.25}$$

and so, conversely:

$$\mathbf{u}'(\mathbf{r}, t) = \frac{1}{(2\pi)^3} \int_{\mathbb{R}^3} \hat{\mathbf{u}}'(\mathbf{K}, t)\, e^{i\mathbf{K}\cdot\mathbf{r}} d^3\mathbf{K} \tag{4.26}$$

The disorderliness of the velocity field over time arises over space as well, and the spatial Fourier transform (4.25) brings about the same remarks as the temporal transform 4.19). Thus, the turbulence specific quantities are distributed over a continuous spectrum. We have just explained that when K exceeds[26] $K_0 \doteqdot 2\pi/\lambda_0$ (according to (4.24)), $\hat{\mathbf{u}}'$ tends towards 0 (the eddies fade away). For that reason, the convergence of the integral (4.26) is possible. Besides, since the eddies cannot have an infinite size, we also get $\hat{\mathbf{u}}'(\mathbf{K} = \mathbf{0}) = 0$. Since velocities are real functions, we also get (first line in eqn (B.39) of Appendix B):

[26] The large values of K correspond, of course, to the small eddies, and the converse.

$$\hat{\mathbf{u}}'(-\mathbf{K}, t) = \hat{\mathbf{u}}'^{*}(\mathbf{K}, t) \tag{4.27}$$

where the star stands for the conjugate complex. We will take advantage of these considerations in Section 4.4.3.

4.3 One-point statistical models

We consider first the Reynolds-averaging operator, and derive its properties. The continuity and momentum equations of a continuous medium and a viscous weakly compressible fluid are averaged, and we introduce the Reynolds stress tensor. Finally, we derive an equation governing the behaviour of the latter, introducing the production, dissipation and pressure-strain correlation tensors.

4.3.1 Reynolds-averaging

The discussion in Section 4.2.2 leaves a fairly pessimistic feeling as to our ability to model the turbulent flows. The latter, indeed, happen to be geometrically quite complex, rapidly varying over time and highly sensitive to the initial conditions (and, it should be added, to the boundary conditions). The so-called DNS (*Direct Numerical Simulation*)[27] approach, which consists in wholly modelling that complexity by solving the high Reynolds number Navier–Stokes equations, is then still a matter for scientific investigation, because it requires a very fine three-dimensional discretization together with very short time steps.[28] Historically, however, the issues due to the turbulence could be addressed through another approach, with the help of one of the difficulties encountered (sensitivity to the initial conditions), that is the Reynolds averaging approach. This idea is closely related to the notion of Boltzmann average as addressed in Section 2.2.1.

The idea consists of considering a given flow (which means a certain geometry, along with preset *average* boundary conditions), then achieving (mentally, if needed) N occurrences of that flow, which we will refer to as flow *instances*. The minute—yet unavoidable—differences in both initial and boundary conditions (boundary irregularities, impurities, etc.) will then result in instances which will differ from each other on details (remember the example of cigarette smoke as mentioned in Section 4.2.2). If one arbitrary point is set in both space and time, then a measure of any field A (velocity, pressure) at both of these points will therefore yield a series of different values according to the flow instance. We will denote these measures as $A^{(i)}$, the superscript$^{(i)}$ referring to a particular instance. An experimental (and consistent with intuition) circumstance subsequently proves helpful: whereas two instances differ on details (i.e. on the structure of the turbulent eddies), on the other hand the *overall* flow remains invariant from one instance to another one.[29] Thus, the smoke plume of a second cigarette is more similar to the smoke plume of a first cigarette. Whereas the wreaths differ, the overall motion, conversely, is globally unvarying. The word 'similar' here means a global resemblance as perceived by 'screwing up' one's eyes. This is how we hit upon the idea of submitting a statistical average of quantity A:

$$\overline{A} \doteqdot \lim_{N\to\infty} \frac{1}{N} \sum_{i=1}^{N} A^{(i)} \tag{4.28}$$

Thus, we define an *averaging* operator (to be interpreted as a probabilistic expectation).[30] Nothing but intuition and experience can then a priori secure the convergence of that series as N increases; a qualitative justification of it, however, will be provided later on.

It must be repeated that the relation (4.28) is defined at one given point in space and time, and so this is a newly defined field, which we will refer to as a *mean field* (according to Reynolds). This field does not depend on the instance of the experiment and happen to be a specific flow feature, a quantity having a deep physical meaning since it can be replicated from one instance to another. On the contrary, each instance is featured by a particular

[27] To be more precise, the DNS technique is generally performed in the Fourier space.

[28] Further details will be provided in Section 4.4.6.

[29] From a statistical point of view, the instances are said to be independent and identically distributed.

[30] That procedure is required in this case. Defining a spatial average as in the elementary theory of continuous media (refer to Chapter 3, Section 3.2.1) would be unreliable because of the too wide range of turbulent fluctuations. In order to be more rigorous, it would be necessary to introduce a probability density of the turbulent quantities and formally build the averages as in the frame of the switching from the Boltzmann equations to those equations governing the continuous media (Section 3.5.1). The turbulent flow equations can then be obtained from an equation governing the transfer of that probability density, which is turned into a non-linear Boltzmann equation. Many details about this topic can be found in Pope (2000).

The following remark should be made too. As explained by Reynolds (1894), for the sake of consistency with the ideas expressed in Chapter 1, the mean velocity should be defined as an average of the density-weighted velocities. By analogy with eqn (1.181):

$$\overline{A} \doteqdot \lim_{N\to\infty} \frac{\sum_{i=1}^{N} \rho^{(i)} A^{(i)}}{\sum_{i=1}^{N} \rho^{(i)}} \tag{4.29}$$

The *Favre average* is then often referred to (Favre, 1989). Within the context of the weakly compressible flows, however, we may consider that that density hardly ever varies from one instance to another, as explained in the next section. The previous formula then comes down to (4.28).

distribution of the turbulent eddies, that is by a particular *fluctuating field* of the quantity A:

$$A'^{(i)} \doteqdot A^{(i)} - \overline{A} \tag{4.30}$$

The field A will henceforth be described as *instantaneous* or *real* (in the following, we will drop the superscript $^{(i)}$ for the sake of clarity). The velocity and pressure (and possibly density) fields of a fluid will subsequently be treated this way. Hence we will get:

$$\begin{aligned} \mathbf{u} &= \overline{\mathbf{u}} + \mathbf{u}' \\ p &= \overline{p} + p' \end{aligned} \tag{4.31}$$

which gives the exact definition of the field $\mathbf{u}'$ as mentioned in Section 4.2.3. Figure 4.6 illustrates the shape of a real velocity field $\mathbf{u}$ and its average $\overline{\mathbf{u}}$. We can already perceive in it the major properties of the mean fields:

- they are smooth, regular, without the tormented aspect of the instantaneous fields. In this respect, they represent a stable dynamics, free from the complexity of the strange attractor;
- they may exhibit recirculations, but the latter can be replicated and then cannot be likened to turbulent eddies, which are sensitive to the initial conditions;
- they may be constant over time, unlike the instantaneous fields which are always variable;
- they may obey an invariance along a given direction and thus occur (for example) in the form of two-dimensional fields, whereas the fluctuating fields (the eddies) are always three-dimensional.

It should be kept in mind that the mean fields are not defined by any spatial or temporal average, but instead by a statistical average (over a large number of instances). If a flow exhibits a time-averaged steady nature, that is if the mean fields are permanent (or hardly variable over time), however, the averages can be defined as temporal, provided that the selected sampling time is sufficiently long with respect to the characteristic time scale of the fluctuating fields,[31] that is the system's Lyapunov time (refer to Section 4.2.2). Likewise, if the fields are spatially hardly variable, the averages can be defined as locally spatial; this is referred to as a homogeneous turbulence.[32]

Since the mean fields can be replicated and are smooth, we will subsequently (except for Sections 4.6.1, 4.6.2 and 4.6.3) try to model them. This approach, which is known the as *one-point statistical average*, is the oldest and most elementary approach, although it gives rise to many issues; nowadays, it is

[31] That time scale characteristic will be provided in Section 4.4.3. It is noteworthy, however, that the presently requisite condition itself a priori involves the notion of mean quantities, which makes the set forth property questionable.

[32] Actually, this condition almost never occurs, because of the required presence of walls and/or interfaces, affecting the spatial distribution of the quantities through the boundary conditions (refer to Section 4.4.6).

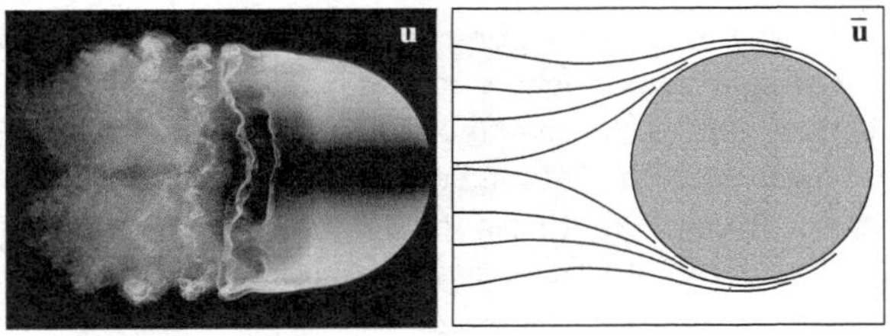

Fig. 4.6 Exemplary turbulent velocity field (Copyright by Onera, French aerospace research center) and shape of its Reynolds average.

still quite useful for numerical modelling in industry and environment. The mean fields, of course, certainly do not satisfy the Navier–Stokes equations, which are non-linear and therefore do not obey the superposition principle. We will then have to apply the Reynolds' averaging operator to the Navier–Stokes equations in order to obtain those equations governing the mean fields. To do this, the mathematical properties of this operator should be set forth (without being demonstrated, because of their intuitive nature).[33] Thus, for all (scalar, vector, tensor) fields A and B, all constants α and β and every spatial index i, we have:

[33] Besides, these properties are partly analogous to those of the averages according to Boltzmann which were discussed in Sections 2.2.1 and 3.5.1.

$$\begin{aligned}\overline{\alpha A + \beta B} &= \alpha \overline{A} + \beta \overline{B} \\ \overline{\frac{\partial A}{\partial t}} &= \frac{\partial \overline{A}}{\partial t} \\ \overline{\frac{\partial A}{\partial x_i}} &= \frac{\partial \overline{A}}{\partial x_i}\end{aligned} \tag{4.32}$$

These properties reveal the linearity of the Reynolds averaging operator, that is the primary result of its definition (4.28). A further property stems from the last of these properties: the averaging operator commutes with all such linear differential operators as divergence and gradient. We furthermore must add the following complementary properties:

$$\begin{aligned}\overline{A'} &= 0 \\ \overline{\overline{A}} &= \overline{A} \\ \overline{\alpha} &= \alpha\end{aligned} \tag{4.33}$$

The first results from the definition of the fluctuating fields as a statistical deviation from the average (4.30), whereas the second means that the Reynolds operator is a projector. Lasly, we must specify an essential counter-property: the average of a product is not equal to the product of the averages, unless one of the involved quantities is already an average:

$$\begin{aligned}\overline{AB} &\neq \overline{A}\,\overline{B} \\ \overline{\overline{A}B} &= \overline{A}\,\overline{B}\end{aligned} \tag{4.34}$$

In particular, the average does not commute with the Lagrangian derivative operator (3.7), which involves an advection term, which is in essence non-linear.

4.3.2 Reynolds equations

The fact that the averaging operator commutes with the linear differential operators has a simple consequence: in the Navier–Stokes' system (3.126), only the advection term in the equation of motion and the second term in the continuity equation (both of them being nonlinear) will be affected by the averaging operator. We will first treat the continuity equation by breaking up both density and velocity into mean and fluctuating parts according to (4.30), then by applying the averaging operator:

$$\frac{\partial \overline{\overline{\rho} + \rho'}}{\partial t} + \mathrm{div}\left[\overline{(\overline{\rho} + \rho')(\overline{\mathbf{u}} + \mathbf{u}')}\right] = 0 \tag{4.35}$$

The non-linear term is treated as follows:

$$\begin{aligned}\overline{(\overline{\rho} + \rho')(\overline{\mathbf{u}} + \mathbf{u}')} &= \overline{\overline{\rho}\,\overline{\mathbf{u}} + \rho'\overline{\mathbf{u}} + \overline{\rho}\mathbf{u}' + \rho'\mathbf{u}'} \\ &= \overline{\rho}\,\overline{\overline{\mathbf{u}}} + \overline{\rho'}\overline{\mathbf{u}} + \overline{\rho}\,\overline{\mathbf{u}'} + \overline{\rho'\mathbf{u}'} \\ &= \overline{\rho}\,\overline{\mathbf{u}} + \overline{\rho'\mathbf{u}'}\end{aligned} \tag{4.36}$$

(we have first used the linearity of the average, then the properties (4.34) and (4.33). Thus, (4.35) becomes

$$\frac{\partial \overline{\rho}}{\partial t} + \mathrm{div}\,(\overline{\rho}\,\overline{\mathbf{u}}) = -\mathrm{div}\left(\overline{\rho'\mathbf{u}'}\right) \tag{4.37}$$

The nearly incompressible nature of the flow, however, substantially restricts the fluctuations of density. Practically, we may then ignore the term in the right-hand side of (4.37) and make $\overline{\rho}$ similar to ρ (which we always will do in the following), and so the continuity equation will remain unchanged (except that it is now concerned with the mean velocity field):

$$\frac{\partial \rho}{\partial t} + \mathrm{div}\,(\rho\overline{\mathbf{u}}) = 0 \tag{4.38}$$

Through a calculation identical to that enabling us to switch from (3.42) to (3.43), equation (4.38) can also be written as:

$$\frac{D\rho}{Dt} = -\rho\mathrm{div}\overline{\mathbf{u}} \tag{4.39}$$

The Lagrangian derivative provided with a capital D denotes a material derivative following the mean motion.[34] As a rule, the following will be defined for every quantity A, by analogy with (3.7):

$$\frac{DA}{Dt} \doteq \frac{\partial A}{\partial t} + \mathbf{grad}\,A \cdot \overline{\mathbf{u}} \tag{4.40}$$

When the flow is really incompressible,[35] since the continuity equation is linear, it is utterly unchanged by the averaging operator:

$$\mathrm{div}\overline{\mathbf{u}} = 0 \tag{4.41}$$

We may note that, in this case, the fluctuating velocity field obeys, by subtraction, the same equation:

$$\mathrm{div}\mathbf{u}' = 0 \tag{4.42}$$

Let us now turn to what the average of the advection term in the Navier–Stokes' equation of motion (3.139) becomes in the general case where density is still assumed to be variable. To that purpose, we will firstly call upon equation (3.45), and so we may write

$$\overline{\rho\frac{d\mathbf{u}}{dt}} = \frac{\partial \rho\overline{\mathbf{u}}}{\partial t} + \mathbf{div}\left(\rho\overline{\mathbf{u} \otimes \mathbf{u}}\right) \tag{4.43}$$

We may then treat the velocity field as in (4.36):[36]

[34] It should be pointed out that this concept has a limited physical, essentially statistical, meaning. We will discuss this idea again in Chapter 6.

[35] With the reservations expressed in Chapter 2 (Section 3.4.2).

[36] We may note the formal analogy of the following with the calculation (3.244).

$$\begin{aligned}\overline{\mathbf{u} \otimes \mathbf{u}} &= \overline{(\overline{\mathbf{u}} + \mathbf{u}') \otimes (\overline{\mathbf{u}} + \mathbf{u}')} \\ &= \overline{\mathbf{u}} \otimes \overline{\mathbf{u}} + \overline{\mathbf{u}' \otimes \mathbf{u}'}\end{aligned} \tag{4.44}$$

Lastly, since the equality (3.44) is linked to the continuity equation, it is still valid for the mean field, and so, ultimately:

$$\begin{aligned}\overline{\rho \frac{d\mathbf{u}}{dt}} &= \frac{\partial \rho \overline{\mathbf{u}}}{\partial t} + \mathbf{div}\,(\rho \overline{\mathbf{u}} \otimes \overline{\mathbf{u}}) + \mathbf{div}\left(\rho \overline{\mathbf{u}' \otimes \mathbf{u}'}\right) \\ &= \rho \frac{\partial \overline{\mathbf{u}}}{\partial t} + \rho \mathbf{grad}\, \overline{\mathbf{u}} \cdot \overline{\mathbf{u}} + \mathbf{div}\,(\rho \mathbf{R})\end{aligned} \tag{4.45}$$

where a new tensor as defined by

$$\mathbf{R} \doteqdot \overline{\mathbf{u}' \otimes \mathbf{u}'} \tag{4.46}$$

is introduced, or else, in the form of Cartesian components:

$$R_{ij} \doteqdot \overline{u'_i u'_j} \tag{4.47}$$

It happens to be the (symmetric) tensor of statistical self-correlations (which are also known as *second-order moments*) of the fluctuating velocity vector, and represent the effect of the velocity fluctuations over the mean flow, which results from the non-linearity of the equations. It can be seen that the quantity

$$\frac{D\overline{\mathbf{u}}}{Dt} \doteqdot \frac{\partial \overline{\mathbf{u}}}{\partial t} + \mathbf{grad}\, \overline{\mathbf{u}} \cdot \overline{\mathbf{u}} \tag{4.48}$$

appears in (4.45) in accordance with the notation (4.40). We have then shown the following identity:

$$\overline{\rho \frac{d\mathbf{u}}{dt}} = \rho \frac{D\overline{\mathbf{u}}}{Dt} + \mathbf{div}\,(\rho \mathbf{R}) \tag{4.49}$$

In order to obtain an equation of motion with respect to the mean velocity field, we just have to multiply the Navier–Stokes equation by ρ, average the result before dividing it again by ρ. Only the inertia term, which is nonlinear, will be modified by that operation. Using (4.49), the Navier–Stokes equations as averaged according to Reynolds—which will now be referred to as the *Reynolds equations*—occur in the following form:

$$\begin{aligned}&\frac{\partial \rho}{\partial t} + \mathrm{div}\,(\rho \overline{\mathbf{u}}) = 0 \\ &\frac{D\overline{\mathbf{u}}}{Dt} = -\frac{1}{\rho} \mathbf{grad}\, \overline{p}^* + \nu \nabla^2 \overline{\mathbf{u}} - \frac{1}{\rho} \mathbf{div}\,(\rho \mathbf{R})\end{aligned} \tag{4.50}$$

for the weakly compressible flows, or

$$\begin{aligned}&\mathrm{div} \overline{\mathbf{u}} = 0 \\ &\frac{D\overline{\mathbf{u}}}{Dt} = -\frac{1}{\rho} \mathbf{grad}\, \overline{p}^* + \nu \nabla^2 \overline{\mathbf{u}} - \mathbf{div}\, \mathbf{R}\end{aligned} \tag{4.51}$$

for the incompressible flows (in the latter case, the new term occurs in the form **div R**).

These equations are in every respect similar to the Navier–Stokes equations, except that they are concerned with the mean fields and include, in the equation of motion, an additional term which we have entered into the right-hand side. Now, returning to the Cauchy equation of motion (3.71), we can see that the same term would appear after the application of the averaging operator:

$$\frac{\partial \overline{\mathbf{u}}}{\partial t} + \mathbf{grad}\,\overline{\mathbf{u}} \cdot \overline{\mathbf{u}} = \frac{1}{\rho}\mathbf{div}\,(\overline{\boldsymbol{\sigma}} - \rho\mathbf{R}) + \mathbf{g} \tag{4.52}$$

The above discussion shows that the new term happens to be an additional stress tensor (it may be considered as such because it is symmetric); **R** will therefore be called the *Reynolds stress tensor*.[37] Thus, the quantity between brackets in (4.52) stands for the total stress tensor,[38] or the total momentum flux (according to the considerations in Section 3.3.3). The physical explanation for these additional stresses is as follows: the turbulent eddies (as represented by the fluctuant velocity vector $\mathbf{u}'$) act in two ways upon the mean flow. To be persuaded of this, let us contemplate a small virtual surface within the fluid, oriented by the unit vector **n**. We can observe the following facts:

[37] Although, strictly speaking, $-\rho\mathbf{R}$ plays that role, according to (4.52).

[38] We also may note the analogy of the definition of **R** with that of the Cauchy's tensor within the context of the Boltzmann approach (eqn (3.250)). Since the Reynolds stresses originate from the average of the inertia term, they are inherently non-linear.

- an eddy whose rotational axis is orthogonal to **n** will exert a 'hammering' effect onto the said surface, which will add further to the pressure which it is subjected to;
- an eddy whose rotational axis is oriented along **n** will induce on the surface a shear which will add further to the viscous molecular friction.

The equation for a scalar (3.61) can be treated likewise, the single term affected by the averaging operator being the advection term. We readily get:

$$\frac{D\overline{B}}{Dt} = \overline{S}_B + \mathrm{div}\left(K_B\mathbf{grad}\overline{B} - \mathbf{Q}^B\right) \tag{4.53}$$

where $\overline{S}_B$ is the average of the source term S_B, and by analogy with (4.46):

$$\mathbf{Q}^B \doteq \overline{B'\mathbf{u}'} = \overline{B'u'_i}\mathbf{e}_i \tag{4.54}$$

This vector represents a turbulent flux of the scalar, which is diffused (mixed) by the turbulent eddies, just like the momentum. The turbulent flux of scalar $\mathbf{Q}^B$ is, versus its molecular flux $\mathbf{q}^B$ (Section 3.3.1, eqn (3.53)) exactly what **R** is versus $\boldsymbol{\sigma}$. We may recall that a scalar is diffused by a vector, whereas a vector is diffused by a second-order tensor (refer to Chapter 2, Section 3.3.3). Thus, it can be observed that the scalar $\overline{B}$ is diffused by the vector $\mathbf{Q}^B$, the vector $\overline{\mathbf{u}}$ being diffused by the tensor **R**.

4.3.3 Reynolds stresses

Just as Cauchy's equation required, in Chapter 3, the utilization of a closure in the form of a behaviour law (Section 3.4.1) expressing $\boldsymbol{\sigma}$, the new stress tensor **R** demands further relations to close the system (4.50). An initial attempt consists in seeking an exact formulation for **R**. In the first place, it is attempted to establish a differential equation over **R**, resulting from the previously available equations. This work will be performed in the particular case of incompressible flows. We will build this equation from several successive steps.

First of all, we may note the following identity, which is based on the equality (3.9):

$$
\begin{aligned}
\mathbf{grad\, u} \cdot \mathbf{u} - \mathbf{grad\, \overline{u}} \cdot \overline{\mathbf{u}} &= \mathbf{div}\,(\mathbf{u} \otimes \mathbf{u} - \overline{\mathbf{u}} \otimes \overline{\mathbf{u}}) \\
&= \mathbf{div}\left(\mathbf{u}' \otimes \overline{\mathbf{u}} + \overline{\mathbf{u}} \otimes \mathbf{u}' + \mathbf{u}' \otimes \mathbf{u}'\right) \\
&= \mathbf{grad\, u}' \cdot \overline{\mathbf{u}} + \mathbf{grad\, \overline{u}} \cdot \mathbf{u}' + \mathbf{div}\left(\mathbf{u}' \otimes \mathbf{u}'\right)
\end{aligned}
\tag{4.55}
$$

(we have used the incompressibility relations, particularly (4.42)). Tensorially multiplying the latter equality on the right-hand side by $\mathbf{u}'$, we get (in an indicial form, to make the calculations clearer):

$$
\begin{aligned}
\left[(\mathbf{grad\, u} \cdot \mathbf{u} - \mathbf{grad\, \overline{u}} \cdot \overline{\mathbf{u}}) \otimes \mathbf{u}'\right]_{ij} =& \frac{\partial u'_i}{\partial x_k}\overline{u}_k u'_j + \frac{\partial \overline{u}_i}{\partial x_k} u'_k u'_j \\
&+ \frac{\partial u'_i u'_k}{\partial x_k} u'_j \\
=& \frac{\partial u'_i u'_j}{\partial x_k}\overline{u}_k - \frac{\partial u'_j}{\partial x_k} u'_i \overline{u}_k + \frac{\partial \overline{u}_i}{\partial x_k} u'_j u'_k \\
&+ \frac{\partial u'_i u'_j u'_k}{\partial x_k} - u'_i u'_k \frac{\partial u'_j}{\partial x_k}
\end{aligned}
\tag{4.56}
$$

Taking the average of the above formula, we soon get, through the already known properties of that operator and the definition (4.47) of the R_{ij}:

$$
\begin{aligned}
\left[\overline{(\mathbf{grad\, u} \cdot \mathbf{u} - \mathbf{grad\, \overline{u}} \cdot \overline{\mathbf{u}}) \otimes \mathbf{u}'}\right]_{ij} =& \frac{\partial R_{ij}}{\partial x_k}\overline{u}_k - \overline{\frac{\partial u'_j}{\partial x_k} u'_i \overline{u}_k} \\
&+ \frac{\partial \overline{u}_i}{\partial x_k} R_{jk} \\
&+ \frac{\partial \overline{u'_i u'_j u'_k}}{\partial x_k} - \overline{u'_i u'_k \frac{\partial u'_j}{\partial x_k}}
\end{aligned}
\tag{4.57}
$$

Let us now do the same work, but tensorially multiplying on the left-hand side the relation (4.55), which initally amounts to performing a permutation of indices i and j:

$$
\begin{aligned}
\left[\overline{\mathbf{u}' \otimes (\mathbf{grad\, u} \cdot \mathbf{u} - \mathbf{grad\, \overline{u}} \cdot \overline{\mathbf{u}})}\right]_{ij} =& \frac{\partial R_{ij}}{\partial x_k}\overline{u}_k - \overline{\frac{\partial u'_i}{\partial x_k} u'_j \overline{u}_k} \\
&+ \frac{\partial \overline{u}_j}{\partial x_k} R_{ik} \\
&+ \frac{\partial \overline{u'_i u'_j u'_k}}{\partial x_k} - \overline{u'_j u'_k \frac{\partial u'_i}{\partial x_k}} \\
=& \overline{\frac{\partial u'_j}{\partial x_k} u'_i \overline{u}_k} + \frac{\partial \overline{u}_j}{\partial x_k} R_{ik} + \overline{\frac{\partial u'_j}{\partial x_k} u'_i u'_k}
\end{aligned}
\tag{4.58}
$$

(we have used the product derivation rule, then (4.42) in an indicial form analogous to (3.48)). Lastly, it is obvious that

$$\overline{\frac{\partial \mathbf{u}'}{\partial t} \otimes \mathbf{u}' + \mathbf{u}' \otimes \frac{\partial \mathbf{u}'}{\partial t}} = \frac{\partial \overline{\mathbf{u}' \otimes \mathbf{u}'}}{\partial t} = \frac{\partial \mathbf{R}}{\partial t} \tag{4.59}$$

Adding up (4.57), (4.58) and (4.59), many terms disappear and we get:

$$\begin{aligned} &\mathbf{u}' \otimes \left(\frac{D\mathbf{u}}{Dt} - \frac{D\overline{\mathbf{u}}}{Dt} \right) + \left(\frac{D\mathbf{u}}{Dt} - \frac{D\overline{\mathbf{u}}}{Dt} \right) \otimes \mathbf{u}' \\ &= \frac{D\mathbf{R}}{Dt} + \left(\frac{\partial \overline{u}_i}{\partial x_k} R_{jk} + \frac{\partial \overline{u}_j}{\partial x_k} R_{ik} + \frac{\partial \overline{u'_i u'_j u'_k}}{\partial x_k} \right) \mathbf{e}_i \otimes \mathbf{e}_j \\ &= \frac{D\mathbf{R}}{Dt} + \mathbf{\Gamma}\mathbf{R} + \mathbf{R}\mathbf{\Gamma}^T + \mathbf{div}\left(\overline{\mathbf{u}' \otimes \mathbf{u}' \otimes \mathbf{u}'} \right) \end{aligned} \tag{4.60}$$

where we have used the Reynolds tensor's symmetry and introduced the gradient tensor of the mean velocity field:

$$\begin{aligned} \mathbf{\Gamma} &\doteqdot \mathbf{grad}\,\overline{\mathbf{u}} \\ &= \frac{\partial \overline{u}_i}{\partial x_j} \mathbf{e}_i \otimes \mathbf{e}_j \end{aligned} \tag{4.61}$$

Furthermore, we can establish the following identity:

$$\begin{aligned} &\mathbf{u}' \otimes \left(\nabla^2 \mathbf{u}' \right) + \left(\nabla^2 \mathbf{u}' \right) \otimes \mathbf{u}' \\ &= \left(u'_i \frac{\partial}{\partial x_k} \frac{\partial u'_j}{\partial x_k} + \frac{\partial}{\partial x_k} \frac{\partial u'_i}{\partial x_k} u'_j \right) \mathbf{e}_i \otimes \mathbf{e}_j \\ &= \left(\frac{\partial}{\partial x_k} \frac{u'_i \partial u'_j}{\partial x_k} - \frac{\partial u'_i}{\partial x_k} \frac{\partial u'_j}{\partial x_k} + \frac{\partial}{\partial x_k} \frac{u'_j \partial u'_i}{\partial x_k} - \frac{\partial u'_j}{\partial x_k} \frac{\partial u'_i}{\partial x_k} \right) \mathbf{e}_i \otimes \mathbf{e}_j \\ &= \left(\frac{\partial^2 u'_i u'_j}{\partial x_k \partial x_k} - 2 \frac{\partial u'_i}{\partial x_k} \frac{\partial u'_j}{\partial x_k} \right) \mathbf{e}_i \otimes \mathbf{e}_j \\ &= \nabla^2 \mathbf{R} - 2 \boldsymbol{\gamma}' \cdot \boldsymbol{\gamma}'^T \end{aligned} \tag{4.62}$$

with

$$\boldsymbol{\gamma}' \doteqdot \mathbf{grad}\,\mathbf{u}' \tag{4.63}$$

We will also need to calculate the following sum:

$$\begin{aligned} \overline{\frac{\partial p'}{\partial x_i} u'_j} + \overline{\frac{\partial p'}{\partial x_j} u'_i} &= \frac{\partial \overline{p' u'_j}}{\partial x_i} - \overline{p' \frac{\partial u'_j}{\partial x_i}} + \frac{\partial \overline{p' u'_i}}{\partial x_j} - \overline{p' \frac{\partial u'_i}{\partial x_j}} \\ &= \frac{\partial}{\partial x_k} \left(\overline{p' u'_i} \delta_{jk} + \overline{p' u'_j} \delta_{ik} \right) - 2 \overline{p' s'_{ij}} \end{aligned} \tag{4.64}$$

Here we have introduced the fluctuating part of the strain rate tensor, that is the symmetric part of (4.63):

$$\mathbf{s}' \doteqdot \boldsymbol{\gamma}'^S = \frac{1}{2}\left(\frac{\partial u'_i}{\partial x_j} + \frac{\partial u'_j}{\partial x_i}\right)\mathbf{e}_i \otimes \mathbf{e}_j \tag{4.65}$$

We have now got all the requisite elements for establishing a differential equation over **R**. The procedure is as follows: the Reynolds momentum equation (4.50) is subtracted from the Navier–Stokes equation (3.123), the result is tensorially multiplied by $\mathbf{u}'$ on both sides, then the results are added up, lastly the Reynolds averaging operator is applied. Using (4.60), (4.62) and (4.64) we come to

$$\frac{D\mathbf{R}}{Dt} = \frac{\partial \mathbf{R}}{\partial t} + \mathbf{gradR} \cdot \bar{\mathbf{u}} = \mathbf{P} - \mathbf{divQ}^{\mathbf{R}} + \Phi - \boldsymbol{\varepsilon} \tag{4.66}$$

with

$$\begin{aligned} \mathbf{P} &\doteqdot -\boldsymbol{\Gamma}\mathbf{R} - \mathbf{R}\boldsymbol{\Gamma}^T = -2\,(\boldsymbol{\Gamma}\mathbf{R})^S \\ \mathbf{Q}^{\mathbf{R}} &\doteqdot \overline{\mathbf{u}' \otimes \mathbf{u}' \otimes \mathbf{u}'} + \frac{2}{\rho}\overline{p'\,(\mathbf{u}' \otimes \mathbf{I}_3)^S} - \nu\mathbf{grad}\,\mathbf{R} \\ \boldsymbol{\Phi} &\doteqdot \frac{2}{\rho}\overline{p'\mathbf{s}'} \\ \varepsilon &\doteqdot 2\nu\overline{\boldsymbol{\gamma}' \cdot \boldsymbol{\gamma}'^T} \end{aligned} \tag{4.67}$$

(in particular, we have used the fact that $(\mathbf{div}\,\mathbf{R}) \otimes \mathbf{u}'$ has a zero average by virtue of the property (4.34), because **R** is already an averaged quantity).

In the first place, we may note that the first term in the diffusion tensor $\mathbf{Q}^{\mathbf{R}}$ is a triple correlation of fluctuating velocity:

$$\check{\mathbf{Q}}^{\mathbf{R}} \doteqdot \overline{\mathbf{u}' \otimes \mathbf{u}' \otimes \mathbf{u}'} \tag{4.68}$$

This term can be rearranged as:

$$\begin{aligned} \mathbf{div}\left(\overline{\mathbf{u}' \otimes \mathbf{u}' \otimes \mathbf{u}'}\right) &= \frac{\partial \overline{u'_i u'_j u'_k}}{\partial x_k}\mathbf{e}_i \otimes \mathbf{e}_j \\ &= \overline{\frac{\partial u'_i u'_j}{\partial x_k}u'_k}\mathbf{e}_i \otimes \mathbf{e}_j + \overline{u'_i u'_j \frac{\partial u'_k}{\partial x_k}}\mathbf{e}_i \otimes \mathbf{e}_j \\ &= \left(\overline{\frac{\partial u'_i}{\partial x_k}u'_k u'_j} + \overline{\frac{\partial u'_j}{\partial x_k}u'_k u'_i}\right)\mathbf{e}_i \otimes \mathbf{e}_j \\ &= 2\overline{\left[\boldsymbol{\gamma}'\,(\mathbf{u}' \otimes \mathbf{u}')\right]}^S \end{aligned} \tag{4.69}$$

(we have used the continuity equation for the fluctuating velocity field (4.42)). In a developed indicial form, (4.67) can also be written as:

$$
\begin{aligned}
\mathbf{P} &= -\left(R_{ik}\frac{\partial \overline{u}_j}{\partial x_k} + R_{jk}\frac{\partial \overline{u}_i}{\partial x_k}\right)\mathbf{e}_i \otimes \mathbf{e}_j \\
\mathbf{Q}^{\mathbf{R}} &= \left(\overline{u'_i u'_j u'_k} + \frac{1}{\rho}\overline{p' u'_i}\delta_{jk} + \frac{1}{\rho}\overline{p' u'_j}\delta_{ik} - \nu\frac{\partial R_{ij}}{\partial x_k}\right)\mathbf{e}_i \otimes \mathbf{e}_j \otimes \mathbf{e}_k \\
\mathbf{\Phi} &= \frac{1}{\rho}\left(\overline{p'\frac{\partial u'_i}{\partial x_j}} + \overline{p'\frac{\partial u'_j}{\partial x_i}}\right) \\
\varepsilon &= 2\nu\overline{\frac{\partial u'_i}{\partial x_k}\frac{\partial u'_j}{\partial x_k}}\mathbf{e}_i \otimes \mathbf{e}_j
\end{aligned}
\tag{4.70}
$$

We will more thoroughly discuss this equation and its meaning in Section 4.5.2. For the moment, we can make the following remark: we have got a tensorial equation which involves new unknown terms, particularly third-order moments (4.68) of the fluctuating velocity field in the right-hand side. The latter stem from the advection term in the equation for $\mathbf{u}'$, just as the Reynolds tensor is a consequence of the advection term in the Navier–Stokes equation. Thus, the non-linearity of the equations leads to the following finding: every attempt to build an equation for the n-th order moments leads to moments of the immediately higher order (on top of further new unknown terms, such as the correlations $\overline{p'u'_i}$ in (4.70)). This is theoretically an infinite hierarchy of equations, which does not enable us to settle the issue. What is more, it turns out that the calculations will become considerably more and more complex in the next steps, provided that one wants to perform them.

Thus, we must drop this purely mathematical quest, interrupt these algebraic calculations and turn to another solution. It will consist of seeking a closure law expressing the $(n+1)$-th order moments versus the n-th order moments, based on heuristic considerations. Depending upon the order n of the mathematical step at which the model maker wants to stop, one refers to a corresponding ordered-closure. Thus, a first-order closure model consists in seeking a relation yielding second-order moments as a function of the first-order moments, that is expressing the Reynolds stresses versus the mean velocities. As we have just explained, the analytical calculation is not suitable for that purpose; we then must utilize physical intuition in order to determine a 'behaviour law' for the Reynolds stresses, as we did in Chapter 2 for the Cauchy stresses.

4.4 First-order closure

In this section we make an attempt to close the Reynolds-averaged momentum equation from the Boussinesq model, introducing the eddy viscosity and turbulent kinetic energy. The equation governing the latter is derived and examined. The Kolmogorov analysis leads to estimations of turbulent scales, then several closure models are proposed for computing the eddy viscosity, in particular the $k-\varepsilon$ model. A generic example is investigated, namely the infinite open-channel steady flow, giving birth to simple models for turbulent boundary conditions.

4.4.1 Boussinesq model

In order to reach the target set by the end of the previous paragraph, it is necessary to understand the effects of the eddies over flow. We have already pointed out the mixing properties of the turbulence, which characterize chaotic processes (Section 4.2.2). Thus, a foreign substance (a passive tracer) initially placed in a limited area of a turbulent fluid flow regime will very quickly be dispersed in the whole fluid. This process looks like a diffusion and is much more rapid and efficient than the molecular diffusion discussed in Section 3.3.1. All the flow specific quantities, indeed, are diffused that way, particularly the momentum (through each of its components). Besides, the eddies happen to be revolving structures and will therefore exert a significant friction against each other (as already evidenced when describing the Reynolds tensor), which in turn brings about an energy dissipation. This phenomenon may be explained otherwise by considering the instantaneous velocity field, which exhibits great velocity variations due to the turbulent oscillations. The rate of strain s is then very high in a turbulent flow and the dissipation is much enhanced,[39] since it is proportional to s^2 (refer to eqn (3.111)). As previously explained, this dissipative nature enables a dynamic system to remain stable on an attractor within the phase space.

[39] Once again, the notion of strain is important for the turbulent mixing, which was mentioned in Section 4.2.2.

Thus, the turbulent eddies feature by a couple of phenomena, namely diffusion and dissipation. We have explained in Section 3.4 that these two processes account for a viscous phenomenon; that is why J. Boussinesq[40] was incited to propose, for **R**, a model which is based on an appropriate notion of viscosity. We have previously highlighted the role of this tensor exerting 'pressure' and friction stresses within the fluid; that is why adopting for **R** a model analogous to that of Stokes (Section 3.4.1) for the Cauchy stress tensor $\boldsymbol{\sigma}$ will appear suitable. From now on, we will refer to incompressible flows, because our arguments will sometimes be rather qualitative, that is applicable to the nearly incompressible flows with an approximation of the same order, whereas the treatment of compressible flows happens to be more complex.[41] The Stokes model (3.107), once averaged according to Reynolds, is written as

[40] Boussinesq (1877) remarkably understood the mechanisms of turbulence through empirical observations.

[41] For the latter, however, please refer to Gatski and Bonnet (2009).

$$-\frac{1}{\rho}\overline{\boldsymbol{\sigma}} = \frac{\overline{p}}{\rho}\mathbf{I}_3 - 2\nu\mathbf{S} \tag{4.71}$$

where **S** is the mean strain rate tensor, as defined by

$$\begin{aligned}\mathbf{S} &\doteqdot \overline{\mathbf{s}} = \boldsymbol{\Gamma}^S \\ &= \frac{1}{2}\left(\frac{\partial \overline{u}_i}{\partial x_j} + \frac{\partial \overline{u}_j}{\partial x_i}\right)\mathbf{e}_i \otimes \mathbf{e}_j\end{aligned} \tag{4.72}$$

Thus, since $-\rho\mathbf{R}$ plays a part analogous to $\overline{\boldsymbol{\sigma}}$ (according to (4.52)), we will write the *Boussinesq model* as

$$\mathbf{R} = \frac{2}{3}k\mathbf{I}_3 - 2\nu_T\mathbf{S} \tag{4.73}$$

(the $\frac{2}{3}$ factor is a matter of convention; its usefulness will become plain further onwards). This equation may be considered as a definition of the quantities k and ν_T. The latter is referred to as (kinematic) *eddy viscosity* and represents the diffusion and dissipation effects of the eddies, whereas the term proportional to k stands for the additional 'pressure' caused by the eddies, that is the hammering force they exert (refer to Section 4.3.2). Sometimes, instead of ν_T, the dynamic eddy viscosity:

$$\mu_T \doteq \rho \nu_T \tag{4.74}$$

is used by analogy with (3.110). If we now form the trace of (4.73), we get:

$$\text{tr}\, \mathbf{R} = 2k \tag{4.75}$$

(because, according to (3.19), the strain rate trace equals the velocity field divergence, which vanishes according to (4.41)). Hence, returning to the definition (4.47) of $\mathbf{R}$, we must define k as follows:

$$k \doteq \frac{1}{2}\overline{u'_i u'_i} = \frac{1}{2}\left|\mathbf{u}'\right|^2 \tag{4.76}$$

Interpreting the quantity k is then easy: it is the kinetic energy of the fluctuating velocity field per unit mass, and we will henceforth refer to it as the *turbulent kinetic energy*. According to its definition, it represents a field. It formally plays, as regards the Reynolds mean velocities, the same role as that played by the internal energy as regards the macroscopic velocities in the Boltzmann theory (for instance, refer to eqn (1.235)). It is clear that it will play a major part in the following, since it represents a measure of turbulence intensity; hence, most of the turbulent quantities are partly related to it, as explained later on. As a first illustration thereof, we can temporarily consider the case of a nearly isotropic turbulence,[42] that is one whose components of the velocity fluctuation vector exhibit identical rms averages which are denoted u'. In this case, the definition (4.76) can be used for writing

[42] The isotropy/anisotropy properties of turbulence will be discussed in Section 4.5.

$$k = \frac{1}{2}\overline{u_x'^2 + u_y'^2 + u_z'^2} = \frac{3}{2}u'^2 \tag{4.77}$$

This formula may be compared with (3.251) (or (2.103)), which highlights the formal relationship of k with the pressure of a perfect gas through (2.103) (more exactly, $2k/3$ should be compared with p/ρ). Just as the molecular viscosity depends on temperature (eqn (3.292)), that is on molecular energy, it is therefore undoubted that ν_T depends on k.

It is noteworthy that pressure generally prevails over turbulent kinetic energy. Indeed, the relation (3.157) can be used to write an estimate of the ratio of the spatial variations in pressure δp to k:

$$\frac{\delta p}{\frac{2}{3}\rho k} \sim \frac{U^2}{2k} \tag{4.78}$$

This is the ratio of the kinetic energy of the mean flow to that of the turbulent motion. Now, it can be felt that the fluctuating velocities are substantially lower than the mean velocities, as confirmed by experience. As explained in

Section 4.4.6, the k/U^2 ratio is then on the order of some thousandths.[43] In quite a lot of cases, the following approximation can then be made:[44]

$$\frac{2}{3}\rho k \ll \delta\overline{p} \tag{4.80}$$

We now have to make some essential remarks about ν_T:

- ν_T representing a feature of turbulence, it is in no way an intrinsic property of the fluid (unlike ν) but it is specific to a given flow, even at a given point in space and time. In other words, it is a field to be determined. This major difference with respect to the molecular viscosity results from the fact that the latter is a symptom of a motion taking place on a very small scale (the mean free path of the molecules, as shown by eqn (3.297)), that is does not depend on those macroscopic quantities characterizing a given flow,[45] whereas the turbulent eddies, on the contrary, have nearly the same size as the dimensions of the overall flow, and so their arrangement, structure and evolution are very sensitive to the specific macroscopic quantities of a particular flow (wall shape, boundary conditions). The dependence of ν_T on k substantiates this intuition. From that viewpoint, the turbulence behaviour may be likened to that of a non-Newtonian fluid (refer to Section 3.4.1). Hence, the model (4.73) is not closed yet, but is a model family whose various forms are still to be disclosed.
- Since turbulent diffusion is much more efficient than molecular diffusion, as already pointed out, we will certainly have

$$\nu \ll \nu_T \tag{4.81}$$

The values of ν_T typically range from 10^{-4} to 10^{-1} $m^2 s^{-1}$, as stated in Section 4.4.5 (for the sake of comparison, we may recall that the kinematic viscosity of water at 20 °C is 10^{-6}, as per Table 3.2). One consequence is that the molecular diffusion is now generally unsignificant. Thus, the exact value of ν has no effect on the mean flow in a turbulent regime. The Reynolds number, which is based on the quantity ν, then exerts no effect on the macroscopic quantities characterizing a given turbulent flow. This result is consistent with Section 3.4.3 of Chapter 3. However, restrictions should be placed on the latter assertion, which only holds true in the presence of a well developed turbulence, that is for a sufficiently high Reynolds number (typically higher than 10^4).

From what was stated above, we can infer that the molecular effects prevail over the normal forces in a fluid, whereas the turbulent effects essentially prevail over the shear stresses. Introducing the Boussinesq model (4.73) into the Reynolds motion equation (4.51), and given the approximations (4.80) and (4.81), we get (unless otherwise specified, the mean Lagrangian derivatives as defined by (4.40) will henceforth be used):

$$\begin{aligned} \frac{D\overline{\mathbf{u}}}{Dt} &= -\frac{1}{\rho}\mathbf{grad}\left(\overline{p}^* + \frac{2}{3}\rho k\right) + \mathbf{div}\left[2\left(\nu + \nu_T\right)\mathbf{S}\right] \\ &\approx -\frac{1}{\rho}\mathbf{grad}\,\overline{p}^* + \mathbf{div}\left(2\nu_m \mathbf{S}\right) \end{aligned} \tag{4.82}$$

[43]One should take care not to make the following error: the relations (2.103) and (4.77) suggest

$$\frac{p}{\frac{2}{3}\rho k} = \frac{\mathring{u}^2}{3u'^2} \tag{4.79}$$

$\mathring{u}$ representing, it must be repeated, the rms velocity of the molecules, which is on the order of magnitude of the speed of sound within the fluid. This relation, however, is erroneous, because we want to compare the variations in pressure, not the absolute pressure. We have already made a similar remark in Section 3.4.3.

[44]Some authors utilize the relationship between these two terms for simply integrating the amount k into pressure. We do not think it is an advisable approach, because the two quantities do not obey the same boundary conditions (refer to Sections 3.4.2 and 4.4.6).

[45]This is how we could determine the molecular viscosity of a rarefied gas by means of the Boltzmann equation (refer to Section 3.5.2). We cannot extend this argument to the case of turbulent viscosity.

with

$$\nu_m \doteqdot \nu + \nu_T = \frac{\mu_m}{\rho} \tag{4.83}$$

We have lumped the two viscosities together in order to lay emphasis on their formal and physical relationship (to some extent), which can only be done within the divergence symbol, because of the spatially variable nature of eddy viscosity. Equation (4.82) is now closed, provided that eddy viscosity can be assessed,[46] which will be done in the next paragraph.

[46] As well as turbulent kinetic energy, when it is desired to keep it in the equation of motion.

Let us make a final remark about the equation we have just obtained. It occurs in a mathematical form which is perfectly identical to the Navier–Stokes equation, and so it should have the same mathematical and physical properties, particularly as regards the stability of the solutions. We then must acknowledge, in accordance with the discussion at the beginning of Section 4.2.2, that its solutions remain stable as long as the ratio of the (destabilizing) inertia forces to the (stabilizing) viscosity forces is not too high. This ratio is estimated in the Navier–Stokes equations by a Reynolds number; here it should be estimated by a Reynolds number which is based on eddy viscosity, by analogy with (3.156):

$$\begin{aligned} \frac{|\mathbf{grad}\,\overline{\mathbf{u}}\cdot\overline{\mathbf{u}}|}{|\mathbf{div}\,[2(\nu+\nu_T)\,\mathbf{S}]|} &\sim \frac{UL}{\nu+\nu_T^*} \\ &\simeq \frac{UL}{\nu_T^*} \doteqdot \overline{Re} \end{aligned} \tag{4.84}$$

where ν_T^* is an order of magnitude of ν_T (we have taken (4.81) into account). The comparatively high values of ν_T suggest that this 'mean flow Reynolds number' is usually moderate. It will be explained, indeed, in Section 4.4.6 that it is on the order of some hundredths, which is below the critical value $Re_{c,2}$ for the appearance of chaotic instabilities. A resulting effect consists in the stability of the mean velocity field, which never becomes chaotic in turn.[47] This is the single qualitative 'evidence' which we can provide as to the existence of the mean quantities. Their very definition as the result of a convergent series (4.28) would be challenged if it happened that they would be subject to the same kind of instabilities as the real velocity field.

[47] The mean velocity field, however, may in turn enter a non-chaotic (periodic) oscillatory regime, which shows itself in the alternating eddying release which can be seen downstream from a solid body (pole of a flapping flag).

The equation of a scalar (4.53) may be closed in the same way, noting that the turbulent flux of the scalar is propitious to such a kind of model as (3.59):

$$\mathbf{Q}^B = -K_T\,\mathbf{grad}\overline{B} \tag{4.85}$$

which is related to (3.59) just as (4.73) is related to (4.71). However, a significant difference exsists: if the molecular diffusion coefficient K_B of the scalar may be quite different from the molecular viscosity ν and vary from one scalar to another, on the other hand the eddy diffusion coefficient K_T is a quantity which does not depend on the scalar, because it represents a process as determined by the turbulence.[48] Then it is a field which it is well-founded to assume is close to the eddy viscosity. Hence we will write

[48] The active scalars (i.e. which affect density), however, are an exception to that rule, since they damp the turbulence. Such is the case with the temperature, even though it may be considered passive when its variations are quite gentle. Refer, for example, to Violeau (2009a).

$$K_T = \frac{\nu_T}{\sigma_T} \tag{4.86}$$

where σ_T is the turbulent Schmidt number, which approximately equals[49] 0.72. With this model, (4.53) takes the following closed form:

$$\frac{D\overline{B}}{Dt} = \overline{S}_B + \text{div}\left(K_m \mathbf{grad}\overline{B}\right) \tag{4.87}$$

to be compared with (4.82), where

$$K_m \doteqdot K_B + \frac{\nu_T}{\sigma_T} \tag{4.88}$$

[49] Refer, for instance, to Violeau (2009a). When the scalar represents the temperature, it is usually referred to as the turbulent Prandtl number. It is noteworthy that it only takes up the presently given constant value when there is no variation of density and for simply sheared flows (refer to Section 4.4.5). Within a more general context, it should be considered that it depends on the non-dimensional mean scalar strain rate, as defined in Section 4.4.2.

4.4.2 Turbulent kinetic energy

In order to suggest models enabling us to assess eddy viscosity, one has to get a better knowledge of the properties of eddies from what was explained in the previous sections. Insofar as ν_T depends on k, let us first determine an equation governing the evolution of that quantity. Given the definition (4.75), it is obtained by taking the half-trace in the transport equation of (4.66):

$$\frac{Dk}{Dt} = P - \breve{\varepsilon} - \text{div}\breve{\mathbf{Q}}^k \tag{4.89}$$

with

$$P \doteqdot -\mathbf{\Gamma} : \mathbf{R} = -\mathbf{S} : \mathbf{R} = -R_{ik}\frac{\partial \overline{u}_i}{\partial x_k} \tag{4.90}$$

$$\breve{\mathbf{Q}}^k \doteqdot \overline{\left(\frac{1}{2}|\mathbf{u}'|^2 + \frac{p'}{\rho}\right)\mathbf{u}'} - \nu\mathbf{grad}\, k = \left(\frac{1}{2}\overline{u'_i u'_i u'_k} + \frac{1}{\rho}\overline{p' u'_k} - \nu\frac{\partial k}{\partial x_k}\right)\mathbf{e}_k$$

$$\breve{\varepsilon} \doteqdot \nu\overline{\boldsymbol{\gamma}' : \boldsymbol{\gamma}'^T} = \nu\overline{\frac{\partial u'_i}{\partial x_k}\frac{\partial u'_i}{\partial x_k}}$$

(the symmetry of $\mathbf{R}$ does provide the first equality).We may already note that the term $\mathbf{\Phi}$ in eqn (4.67) has disappeared owing to the continuity equation of the fluctuating velocity field (4.42) (we will go back over that circumstance in Section 4.5.2). Each of the terms in equation (4.90) is the half-trace of the corresponding term in (4.67); in particular, we have:

$$P = \frac{1}{2}\text{tr}\,\mathbf{P} \tag{4.91}$$

(in a similar way to (4.75)).

The following mathematical modification is commonly made:

$$\begin{aligned}
\frac{\partial^2 k}{\partial x_k \partial x_k} - \frac{\breve{\varepsilon}}{\nu} &= \overline{\frac{1}{2}\frac{\partial^2 u'_i u'_i}{\partial x_k \partial x_k} - \frac{\partial u'_i}{\partial x_k}\frac{\partial u'_i}{\partial x_k}} \\
&= \overline{\frac{\partial}{\partial x_k}\left(u'_i\frac{\partial u'_i}{\partial x_k}\right) - \frac{\partial u'_i}{\partial x_k}\frac{\partial u'_i}{\partial x_k}} \\
&= \overline{u'_i\frac{\partial}{\partial x_k}\frac{\partial u'_i}{\partial x_k} + u'_i\frac{\partial}{\partial x_k}\frac{\partial u'_k}{\partial x_i}} \\
&= \overline{2u'_i\frac{\partial s'_{ik}}{\partial x_k}}
\end{aligned} \tag{4.92}$$

(the second term on the third line, which was artificially added, is identically zero by virtue of the continuity equation). Next:

$$
\begin{aligned}
2\overline{u'_i \frac{\partial s'_{ik}}{\partial x_k}} &= 2\overline{\frac{\partial u'_i s'_{ik}}{\partial x_k} - \frac{\partial u'_i}{\partial x_k} s'_{ik}} \\
&= 2\overline{\frac{\partial u'_i s'_{ik}}{\partial x_k} - s'_{ik} s'_{ik}} \\
&= \frac{\partial 2\overline{s'_{ki} u'_i}}{\partial x_k} - \overline{s'^2}
\end{aligned}
\tag{4.93}
$$

with, in accordance with notation (3.29):

$$
\begin{aligned}
s' &\doteqdot \sqrt{2\mathbf{s}' : \mathbf{s}'} \\
&= \sqrt{2 s'_{ij} s'_{ij}}
\end{aligned}
\tag{4.94}
$$

which stands for the fluctuating strain rate.[50] Hence, the elements in (4.90) may be rearranged as

[50] Note that it is not the fluctuating part of s, contrary to what is suggested by the notation.

$$
\begin{aligned}
\breve{\mathbf{Q}}^k \rightarrow \mathbf{Q}^k &\doteqdot \overline{\left(\frac{1}{2}|\mathbf{u}'|^2 + \frac{p'}{\rho}\right)\mathbf{u}'} - 2\nu\overline{\mathbf{s}'\mathbf{u}'} \\
&= \left(\frac{1}{2}\overline{u'_i u'_i u'_k} + \frac{1}{\rho}\overline{p' u'_k} - 2\nu\overline{u'_i s'_{ik}}\right)\mathbf{e}_k \\
\breve{\varepsilon} \rightarrow \varepsilon &\doteqdot \nu\overline{s'^2}
\end{aligned}
\tag{4.95}
$$

Then we get

$$
\frac{Dk}{Dt} = P - \varepsilon - \operatorname{div}\mathbf{Q}^k \tag{4.96}
$$

It can already be pointed out that the quantities $\breve{\varepsilon}$ and ε are numerically very close, since the difference does not exceed a few percentiles. Some authors sometimes mix them up, but we have preferred distinguishing them for reasons which will become clear in the following.

Equation (4.96) is formally analogous to the transport-diffusion equation of a scalar (3.53) (or (4.87)), with source terms P and $-\varepsilon$. Vector $\mathbf{Q}^k$ serves as the flux of quantity k, and its definition clearly reveals that it represents a transport of both kinetic and potential energies by the eddies and the molecular viscosity. The term P, provided with the Boussinesq model (4.73), is reduced to

$$
\begin{aligned}
P &= -\mathbf{S} : \left(\frac{2}{3}k\mathbf{I}_3 - 2\nu_T\mathbf{S}\right) \\
&= -\frac{2}{3}k \operatorname{tr}\mathbf{S} + 2\nu_T\mathbf{S} : \mathbf{S} \\
&= \nu_T S^2
\end{aligned}
\tag{4.97}
$$

(we have used the continuity equation). That formula involves the mean scalar strain rate[51] S:

[51] Which is not the scalar strain rate average, but the norm of the mean strain rate tensor.

$$S \doteqdot \sqrt{2\mathbf{S} : \mathbf{S}} \\ = \sqrt{2S_{ij}S_{ij}} \tag{4.98}$$

Thus, P is positive with that model;[52] hence it is a kinetic energy input. We will call it *production of turbulent kinetic energy*. It can easily be interpreted: the mean velocity gradients (which determine S) account for instabililities which give rise to the largest eddies, imparting energy to them. These gradients are necessarily present, if only because of the nearly inevitable existence of solid walls determining a condition on velocity. For instance, just think of a train arriving at a platform: the spatial variations of mean velocity as caused by the train motion generate an intense eddying agitation. It is attractive, however, to consider things from another viewpoint taking into account the affinity of (4.97) and (3.111):[53] since the Reynolds equation has the same form as the Navier–Stokes equation, then the energy which is dissipated by the mean flow will naturally take the same form as that mentioned in Chapter 3, as given by (3.89) and (3.111), but from now on eddy viscosity comes on top of molecular viscosity. If we ignore the energy diffusion term, we get:

[52] We must emphasize that it is a consequence of the Boussinesq model (4.73). Practically, it happens that P be locally or temporarily negative. This is one of the weak points of the Boussinesq model; we will review them at the end of Sections 4.4.4 and 4.5.1.

[53] Or, more generally, that of the first line in (4.90) and (3.85).

$$\frac{de_k(\overline{\mathbf{u}})}{dt} = -(\nu + \nu_T)S^2 \\ = -P - \nu S^2 \tag{4.99}$$

By the way, we may note that equation (4.99) can be integrated over the fluid volume to yield, through the rule (3.50) and the definition (4.83), the following relation:

$$\frac{dE_k(\overline{\mathbf{u}})}{dt} = \frac{d}{dt}\int_\Omega \rho e_k(\overline{\mathbf{u}})\,d\Omega \\ = \int_\Omega \rho\frac{de_k(\overline{\mathbf{u}})}{dt}d\Omega \\ = -2\int_\Omega \rho\nu_m S^2 d\Omega \tag{4.100}$$

Furthermore, the mean total kinetic energy is written as

$$\overline{e_k(\mathbf{u})} = \frac{1}{2}\overline{\mathbf{u}\cdot\mathbf{u}} \\ = \frac{1}{2}\overline{(\overline{\mathbf{u}}+\mathbf{u}')\cdot(\overline{\mathbf{u}}+\mathbf{u}')} \\ = \frac{1}{2}\overline{\overline{\mathbf{u}}\cdot\overline{\mathbf{u}}+\overline{\mathbf{u}}\cdot\mathbf{u}'+\mathbf{u}'\cdot\overline{\mathbf{u}}+\mathbf{u}'\cdot\mathbf{u}'} \\ = e_k(\overline{\mathbf{u}}) + k \tag{4.101}$$

(the calculation is analogous to (4.44)). That means that a large part of the kinetic energy is lost by the mean flow transferred into the fluctuating flow. Such is the role of the quantity P, which then stands for an energy flux from the mean field to the eddies (the other part of eqn (4.99), namely νS^2, represents an energy which is dissipated through molecular friction by the mean flow and, by comparison, is very low). Now, from Section 4.2.3 we know that the decaying eddies generate a set of smaller eddies. This way, the 'mother' structures transfer some part of their energy to the 'daughter' structures; hence, this is what is commonly known as an *energy cascade*, the energy P taken from the mean field being passed on 'from mother to daughter' down to the Kolmogorov scale at which it is ultimately dissipated. Such dissipation is expressed in equation (4.95) by the term $-\varepsilon$, which is inherently negative and is referred to as the *turbulent kinetic energy dissipation rate*.[54] One can directly verify that the mean flow and the eddies jointly dissipate the whole energy, whose actual dissipation rate is given by $f = \nu s^2$:

[54] $\breve{\varepsilon}$ is then often called a 'pseudo-dissipation' (Pope, 2000).

$$\begin{aligned} \overline{s^2} &= \overline{\left(S_{ij} + s'_{ij}\right)\left(S_{ij} + s'_{ij}\right)} \\ &= \overline{\left(S_{ij}S_{ij} + 2S_{ij}s'_{ij} + s'_{ij}s'_{ij}\right)} \\ &= S^2 + s'^2 \end{aligned} \tag{4.102}$$

whence, owing to (4.95),

$$\begin{aligned} \overline{f} = \nu\overline{s^2} &= \nu S^2 + \varepsilon \\ &\approx \varepsilon \end{aligned} \tag{4.103}$$

We now will turn to the particular case of a homogeneous turbulence.[55] In this case, the diffusion term vanishes and eqn (4.96) is simply reduced to

[55] In the meaning as defined in Section 4.3.1, with the relevant restrictions. The next result, however, could be derived from the integration of (4.96) over the whole fluid, the diffusion term then coming down to a boundary integral which vanishes pursuant to the boundary conditions.

$$\frac{Dk}{Dt} = P - \varepsilon \tag{4.104}$$

If, additionally, the turbulence is statistically permanent, then we will get

$$P = \varepsilon \tag{4.105}$$

The production and dissipation terms are then in equilibrium in this ideal case. We are therefore in exactly the same situation as the sustained pendulum discussed in Section 4.2.1, and the equilibrium which we have just demonstrated is analogous to equation (4.3). The turbulent eddy system then occurs in the form of a dramatically complex pendulum whose mechanical behaviour can nevertheless be described, on an average, by simple laws. Let us imagine an illustrative case of a closed vessel which is filled with water being stirred and imparting energy to the whole unit. The velocity differences from one place to the other, particularly resulting from the presence of walls, cause turbulent kinetic energy to be generated, and the fluid soon enters a turbulent regime if the Reynolds number is high enough. The energy cascade then starts running, transferring an energy flux $P = \varepsilon$ towards the smallest eddies, so that they can dissipate the whole input energy. The higher the Reynolds number is, the greater the energy (4.97) injected into the eddy system is. The Kolmogorov cascade is then extended towards the high wave numbers (small eddies) down to a scale which is sufficiently small for the production to be balanced by dissipation (4.103). If the motion is halted, since the production of energy is terminated, then dissipation quickly brings the fluid back to rest like a damped simple pendulum (the attractor then suddenly becomes punctual).

We may note that although the equilibrium is only true in the idealized case which we have mentioned (4.105), nevertheless it is approximately true most of the time,[56] and so it may be written that $P \sim \varepsilon$. Thus, returning to the definitions (4.97) and (4.95) of these two quantities, we get

$$\nu_T S^2 \sim \nu \overline{s'^2} \tag{4.106}$$

Now, the fluctuating velocity field obviously exhibits much larger variations than the mean field. Thus, S is quite small versus s', and the result (4.81) can be found again.

Let us summarize what we have learnt from the Boussinesq model. For model designers adopting a mean field-based approach, intentionally ignoring the eddies in the representation of the fields is possible provided that eddy viscosity is added to molecular viscosity (which is nearly tantamount to substituting the former for the latter, taking their respective orders of magnitude into account). In other words, the purpose of eddy viscosity is taking from the mean field, in spite of its smooth nature, the dissipation which is actually exerted by the genuine velocity field which, as to it, is quite varying.

[56]The following circumstance should also be noticed. The production-dissipation equilibrium could not be explained by an analysis of magnitudes if the eddies had a single size and a single velocity; the spectral nature of turbulence does allow such an equilibrium, as explained in the next section.

4.4.3 Kolmogorov analysis

We have learnt from Section 4.2.3 that the turbulent fluctuations were distributed over a spectrum, and so the same property should characterize all the turbulent structure-related quantities. We presently know two of them, namely k and ε. Kolmogorov's work (1941, one can refer to its English translation published in 1991, or to the paper published in 1962)[57] is based on the assumption that these two quantities are sufficient to characterize turbulence at a given point in space and time.[58] For isotropy reasons, as a first approach, only the norm K in the wave vector $\mathbf{K}$ will here be of interest. We will then attempt to determine a formulation in such a form as

$$\begin{aligned} k &= \int_0^{+\infty} E(K)\,dK \\ \varepsilon &= \int_0^{+\infty} D(K)\,dK \end{aligned} \tag{4.107}$$

where E and D are called the energy and dissipation spectra, respectively.[59] From the definitions of k and ε, we will try to relate their respective spectra to that of the fluctuating velocity field as given by (4.26). Let us first turn to kinetic energy, writing

$$\frac{1}{2}u'_i u'_i = \frac{1}{(2\pi)^6}\int_{\mathbb{R}^3}\int_{\mathbb{R}^3}\frac{1}{2}\hat{u}'_i(\mathbf{K},t)\,\hat{u}'_i\left(\mathbf{K}',t\right)e^{i(\mathbf{K}+\mathbf{K}')\cdot\mathbf{r}}d^3\mathbf{K}d^3\mathbf{K}' \tag{4.108}$$

According to (4.76), the spectrum of k is then obtained by averaging that quantity. We will turn to the case of a homogeneous turbulence, and so the averaging operator can be viewed as a space integral:

[57]Obhukov (1962) and other renowned scientists took part, either independently or not, in the elaboration of these foundations.

[58]ν_T depends on k and P has the same order of magnitude as ε, and so we ignore them in this approach. We may note that only considering the quantities k and ε is theoretically only valid on the assumption of an isotropic turbulence, i.e. whose eddies do not have any privileged direction. As explained later on in this chapter, particularly in Section 4.5.1 and 4.6.1, this hypothesis is often found to be at fault for the biggest eddies.

[59]Spectra, like all the other quantities, a priori are fields. We keep, however, ignoring their explicit dependence on time and space, so much the more as we will make an hypothesis of spatial homogeousness in the following. Besides, we believe it is impossible to confuse the subscript integer i and the unitary complex i.

$$k = \frac{1}{2(2\pi)^6 V} \int_{\Omega} \int_{\mathbb{R}^3} \int_{\mathbb{R}^3} \hat{u}'_i(\mathbf{K}, t)\, \hat{u}'_i(\mathbf{K}', t)\, e^{i(\mathbf{K}+\mathbf{K}')\cdot\mathbf{r}} d^3\mathbf{K} d^3\mathbf{K}' d^3\mathbf{r} \tag{4.109}$$

where V denotes the volume of the domain Ω, which will be assumed to be large enough to be likened to the whole space $\mathbb{R}^3$. The integral of $\exp[i(\mathbf{K}+\mathbf{K}')\cdot\mathbf{r}]\,d^3\mathbf{r}$ over $\mathbb{R}^3$ then vanishes, except for $\mathbf{K}+\mathbf{K}'=\mathbf{0}$, in which case it equals V. Thus, we get an integral which is reduced to a single wave vector:

$$\begin{aligned} k &= \frac{1}{2(2\pi)^6} \int_{\mathbb{R}^3} \hat{u}'_i(\mathbf{K}, t)\, \hat{u}'_i(-\mathbf{K}, t)\, d^3\mathbf{K} \\ &= \frac{1}{2(2\pi)^6} \int_{\mathbb{R}^3} \left|\hat{\mathbf{u}}'(\mathbf{K}, t)\right|^2 d^3\mathbf{K} \end{aligned} \tag{4.110}$$

(we have used the relation (4.27)).

The dissipation spectrum is treated in a similar way, but working with the pseudo-dissipation $\breve{\varepsilon}$, as given by (4.90), is more convenient. We know that the derivation, in the Fourier space, results in a simple multiplication times the spectrum variable (refer to eqn (B.53)), whence

$$\frac{\partial u'_i}{\partial x_j} = \frac{1}{(2\pi)^3} \int_{\mathbb{R}^3} i K_j \hat{u}'_i(\mathbf{K}, t)\, e^{i\mathbf{K}\cdot\mathbf{r}} d^3\mathbf{K} \tag{4.111}$$

Multiplying that relation by itself, and subsequently adding over the subscripts i and j, we readily get

$$\frac{\partial u'_i}{\partial x_j}\frac{\partial u'_i}{\partial x_j} = -\frac{1}{(2\pi)^6} \int_{\mathbb{R}^3} \int_{\mathbb{R}^3} K_j K'_j \hat{u}'_i(\mathbf{K}, t)\, \hat{u}'_i(\mathbf{K}', t)\, e^{i(\mathbf{K}+\mathbf{K}')\cdot\mathbf{r}} d^3\mathbf{K} d^3\mathbf{K}' \tag{4.112}$$

Such an averaging procedure as that we have used for kinetic energy then yields

$$\varepsilon \approx \breve{\varepsilon} = \frac{\nu}{(2\pi)^6} \int_{\mathbb{R}^3} K^2 \left|\hat{\mathbf{u}}'(\mathbf{K}, t)\right|^2 d^3\mathbf{K} \tag{4.113}$$

The E and D spectra as defined by (4.107) are obtained by integrating (4.110) and (4.113) over all the possible directions of the wave vector, that is over a unit sphere. Without even carrying on with the integration, it becomes manifest that they are related by

$$D(K) = 2\nu K^2 E(K) \tag{4.114}$$

Some first information emerges from that relation: because of the small value of the viscosity of ordinary fluids, dissipation only gets significant values for high values of K, that is they occur in the small eddies, which corroborates the analysis made in Section 4.4.2. From intuitive arguments, we will try to guess the shape of these two spectra:

- Both of them verify the properties of $\hat{\mathbf{u}}'$ as set out in Section 4.2.3, in particular they vanish upon $K = 0$ and when K tends towards infinity.
- Near infinity, E tends more quickly towards zero than D, because of (4.114), in order to provide for the convergence of the integrals.

- Both of them are inherently positive.
- We know from experience that they are continuous, in accordance with the discussion in Section 4.2.2.

Thus, both spectra should have a maximum,[60] that of E coming before that of D (i.e. for a smaller value of K). We may check the latter result through a calculation by deriving (4.114):

$$\frac{dD}{dK} = 2\nu K \left(2E + K\frac{dE}{dK}\right) \tag{4.115}$$

Thus, when E reaches its maximum, the derivative of D equals $4\nu K E$ which is strictly positive, and then D is still growing. Figure 4.10 (*a*) (Section 4.6.2) illustrates the shape of these two spectra; we can see that the large eddies carry most of kinetic energy, whereas the small ones dissipate it, as we found on several occasions. The dissipation peak is carried by a wave number corresponding to a scale which is very similar to the Kolmogorov scale $\lambda_0 = 2\pi/K_0$ (refer to Section 4.2.3), whereas the peak of energy is carried by eddies the size of which will be denoted as $L_t \doteqdot 2\pi/K_t$ (refer to (4.24)) and is known as the *integral scale*, corresponding to the characteristic size of the largest eddies.

Kolmogorov has formulated from the above a number of hypotheses which we summarize in the following quite simplified way:

- everything connected to the large eddies cannot depend on molecular viscosity,[61] since it is peculiar to dissipation, which is carried by the smallest eddies;
- everything connected to the small eddies cannot depend on k, since that quantity is peculiar to the largest eddies;
- all the eddies depend on ε, which stands for the energy flux from the larger to the smaller eddies. This flux does not depend on the eddy scale.[62]

The independence of ε with respect to the eddy size is a quite essential feature. It is, indeed, a quantity which characterizes the turbulence of the relevant flow at one given point in time and space. These assertions are obviously based on the principle that the large and small structures are uncorrelated, which may only hold true if the turbulence is sufficiently 'developed'; we will subsequently provide a sufficient condition for that hypothesis to be valid.

Let us now make out the list of available independent parameters for describing turbulence at one given point in time and space: we must in the first place consider the fluid parameters, namely ρ and ν, then those locally characterizing turbulence, that is k and ε. Density, which is the single parameter depending on the mass unit, should be excluded from our reasoning. We come to the conclusion that every unknown quantity (independent of the mass unit) characterizing turbulence at one point in space should only depend on the parameters k, ε and ν. Considering Kolmogorov's hypotheses, we may write, for instance, that the Kolmogorov scale λ_0, characterizing the small eddies, may only depend on ε and ν. Now, the single combination of these two quantities which has the dimension of a length is $\left(\nu^3/\varepsilon\right)^{1/4}$, which gives a plausible magnitude of λ_0. More rigorously, we may resort to the Vaschy–

[60] It is a natural economy principle which suggests they have only one, which assumption is substantiated by experience.

[61] This principle is consistent with the idea that the Reynolds number does not play any role when it is very high. More precisely, *Re* only affects the small turbulent structures.

[62] One could think that the dissipation rate probably only influences the small scales, since they carry most of the dissipation. By virtue of (4.105), however, ε is in equilibrium with P, characterizing the large eddies.

Buckingham theorem as introduced in Section 3.4.3 and state that the problem is featured by $n - m = 3 - 2 = 1$ constant non-dimensional number, that is $\lambda_0^4 \nu^3/\varepsilon$. Through that procedure, we readily get the orders of magnitude of those scales characterizing the small eddies (length, time, velocity):

$$\begin{aligned}
\lambda_0 &\sim \left(\frac{\nu^3}{\varepsilon}\right)^{1/4} \\
\tau_0 &\sim \left(\frac{\nu}{\varepsilon}\right)^{1/2} \\
u_0 &\sim (\varepsilon\nu)^{1/4}
\end{aligned} \tag{4.116}$$

where τ_0 is the small eddies' time scale. We may note that the third equation in (4.116) results from the first two equations, or else from the first one together with (4.23). For the large eddies, an analogous argument will yield

$$\begin{aligned}
L_t &\sim \frac{k^{3/2}}{\varepsilon} \\
\tau &\sim \frac{k}{\varepsilon} \\
u_t &\sim k^{1/2}
\end{aligned} \tag{4.117}$$

where L_t, τ and u_t are the length, time and velocity scales of large eddies (τ is sometimes called the *turnover time scale*). We unsurprisingly find that the characteristic velocity of the large fluctuating scales is the square root of k, which is consistent with the fact that they carry most of the turbulent kinetic energy. We can see too that the small eddies are even larger and slower since the fluid is viscous, which fits intuition. The scale ratio of the large structures to the small ones can be assessed by

$$\begin{aligned}
\frac{L_t}{\lambda_0} &\sim \left(\frac{k^2}{\nu\varepsilon}\right)^{3/4} \sim \left(\frac{L_t u_t}{\nu}\right)^{3/4} \doteqdot Re_t^{3/4} \\
\frac{\tau}{\tau_0} &\sim \left(\frac{k^2}{\nu\varepsilon}\right)^{1/2} = Re_t^{1/2}
\end{aligned} \tag{4.118}$$

The *turbulent Reynolds number* Re_t represents the large scales (actually, it is the Re_λ of eqn (4.22), for $\lambda = L_t$). Since the large eddies get their energy from the mean flow, it is, indeed, an increasing function of the overall Reynolds number Re. Thus, if the latter is high enough, the turbulent spectrum is very broad and Kolmogorov's hypotheses are valid. Then there is a wide range, the so-called *inertial range*, of intermediate eddies between the larger and the smaller ones. In that spectrum domain, the eddies merely transfer energy $P \sim \varepsilon$ in a relative equilibrium. Since these eddies are both large-sized before λ_0 and small-sized before L_t, they satisfy both Kolmogorov's hypotheses, and then the dimensional analysis reveals $n = 3$ parameters (E, ε and the wave number K) out of which $m = 2$ are independent (two physical units), which gives a general law for the spectrum E as a function of ε and K:

$$K_t \ll K \ll K_0,\ E(K) = c_K \varepsilon^{2/3} K^{-5/3}$$
$$D(K) = 2c_K \nu \varepsilon^{2/3} K^{1/3} \tag{4.119}$$

(the second equality results from (4.114)). These laws are very well verified by experience, with $c_K = 1.5$ (Kolmogorov constant).

The time and velocity scales of the inertial range eddies may also be related through that procedure to the eddy size $\lambda = 2\pi/K$:

$$\lambda_0 \ll \lambda \ll L_t,\ \tau(\lambda) \sim \left(\frac{\lambda^2}{\varepsilon}\right)^{1/3}$$
$$u(\lambda) \sim (\varepsilon\lambda)^{1/3} \tag{4.120}$$

with, of course

$$u(\lambda) \sim \frac{\lambda}{\tau(\lambda)} \tag{4.121}$$

We may note that these formulas not unexpectedly yield $\tau(L_t) \sim \tau$, $\tau(\lambda_0) \sim \tau_0$, $u(L_t) \sim u_t$ and $u(\lambda_0) \sim u_0$ [63] once again. The formulas (4.120) indicate that the eddies are even slower[64] since they are small, which is still consistent with what we have said so far.

[63] Although neither largest nor smallest eddies are within the inertial range. This is a necessary result of dimensional considerations; there is no evidence, however, that the proportionality constants in the equation (4.120) are the same as those in equations (4.116) and (4.117).

Let us now return to the remarks made by the end of Section 4.4.2, particularly about the production-dissipation equilibrium (4.105). We can now assess the orders of magnitude of these two quantities, knowing that the first one is carried by the large-scaled structures and the second one by the small ones. From the definition (4.90) and using (4.47), as well as the model (4.73), we get

$$P = -\overline{u_i' u_k'} S_{ik} \sim k \frac{U}{L} \tag{4.122}$$

where U is an order of magnitude of the mean velocity and L is a characteristic length scale of the mean flow, as in Section 3.4.3. With (4.117), the production-to-dissipation ratio is now estimated by

[64] This means their velocity increases with λ. We may notice, however, that their characteristic time also increases, although less quickly, with λ.

$$\frac{P}{\varepsilon} \sim \frac{k}{\varepsilon}\frac{U}{L} \sim \frac{\tau}{T} \tag{4.123}$$

where T is a characteristic time scale of the mean flow. Intuition confirms that the product of these two ratios is close to one, which accounts for the estimate $P \sim \varepsilon$. We may note that if we had used one velocity scale and one spatial scale to describe P and ε, then we would have come to an absurdity (namely $P/\varepsilon \sim 1/Re \to 0$), which highlights the significance of the spectral nature of turbulence. We must point out the paradoxical nature of the relation $P \sim \varepsilon$, which means that the dissipation does not depend on the value of viscosity, in spite of the fact that its definition is based on ν. The above analysis sheds light on that seeming paradox: when the viscosity is smaller, the small turbulent scales become smaller, that is the fluctuating velocity gradients s' are higher:

$$s' \sim \frac{u_0}{\lambda_0} \propto \frac{1}{\sqrt{\nu}} \tag{4.124}$$

The eddy dislocation process then keeps the $\varepsilon = \nu\overline{s'^2}$ product constant. In short, the turbulence gets adapted and generates such small structures that the amount of energy P from the mean velocity gradients is dissipated therein.

Through the elements we have just determined, we can have some notion of the fractal nature of turbulence (Frisch, 1995). We may recall, indeed, that as explained in Section 4.2.2, the fractality of the attractor in a chaotic system should be found again in the actual geometry of the system being investigated. We have explained in Section 4.2.3 that this aspect shows itself in the presence of a spectrum of eddies which we have just assessed the features of, which will enable us to estimate the fractal dimension d of a turbulent flow. We know, indeed, that this dimension is calculated by counting the sub-units of the object being studied according to their sizes (refer to Section 4.2.2). Denoting as $dp(\lambda)$ the proportion of eddies whose sizes range from λ and $\lambda + d\lambda$, we can state that the energy they contain equals

$$E(K)\,dK = dp(\lambda)\,\frac{u(\lambda)^2}{2} \tag{4.125}$$

Let us now turn to a large eddy, containing eddies of each size $\lambda \leqslant L_t$ according to the distribution law $p(\lambda)$. Obviously, the rarer the eddies are as their sizes decrease, the smaller the fractal dimension is. A short reflection, analogous to the discussion in Section 4.2.2, shows that we have the following relation:

$$p(\lambda) \propto \left(\frac{\lambda}{L_t}\right)^{3-d} \tag{4.126}$$

Hence, if $d = 3$, the turbulence 'fills up' the space, and $p(\lambda)$ is constant. Deriving (4.126) and associating the result with (4.125) and (4.24), we find

$$E(K) \propto \left(\frac{\lambda}{L_t}\right)^{3-d} \lambda u(\lambda)^2 \tag{4.127}$$

For further thought, let us now estimate the energy flux which is transmitted by a λ-sized eddy lying in the inertial range. The latter releases its whole energy within an infinitesimal time $d\tau(\lambda)$, which yields

$$\varepsilon \sim \frac{E(K)\,dK}{d\tau(\lambda)} \tag{4.128}$$

Time $d\tau(\lambda)$ is linked to the scale variation $d\lambda$ by a dependence analogous to (4.121), that is

$$d\tau(\lambda) \sim \frac{d\lambda}{u(\lambda)} \tag{4.129}$$

Combining (4.127), (4.128), (4.129) and invoking (4.24), we get

$$\varepsilon \sim \left(\frac{\lambda}{L_t}\right)^{3-d} \frac{u(\lambda)^3}{\lambda} \tag{4.130}$$

In particular, for $\lambda = L_t$, we have

$$\varepsilon \sim \frac{u_t^3}{L_t} \tag{4.131}$$

in accordance with (4.117). The latter equation can be used to rearrange (4.130) as

$$u(\lambda) \sim \left(\frac{\lambda}{L_t}\right)^{\frac{d-2}{3}} u_0 \qquad (4.132)$$

That relation being provided, (4.127) becomes

$$E(K) \propto \varepsilon^{\frac{2}{3}} L_t^{\frac{d-3}{3}} K^{\frac{d-8}{3}} \qquad (4.133)$$

Formulas (4.132) and (4.133) are in line with (4.119) and (4.120) if $d = 3$. The latest measurements, however, show that d is not necessarily universal and that its probable values are closer to 2.8 (Frisch, 1995). These experimental results demonstrate that the Kolmogorov theory is not perfect and that it is merely one stage in the understanding of turbulence by physicists. Throughout the remainder of this book, however, we will, failing anything better, consider it is valid.

For completing this section, let us briefly turn to the question of the temporal correlation of the fluctuating velocities. Velocity at one given point in space and time depends, via the equations of motion, on the velocity at the time points before. The chaotic nature of the velocity field, however, makes the velocity signal at one point $\mathbf{u}(t)$ so rapidly variable that the flow loses track of its past fairly early.[65] The velocity deviations at two different time points should then statistically increase as the time between them becomes longer. One can get some idea of that effect by considering the velocity field temporal autocorrelation tensor linked to a fluid particle[66] at two given time points t' and t'', which necessarily is a function of time deviation $t'' - t'$, by invariance with respect to the time origin (refer to Chapter 1):

$$\mathbf{D}_u\left(t'' - t'\right) \doteqdot \overline{[\mathbf{u}(t'') - \mathbf{u}(t')] \otimes [\mathbf{u}(t'') - \mathbf{u}(t')]} \qquad (4.134)$$

This definition may be compared with (2.311). The time τ of the large eddies here plays the part of the Lyapunov time as mentioned at the end of Section 4.2.2, and we may state that the velocity signal does not exhibit any autocorrelation for a sufficiently large deviation $t'' - t'$ before τ. Within the inertial range,[67] on the contrary, $\mathbf{D}_u$ will not be equal to zero. Kolmogorov's hypotheses make it possible to seek a dependence with respect to both ε and increment $t'' - t'$. We may note too that if the direction of time is reversed, the velocities change sign, which leaves $\mathbf{D}_u$ unchanged. One then has to seek $\mathbf{D}_u$, which has the dimension of the square of a velocity, like a function of ε and $\left|t'' - t'\right|$. Thus, there is such a dimensionless tensor $\mathbf{D}$, depending on the fluid particle being considered, as:

$$\tau_0 \ll \left|t'' - t'\right| \ll \tau, \quad \mathbf{D}_u\left(t'' - t'\right) = \varepsilon \left|t'' - t'\right| \mathbf{D} \qquad (4.135)$$

The 'memory' of the signal $\mathbf{u}$ is then proportional to the square root of the time increment t. Not surprisingly, it can be found that the more dissipative the system is, the less it remembers its past for long, which is consistent with the fact that it is all the more sensitive to forgetting the initial conditions. Under an isotropic turbulence, that is in the absence of any privileged direction (particularly, far off the walls), the tensor $\mathbf{D}$ must be proportional to the identity, which yields

[65] This is an instance of forgotten initial conditions as mentioned in Section 4.2.2. This phenomenon may also be compared with the erratic motion of a particle under the action of a collection of molecules (Section 2.5.3).

[66] This is a Lagrangian concept, the dependence on time partly resulting from the particle motion (refer, for example, to Section 2.2.1).

[67] The 'inertial range' should here be understood as the range of times $t'' - t'$ corresponding to scales $\tau(\lambda)$ which are neither too small nor too large.

$$\tau_0 \ll t'' - t' \ll \tau, \quad \mathbf{D}_u\left(t'' - t'\right) = C_0\varepsilon\left|t'' - t'\right|\mathbf{I}_3 \tag{4.136}$$

This law is in agreement with equation (2.313) of Chapter 1. To some extent, the turbulence may be regarded as a stochastic process governed by those laws formulated in Section 2.5.3, which will advantageously be used in Section 4.6.3. The value of the dimensionless constant C_0 is a still unsettled question; Pope (2000) suggests $C_0 = 2.1$, but he says that higher values (probably up to 6) are not impossible.

4.4.4 Closures for eddy viscosity

We now have all the required elements for estimating the eddy viscosity. We may firstly note that its diffusion role illustrates the large turbulent structures, which are the most efficient ones in that process. Thus, we know from the Kolmogorov hypotheses as set out in the previous section that ν_T can only depend on k and ε, which yields, thanks to the results in Section 3.4.3, $\nu_T \sim k^2/\varepsilon$. There must then be such a constant C_μ that:

$$\nu_T = C_\mu \frac{k^2}{\varepsilon} \tag{4.137}$$

The value of C_μ is determined from experience and, for most authors, equals[68] 0.09. In such a way, if we were able to calculate k and ε as a function of space and time, we would be successful. In addition, we may note that the relations (4.117) can be used to write (4.137) as

$$\nu_T \sim L_t k^{1/2} \tag{4.138}$$

There is nothing surprising about that relation. As in the kinetic theory of gases (Section 3.5.2, eqn (3.296)), viscosity happens to be the product of a length and a velocity which characterize the agitation motion determining it, as given here by the third equation of (4.117). Hence, the macroscopic size of the large eddies gives ν_T a value which expectedly is much higher that ν. Combining (4.137), (4.138), (4.97) and (4.105), we get

$$\nu_T \sim L_t^2 S \tag{4.139}$$

There still is an unknown, but having returned it to the large scale size is an already major advance, because it is a rather intuitive quantity which lends itself to observation.[69] Like Prandtl (1925) and Smagorinsky (1963),[70] we will assume there is such a local length $L_m(x_i, t)$ that the above relation will become exact:

$$\nu_T = L_m^2 S \tag{4.140}$$

L_m is on the order of magnitude of L_t and is known as *mixing length*. The model (4.140) is named after it, which is advantageously explicit. It provides the eddy viscosity field as a function of the mean strain rate, and is an increasing function of it, as it should be (we may recall that the strain generates eddies via the term P). Moreover, this model requires details on L_m, an unknown quantity which is liable to vary according to the flows, space and

[68] The fact that it is only given with one significant figure is ... significant: the model (4.137) is rather rough, since the very notion of turbulent viscosity is based on an assumed isotropy of the eddies. We will come back to that issue in Section 4.5.1. By the way, we notice that with that model the turbulent Reynolds number Re_t, as defined by (4.118), is nothing else but the ratio $\nu_T/\nu = \mathrm{Re}/\overline{\mathrm{Re}}$, where $\overline{\mathrm{Re}}$ is defined by (4.84). Its high value is in accordance with eqn (4.81).

[69] The size of the large turbulent eddies is typically on the order of one tenth of characteristic size of the flow as a whole.

[70] It should be pointed out that Prandtl made that hypothesis within the frame of simply sheared flows (refer to Section 4.4.5). This formulation was imagined by Smagorinsky for extending the idea to the case of any flow; it should be specified, however, that his work was initially conducted within the context of the large eddy simulation (Section 4.6.1).

time. It can be observed that this model, as expected, leads to an eddy viscosity varying in space and time, as requested.

An important remark should be made here. In spite of appearance, equation (4.138) only has a limited similarity to the molecular model (3.296). We, indeed, have laid emphasis, in Section 3.5.2 on the fact that the molecular and mesoscopic scales are sufficiently distinct to let us assume that the molecular viscosity is independent of the fluid velocity. On the other hand, since the respective scales of mean flow and turbulence are similar, it may not be considered that the eddy viscosity is independent of mean flow velocity. Equation (4.140), on the contrary, shows that the eddy viscosity should be proportional to the latter. Hence, the Boussinesq model (4.73) shows that the Reynolds shear stresses are quadratic functions of velocity:

$$\forall i \neq j \quad R_{ij} = -2L_m^2 S S_{ij} \propto U^2 \tag{4.141}$$

All things considered, this result is dimensionally in agreement with their definition (4.46). This is a major difference from the Cauchy stresses for a Newtonian viscous fluid, whose sheared components (second line in (3.107)) are simply proportional to velocity.[71] We can infer that all the energy transfer processes caused by turbulence depend in a cubic way on velocities. This finding is in line with the observation we made in Section 3.4.3, in which we discussed the dependence of forces on velocities. As mentioned in that section, this result is due to the fact that eddy viscosity is not constant, unlike molecular viscosity, but proportional to velocity. This is a general case in which the linear (molecular) friction forces can be neglected, compared to the (turbulent) quadratic forces,[72] which come within the frame of the discussion in Section 2.5.2, with the theory of mirror systems. We will come back to this in Chapter 5 (Section 6.5.2) when dealing with turbulence modelling through the SPH method.

The mixing length model is based on so strong hypotheses that it can only be valid in simple cases. This is because we have assumed the turbulent equilibrium $P = \varepsilon$, which only holds true in an approximate way. Let us consider the particular case of a flow whose mean quantities (according to Reynolds) vary over time with a characteristic time[73] of the same order of magnitude as the time scale of the large eddies, that is.[74]

$$\tau = \frac{k}{\varepsilon} \tag{4.142}$$

In this case, the equilibrium $P = \varepsilon$ is much challenged, because some time is needed for the energy cascade to be set up, and so an energy input P does not instantly induce a dissipation ε of the same order of magnitude; that equilibrium is only established on completion of a relaxation time on the order of τ. Likewise, production only generates the energy k after an equivalent characteristic time, as illustrated in the kinetic energy equation (4.96). Lastly, we may recall that the original of the equation of motion (4.82) includes kinetic energy, which in some cases is not negligible as compared to pressure, into the pressure term. For all these reasons, the mixing length model should in many cases, particularly when the mean flow varies comparatively rapidly, give way to more sophisticated models.

[71] It may be objected that the Cauchy stresses exhibit a quadratic shape in (3.250) too. The difference stems from the fact that, unlike the molecular motion, the turbulent motions cannot be disconnected from the mean flow, because their spatio-temporal scales are too similar.

[72] Reynolds (1883), in his basic paper, perfectly understood the link between the occurrence of the turbulence and the quadratic behaviour of the energy losses in a fluid.

[73] It should be remembered that this time scales as S^{-1}, according to Section 3.2.2.

[74] According to (4.117), we may only state that the following equality should be substituted for by the 'order of magnitude' symbol $\sim$. There is nothing, however, preventing us from defining τ that way, what we will do in the following.

A more suitable model would consist in solving the turbulent kinetic energy equation (4.96). Now, this equation is not closed, because it includes the quantities $\mathbf{Q}^k$ and ε. However, we have established a link between that equation and that of a scalar, and highlighted the role of spatial flux of energy as played by $\mathbf{Q}^k$ (eqn (4.95)); then, by analogy with the model (4.85), it seems it is advisable to model this flux as follows:

$$\mathbf{Q}^k = -\nu_k \, \mathbf{grad} \, k \tag{4.143}$$

The diffusion coefficient ν_k models the diffusion as caused by the molecular viscosity (which explicitly appears in (4.95)) and the turbulent processes, which we have previously modelled through eddy viscosity. It should then be written as

$$\nu_k \doteqdot \nu + \frac{\nu_T}{\sigma_k} \tag{4.144}$$

By analogy with (4.88), the dimensionless coefficient σ_k illustrates the fact that the eddy mixing processes probably do not exert quantitatively equal actions over momentum and energy. σ_k, however, is probably on the order of magnitude of one, and most authors agree to give it the value $\sigma_k = 1.0$. Being provided with that model, the transport-diffusion equation of k becomes

$$\frac{Dk}{Dt} = P - \varepsilon + \mathrm{div} \, (\nu_k \, \mathrm{grad} \;\, k) \tag{4.145}$$

In order to close that model, dissipation is still to be estimated. A preliminary idea consists in temporarily coming back to the equilibrium $P = \varepsilon$ which, along with (4.97), (4.137) and (4.140), yields

$$\varepsilon = C_\mu^{3/4} \frac{k^{3/2}}{L_m} \tag{4.146}$$

which relation is consistent with the estimate (4.117). Equations (4.145) and (4.146) theoretically enable us to calculate k, what subsequently yields ν_T through the model (4.137), which, with (4.146) is now written as

$$\nu_T = C_\mu^{1/4} k^{1/2} L_m \tag{4.147}$$

This model, which we will henceforth refer to as $k - L_m$, is also known as a *one-equation* model because it involves an additional differential equation to reach our goal. It must therefore be accompanied by boundary conditions as regards k; details will be provided in Section 4.4.6. Despite its relative complexity compared to the mixing length model, it is much more suitable for modelling the quickly varying flows. It is still based, however, on a number of approximations, and above all once again utilizes the notion of mixing length, with the previously set forth restrictions.

In order to palliate its drawbacks, Launder and Spalding (1974) have suggested determining the dissipation value by solving an equation analogous to the energy equation, in order to achieve what is commonly known as the $k - \varepsilon$ model. Since the exact equation with respect to ε is too complex and involves terms whose physical meaning can only be elucidated in the Fourier space, these authors have proposed considering a heuristic equation taking a form similar to (4.145):

$$\frac{D\varepsilon}{Dt} = \frac{\varepsilon}{k}\left(C_{\varepsilon 1}P - C_{\varepsilon 2}\varepsilon\right) + \text{div}\left(\nu_\varepsilon \,\mathbf{grad}\ \varepsilon\right) \tag{4.148}$$

with

$$\nu_\varepsilon \doteqdot \nu + \frac{\nu_T}{\sigma_\varepsilon} \tag{4.149}$$

and some constants determined from semi-empirical considerations[75] as given by Table 4.1. We may note that equation (4.148) explicitly involves the relaxation time required for the action of production, corresponding to the large eddies' characteristic time $\tau = k/\varepsilon$. Once again, this equation should be provided with boundary conditions which are discussed later.

[75]The models based on the renormalization Groups (RNG), however, make it possible to calculate, from a rather complex spectral formalism, the values of these 'constants', as well as other previously mentioned constants (for example, refer to Yakhot and Orszag, 1986). That theory also suggests the addition of a term into the dissipation equation.

Table 4.1 Coefficients of the main two-equation turbulence closure models.

$k-\varepsilon$	$C_{\varepsilon 1}$	$C_{\varepsilon 2}$	σ_ε
	1.44	1.92	1.3
$k-\omega$	$C_{\omega 1}$	$C_{\omega 2}$	σ_ω
	$\frac{5}{9}$	$\frac{3}{40}$	2

The $k-\varepsilon$ model is now a standard in both industrial and environmental modelling practice. There are many variants, such as the Mellor and Yamada model (1974). As an example of the drawbacks of this approach, it should be mentioned that the size of the large eddies, as given by the estimate (4.117), is liable to take arbitrary values, whereas simple physical considerations show that it would have to be bounded, *a minima* by the size L of the flow, and even by the distance from the point being considered to the nearest wall. To remedy this problem, Yap (1987) suggests increasing ε by reducing the very dissipation term in the dissipation equation, which amounts to modifying the coefficient $C_{\varepsilon 2}$ as follows:

$$C_{\varepsilon 2,Yap} = C_{\varepsilon 2} - \max\left[0;\, 0.83\left(\frac{L_t}{L} - 1\right)\left(\frac{L_t}{L}\right)^2\right] \tag{4.150}$$

Yap suggests choosing $L = \kappa z / C_\mu^{3/4}$, z being the distance to the closest wall. The presence of the constant κ (as given by eqn (4.179)) will be substantiated in Section 4.4.5.

The best known and mostly used variant of the $k-\varepsilon$ model, however, is the $k-\omega$ model (Wilcox, 1988). The latter consists of replacing the heuristic equation (4.148) with an analogous equation with respect to the large eddies' frequency ω as defined by

$$\omega \doteqdot \frac{\varepsilon}{C_\mu k} \tag{4.151}$$

From (4.137) onwards, the eddy viscosity can be calculated through

$$\nu_T = \frac{k}{\omega} \tag{4.152}$$

The transport equation of that quantity, in the simplest variant of the model, is written as:

$$\frac{D\omega}{Dt} = C_{\omega 1}\frac{\omega}{k}P - C_{\omega 2}\omega^2 + \text{div}\left(\nu_\omega \,\mathbf{grad}\ \omega\right) \tag{4.153}$$

with

$$\nu_\omega \doteqdot \nu + \frac{\nu_T}{\sigma_\omega} \tag{4.154}$$

and constants as given by Table 4.1.

The *two-equation models*, in the above set out forms, are found to be defective in close proximity to a wall, in the so-called viscous sublayer (refer to Section 4.4.6). These gaps are usually bridged through a local change in the coefficients of Table 4.1 (for example, refer to Craft et al., 2002).

Before completing this section, a remark should be made about one of the major defects in the two-equation models we have just disclosed. This drawback directly results from the shape of the Reynolds tensor as estimated by the Boussinesq model (4.73). It can be seen that the dependence on the strain tensor **S** is linear. Now, if this kind of model is suitable for the molecular Cauchy stress tensor, it is only coarsely applicable to **R**, for reasons which will be elucidated in Section 4.5.1. We are going to explain several evidently erroneous consequences of it. To that purpose, let us rearrange the Boussinesq model in a somewhat different form, introducing the non-dimensional deviator of **R**:

$$\begin{aligned}\mathbf{a} &\doteq \frac{1}{k}\mathbf{R}^D \\ &= \frac{1}{k}\left(\mathbf{R} - \frac{2}{3}k\mathbf{I}_3\right) \\ &= -\frac{2\nu_T}{k}\mathbf{S} \\ &= -2C_\mu \tau \mathbf{S}\end{aligned} \tag{4.155}$$

(we have used (4.73), (4.75), (4.137) and (4.142), as well as the definition (A.23) from Appendix A with $n = 3$). Tensor **a** stands for the portion of **R** which is not proportional to the identity tensor, that is the anisotropic part of the Reynolds stresses, and is then referred to as the *anisotropy tensor*.[76] The preceding calculation states that the Boussinesq model assumes an anisotropy which is everywhere proportional to the strain. Now, as a dimensionless tensor, **a** is certainly norm-bounded. Thus, **a** cannot be proportional to **S** (which cannot be bounded for any reason), which seems to be at odds with (4.155), unless τ decreases in $1/S$. Another effect of (4.155) is that the production is written, according to (4.97), as:

$$\begin{aligned}P &= -\mathbf{S} : k\left(\mathbf{a} + \frac{2}{3}\mathbf{I}_3\right) \\ &= -k\mathbf{S} : \mathbf{a} - \frac{2}{3}k \operatorname{tr} \mathbf{S} \\ &= -k\mathbf{S} : \mathbf{a} \\ &\leqslant kS\sqrt{\frac{II_\mathbf{a}}{2}}\end{aligned} \tag{4.156}$$

where $II_\mathbf{a}$ is the second-order invariant of tensor **a**, defined as in (3.26). Once again, incompressibility was used in the calculation (4.156), while the latest inequality results from the Cauchy–Schwartz theorem. Since invariant $II_\mathbf{a}$ is necessarily bounded, the relation (4.156) shows that production probably behaves at most as a linear function of the strain when the latter is large,

[76] Its properties will be disclosed in Section 4.5.1.

which is at variance with the quadratic dependence (4.97). For larger strains (in close vicinity to a stagnation point immediately upstream of an obstacle, for instance), the standard $k - \varepsilon$ model will then yield too high values of k, and hence of ν_T. In order to palliate this drawback, Guimet and Laurence (2002) have put forward the following argument. Turning back to the definition (4.97) of production, and using the equilibrium (4.105) and the relation (4.137), another formulation of P can be found:

$$P = \sqrt{C_\mu} k S \tag{4.157}$$

In order to avoid overestimating k, we then may contemplate selecting, out of (4.97) and (4.157), the smaller one, which restricts the production of energy to a linear behaviour when the rate of strain S is high. We then achieve a semiquadratic, semilinear model for production:

$$P = \min\left(1, \sqrt{C_\mu} \tau S\right) \sqrt{C_\mu} k S \tag{4.158}$$

We may note that this procedure amounts to increasing the invariant $II_{\mathbf{a}}$ by the value it takes for the most naturally and definitely anisotropic flow, namely the case of a simple shear flow, that is when velocity is directed along one direction. We will show, indeed, in the next section (eqn (4.206)) that in such circumstances this invariant then equals $2C_\mu$, a limiting value ascribed to it by (4.158) for large strains. This model again provides the quadratic dependence (4.97) for small strains, whereas it becomes linear when S is high, in accordance with the preceding comments.

Lastly, it should be pointed out that equation (4.156) suggests that the production may be negative, if tensors $-\mathbf{S}$ and $\mathbf{a}$ are directed along quite different specific directions.[77] Such is the case of some flows which are either rapidly oscillating or highly variable in space.[78] The Reynolds stress equation (4.66) suggests in that respect that the latter follow the average flow strain with a delay in the order of τ, just like kinetic energy. Thus, if the average field varies over time with a charactetistic time S^{-1} in the order of τ, the specific directions of $\mathbf{R}$ (hence of $\mathbf{a}$) will possibly be in phase opposition with those of $-\mathbf{S}$. Revell et al. (2006) suggest remedying that drawback by calculating P (or rather the quantity $C_{as} \doteqdot \mathbf{S} : \mathbf{a}/S$) using an additional equation, which results in a three-equation model ($k - \varepsilon - C_{as}$ model). We will not disclose this model in this book; we let inquiring readers discover on their own what is one of the latest advances (at the time this page is being written) as regards the turbulent flow modelling through Reynolds-averaging type methods.

[77] Being symmetric tensors, both are diagonalizable in orthonormal basis, but there is no reason why they would coincide, because the tensors do not necessarily commute.

[78] That means the average quantities vary over time (respectively in space) with a characteristic time (respectivement spatial) scale which is close to that of the large turbulent scales.

4.4.5 Generic case study: the infinite open channel

In order to determine the boundary conditions of those turbulent models we have just introduced, we will start from those conditions previously discussed in Section 3.4.2 in the general case of the Navier–Stokes equations. Through a mere application of the averaging operator, the free surface conditions (3.136) and (3.143) will, of course, yield[79]

[79] The second following condition assumes that $\mathbf{n}$ is constant. This hypothesis is challenged for a solid body which is liable to oscillate through eddying agitation. The conditions under which the motion of a free solid would be affected by the turbulent eddies are not considered here.

$$[\overline{p}] = 0$$
$$[\overline{\tau}\cdot\mathbf{n}] = 0 \qquad (4.159)$$
$$\overline{\mathbf{u}}\,(\mathbf{r} = \mathbf{r}_{wall}) = \mathbf{u}_{wall}\,(\mathbf{r}_{wall})$$

(the brackets, as in Section 3.4.2, standing for the discontinuity of the quantity they encompass when passing through the relevant interface, like a free surface). Subtracting the new wall condition from the original condition (3.143), we find out that there is no velocity fluctuation on the walls:

$$\mathbf{u}'\,(\mathbf{r} = \mathbf{r}_{wall}) = \mathbf{0} \qquad (4.160)$$

(and no Reynolds stresses either). These conditions, however, are practically inapplicable, since the flow is seldom modelled up to the wall under a turbulent regime. There are a couple of reasons: first, as subsequently explained in this section, the velocity field exhibits a strong gradient near a solid wall, which would numerically imply a very fine discretization; next, the very notion of wall is quite often ambiguous, particularly for the natural flows (in the environment) whose edges exhibit significant irregularities (just think of a river bed).

Going further into these matters, it is convenient to turn to the peculiar case of a open channel with an elevation H, having a bed with a gentle slope angle θ, an 'infinite' length and submitted to a longitudinally uniform and steady regime. We will utilize the notations from Fig. 4.7, with an origin of the axis z being located at the bottom level (note that the axis z is not vertical but orthogonal to the bed). One feature of such a flow is that it is invariant along the perpendicular direction (denoted as y) to that figure, since all the average quantities are invariant. We mean by the phrase 'longitudinally invariant' that the average quantities do not vary along the axis x, which is made possible by the theoretically infinite nature of the bed. Besides, the *mean* velocity field is in every point aligned with x. Thus, the velocity field is written as $\overline{\mathbf{u}} = \overline{u}\,(x, z)\,\mathbf{e}_x$ and the boundary conditions (4.159) are written as[80]

[80]The second condition, regarding the surface friction, is written here in the absence of wind (no surface friction).

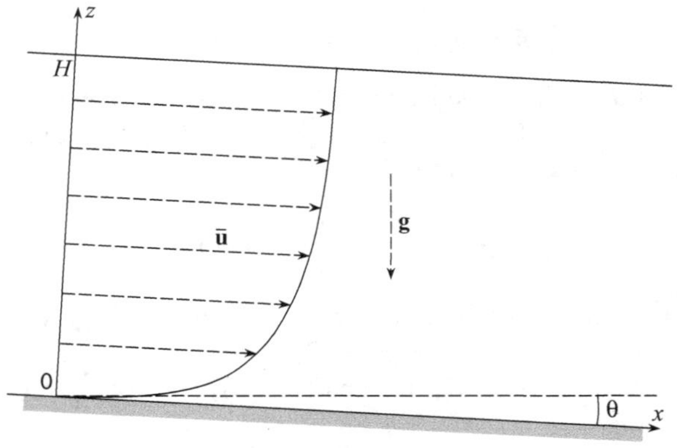

Fig. 4.7 Free surface infinite channel under steady uniform regime.

$$\begin{aligned} \overline{p}\,(z = H) &= \overline{p}_{atm} \\ \frac{\partial \overline{u}}{\partial z}\,(z = H) &= 0 \\ \overline{u}\,(z = 0) &= 0 \end{aligned} \tag{4.161}$$

where $\overline{p}_{atm}$ denotes the average atmospheric pressure, which is supposedly constant. The continuity equation is reduced to

$$\frac{\partial \overline{u}}{\partial x} = 0 \tag{4.162}$$

which reveals that $\overline{\mathbf{u}} = \overline{u}\,(z)\,\mathbf{e}_x$. The strain rate tensor is then written as:[81]

[81] Such a flow is said to be *simply sheared.*

$$\mathbf{S} = \frac{1}{2}\frac{\partial \overline{u}}{\partial z}\begin{pmatrix} 0 & 0 & 1 \\ 0 & 0 & 0 \\ 1 & 0 & 0 \end{pmatrix} \tag{4.163}$$

and its norm simply equals

$$S = \left|\frac{\partial \overline{u}}{\partial z}\right| \tag{4.164}$$

Because of the hypotheses, projecting the Reynolds equation (4.82) onto the (x, z) the plane yields two scalar equations for the motion, using the definition (3.138) of the dynamic pressure:

$$\begin{aligned} 0 &= -\frac{1}{\rho}\frac{\partial \overline{p}}{\partial x} + \frac{\partial}{\partial z}\left[(\nu + \nu_T)\frac{\partial \overline{u}}{\partial z}\right] + g\sin\theta \\ 0 &= -\frac{1}{\rho}\frac{\partial \overline{p}}{\partial z} - g\cos\theta \end{aligned} \tag{4.165}$$

It can be observed that the inertia terms have disappeared. Nevertheless, the equations are not linear, because of the dependence of ν_T on the velocities (eqn (4.140), for instance). We may recall in that respect that the eddy viscosity ν_T models, through the component R_{xz} of the Reynolds stress tensor, the effect of the eddies on the average flow, which effect is inertia, hence not linear. In view of the first condition in (4.161), the second equation of (4.165) gives a hydrostatic pressure profile which is analogous to (3.135):

$$\overline{p} = \overline{p}_{atm} + \rho g\,(H - z)\cos\theta \tag{4.166}$$

and so the pressure does not depend on x (once again, we can observe the longitudinal invariance).[82] The first equation then yields, integrating first over z, and using the second condition of (4.161):

[82] That would be untrue in a closed pipe where, since there is no free surface, that invariance could not be ensured for pressure. On the contrary, a longitudinal pressure gradient can be used, in this case, as a driving term in addition to the possible action of gravity.

$$(\nu + \nu_T)\frac{\partial \overline{u}}{\partial z} = \frac{\tau_b}{\rho} - gz\sin\theta \tag{4.167}$$

The left-hand side (multiplied by ρ) represents the $\overline{\sigma}_{xz} - \rho R_{xz}$ component of the total stress tensor (refer to eqn (4.52)), which shows that the integration constant τ_b is the shear stress at the bed ($z = 0$). The common practice then consists in putting

$$u_* \doteqdot \sqrt{\frac{\tau_b}{\rho}} \tag{4.168}$$

That quantity is known as *shear velocity*, and a physical interpretation of it will subsequently be provided. In the absence of wind, the stress vanishes at the surface ($z = H$), according to (4.159), and equation (4.167) is reduced to

$$(\nu + \nu_T)\frac{\partial \overline{u}}{\partial z} = u_*^2\left(1 - \frac{z}{H}\right) \tag{4.169}$$

Once again, we observe the quadratic nature of the Reynolds stresses which was mentioned in Section 4.4.4. Equation (4.167) also yields the shear velocity as

$$u_* = \sqrt{gH\sin\theta} \tag{4.170}$$

u_* is, of course, an increasing function of slope;[83] this is (like the more general relation (4.167)) an equilibrium between the channel slope, which is the flow driving term, and the friction exerted onto the bed, which is a resistance-like term allowing a steady regime equilibrium.[84] In the case of a laminar flow, that is with a Reynolds number below $Re_{c,1}$ (refer to Section 4.2.2), $\overline{u}$ is reduced to the actual velocity field u and the eddy viscosity equals zero. Equation (4.169) is then immediately integrated and yields, taking the third condition of (4.161) into account:

$$u = \frac{u_*^2}{\nu} z\left(1 - \frac{z}{2H}\right) \tag{4.171}$$

The velocity profile is then parabolic. The maximum velocity $U_{\max} = u\,(z = H)$ is given, in view of (4.170), by

$$U_{\max} = \frac{gH^2\sin\theta}{2\nu} \tag{4.172}$$

and the Reynolds number as built on $U_{\max}$ and H equals

$$Re = \frac{gH^3\sin\theta}{2\nu^2} \tag{4.173}$$

The circumstances governing the establishment of such a flow, however, quite seldom occur, the condition $Re < Re_{c,1}$ corresponding to a slope $\tan\theta \approx \sin\theta < 4 \times 10^{-7}$ if the channel is filled with water to a height $H = 10$ cm.

In the turbulent case, which is by far the most frequent one, the equilibrium equation (4.169) does not have an exact analytical solution everywhere; we then must make a distinction between several cases. We first invoke the mixing length model (4.140), which is valid here owing to the steady regime hypotheses. With (4.164), we get:[85]

$$\nu_T = L_m^2 \frac{\partial \overline{u}}{\partial z} \tag{4.174}$$

This is where the mixing length model reveals its true difficulty, because L_m is unknown. It can be stated, however, according to Section 4.4.4, that it has the same order of magnitude as the size of the large eddies. Now, the

[83] Once again, this formula is only valid in the absence of a horizontal pressure gradient, and so it is useless in a closed pipe. Later on, we will discuss a general method for assessing the friction velocity.

[84] This is an old common topic in fluid mechanics, where the notion of equilibrium plays a prevailing role: just think of a body falling in a viscous fluid, or else the production-dissipation balance which is at the core of the turbulence mechanisms.

[85] The velocity is assumed to be increasing along z, which is both in accordance with intuition and right.

condition (4.160) tells us that the eddies fade at the bottom. A first attempt[86] then consists in making a Taylor series expansion yielding $L_m = \kappa z$, κ being a constant which is unknown for the time being. This model, however, is still too simplistic, because of the free surface effect. A mere observation of a river surface shows, indeed, that the vertical fluctuations of velocity vanish at it,[87] which may be explained by the great difference of density between water and air, since gravity inhibits every incipient vertical fluctuation.[88] Thus, we may consider that the mixing length equals zero at the surface. The κz term which we have just established should then be multiplied by a function vanishing in $z = H$. We know from experience (refer to Nezu and Nakagawa, 1993) that satisfactory results are achieved, in the presently discussed case, with the following form:

$$L_m = \kappa z \sqrt{1 - \frac{z}{H}} \tag{4.175}$$

In close vicinity to the bottom, within sufficiently small distances, formula (4.175) shows that the eddy viscosity (4.174) tends towards zero. Thus, there is a very thin layer, which is known as the *viscous sublayer*, near the channel bottom, where the molecular effects prevail over the turbulent fluctuations. We then have a profile which is analogous to (4.171), which can be made even simpler by dropping the quadratic term. The characteristic thickness e_ν of that layer depends on both u_* and ν, and the considerations about dimensions in Section 3.4.3 show it is written as

$$e_\nu \sim 8 \frac{\nu}{u_*} \tag{4.176}$$

(the factor 8 should be considered as a piece of experimental data). Equation (3.160) of Section 3.4.2 shows that e_ν is expected on the order of magnitude of L_ν.

Beyond the viscous sublayer (typically for $z > 100\frac{\nu}{u_*}$), on the contrary, the turbulent forces ρR_{xz} do prevail over the molecular forces $\overline{\sigma}_{xz}$, which amounts to saying that $\nu \ll \nu_T$ in the greater part of the flow (in accordance with eqn (4.81)). The equilibrium equation (4.169) and the models (4.174) and (4.175) then give the very simple relation

$$\frac{\partial \overline{u}}{\partial z} = \frac{u_*}{\kappa z} \tag{4.177}$$

For building a solution, we still need a boundary condition. Unfortunately, since we have only made valid assumptions from a small distance to the wall, we may not invoke the condition (4.161). We then have to introduce a further empirical constant to reach the following solution:

$$\overline{u} = u_* \left(\frac{1}{\kappa} \ln \frac{z u_*}{\nu} + C_\nu \right) \tag{4.178}$$

Practice yields

$$\begin{aligned} \kappa &= 0.41 \\ C_\nu &= 5.2 \end{aligned} \tag{4.179}$$

[86] Proposed by Prandtl.

[87] Here we ignore the gravity waves (swell).

[88] This process is analogous to the gravitational damping within the stratified turbulent flows. We may note in that respect that this phenomenon makes the velocity fluctuations very anisotropic near the surface, which echoes the remarks made in the preceding section and the next ones as regards the validity of the Boussinesq model. Such are the limits of the notion of turbulent viscosity at the free surface. This trouble, however, does not affect the arguments in this section, since only vertical diffusion is of interest here, the flow being horizontally unvarying.

κ is referred to as the von Karman constant.[89] The logarithmic profile (4.178) exhibits a velocity distribution which is much more homogeneous along the vertical direction than the laminar parabola (4.171), which reveals the significant diffusion (here, of momentum) being exerted by the turbulence.

At this stage, a remark about the viscous sublayer should be made without delay. Formula (4.176) suggests it is very thin (typically less than one millimetre or even one or two orders of magnitude less), and so its pertinence should be questioned, especially in a natural environment. The cemented wall of an artificial channel, and even more the natural bed of a river, bear asperities whose characteristic size may largely exceed e_ν on their surfaces. In such a case, the viscous sublayer is concealed by the asperities and we must challenge our arguments, because the boundary condition at the wall becomes partly meaningless since it is determined by these asperities which may have quite intricate and irregular shapes. It may be considered that the roughness of a wall induces a greater friction, that is more turbulent fluctuations near the wall. As such, based on the discussions in Section 4.4.1, it is advisable to model that effect by adding some sort of 'roughness viscosity', here being denoted as ν_R, to the molecular viscosity. From the dimensional considerations in Section 3.4.3, such a 'roughness viscosity' can be linked to the characteristic size k_s of the asperities (which is referred to as the wall *roughness*) and to u_* by the following relation:

$$\nu_R = \alpha_R k_s u_* \tag{4.180}$$

with a constant $\alpha_R = 0.26$, as provided by experience. The profile (4.178) should then be changed by adding ν_R to ν:

$$\overline{u} = u_* \left(\frac{1}{\kappa} \ln \frac{z u_*}{\nu + \alpha_R k_s u_*} + C_\nu \right) \tag{4.181}$$

The effect of the asperities can be assessed through the ratio of these two viscosities:

$$Re_* \doteq \frac{k_s u_*}{\nu} \sim \frac{\nu_R}{\nu} \sim \frac{k_s}{e_\nu} \tag{4.182}$$

Three regimes can be differentiated by means of that *shear Reynolds number*, namely:

- $Re_* < 5$: smooth regime, in which the molecular viscosity near the wall prevails over the effects of the asperities, that is $\nu_R \ll \nu$. The viscous sublayer is substantially thicker than the asperities. The profile (4.181) once again gives the *smooth velocity profile* (4.178);
- $5 < Re_* < 70$: mixed regime, in which the effects of the asperities are of the same order of magnitude as those of molecular viscosity, that is $\nu_R \sim \nu$. The regime (4.181) remains unchanged;
- $70 < Re_*$: rough regime, in which the effects of the asperities prevail over molecular viscosity near the wall, that is $\nu \ll \nu_R$. The viscous sublayer is concealed by the asperities. The profile (4.181) is simplified and gives a *rough velocity profile*:

$$\overline{u} = u_* \left(\frac{1}{\kappa} \ln \frac{z}{k_s} + C_* \right) \tag{4.183}$$

[89] In the literature, scientists are not unanimous about the value of the constant which here we call C_ν. Values typically ranging from 5.0 to 5.5 are reported.

with

$$C_* \doteqdot C_\nu - \frac{1}{\kappa}\ln \alpha_R \approx 8.5 \tag{4.184}$$

Whatever regime may be achieved, the formulas (4.174), (4.175) and (4.177) yield the eddy viscosity as

$$\nu_T = \kappa u_* z\left(1 - \frac{z}{H}\right) \tag{4.185}$$

Even in the simple case being dealt with, the equations (4.145) and (4.148) in the $k - \varepsilon$ model do not have any analytical solution; these two quantities, however, can be estimated from what was explained above by recovering the production-dissipation equilibrium:

$$\begin{aligned}\varepsilon = P &= \nu_T S^2 \\ &= \frac{u_*^3}{\kappa z}\left(1 - \frac{z}{H}\right)\end{aligned} \tag{4.186}$$

(we have taken advantage of (4.164), (4.177) and (4.185)). The dimensional relation (4.137) then yields k:

$$\begin{aligned}k &= \sqrt{\frac{\nu_T \varepsilon}{C_\mu}} \\ &= \frac{u_*^2}{\sqrt{C_\mu}}\left(1 - \frac{z}{H}\right)\end{aligned} \tag{4.187}$$

Thus, the shear velocity is in the form of an order of magnitude of the velocity $\sqrt{k}$ of the large eddies near the bottom. The above results may also be rearranged in a dimensionless form:

$$\begin{aligned}k^+ &\doteqdot \frac{k}{u_*^2} = \frac{1 - z^+}{\sqrt{C_\mu}} \\ \varepsilon^+ &\doteqdot \frac{\varepsilon H}{u_*^3} = \frac{1 - z^+}{\kappa z^+} \\ \nu_T^+ &\doteqdot \frac{\nu_T}{u_* H} = \kappa z^+\left(1 - z^+\right)\end{aligned} \tag{4.188}$$

with[90] $z^+ \doteqdot z/H$. The formulas (4.182), (4.185), (4.186) and (4.187) are fairly well confirmed in practice, particularly the velocity and energy dissipation profiles (refer to Nezu and Nakagawa, 1993). Some exemplary comparisons can be found in Chapter 7, Section 7.2.3.

[90] It should be pointed out that the notation z^+ being used is not universal. In most of the publications about turbulence, it is rather used for denoting $z u_*/\nu$.

The flow whose theory has just been briefly described will be modelled by means of the SPH method in Chapter 7 (Section 7.2.3).

4.4.6 Turbulent boundary conditions

The behaviour which we have just highlighted may be considered as comparatively general near a wall which is not too curved, with, however, a rather coarse approximation in the case of rapidly varying flows. It will then enable us to define wall conditions for the turbulent quantities, within a rather general frame. Thus, in a rough regime, it will be considered that within a small distance δ from a wall, the conditions on the mean velocity, the kinetic energy and its dissipation rate can be specified[91] by analogy with (4.183), (4.186) and (4.187):

[91] Setting the velocity is one alternative. Some authors, however, recommend, instead, to set the stress value at the wall, i.e. the momentum flux.

$$\begin{aligned}\overline{\mathbf{u}}\,(z=\delta) &= u_*\left(\frac{1}{\kappa}\ln\frac{\delta}{k_s}+C_*\right)\mathbf{t}\\ k\,(z=\delta) &= \frac{u_*^2}{\sqrt{C_\mu}}\\ \varepsilon\,(z=\delta) &= \frac{|u_*|^3}{\kappa\delta}\end{aligned} \tag{4.189}$$

where $\mathbf{t}$ is the wall-tangential unit vector. We have ignored the terms in $1-\delta/H$, in view of the smallness of δ, which may be arbitrarily chosen, but satisfying the condition $e_\nu \ll \delta \ll H$, H being, in the general case, a characteristic dimension of the overall flow. Numerically, the values given by (4.189) at each point of a wall[92] are prescribed, which amounts to likening the calculation point located on the wall to a virtual point located within the flow and at a distance δ from the wall, in the perpendicular direction to the latter.

[92] The formulas (4.189) are called *wall functions*.

Lastly, these conditions imply an estimated shear velocity, which is by no means constant as a rule, whether in time or space. For that reason, in many calculation softwares, the mean velocity $\overline{\mathbf{u}}$ is estimated at a distance δ' from the wall which exceeds δ, and u_* is subsequently derived from it by means of (4.183):

$$u_* = \frac{\overline{\mathbf{u}}\left(z=\delta'\right)\cdot\mathbf{t}}{\frac{1}{\kappa}\ln\frac{\delta'}{k_s}+C_*} \tag{4.190}$$

The shear velocity then has the sign of $\overline{\mathbf{u}}\left(z=\delta'\right)\cdot\mathbf{t}$, and the velocity $\overline{\mathbf{u}}\,(z=\delta)$ can therefore be suitably directed; thus, it may be negative in some cases. That is why u_* occurs as an absolute value in the condition with respect to ε in (4.189), because dissipation is inherently positive. It should be pointed out, however, that limitations are encountered in that approach in the case of a comparatively variable flow in space and/or time.

The question of turbulent conditions at a free surface is still partly unsettled. Imposing a Von Neumann condition for turbulent kinetic energy looks logical. That option is dictated by the model (4.143), given the absence of any energy flux at the surface:

$$\frac{\partial k}{\partial \mathbf{n}_s} = 0 \tag{4.191}$$

with, as previously stated, $\partial A/\partial\mathbf{n} \doteqdot \mathbf{grad}\ A\cdot\mathbf{n}$, and $\mathbf{n}_s$ standing for the unit vector normal to the free surface at the point being contemplated. Nezu and Nakagawa (1993) recommend relating the dissipation at the surface to the value of the turbulent energy, due to dimensional considerations:

$$\varepsilon_{surf} = \frac{k_{surf}^{3/2}}{\alpha_\varepsilon H} \tag{4.192}$$

where H is the water depth, and with $\alpha_\varepsilon = 0.18$. Clearly enough, that condition is hardly generalizable, since it is based on the water depth concept, which is often ill-defined for a complex free surface flow, as in the case of a wave breaking.

To complete this section, we will estimate several orders of magnitude. First, u_* may be related to the defined averaged velocity U[93] by

$$U \doteqdot \frac{1}{H}\int_0^H \overline{u}(z)\, dz \tag{4.193}$$

[93] We have used the rough profile (4.183), and considered that the integration can be made from $z = 0$ without significantly changing the result, although that profile is strictly exclusively valid for $z > k_s$.

With the velocity profile (4.183), that yields

$$U = u_* \left(\frac{1}{\kappa} \ln \frac{H}{ek_s} + C_*\right) \tag{4.194}$$

where e is the basis of the natural logarithm. Given the usual values of water depth and roughness, the shear velocity then typically ranges from 3 to 8% of U. The maximum velocity value is reached at the surface and equals

$$U_{\max} = \overline{u}(H) = u_* \left(\frac{1}{\kappa} \ln \frac{H}{k_s} + C_*\right) \tag{4.195}$$

The $U_{\max}/U$ ratio depends on both roughness and depth and is typically around 1.2. By comparison, the definition (4.193) gives, for the laminar profile (4.171):

$$U_{\max} = \frac{3}{2} U \tag{4.196}$$

((4.171) and (4.172) were used). The parabolic laminar profile is then much more heterogeneous than the logarithmic turbulent profile.

Let us come back to the turbulent case. The formulas (4.186) and (4.187) then yield

$$\begin{aligned} L_t &\sim \frac{\kappa z}{C_\mu^{3/4}} \sqrt{1 - \frac{z}{H}} \\ &= \frac{L_m}{C_\mu^{3/4}} \end{aligned} \tag{4.197}$$

((4.175) was used). We unsurprisingly find that the mixing length provides an order of magnitude of the size of the large eddies. As regards the turbulent Reynolds number, it is written, according to (4.118) and (4.175) as:

$$\begin{aligned} Re_t &= \frac{L_t u_t}{\nu} \\ &\sim \kappa Re \frac{u_*}{U} \frac{z}{H} \sqrt{1 - \frac{z}{H}} \end{aligned} \tag{4.198}$$

where the ordinary Reynolds number is here given by

$$Re = \frac{UH}{\nu} \tag{4.199}$$

In frequent circumstances (natural, industrial or experimental flows), Re_t takes values ranging from 10^4 to 10^8, which, by virtue of (4.118), generates small eddies which are 10^3 to 10^6 times smaller than the large ones. The inertial range is then quite well developed. The formulas (4.116) and (4.186) provide (dimensionless) time and space scales of the small eddies in the following form:

$$\begin{aligned}\frac{\lambda_0}{H} &= \left(\frac{\kappa}{Re_*^3}\frac{z/H}{1-\frac{z}{H}}\right)^{1/4} \sim \frac{1}{Re_*^{3/4}}\\ \frac{\tau_0 u_*}{H} &= \left(\frac{\kappa}{Re_*}\frac{z/H}{1-\frac{z}{H}}\right)^{1/2} \sim \frac{1}{Re_*^{1/2}}\end{aligned} \tag{4.200}$$

The Kolmogorov scale then exhibits the same ratio to the water depth as to the large eddies (which means that the size of the latter does not vary as a function of the Reynolds number, which size equals L_m according to eqn (4.197)). Given the three-dimensionality of the turbulent structures, we may find out that the required spatio-temporal resolution increases as $\left(Re_*^{3/4}\right)^3 \times Re_*^{1/2} = Re_*^{9/8}$. Thus, the required resolution for a direct numerical (DNS, refer to the beginning of Section 4.3.1) is inhibitory even in this simple case.[94]

Lastly, let us turn to the mean flow Reynolds number as defined by (4.84). In order to determine an order of magnitude of it, we will temporarily return to the case of the H depth channel as defined in Section 4.4.5. It is provided by

$$\begin{aligned}\overline{Re} &= \frac{\overline{u}(z=H)\,H}{\nu_T(z=H/2)}\\ &= \frac{4}{\kappa^2}\left(\ln\frac{H}{k_s}+C_*\right)\end{aligned} \tag{4.201}$$

With the natural values of roughness,[95] this number is approximately 300, which is substantially below the critical threshold $Re_{c,2}$ for the onset of chaos, which equals about 3000 for an open channel (refer to Section 4.2.2). Although it is a rough estimate, it substantiates the mean flow stability, in accordance with the comment we made by the end of Section 4.4.1.

We now can quantitatively compare kinetic energy and pressure in this case in order to demonstrate that the approximation (4.80) is justified. We may recall that such an approximation consists of ignoring $\frac{2}{3}\rho k$ before $\overline{p}$ in the mean field equation of motion. In the case of the infinite channel, the profiles (4.166) and (4.187), together with the equilibrium (4.170), make it possible to write

$$\begin{aligned}\frac{\frac{2}{3}\rho k}{\overline{p}-\overline{p}_{atm}} &= \frac{2}{3\sqrt{C_\mu}}\frac{u_*^2}{gH\cos\theta}\\ &= \frac{2}{3\sqrt{C_\mu}}\tan\theta\\ &\approx 2.22\tan\theta\end{aligned} \tag{4.202}$$

[94] A the time these lines are being written, the literature mentions a DNS of up to $Re_* = 2000$ in a channel (Hoyas and Jimenez, 2006).

[95] Because of the logarithmic dependence, the exact value of the water depth-to-roughness ratio only slightly affects the result.

Our hypothesis then holds true as long as the bed slope is not too steep; for ordinary values ($\tan\theta \sim 10^{-5}$ to 10^{-2}), the ratio (4.202) hardly exceeds 2%. For extending our assertion to the more general case of an undiscriminated flow (with a free surface), one can take advantage of (4.202). Since the u_*/U ratio ranges from 3 to 8% in most cases, and likening $\cos\theta$ to unit, it is inferred that the estimate (4.202) is reduced to

$$\frac{\frac{2}{3}\rho k}{\overline{p} - \overline{p}_{atm}} < 0.01 Fr^2 \tag{4.203}$$

where Fr is the Froude number, as defined by (3.154). Thus, we may state that the turbulent kinetic energy only takes significant values for Froude numbers above 10, which is still fairly rare. The above approximation can be further strengthened by writing, owing to (4.187):

$$\frac{k}{U^2} \sim \left(\frac{u_*}{U}\right)^2 \tag{4.204}$$

The estimate of u_* as it was made above provides values on the order of several thousandths for that ratio. The relation (4.78) can be used to conclude. The above consideration will, of course, still be restricted to the free surface flows, but gives a fairly good idea of the ratio to be found within a more general frame.

Because of needs to be elucidated later on, we lastly calculate an non-dimensionalized form of the scalar strain rate, for a simple shear flow, owing to (4.164), (4.177), (4.186) and (4.187):

$$\tau S = \frac{k}{\varepsilon}\left|\frac{\partial \overline{u}}{\partial z}\right| = \frac{1}{\sqrt{C_\mu}} = 3.33 \tag{4.205}$$

That formula gives the anisotropy invariant as mentioned at the end of Section 4.4.4:

$$\begin{aligned} II_{\mathbf{a}} &\doteqdot a_{ij}a_{ij} \\ &= 2C_\mu^2 \tau^2 S^2 \\ &= 2C_\mu \end{aligned} \tag{4.206}$$

4.5 Advanced averaged models

We investigate now more sophisticated averaged models. By reviewing the deficiencies of first-order models, we introduce the concept of Lumley realizability of turbulent models. As a second step, we present closure equations for the second-order statistical turbulent models. The particular case of weak-equilibrium turbulence finally drives us to explicit algebraic models for Reynolds stress closure.

4.5.1 First-order model deficiencies

The first-order closure models which we have discussed in Section 4.4 exhibit deficiencies which are primarily linked to the utilization of the Boussinesq model (4.73), which is equivalent to the Stokes model for the Cauchy stress tensor. We have previously explained several effects of that approach in the assessment of the turbulent kinetic energy production term and mentioned some of the available means to remedy these errors; there are, however, additional shortcomings in the Boussinesq model; we are going to discuss the main ones. Since the trace of $\mathbf{R}$ is known through the definition of k, the following reasonings will deal with the anisotropy tensor $\mathbf{a}$ as defined by (4.155):

- We explained in Section 3.4.1 that thanks to the isotropic nature of a fluid, the Cauchy stress deviator $\boldsymbol{\sigma}^D$ only depends on $\mathbf{s}$, because a pure rotation does not then generate any friction. Now there is no reason why the turbulent eddies would obey such an isotropic property, since they are influenced by the nearby solid walls. This is due to their macroscopic size, preventing us thinking that their properties are rotationally invariant, unlike the molecular motion in the Boltzmann's approach discussed in Chapters 2 and 3. We have previously explained that the size of the eddies varies as they come nearer to a wall (Section 4.4.5); the velocity fluctuations orthogonal to that wall happen to be more influenced, indeed, as one comes nearer to it, which results in an anisotropy of the fluctuating velocities, that is of the Reynolds stresses. Thus, the deviator of $\mathbf{R}$ probably depends not only on the symmetric part $\mathbf{S}$ of the velocity gradients (which dependence is already included in the Boussinesq model (4.73)), but also on its asymmetric part $\boldsymbol{\Omega}$, representing the rotation of the mean field (or mean vorticity tensor), as defined by

$$\begin{aligned}\boldsymbol{\Omega} &\doteq \overline{\boldsymbol{\omega}} \\ &= \frac{1}{2}\left(\frac{\partial \overline{u}_i}{\partial x_j} - \frac{\partial \overline{u}_j}{\partial x_i}\right)\mathbf{e}_i \otimes \mathbf{e}_j\end{aligned} \tag{4.207}$$

  where $\boldsymbol{\omega}$ is defined by (3.30).
- That anisotropic property produces other effects: the linear relation established by the Boussinesq model is not written in a general form, and would actually have to involve a fourth-order eddy viscosity tensor in order to substitute such a formula as

$$a_{ij} = -\frac{2}{k}\nu_{T,ijkl}S_{lk} \tag{4.208}$$

  for (4.155). This would give not 1, but 81 viscosity coefficients (with, as it should be acknowledged, symmetry properties which would reduce the number of independent coefficients). The Boussinesq model, pursuant to (4.155), corresponds to coefficients $\nu_{T,ijkl} = \nu_T\delta_{il}\delta_{jk}$, which is much too simplistic for such a complex 'medium' as a turbulent flow.
- The Taylor series expansion (3.102) is only valid because the molecular viscous forces within an ordinary fluid are quite moderate due to the low value of molecular viscosity. On the other hand, eddy viscosity is a sufficiently large quantity for the turbulent shearing to be usually at least of the same order of magnitude as the turbulent kinetic energy term in the Boussinesq model. For instance, let us consider the infinite channel case that will be described in Section 4.4.5: with the formulas (4.164), (4.177),

(4.185) and (4.187), the ratio of the first two terms in the Boussinesq model is written as

$$\frac{|2\nu_T \mathbf{S}|}{\left|\frac{2}{3}k\mathbf{I}_3\right|} = 3\sqrt{C_\mu} \approx 1 \tag{4.209}$$

Thus, additional terms are expected, at least of the second order either in $\mathbf{S}$ or $\mathbf{\Omega}$. For instance, terms in such forms as $\mathbf{S}^2$ and $\mathbf{S\Omega}$ would probably appear.

All three above comments show that the Boussinesq model is not sufficient for a fine approach to turbulence. Before attempting to address this issue, another essential remark has to be made. The discussion at the end of Section 4.4.4 suggests that further investigations into the mathematical relations which $\mathbf{R}$ would necessarily have to verify are useful. Thus, we may state that its diagonal values $R_{ij} = \overline{u'_i u'_j}$ are positive. Besides, according to (4.47), and in so far as the Reynolds stresses happen to be statistical averages, we can see they should verify the Cauchy–Schwartz theorem; this is written here as $\overline{u'_i u'_j}^2 \leqslant \overline{u'^2_i}\,\overline{u'^2_j}$. Thus, $\mathbf{R}$ should theoretically satisfy both following properties:

$$\begin{aligned} &\forall i, j \quad R_{ij}^2 \leqslant R_{ii} R_{jj} \\ &\forall i, \quad R_{ii} \geqslant 0 \end{aligned} \tag{4.210}$$

(there is no summation over i and j, as well as in the preceding relations). These properties directly result from the mathematical definition of $\mathbf{R}$ as a statistical autocorrelation tensor. It is well known (refer to Appendix A, eqn (A.32)) that the tensor product of two vectors is a tensor. $\mathbf{R}$ is then transformed like the vector filed $\mathbf{u}'$ which determines it; its properties (4.210) are then independent of the coordinate system and then have a strong physical meaning.[96] As a consequence, they can probably be expressed as a function of k and the invariants of $\mathbf{a}$. The latter, which are defined as in (3.26) (Section 3.2.2) are two in number, because $\mathbf{a}$ has a zero trace, like any deviator. For the sake of convenience, we will note:

$$\begin{aligned} \eta_{\mathbf{a}} &\doteqdot \left(\frac{II_{\mathbf{a}}}{24}\right)^{1/2} = \left(\frac{\operatorname{tr}\mathbf{a}^2}{24}\right)^{1/2} \\ \xi_{\mathbf{a}} &\doteqdot \left(\frac{III_{\mathbf{a}}}{48}\right)^{1/3} = \left(\frac{\operatorname{tr}\mathbf{a}^3}{48}\right)^{1/3} \end{aligned} \tag{4.211}$$

Simple calculations (Pope, 2000) show that the inequalities (4.210) are satisfied if and only if these two invariants verify the following relations:

$$|\xi_{\mathbf{a}}| \leqslant \eta_{\mathbf{a}} \leqslant \sqrt{\frac{1}{27} + 2\xi_{\mathbf{a}}^3} \tag{4.212}$$

On the $(\eta_{\mathbf{a}}, \xi_{\mathbf{a}})$ plane, these relations mean that one must refer to the inside of an area of the so-called *Lumley triangle* (1978), which is shown[97] in Fig. 4.8. Since the tensor $\mathbf{R}$ is symmetric, it is diagonalisable[98] in an orthonormal basis, and its eigenvalues are the squares of the *main velocity fluctuations* u', v' and w', which are defined as the rms values of the fluctuations of the velocity components. Tensor $\mathbf{a}$ provides a measure of the relative differences between

[96] Once again, in every point in time and space. It should be pointed out that this result holds true provided that the coordinate system is orthonormal, which we had not mentioned in the definition of $\mathbf{R}$. We must also specify that the latter is invariant under a Galilean transformation, like the Cauchy tensor (refer to Section 3.4.2).

[97] That anisotropy map is analogous to the Von Mises or Tresca criteria in the theory of continuous media. We will express it here using Pope's notations (2000).

[98] Once again, its diagonalization properties (particularly its eigendirections) may depend on time and space.

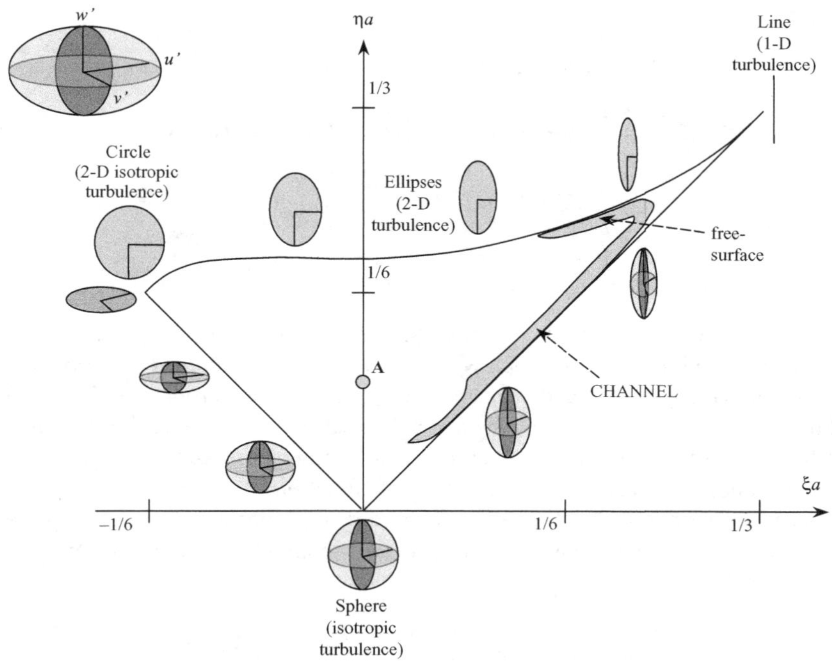

Fig. 4.8 The Lumley triangle. The dimensions of the ellipsoids correspond to the rms values of the velocity component fluctuations. Point A indicates the predictions of the Boussinesq model in the case of a permanent open channel, whereas the DNS (after Kim et al., 1987) are represented by the grey pattern.

these three quantities: when they are equal, **a** is zero and the turbulence is isotropic (ideal case). The various assumable cases are indicated in Fig. 4.8. The upper boundary of the flow domain corresponds to a turbulence with two-dimensional fluctuations, occurring at an interface between two fluids having quite different densities, as happens with a free surface.

A turbulence model which satisfies the Lumley conditions (4.210) is said to be *realizable*, since it predicts Reynolds stresses which are compatible with the mathematical definition of this tensor. The Boussinesq model, for instance, gives $\xi_{\mathbf{a}} = 0$ and $\eta_{\mathbf{a}} = C_\mu \tau S/\sqrt{12}$ with (4.155). As regards the channel being studied in Section 4.4.5, equation (4.205) yields $\eta_{\mathbf{a}} = C_\mu/\sqrt{12} = 7.5 \times 10^{-3} \leqslant 1/\sqrt{27} = 0.19$. Thus, the model is realizable in this very case, but there is no evidence that some flows would not give rise to higher values of $\eta_{\mathbf{a}}$. Besides, it was experimentally demonstrated that the anisotropy as predicted by the Boussinesq model is erroneous even in this simple case (refer to Fig. 4.8).

4.5.2 Second order closure

In order to put forward models which (at least partially) satisfy the requirements of the previous paragraph, let us momentarily return to those equations governing **R**, namely (4.66) and (4.67). We may recall that they occur in the form

$$\frac{D\mathbf{R}}{Dt} = \frac{\partial \mathbf{R}}{\partial t} + \mathbf{grad}\ \mathbf{R} \cdot \overline{\mathbf{u}} = \mathbf{P} - \mathbf{div}\ \mathbf{Q}^{\mathbf{R}} + \boldsymbol{\Phi} - \boldsymbol{\varepsilon} \tag{4.213}$$

with

$$\begin{aligned}
\mathbf{P} &\doteq -\boldsymbol{\Gamma}\mathbf{R} - \mathbf{R}\boldsymbol{\Gamma}^T \\
\mathbf{Q}^{\mathbf{R}} &\doteq \overline{\mathbf{u}' \otimes \mathbf{u}' \otimes \mathbf{u}'} + \frac{2}{\rho}\overline{p'\left(\mathbf{u}' \otimes \mathbf{I}_3\right)^S} - \nu \mathbf{grad}\,\mathbf{R} \\
\boldsymbol{\Phi} &\doteq \frac{2}{\rho}\overline{p'\mathbf{s}'} \\
\boldsymbol{\varepsilon} &\doteq 2\nu\overline{\boldsymbol{\gamma}' \cdot \boldsymbol{\gamma}'^T}
\end{aligned} \tag{4.214}$$

This system of equations is not closed, due to many unknown quantities. By analogy with the study which we have conducted in Section 4.4.4 about the equation with respect to k, the tensor $\mathbf{P}$ clearly stands for a production (as in the kinetic energy equation), whereas $\boldsymbol{\varepsilon}$ is a dissipation tensor and $\mathbf{Q}^{\mathbf{R}}$ a flux (it is a third-order tensor). Lastly, $\boldsymbol{\Phi}$ is referred to as the *pressure-strain correlation tensor*, and we will subsequently explain its function.

Let us now go into further details as regards each of these terms. Using the notations which we introduced in the previous section, the production tensor is also written as

$$\begin{aligned}
\mathbf{P} &= -\left(\mathbf{S} + \boldsymbol{\Omega}\right) k \left(\mathbf{a} + \frac{2}{3}\mathbf{I}_3\right) - k\left(\mathbf{a} + \frac{2}{3}\mathbf{I}_3\right)\left(\mathbf{S} - \boldsymbol{\Omega}\right) \\
&= -k\left(\mathbf{Sa} + \mathbf{aS} + \boldsymbol{\Omega}\mathbf{a} - \mathbf{a}\Omega + \frac{4}{3}\mathbf{S}\right)
\end{aligned} \tag{4.215}$$

It is perfectly determined by the mean quantities and the tensor $\mathbf{R}$ being the unknown in the equation being considered; it is accordingly explicit and does not require any closure model.

The dissipation tensor represents the energy dissipated by friction, which we have explained in Section 4.4.2 it is exerted by the smallest turbulent eddies. Now, pursuant to the Kolmogorov cascade, as they break down the eddies do not keep track of their anisotropy, if any, and so the small eddies can be considered as isotropic.[99] It is then all the same with $\boldsymbol{\varepsilon}$, and we will write to a very good approximation

[99] This is an effect of the term $\boldsymbol{\Phi}$ being considered below.

$$\boldsymbol{\varepsilon} = \frac{2}{3}\breve{\varepsilon}\mathbf{I}_3 \approx \frac{2}{3}\varepsilon\mathbf{I}_3 \tag{4.216}$$

and so the half-trace of this term provides the scalar dissipation ε.

The flux $\mathbf{Q}^{\mathbf{R}}$ of Reynolds stresses may be treated like the diffusion term of k (4.143). Only the first two terms (characterizing the turbulent diffusion) should be modelled, the third one standing for an explicit molecular diffusion. Thus, a diffusion coefficient containing a molecular part and a turbulent part, by analogy with (4.144), is contemplated. However, because of the anisotropic nature of the large eddies, accounting for the mixing processes, we will here introduce an eddy viscosity tensor being substituted for a scalar eddy viscosity, which looks more like the general form (4.208). Thus, Daly and Harlow (1970) suggest putting

$$\mathbf{Q}^{\mathbf{R}} = -\,(\mathbf{grad}\ \mathbf{R}) \cdot \boldsymbol{\nu}_R = -\nu_{kl} \frac{\partial R_{ij}}{\partial x_k} \tag{4.217}$$

grad R is a third-order tensor, whereas $\boldsymbol{\nu}_R$ is a second-order tensor as given by

$$\boldsymbol{\nu}_R \doteqdot \nu \mathbf{I}_3 + C_s \frac{k}{\varepsilon} \mathbf{R} \tag{4.218}$$

The second term in that model stands for the turbulent diffusion, which was written by analogy with (4.137) but in a tensor form, **R** being the single known tensor having the same dimension as k. Launder (1990) then suggests[100] $C_s = 0.22$. In the form of components, the models (4.217) and (4.218) are read as

[100] We may note, however, that with that model the half trace of the diffusion term does not give exactly the same diffusion term of k as that modelled in (4.145).

$$\mathbf{Q}^{\mathbf{R}} = -\left(\nu\delta_{kl} + C_s \frac{k}{\varepsilon} R_{kl}\right) \frac{\partial R_{ij}}{\partial x_l} \mathbf{e}_i \otimes \mathbf{e}_j \otimes \mathbf{e}_k \tag{4.219}$$

We still have to consider the tensor $\mathbf{\Phi}$, which is essential in the next steps of our discussion. As explained in Section 4.4.2, the latter does not leave any 'trace' in the energy equation, which means that it takes part in the redistribution of the components of **R** without modifying its trace, that is the energy. For a better understanding of the role of this tensor, we must come back to the channel case of Section 4.4.5, but in a possible unsteady flow regime and forgetting the diffusion terms (whose action only consists in redistributing each stress in space during the time). Adopting the model (4.216) for dissipation, the six equations (4.213) are significantly simplified, and only three of them remain different from zero:

$$\begin{aligned}
\frac{DR_{xx}}{Dt} &= -2R_{xx}\frac{\partial \bar{u}}{\partial z} + \Phi_{xx} - \frac{2}{3}\varepsilon \\
\frac{DR_{xz}}{Dt} &= \ \ -R_{zz}\frac{\partial \bar{u}}{\partial z} + \Phi_{xz} - \frac{2}{3}\varepsilon \\
\frac{DR_{zz}}{Dt} &= \qquad\qquad\quad \Phi_{zz} - \frac{2}{3}\varepsilon
\end{aligned} \tag{4.220}$$

We find out that the production does not influence $R_{zz} = \overline{u_z'^2}$; intuition however suggests (which is experimentally substantiated) that the fluctuations of vertical velocity do exist. Thus, the term Φ_{zz} actually accounts for the occurrence of such fluctuations. Since $\mathbf{\Phi}$ leaves no trace, this term is opposed to Φ_{xx}, and we come to the conclusion that the tensor $\mathbf{\Phi}$ has drawn energy from the fluctuations along x and subsequently restored it into the fluctuations along z. This tensor then induces a 're-isotropization' of the fluctuations, that is makes the tensor **R** as isotropic as possible, in other words reduces its deviator. That is why Rotta (1951) proposed a simple model of *return to isotropy* representing $\mathbf{\Phi}$ by a term which is based on a relaxation time on the order of magnitude of the turbulent time scale τ:

$$\mathbf{\Phi} = -C_R \frac{\varepsilon}{k} \mathbf{R}^D = -C_R \varepsilon \mathbf{a} \tag{4.221}$$

It can be found that the processes included in the pressure-strain correlation term $\mathbf{\Phi}$ are more complex than is suggested by this simple model, since the return to isotropy does not only involve the tensor $\mathbf{R}$. From rather complex tensorial considerations beyond the scope of this book,[101] Launder et al. (1975) have come up with the following more general model, which is referred to as LRR-QI:

$$\mathbf{\Phi} = -C_R \varepsilon \mathbf{a} - C_P \mathbf{P}^D - C_D \mathbf{D}^D - C_S k \mathbf{S} \tag{4.222}$$

where $\mathbf{D}$ is a new tensor, looking like $\mathbf{P}$ and defined by

$$\begin{aligned} \mathbf{D} &\doteqdot -\mathbf{R}\mathbf{\Gamma} - \mathbf{\Gamma}^T \mathbf{R} = -2\left(\mathbf{R}\mathbf{\Gamma}\right)^S \\ &= -k\left(\mathbf{S}\mathbf{a} + \mathbf{a}\mathbf{S} - \mathbf{\Omega}\mathbf{a} + \mathbf{a}\mathbf{\Omega} + \frac{4}{3}\mathbf{S}\right) \end{aligned} \tag{4.223}$$

which may be compared to (4.215). The deviators occurring in the model (4.222) are given, according to the definitions of k and P, by

$$\begin{aligned} \mathbf{R}^D &\doteqdot \mathbf{a} = \frac{1}{k}\left(\mathbf{R} - \frac{2}{3}k\mathbf{I}_3\right) \\ \mathbf{P}^D &\doteqdot \mathbf{P} - \frac{2}{3}P\mathbf{I}_3 \\ \mathbf{D}^D &\doteqdot \mathbf{D} - \frac{2}{3}P\mathbf{I}_3 \end{aligned} \tag{4.224}$$

(the first of these formulas is identical to (4.155), and we have resorted to (4.91)). The constants in the model (4.222) equal

$$\begin{aligned} C_R &= 1.5 \\ C_P &= \frac{C+8}{11} \\ C_D &= \frac{8C-2}{11} \\ C_S &= \frac{60C-4}{55} \end{aligned} \tag{4.225}$$

C_R is known as the Rotta constant. The coefficient C, which is calibrated either on experimental simulations or Direct Simulations (DNS), is discussed in many papers. The generally adopted values are $C = 0.4$ (Launder et al., 1975) and $C = 5/9$ (Lumley, 1978). The corresponding values of the coefficients (4.225) are given by Table 4.2. As a rule, there is no general agreement about the values of the coefficients appearing here; Launder et al. (1975) therefore propose a simplified set denoted as LRR-IP, whose values occur in the same table.[102]

[101] This model is based on a Poisson equation which is satisfied by the pressure fluctuations and is close to (3.187), whose solution is analogous to (3.191).

[102] As with the two-equation models which are set out in Section 4.4.4, the present coefficients are not valid in a viscous sublayer, where corrections are necessary to take the effects of Re_* into account.

Table 4.2 Coefficients of several second-order turbulent closure models (LRR-QI-1: Launder et al., 1975; LRR-QI-2: Lumley, 1978).

Model	C_R	C_P	C_D	C_S
LRR-IP	1.8	$\frac{3}{5}$	0	0
LRR-QI-1	1.5	0.764	0.109	0.364
LRR-QI-2	1.5	$\frac{7}{9}$	$\frac{2}{9}$	$\frac{8}{15}$

Anyway, the equations (4.213) and (4.214), being provided with the models (4.216), (4.219) and (4.222), get a new form

$$\begin{aligned}\frac{D\mathbf{R}}{Dt} &= \mathbf{P} + \mathbf{div}\left[(\mathbf{grad}\ \mathbf{R})\,\nu_R\right] - C_R\varepsilon\mathbf{a} - C_P\mathbf{P}^D \\ &\quad - C_D\mathbf{D}^D - C_S k\mathbf{S} - \frac{2}{3}\varepsilon\mathbf{I}_3\end{aligned} \tag{4.226}$$

With the formulas (4.215), (4.218) and (4.223) and the definition (4.155) of **a**, these six equations[103] are not yet fully closed, because the scalar dissipation rate ε plays a part (k is not a further unknown, since it is the half trace of **R**). However, supplementing them with the equation (4.148) governing the evolution of ε,[104] they make up a closed system of seven equations. The solution **R** is directly used for solving the Reynolds equations (4.51). The $R_{ij} - \varepsilon$ model is then referred to; it is a second-order closure model, in accordance with the nomenclature being mentioned at the end of Section 4.3.3.

[103] We may recall that, as a symmetric tensor, **R** only has six independent components.

[104] Some authors then prefer to model the diffusion of ε by means of a diffusion tensor analogous to (4.218) instead of the scalar turbulent viscosity model of the standard equation (4.148).

There are many variants to this kind of model[105] on top of the three 'standard' versions which are discussed here; all of them exhibit great qualities. In particular, they explain the consequences of the anisotropy of the large turbulent structures, notably thanks to the exactness of the production tensor. The production term in the equation of ε is also accurately represented[106] by the formula (4.90), which makes it possible to avoid the difficulties of the $k - \varepsilon$ models which were spoken of at the end of Section 4.4.4. In addition, these models generally satisfy the realizability conditions (Section 4.5.1). Thus, they manage to rather closely represent flows having features which cannot be replicated with the $k - \varepsilon$ model, and then are quite useful in industry (for instance, refer to Rodi, 2000).

[105] A good review of them is made by Pope (2000).

[106] With the restrictions as previously expressed in Section 4.4.4 as regards the validity of that equation.

The second-order models, however, are complex and imply appropriate boundary conditions which we will not discuss here. However, we are going to use the arguments which were previously put forward and build, in the next section, a class of models which will be markedly simpler and will contain a good deal of physical information.

4.5.3 Explicit algebraic models

In this section, we will come back to the first-order models, attempting to settle some of those issues raised by the Boussinesq model, without reaching the complexity level of the second-order models.

Taking advantage of the first three observations of Section 4.5.1, we could try to explicitly determine the Reynolds stresses **R** in a more general and accurate way as the simple Boussinesq model, through similar arguments as those in Section 3.4.1. The natural properties of the tensor **R** (or, rather, of **a**), however, allow us to operate more simply. Since **a** is a traceless symmetric tensor, it should be sought, indeed, in the vector space of the three-dimensional second-order tensors satisfying these properties. Gatski and Speziale (1993) show that such a space is ten-dimensional, but in the following we will restrict our arguments to the case of the statistically invariant flow along one direction (which we will denote here as $x_2 \equiv y$) and whose mean velocity field $(\overline{u}_x, \overline{u}_z) \equiv (\overline{u}, \overline{w})$ is located within the plane which is orthogonal to it $(x_1, x_3) \equiv (x, z)$. That may be called a 'two-dimensional' flow, since the mean quantities are two-dimensional too, although the turbulent fluctuations are definitely three-dimensional. Examples of flows within that scope are provided by the case of a unidirectional wave, as well as that of a sufficiently broad channel (as in Section 4.4.5). Under these assumptions, the tensors **S** and **Ω** are written as

$$\mathbf{S} = \begin{pmatrix} \dfrac{\partial \overline{u}}{\partial x} & 0 & \dfrac{1}{2}\left(\dfrac{\partial \overline{u}}{\partial z} + \dfrac{\partial \overline{w}}{\partial x}\right) \\ 0 & 0 & 0 \\ \dfrac{1}{2}\left(\dfrac{\partial \overline{u}}{\partial z} + \dfrac{\partial \overline{w}}{\partial x}\right) & 0 & \dfrac{\partial \overline{w}}{\partial z} \end{pmatrix}.$$

$$\boldsymbol{\Omega} = \begin{pmatrix} 0 & 0 & \dfrac{1}{2}\left(\dfrac{\partial \overline{u}}{\partial z} - \dfrac{\partial \overline{w}}{\partial x}\right) \\ 0 & 0 & 0 \\ \dfrac{1}{2}\left(\dfrac{\partial \overline{u}}{\partial z} - \dfrac{\partial \overline{w}}{\partial x}\right) & 0 & 0 \end{pmatrix} \tag{4.227}$$

or else

$$\mathbf{S} = \begin{pmatrix} S_{xx} & 0 & S_{xz} \\ 0 & 0 & 0 \\ S_{xz} & 0 & -S_{xx} \end{pmatrix}$$

$$\boldsymbol{\Omega} = \Omega_{xz} \begin{pmatrix} 0 & 0 & 1 \\ 0 & 0 & 0 \\ -1 & 0 & 0 \end{pmatrix} \tag{4.228}$$

We may also define, by analogy with (3.31):

$$\begin{aligned} \Omega &\doteqdot \sqrt{2\boldsymbol{\Omega} : \boldsymbol{\Omega}} \\ &= \sqrt{2\Omega_{ij}\Omega_{ij}} \\ &= 2\left|\Omega_{xz}\right| \end{aligned} \tag{4.229}$$

which represents the mean scalar vorticity. By the way, we may note that in such a simple shear flow as that of the infinite channel, that is when the velocity is aligned along x and only depends on z, we have $\Omega = S = |\partial \overline{u}/\partial z|$.

The anisotropy tensor also gets a simpler form. If the direction of axis y is inverted, then it can be seen that the fluctuations u'_y change their signs, which also changes the signs of the quantities $a_{xy} = \overline{u'_x u'_y}$ and $a_{yz} = \overline{u'_y u'_z}$. Now, the physics of turbulence is quite obviously unchanged by that arbitrary reversal of the axes,[107] and so we have $a_{xy} = a_{yz} = 0$. The only terms which play a role are then a_{xx}, a_{xz} and a_{zz}, as well as $a_{yy} = -a_{xx} - a_{zz}$. The tensor **a** is then reduced to three degrees of freedom and should be sought in a three-dimensional space; we will try to break it down on a basis of that space. We already know a traceless symmetric tensor, namely **S**. Besides, we have explained in Section 4.5.1 that a general analytical formula for **a** would probably *a minima* involve quadratic terms in such forms as $\mathbf{S\Omega}$ and $\mathbf{S}^2$. A brief calculation will yield, under our assumptions:

[107] It is a symmetry-related invariance, which was not dealt with in Chapter 1. It does not provide, however, any further information as regards Hamiltonian mechanics.

$$\mathbf{S\Omega} = \frac{\Omega}{2}\begin{pmatrix} -\frac{1}{2}\left(\frac{\partial \overline{u}}{\partial z} + \frac{\partial \overline{w}}{\partial x}\right) & 0 & \frac{\partial \overline{u}}{\partial x} \\ 0 & 0 & 0 \\ -\frac{\partial \overline{w}}{\partial z} & 0 & \frac{1}{2}\left(\frac{\partial \overline{u}}{\partial z} + \frac{\partial \overline{w}}{\partial x}\right) \end{pmatrix} = -\mathbf{\Omega} S \tag{4.230}$$

Thus, this is a traceless tensor and the continuity equation shows it is symmetric.[108] As regards $\mathbf{S}^2$, as the square of a symmetric tensor, it has the same property. In order to turn it into a traceless tensor, considering its deviator is sufficient. Thus, **a** will be sought in the form

[108] In the general case of a flow varying along all three directions, that result would be wrong; however, the tensor $\mathbf{S\Omega} - \mathbf{\Omega S}$, which complies with the prerequisites, could be considered. In the relevant case, the latter simply equals $2\mathbf{S\Omega}$.

$$\mathbf{a} = C'_1\mathbf{S}^* + C'_2\mathbf{S}^*\Omega^* + C'_3\left(\mathbf{S}^{*2}\right)^D \tag{4.231}$$

where the C'_is are so far undetermined coefficients. Since **a** is dimensionless, we have non-dimensionalized the tensors **S** and $\mathbf{\Omega}$ as

$$\begin{aligned} \mathbf{S}^* &\doteq \tau\mathbf{S} \\ \mathbf{\Omega}^* &\doteq \tau\mathbf{\Omega} \end{aligned} \tag{4.232}$$

The orthogonality of the basis provided by the tensors $\mathbf{S}^*$, $\mathbf{S}^*\Omega^*$ and $\left(\mathbf{S}^{*2}\right)^D$ is a major result. It is one of the consequences of the symmetric and antisymmetry properties of $\mathbf{S}^*$ and $\mathbf{\Omega}^*$. By way of example, let us make the following calculation, which is based upon the trace associativity:

$$\begin{aligned} \mathbf{S}^* : \left(\mathbf{S}^*\mathbf{\Omega}^*\right) &= \mathrm{tr}\left(\mathbf{S}^{*2}\mathbf{\Omega}^*\right) \\ &= \mathrm{tr}\left[\left(\mathbf{S}^{*2}\mathbf{\Omega}^*\right)^T\right] \\ &= \mathrm{tr}\left[\mathbf{\Omega}^{*T}\left(\mathbf{S}^{*2}\right)^T\right] \\ &= \mathrm{tr}\left(-\mathbf{\Omega}^*\mathbf{S}^{*2}\right) \\ &= -\mathrm{tr}\left(\mathbf{S}^{*2}\mathbf{\Omega}^*\right) \\ &= 0 \end{aligned} \tag{4.233}$$

The powers of $\mathbf{S}^*$ of the third or higher order powers are naturally not taken into account in (4.231). This is because the problem we face may be regarded as two-dimensional, with respect to the mean velocities, as indicated by the formulas (4.228). That being the case, the Cayley–Hamilton theorem states that

$$\mathbf{S}^{*2} - \left(\mathrm{tr}\,\mathbf{S}^*\right)\mathbf{S}^* + \left(\det \mathbf{S}^*\right)\mathbf{I}_2 = \mathbf{0} \tag{4.234}$$

(refer to the three-dimensional case (3.27) in Section 3.2.2). Multiplying that relation by $\mathbf{S}^*$, we may state that $\mathbf{S}^{*3}$ (and by mathematical induction all the higher-order powers) depend on $\mathbf{S}^*$ and $\mathbf{S}^{*2}$. Though similar considerations, we can show that no other combination of $\mathbf{S}^*$ and $\boldsymbol{\Omega}^*$ is required.

The realizability of the model (4.231) is discussed by Pope (2000). Violeau (2009a) also demonstrates that the first realizability condition in (4.212) is always satisfied. Due to the orthogonality of the working base, indeed, we can write, after making some simple calculations:

$$\begin{aligned}
\mathrm{tr}\,\mathbf{a}^2 &= \left(C_1'^2 + \frac{C_2'^2}{4}\Omega^{*2} + C_3'^2\frac{S^{*2}}{12}\right)\frac{S^{*2}}{2} \\
\mathrm{tr}\,\mathbf{a}^3 &= \left(C_1'^2 + \frac{C_2'^2}{4}\Omega^{*2} - C_3'^2\frac{S^{*2}}{36}\right)C_3'\frac{S^{*4}}{8}
\end{aligned} \tag{4.235}$$

with the following definitions:

$$\begin{aligned}
S^* &\doteqdot \tau S = \sqrt{2\mathbf{S}^* : \mathbf{S}^*} \\
\Omega^* &\doteqdot \tau\Omega = \sqrt{2\boldsymbol{\Omega}^* : \boldsymbol{\Omega}^*}
\end{aligned} \tag{4.236}$$

Lengthy algebraic calculations then yield the following result:

$$\begin{aligned}
\eta_{\mathbf{a}}^6 - \xi_{\mathbf{a}}^6 &= \left(\mathrm{tr}\,\mathbf{a}^2\right)^3 - 6\left(\mathrm{tr}\,\mathbf{a}^3\right)^2 \\
&= \left(C_1'^2 + \frac{C_2'^2}{4}\Omega^{*2}\right)\left(C_1'^2 + \frac{C_2'^2}{4}\Omega^{*2} + C_3'^2\frac{S^{*2}}{4}\right)^2\frac{S^{*6}}{8} \geqslant 0
\end{aligned} \tag{4.237}$$

which then secures $|\xi_{\mathbf{a}}| \leqslant \eta_{\mathbf{a}}$ for all sets of coefficients C_i'.

In order to determine the latter, we will resort to a method which was proposed by Pope (1975) as a variant of Rodi's idea (1976). It consists of seeking approximate solutions of equation (4.226) making up the second-order models, making a simplifying hypothesis for a reduction to an equation in which the partial derivatives of $\mathbf{R}$ disappear. To that purpose, we are going to assume that the time evolution of Reynolds stresses is basically caused by the evolution of energy rather than that of anisotropy, which amounts to stating that $D\mathbf{a}/Dt$ is nearly zero. Thus, with (4.155) we get[109]

[109] In this paragraph, the identity matrix $\mathbf{I}_3$ is definitely a three-dimensional one, even though this problem is two-dimensional on an average, because the turbulence remains three-dimensional, exhibiting transverse fluctuations to the main directions of the velocity field.

$$\begin{aligned}\frac{D\mathbf{R}}{Dt} &= \frac{D}{Dt}\left[k\left(\mathbf{a}+\frac{2}{3}\mathbf{I}_3\right)\right]\\ &= \frac{Dk}{Dt}\frac{\mathbf{R}}{k}+k\frac{D\mathbf{a}}{Dt}\\ &\approx \frac{Dk}{Dt}\frac{\mathbf{R}}{k}\end{aligned} \tag{4.238}$$

which means that the relative rate of change of each component in $\mathbf{R}$ is identical to the others, which is consistent with the initially assumed constant anisotropy (this is the so-called *weak equilibrium*). This kind of equilibrium is ultimately similar to the hypothesis $P = \varepsilon$ which we have already come upon when preparing much simpler models; the presently achieved equilibrium, however, is more sophisticated, especially because it is pertinent to tensorial quantities. We will make the same hypothesis for the diffusion term, which is written as:

$$\mathbf{div}\ \mathbf{Q}^{\mathbf{R}} \approx \left(\mathrm{div}\mathbf{Q}^k\right)\frac{\mathbf{R}}{k} \tag{4.239}$$

Combining the latter two relations, the tensor transport equation (4.226) becomes

$$\begin{aligned}&\mathbf{P} - C_R\varepsilon\mathbf{a} - C_P\mathbf{P}^D - C_D\mathbf{D}^D - C_S k\mathbf{S} - \frac{2}{3}\varepsilon\mathbf{I}_3\\ &= \left(\frac{Dk}{Dt} - \mathrm{div}\mathbf{Q}^k\right)\frac{\mathbf{R}}{k}\\ &= (P-\varepsilon)\frac{\mathbf{R}}{k}\end{aligned} \tag{4.240}$$

the latter equality resulting from the energy equation (4.96). Turning back to the definitions (4.155), (4.215) and (4.223) of $\mathbf{a}$, $\mathbf{P}$ and $\mathbf{D}$, and after some algebraic calculations (especially utilizing (4.224)), that equation is rearranged in the form of a linear implicit relation[110] in $\mathbf{a}$ with coefficients which are functions of the non-dimensional velocity gradients:

$$\begin{aligned}&C_0\mathbf{a} + C_1\left(\mathbf{S}^*\mathbf{a} + \mathbf{a}\mathbf{S}^*\right) + C_2\left(\boldsymbol{\Omega}^* a - \mathbf{a}\Omega^*\right)\\ &+C_3\mathbf{S}^* + C_4\mathrm{tr}\left(\mathbf{a}\mathbf{S}^*\right)\mathbf{I}_3 = \mathbf{0}\end{aligned} \tag{4.241}$$

where the following has be defined

$$\begin{aligned}P^* &\doteqdot \frac{P}{\varepsilon}\\ C_0 &\doteqdot P^* + C_R - 1\\ C_1 &\doteqdot 1 - C_P - C_D = \frac{5-9C}{11}\\ C_2 &\doteqdot 1 - C_P + C_D = \frac{7C+1}{11}\\ C_3 &\doteqdot \frac{4}{3}C_1 + C_S = \frac{8}{15}\\ C_4 &\doteqdot -\frac{2}{3}C_1 = \frac{18C+10}{33}\end{aligned} \tag{4.242}$$

[110] Actually nearly-linear, because the coefficients C_i depend via (4.242) on $P^* = -\mathrm{tr}(\mathbf{a}\mathbf{S}^*)$, as evidenced by (4.243). Here the equation is linearized by temporarily assuming that P^* is constant.

In order to get equation (4.241), we have used the relation

$$k \operatorname{tr}(\mathbf{Sa}) = k\mathbf{S} : \mathbf{a} = -P \tag{4.243}$$

(in which the third line of (4.156) occurs). We also have utilized the following calculation:

$$\begin{aligned}
\operatorname{tr}(\boldsymbol{\Omega}\mathbf{a}) &= \operatorname{tr}\left[(\boldsymbol{\Omega}\mathbf{a})^T\right] \\
&= \operatorname{tr}\left(\mathbf{a}^T\boldsymbol{\Omega}^T\right) \\
&= \operatorname{tr}\left[\mathbf{a}(-\boldsymbol{\Omega})\right] \\
&= -\operatorname{tr}(\boldsymbol{\Omega}\mathbf{a}) \\
&= 0
\end{aligned} \tag{4.244}$$

Substituting the form being sought for **a**, that is (4.231), into the behaviour equation (4.241), we get a system of three equations whose three unknowns are C_1, C_2 and C_3. Solving it is a lengthy task, but owing to simple properties[111] of tensors **S** and **Ω**, many terms are simplified and the solution is ultimately reduced to

[111] Properties especially, resulting from the two-dimensional frame in which we are placed.

$$\begin{aligned}
C_1' &= -2C_\mu \\
C_2' &= -4B_2C_\mu \\
C_3' &= -2B_1C_\mu
\end{aligned} \tag{4.245}$$

with

$$\begin{aligned}
B_1 &\doteqdot \frac{C_1}{C_0} \\
B_2 &\doteqdot \frac{C_2}{C_0} \\
B_3 &\doteqdot \frac{C_3}{C_0} \\
C_\mu &\doteqdot \frac{B_3/2}{\frac{1}{3}B_1^2S^{*2} - B_2^2\Omega^{*2} - 1}
\end{aligned} \tag{4.246}$$

Given these coefficients, and turning back to the dimensional quantities, we may write the Reynolds tensor as

$$\mathbf{R} = \frac{2}{3}k\mathbf{I}_3 - 2C_\mu\frac{k^2}{\varepsilon}\left[\mathbf{S} + 2B_2\frac{k}{\varepsilon}\mathbf{S}\boldsymbol{\Omega} + B_1\frac{k}{\varepsilon}\left(\mathbf{S}^2\right)^D\right] \tag{4.247}$$

The Boussinesq model (4.73) provided with the model (4.137) for estimating the tubulent viscosity can be recognized in the first two terms, except that the coefficient C_μ, which is presently given by (4.246), is no longer a constant but is a function of S^*, Ω^* and P^* (through the agency of C_0, which is given by (4.242)). Thus, we have determined an extension of the Boussinesq model containing the expected non-linear, vorticity-related terms; this is what is known as an *explicit algebraic Reynolds stress model*.

In the particular case where $C = 5/9$ (refer to Table 4.2, Section 4.5.2), the coefficient C_1 vanishes and we get the Wallin and Johansson model (2000):

$$\mathbf{R} = \frac{2}{3}k\mathbf{I}_3 - 2C_\mu \frac{k^2}{\varepsilon}\left(\mathbf{S} + 2B_2\frac{k}{\varepsilon}\mathbf{S\Omega}\right) \tag{4.248}$$

with, taking (4.242) into account:

$$\begin{aligned} C_0 &= P^* + B_0 \\ B_0 &= C_R - 1 \\ B_2 &= \frac{4}{9C_0} \\ B_3 &= \frac{8}{15C_0} \\ C_\mu &= -\frac{B_3/2}{B_2^2\Omega^{*2} + 1} \end{aligned} \tag{4.249}$$

By means of the asymmetry relation (4.230), the non-linear model (4.247) may also be formulated as

$$\mathbf{a} = -2C_\mu\left(\mathbf{I}_3 + 2B_2\tau\mathbf{\Omega}\right)\tau\mathbf{S} \tag{4.250}$$

which corresponds to the general model (4.208) for the eddy viscosity tensor, with from now on

$$\nu_{T,ijkl} = C_\mu \frac{k^2}{\varepsilon}\left(\delta_{il}\delta_{jk} + 2B_2\tau\Omega_{ik}\delta_{jl}\right) \tag{4.251}$$

Once again, it could be reduced to a second-order viscous tensor in which (4.208) would be substituted for by

$$a_{ij} = -\frac{2}{k}\nu_{T,ik}S_{kj} \tag{4.252}$$

where

$$\nu_{T,ik} = C_\mu \frac{k^2}{\varepsilon}\left(\delta_{ik} + 2B_2\tau\Omega_{ik}\right) \tag{4.253}$$

has been defined. The first term duly corresponds to the Boussinesq linear isotropic model.

We may recall that the production term implicitly occurs in the coefficients of this model, via (4.242), and then is still to be calculated. To that purpose, let us come back to the definition (4.90) of P and write, based on (4.248):

$$\begin{aligned} P &= -\mathbf{S} : \mathbf{R} \\ &= -\mathbf{S} : \left[\frac{2}{3}k\mathbf{I}_3 - 2C_\mu\frac{k^2}{\varepsilon}\left(\mathbf{S} + 2B_2\frac{k}{\varepsilon}\mathbf{S\Omega}\right)\right] \\ &= 2C_\mu\left(S^*, \Omega^*, P^*\right)\frac{k^2}{\varepsilon}S^2 \end{aligned} \tag{4.254}$$

(we have taken advantage of the tensor basis orthogonality). The resulting relation is analogous to (4.97) as combined with (4.137), but with a coefficient C_μ which depends on the very production, as we have just explained it. Thus, the term $\mathbf{S\Omega}$ does not dissipate the mean energy nor, which comes to the same thing, produce any turbulent kinetic energy. Combining (4.249) with (4.254), we get a third-order polynomial equation in P^* with coefficients which are functions of S^* and Ω^*:

$$P^{*3} + 2B_0 P^{*2} + \left(\frac{B_3}{2} S^{*2} + B_2^2 \Omega^{*2} + B_0^2\right) P^* + \frac{B_0 B_3}{2} S^{*2} = 0 \quad (4.255)$$

We directly give its single positive solution (Girimaji, 1995):

$$P^* = \begin{cases} -\frac{2}{3} B_0 + \frac{4}{9}\left(P_1 + \sqrt{P_2}\right)^{1/3} \\ \quad + \frac{4}{9} \operatorname{sign}\left(P_1 - \sqrt{P_2}\right) \left|P_1 - \sqrt{P_2}\right|^{1/3} & \text{if } P_2 \geqslant 0 \\ -\frac{2}{3} B_0 + \frac{4}{9}\left(P_1^2 - P_2\right)^{1/6} \\ \times \cos\left(\frac{1}{3} \cos^{-1} \frac{P_1}{\sqrt{P_1^2 - P_2}}\right) & \text{if } P_2 < 0 \end{cases} \quad (4.256)$$

with

$$\begin{aligned} P_1 &\doteqdot \frac{9B_0}{4}\left(\frac{3B_0^2}{16} + \frac{9}{40} S^{*2} - \frac{1}{3} \Omega^{*2}\right) \\ P_2 &\doteqdot P_1^2 - \left(\frac{9B_0^2}{16} + \frac{9}{20} S^{*2} + \frac{1}{3} \Omega^{*2}\right)^3 \end{aligned} \quad (4.257)$$

The production then presently depends not merely on the scalar strain rate, but also on the mean scalar vorticity. Figure 4.9 (*a*) illustrates the variation of the dimensionless production $P^* = P/\varepsilon$ with respect to S^* and Ω^*. It can be seen that in the case of a simple shear flow (as it is with the infinite channel), the graphs corresponding to $S^* = \Omega^*$ behave quadratically (as with the Boussinesq model (4.97)) for small strain values, then looks more like the semi-linear model (4.158) upon large strains. Thus, this model addresses one of the issues mentioned in Section 4.5.1; on the contrary, it always results in positive production values, and then does not cover the theoretical cases which can be addressed using the three-equation model mentioned at the end of the same section. Besides, we can see that the production of energy is reduced by the vorticity of the mean flow, which is accounted for by the occurrence of inertial forces counteracting the development of eddies and stabilizing the flow.[112]

[112] This is a well-known phenomenon, analogous to the gravitational damping which is a specific feature of the density-stratified flows.

In the infinite channel case (with $S^* = 3.33$ according to (4.205)), the Wallin and Johansson model yields $P^* = 0.982$, which is near equilibrium $P = \varepsilon$ (*i.e.* $P^* = 1$) which we have used many a time (particularly in Section 4.4.5 for analytically treating the flow in the said channel). The C_μ function is illustrated in Fig. 4.9 (*b*) with respect to S^* and Ω^*. In the infinite channel case, its value

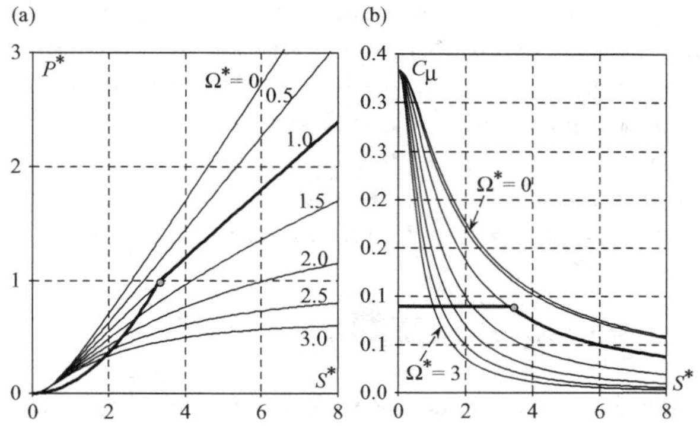

Fig. 4.9 Wallin and Johansson model: non-dimensionalized production of turbulent kinetic energy (*a*) and C_μ function (*b*) with respect to the mean strain rate and vorticity. Each curve corresponds to a value of vorticity; the values on the graph (*b*) are the same as on (*a*). The grey point on each graph stands for the case of a permanent infinite channel, and the thick curve illustrates the Guimet and Laurence model (2002).

is found to be 0.0897 (which is close to the 0.09 value being used within the frame of the standard $k - \varepsilon$ model).

The model we have just disclosed is simple and smart because it consists of specifying the eddy viscosity in the form of a tensor as given by (4.251). More generally, this model explicitly gives the Reynolds tensor **R** via (4.248), in order to solve the Reynolds equations (4.51). Since all the quantities are based upon k and ε (notably S^* and Ω^*), it necessitates using the equations of the standard $k - \varepsilon$ model,[113] except that the production is presently (better) estimated by (4.256) and (4.257). For that reason, a *non-linear* $k - \varepsilon$ *model* is often referred to. The production calculation step is the most complex one in it—that is the most time consuming in software terms; it should be pointed out, however, that the main advantage of this model lies in properly predicted production. In the case of a flow upstream of a body, for instance, we have previously mentioned the high strain. Thus, the standard $k - \varepsilon$ model will overrate the kinetic energy, providing too high eddy viscosity values near the stagnation point. That shortcoming significantly affects the flow shape in the wake immediately downstream of the solid, as already explained at the end of Section 4.4.4 (Guimet and Laurence, 2002). These problems can be solved almost as properly as the second-order models, at a lesser cost, by the explicit algebraic models. Furthermore, it is noteworthy that the absence of differential equations governing the Reynolds stresses makes it possible to get free of the tricky issue of the boundary conditions.

[113] In this case, energy k cannot be calculated, as in the case of the second-order models being described in Section 4.5.2, by taking the half trace of the Reynolds tensor (4.248), because that operation would, by construction, result in a tautology.

The boundary conditions relative to k and ε in the non-linear $k - \varepsilon$ model are left unchanged with respect to those in the standard model; these conditions were explained in Section 4.4.4. This is because the ideal case of the infinite channel is not affected by the presently introduced non-linear terms, as we have just explained.

It is noteworthy that some authors have derived explicit algebraic models for the Reynolds stresses from the Boltzmann equations, applying them in such a treatment as we did for the continua in Chapter 3, Section 3.5.2. The calculations, however, are excessively complex because of the required estimate of the higher than first-order terms upon the estimate of the collision integral in the Boltzmann equation. In addition, it should be pointed out that these models usually result in forms which are analogous to the presently disclosed models, but provided with constant coefficients which generally do not properly reflect all the physical processes pertaining to an efficient turbulence model. The work by Chen et al. (2004), for instance, does not take the effect of the pressure gradient into account, and so the pressure-strains correlations tensor $\mathbf{\Phi}$ plays no part in it, whereas its physical role is crucial, as previously explained.

Similar considerations may be formulated in order to set up algebraic closures for a turbulent scalar flux $\mathbf{Q}^B$, as defined by (4.54), improving the simplistic model (4.85) (for example, refer to Violeau, 2009a).

4.6 Non-averaged models

To conclude this overview of the main models of turbulence, we give a few features regarding non-averaged models, starting with large eddy simulation (LES). The α model, less popular but based on attractive theoretical grounds, is then briefly presented. Lastly, we consider stochastic models, on the basis of the Langevin-type equations introduced at the end of Chapter 2.

4.6.1 Large eddy simulation

The Reynolds average approach which this chapter almost exclusively deals with is attractive because it represents the motion with steady, smooth fields. However, it is disadvantageous in that it requires resorting to more or less complex closure models, with the obvious consequence that adequate models are obtained at the cost of more complexity. Furthermore, the mean fields are often not sufficient for engineering purposes, where some information about fluctuating quantities may be useful in order to address a number of practical problems. Thus, the solid structures in contact with fluids (vehicles, industrial pipes, etc.) are subject to vibration which is directly related to the behaviour of the largest adjoining turbulent eddies. The Kolmogorov analysis which was disclosed in Section 4.4.3 makes it possible, indeed, to assess the characteristic scales via quantities (kinetic energy and dissipation rate) which can be fairly accurately determined through most of the previously discussed models. The anisotropic nature of the large structures, however, do not come under the first-order models, and the second-order models only take their effects into account as regards their influence over the mean quantities.

Besides, we mentioned in Section 4.3.1 the practical impossibility of accurately modelling all the turbulent structures by means of a thorough numerical simulation of the Navier–Stokes equations because of the very fine spatio-temporal resolution which such an approach (DNS) would require (also refer to Section 4.4.6). In spite of the increasing calculation capacity of computers, an intermediate method between the DNS and the mean Reynolds analysis was developed in the 1980s (although its foundations date back to 1960); this method offers advantages since it is more precise than the latter without featuring the same impracticable computational time and storage capacity. The aim is to simulate the larger turbulent structures by means of the resolution made possible by the selected discretization, while using closure equations for modelling the effects of the smaller structures, which are beyond the reach of the spatial resolution. This is known as LES (*large eddy simulation*), and we will review its broad outlines, focussing on the weakly compressible flows, as we have done since the beginning of this chapter.

In order to only keep, out of the velocity and pressure signals, those variations corresponding to the largest structures, each field A is associated with a filtered field $\tilde{A}$:

$$\tilde{A}(\mathbf{r},t) \doteqdot \int_{\Omega} A\left(\mathbf{r}',t\right) G_{\Delta}\left(\mathbf{r},\mathbf{r}'\right) d^3\mathbf{r}' \tag{4.258}$$

where Ω is the fluid domain being considered. The *filter* function G_{Δ} may take the shape either of a Gaussian with a characteristic 'size' Δ or of a rectangular function with a width Δ. As a rule, Δ is a filter scale which represents the typical size of the smallest modelled structures, and then is on the order of magnitude of the spatial resolution in the calculation being contemplated.[114] It generally has, for practical applications, a compact support with a size Δ, which we will assume in the following. The filter function should also satisfy the following normalization condition:

$$\forall \mathbf{r}, \quad \int_{\Omega} G_{\Delta}\left(\mathbf{r},\mathbf{r}'\right) d^3\mathbf{r}' = 1 \tag{4.259}$$

From now on, we will consider the more general case of a filter which only depends on distance:

$$G_{\Delta}\left(\mathbf{r},\mathbf{r}'\right) = G_{\Delta}\left(\left|\mathbf{r}-\mathbf{r}'\right|\right) \tag{4.260}$$

Temporarily reasoning in one dimension, the following exemplary filter can be used:

$$G_{\Delta}(r) = \frac{1}{\pi r}\sin\frac{K_{\Delta} r}{2} \tag{4.261}$$

with

$$K_{\Delta} \doteqdot \frac{2\pi}{\Delta} \tag{4.262}$$

More generally, the filter has the dimension of the inverse of a volume.

With the property (4.260), the filter operator (4.258) is in the form of a convolution. Hence, the Fourier transforms of both original and filtered fields are related by

$$\widehat{\tilde{A}}(\mathbf{K},t) = \widehat{A}(\mathbf{K},t)\,\widehat{G}_{\Delta}(\mathbf{K}) \tag{4.263}$$

where $\widehat{G}_{\Delta}$ is the filter transform (Appendix B, eqn (B.10)). With the option (4.261), it is a *gate function* which only depends on the norm K of the wave vector $\mathbf{K}$ (refer to eqn (B.66)):

$$\begin{aligned} \widehat{G}_{\Delta}(K) &= 1 \quad \text{if} \quad |K| \leqslant \frac{K_{\Delta}}{2} \\ &= 0 \quad \text{else} \end{aligned} \tag{4.264}$$

Thus, the relation (4.263) shows that the filter (4.261) deletes all the oscillatory modes whose wave number exceeds the threshold K_{Δ}. As a rule, that reasoning shows that G_{Δ} behaves like a low-pass filter, only keeping the small

[114] For practical applications, it should be pointed out that the characteristic time of the smallest simulated structures is also made longer with respect to the time scale of the smallest eddies physically existing within the flow; thus, the selected time step can be larger than in DNS.

wave numbers, that is the largest turbulent structures. Figures 7.12 and 8.16 (Chapters 7 and 8) illustrate examples of a filtered velocity field.

The residual field is defined as the difference between the real field and the filtered field:

$$A'' \doteqdot A - \tilde{A} \tag{4.265}$$

That relation may be likened to (4.30), except for an essential difference: unlike the first property of (4.33), the filtered value of a residual field is non-zero:

$$\widetilde{A''} \neq 0 \tag{4.266}$$

Instead, the linearity properties (4.32) are satisfied by the filter operator, as can readily be seen from its definition (4.258). In particular, if the property of commutation with the partial time derivative operator is obvious, the commutation with the spatial derivatives needs to be demonstrated. Let us start with observing the following relation:

$$\frac{\partial G_\Delta\left(|\mathbf{r}-\mathbf{r}'|\right)}{\partial x_i} = -\frac{\partial G_\Delta\left(|\mathbf{r}-\mathbf{r}'|\right)}{\partial x_i'} \tag{4.267}$$

where x_i and x_i' denote the components of $\mathbf{r}$ and $\mathbf{r}'$vectors, respectively. Using that result, we refer, as in Section 4.4.3, to the case in which the fluid domain is extended to the whole space, and write (dropping the time dependency for the sake of simplicity):

$$\begin{aligned}\frac{\partial \tilde{A}(\mathbf{r})}{\partial x_i} &= -\int_{\mathbb{R}^3} A(\mathbf{r}') \frac{\partial G_\Delta\left(|\mathbf{r}-\mathbf{r}'|\right)}{\partial x_i'} d^3\mathbf{r}' \\ &= -\int_{\mathbb{R}^3} \left[\frac{\partial A(\mathbf{r}') G_\Delta\left(|\mathbf{r}-\mathbf{r}'|\right)}{\partial x_i'} - \frac{\partial A(\mathbf{r}')}{\partial x_i'} G_\Delta\left(|\mathbf{r}-\mathbf{r}'|\right)\right] d^3\mathbf{r}' \\ &= -\oint A(\mathbf{r}') G_\Delta\left(|\mathbf{r}-\mathbf{r}'|\right) \mathbf{n}' d\Gamma' \\ &\quad + \int_{\mathbb{R}^3} \frac{\partial A(\mathbf{r}')}{\partial x_i'} G_\Delta\left(|\mathbf{r}-\mathbf{r}'|\right) d^3\mathbf{r}' \\ &= \widetilde{\frac{\partial A}{\partial x_i}}(\mathbf{r})\end{aligned} \tag{4.268}$$

(the boundary integral is zero because the filter has a compact support).[115]

That commutation property, of course, covers all the spatial differentiation operators. Thus, for an incompressible fluid, the filtered velocity field obeys the standard continuity equation:[116]

$$\mathrm{div}\tilde{\mathbf{u}} = 0 \tag{4.269}$$

The weakly compressible continuity equation, through a treatment analogous to the calculation (4.35)–(4.38), yields the following filtered form:

$$\frac{\partial \rho}{\partial t} + \mathrm{div}\left(\rho\tilde{\mathbf{u}}\right) = 0 \tag{4.270}$$

[115] One may have noticed that the hypothesis of a filter only depending on distance is crucial in this demonstration, via the the equality (4.267).

[116] We hope that the notation $\tilde{\mathbf{u}}$ for filtered velocity will not allow confusion with the molecule velocity of Chapters 2 and 3.

Through a suitable treatment, we get an equation similar to (4.39):

$$\frac{\tilde{D}\rho}{Dt} = -\rho \mathrm{div}\tilde{\mathbf{u}} \tag{4.271}$$

the symbol $\tilde{D}$ meaning that the derivation takes places along the filtered velocity field:

$$\frac{\tilde{D}A}{Dt} \doteqdot \frac{\partial A}{\partial t} + \mathbf{grad}A \cdot \tilde{\mathbf{u}} \tag{4.272}$$

As in the Reynolds average treatment being carried out in Section 4.3.3, the advection term, because of its non-linearity, will only be modified by the application of the filter, and generate an additional term on the right-hand side of the equation of motion (by analogy with the calculations (4.43) through (4.45)). This term is the divergence of a new stress tensor which is known as the *residual tensor*:

$$\boldsymbol{\tau}_R \doteqdot \widetilde{\mathbf{u} \otimes \mathbf{u}} - \tilde{\mathbf{u}} \otimes \tilde{\mathbf{u}} \tag{4.273}$$

Because of the counter-property (4.266), this tensor, unlike $\mathbf{R}$, cannot be reduced to such a simple expression as (4.46). The equation of motion of the filtered fields is then written as

$$\frac{\tilde{D}\tilde{\mathbf{u}}}{Dt} = \frac{\partial \tilde{\mathbf{u}}}{\partial t} + \mathbf{grad}\ \tilde{\mathbf{u}} \cdot \tilde{\mathbf{u}} = -\frac{1}{\rho}\mathbf{grad}\ \tilde{p}^* + \nu \nabla^2 \tilde{\mathbf{u}} - \mathbf{div}\ \boldsymbol{\tau}_R \tag{4.274}$$

The tensor $\boldsymbol{\tau}_R$ represents the impact of the residual structures on the filtered field, and the aim is to model it. It then appears that we are back to the issues raised by the Reynolds-averaging approach; one circumstance, however, will help us. We have explained in Section 4.2.3 that when the eddies decay in the Kolmogorov cascade process, they gradually recover isotropy, and so if Δ is small enough,[117] the residual structures may be regarded as isotropic. Let us put Δ into the inertial range.[118] Since it is unlikely that the modelling specialist will define Δ as smaller than the Kolmogorov scale,[119] that amounts to prescribing that Δ be sufficiently smaller than the size of the large eddies L_t, that is taking (4.117) and (4.197) into account:

$$\Delta \ll \frac{k^{3/2}}{\varepsilon} = \frac{L_m}{C_\mu^{3/4}} \tag{4.275}$$

Section 4.4.2 then states that these eddies transfer the energy towards the smaller structures without any effect of the large eddies on the process. Furthermore, the relative smallness of Δ gives the residual tensor values which are substantially lower than the Reynolds stresses and so, on initial examination, the non-linear contributions of the strain can be ignored (refer to Section 4.5.1). Thus, on the scale of eddies with sizes similar to Δ, this is a comparatively moderate, isotropic and quasi-equilibrium turbulence, which allows us to invoke a model analogous to that of Boussinesq (4.73), provided with a 'mixing length' type closure (eqn (4.140)), namely:

$$\boldsymbol{\tau}_R = \frac{2}{3}k_R \mathbf{I}_3 - 2\nu_s \tilde{\mathbf{s}} \tag{4.276}$$

[117] As explained below, it cannot be too small.

[118] It may be recalled it is the broad range of medium-sized eddies, between the smallest and the largest ones (refer to Section 4.4.3).

[119] That would amount, indeed, to implementing the DNS.

with

$$\nu_s = L_s^2 \tilde{s} \tag{4.277}$$

The quantity ν_s is often referred to as *subgrid viscosity*,[120] whereas k_R stands for the *residual kinetic energy*, $\tilde{\mathbf{s}}$ is the filtered strain tensor and $\tilde{s}$ is its second-order invariant (*filtered scalar strain rate*), as defined by

$$\begin{aligned} k_R &\doteqdot \frac{1}{2}\mathrm{tr}\,\boldsymbol{\tau}_R \\ \tilde{\mathbf{s}} &\doteqdot \frac{1}{2}\left(\frac{\partial \tilde{u}_i}{\partial x_j} + \frac{\partial \tilde{u}_j}{\partial x_i}\right)\mathbf{e}_i \otimes \mathbf{e}_j \\ \tilde{s} &\doteqdot \sqrt{2\tilde{\mathbf{s}} : \tilde{\mathbf{s}}} \end{aligned} \tag{4.278}$$

[120] That wording refers to the mesh methods. In the second Part of the book, dealing with the SPH method, we will obviously speak of *sub-particle viscosity*.

The equation of motion (4.274) then takes a form analogous to (4.82):[121]

$$\frac{\tilde{D}\tilde{\mathbf{u}}}{Dt} = -\frac{1}{\rho}\mathbf{grad}\,\tilde{p}^* - \mathbf{div}\left(\nu_M \tilde{\mathbf{s}}\right) \tag{4.279}$$

[121] As in (4.82), we assume that the residual energy is negligible when compared with the filtered pressure.

with

$$\nu_M \doteqdot \nu + \nu_s \tag{4.280}$$

Comparing (4.279) with (4.82), we find that as compared with those models based upon the Reynolds averages, the subgrid viscosity supersedes the eddy viscosity. It represents the dissipative and diffusive effect being exerted by the unsolved (residual) structures, that is the sizes of which are smaller than Δ. Thus, it seems we may justifiably seek the length L_s, representing a numerical mixing scale, or a 'sub-grid' scale which characterizes these structures, in the following form

$$L_s = c_S \Delta \tag{4.281}$$

(Smagorinsky, 1963). In order to determine the Smagorinsky constant c_S, we will conduct some energy-oriented reasonings similar to those which were developed for treating the first-order Reynolds models, but without getting deep into the details (we will primarily rely on Section 4.4.2). By analogy with (4.99), the energy dissipated by the viscous term in equation (4.279) equals

$$\frac{de_k\left(\tilde{\mathbf{u}}\right)}{dt} = -\left(\nu + \nu_s\right)\tilde{s}^2 \tag{4.282}$$

The molecular dissipation is patently negligible when compared with the dissipation induced by the residual scales, because Δ is large enough in the face of the Kolmogorov scale. Energy is then essentially dissipated by the smallest solved scales, which typically have a size L_s. Inasmuch as that scale lies within the inertial range, these eddies, as previously noted, are subject to a production-dissipation equilibrium owing to which it can be stated that this dissipated energy is, on an average (in the Reynolds sense), nearly equal to ε:

$$\varepsilon \approx \overline{\nu_s \tilde{s}^2} = L_s^2 \overline{\tilde{s}^3} \tag{4.283}$$

(using (4.277)). The energy lost due to molecular viscosity and occurring in (4.282) is featured, on an average, by a spectrum:[122]

$$\tilde{\varepsilon} \doteqdot \nu \overline{\tilde{s}^2} = \int_0^{+\infty} \tilde{D}(K)\, dK \tag{4.284}$$

[122] As with $\tilde{s}$, and contrary to what is suggested by the notations being adopted, $\tilde{\varepsilon}$ is not the result of the filtering of ε. The same remark can be made about the $\tilde{D}(K)$ spectrum. This is an effect of the non-linearity of the formulas relating these quantities to the velocity field.

We will attempt to relate that spectrum to that of the true fluctuating velocity field, always looking at the case in which the fluid domain encompasses the whole space. By analogy with (4.113), it can then be written

$$\begin{aligned}
\tilde{D}(K) =& \frac{\nu K^2}{(2\pi)^6} \left| \widehat{\tilde{\mathbf{u}}}(\mathbf{K}, t) \right|^2 \\
=& \frac{\nu K^2}{(2\pi)^6} \left| \int_{\mathbb{R}^3} \int_{\mathbb{R}^3} \mathbf{u}(\mathbf{r}', t)\, G_\Delta(|\mathbf{r} - \mathbf{r}'|)\, e^{-i\mathbf{K}\cdot\mathbf{r}} d^3\mathbf{r}' d^3\mathbf{r} \right|^2 \\
=& \frac{\nu K^2}{(2\pi)^6} \int_{\mathbb{R}^3} \int_{\mathbb{R}^3} \int_{\mathbb{R}^3} \int_{\mathbb{R}^3} \mathbf{u}(\mathbf{r}', t) \cdot \mathbf{u}(\mathbf{r}''', t)\, G_\Delta(|\mathbf{r} - \mathbf{r}'|) \\
& \times G_\Delta(|\mathbf{r}'' - \mathbf{r}'''|)\, e^{i\mathbf{K}\cdot(\mathbf{r}'' - \mathbf{r})} d^3\mathbf{r} d^3\mathbf{r}' d^3\mathbf{r}'' d^3\mathbf{r}'''
\end{aligned} \tag{4.285}$$

(we have used the definitions (4.25) and (4.258), then expressed the squared norm of a complex z as zz^*). Since the terms in G_Δ are the only terms depending on the variables $\mathbf{r}$ and $\mathbf{r}''$, it is possible to isolate them, and afterwards to split up the exponential into three contributions to get:

$$\begin{aligned}
\tilde{D}(K) =& \frac{\nu K^2}{(2\pi)^6} \int_{\mathbb{R}^3} \int_{\mathbb{R}^3} \mathbf{u}(\mathbf{r}', t) \cdot \mathbf{u}(\mathbf{r}''', t)\, e^{i\mathbf{K}\cdot(\mathbf{r}''' - \mathbf{r}')} \\
& \times \int_{\mathbb{R}^3} G_\Delta(|\mathbf{r} - \mathbf{r}'|)\, e^{-i\mathbf{K}\cdot(\mathbf{r} - \mathbf{r}')} \\
& \times \int_{\mathbb{R}^3} G_\Delta(|\mathbf{r}''' - \mathbf{r}''|)\, e^{-i\mathbf{K}\cdot(\mathbf{r}''' - \mathbf{r}'')} d^3\mathbf{r} d^3\mathbf{r}'' d^3\mathbf{r}' d^3\mathbf{r}'''
\end{aligned} \tag{4.286}$$

In the last two integrals, the respective changes $\mathbf{r} - \mathbf{r}' \to \mathbf{r}$ and $\mathbf{r}''' - \mathbf{r}'' \to \mathbf{r}''$ of variables can be made and the Fourier transform of the filter $\widehat{G}_\Delta$ appears twice. Hence:

$$\begin{aligned}
\tilde{D}(K) =& \widehat{G}_\Delta(K)^2 \frac{\nu K^2}{(2\pi)^6} \int_{\mathbb{R}^3} \int_{\mathbb{R}^3} \mathbf{u}(\mathbf{r}', t)\, e^{-i\mathbf{K}\cdot\mathbf{r}'} \\
& \cdot \left[\mathbf{u}(\mathbf{r}''', t)\, e^{-i\mathbf{K}\cdot\mathbf{r}'''} \right]^* d^3\mathbf{r}' d^3\mathbf{r}''' \\
=& \widehat{G}_\Delta(K)^2 \frac{\nu K^2}{(2\pi)^6} \left| \hat{\mathbf{u}}(\mathbf{K}, t) \right|^2 \\
=& \widehat{G}_\Delta(K)^2 D(K)
\end{aligned} \tag{4.287}$$

Equation (4.284) then yields

$$\begin{aligned}\overline{\tilde{s}^2} &= \frac{1}{\nu}\int_0^{+\infty} \tilde{D}(K)\,dK \\ &= \frac{1}{\nu}\int_0^{+\infty} \widehat{G}_\Delta(K)^2 D(K)\,dK\end{aligned} \tag{4.288}$$

Since the important scales for our arguments lie in the inertial range, we can approach the previous integral by means of the Kolmogorov spectrum (4.119), which yields:

$$\overline{\tilde{s}^2} = 2c_K\varepsilon^{2/3}\int_0^{+\infty} \widehat{G}_\Delta(K)^2 K^{1/3}\,dK \tag{4.289}$$

The integral only depends on the selected filter and its width Δ. As regards the filter (4.261), it is $3(\pi/\Delta)^{4/3}/4$. With such an option, recombining (4.281), (4.283) and (4.289), we ultimately find:

$$c_S = \frac{1}{\pi}\left(\frac{2}{3c_K}\right)^{3/4}\left(\frac{\overline{\tilde{s}^2}^{1/2}}{\overline{\tilde{s}^3}^{1/3}}\right)^{3/2} \tag{4.290}$$

Lilly (1992) has discussed that model and made the hypothesis that the last term inside parentheses[123] is equal to one, which gives

[123] In the jargon of statisticians, it is the inverse cube of a quantity which is referred to as the spectrum 'asymmetry'.

$$c_S \approx 0.17 \tag{4.291}$$

Other filters provide values of the same order of magnitude, which are still often used in large eddy simulation, although there are more sophisticated models (in particular, refer to Pope, 2000).

Equations (4.269) and (4.279), fitted with the models (4.277) and (4.281), are the standard model of LES. Its simplicity is beneficial, since the equation ultimately comes down, at least formally, to the mixing length model. This is why it is more and more used in industry and has recently been applied in environmental science, but it requires a rather intensive spatio-temporal discretization. Above all, it can theoretically only work in three dimensions, because the simulated structures are three-dimensional. It is sometimes used, however, in two dimensions, and the comparisons of two-dimensional[124] with three-dimensional give suitable results, although the large structures are too marked in two dimensions because of the lack of energy transfer over the third dimension. The two-dimensional LES, however, may to some extent be considered an alternative to the conventional mixing length model, being advantageous in that the mixing length is known and constant.

[124] In this textbook, we only deal with three-dimensional turbulence. Two-dimensional turbulence, however, can exist in some special cases, such as large-scale atmospheric flows. It is noteworthy that its properties are significantly different than those of three-dimensional turbulence (see e.g. Frisch, 1995, as well as the end of Section 6.5.1).

One of the difficulties with the LES lies in that, for many applications, it needs to have access to the mean quantities after the calculation. In the general case, they are obtained through several simulations from slightly different initial conditions before carrying out a statistical averaging analogous to the Reynolds average (4.28). That approach, however, is lengthy and costly. In the case of a statistically steady flow, however, an averaging of all the quantities can be made over time.

It should also be pointed out that the initial conditions should be treated in a specific way in order to represent a satisfactory spatial distribution of the velocity fluctuations. Several methods can be used for preparing a 'synthetic turbulence' as an adequate starting point. A further difficulty stems from the choice of the spatial discretization size Δ, whose condition (4.275) shows that it depends on both space and time via the quantities k and ε. Sometimes, the modelling specialists make a preliminary calculation based upon the $k-\varepsilon$ model (Section 4.4.4) and resting on a comparatively coarse discretization, before deriving therefrom a discretization which is suitable for the LES.

4.6.2 The α model

The issue of the existence, unicity and regularity of the solutions to the Navier–Stokes equations has fueled considerable controversy and has been the subject of lots of works. The problem is not settled yet and is one of the major open questions in theoretical physics, since it calls into question the validity of most of the topics being discussed in this chapter. It appears indeed that due to the lack of certainty as regards the regularity of the velocity fields for a fluid in a turbulent regime, nobody really knows whether the viscous term sets a lower limit for the turbulent fluctuations of velocity. As explained in Section 4.2.3, the inertia term is at the origin of those instabilities, and so one may imagine that a formal modification of the advection velocity could provide part of the answer to that question. Among the major advances as proposed to the physicist community on the matter is the Leray's theorem for regularization (1934), which states that the existence, unicity and regularity of the velocity field can be secured provided that a *regularized* velocity is substituted for the advection velocity, which makes it possible to define a regularized operator of Lagrangian derivative:

$$\frac{\breve{D}A}{Dt} \doteq \frac{\partial A}{\partial t} + \mathbf{grad}A \cdot \breve{\mathbf{u}} \tag{4.292}$$

where $\breve{\mathbf{u}}$ is a regularized velocity related to 'true' velocity by a so far unknown linear operator as denoted by O:

$$\mathbf{u} \doteq O\breve{\mathbf{u}} \tag{4.293}$$

If we now decide to substitute the regularized advection for the standard advection, the Navier–Stokes equation of motion (3.139) then becomes

$$\frac{\breve{D}\mathbf{u}}{Dt} = \frac{\partial \mathbf{u}}{\partial t} + \mathbf{grad}\ \mathbf{u} \cdot \breve{\mathbf{u}} = -\frac{1}{\rho}\mathbf{grad}\ p^* + \nu\nabla^2\mathbf{u} \tag{4.294}$$

Holm (1999) has demonstrated that the continuous variational method as set out in Section 3.6 makes it possible to build the regularized equation (4.294) from the following amended Lagrangian:[125]

$$\breve{\mathscr{L}}(\mathbf{r}, \mathbf{u}) = \frac{1}{2}\mathbf{u} \cdot \breve{\mathbf{u}} + \mathbf{g} \cdot \mathbf{r} \tag{4.295}$$

[125] The formulation (4.295) suggests an affinity with the mirror systems discussed in Section 2.5.2 of Chapter 2 which, moreover, could have been extended to the continuous frame owing to the content of Section 3.6 in Chapter 3. We will make use thereof in a discrete framework in Section 6.5.2 (Chapter 6).

This Lagrangian supersedes (3.342), provided that it is considered that the mesoscopic fluid particles are now moving at the velocity $\breve{\mathbf{u}}$:

$$\breve{\mathbf{u}} = \frac{d\mathbf{r}}{dt} \tag{4.296}$$

Holm also explains how this approach enables us to contemplate a smart modelling of turbulence, considering, in order to define (4.293), the Helmoltz operator:

$$O \doteqdot \mathbf{I}_3 - \alpha^2 \nabla^2 \tag{4.297}$$

where α is a distance whose meaning will become clear later on. The regularized velocity is now defined by

$$\breve{\mathbf{u}} \doteqdot \left(\mathbf{I}_3 - \alpha^2 \nabla^2\right)^{-1} \mathbf{u} \tag{4.298}$$

As with the filter (4.258), it may be noticed that the operator commutes with the partial temporal and spatial derivatives. In particular, we have

$$\begin{aligned} \mathrm{div}\mathbf{u} &= \mathrm{div}\left(O\breve{\mathbf{u}}\right) \\ &= O\mathrm{div}\breve{\mathbf{u}} \end{aligned} \tag{4.299}$$

We therefore come to the conclusion that the regularized field obeys an continuity equation which is analogous to the initial field:

$$\mathrm{div}\breve{\mathbf{u}} = 0 \tag{4.300}$$

A calculation identical to (4.35)–(4.39) will yield, in the weakly compressible case:

$$\frac{\breve{D}\rho}{Dt} = -\rho\mathrm{div}\breve{\mathbf{u}} \tag{4.301}$$

It is noteworthy that the regularized velocity shall duly be used in the continuity equation, jointly with the regularized equation of motion (4.294). This is because the conservation of mass as established in Section 3.2.3 relies on the analysis of motion of the mesoscopic particles making up the fluid. Now, equation (4.296) indicates that the Lagrangian derivative of the position presently equals $\breve{\mathbf{u}}$.

For the same reason, in the state vector $\mathbf{y}(\mathbf{r}, t)$ of a fluid particle, as given by (3.366), the relevant velocity should then be $\breve{\mathbf{u}}$. The definition (4.295) of the Lagrangian then shows that the vector $\partial e/\partial \mathbf{y}$ as defined by (3.367) is left unchanged. Since the equation of motion (4.294) is about $\mathbf{u}$, the calculation (3.369) is still valid if one ignores the dissipative forces, but the velocity is not the same in the two terms of the integral:

$$\frac{dH}{dt} = \int_{\Omega(t)} \left(-\mathbf{u} \cdot \mathbf{grad}\, p^* - p^* \mathrm{div}\breve{\mathbf{u}}\right) d\Omega \tag{4.302}$$

Thus, the energy is not conserved: the velocity field is subject to dissipative forces. More specifically, the energy built on the Lagrangian form (4.295) is conserved, as expected. The traditional energy, being defined based on the square of the velocity field, however, is not conserved, the regularized velocity field giving up some energy to the real velocity field.[126] For getting more deeply into that process, we have to modify the equation governing **u** so that the advection involves **u** itself. To that purpose, we just have to invoke the transformation (3.9) and (4.300) and write

$$\mathbf{grad}\,\mathbf{u}\cdot\breve{\mathbf{u}} - \mathbf{grad}\,\mathbf{u}\cdot\mathbf{u} = \mathbf{div}\left(\mathbf{u}\otimes\breve{\mathbf{u}} - \mathbf{u}\otimes\mathbf{u}\right) \tag{4.303}$$

Thus, the equation of motion (4.294) yields

$$\frac{d\mathbf{u}}{dt} = \frac{\partial \mathbf{u}}{\partial t} + \mathbf{grad}\,\mathbf{u}\cdot\mathbf{u} = -\frac{1}{\rho}\mathbf{grad}\,p^* + \nu\nabla^2\mathbf{u} - \mathbf{div}\,\boldsymbol{\tau}_\alpha \tag{4.304}$$

with

$$\boldsymbol{\tau}_\alpha \doteq \mathbf{u}\otimes\breve{\mathbf{u}} - \mathbf{u}\otimes\mathbf{u} \tag{4.305}$$

This quantity represents a further stress tensor, whose formal affinity with (4.273) will have been noticed. With the operator (4.298), (4.305) is also written as

$$\begin{aligned}\boldsymbol{\tau}_\alpha &= \mathbf{u}\otimes\left(\breve{\mathbf{u}} - \mathbf{u}\right)\\ &= \alpha^2\breve{\mathbf{u}}\otimes\left(\nabla^2\breve{\mathbf{u}}\right)\end{aligned} \tag{4.306}$$

Once again, we can find, in the definition (4.306), the quadratic nature of the turbulent stresses which were mentioned in Section 4.4.4.

Equation (4.304) is often referred to either as the *Navier–Stokes-α equation*, or as the *viscous Camassa–Holm equation*. It may have been noticed that it involves a standard Lagrangian temporal derivative, which is no longer based upon an amended velocity field (such as $\overline{\mathbf{u}}$, $\tilde{\mathbf{u}}$ or $\breve{\mathbf{u}}$). For interpreting the meaning of the tensor (4.305), we may introduce the strain rate tensor of the regularized field (denoted as $\breve{\mathbf{s}}$) and its second-order invariant:

$$\begin{aligned}\breve{\mathbf{s}} &\doteq \frac{1}{2}\left(\frac{\partial \breve{u}_i}{\partial x_j} + \frac{\partial \breve{u}_j}{\partial x_i}\right)\mathbf{e}_i\otimes\mathbf{e}_j\\ \breve{s} &\doteq \sqrt{2\breve{\mathbf{s}}:\breve{\mathbf{s}}}\end{aligned} \tag{4.307}$$

If can then be noticed that (4.306) allows us to make the following estimate:

$$\begin{aligned}|\boldsymbol{\tau}_\alpha| &\sim \left(\frac{\alpha U}{L}\right)^2\\ &\sim \left(\alpha^2\breve{s}\right)|\breve{\mathbf{s}}|\end{aligned} \tag{4.308}$$

which is similar to the viscous part of (4.276). Thus, we can define a regularized viscosity:

$$\breve{\nu} \sim \alpha^2\breve{s} \tag{4.309}$$

Through a brief analogy with (4.277), we can find that α serves as a numerical mixing length,[127] just like $L_s \sim \Delta$ in large scale simulation. The regularization procedure is then something like a form of LES, filtering the quickest oscillatory modes.[128]

[126] In order to echo the previous margin note, it is a process which could be described within the context of the dissipative 'mirror' systems (refer to Chapter 2, Section 2.5.2).

[127] α should, of course, be distinguished from the kinetic coefficients of Section 2.4.2.

[128] Details about that affinity can be found in Foias et al. (2001), as well as in Guermond et al. (2003). The latter also analyze other formulations of the momentum equation with respect to invariance under a change of reference frame.

For a deeper insight into the effects of that regularization, let us now turn to the kinetic energy. The Lagrangian (4.295) suggests that kinetic energy will henceforth be written as

$$E_k = \int_\Omega \frac{1}{2}\rho \mathbf{u} \cdot \breve{\mathbf{u}}\, d\Omega \tag{4.310}$$

With the Helmholtz operator (4.297), an integration by parts will yield

$$\begin{aligned} E_k &= \int_\Omega \frac{1}{2}\rho \left[\left(\mathbf{I}_3 - \alpha^2\nabla^2\right)\breve{\mathbf{u}}\right] \cdot \breve{\mathbf{u}}\, d\Omega \\ &= \int_\Omega \frac{1}{2}\rho \left[\breve{u}^2 - \left(\alpha^2\nabla^2\breve{\mathbf{u}}\right)\cdot\breve{\mathbf{u}}\right] d\Omega \\ &= \int_\Omega \frac{1}{2}\rho\breve{u}^2 d\Omega - \oint_{\partial\Omega} \frac{1}{2}\rho \left[\left(\mathbf{grad}\, \breve{\mathbf{u}}\right)\breve{\mathbf{u}}\right]\cdot \mathbf{n}\, d\Gamma \\ &\quad + \int_\Omega \frac{1}{2}\rho\alpha^2 \left(\mathbf{grad}\, \breve{\mathbf{u}}\right)^2 d\Omega \end{aligned} \tag{4.311}$$

The second integral (4.311) extends to the boundary of the fluid domain. If the latter extends to infinity[129] (as in Section 4.6.1), it vanishes and yields:

[129] If the fluid is bounded, Shkoller (1999) shows that the second term in (4.312) should be replaced by $4\breve{\mathbf{s}} : \breve{\mathbf{s}} = 2\breve{s}^2$.

$$E_k = \int_\Omega \frac{1}{2}\rho \left[\breve{u}^2 + \alpha^2 \left(\mathbf{grad}\, \breve{\mathbf{u}}\right)^2\right] d\Omega \tag{4.312}$$

This formula (which is still a positive definite quadratic form) shows that the regularization term increases energy. The same result will obviously apply to the turbulent kinetic energy, whose spectrum (as introduced in Section 4.4.3) is then affected. In the inertial range, for such wave numbers K as $\alpha K \ll 1$, the term in $\left(\mathbf{grad}\, \breve{\mathbf{u}}\right)^2$ is negligible with respect to the standard term, and the spectrum is unchanged when compared to the Kolmogorov spectrum as given by (4.119). Moreover, for the wave numbers satisfying $1 \ll \alpha K$ (i.e. for those turbulent eddies with a sufficiently small size when compared to α), the first term of (4.312), indeed, is negligible when compared to the second. Thus, a cut-off wave number is defined

$$K_\alpha \doteqdot \frac{2\pi}{\alpha} \tag{4.313}$$

The velocity field regularization amounts to only modelling the largest eddies, that is the small wave numbers ($K \ll K_\alpha$). As regards the large wave numbers, they are distributed over a spectrum. Our method in Section 4.4.3 for determining the Kolmogorov spectrum may be repeated here, to within one detail: at such a scale, the energy flux characterizing the energy cascade concerns the quantity $\left(\mathbf{grad}\, \breve{\mathbf{u}}\right)^2$ (and not $\breve{u}^2$) and then is measured in s^{-3} (not longer in $m^2 s^{-3}$). We will denote it as η, and the Kolmogorov hypotheses make it possible to seek the regularized spectrum $\breve{E}$ as being a function of η and K (whereas the spectrum E of Section 4.4.3 depends on ε and K). We get:

$$\begin{aligned} K_\alpha \ll K \ll K_0,\ \breve{E}(K) &= c_K \eta^{2/3} K^{-3} \\ \breve{D}(K) &= 2c_K \nu \eta^{2/3} K^{-1} \end{aligned} \tag{4.314}$$

We may attempt to unify the spectrum (4.314) with the Kolmogorov spectrum by reasoning about the fluctuating velocities of the regularized velocity field. This is because the Fourier transform of the standard fluctuating velocity is written as:

$$\begin{aligned}\hat{\mathbf{u}}'(\mathbf{K},t) &= \left(\mathbf{I}_3 - \widehat{\alpha^2\nabla^2}\right)\breve{\mathbf{u}}' \\ &= \left(1+\alpha^2K^2\right)\widehat{\breve{\mathbf{u}}'}(\mathbf{K},t)\end{aligned} \tag{4.315}$$

(we have used the property (B.42) of Appendix B). The turbulent kinetic energy related to the regularized field is written, by analogy with (4.110) and (4.310), as:

$$\breve{k} = \frac{1}{2(2\pi)^6}\int_{\mathbb{R}^3}\hat{\mathbf{u}}'(\mathbf{K},t)\cdot\widehat{\breve{\mathbf{u}}'}(\mathbf{K},t)\,d^3\mathbf{K} \tag{4.316}$$

Introducing (4.315) thereinto, we can see that the standard and regularized energy spectra are connected by

$$\breve{E}(K) = \frac{E(K)}{1+\alpha^2K^2} \tag{4.317}$$

It would be wrong not to go more deeply into the estimate of $\breve{E}$, because the regularization, as we have just explained it, also affects the energy flux supplied to the Kolmogorov cascade. For quantifying this effect, we initally write that the time $\breve{\tau}_\lambda$ characterizing a turbulent structure with a scale $\lambda = 2\pi/K$ in the regularized spectrum is related to its characteristic velocity $\breve{u}_\lambda$ by

$$\breve{\tau}_\lambda \sim \frac{1}{\breve{u}_\lambda K} \sim \frac{1}{\sqrt{\breve{E}(K)K}} \sim \sqrt{\frac{1+\alpha^2K^2}{E(K)\,K}} \tag{4.318}$$

We subsequently notice that $\breve{\tau}_\lambda$ stands for the lifetime of an eddy, and so we may write the energy flux it transfers to the immediately smaller structures as

$$\breve{\varepsilon} \sim \frac{u_\lambda^2}{\breve{\tau}_\lambda} \sim u_\lambda^2\sqrt{\frac{E(K)\,K}{1+\alpha^2K^2}} \tag{4.319}$$

In the specific case where $\alpha = 0$, the Kolmogorov spectrum should be found again, which requires

$$\varepsilon = \breve{\varepsilon}\sqrt{1+\alpha^2K^2} \tag{4.320}$$

Combining (4.317) and (4.320), and utilizing (4.119), we ultimately get:

$$E(K) \sim \left[\frac{\breve{\varepsilon}^2}{\left(1+\alpha^2K^2\right)^2K^5}\right]^{1/3} \tag{4.321}$$

We can lastly invoke the relation (4.114) to come to the regularized Kolmogorov spectra:

$$K_t \ll K \ll K_0, \ \breve{E}(K) = c_K \left[\frac{\breve{\varepsilon}^2}{\left(1+\alpha^2 K^2\right)^2 K^5} \right]^{1/3}$$
$$\breve{D}(K) = 2 c_K \nu \left[\frac{\breve{\varepsilon}^2 K}{\left(1+\alpha^2 K^2\right)^2} \right]^{1/3} \tag{4.322}$$

For smaller wavenumbers ($K \ll K_\alpha$), we come across the usual spectrum (4.119), whereas the large wavenumbers are featured by (4.314), with

$$\eta = \frac{\breve{\varepsilon}}{\alpha^2} \tag{4.323}$$

as was suggested by the initial reasoning.

Thus, the smoothing carried out by the Helmoltz operator results in deleting the small structures from the spectrum, as substantiated by the decrease in K^{-3} which is valid for $K \gg K_\alpha$. The spectrum is then considerably restricted to the large wave numbers (refer to Fig. 4.10 (*b*)), which smoothes the velocity field, settling the problem set out at the beginning of this section, but also allowing us to apply a coarser time discretization upon a numerical modelling, as in LES.[130] The smoothing caused by the Helmoltz operator, however, is less abrupt than the smoothing implemented by the LES filter as proposed by the formula (4.261), for example.

[130] Refer to Foias et al. (2001) for further details on that matter.

It can be seen that the regularization also results in an energy flux which varies depending on the scale, since the smoothed structures transfer more energy in order to compensate for the spectrum narrowing. The choice of the length α will, of course, be left up to the user (just like the length Δ in LES), but should satisfy

$$\lambda_0 \ll \alpha \ll L_t \tag{4.324}$$

where the turbulence scales are given by (4.116) and (4.117). If the first condition is not satisfied (α is too small), the model takes all the turbulent structures into account and is then reduced to a DNS, that is to solving the ordinary Navier–Stokes equations, the Helmoltz operator coinciding with identity. Moreover, the second condition in (4.324) is essential, because choosing too

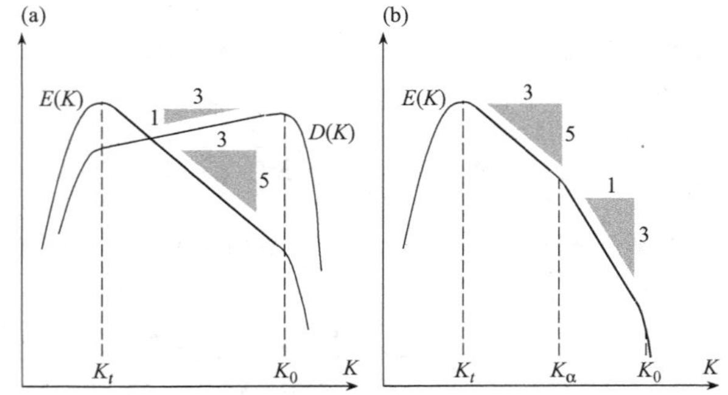

Fig. 4.10 The Kolmogorov spectra (*a*) and the truncated kinetic energy spectrum for the α model (*b*).

high an α would be physically meaningless with respect to being necessarily bounded. With (4.117), it is also written as

$$\alpha \ll \min \frac{k^{3/2}}{\varepsilon} \tag{4.325}$$

The generic case of the channel as discussed in Section 4.4.5 gives, with (4.197):

$$\frac{\alpha}{H} \ll \frac{L_m}{H} \sim \kappa\sqrt{2} \approx 0.58 \tag{4.326}$$

Foias et al. (2001) develop interesting considerations about the anisotropy and realizability of that model, whereas Chen et al. (1999) apply it to the steady infinite channel case dealt with in Section 4.4.5, providing more accurate velocity profiles than the logarithmic law (4.181).

4.6.3 Stochastic models

To complete this brief overview of the best known and most used turbulent models, we will turn to the stochastic models, which are very promising (particularly in the Lagrangian formalisms) because they make it possible to model a large part of turbulent physics at little cost (one may refer in the first place to Pope, 1994; 2000).

The initial paragraphs in this chapter have described turbulence as a process which may be viewed as a random signal, though it obeys some rules. The velocity of a fluid particle may be regarded as consisting of a steady part (the mean field as understood by Reynolds) and a fluctuating part, which is tantamount to a stochastic signal. As we have seen it, that process may be formally likened, to some extent, to the molecular motion. The fluctuating velocity field could then be regarded as a rapidly varying stochastic signal affecting the motion of a fluid particle like Brownian motion.[131] On the basis of the arguments expressed in Section 2.5.3, we will then write that the velocity of a fluid particle varies, during an infinitesimal time dt, by a term which is proportional to dt and represents the internal (pressure), external (gravity) and dissipative (viscous friction) forces, according to equation (2.285), complemented by a further term which is proportional to $\sqrt{dt}$ and represents the stochastic signal (2.281)):

$$d\mathbf{u} = \left[-\frac{1}{\rho}\mathbf{grad}\,\overline{p}^{*} - \mathbf{G}\left(\mathbf{u} - \overline{\mathbf{u}}\right)\right] dt + \sqrt{Z dt}\boldsymbol{\xi} \tag{4.327}$$

where $\boldsymbol{\xi}(\mathbf{r}, t)$ is a random vector field with components being independent standard normal distributions. The symbol $d\mathbf{u}$ here denotes a variation of $\mathbf{u}$ during dt following the instantaneous motion of a particle rather than its mean motion (the symbol $D\mathbf{u}$ being, it will remembered, used solely with that concept). Compared with Section 2.5.3, it will have been noticed that the fluctuating velocity $\mathbf{u} - \overline{\mathbf{u}}$ does determine the friction force.[132] Moreover, the Reynolds average is substituted for the Boltzmann average.

Thus, we will *mutatis mutandis* apply the results of Section 2.5.3, formally switching from the molecular context to the turbulent context, which, in

[131] Pope (2000) reviews the stochastic models being dealt with in this Section on the basis of the probability density of the turbulent velocity (refer to the margin note of Section 4.3.1). The equation governing the transport of that probability density is then similar to the Fokker–Planck equation (refer to Section 2.5.3).

[132] This is because, in Section 2.5.3, the calculations were made in the fluid's local inertial frame of reference.

particular, will induce changes in the specific time scales of the stochastic signal. The two characteristic time scales are that of the agitation motion which is modelled by the stochastic signal (in this case the smallest eddies, formally playing the part of τ_m) and that of the mean motion (in this case being at least equal to the time scale of the largest eddies, here formally playing the part of T_R). Thus, we shift from the molecular formalism of Section 2.5.3 to the turbulence formalism through the following transformations:

$$\begin{aligned} \langle A \rangle &\longleftarrow \overline{A} \\ \tau_m &\longleftarrow \tau_0 \\ T_R &\longleftarrow \tau \\ E_{int} &\longleftarrow k \end{aligned} \tag{4.328}$$

where τ_0 is the Kolmogorov time scale and τ is the time scale for the large eddies; these scales were introduced in (4.116) and (4.117). According to equation (2.274), we must, upon a numerical interpretation of equation (4.327), prescribe the following condition for the time step dt:

$$\left(\frac{\nu}{\varepsilon}\right)^{1/2} \sim \tau_0 \ll dt \ll \tau \sim \frac{k}{\varepsilon} \tag{4.329}$$

In accordance with equations (2.275) and (2.284), we prescribe the following conditions for the quantities $\mathbf{G}$ and $\boldsymbol{\xi}$:

$$\begin{aligned} \overline{\mathbf{G}} &= \mathbf{G} = \mathbf{G}^T \\ \overline{\boldsymbol{\xi}} &= \mathbf{0} \\ \overline{\boldsymbol{\xi} \otimes \boldsymbol{\xi}} &= \mathbf{I}_3 \\ \overline{\boldsymbol{\xi} \otimes \mathbf{u}} &= \mathbf{0} \end{aligned} \tag{4.330}$$

We should recall that the symmetry of $\mathbf{G}$ is based on the Onsager theorem (Section 2.4.1). We add the further property $\overline{\mathbf{G}} = \mathbf{G}$, which is not mentioned in Section 2.5.3. It should be pointed out that the second and third properties (4.330) are centralization and statistical normalization conditions, whereas the last one prescribes the independence of the stochastic process $\boldsymbol{\xi}$ with respect to the velocity field, according to equation (2.316) in Section 2.5.3. We should note that it immediately implies

$$\begin{aligned} \overline{\boldsymbol{\xi} \otimes \mathbf{u}'} &= \overline{\boldsymbol{\xi} \otimes \mathbf{u}} - \overline{\boldsymbol{\xi}} \otimes \overline{\mathbf{u}} \\ &= \mathbf{0} \end{aligned} \tag{4.331}$$

(we have used the properties (4.33) of the average operator).

We are going to discuss the statistical properties of the stochastic model we have just set up, firstly disclosing an important property enabling us to make its content clearer in terms of temporal correlations. For a time increment dt satisfying the conditions (4.329), and writing $\mathbf{u}\left(t''\right) - \mathbf{u}\left(t'\right) = d\mathbf{u}$, the model (4.327) allows us to write the velocity correlation tensor as defined in (4.134) in a simple form. Only keeping[133] the first-order terms in dt and taking advantage of the properties (4.330), we actually get:

[133]For that reason, we must use a sufficiently small time increment dt, which is a stronger restriction than the physical condition (4.329).

$$\begin{aligned}\mathbf{D}_u\,(dt) &= \overline{d\mathbf{u}\otimes d\mathbf{u}}\\ &= Zdt\overline{\boldsymbol{\xi}\otimes\boldsymbol{\xi}}\\ &= Zdt\mathbf{I}_3\end{aligned} \tag{4.332}$$

This law is in accordance with (2.313) and (4.136), which suggests

$$Z = C_0\varepsilon \tag{4.333}$$

Since the dissipation of turbulent kinetic energy per unit mass is defined by ε, this result agrees with equation (2.318), the dissipation being closely related to the fluctuations. The cause of the C_0 factor will become clear later on. We will write, instead of (4.327):

$$d\mathbf{u} = \left(-\frac{1}{\rho}\mathbf{grad}\,\overline{p}^* - \mathbf{G}\mathbf{u}'\right)dt + \sqrt{C_0\varepsilon dt}\boldsymbol{\xi} \tag{4.334}$$

with, as usual, $\mathbf{u}' \doteqdot \mathbf{u} - \overline{\mathbf{u}}$.

Let us now turn to the behaviour of the statistical velocity moments as provided in the model being considered. The averaging of equation (4.334) yields, due to the conditions (4.330):

$$\frac{\overline{d\mathbf{u}}}{dt} = -\frac{1}{\rho}\mathbf{grad}\,\overline{p}^* \tag{4.335}$$

However, the relation (4.49) with a constant ρ gives

$$\frac{\overline{d\mathbf{u}}}{dt} = \frac{D\overline{\mathbf{u}}}{Dt} + \mathbf{div}\,\mathbf{R} \tag{4.336}$$

Making the latter two equations equivalent, we find

$$\frac{D\overline{\mathbf{u}}}{Dt} = -\frac{1}{\rho}\mathbf{grad}\,\overline{p}^* - \mathbf{div}\,\mathbf{R} \tag{4.337}$$

The incompressible Reynolds equation (4.51) is then satisfied[134] (on a first-order statistical average) by the model (4.334). It might be said that (4.334) implicitly and stochastically includes the Reynolds equation of motion.

To further investigate the behaviour of that model, we then have to look at the statistical evolution it provides to the Reynolds stress tensor $\mathbf{R} \doteqdot \overline{\mathbf{u}'\otimes\mathbf{u}'}$. To that purpose, we might believe that referring to the derivations in Section 4.5.2 is enough; here, however, that reasoning is partially wrong. It should be kept in mind, indeed, that the velocity field as defined by (4.334) is theoretically underivable, and so we have to proceed otherwise.[135] We will write that the increment of $\mathbf{R}$ between two time points which are separate by dt (following the mean velocity field) is given by

$$\begin{aligned}D\mathbf{R} &= \overline{\mathbf{u}'\otimes\mathbf{u}'}\,(t+dt) - \overline{\mathbf{u}'\otimes\mathbf{u}'}\,(t)\\ &= \overline{(\mathbf{u}'+D\mathbf{u}')\otimes(\mathbf{u}'+D\mathbf{u}')} - \overline{\mathbf{u}'\otimes\mathbf{u}'}\\ &= 2\overline{\mathbf{u}'\otimes D\mathbf{u}'}^S + \overline{D\mathbf{u}'\otimes D\mathbf{u}'}\end{aligned} \tag{4.338}$$

[134] The molecular viscosity term is not included in it, but it may, of course, be added into the generic model (4.334). Here we take advantage of the fact that turbulent diffusion largely prevails over molecular diffusion, to ignore the latter just for the sake of simplicity.

[135] We have faced the same problem in Section 2.5.3 upon the derivation of the Fokker–Planck equation.

Subsequently, from the definition of the instantaneous (d/dt) and mean (D/Dt) Lagrangian derivatives, we get:

$$\begin{aligned}\frac{D\mathbf{u}'}{Dt} &= \frac{\partial \mathbf{u}'}{\partial t} + \mathbf{grad}\ \mathbf{u}' \cdot \overline{\mathbf{u}} \\ &= \frac{\partial \mathbf{u}'}{\partial t} + \mathbf{grad}\ \mathbf{u}' \cdot \mathbf{u} - \mathbf{grad}\ \mathbf{u}' \cdot \mathbf{u}' \\ &= \frac{d\mathbf{u}'}{dt} - \boldsymbol{\gamma}' \cdot \mathbf{u}'\end{aligned} \tag{4.339}$$

where $\boldsymbol{\gamma}'$ is defined by (4.63). Through the relation (A.43) in Appendix A, it follows

$$\mathbf{u}' \otimes \frac{D\mathbf{u}'}{Dt} = \mathbf{u}' \otimes \frac{d\mathbf{u}'}{dt} - \boldsymbol{\gamma}' \left(\mathbf{u}' \otimes \mathbf{u}'\right) \tag{4.340}$$

Consequently, invoking (4.69) and the definition (4.68), (4.340) yields

$$2\overline{\left(\mathbf{u}' \otimes \frac{D\mathbf{u}'}{Dt}\right)^S} = 2\overline{\left(\mathbf{u}' \otimes \frac{d\mathbf{u}'}{dt}\right)^S} - \mathbf{div}\ \check{\mathbf{Q}}^{\mathbf{R}} \tag{4.341}$$

Furthermore, re-using (4.339) and still only keeping the first-order terms in dt, we may write

$$\overline{D\mathbf{u}' \otimes D\mathbf{u}'} = \overline{d\mathbf{u}' \otimes d\mathbf{u}'}$$

(4.338) is then reduced to the following form

$$\frac{D\mathbf{R}}{Dt} = -\mathbf{div}\ \check{\mathbf{Q}}^{\mathbf{R}} + 2\overline{\left(\mathbf{u}' \otimes \frac{d\mathbf{u}'}{dt}\right)^S} + \frac{\overline{d\mathbf{u}' \otimes d\mathbf{u}'}}{dt} \tag{4.342}$$

Compared with such a reasoning as (4.44), we now find that a new term appears, since $d\mathbf{u}' \otimes d\mathbf{u}'$ has a first-order term in dt via the last term in the initial equation (4.334). The increment of the $d\mathbf{u}'$ fluctuation is now to be related to the fluctuation of the $(d\mathbf{u})'$ increment by writing[136]

[136] As is by now well established, these two operators will obviously not commute.

$$\begin{aligned}(d\mathbf{u})' &= d\mathbf{u} - \overline{d\mathbf{u}} \\ &= d\mathbf{u}' + d\overline{\mathbf{u}} - \overline{d\mathbf{u}}\end{aligned} \tag{4.343}$$

Let us now calculate the following term:

$$\begin{aligned}\overline{\mathbf{u}' \otimes \frac{d\overline{\mathbf{u}}}{dt}} &= \overline{\mathbf{u}' \otimes \frac{\partial \overline{\mathbf{u}}}{\partial t}} + \overline{\mathbf{u}' \otimes (\mathbf{grad}\ \overline{\mathbf{u}} \cdot \mathbf{u})} \\ &= \overline{\mathbf{u}'} \otimes \frac{\partial \overline{\mathbf{u}}}{\partial t} + \overline{\mathbf{u}' \otimes (\boldsymbol{\Gamma} \cdot \mathbf{u})} \\ &= \overline{\boldsymbol{\Gamma}\,(\mathbf{u}' \otimes \mathbf{u})}\end{aligned} \tag{4.344}$$

(we have used the definitions of a Lagrangian derivative and the mean velocity gradient $\boldsymbol{\Gamma}$, as well as the properties (4.32) and (4.33) of the averaging operator). Invoking the same properties of the average, we also have

$$\begin{aligned}\overline{\boldsymbol{\Gamma}\,(\mathbf{u}' \otimes \mathbf{u})} &= \overline{\boldsymbol{\Gamma}\,(\mathbf{u}' \otimes \mathbf{u}')} + \overline{\boldsymbol{\Gamma}\,(\overline{\mathbf{u}} \otimes \mathbf{u}')} \\ &= \boldsymbol{\Gamma}\mathbf{R}\end{aligned} \tag{4.345}$$

and

$$\overline{\mathbf{u}' \otimes \overline{\frac{d\mathbf{u}}{dt}}} = \overline{\mathbf{u}'} \otimes \overline{\frac{d\mathbf{u}}{dt}} = \mathbf{0} \tag{4.346}$$

Thus, with (4.343), (4.344), (4.345) and (4.346) we may write

$$\overline{\mathbf{u}' \otimes \frac{d\mathbf{u}'}{dt}} = \overline{\mathbf{u}' \otimes \frac{(d\mathbf{u})'}{dt}} - \boldsymbol{\Gamma}\mathbf{R} \tag{4.347}$$

Equation (4.342) then becomes, due to the definition (4.67) of $\mathbf{P}$:

$$\frac{D\mathbf{R}}{Dt} = \mathbf{P} - \mathbf{div}\,\check{\mathbf{Q}}^{\mathrm{R}} + 2\overline{\left(\mathbf{u}' \otimes \frac{(d\mathbf{u})'}{dt}\right)}^{S} + \frac{\overline{d\mathbf{u}' \otimes d\mathbf{u}'}}{dt} \tag{4.348}$$

We still have to calculate the last two terms of that equation. The quantity $(d\mathbf{u})'$ is derived from the initial model (4.334):

$$(d\mathbf{u})' = -\mathbf{G}\mathbf{u}'dt + \sqrt{C_0\varepsilon dt}\,\boldsymbol{\xi} \tag{4.349}$$

We have in particular, with (4.331) and the definition of $\mathbf{R}$:

$$\begin{aligned}\overline{\mathbf{u}' \otimes \frac{(d\mathbf{u})'}{dt}} &= -\overline{\mathbf{u}' \otimes \mathbf{G}\mathbf{u}'} + \sqrt{\frac{C_0\varepsilon}{dt}}\,\overline{\mathbf{u}' \otimes \boldsymbol{\xi}} \\ &= -\mathbf{G}\mathbf{R}\end{aligned} \tag{4.350}$$

To calculate the last term in (4.348), we will write the following relation:

$$\begin{aligned}\frac{d\overline{\mathbf{u}}}{dt} &= \frac{\partial\overline{\mathbf{u}}}{\partial t} + \mathbf{grad}\,\overline{\mathbf{u}}\cdot\mathbf{u} \\ &= \frac{\partial\overline{\mathbf{u}}}{\partial t} + \mathbf{grad}\,\overline{\mathbf{u}}\cdot\overline{\mathbf{u}} + \mathbf{grad}\,\overline{\mathbf{u}}\cdot\mathbf{u}' \\ &= \frac{D\overline{\mathbf{u}}}{Dt} + \boldsymbol{\Gamma}\cdot\mathbf{u}' \\ &= \overline{\frac{d\mathbf{u}}{dt}} - \mathbf{div}\,\mathbf{R} + \boldsymbol{\Gamma}\cdot\mathbf{u}'\end{aligned} \tag{4.351}$$

(we have utilized (4.336)). We then have

$$\begin{aligned}d\mathbf{u}' &= d\mathbf{u} - d\overline{\mathbf{u}} \\ &= (d\mathbf{u})' + \left(\mathbf{div}\,\mathbf{R} - \boldsymbol{\Gamma}\cdot\mathbf{u}'\right)dt\end{aligned} \tag{4.352}$$

In the above expression, only $(d\mathbf{u})'$ has a term in $\sqrt{dt}$ (according to equation (4.349)), and so we have, once again to the first order:

$$d\mathbf{u}' \otimes d\mathbf{u}' = (d\mathbf{u})' \otimes (d\mathbf{u})' \tag{4.353}$$

Lastly, with the normalization property (4.330), it follows

$$\begin{aligned}\overline{d\mathbf{u}'\otimes d\mathbf{u}'} &= C_0\varepsilon dt\overline{\boldsymbol{\xi}\otimes\boldsymbol{\xi}}\\ &= C_0\varepsilon dt\mathbf{I}_3\end{aligned} \tag{4.354}$$

which may be compared with (4.332). Let us now come back to the transport equation (4.348); the formulas (4.350) and 4.354) ultimately yield:

$$\frac{D\mathbf{R}}{Dt} = \mathbf{P} - \mathbf{div}\,\check{\mathbf{Q}}^{\mathbf{R}} - 2\left(\mathbf{G}\mathbf{R}^S\right) + C_0\varepsilon\mathbf{I}_3 \tag{4.355}$$

That equation is perfectly identical[137] to equation (4.66), provided that the following is prescribed

$$-2\left(\mathbf{G}\mathbf{R}^S\right) + C_0\varepsilon\mathbf{I}_3 = \boldsymbol{\Phi} - \boldsymbol{\varepsilon} \tag{4.356}$$

With the LRR-QI (or LRR-IP) model for $\boldsymbol{\Phi}$ (eqn (4.222)), supplemented by the isotropic dissipation model (4.216), that is tantamount to the following equality, taking the definitions (4.67) and (4.223) into account:

$$\begin{aligned}-2\left(\mathbf{G}\mathbf{R}\right)^S + C_0\varepsilon\mathbf{I}_3 &= -\,C_R\varepsilon\mathbf{a} - C_P\mathbf{P}^D - C_D\mathbf{D}^D\\ &\quad -\,C_S k\mathbf{S} - \frac{2}{3}\varepsilon\mathbf{I}_3\\ &= -\,C_R\frac{\varepsilon}{k}\mathbf{R} + 2C_P\left(\boldsymbol{\Gamma}\mathbf{R}\right)^S + 2C_D\left(\boldsymbol{\Gamma}^T\mathbf{R}\right)^S\\ &\quad -\,2C_S k\boldsymbol{\Gamma}^S + \frac{2}{3}\left[\left(C_P + C_D\right)P^* + C_R - 1\right]\varepsilon\mathbf{I}_3\end{aligned} \tag{4.357}$$

for the LRR-IP model ($C_D = C_S = 0$), the following option is suitable:

$$\begin{aligned}\mathbf{G} &= \frac{C_R}{2}\frac{\varepsilon}{k}\mathbf{I}_3 - C_P\boldsymbol{\Gamma}\\ C_0 &= \frac{2}{3}\left(C_P P^* + C_R - 1\right)\end{aligned} \tag{4.358}$$

Due to that option, the stochastic model (4.334) is written as follows for the velocity increment:

$$\begin{aligned}d\mathbf{u} &= \left(-\frac{1}{\rho}\mathbf{grad}\,\overline{p}^* - \frac{C_R}{2}\frac{\varepsilon}{k}\mathbf{u}' + C_P\boldsymbol{\Gamma}\mathbf{u}'\right)dt\\ &\quad + \sqrt{\frac{2}{3}\left[C_P P + \left(C_R - 1\right)\varepsilon\right]dt}\,\boldsymbol{\xi}\end{aligned} \tag{4.359}$$

This is the *generalized Langevin model.* We have just explained that it is statistically compatible with the (exact) Reynolds equation for the first-order moments and with the (closed, i.e. approximate) equation of the LRR-IP model for the second-order moments. We may recall that the constants in the latter model are provided by Table 4.2. Due to their positive nature, C_0 is positive if $P^* \doteqdot P/\varepsilon$ is positive, which is satisfied with the conventional models (4.97) and (4.158).[138]

[137] Provided that the flux $\mathbf{Q}^{\mathbf{R}}$ is considered as being restricted to its part $\check{\mathbf{Q}}^{\mathbf{R}}$. Pope (2000) argues that this is a relevant approximation. Besides, it may be pointed out that with a Gaussian fluctuating signal, the third-order moments are null, and the definition (4.68) then indicates that the flux $\check{\mathbf{Q}}^{\mathbf{R}}$ of the Reynolds stresses is null. Strictly speaking, this model represents a homogeneous turbulence.

[138] It should be pointed out, however, that actually this may not always be true (refer to the end of Section 4.4.4).

Moreover, other second-order closure models could be invoked[139] adequately specifying the tensor $\mathbf{G}$. The *simplified Langevin model*, for instance, consists of specifying $C_P = C_D = C_S = 0$, that is

$$\begin{aligned} \mathbf{G} &= \frac{C_R}{2}\frac{\varepsilon}{k}\mathbf{I}_3 \\ C_0 &= \frac{2}{3}(C_R - 1) \end{aligned} \tag{4.360}$$

It is therefore compatible with a closure of the Reynolds stress equation based upon the simplified Rotta model (4.221).

An important remark should be made here. The fluctuation-dissipation theorem (2.300) is theoretically applicable to this case, $C_0\varepsilon$ playing the part of Z (eqn (4.333)) and $\tau = k/\varepsilon$ playing the part of T_R, according to (4.328). With the second line of (4.360), that yields

$$\begin{aligned} \mathbf{G} &= \frac{3C_0}{4}\frac{\varepsilon}{k}\mathbf{I}_3 \\ &= \frac{C_R - 1}{2}\frac{\varepsilon}{k}\mathbf{I}_3 \end{aligned} \tag{4.361}$$

Compared with the first line in (4.360), the presence of the factor $C_R - 1$ instead of C_R is caused by the fact that the fluctuation-dissipation theorem rests on the assumption of a particle in thermodynamic equilibrium with the surrounding particles. In this case, this amounts to assuming that k is constant, that is $Dk/Dt = 0$, whereas equation (4.104) yields $Dk/Dt = P - \varepsilon$. The presence of the gradient $\mathbf{\Gamma}$ in the more general case of equation (4.358) is caused by the non-linearity of the turbulence, whereas the molecular arguments in Section 2.5.3 rely on a linear theory of friction.[140] Moreover, it should be kept in mind that most of the arguments in Section 2.5.3 are based upon the assumption of constant external parameters, unlike the pressure and friction forces in this case, because $\overline{p}^*$ is variable, as well as $\mathbf{G}$. Anyway, equation (4.360) shows that the quantity $C_R\tau/2$ (rather than τ) does play the formal part of T_R.

So, with the simplified model (4.360), the velocity increment is given by

$$d\mathbf{u} = \left(-\frac{1}{\rho}\mathbf{grad}\,\overline{p}^* - \frac{C_R}{2}\frac{\varepsilon}{k}\mathbf{u}'\right)dt + \sqrt{\frac{2}{3}(C_R - 1)\,\varepsilon dt}\boldsymbol{\xi} \tag{4.362}$$

Contrary to what is suggested by a superficial examination, this model is duly non-linear, because ε and k implicitly depend on the mean velocity field.

The expected benefits from these models are obvious: using the explicit equation (4.359) (or (4.362)), a complex physics can be explained at a low cost. In addition, as with the explicit algebraic models (refer to Section 4.5.3), there is no need to use boundary conditions for the second-order moments. These conclusions, however, are to be qualified by means of a couple of remarks:

- The physical content of this model is only valid on an average; so, it looks like an LES type model (refer to the preceding two sections) with the same remarks about the interpretation of the quantity $\mathbf{u}$. An objection is that the mean velocity $\overline{\mathbf{u}}$ occurs in equation (4.359) through the quantities

139 The LRR-QI model, which is more general, is troublesome: adding the term $C_D\mathbf{\Gamma}^T$ to the definition (4.358) of $\mathbf{G}$ does allow us to get the term $2C_D\left(\mathbf{\Gamma}^T\mathbf{R}\right)^S$ of (4.357), but the term $-2C_Sk\mathbf{\Gamma}^S$ cannot be recovered through that procedure.

140 We may also recall the hypothesis of an erratic isotropic motion of the molecules, which was defeated as regards the turbulent fluctuations.

$\mathbf{u}' = \mathbf{u} - \overline{\mathbf{u}}$, but this is more a problem than an adavantage, because $\overline{\mathbf{u}}$ (and $\overline{p}$) are still to be calculated, which may be an awkward task. Using the Reynolds equation to that purpose would result in wasting the whole benefit from the stochastic model, whose virtue lies in the estimate of a turbulent velocity field which implicitly satisfies the Reynolds equations without solving them.

- The model requires knowledge of the quantities k, ε and P, which is possible through the (either conventional or non-linear) $k - \varepsilon$ models. It would be naïve, however, to believe that k and $P = -\mathbf{R} : \mathbf{S}$ can be determined otherwise, from the very model (4.359). We may recall, indeed, that the equation for $\mathbf{R}$ is only implicitly contained by the model, since equation (4.355) is not solved. The same applies[141] to the equation for k. Here, one of the advantages of the second-order models is lost (refer to Section 4.5.2).
- Lastly, we may notice that if we assume, just for a moment, that the production-dissipation equilibrium is valid $P^* = 1$, then the quantity C_0 as given by (4.358) becomes

$$C_0 = \frac{2}{3}\left(C_P + C_R - 1\right) \tag{4.363}$$

With the constants of the LRR-IP model (Table 4.2, Section 4.5.2), we find $C_0 = 0.93$, which is lower than the 2.1 value mentioned at the end of Section 4.4.3, but is still a suitable order of magnitude. It is recommended, however, to keep (4.358).

The stochastic models also find many applications in the field of effluent dilution in natural or industrial environments. This approach is sometimes preferable to solving a transport-diffusion equation of a scalar (eqn (4.87)) when the amount of material being transported occurs in the form of a collection of small bodies (corrosion or combustion products, sand grains, etc.). The review by Peirano et al. (2006) may be referred to about that matter. A practical application to a natural environment can be found in Issa et al. (2009b).

[141] In line with the reservations expressed above as regards the approximation of the diffusion term.

The SPH method in hydraulics

Part II

Principles of the SPH method

5

5.1 Introduction

The SPH (Smoothed Particle Hydrodynamics) numerical method was developed in the late 1970s for treating the behaviour of the systems existing in Astrophysics, for which the (Eulerian) mesh-based methods are found to be quite unsuitable due to the lack of defined boundaries (Lucy, 1977; Gingold and Monaghan, 1977). After being subsequently tailored to the modelling of elastic media, particularly for the calculation of the very large strains and the fracture phenomenon (see e.g. Campbell et al., 2000; Maurel, 2008), it was repeatedly improved before being applied, as of the early 1990s, first to weakly compressible hydrodynamics, then to strongly compressible (Sigalotti et al., 2009) and truly incompressible (Cummins and Rudman, 1999) hydrodynamics. SPH is now the most popular meshfree method in scientific literature and is close to the MPS (moving particle semi-implicit method, refer e.g. to Gotoh et al., 2005) and DPD (dissipative particle dynamics, refer among others to Español, 1995) methods. In addition, it has given rise to such interesting variants as remeshed SPH (Chaniotis et al., 2002) and PVFM (particle finite volume method, refer e.g. to Nestor and Quinlan, 2009). This presentation will only deal with the standard SPH method, although some of the concepts we will discuss can be extended to these closely related methods[1].

[1] It is interesting to mention another class of numerical particle-based methods, which could be labelled as semi-Lagrangian. These are the *Lattice-Boltzmann* type methods in which the particles can freely move over a fixed network while being subject to mutual collisions. The Boltzmann theory discussed in Chapter 2 is at the core of this approach.

Among the main attractions of SPH is its ability to predict highly strained motions without utilizing any computational mesh or grid, on the basis of a collection of macroscopic particles. It is based on the use of a ‘kernel’ function which is suitable for representing the derivatives of continuous fields in a discrete form. This formalism is achieved on completion of two successive, respectively continuous and discrete, interpolations. The resulting equations then consist of discrete mechanical systems in which the interactions between particles are expressed by fluxes which depend, in particular, on their mutual distances and their mechanical and thermodynamic features. One of the most attractive advantages of that approach is the consistency of that method with both Lagrangian and Hamiltonian mechanics as set out in Chapter 1, particularly in terms of conservativity. In addition, the development of the large system modelling for astrophysics allows us to extend that notion to the discrete treatment of time, making SPH a fully conservative method, provided that some care is exercised.

SPH has been somewhat distrusted by some theoricists and modelling specialists, because of the relatively 'experimental' nature of its initial developments. Nowadays, it is regarded as a quite reliable method due to further improvements which provided it with a number of properties which are deemed essential to the trustworthiness of a numerical method. In particular, tools provide for the consistency of the SPH interpolations under some conditions. Recent papers also substantiate its properties of convergence (Di Lisio et al., 1998; Ben Moussa, 2006), although some gaps are still to be bridged in that respect, and further investigations into that method have therefore become excessively appealing.

The first stage in the disclosure of the SPH method will consist in explaining the continuous and discrete interpolations and their main properties, then we will introduce some of the most commonly used kernels before reviewing the renormalization schemes. This will lead to the notion of discrete operators, which will make it possible to establish discrete forms of the continuity and Euler equations as built in Chapter 3 in a continuous formalism. The link to Chapter 1 (Lagrange and Hamilton equations) will then be set up through a variational method which will enable us to build the usual conservation laws for the SPH method, within the context of a time being considered as a continuous variable. The requisite time discretization will then lead us to the notion of temporal integrators, particularly symplectic integrators (based on the Hamilton equations), together with their exact conservation properties. These integrators will be illustrated by means of the simple pendulum case which was briefly discussed in Chapter 1. We will complete this presentation with an analysis of the numerical stability of the equations in the SPH method.

5.2 Lagrangian interpolation

We begin our analysis of the SPH method with the continuous interpolation process and its properties. Interpolating kernel examples are presented and analysed. The discrete (particle-based) interpolation is then considered, and the question of its numerical accuracy is addressed. This leads us to the concept of discrete differential operators. We finally review some kernel renormalization techniques.

5.2.1 Kernel and continuous interpolation

One of the major issues with respect to numerical modelling consists of actually switching from a continuous formalism to a discrete set for modelling the derivation operators occurring in the equations of motion (e.g. advection, diffusion). Within the frame of the SPH method, a two-step approach should be adopted; we will introduce a continuous interpolation scheme in this and the next sections[2] before dealing with the (discrete) particle-based discretization.

Let us start with writing an arbitrary scalar field A as a spatial convolution product with the Dirac distribution δ (refer to Appendix B, eqn (B.56)):

[2]There are lots of scientific papers about the subject of this section. For example, refer to Oger et al. (2007), as well as Marongiu (2007).

$$
\begin{aligned}
A(\mathbf{r},t) &= (A*\delta)(\mathbf{r},t) \\
&= \int_{\Omega} A\left(\mathbf{r}',t\right)\delta\left(\mathbf{r}-\mathbf{r}'\right)d^n\mathbf{r}'
\end{aligned} \tag{5.1}
$$

where Ω denotes the whole continuous medium. Having regard to 2- and 3-dimensional applications, we denote the dimension of the problem as n, without loss of generality (the dimension $n = 1$ will also be briefly discussed in the next paragraph). For practical reasons, the Dirac distribution should be approached through an interpolation kernel, here being denoted as $w_h\left(\mathbf{r}-\mathbf{r}'\right)$, which is a regular function and has non-zero values on a compact or infinite-sized support. The size of that support, when finite, is on the order of magnitude of a length parameter h a precise definition of which will be given in Section 5.2.2. This kernel may, for instance, be either a Gaussian function (in such a case, it continuously tends towards zero at infinity) or a Gaussian-like function with a compact support (in such a case, it equals zero at a finite distance). We will denote the $\mathbf{r}$ point-centred domain as $\Omega_{\mathbf{r}}$, w_h getting non-zero values at that point. We then will write[3]

$$
\begin{aligned}
A(\mathbf{r},t) \approx [A]_c(\mathbf{r},t) &\doteqdot \int_{\Omega_{\mathbf{r}}} A\left(\mathbf{r}',t\right) w_h\left(\mathbf{r}-\mathbf{r}'\right)d^n\mathbf{r}' \\
&= (A*w_h)(\mathbf{r},t)
\end{aligned} \tag{5.2}
$$

which defines an interpolated field[4] being denoted as $[A]_c$. Considering the definition of the kernel support $\Omega_{\mathbf{r}}$, the integration can be extended to Ω, and (5.1) then shows that $[A]_c(\mathbf{r},t)$ is close to $A(\mathbf{r},t)$. In order to determine the accuracy[5] of the approximation (5.2), we now write the field A (assumed to be regular enough) in the form of a second-order Taylor series expansion about the point $\mathbf{r}$, putting $\tilde{\mathbf{r}} \doteqdot \mathbf{r}-\mathbf{r}'$:

$$
A\left(\mathbf{r}',t\right) = A(\mathbf{r},t) - \frac{\partial A}{\partial \mathbf{r}}\cdot\tilde{\mathbf{r}} + \frac{1}{2}\tilde{\mathbf{r}}^T\frac{\partial^2 A}{\partial \mathbf{r}^2}\tilde{\mathbf{r}} + O\left(|\tilde{\mathbf{r}}|^3\right) \tag{5.3}
$$

where the derivatives are considered at points $\mathbf{r}$ in space and t in time. Then, substituting that formulation into (5.2):

$$
\begin{aligned}
[A]_c(\mathbf{r},t) &= A(\mathbf{r},t)\int_{\Omega_0} w_h\left(\tilde{\mathbf{r}}\right)d^n\mathbf{r}' - \frac{\partial A}{\partial \mathbf{r}}\cdot\int_{\Omega_0}\tilde{\mathbf{r}}w_h\left(\tilde{\mathbf{r}}\right)d^n\mathbf{r}' \\
&+ \frac{1}{2}\frac{\partial^2 A}{\partial \mathbf{r}^2} : \int_{\Omega_0}\tilde{\mathbf{r}}\otimes\tilde{\mathbf{r}}w_h\left(\tilde{\mathbf{r}}\right)d^n\mathbf{r}' + \int_{\Omega_0} O\left(|\tilde{\mathbf{r}}|^3\right)w_h\left(\tilde{\mathbf{r}}\right)d^n\mathbf{r}'
\end{aligned} \tag{5.4}
$$

where $\Omega_{\mathbf{0}}$ is a copy of $\Omega_{\mathbf{r}}$ translated about the origin vector. Noting that $d^n\mathbf{r}' = d^n\tilde{\mathbf{r}}$, for the approximation $[A]_c \approx A$ to be accurate to the first order, the following two conditions should be satisfied:

$$
\int_{\Omega_0} w_h\left(\tilde{\mathbf{r}}\right)d^n\tilde{\mathbf{r}} = 1 \tag{5.5}
$$

and

$$
\int_{\Omega_0} w_h\left(\tilde{\mathbf{r}}\right)\tilde{\mathbf{r}}\,d^n\tilde{\mathbf{r}} = \mathbf{0} \tag{5.6}
$$

[3]The subscript $_c$ in the notation $[A]_c$ refers to a continuous interpolation, unlike what will be done in Section 5.2.3.

[4]We may notice that the definition (5.2) is closely related to the formula (2.3) giving $n(\mathbf{r},t)$ in the Boltzmann theory as set out in Chapter 2. The MPS (moving particle semi-implicit) method, which is closely related to SPH, takes advantage of that affinity to define the density of particles per unit volume (refer, for instance, to Gotoh et al., 2005).

In addition, a formal analogy can be found between (5.2) and the definition (4.258) of the filtered fields for the treatment of turbulence by the large eddy simulation (Section 4.6.1). This link will become clearer in Section 6.5.1.

[5]The following is valid for a point lying far enough from the boundary of the fluid domain (refer to Fig. 5.1). Section 5.2.5 will provide further information in this respect.

The condition (5.5) stipulates that the kernel average (its zeroth-order moment) should be equal to one (like the Dirac distribution), which is a normalization condition about which we will assume it will always be verified in the rest of this paragraph. Under that hypothesis, and noting that $w_h\left(\tilde{\mathbf{r}}\right) d^n\mathbf{r}'$ is substantially equal to one, the last integral in (5.4) is a $O\left(h^3\right)$, the last but one is a $O\left(h^2\right)$. As regards the condition (5.6), it stipulates that the kernel first-order moment should be zero and is far from trivial. It can be contemplated again, however, where $\Omega_\mathbf{0}$ is central-symmetrically invariant[6] and where the kernel is even, that is if verifies

[6] See margin note 5.

$$\forall\tilde{\mathbf{r}},\quad w_h\left(-\tilde{\mathbf{r}}\right) = w_h\left(\tilde{\mathbf{r}}\right) \tag{5.7}$$

We may incidentally note that the condition (5.7) immediately leads to

$$\nabla w_h\left(-\tilde{\mathbf{r}}\right) = -\nabla w_h\left(\tilde{\mathbf{r}}\right) \tag{5.8}$$

where ∇w_h denotes the kernel gradient. Under that condition, making the variable change $\tilde{\mathbf{r}}' \doteqdot -\tilde{\mathbf{r}}$ in the first integral of (5.6) is helpful. The element of $n-$integration volume is then left unchanged ($d^n\tilde{\mathbf{r}}' = d^n\tilde{\mathbf{r}}$); thus, using equation (5.7), we get:

$$\begin{aligned}\int_{\Omega_\mathbf{0}} w_h\left(\tilde{\mathbf{r}}\right)\tilde{\mathbf{r}} d^n\tilde{\mathbf{r}} &= \int_{\Omega_\mathbf{0}} w_h\left(-\tilde{\mathbf{r}}\right)\tilde{\mathbf{r}}\, d^n\tilde{\mathbf{r}} \\ &= -\int_{\bar{\Omega}_\mathbf{0}} w_h\left(\tilde{\mathbf{r}}'\right)\tilde{\mathbf{r}}'\, d^n\tilde{\mathbf{r}}'\end{aligned} \tag{5.9}$$

where $\bar{\Omega}_\mathbf{0}$ is the mirror image of $\Omega_\mathbf{0}$ under central symmetry with respect to the origin $\mathbf{0}$. If $\Omega_\mathbf{0}$ is symmetrically invariant (i.e. if $\bar{\Omega}_\mathbf{0} = \Omega_\mathbf{0}$), the integral then equals its opposite, that is it is zero, and the condition (5.6) is satisfied.

A Taylor series expansion to a higher order than (5.3) would not give rise to more difficulties and would unsurprisingly lead to zero conditions of the m-th-order moments for a consistency with the corresponding order. Under the hypothesis (5.7), the kernel moments of odd orders are identically zero, for the same reasons as above. More generally, we have:

$$\int_{\Omega} F\left(\tilde{r}\right) w_h\left(\tilde{\mathbf{r}}\right) \underbrace{\tilde{\mathbf{r}}\otimes\ldots\otimes\tilde{\mathbf{r}}}_{2m+1} d^n\tilde{\mathbf{r}} = \mathbf{0} \tag{5.10}$$

for every natural integer m and every function F of the norm $\tilde{r} = \left|\tilde{\mathbf{r}}\right|$ of $\tilde{\mathbf{r}}$, because F is then even with respect to the vector $\tilde{\mathbf{r}}$. On the other hand, this procedure has no effect on the moments of even order of w_h, especially on the last but one integral of (5.4). The second-order moment, for instance, which will sometimes be useful in the following, is a second-order tensor which is denoted here as

$$\mathbf{M}_{2,w} \doteqdot \int_{\Omega_\mathbf{0}} w_h\left(\tilde{\mathbf{r}}\right)\tilde{\mathbf{r}}\otimes\tilde{\mathbf{r}}\, d^n\tilde{\mathbf{r}} \tag{5.11}$$

(its components will be calculated in the next section). Of course, it can only be zero if the kernel gets negative values over some part of its support, which is troublesome from a physical point of view, as explained later on. Thus, the property (5.7), which will from now on presumably be true, only provides the consistency of continuous interpolation to the first order in h.

Owing to the results we have just come to, (5.4) is reduced to

$$[A]_c(\mathbf{r},t) = A(\mathbf{r},t) + \frac{1}{2}\frac{\partial^2 A}{\partial \mathbf{r}^2} : \mathbf{M}_{2,w} + O\left(h^4\right) \qquad (5.12)$$
$$= A(\mathbf{r},t) + O\left(h^2\right)$$

We are going to take a closer look at the way the continuous interpolation procedure we have introduced behaves with respect to the first-order derivatives of a scalar field. To that purpose, let us apply the approximation (5.2) to the field **grad** A:

$$[\mathbf{grad}\, A]_c(\mathbf{r},t) = \int_{\Omega_\mathbf{r}} \frac{\partial A(\mathbf{r}',t)}{\partial \mathbf{r}'} w_h(\mathbf{r}-\mathbf{r}')\, d^n\mathbf{r}'$$
$$= \int_{\Omega_\mathbf{r}} \frac{\partial}{\partial \mathbf{r}'}\left[A(\mathbf{r}',t)\, w_h(\mathbf{r}-\mathbf{r}')\right] d^n\mathbf{r}' \qquad (5.13)$$
$$- \int_{\Omega_\mathbf{r}} A(\mathbf{r}',t)\, \frac{\partial w_h(\mathbf{r}-\mathbf{r}')}{\partial \mathbf{r}'} d^n\mathbf{r}'$$

(we have written the gradients like vectorial derivatives to lay emphasis on the derivation variable $\mathbf{r}'$). We may subsequently notice the following identity:

$$\frac{\partial w_h(\mathbf{r}-\mathbf{r}')}{\partial \mathbf{r}'} = -\frac{\partial w_h(\mathbf{r}-\mathbf{r}')}{\partial \mathbf{r}} \qquad (5.14)$$

Besides, the first integral of (5.13) may be turned into a surface integral through the Gauss theorem (refer to Appendix A), whereas the property (5.14) allows us to change the sign of the second integral and yield:

$$[\mathbf{grad}\, A]_c(\mathbf{r},t) = \oint_{\partial\Omega_\mathbf{r}} A(\mathbf{r}',t)\, w_h(\mathbf{r}-\mathbf{r}')\, \mathbf{n}(\mathbf{r}')\, d^{n-1}\Gamma$$
$$+ \int_{\Omega_\mathbf{r}} A(\mathbf{r}',t)\, \frac{\partial w_h(\mathbf{r}-\mathbf{r}')}{\partial \mathbf{r}} d^n\mathbf{r}' \qquad (5.15)$$

$\mathbf{n}(\mathbf{r}')$ being the unit vector normal to $\partial\Omega_\mathbf{r}$ at point $\mathbf{r}'$ and directed towards the outside (the first integration is made on the element of $(n-1)$ −surface $d^{n-1}\Gamma$). If the kernel has an infinite support (e.g. Gaussian), the surface integral only certainly vanishes in the case of a fluid which is infinite in all directions (which almost never occurs), or even when the field A has a compact support (which seldom occurs). Moreover, with a kernel having a compact support, the surface integral is reduced to the intersection of $\Omega_\mathbf{r}$ with $\partial\Omega$, that is $\partial\Omega \cap \Omega_\mathbf{r}$ (refer to Fig. 5.1). The same applies to the volume integral. Once again as regards a compact support kernel, if the calculation point $\mathbf{r}$ is sufficiently far off $\partial\Omega$ (Fig. 5.1), the boundary integral vanishes and $\Omega \cap \Omega_\mathbf{r} = \Omega_\mathbf{r}$, which ultimately yields

$$[\mathbf{grad}\, A]_c(\mathbf{r},t) = \int_{\Omega_\mathbf{r}} A(\mathbf{r}',t)\, \nabla w_h(\mathbf{r}-\mathbf{r}')\, d^n\mathbf{r}' \qquad (5.16)$$

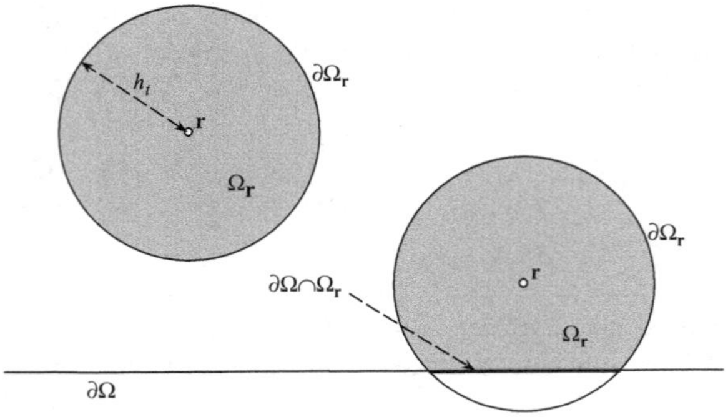

Fig. 5.1 Continuous approximation of a field by means of a compact support kernel.

Formula (5.16) defines a continuously interpolated gradient operator. We may note that the interpolation (5.16) is identical to that of the initial function (5.2), but using the kernel gradient as an interpolation function. That relation could be directly obtained deriving the original formula (5.2), on the basis of the previously mentioned properties:

$$\begin{aligned}\mathbf{grad}\,[A]_c\,(\mathbf{r},t) &= \frac{\partial}{\partial \mathbf{r}}\int_{\Omega_{\mathbf{r}}} A\left(\mathbf{r}',t\right) w_h\left(\mathbf{r}-\mathbf{r}'\right) d^n\mathbf{r}' \\ &= \int_{\Omega_{\mathbf{r}}} A\left(\mathbf{r}',t\right) \frac{\partial}{\partial \mathbf{r}} w_h\left(\mathbf{r}-\mathbf{r}'\right) d^n\mathbf{r}' \\ &= -\int_{\Omega_{\mathbf{r}}} A\left(\mathbf{r}',t\right) \frac{\partial}{\partial \mathbf{r}'} w_h\left(\mathbf{r}-\mathbf{r}'\right) d^n\mathbf{r}' \\ &= \int_{\Omega_{\mathbf{r}}} \frac{\partial A\left(\mathbf{r}',t\right)}{\partial \mathbf{r}'} w_h\left(\mathbf{r}-\mathbf{r}'\right) d^n\mathbf{r}' \\ &= [\mathbf{grad}\,A]_c\,(\mathbf{r},t)\end{aligned} \tag{5.17}$$

which means that the **grad** and $[\cdot]_c$ operators commute.

Applying the gradient operator to (5.12), we immediately get:

$$\mathbf{grad}\,[A]_c\,(\mathbf{r},t) = (\mathbf{grad}\,A)\,(\mathbf{r},t) + O\left(h^2\right) \tag{5.18}$$

Another demonstration of that result can be achieved through the previously adopted procedure, that is inserting (5.3) into (5.16):

$$\begin{aligned}[\mathbf{grad}\,A]_c\,(\mathbf{r},t) = {} & A\,(\mathbf{r},t) \int_{\Omega_{\mathbf{r}}} \nabla w_h\left(\tilde{\mathbf{r}}\right) d^n\mathbf{r}' \\ & - \left[\int_{\Omega_{\mathbf{r}}} \nabla w_h\left(\tilde{\mathbf{r}}\right) \otimes \tilde{\mathbf{r}}\, d^n\mathbf{r}'\right] \frac{\partial A}{\partial \mathbf{r}} \\ & + \frac{1}{2}\left[\int_{\Omega_{\mathbf{r}}} \nabla w_h\left(\tilde{\mathbf{r}}\right) \otimes \tilde{\mathbf{r}} \otimes \tilde{\mathbf{r}}\, d^n\mathbf{r}'\right] : \frac{\partial^2 A}{\partial \mathbf{r}^2} + O\left(h^2\right)\end{aligned} \tag{5.19}$$

(the integral giving the error is an $O\left(h^2\right)$ because $\nabla w_h\left(\tilde{\mathbf{r}}\right) d^n\mathbf{r}'$ is on the order of $1/h$). The conditions to be fulfilled for a second-order accuracy of the $[\mathbf{grad}\,A]_c \approx \mathbf{grad}\,A$ approximation are then

$$\int_{\Omega_0} \nabla w_h\left(\tilde{\mathbf{r}}\right) d^n\tilde{\mathbf{r}} = \mathbf{0}$$
$$\int_{\Omega_0} \nabla w_h\left(\tilde{\mathbf{r}}\right) \otimes \tilde{\mathbf{r}}\, d^n\tilde{\mathbf{r}} = -\mathbf{I}_n \tag{5.20}$$
$$\int_{\Omega_0} \nabla w_h\left(\tilde{\mathbf{r}}\right) \otimes \tilde{\mathbf{r}} \otimes \tilde{\mathbf{r}}\, d^n\tilde{\mathbf{r}} = \mathbf{0}$$

where $\mathbf{I}_n$ is the n-dimensional identity tensor,[7] characterizing the problem. The first condition is satisfied with the property (5.8). This is because the change of variable $\tilde{\mathbf{r}}' = -\tilde{\mathbf{r}}$ gives

[7] We may recall that we are contemplating the general case of an arbitrary dimension, contrary to the first section of the book in which, for physical reasons, we were practically limited to $n = 3$.

$$\begin{aligned}\int_{\Omega_0} \nabla w_h\left(\tilde{\mathbf{r}}\right) d^n\tilde{\mathbf{r}} &= -\int_{\Omega_0} \nabla w_h\left(-\tilde{\mathbf{r}}\right) d^n\tilde{\mathbf{r}} \\ &= -\int_{\Omega_0} \nabla w_h\left(\tilde{\mathbf{r}}'\right) d^n\tilde{\mathbf{r}}' = \mathbf{0}\end{aligned} \tag{5.21}$$

And, likewise:

$$\int_{\Omega_0} \nabla w_h\left(\tilde{\mathbf{r}}\right) \otimes \tilde{\mathbf{r}} \otimes \tilde{\mathbf{r}}\, d^n\tilde{\mathbf{r}} = -\int_{\Omega_0} \nabla w_h\left(\tilde{\mathbf{r}}\right) \otimes \tilde{\mathbf{r}} \otimes \tilde{\mathbf{r}}\, d^n\tilde{\mathbf{r}} = \mathbf{0} \tag{5.22}$$

Conversely, this procedure does not make it possible to draw a conclusion about the second integral of (5.20). We may, however, modify it and write:

$$\begin{aligned}\int_{\Omega_0} \nabla w_h\left(\tilde{\mathbf{r}}\right) \otimes \tilde{\mathbf{r}}\, d^n\tilde{\mathbf{r}} &= \int_{\Omega_0} \left[\frac{\partial}{\partial \tilde{\mathbf{r}}}\left[w_h\left(\tilde{\mathbf{r}}\right)\tilde{\mathbf{r}}\right] - w_h\left(\tilde{\mathbf{r}}\right)\left(\frac{\partial \tilde{\mathbf{r}}}{\partial \tilde{\mathbf{r}}}\right)^T\right] d^n\tilde{\mathbf{r}} \\ &= \oint_{\partial\Omega_0} w_h\left(\tilde{\mathbf{r}}\right)\tilde{\mathbf{r}} \otimes \mathbf{n}\left(\tilde{\mathbf{r}}\right) d^{n-1}\Gamma \\ &\quad - \left(\int_{\Omega_0} w_h\left(\tilde{\mathbf{r}}\right) d^n\tilde{\mathbf{r}}\right)\mathbf{I}_n\end{aligned} \tag{5.23}$$

(once again, we have used the Gauss theorem and the fact that $\partial\tilde{\mathbf{r}}/\partial\tilde{\mathbf{r}} = \mathbf{I}_n$). The surface integral is zero in this calculation because w_h vanishes at the boundaries of its support if the latter does not intersect $\partial\Omega$, whereas the property (5.5) ultimately provides for the fulfillment of the second condition of (5.20). Lastly, the estimate (5.19) is reduced to (5.18).

The m-th-order consistency of the interpolation of gradient (5.16) would, of course, imply the nullity of the corresponding order moments, the definitions of which are analogous to (5.20). A procedure like the one above would show that, under the selected conditions, all the even-order moments of the kernel gradient equal zero, and furthermore that

$$\int_{\Omega} F\left(\tilde{r}\right) \nabla w_h\left(\tilde{\mathbf{r}}\right) \otimes \underbrace{\tilde{\mathbf{r}} \otimes \ldots \otimes \tilde{\mathbf{r}}}_{2m} d^n\tilde{\mathbf{r}} = \mathbf{0} \tag{5.24}$$

for every natural integer m and every function F (which may be compared with (5.10)). On the contrary, nothing can a priori be stated about the odd-order moments of ∇w_h. An integration by parts similar to (5.23) would lead, for instance, for the third-order moment of ∇w_h at a point within the domain, to

$$\int_{\Omega_0} \nabla w_h\left(\tilde{\mathbf{r}}\right) \otimes \tilde{\mathbf{r}} \otimes \tilde{\mathbf{r}} \otimes \tilde{\mathbf{r}}\, d^n\tilde{\mathbf{r}} = -\left[\int_{\Omega_0} w_h\left(\tilde{\mathbf{r}}\right) \tilde{\mathbf{r}} \otimes \tilde{\mathbf{r}}\, d^n\tilde{\mathbf{r}}\right] \otimes \mathbf{I}_n \qquad (5.25)$$

Now, the right-hand integral is the second-order moment of $w_h$which, as previously explained, is not necessarily zero.

To sum up our calculations, we can state that the interpolation method based upon the use of a kernel is consistent at the first order inclusively for both scalar fields and assessment of scalar gradients.[8] That property rests on the conditions (5.5) and (5.7) (which yields (5.8)), and then holds true for all such points as $\partial\Omega \cap \Omega_{\mathbf{r}} = \varnothing$ within the domain.

[8] These two results are coherent with each other, since every scalar may be considered as a component of the gradient of another scalar.

Similar arguments can be developed for the vector fields $\mathbf{A}$. In particular, we get:

$$\mathbf{A}(\mathbf{r}, t) \approx [\mathbf{A}]_c(\mathbf{r}, t) \doteqdot \int_{\Omega_{\mathbf{r}}} \mathbf{A}\left(\mathbf{r}', t\right) w_h\left(\mathbf{r} - \mathbf{r}'\right) d^n\mathbf{r}' \qquad (5.26)$$

with a first-order consistency, subject to the conditions (5.5) and (5.7) for a point which is sufficiently far off the boundary. This result does not require any demonstration, since the above demonstrated properties remain true for each component of the field $\mathbf{A}$. We also may write, under the same conditions as previously, a continuously interpolated divergence operator:

$$\begin{aligned}
[\mathrm{div}\mathbf{A}]_c(\mathbf{r}, t) &= \int_{\Omega_{\mathbf{r}}} \frac{\partial}{\partial \mathbf{r}'} \cdot \mathbf{A}\left(\mathbf{r}', t\right) w_h\left(\mathbf{r} - \mathbf{r}'\right) d^n\mathbf{r}' \\
&= \int_{\Omega_{\mathbf{r}}} \frac{\partial}{\partial \mathbf{r}'} \cdot \left[\mathbf{A}\left(\mathbf{r}', t\right) w_h\left(\mathbf{r} - \mathbf{r}'\right)\right] d^n\mathbf{r}' \\
&\quad - \int_{\Omega_{\mathbf{r}}} \mathbf{A}\left(\mathbf{r}', t\right) \cdot \frac{\partial w_h\left(\mathbf{r} - \mathbf{r}'\right)}{\partial \mathbf{r}'} d^n\mathbf{r}'
\end{aligned} \qquad (5.27)$$

The first integral vanishes pursuant to the Gauss theorem and the second one is antisymmetric due to the change of $\mathbf{r}'$ into $\mathbf{r}$, so that[9]

[9] Developing that procedure, it would be shown that the div and $[\cdot]_c$ operators commute.

$$[\mathrm{div}\mathbf{A}]_c(\mathbf{r}, t) = \int_{\Omega_{\mathbf{r}}} \mathbf{A}\left(\mathbf{r}', t\right) \cdot \nabla w_h\left(\mathbf{r} - \mathbf{r}'\right) d^n\mathbf{r}' \qquad (5.28)$$

and

$$\begin{aligned}
[\mathrm{div}\mathbf{A}]_c(\mathbf{r}, t) &= \mathbf{A}(\mathbf{r}, t) \cdot \int_{\Omega_{\mathbf{r}}} \nabla w_h\left(\tilde{\mathbf{r}}\right) d^n\mathbf{r}' \\
&\quad - \left[\int_{\Omega_{\mathbf{r}}} \nabla w_h\left(\tilde{\mathbf{r}}\right) \otimes \tilde{\mathbf{r}}\, d^n\mathbf{r}'\right] : \frac{\partial \mathbf{A}}{\partial \mathbf{r}} \\
&\quad + \frac{1}{2}\left[\int_{\Omega_{\mathbf{r}}} \nabla w_h\left(\tilde{\mathbf{r}}\right) \otimes \tilde{\mathbf{r}} \otimes \tilde{\mathbf{r}}\, d^n\mathbf{r}'\right] \vdots \frac{\partial^2 \mathbf{A}}{\partial \mathbf{r}^2} + O\left(h^2\right)
\end{aligned} \qquad (5.29)$$

($\partial^2\mathbf{A}/\partial\mathbf{r}^2$ is a third-order tensor, here being contracted over its three subscripts with the tensor occurring under the integral; refer to Appendix A for further details about the tensor calculations). The conditions (5.20) then secures the first-order consistency as for the gradients.

A few additional remarks of a general nature are still to be made. The whole preceding discussion could readily be applied to the tensor fields and continuous differential operators of any order. It may be noticed that a linear operator remains linear after being interpolated that way. Lastly, it is noteworthy before carrying on with this chapter to mention a crucial property of the interpolated gradient and divergence operators. On the ground of the antisymmetry property (5.8), we can indeed write the scalar product of an arbitrary scalar field A times the interpolated divergence of a vector field $\mathbf{B}$, on the basis of equation (3.194) in Section 3.4.4:

$$\begin{aligned}(A, [\mathrm{div}\mathbf{B}]_c) &= \int_\Omega A(\mathbf{r}, t) \\ &\times \int_\Omega \mathbf{B}(\mathbf{r}', t) \cdot \nabla w_h(\mathbf{r} - \mathbf{r}')\, d^n\mathbf{r}' d^n\mathbf{r} \\ &= -\int_\Omega \mathbf{B}(\mathbf{r}', t) \cdot \int_\Omega A(\mathbf{r}, t) \nabla w_h(\mathbf{r}' - \mathbf{r})\, d^n\mathbf{r} d^n\mathbf{r}' \\ &= -\left\langle [\mathbf{grad}\, A]_c, \mathbf{B} \right\rangle\end{aligned} \tag{5.30}$$

Thus, the interpolated operators are skew-adjoint, like the standard operators (eqn (3.198)).

5.2.2 Kernel examples

The simplest way to secure the symmetry condition (5.7) is by making the kernel a function of the distance between spatial points:

$$w_h(\tilde{\mathbf{r}}) = w_h(|\tilde{\mathbf{r}}|) = w_h(\tilde{r}) \tag{5.31}$$

The continuous approximations (5.2), (5.16), (5.26) and (5.28) are now written, in view of the results of the previous section, as:

$$\begin{aligned}[A]_c(\mathbf{r}, t) &= \int_{\Omega_\mathbf{r}} A(\mathbf{r}', t)\, w_h(|\mathbf{r} - \mathbf{r}'|)\, d^n\mathbf{r}' \\ &= A(\mathbf{r}, t) + O\left(h^2\right) \\ [\mathbf{grad}\, A]_c(\mathbf{r}, t) &= \int_{\Omega_\mathbf{r}} A(\mathbf{r}', t) \nabla w_h(|\mathbf{r} - \mathbf{r}'|)\, d^n\mathbf{r}' \\ &= (\mathbf{grad}\, A)(\mathbf{r}, t) + O\left(h^2\right) \\ [\mathbf{A}]_c(\mathbf{r}, t) &= \int_{\Omega_\mathbf{r}} \mathbf{A}(\mathbf{r}', t)\, w_h(|\mathbf{r} - \mathbf{r}'|)\, d^n\mathbf{r}' \\ &= \mathbf{A}(\mathbf{r}, t) + O\left(h^2\right) \\ [\mathrm{div}\mathbf{A}]_c(\mathbf{r}, t) &= \int_{\Omega_\mathbf{r}} \mathbf{A}(\mathbf{r}', t) \cdot \nabla w_h(|\mathbf{r} - \mathbf{r}'|)\, d^n\mathbf{r}' \\ &= (\mathrm{div}\mathbf{A})(\mathbf{r}, t) + O\left(h^2\right)\end{aligned} \tag{5.32}$$

Through the hypothesis (5.31), we immediately find

$$\begin{aligned}\nabla w_h\left(\tilde{\mathbf{r}}\right) &= w_h'\left(\tilde{r}\right)\frac{\tilde{\mathbf{r}}}{\tilde{r}}\\ &= -w_h'\left(\left|-\tilde{\mathbf{r}}\right|\right)\frac{-\tilde{\mathbf{r}}}{\left|-\tilde{\mathbf{r}}\right|}\\ &= -\nabla w_h\left(-\tilde{\mathbf{r}}\right)\end{aligned} \tag{5.33}$$

where $w_h'\left(\tilde{r}\right)$ is the derivative of $w_h\left(\tilde{r}\right)$, being considered as a function of the scalar variable $\tilde{r}$. As expected, the antisymmetry requirement of the gradient (5.8) is then systematically satisfied. The property (5.24) is now written as

$$\int_{\Omega} F\left(\tilde{r}\right) w_h'\left(\tilde{r}\right) \otimes \underbrace{\tilde{\mathbf{r}} \otimes \ldots \otimes \tilde{\mathbf{r}}}_{2m+1} d^n\tilde{\mathbf{r}} = \mathbf{0} \tag{5.34}$$

As a consequence, the kernel support is also rotation invariant, that is occurring in the form of a $n-$sphere (n being, by definition, the dimension of the problem being addressed). We will explain in Section 5.3.3 a significant effect of this option. When the support is compact, the radius of that sphere will be denoted as h_t. In order to encompass the case of an infinite support in our discussion, however, we introduce[10] a *smoothing length* denoted h, characterizing the dependence of w_h with respect to $\tilde{r} = \left|\tilde{\mathbf{r}}\right|$. More specifically, wanting to write the kernel in a dimensionless form, we arbitrarily set h before making the following change of variable:

$$q \doteqdot \frac{\left|\tilde{\mathbf{r}}\right|}{h} = \frac{\left|\mathbf{r}-\mathbf{r}'\right|}{h} \tag{5.35}$$

Furthermore, the discussion in the previous section (e.g. eqn (5.2)) show that w_h is homogeneous[11] to a length raised to the $(-n)$-th power. We then define the kernel as follows:

$$w_h\left(\tilde{r}\right) = \frac{\alpha_{w,n}}{h^n} f\left(q\right) \tag{5.36}$$

where f is a function defined over $\mathbb{R}^+$, which is positive,[12] sufficiently regular, that is at least once continuously derivable, and $\alpha_{w,n}$ is a dimensionless constant depending on the kernel and the space dimension, which is calculated in such a way that the normalization condition (5.5) is met, which is now written as

$$\alpha_{w,n} \int_{\Omega_{\mathbf{q}}} f\left(q\right) d^n\mathbf{q} = 1 \tag{5.37}$$

where $\mathbf{q}$ is the vector $\tilde{\mathbf{r}}/h$ (with a norm q) and $\Omega_{\mathbf{q}}$ stands for the domain defined by all possible values of $\mathbf{q}$. We may notice that the positive nature of f makes $\alpha_{w,n}$ positive too. The domain $\Omega_{\mathbf{q}}$ is a $n-$sphere, whose $n-$volume element equals $d^n\mathbf{q} = q^{n-1}dq\,d\omega$, $d\omega$ being the infinitesimal $n-$angle[13] and so the condition (5.37) may also be written as

$$\alpha_{w,n} = \left[S_n \int_0^{R_f} f\left(q\right) q^{n-1} dq\right]^{-1} \tag{5.38}$$

[10] In this book, we consider a constant smoothing length h, in view of the nearly incompressible flows, for which the inter-particular distance then only slightly varies (refer to Section 5.2.3). The effect of a variable h was investigated by Bonet and Rodriguez-Paz (2005), among others.

[11] Whereas ∇w_h (or w_h') is in $r^{-(n+1)}$.

[12] Its positive nature is a priori unnecessary at this stage of our discussion, but it will later on be found that it is the best physical option.

[13] It is an ordinary angle if $n = 2$, a solid angle if $n = 3$.

where R_f is the (possibly infinite) radius of the $n-$sphere making up the support of the function f, and S_n is the surface area of the $n-$sphere of unit radius, that is

$$S_n \doteq \frac{2\pi^{n/2}}{\Gamma\left(\frac{n}{2}\right)} \tag{5.39}$$

Γ being the Euler function of the second kind (refer to Kendall 1961). In particular, we get

$$\begin{aligned} S_1 &= 2 \\ S_2 &= 2\pi \\ S_3 &= 4\pi \end{aligned} \tag{5.40}$$

As regards a kernel with a compact support of radius h_t, we obviously get:

$$R_f = \frac{h_t}{h} \tag{5.41}$$

The gradient of w_h will henceforth be expressed with respect to the derivative $f'$of f:

$$\begin{aligned} \nabla w_h(\tilde{r}) &= \frac{\alpha_{w,n}}{h^n}\frac{\partial f(q)}{\partial \mathbf{r}} \\ &= \frac{\alpha_{w,n}}{h^n}\frac{df(q)}{dq}\frac{1}{h}\frac{\partial\left|\mathbf{r}-\mathbf{r}'\right|}{\partial \mathbf{r}} \\ &= \frac{\alpha_{w,n}}{h^{n+1}}f'(q)\frac{\tilde{\mathbf{r}}}{\tilde{r}} \end{aligned} \tag{5.42}$$

which gives a relation between the derivatives of the dimensional and dimensionless kernels, along the lines of (5.36):

$$w_h'(\tilde{r}) = \frac{\alpha_{w,n}}{h^{n+1}}f'(q) \tag{5.43}$$

The higher order derivatives obey an identical law:

$$w_h^{(k)}(\tilde{r}) = \frac{\alpha_{w,n}}{h^{n+k}}f^{(k)}(q) \tag{5.44}$$

There are many examples of kernels and we do not intend to review all of them; we will just refer to some of the most popular ones.[14] The most natural choice, as mentioned at the beginning of the previous section, is provided by a Gaussian function $f = f_G$:

[14] A more thorough review can be found in Liu and Liu (2003).

$$\begin{aligned} f_G(q) &\doteq e^{-q^2} \\ f_G'(q) &= -2qe^{-q^2} \end{aligned} \tag{5.45}$$

In this case, the support is infinite. The normalization constants $\alpha_{w,n} = \alpha_{G,n}$ equal, according to (5.38) (with $R_f = R_G = +\infty$):

$$\alpha_{G,n} = \frac{\Gamma\left(\frac{n}{2}\right)}{2\pi^{n/2}\int_0^{+\infty} e^{-q^2} q^{n-1} dq} = \frac{1}{\pi^{n/2}} \tag{5.46}$$

We will subsequently find that one can conveniently work on kernels with compact supports, frequent examples of which are given here, namely the *B-splines* (refer, for instance, to Monaghan, 1985a). An M-degree B-spline is a piecewise polynomial function of degree M, defined over $\mathbb{R}^+$ by

$$f_M(q) \doteqdot \frac{1}{M!}\sum_{k=0}^{M+1} (-1)^k \binom{M+1}{k} \times \max\left(0; q + \frac{M+1}{2} - k\right)^M \tag{5.47}$$

where the brackets stand for the binomial coefficients. The size of its support is $(M+1)/2$. It was demonstrated (De Boor, 2001) that f_M is the most regular piecewise polynomial function for a support with such a size. The B-spline of order $M = 3$ (very commonly used in the literature dealing with the SPH method, though it is rather inaccurate) is of degree 3 and has a support $R_f = R_3 = 2$, and is written as

$$f_3(q) = \begin{cases} 1 - \frac{3}{2}q^2 + \frac{3}{4}q^3 & 0 \le q \le 1 \\ \frac{1}{4}(2-q)^3 & 1 < q \le 2 \\ 0 & 2 < q \end{cases} \tag{5.48}$$

The normalization condition provides the constants $\alpha_{w,n} = \alpha_{M,n}$ in the form[15]

$$\alpha_{3,1} = \frac{2}{3}, \quad \alpha_{3,2} = \frac{10}{7\pi}, \quad \alpha_{3,3} = \frac{1}{\pi} \tag{5.49}$$

Its derivative is given by

$$f_3'(q) = \begin{cases} -3q + \frac{9}{4}q^2 & 0 \le q \le 1 \\ -\frac{3}{4}(2-q)^2 & 1 < q \le 2 \\ 0 & 2 < q \end{cases} \tag{5.50}$$

The B-spline of order $M = 4$ (of degree 4 and with a support $R_f = R_4 = 5/2$) is provided by

[15] The first sub-script of $\alpha_{M,n}$ here refers to the order M of the spline; the second one refers to the spatial dimension n.

$$f_4(q) = \begin{cases} \left(\frac{5}{2}-q\right)^4 - 5\left(\frac{3}{2}-q\right)^4 + 10\left(\frac{1}{2}-q\right)^4 & 0 \le q \le 0.5 \\ \left(\frac{5}{2}-q\right)^4 - 5\left(\frac{3}{2}-q\right)^4 & 0.5 < q \le 1.5 \\ \left(\frac{5}{2}-q\right)^4 & 1.5 < q \le 2.5 \\ 0 & 2.5 < q \end{cases} \tag{5.51}$$

The normalization condition yields

$$\alpha_{4,1} = \frac{1}{24} \tag{5.52}$$
$$\alpha_{4,2} = \frac{96}{1199\pi}$$
$$\alpha_{4,3} = \frac{1}{20\pi}$$

Its derivative is given by

$$f_4'(q) = -4\begin{cases} \left(\frac{5}{2}-q\right)^3 - 5\left(\frac{3}{2}-q\right)^3 + 10\left(\frac{1}{2}-q\right)^3 & 0 \le q \le 0.5 \\ \left(\frac{5}{2}-q\right)^3 - 5\left(\frac{3}{2}-q\right)^3 & 0.5 < q \le 1.5 \\ \left(\frac{5}{2}-q\right)^3 & 1.5 < q \le 2.5 \\ 0 & 2.5 < q \end{cases} \tag{5.53}$$

We will also consider the B-spline of order $M = 5$, with a support $R_f = R_5 = 3$:

$$f_5(q) = \begin{cases} (3-q)^5 - 6(2-q)^5 + 15(1-q)^5 & 0 \le q \le 1 \\ (3-q)^5 - 6(2-q)^5 & 1 < q \le 2 \\ (3-q)^5 & 2 < q \le 3 \\ 0 & 3 < q \end{cases} \tag{5.54}$$

with

$$\alpha_{5,1} = \frac{1}{120}$$
$$\alpha_{5,2} = \frac{7}{478\pi} \tag{5.55}$$
$$\alpha_{5,3} = \frac{1}{120\pi}$$

and

$$f_5'(q) = -5\begin{cases} (3-q)^4 - 6(2-q)^4 + 15(1-q)^4 & 0 \le q \le 1 \\ (3-q)^4 - 6(2-q)^4 & 1 < q \le 2 \\ (3-q)^4 & 2 < q \le 3 \\ 0 & 3 < q \end{cases} \tag{5.56}$$

Lastly, we may mention the Wendland kernels (1995), which are advantageously uniform (i.e. based upon one algebraic equation) while having compact supports. Here we provide only one, of order 4, with $R_f = R_W = 2$:

$$f_W(q) = \begin{cases} \left(1 - \frac{q}{2}\right)^4 (1 + 2q) & 0 \le q \le 2 \\ 0 & 2 < q \end{cases} \tag{5.57}$$

The normalization constants equal

$$\alpha_{W,1} = \frac{3}{4}$$
$$\alpha_{W,2} = \frac{7}{4\pi} \tag{5.58}$$
$$\alpha_{W,3} = \frac{21}{16\pi}$$

and

$$f'_W(q) = \begin{cases} -5q\left(1 - \frac{q}{2}\right)^3 & 0 \le q \le 2 \\ 0 & 2 < q \end{cases} \tag{5.59}$$

The $\alpha_{w,n} f(q)$ and $\alpha_{w,n} f'(q)$ functions are illustrated in Fig. 5.2, for $n = 2$ and $n = 3$. As can be seen, they get their maximum at zero and are decreasing, whereas their derivatives (which are consequently negative) get a minimum before increasing towards zero. Determining the behaviour of f' near zero is interesting. We find, for the selected three examples:

$$f_G(q) \underset{q \to 0}{\sim} 1 - q^2 \tag{5.60}$$

$$f_3(q) \underset{q \to 0}{\sim} 1 - \frac{3}{2}q^2$$

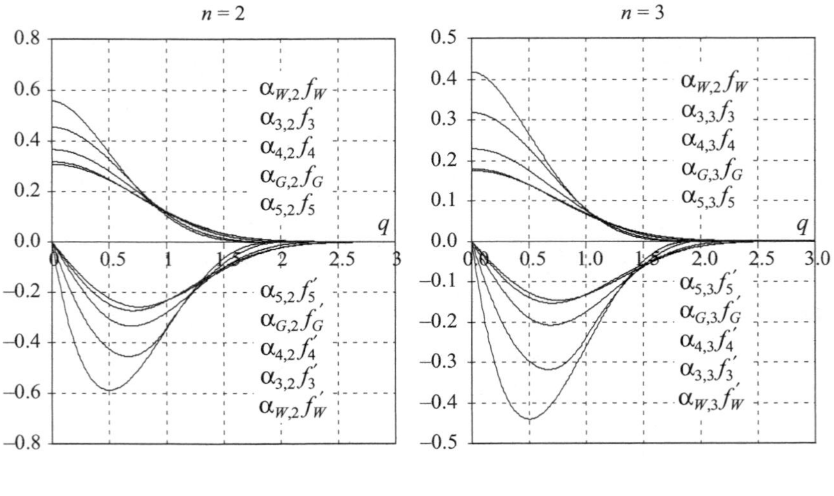

Fig. 5.2 Graphs of the kernels used in the present section and of their derivatives. For each value of the dimension (n) (=2 and 3), the kernels are positive, while their derivatives are negative. The curves are displayed in the same order as in the legend.

$$f_4(q) \underset{q\to 0}{\sim} \frac{115}{8} - 15q^2$$

$$f_5(q) \underset{q\to 0}{\sim} 66 - 60q^2$$

$$f_W(q) \underset{q\to 0}{\sim} 16 - 40q^2$$

We can see that all the first derivatives $f'(0)$ are zero, whereas the second derivatives $f''(0)$ are strictly negative. These kernels are then 'bell-shaped'. We may note that the spline of order 5 looks much like the Gaussian kernel.

We are now going to calculate the components of the second-order moment of the kernel as given by (5.11), a tensor whose components equal

$$M_{2,w,ij} = \int_{\Omega_0} w_h(\tilde{r})\, \tilde{x}_i \tilde{x}_j d^n\tilde{\mathbf{r}} \tag{5.61}$$

where the $\tilde{x}_i$ are the components of $\tilde{\mathbf{r}}$. As a rule, we can first estimate

$$\int_{\Omega_0} F(\tilde{r})\, \tilde{x}_i \tilde{x}_j d^n\tilde{\mathbf{r}} \tag{5.62}$$

where F is an arbitrary function. Changing $\tilde{x}_i$ into $-\tilde{x}_i$ will, of course, leave the integral unchanged, since the sphere is invariant under all the reflections around planes containing the origin. Thus, only those integrals of (5.61) corresponding to $i = j$ do not vanish. An isotropic argument shows that the latter (which are n in number) are equal, and their sum equals the integral of $F(\tilde{r})\,\tilde{r}^2$, which gives:

$$\int_{\Omega_0} F(\tilde{r})\, \tilde{\mathbf{r}} \otimes \tilde{\mathbf{r}}\, d^n\tilde{\mathbf{r}} = \frac{1}{n}\left[\int_{\Omega_0} F(\tilde{r})\, \tilde{r}^2 d^n\tilde{\mathbf{r}}\right] \mathbf{I}_n \tag{5.63}$$

In particular:

$$\mathbf{M}_{2,w} = \frac{M_{2,w}}{n} \mathbf{I}_n \tag{5.64}$$

with

$$\begin{aligned} M_{2,w} &\doteqdot \int_{\Omega_0} w_h(\tilde{r})\, \tilde{r}^2 d^n\tilde{\mathbf{r}} \\ &= C_{w,2} h^2 \end{aligned} \tag{5.65}$$

and

$$C_{w,2} \doteqdot \alpha_{w,n} S_n \int_0^{R_f} f(q)\, q^{n+1} dq \tag{5.66}$$

From the above, the estimation (5.12) may be written as

$$\begin{aligned} [A]_c(\mathbf{r}, t) &= A(\mathbf{r}, t) + \frac{C_{w,2}}{2n} \mathrm{tr}\left(\frac{\partial^2 A}{\partial \mathbf{r}^2}\right) h^2 + O\left(h^4\right) \\ &= A(\mathbf{r}, t) + \frac{C_{w,2}}{2n} \nabla^2 A(\mathbf{r}, t)\, h^2 + O\left(h^4\right) \end{aligned} \tag{5.67}$$

We will give the current value for the Gaussian kernel (5.45). Due to (5.46), we get:

$$C_{G,2} = \alpha_{G,n} S_n \int_0^{+\infty} e^{-q^2} q^{n+1} dq = \frac{\Gamma\left(\frac{n}{2}+1\right)}{\Gamma\left(\frac{n}{2}\right)} = \frac{n}{2} \tag{5.68}$$

that is with (5.64):

$$\mathbf{M}_{2,G} = \frac{h^2}{2}\mathbf{I}_n \tag{5.69}$$

Lastly, we will sometimes resort in the following (particularly in Section 5.4.4) to the Fourier transform of the kernel (refer to Appendix B):

$$\widehat{w_h}(\mathbf{K}) \doteq \int_{\mathbb{R}^n} w(\tilde{r}) \exp\left(-i\mathbf{K}\cdot\tilde{\mathbf{r}}\right) d^n\tilde{\mathbf{r}} \tag{5.70}$$

where $\mathbf{K}$ is an arbitrary wave vector. In the dimension $n = 1$, the transform of $w_h(\tilde{r})$ will obviously become identical to that of the dimensionless kernel $f(q)$:

$$\widehat{w_h}(K) = \alpha_{w,1}\widehat{f}\left(K^+\right) \tag{5.71}$$

with[16]

$$\widehat{f}\left(K^+\right) = 2\int_0^{+\infty} f(q)\cos\left(K^+ q\right) dq \tag{5.72}$$

and

$$K^+ \doteq hK \tag{5.73}$$

The transforms $\widehat{w_h}$ and $\widehat{f}$ are dimensionless. The B-spline transform (5.47) is provided by

$$\widehat{f_M}\left(K^+\right) = \frac{1}{\alpha_{M,1}}\left(\frac{2}{K^+}\sin\frac{K^+}{2}\right)^{M+1} \tag{5.74}$$

with the limiting value at the origin $\widehat{f_M}(0) = 1$ (refer to Appendix B, eqn (B.64)). The Fourier transform of the Wendland kernel (5.57) is given by

$$\widehat{f_W}(K) = \frac{30}{K^{+6}}\left(K^{+2} + \frac{1}{2}K^+\sin 2K^+ - 2\sin^2 K^+\right) \tag{5.75}$$

with the limiting value at the origin $\widehat{f_W}(0) = 4/3$.

As regards an arbitrary isotropic kernel $w_h(\tilde{r})$ of any dimension, it is well-known that the Fourier transform is only a function of the norm K of the wave vector (refer to Appendix B):

$$\widehat{w_h}(\mathbf{K}) = \widehat{w_h}(K) \tag{5.76}$$

[16] The factor 2 in (5.72) corresponds to S_1 (refer to equation (5.40)). This is because f is only defined over $\mathbb{R}^+$, whereas this integration should be made over $\mathbb{R}$, by analogy with (5.38). It is presently considered that f is even, in order to remain consistent with the fact that the kernel is a function of the distance between two points. In addition, that property enables us to calculate the Fourier transforms by substituting $\cos\left(K^+q\right)$ for the term $\exp\left(-iK^+q\right)$, which therefore turns them into real functions.

There is a general relation linking the multi-dimensional Fourier transform of a radial function $w_h(\tilde{r})$ to the transform of the function of a real variable $f(q)$ (refer to eqn (B.46) of Appendix B). It only seldom leads, however, to simple algebraic results. Nevertheless, in the case of a Gaussian kernel, the Fourier transform of a Gaussian being a Gaussian (eqn (B.51)) is of great help. We may write, for an arbitrary dimension n, using (5.46):

$$\begin{aligned} \widehat{w_h}(K) &= \frac{\alpha_{G,n}}{h^n} \widehat{\exp(-q^2)}(K) \\ &= \frac{\alpha_{G,n}}{h^n} \left(\sqrt{\pi}h\right)^n \exp\left[-\frac{(Kh)^2}{4}\right] \\ &= \exp\left[-\left(\frac{K^+}{2}\right)^2\right] \end{aligned} \tag{5.77}$$

For $n = 1$, the relation (5.71) yields, with (5.46):

$$\widehat{f_G}\left(K^+\right) = \frac{\widehat{w_h}(K)}{\alpha_{G,1}} = \sqrt{\pi} \exp\left[-\left(\frac{K^+}{2}\right)^2\right] \tag{5.78}$$

In the case of a Gaussian kernel, regardless of the dimension, we may note that $\widehat{w_h}(K)$ is, like $w_h(\tilde{r})$, positive and decreasing. Furthermore, it is lower than one. These conditions, which will be efficiently utilized in Section 5.4.4, are far from insignificant. It can readily be verified that (5.75) satisfies them too, whereas the sign and the growth of (5.74) remain undetermined. The functions $\alpha_{w,1}\widehat{f}\left(K^+\right)$ are illustrated in Fig. 5.3 and show that the splines

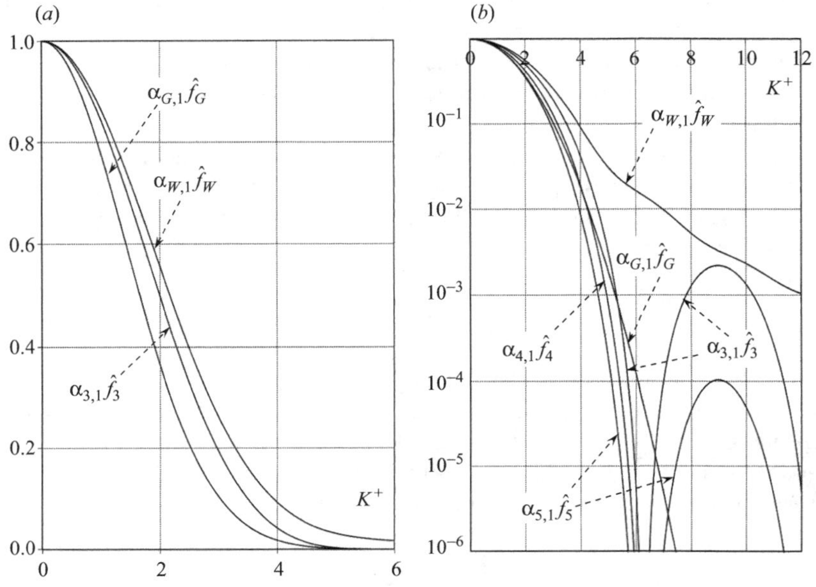

Fig. 5.3 Graphs of the Fourier transforms of the 1-dimensional kernels introduced in this paragraph, on linear (*a*) and logarithmic (*b*) scales. The B-splines of orders 4 and 5 are not illustrated on the graph (*a*).

do have decreasing Fourier transforms near the origin before being subject to oscillations for sufficiently high values of the dimensionless wave number K^+. Additional information about multi-dimensional kernel Fourier transforms can be found in Dehnen and Aly (2012) and Violeau and Leroy (2014).

5.2.3 Particles and discrete interpolation

The basic idea in the SPH method consists of modelling a continuous medium, considering it as a collection of particles as defined in Chapters 1 and 2. From a dynamic point of view, the particles then behave like material points (refer to Section 1.3.1) but actually are macroscopic entities,[17] bearing kinematic (velocity $\mathbf{u}_a$, etc.) and thermodynamic (pressure p_a, etc.) quantities A_a defined as in Chapter 2, and moving over time. In particular, the position $\mathbf{r}_a$ will be defined as the centre of inertia of a set of molecules having a fixed mass m_a,[18] whereas the volume V_a and the density ρ_a are linked by the second line in (1.254), which we recall here:

$$V_a = \frac{m_a}{\rho_a} \tag{5.79}$$

Since we are dealing with weakly compressible flows, it may be considered that each particle is featured by a rather stable specific size or diameter. Thus, it may be considered that the particles volumes are very close to a mean value[19] V_p. The mean diameter δr of a particle can then be introduced:

$$\delta r \doteq V_p^{1/n} \tag{5.80}$$

If the particles lie, upon the initial point in time, at the vertices of a rectangular grid, then δr also stands for the initial interparticular distance, that is the grid size. This item is usually chosen first, and the formulas (5.80) and (5.79) are used for deriving the volume, then the mass of the particles, which in this case is constant for each particle, if the density is constant as well. In the case of a flow involving several fluids, the particle masses are constant for each fluid.

The $h/\delta r$ ratio, where h is the smoothing length of the kernel defined in the previous section, clearly plays a major part. It is also recommended to give that parameter values order unity. We then will provide the following order of magnitude:

$$\frac{h}{\delta r} \sim 1.0 \text{ to } 2.0 \tag{5.81}$$

Now that we are back to the Lagrangian formalism of Chapters 1 and 2, we may state that every quantity characterizing a given particle only formally depends on time (refer to Section 2.2.1). As a particle a moves about, an arbitrary quantity[20] A_a which is assigned to it will be subject over time to a variation rate as given by its Lagrangian derivative:

$$\dot{A}_a \doteq \frac{dA_a(t)}{dt} \tag{5.82}$$

[17] Although there is normally nothing wrong with considering mesoscopic scale particles.

[18] In the two-dimensional simulations, like all the additive quantities, mass is measured per unit length according to the invariance direction of the problem.

[19] This assumption is consistent with that of a constant smoothing length, which was expressed in the preceding section. It should be kept in mind, however, that it is an hypothesis which other authors drop for treating flows with a substantially variable density.

[20] The next equation remains valid, for example, for vector fields.

Thus, the velocity of a particle is given by the derivative of the position according to the natural equation (1.2):

$$\forall a, \; \frac{d\mathbf{r}_a}{dt} = \mathbf{u}_a \tag{5.83}$$

Compared with an Eulerian approach, such a concept is advantageous in that it does not use any advection term occurring in the definition (3.7), so that it is not necessary to pay attention to their discretization, which is all the less convenient because they are nonlinear.

In order to discretize the other operators occurring in the mechanical equations, we start from the continuous approximation as described in the preceding paragraph, then we approximate the integrals through discrete sums. To do this, it should be pointed out that the integral of an arbitrary function C can be approximated through a Riemann sum:

$$\int_{\Omega_\mathbf{r}} C\left(\mathbf{r}', t\right) d^n\mathbf{r}' \approx \sum_b C\left(\mathbf{r}_b, t\right) V_b \tag{5.84}$$

Here the summation points are the particle positions $\mathbf{r}_b$ (playing the part of the continuous variable $\mathbf{r}'$ in the integral), whereas the volume V_b of each particule is substituted for the integration volume $d^n\mathbf{r}'$. We are now going to apply this method to the continuous approximation (5.32), that is putting

$$C\left(\mathbf{r}'\right) \doteqdot A\left(\mathbf{r}'\right) w_h\left(\left|\mathbf{r} - \mathbf{r}'\right|\right) \tag{5.85}$$

We get, for a scalar field:

$$\begin{aligned} [A]_c\left(\mathbf{r}, t\right) &= \int_{\Omega_\mathbf{r}} A\left(\mathbf{r}', t\right) w_h\left(\mathbf{r} - \mathbf{r}'\right) d^n\mathbf{r}' \\ &\approx \sum_b A\left(\mathbf{r}_b, t\right) w_h\left(\left|\mathbf{r} - \mathbf{r}_b\right|\right) V_b \doteqdot [A]_d\left(\mathbf{r}, t\right) \end{aligned} \tag{5.86}$$

which defines a discrete interpolation of the field, being denoted as $[A]_d$. The discrete sum covers all the particles. If the kernel support is compact, however, that sum is reduced to the particles b lying around the selected point $\mathbf{r}$ in a $n-$sphere of radius h_t (refer to Fig. 5.4). Thus, each particle interacts with a

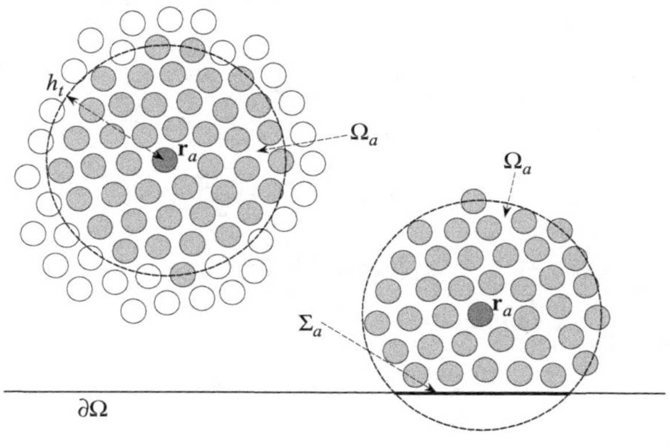

Fig. 5.4 Discrete approximation of a field by means of a compact support kernel and a collection of particles.

finite number N_n of adjacent particles, which can be estimated at $V_n h_t^n / V_p$, $V_n = S_n / n$ being the volume of the unit n−sphere. In view of (5.39), (5.41) and (5.80), N_n can be estimated as

$$N_n \approx \frac{2\pi^{n/2}}{n\Gamma\left(\frac{n}{2}\right)} \left(\frac{R_f h}{\delta r} \right)^n \tag{5.87}$$

With those values of R_f as provided in the previous section, and due to the order of magnitude (5.81), N_2 ranges from 30 to 50, N_3 ranges from 80 to 200. These values are significantly higher than the number of elements adjacent to an element in the frame of the mesh methods, which is a weakness in the SPH method. An examplary discrete interpolation can be seen in Fig. 7.1 (Chapter 7).

We may note that if the kernel is equated to the Dirac distribution, as in the beginning of Section 5.2.1 (eqn (5.1)), the formula (5.86) becomes

$$[A]_d (\mathbf{r}, t) \approx \sum_b V_b A (\mathbf{r}_b, t)\, \delta (|\mathbf{r} - \mathbf{r}_b|) \tag{5.88}$$

which is tantamount to considering that the field A is 'focused' on the centres of inertia of the particles.

The discrete operator (5.86) may now be applied to the vectors and differential operators and yields

$$\begin{aligned}
[A]_d (\mathbf{r}, t) &= \sum_b V_b A (\mathbf{r}_b, t)\, w_h (|\mathbf{r} - \mathbf{r}_b|) \approx A (\mathbf{r}, t) \\
[\mathbf{grad}\ A]_d (\mathbf{r}, t) &= \sum_b V_b A (\mathbf{r}_b, t)\, \nabla w_h (|\mathbf{r} - \mathbf{r}_b|) \approx (\mathbf{grad}\ A) (\mathbf{r}, t) \\
[\mathbf{A}]_d (\mathbf{r}, t) &= \sum_b V_b \mathbf{A} (\mathbf{r}_b, t)\, w_h (|\mathbf{r} - \mathbf{r}_b|) \approx A (\mathbf{r}, t) \\
[\mathrm{div}\mathbf{A}]_d (\mathbf{r}, t) &= \sum_b V_b \mathbf{A} (\mathbf{r}_b, t) \cdot \nabla w_h (|\mathbf{r} - \mathbf{r}_b|) \approx (\mathrm{div}\mathbf{A}) (\mathbf{r}, t)
\end{aligned} \tag{5.89}$$

We may note right now that all the above equations are linked: the third one is a vector form of the first one, whereas the second and fourth are achieved though a termwise derivation of the first and third. This is because the discretely interpolated fields are still functions of the continuous variable $\mathbf{r}$. For example:

$$\begin{aligned}
\mathbf{grad} ([A]_d) (\mathbf{r}, t) &= \sum_b V_b A (\mathbf{r}_b, t) \frac{w_h (|\mathbf{r} - \mathbf{r}_b|)}{\partial \mathbf{r}} \\
&= \sum_b V_b A (\mathbf{r}_b, t)\, \nabla w_h (|\mathbf{r} - \mathbf{r}_b|) \\
&= [\mathbf{grad}\ A]_d (\mathbf{r}, t)
\end{aligned} \tag{5.90}$$

Thus, we have two commutation relations:

$$\begin{aligned}
\mathbf{grad}\, [A]_d &= [\mathbf{grad}\ A]_d \\
\mathrm{div}\, [\mathbf{A}]_d &= [\mathrm{div}\mathbf{A}]_d
\end{aligned} \tag{5.91}$$

The first of these two relations may be compared with (5.17), in a continuous viewpoint.

It should be pointed out that we are now faced with two approximations: first the continuous interpolator considered in Section 5.2.1, second the discrete interpolator which we have just introduced. We have seen that the continuous interpolation operator $[\cdot]_c$ as defined by (5.2) approaches a field within an error margin of h^2 (eqns (5.12) and (5.18)). Estimating the error made by the discrete operator $[\cdot]_d$ is more complicated because it depends on the particle positions. When the latter are evenly distributed (for instance in a Cartesian grid), there is a precise answer to that problem which we are about to provide. If, conversely, the particles lie in positions which can said to be random positions, the discrete interpolation operator is similar to a Monte Carlo type statistical evaluation. Theoretical means are then available to provide an estimation of the mean error being made (for instance, refer to Pope, 2000). The particles, however, usually exhibit a more disorderly configuration than a Cartesian grid, but which is much more orderly than a random cloud. It then seems that neither of these two approaches is quite suitable. Another issue arises, in the frame of the SPH method: the particles experience a motion prescribed by discrete deterministic equations which will be given in Section 5.3 and which approximates to the most possible extent the equations of the real fluids being discussed in Chapter 3. So, the particle disorder is fully determined by these discrete equations. If this disorder can be measured by some quantities, the latter are then weak but non-zero and vary over space and time. This creates a significant complication in the assessment of the accuracy of the discrete interpolation, which remains an open problem (refer, in particular, to Monaghan, 1985b).

Let us get into the issue of the accuracy of the discrete interpolator in an elementary case. Since the configuration assumed by the whole set of particles is generally fairly regular, particularly as regards a weakly compressible flow, it is advisable to consider in the first place the ideal case in which the particles cover an infinite regular Cartesian grid with an arbitrary dimension n. For an arbitrary (scalar, vector or tensor) field[21] $C(\mathbf{r}')$, we define the discrete error being made taking on the relation (5.84) as follows:

[21] To make the notations easier to read, we will henceforth ignore the explicit time-dependence.

$$E_d[C(\cdot)] \doteqdot \sum_b C(\mathbf{r}_b)\, V_b - \int_{\Omega_\mathbf{r}} C(\mathbf{r}')\, d^n\mathbf{r}' \tag{5.92}$$

where the dot in the notation $C(\cdot)$ stands for the spatial variable (continuous $\mathbf{r}'$ or discrete $\mathbf{r}_b$ variable). Since the grid is regular, all the volumes V_b equal V_p (eqn (5.80)). Besides, the $\mathbf{r}_b$ coordinates are multiples of the spacing δr. A fundamental wavenumber can then be defined:

$$K_{\delta r} \doteqdot \frac{2\pi}{\delta r} \tag{5.93}$$

and the Poisson formula can be applied (refer to Appendix B, eqn (B.54) as well as Stein and Weiss, 1971), here being written as:

$$\sum_{\{m_i\}\in\mathbb{Z}^n} C(m_1\delta r, ..., m_n\delta r) = \frac{1}{\delta r^n} \sum_{\{k_i\}\in\mathbb{Z}^n} \widehat{C}(k_1 K_{\delta r}, ..., k_n K_{\delta r}) \tag{5.94}$$

where $\widehat{C}$ is the space Fourier transform of the function C. The $\{m_i\}$ range over all the relative integers in the expression (5.94), and the left-hand side then equals the discrete sum of (5.92):

$$V_p \sum_b C(\mathbf{r}_b) = \sum_{\{k_i\} \in \mathbb{Z}^n} \widehat{C}(k_1 K_{\delta r}, ..., k_n K_{\delta r}) \tag{5.95}$$

Furthermore, the integral $C(\mathbf{r}')$ equals the value of the Fourier transform at the origin (refer to eqn (B.2)):

$$\int_{\Omega_{\mathbf{r}}} C(\mathbf{r}') \, d^n \mathbf{r}' = \widehat{C}(\mathbf{0}) \tag{5.96}$$

Combining (5.95) and (5.96), the discrete error (5.92) is written as

$$E_d[C(\cdot)] = \sum_{\{k_i\} \in \mathbb{Z}^n \setminus \{0\}} \widehat{C}(k_1 K_{\delta r}, ..., k_n K_{\delta r}) \tag{5.97}$$

where at least one of the integers $\{k_i\}$ is non-zero. Let us now return to the case (5.85), in order to assess the error in the discrete SPH interpolation. To do this, we temporarily make a change of origin:

$$E_d[A(\cdot) w_h(|\mathbf{r} - \cdot|)] = E_d[A(\mathbf{r} - \cdot) w_h(|\cdot|)] \tag{5.98}$$

(which amounts to performing a translation of the system, without any effect onto the error). We also may note that the function $C(\mathbf{r}')$ vanishes, like $w_h(|\mathbf{r} - \mathbf{r}'|)$, for $|\mathbf{r} - \mathbf{r}'| \geqslant h_t$ (refer to Section 5.2.1). Since the integration takes place over a comparatively small domain, it is probably appropriate to expand the first-order Taylor series function $A(\mathbf{r}')$ in a manner similar to (5.3):

$$A(\mathbf{r} - \cdot) = A(\mathbf{r}) - \frac{\partial A}{\partial \mathbf{r}} \cdot (\cdot) + \frac{1}{2} (\cdot)^T \frac{\partial^2 A}{\partial \mathbf{r}^2} (\cdot) + O\left(|\cdot|^3\right) \tag{5.99}$$

where the derivatives are considered at the point $\mathbf{r}$. Inserting (5.99) into (5.98), we find

$$\begin{aligned} E_d[A(\cdot) w_h(|\mathbf{r} - \cdot|)] &= A(\mathbf{r}) E_d[w_h(|\cdot|)] - \frac{\partial A}{\partial \mathbf{r}} \cdot E_d[w_h(|\cdot|)(\cdot)] \\ &+ \frac{1}{2} \frac{\partial^2 A}{\partial \mathbf{r}^2} : E_d[w_h(|\cdot|)(\cdot) \otimes (\cdot)] \\ &+ E_d\left[O\left(|\cdot|^3\right) w_h(|\cdot|)\right] \end{aligned} \tag{5.100}$$

Since the function $\tilde{\mathbf{r}} w_h(|\tilde{\mathbf{r}}|)$ is odd, its integral is zero (refer to Section 5.2.1). It is all the same with the corresponding discrete sum, because of the regular nature of the particle network. We then may state that $E_d[(\cdot) w_h(|\cdot|)] = \mathbf{0}$, which is substantiated by the relation (5.97), since an odd real function features a Fourier transform whose real part is zero (refer to Appendix B). Furthermore, the last term in (5.100) is at most of the order of h^3. As regards the other quantities occurring in that development, we resort to equation (5.97) to assess them. In particular, we note the following relation:

$$\widehat{w_h(|\mathbf{r}|)\, \mathbf{r} \otimes \mathbf{r}} = -\frac{\partial^2 \widehat{w_h}}{\partial K^2} \mathbf{e}_K \otimes \mathbf{e}_K \tag{5.101}$$

with

$$\mathbf{e}_K \doteqdot \frac{\mathbf{K}}{K} \tag{5.102}$$

With the above arguments, we get:

$$\begin{aligned}
E_d \left[w_h \left(|\cdot| \right) \right] &= \sum_{\{k_i\} \in \mathbb{Z}^n \setminus \{0\}} \widehat{w_h} \left(\overline{k} K_{\delta r} \right) \\
E_d \left[w_h \left(|\cdot| \right) (\cdot) \right] &= \mathbf{0} \\
E_d \left[w_h \left(|\cdot| \right) (\cdot) \otimes (\cdot) \right] &= - \sum_{\{k_i\} \in \mathbb{Z}^n \setminus \{0\}} \frac{\partial^2 \widehat{w_h}}{\partial K^2} \left(\overline{k} K_{\delta r} \right) \mathbf{e}_k \otimes \mathbf{e}_k
\end{aligned} \tag{5.103}$$

We here have used the fact that the kernel Fourier transform is a function of the wavenumber norm (Section 5.2.2, eqn (5.76)), and defined

$$\begin{aligned}
\overline{k} &\doteqdot \sqrt{\sum_i k_i^2} \\
\mathbf{e}_k &\doteqdot \frac{1}{\overline{k}} \left(k_1, \ldots, k_n \right)
\end{aligned} \tag{5.104}$$

where the $\{k_i\}$, as it should be remembered, are integers not all of which are zero. With (5.103), the estimation (5.100) now gives

$$\begin{aligned}
E_d \left[A (\cdot) \, w_h \left(|\mathbf{r} - \cdot| \right) \right] = A (\mathbf{r}) & \sum_{\{k_i\} \in \mathbb{Z}^n \setminus \{0\}} \widehat{w_h} \left(\overline{k} K_{\delta r} \right) \\
& - \frac{1}{2} \sum_{\{k_i\} \in \mathbb{Z}^n \setminus \{0\}} \left[\left(\mathbf{e}_k^T \frac{\partial^2 A}{\partial \mathbf{r}^2} \mathbf{e}_k \right) \frac{\partial^2 \widehat{w_h}}{\partial K^2} \left(\overline{k} K_{\delta r} \right) \right] \\
& + O \left(h^3 \right)
\end{aligned} \tag{5.105}$$

Section 5.2.2 explains that $\widehat{w_h}$ is a rapidly decreasing function. According to Fig. 5.3, it may be considered that $\widehat{w_h} \left(K^+ \right)$ is negligible (about 0.01) as soon as $K^+ = hK$ is higher than 10. Thus, only the $\{k_i\}$ such as $h\overline{k}K_{\delta r} < 10$ can be kept in the estimation (5.105), that is with (5.93):

$$\overline{k} < \frac{10}{2\pi} \frac{\delta r}{h} \sim 1 \tag{5.106}$$

(the estimation (5.81) was used). Thus, it is sufficient, as an initial approximation, to only keep the $\{k_i\}$ one of which equals 1 whereas all of the others are zero, which corresponds to $\overline{k} = 1$. Thus, there are n possible choices, which ultimately makes it possible to write

$$E_d\left[A(\cdot)\,w_h(|\mathbf{r}-\cdot|)\right] = A(\mathbf{r})\,n\widehat{w_h}(K_{\delta r}) - \frac{1}{2}\nabla^2 A(\mathbf{r})\,\frac{\partial^2 \widehat{w_h}}{\partial K^2}(K_{\delta r}) + O\left(h^3\right) \tag{5.107}$$

Restricting ourselves to a first-order development, we can simplify that formula and ultimately write

$$\frac{E_d\left[A(\cdot)\,w_h(|\mathbf{r}-\cdot|)\right]}{A(\mathbf{r})} \approx n\widehat{w_h}\left(K^+ = \frac{2\pi h}{\delta r} \sim 10\right) + O\left(h^2\right) \tag{5.108}$$

With the kernels as disclosed in Section 5.2.2, the relative error in the discrete interpolation is then in the order of magnitude of a few per cent. We may recall, however, that this result is valid for a Cartesian regular arrangement of the particles and is subject to an approximation of the order of h^2.

It is noteworthy that the kernel Fourier transform plays a crucial role in the error being made, as confirmed by several authors (in particular, refer to Monaghan, 2005).We now must make an important remark about the above result. The higher $h/\delta r$, the lower $\widehat{w_h}(2\pi h/\delta r)$ if the kernel has a decreasing Fourier transform (we may recall that such is the case with the Gaussian (5.45) and Wendland (5.57) kernels). Moreover, the higher h, the higher the term $O\left(h^2\right)$. Thus, the smoothing length produces two contradictory effects. In order to clarify this point, we can accurately calculate the discrete error in the one-dimensional case ($n = 1$). The integration domain $\Omega_{\mathbf{r}}$, which is centred on the point x (playing the part of $\mathbf{r}$ in one dimension) is then reduced to the segment $[x - h_t; x + h_t]$. Still in the particular case in which the particles are equidistant, and if no other particle coincides with the ends of the domain, the second Euler–MacLaurin formula (refer, e.g. to Abramovitz and Stegun, 1972) can be used to write the error (5.92) as

$$E_d\left[C(\cdot)\right] = \sum_{k=1}^{+\infty} \frac{\left(1-2^{1-2k}\right)B_{2k}}{(2k)!}\delta r^{2k} \times \left[C^{(2k-1)}(x+h_t) - C^{(2k-1)}(x-h_t)\right] \tag{5.109}$$

(Quinlan et al., 2006). In this equation, the B_k stand for the Bernoulli numbers,[22] and $C^{(2k-1)}$ designates the $(2k-1)$-th-order derivative in the function C. Before applying the above formula to the case of the discrete SPH interpolation, that is defining C by (5.85), we already may note that, since the kernel derivatives are inversely proportional to $h^{n+2k-1} = h^{2k}$ (refer to eqn (5.44)), each of the terms in the sum (5.109) is proportional to $(\delta r/h)^{2k-1}$, and then is of a lower order than the previous one. Furthermore, some of the $C^{(2k-1)}(x \pm h_t)$ terms are obviously zero, because the kernel and its first derivatives are zero at the ends of the integration domain. The first non-zero term in the sum (5.109) appears when the integer $2k-1$ is large enough for the $(2k-1)$ −th kernel derivative not to vanish in h_t. We denote as β_w the smallest integer verifying that condition, that is the smallest odd integer such that $w_h^{(\beta_w)}(h_t) \neq 0$. For the Gaussian kernel, the M-order B-splines and the Wendland kernel (refer to Section 5.2.2), these integers equal, respectively

[22] We may recall that these numbers are defined as the polynomial development coefficients for the sum of the integral powers of the first integers:

$$\sum_{k=1}^{n} k^m = \frac{1}{m+1}\sum_{k=0}^{m}\binom{m+1}{k} \times B_k n^{m+1-k}$$

Except for B_1, the B_k are zero for odd k. The first values in that series are $B_0 = 1$, $B_1 = 1/2$, $B_2 = 1/6$, $B_4 = -1/30$, $B_6 = 1/42$, $B_8 = -1/30$, $B_{10} = 5/66$.

$$
\begin{aligned}
\beta_G &= +\infty \\
\beta_M &= M+1 \quad \text{if} \quad M \quad \text{even} \\
&= M \qquad\ \ \text{if} \quad M \quad \text{odd} \\
\beta_W &= 5
\end{aligned}
\tag{5.110}
$$

The first non-zero term in the sum (5.109), that is the prevailing term, then corresponds to $2k-1=\beta_w$. The quantities $C^{(\beta_w)}(x \pm h_t)$ are then proportional to $w_h^{(\beta_w)}(h_t)$, and (5.109) can be approached by[23]

$$
\begin{aligned}
E_d\left[A(\cdot)\, w_h(|x-\cdot|)\right] &= \frac{\left(1-2^{-\beta_w}\right) B_{\beta_w+1}}{(\beta_w+1)!}\delta r^{\beta_w+1} w_h^{(\beta_w)}(h_t) \\
&\quad \times \left[A(x+h_t)-A(x-h_t)\right] + O\left(\delta r^{\beta_w+3}\right)
\end{aligned}
\tag{5.111}
$$

[23]The Gaussian kernel case clearly cannot be assessed as follows, because both its support and β_G are infinite.

The term between square brackets can still be developed:

$$
A(x+h_t)-A(x-h_t) = 2\frac{\partial A}{\partial x}h_t + O\left(h^3\right) \tag{5.112}
$$

Lastly, (5.111) can be simplified using the dimensionless kernel as defined by (5.36) and the relation (5.44):

$$
\begin{aligned}
E_d\left[A(\cdot)\, w_h(|x-\cdot|)\right] &= \frac{2\left(1-2^{-\beta_w}\right) B_{\beta_w+1}}{(\beta_w+1)!}\alpha_{w,n} f^{(\beta_w)}(1) \\
&\quad \times \left(\frac{\delta r}{h}\right)^{\beta_w+1}\left[\frac{\partial A}{\partial x}h_t + O\left(h^3\right)\right] + O\left[\left(\frac{\delta r}{h}\right)^{\beta_w+3}\right] \\
&= O\left[\left(\frac{\delta r}{h}\right)^{\beta_w+1}\right]
\end{aligned}
\tag{5.113}
$$

Let us now estimate the error being made when likening the discrete and continuous interpolations. We may rearrange (5.92) as

$$
[A]_d(\mathbf{r},t) - [A]_c(\mathbf{r},t) = E_d\left[A(\cdot,t)\, w_h(|\mathbf{r}-\cdot|)\right] \tag{5.114}
$$

With (5.12) and (5.113), that yields

$$
\begin{aligned}
[A]_d(\mathbf{r},t) &= [A]_c(\mathbf{r},t) + O\left[\left(\frac{\delta r}{h}\right)^{\beta_w+1}\right] \\
&= A(\mathbf{r},t) + O\left(h^2\right) + O\left[\left(\frac{\delta r}{h}\right)^{\beta_w+1}\right]
\end{aligned}
\tag{5.115}
$$

This confirms that the SPH interpolation is accompanied by an error comprising two terms, so we can find once again the two contradictory effects of the smoothing length. The first error, in h^2, stems from the continuous interpolation and is an increasing function of h, whereas the second error, in $(\delta r/h)^{\beta_w+1}$, is a decreasing function of it, which results from the discrete interpolator. Thus, contrary to the continuous interpolation as described in Section 5.2.1, the discrete approximation is not even accurate at the zeroth order. This result may be generalized to the case of a collection of arbitrarily arranged particles (refer to Quinlan et al., 2006).[24]

[24] Pope (2000) achieves a qualitatively similar outcome for randomly arranged particles.

This phenomenon can be explained as follows: the higher $h/\delta r$, the larger the number of particles involved in the discrete sum, which tends to reduce the statistical inaccuracy of the discrete estimation; on the other hand, the higher h, the broader the integration domain, which disrupts the estimation of the interpolated quantity at the point of interest and causes all the greater errors (the *bias*). It can be inferred from the above that the convergence of the discrete interpolation is subject to two conditions,[25] namely h should tend towards zero whereas $h/\delta r$ should tend towards infinity. The particle size shall then obviously be as small as possible. The expression (5.115) also shows that there should be an optimal value of the $h/\delta r$ ratio; that value, however, is not universal (it depends of the particle arrangements as well as on the interpolated quantity A).

[25] This is one of the least attractive features of the SPH methods, as compared with the conventional mesh-based methods.

5.2.4 SPH operators

Let us now apply the relations (5.89) to the case in which the estimation point $\mathbf{r}$ coincides with the position of a particle a. With respect to the continuous interpolation of Section 5.2.1, that amounts to an integration over the volume Ω_a encompassing all the points interacting with particle a (Fig. 5.4). We get, by means of the usual notation[26] $A_b = A(\mathbf{r}_b, t)$:

[26] The quantities A_b and $\mathbf{A}_b$ then implicitly depend on time t.

$$\begin{aligned}
[A]_d(\mathbf{r}_a, t) &= \sum_b V_b A_b w_h(r_{ab}) \\
[\mathbf{grad}\, A]_d(\mathbf{r}_a, t) &= \sum_b V_b A_b \nabla_a w_h(r_{ab}) \\
[\mathbf{A}]_d(\mathbf{r}_a, t) &= \sum_b V_b \mathbf{A}_b w_h(r_{ab}) \\
[\mathrm{div}\mathbf{A}]_d(\mathbf{r}_a, t) &= \sum_b V_b \mathbf{A}_b \cdot \nabla_a w_h(r_{ab})
\end{aligned} \tag{5.116}$$

where we have noted $r_{ab} \doteqdot |\mathbf{r}_{ab}|$ with $\mathbf{r}_{ab} \doteqdot \mathbf{r}_a - \mathbf{r}_b$ (as in Chapter 1, Section 1.3.2), and

$$\begin{aligned}
\nabla_a w_h(r_{ab}) &\doteqdot \frac{\partial w_h(r_{ab})}{\partial \mathbf{r}_a} \\
&= -\frac{\partial w_h(r_{ab})}{\partial \mathbf{r}_b} \\
&= -\nabla_b w_h(r_{ab})
\end{aligned} \tag{5.117}$$

The second line in the above calculation is based on the kernel symmetry property (5.8). Noting $\mathbf{e}_{ab} \doteqdot \mathbf{r}_{ab}/r_{ab}$, as in Chapter 1 (Section 1.3.3), the relation (5.33) also shows that

$$\nabla_a w_h(r_{ab}) = w_h'(r_{ab})\,\mathbf{e}_{ab} \tag{5.118}$$

Through (5.116), we have defined linear operators transforming a discrete field $\{A_a\}$ into another discrete field $\{[A]_d(\mathbf{r}_a, t)\}$, etc. The discrete operators approximating[27] the identity operator will be denoted as J (for the scalar fields) and $\mathbf{J}$ (for the vector fields), whereas we will denote as $\mathbf{G}$ and D the discrete operators approaching the continuous operators **grad** and div respectively, which, using (5.118), will yield:[28]

$$
\begin{aligned}
J\{A_a\} &\doteqdot \left\{\sum_b V_b A_b w_{ab}\right\} \approx \{A_a\} \\
\mathbf{G}\{A_a\} &\doteqdot \left\{\sum_b V_b A_b w'_{ab}\mathbf{e}_{ab}\right\} \approx \left\{(\mathbf{grad}\ A)_a\right\} \\
\mathbf{J}\{\mathbf{A}_a\} &\doteqdot \left\{\sum_b V_b \mathbf{A}_b w_{ab}\right\} \approx \{\mathbf{A}_a\} \\
D\{\mathbf{A}_a\} &\doteqdot \left\{\sum_b V_b \mathbf{A}_b \cdot w'_{ab}\mathbf{e}_{ab}\right\} \approx \left\{(\mathrm{div}\mathbf{A})_a\right\}
\end{aligned}
\qquad (5.119)
$$

The braces $\{\cdot\}$ in these relations, as in Chapters 1 and 2, symbolize a discrete set of values the discrete subscript of which is the free label a (it is appropriate to ensure that the free label be identical on both sides when handling such an equality). We have also introduced the notations[29]

$$
\begin{aligned}
w_{ab} &= w_h(r_{ab}) = w_{ba} \\
w'_{ab} &= w'_h(r_{ab}) = w'_{ba}
\end{aligned}
\qquad (5.120)
$$

(the symmetry relations stem from the hypothesis (5.7) and are consistent with (5.117) and (5.118)). We, of course, get from them:

$$
w'_{ab}\mathbf{e}_{ab} = -w'_{ba}\mathbf{e}_{ba} \qquad (5.121)
$$

which is equivalent to (5.8). In view of (5.36) and (5.42), the definitions (5.120) could be rearranged as

$$
\begin{aligned}
w_{ab} &= \frac{\alpha_{w,n}}{h^n} f(q_{ab}) \\
w'_{ab} &= \frac{\alpha_{w,n}}{h^{n+1}} f'(q_{ab})
\end{aligned}
\qquad (5.122)
$$

with $q_{ab} \doteqdot r_{ab}/h$. For the sake of the simplicity of notations, we will quite often use the values of the discrete fields $J\{A_a\}$, and so on at the point being occupied by a given particle a.[30] We then will note

$$
\begin{aligned}
J_a\{A_b\} &= \sum_b V_b A_b w_{ab} \\
\mathbf{G}_a\{A_b\} &= \sum_b V_b A_b w'_{ab}\mathbf{e}_{ab}
\end{aligned}
\qquad (5.123)
$$

[27] One may notice that it is an actual approximation, $J\{A_a\}$ usually not being equal to $\{A_a\}$, as explained in the preceding section. The estimation of a function at the point being occupied by a particle by the discrete SPH operator is then not equal to the value of that function at that point.

[28] As pointed out in Section 5.2.1, this is still possible for the higher-ordered tensors or operators. Examples will be provided a bit later on.

[29] The remark made in Chapter 1 about the ambiguity of some notations has to be repeated here: whereas A_{ab} stands for the difference $A_a - A_b$, w_{ab} does not happen to be the difference between two terms.

[30] We will sometimes talk about 'the operator $\mathbf{G}_a$', although it is misleading. More precisely, it is the value of the field resulting from the application of the operator $\mathbf{G}$ to the field $\{A_b\}$, a value corresponding to the particle a.

$$\mathbf{J}_a\{\mathbf{A}_b\} = \sum_b V_b \mathbf{A}_b w_{ab}$$
$$D_a\{\mathbf{A}_b\} = \sum_b V_b \mathbf{A}_b \cdot w'_{ab}\mathbf{e}_{ab}$$

which amounts to defining $J_a\{A_b\} \doteqdot [A]_d(\mathbf{r}_a, t)$, and so on.

Formulas (5.123) are not the only ones which make it possible to approximate either the gradient or the divergence of a field through the procedure being introduced. To be convinced of it, let us have a look, for instance, at the following identities, which are valid for every real k (refer to Appendix A, particularly eqn (A.65)):

$$\begin{aligned} \mathbf{grad}\, A &= \rho^k \mathbf{grad}\frac{A}{\rho^k} + \frac{A}{\rho^k}\mathbf{grad}\left(\rho^k\right) \\ \mathrm{div}\mathbf{A} &= \rho^k \mathrm{div}\frac{\mathbf{A}}{\rho^k} + \frac{\mathbf{A}}{\rho^k}\cdot\mathbf{grad}\left(\rho^k\right) \end{aligned} \tag{5.124}$$

Let us now rearrange these two expressions at the point being occupied by the particle a, approximating the continuous operators in the right-hand sides of (5.124) by means of the just defined discrete operators $\mathbf{G}$ and D. We get

$$\begin{aligned} (\mathbf{grad}\, A)_a &\approx \rho_a^k \mathbf{G}_a\left\{\left(\frac{A}{\rho^k}\right)_b\right\} + \frac{A_a}{\rho_a^k}\mathbf{G}_a\left\{\left(\rho^k\right)_b\right\} \\ (\mathrm{div}\mathbf{A})_a &\approx \rho_a^k D_a\left\{\left(\frac{\mathbf{A}}{\rho^k}\right)_b\right\} + \frac{\mathbf{A}_a}{\rho_a^k}\cdot\mathbf{G}_a\left\{\left(\rho^k\right)_b\right\} \end{aligned} \tag{5.125}$$

(Monaghan, 1992). Using the definitions (5.123) and taking advantage[31] of the relation (5.79), we soon find other discrete approximations of the continuous operators:

[31] We could have made this work earlier, using the continuous operators (5.32).

$$\begin{aligned} \mathbf{G}_a^k\{A_b\} &\doteqdot \sum_b V_b \frac{\rho_b^{2k}A_a + \rho_a^{2k}A_b}{(\rho_a\rho_b)^k} w'_{ab}\mathbf{e}_{ab} \approx (\mathbf{grad}\, A)_a \\ D_a^k\{\mathbf{A}_b\} &\doteqdot \sum_b V_b \frac{\rho_b^{2k}\mathbf{A}_a + \rho_a^{2k}\mathbf{A}_b}{(\rho_a\rho_b)^k}\cdot w'_{ab}\mathbf{e}_{ab} \approx (\mathrm{div}\mathbf{A})_a \end{aligned} \tag{5.126}$$

These relations define two infinite sets of 'gradient' and 'divergence' discrete operators, which are denoted as $\mathbf{G}_a^k$ and D_a^k. For $k = 0$, for example, we get

$$\begin{aligned} \mathbf{G}_a^0\{A_b\} &= \sum_b V_b\left(A_a + A_b\right) w'_{ab}\mathbf{e}_{ab} \\ D_a^0\{\mathbf{A}_b\} &= \sum_b V_b\left(\mathbf{A}_a + \mathbf{A}_b\right)\cdot w'_{ab}\mathbf{e}_{ab} \end{aligned} \tag{5.127}$$

and for $k = 1$:

$$\begin{aligned} \mathbf{G}_a^1\{A_b\} &= \rho_a \sum_b m_b \left(\frac{A_a}{\rho_a^2} + \frac{A_b}{\rho_b^2}\right) w'_{ab}\mathbf{e}_{ab} \\ D_a^1\{\mathbf{A}_b\} &= \rho_a \sum_b m_b \left(\frac{\mathbf{A}_a}{\rho_a^2} + \frac{\mathbf{A}_b}{\rho_b^2}\right) \cdot w'_{ab}\mathbf{e}_{ab} \end{aligned} \tag{5.128}$$

We can proceed differently by writing a variant of (5.124):

$$\begin{aligned} \mathbf{grad}\, A &= \frac{1}{\rho^k}\mathbf{grad}\left(\rho^k A\right) - \frac{A}{\rho^k}\mathbf{grad}\left(\rho^k\right) \\ \mathrm{div}\mathbf{A} &= \frac{1}{\rho^k}\mathrm{div}\left(\rho^k \mathbf{A}\right) - \frac{\mathbf{A}}{\rho^k}\cdot \mathbf{grad}\left(\rho^k\right) \end{aligned} \tag{5.129}$$

Following the same procedure, we then can find another series of operators:

$$\begin{aligned} \tilde{\mathbf{G}}_a^k\{A_b\} &\doteqdot -\frac{1}{\rho_a^{2k}} \sum_b V_b \left(\rho_a\rho_b\right)^k A_{ab} w'_{ab}\mathbf{e}_{ab} \approx \left(\mathbf{grad}\, A\right)_a \\ \tilde{D}_a^k\{\mathbf{A}_b\} &\doteqdot -\frac{1}{\rho_a^{2k}} \sum_b V_b \left(\rho_a\rho_b\right)^k \mathbf{A}_{ab}\cdot w'_{ab}\mathbf{e}_{ab} \approx \left(\mathrm{div}\mathbf{A}\right)_a \end{aligned} \tag{5.130}$$

with, as in Chapter 1 and 2, $A_{ab} \doteqdot A_a - A_b$ and $\mathbf{A}_{ab} \doteqdot \mathbf{A}_a - \mathbf{A}_b$. For $k = 0$, that yields

$$\begin{aligned} \tilde{\mathbf{G}}_a^0\{A_b\} &= -\sum_b V_b A_{ab} w'_{ab}\mathbf{e}_{ab} \\ \tilde{D}_a^0\{\mathbf{A}_b\} &= -\sum_b V_b \mathbf{A}_{ab}\cdot w'_{ab}\mathbf{e}_{ab} \end{aligned} \tag{5.131}$$

and for $k = 1$:

$$\begin{aligned} \tilde{\mathbf{G}}_a^1\{A_b\} &= -\frac{1}{\rho_a} \sum_b m_b A_{ab} w'_{ab}\mathbf{e}_{ab} \\ \tilde{D}_a^1\{\mathbf{A}_b\} &= -\frac{1}{\rho_a} \sum_b m_b \mathbf{A}_{ab}\cdot w'_{ab}\mathbf{e}_{ab} \end{aligned} \tag{5.132}$$

Two remarks are immediately to be made about these proposed forms of discrete operators: on the one hand, all of them consist of linear operators. Moreover, when we apply them to a constant field, the forms (5.130) provide discrete gradients and divergences which are strictly zero, as expected (since, in this case, $A_{ab} = 0$). Through linearity, this property is reduced to the following formulas:

$$\begin{aligned} \tilde{\mathbf{G}}_a^k\{1\} &= \mathbf{0} \\ \tilde{D}_a^k\{\mathbf{1}\} &= 0 \end{aligned} \tag{5.133}$$

The forms (5.123) and (5.126), on the other hand, do not have this elementary property. It is also vital to remember that the basic discrete interpolation corresponding to the first line in (5.123) does a priori not verify $J_a\{1\} = 1$, in accordance with the calculations in Section 5.2.3. Among the anti-properties of the discrete operators, we then will mention the following:

$$\begin{aligned} J_a\{1\} &\neq 1 \\ \mathbf{G}_a\{1\} &\neq \mathbf{0} \\ \mathbf{G}_a^k\{1\} &\neq \mathbf{0} \\ D_a\{\mathbf{1}\} &\neq 0 \\ D_a^k\{\mathbf{1}\} &\neq 0 \end{aligned} \tag{5.134}$$

We will deal again with these ideas in Section 5.2.5, and try to remedy them.

As previously explained, these arguments can be readily applied to the higher-ordered tensors or operators. Thus, for example, we may write approximations of the gradient tensor of a vector $\mathbf{A}$ and of the divergence vector of a second-order tensor in the following forms, which are analogous to (5.127) and (5.131), respectively:

$$\begin{aligned} \mathbf{G}_a^0\{\mathbf{A}_b\} &= \sum_b V_b\left(\mathbf{A}_a + \mathbf{A}_b\right) \otimes w'_{ab}\mathbf{e}_{ab} \approx (\mathbf{grad\ A})_a \\ \mathbf{D}_a^0\{\mathbf{A}_b\} &= \sum_b V_b\left(\mathbf{A}_a + \mathbf{A}_b\right) w'_{ab}\mathbf{e}_{ab} \approx (\mathbf{div\ A})_a \end{aligned} \tag{5.135}$$

and

$$\begin{aligned} \tilde{\mathbf{G}}_a^0\{\mathbf{A}_b\} &= -\sum_b V_b\mathbf{A}_{ab} \otimes w'_{ab}\mathbf{e}_{ab} \approx (\mathbf{grad\ A})_a \\ \tilde{\mathbf{D}}_a^0\{\mathbf{A}_b\} &= -\sum_b V_b\mathbf{A}_{ab} w'_{ab}\mathbf{e}_{ab} \approx (\mathbf{div\ A})_a \end{aligned} \tag{5.136}$$

Making a choice among the proposed formulas (5.127), (5.128), (5.131) and (5.132) (and other possible ones) is not inconsequential, as explained in the following. Each of these operators has, indeed (unlike the original formulas (5.119)), conspicuous properties of symmetry and antisymmetry with respect to the a and b subscripts; using one instead of the other is therefore suitable depending on the circumstances. More precisely, pursuant to the law (5.121) the $\mathbf{G}_a^k$ and D_a^k operators are antisymmetric, whereas $\tilde{\mathbf{G}}_a^k$ and $\tilde{D}_a^k$ are symmetric. That means that those terms occurring under the summation symbols consist of symmetric or antisymmetric expressions with respect to the a and b subscripts.

The series of operators which we have used can be expanded to infinity. Starting from the relation (A.65) in Appendix A, for instance, we can set up the following approximation:

$$\begin{aligned} (B\mathrm{div}\mathbf{A})_a &= \left[\mathrm{div}\,(B\mathbf{A}) - \mathbf{A}\cdot\mathbf{grad}\ B\right]_a \\ &\approx D_a\left\{(B\mathbf{A})_b\right\} - \mathbf{A}_a\cdot\mathbf{G}_a\{B_b\} \\ &= -\sum_b V_b B_b \mathbf{A}_{ab}\cdot w'_{ab}\mathbf{e}_{ab} \end{aligned} \tag{5.137}$$

To complete this section, we are going to disclose a crucial property of some of the just introduced operators. We may initially observe that such kinds of sets as $\{A_b\}$ and $\{\mathbf{A}_b\}$ are two finite-dimensional vector spaces (the dimension is equal to N_p, namely the number of particles being involved), on which we can construct two vector products and their relevant norms, in line with the continuous formalism (3.194) and using the discrete integration[32] (5.84):

$$\begin{aligned} \langle \{\mathbf{A}_b\}, \{\mathbf{B}_b\} \rangle &= \sum_a V_a \mathbf{A}_a \cdot \mathbf{B}_a \qquad \| \{\mathbf{A}_b\} \|^2 = \sum_a V_a |\mathbf{A}_a|^2 \\ (\{A_b\}, \{B_b\}) &= \sum_a V_a A_a B_a \qquad \| \{A_b\} \|^2 = \sum_a V_a A_a^2 \end{aligned} \tag{5.138}$$

[32] It can be pointed out that the following notation slightly differs from that in Section 1.4.2, the volumes V_a now appearing whereas they are absent in eqn (1.97).

We are now trying to estimate the scalar product of the discrete fields $\mathbf{G}_a^k \{A_b\}$ and $\{\mathbf{B}_a\}$, for two arbitrary quantities A and $\mathbf{B}$. To that purpose, we will perform an exchange of the dummy subscripts a and b in the following expression:

$$\begin{aligned} \left\langle \mathbf{G}_a^k \{A_b\}, \{\mathbf{B}_a\} \right\rangle &= \sum_{a,b} m_b m_a \frac{\rho_b^{2k} A_a + \rho_a^{2k} A_b}{(\rho_a \rho_b)^{k+1}} \mathbf{B}_a \cdot w'_{ab} \mathbf{e}_{ab} \\ &= -\sum_{a,b} m_a m_b \frac{\rho_b^{2k} A_a + \rho_a^{2k} A_b}{(\rho_a \rho_b)^{k+1}} \mathbf{B}_b \cdot w'_{ab} \mathbf{e}_{ab} \\ &= \frac{1}{2} \sum_{a,b} m_a m_b \frac{\rho_b^{2k} A_a + \rho_a^{2k} A_b}{(\rho_a \rho_b)^{k+1}} \mathbf{B}_{ab} \cdot w'_{ab} \mathbf{e}_{ab} \end{aligned} \tag{5.139}$$

(the last line is the average of the preceding two lines). Following the same procedure, we may write

$$\begin{aligned} \left(\{A_a\}, \tilde{D}_a^k \{\mathbf{B}_b\} \right) &= -\sum_{a,b} m_b m_a \frac{\rho_b^{2k} A_a}{(\rho_a \rho_b)^{k+1}} \mathbf{B}_{ab} \cdot w'_{ab} \mathbf{e}_{ab} \\ &= -\frac{1}{2} \sum_{a,b} m_b m_a \frac{\rho_b^{2k} A_a + \rho_a^{2k} A_b}{(\rho_a \rho_b)^{k+1}} \mathbf{B}_{ab} \cdot w'_{ab} \mathbf{e}_{ab} \end{aligned} \tag{5.140}$$

which is identical to (5.139), but with a minus sign. Combining these two relations, we then get

$$\left(\{A_a\}, \tilde{D}_a^k \{\mathbf{B}_b\} \right) = -\langle \mathbf{G}_a^k \{A_b\}, \{\mathbf{B}_a\} \rangle \tag{5.141}$$

This equation means that the $-\mathbf{G}^k$ and $\tilde{D}^k$ operators are adjoint, just like the $-\mathbf{grad}$ and div continuous operators (eqn (3.198)) or the interpolated operators (eqn (5.30)). It could likewise be shown that $-\tilde{\mathbf{G}}^k$ and D^k mutually verify the same property. Thus, the discrete operators are skew-adjoint, like the continuous ones (see the end of Section 5.2.1). These results are vital for the consistency of the SPH method, as explained on several occasions in the following.

The gradient of the kernel gradient will advantageously be calculated before carrying on; it is the gradient of the $w'_{ab}\mathbf{e}_{ab}$ quantity with respect to the coordinates of one of the two particles, that is the second-order tensor as defined by $\mathbf{W}_{ab} \doteqdot \nabla_b \left(w'_{ab} \mathbf{e}_{ab} \right)$. We immediately note that

$$\begin{aligned}\frac{\partial \mathbf{e}_{ab}}{\partial \mathbf{r}_b} &= \frac{\partial}{\partial \mathbf{r}_b}\left(\frac{\mathbf{r}_{ab}}{r_{ab}}\right) \\ &= \frac{1}{r_{ab}}\frac{\partial \mathbf{r}_{ab}}{\partial \mathbf{r}_b} - \frac{1}{r_{ab}^2}\frac{\partial r_{ab}}{\partial \mathbf{r}_b} \otimes \mathbf{r}_{ab} \\ &= \frac{1}{r_{ab}}\left(\mathbf{e}_{ab} \otimes \mathbf{e}_{ab} - \mathbf{I}_n\right)\end{aligned} \tag{5.142}$$

Then, introducing the second derivative w_h'' of the w_h function, and writing $w_{ab}'' \doteqdot w_h''(r_{ab})$, we get

$$\begin{aligned}\frac{\partial w_{ab}'}{\partial \mathbf{r}_b} &= w_{ab}''\frac{\partial r_{ab}}{\partial \mathbf{r}_b} \\ &= -w_{ab}''\mathbf{e}_{ab}\end{aligned} \tag{5.143}$$

Combining (5.142) and (5.143), we ultimately find:

$$\begin{aligned}\mathbf{W}_{ab} &\doteqdot \frac{\partial w_{ab}'\mathbf{e}_{ab}}{\partial \mathbf{r}_b} \\ &= \frac{\partial w_{ab}'}{\partial \mathbf{r}_b} \otimes \mathbf{e}_{ab} + w_{ab}'\frac{\partial \mathbf{e}_{ab}}{\partial \mathbf{r}_b} \\ &= \left(\frac{w_{ab}'}{r_{ab}} - w_{ab}''\right)\mathbf{e}_{ab} \otimes \mathbf{e}_{ab} - \frac{w_{ab}'}{r_{ab}}\mathbf{I}_n \\ &= \mathbf{W}_{ba} = -\frac{\partial w_{ab}'\mathbf{e}_{ab}}{\partial \mathbf{r}_a}\end{aligned} \tag{5.144}$$

These relations will be useful in Sections 5.4.4 and 6.3.1.

5.2.5 Renormalization

We now will put forward a few guidelines to enhance the SPH interpolation operators. For further details about this topic, please refer, for instance, to Bonet and Lok (1999), Vila (1999), Belytschko and Xiao (2002), as well as Randles and Libersky (1996) and Oger et al. (2007). In order to outline the objective of what will be said, we may recall that the continuous interpolation introduced in Section 5.2.1 was analyzed in the case where the interpolation point lies sufficiently far off the edges of the domain to prevent the integration support from crossing the boundary (free surface or solid wall). Otherwise, that is if the particle being considered lies near a wall or a free surface, the condition (5.5) is not achieved, and so the estimation (5.12) can be altered as follows:

$$[A]_c(\mathbf{r}, t) = A(\mathbf{r}, t)\int_{\Omega_\mathbf{r}} w_h(\tilde{r})\, d^n\mathbf{r}' + O\left(h^2\right) \tag{5.145}$$

The estimation is then only close to the interpolated function provided that a continuous interpolation operator is defined as different from (5.2), being noted $[A]_c^\gamma$ and defined by

$$A(\mathbf{r},t) \approx [A]_c^\gamma(\mathbf{r},t) \doteqdot \frac{[A]_c(\mathbf{r},t)}{\gamma(\mathbf{r})} \tag{5.146}$$

with

$$\gamma(\mathbf{r}) \doteqdot [1]_c(\mathbf{r},t) = \int_{\Omega_\mathbf{r}} w_h\left(\left|\mathbf{r}-\mathbf{r}'\right|\right) d^n\mathbf{r}' \tag{5.147}$$

Far off the boundary of the domain, $\gamma(\mathbf{r}) = 1$ (eqn (5.5)) and the conventional interpolation (5.2), that is $[A]_c = [A]_c^\gamma$ is found again, whereas $\gamma(\mathbf{r})$ is lower than 1 near the boundary. We may note that the definition (5.146) is formally identical to the original interpolation (5.2), provided that the kernel is corrected by

$$\tilde{w}_h\left(\mathbf{r},\mathbf{r}'\right) \doteqdot \frac{w_h\left(\left|\mathbf{r}-\mathbf{r}'\right|\right)}{\gamma(\mathbf{r})} > w_h\left(\mathbf{r}-\mathbf{r}'\right) \tag{5.148}$$

It can be seen that the drawback of this formalism is that the corrected kernel $\tilde{w}_h$ is no longer a mere function of the norm $\left|\mathbf{r}-\mathbf{r}'\right|$, and is not even any longer a function of $\mathbf{r}-\mathbf{r}'$. As a result, most of the estimations of order of accuracy announced in Section 5.2.1 prove deficient.

The discrete interpolation can be altered in the same way. By analogy with (5.146), the altered discrete interpolator J_a^γ is defined by:

$$A(\mathbf{r}_a,t) \approx J_a^\gamma\{A_b\} \doteqdot \frac{J_a\{A_b\}}{\gamma_a} \tag{5.149}$$

with

$$\gamma_a \doteqdot J_a\{1\} = \sum_b V_b w_{ab} \tag{5.150}$$

which obeys the following approximation:

$$\gamma_a \approx \int_\Omega w_h\left(\left|\mathbf{r}_a-\mathbf{r}'\right|\right) d^n\mathbf{r}' = \gamma(\mathbf{r}_a) \tag{5.151}$$

(Shepard, 1968). We may note that γ_a is close to 1 for the particles within the domain, but there is no reason for it to equal 1, because of the inequality of the first line of (5.134), that is $J_a\{1\} \neq 1$. Equation (5.149) supersedes the first line of (5.123). This new operator is advantageous, since it satisfies

$$\begin{aligned} J_a^\gamma\{1\} &= \frac{1}{\gamma_a}\sum_b V_b w_{ab} \\ &= 1 \end{aligned} \tag{5.152}$$

Through linearity, this property holds true for every constant function. The operator J_a^γ is then more consistent than J_a, providing for the consistency of the zero-order discrete interpolation. The new discrete interpolators amount to modifying the kernel as follows:

$$\tilde{w}_{ab} \doteqdot \frac{w_{ab}}{\gamma_a} \neq \tilde{w}_{ba} \tag{5.153}$$

As in the case of the continuous interpolation, the symmetry properties as set out in Section 5.2.4 (eqns (5.120)) are lost. A vectorial operator $\mathbf{J}_a^\gamma$ can, of course, be defined along the same lines, this vector being a corrected form of the operator as given by the third line in (5.123).

Through a similar procedure, it may be required that the corrected operator be more efficient, for example that it allows the accurate interpolation of a polynomial of a given order. For further details in that respect, refer to Bonet and Lok (1999) for a first-order consistency, to Liu et al. (1996) for an arbitrary order.

Gradient operators can be defined from (5.149). It should be pointed out that the derivative of γ_a with respect to $\mathbf{r}_a$ is written as

$$\begin{aligned} \boldsymbol{\gamma}'_a &\doteq \frac{\partial \gamma_a}{\partial \mathbf{r}_a} \\ &= \sum_b V_b w'_{ab} \mathbf{e}_{ab} \\ &= \mathbf{G}_a \{1\} \end{aligned} \tag{5.154}$$

where $\mathbf{G}_a$ is the standard discrete gradient operator as defined in the second line of (5.123). The approximation (5.151) of γ_a makes it possible to provide a continuous approximation[33] of $\boldsymbol{\gamma}'_a$: this vector is simply tantamount to the kernel integral over the part of the wall being encompassed within the influence domain Ω_a of the particle a, that is $\Sigma_a = \Omega_a \cap \partial\Omega$ (refer to Fig. 5.4, which shall be compared with Fig. 5.1):

[33] Just as γ_a does not exactly equal 1 for the internal particles, $\boldsymbol{\gamma}'_a$ is not strictly zero, because $\mathbf{G}_a \{1\} \neq 0$ (refer to Section 5.2.3, eqn 5.135). It is definitely non-zero for the particles lying near the boundary.

$$\boldsymbol{\gamma}'_a \approx \int_{\Sigma_a} w_h \left(|\mathbf{r}' - \mathbf{r}_a|\right) \mathbf{n}\left(\mathbf{r}'\right) d^{n-1}\Gamma\left(\mathbf{r}'\right) \tag{5.155}$$

($\mathbf{n}$ here denotes the inward, locally wall-normal unit vector). Deriving (5.149) termwise, we can now define a corrected gradient operator, being denoted as $\mathbf{G}_a^{\gamma}$:

$$\begin{aligned} \mathbf{G}_a^{\gamma} \{A_b\} &\doteq \frac{\partial J_a^{\gamma} \{A_b\}}{\partial \mathbf{r}_a} \\ &= \frac{1}{\gamma_a} \sum_b V_b A_b w'_{ab} \mathbf{e}_{ab} - \frac{1}{\gamma_a^2} \left(\sum_c V_c A_c w_{ac} \right) \boldsymbol{\gamma}'_a \\ &= \frac{1}{\gamma_a} \sum_b V_b \left[A_b - \frac{J_a \{A_c\}}{\gamma_a} \right] w'_{ab} \mathbf{e}_{ab} \\ &= \frac{1}{\gamma_a} \left(\mathbf{G}_a \{A_b\} - J_a^{\gamma} \{A_b\} \mathbf{G}_a \{1\} \right) \end{aligned} \tag{5.156}$$

which corrects the second line in (5.123). Applying this operator to the constant function equal to one, using (5.152), we find

$$\begin{aligned} \mathbf{G}_a^{\gamma} \{1\} &= \frac{1}{\gamma_a} \left(\mathbf{G}_a \{1\} - J_a^{\gamma} \{1\} \mathbf{G}_a \{1\} \right) \\ &= \mathbf{0} \end{aligned} \tag{5.157}$$

The $\mathbf{G}_a^{\gamma}$ operator then does not have the main defect of $\mathbf{G}_a$, as given by the second line in (5.134).[34] This is known as a *renormalized* operator. This

[34] The $\tilde{\mathbf{G}}_a^k$ operators as defined by (5.130) satisfy that property. The developments in the following section will show, however, that the $\mathbf{G}_a^k$ operators are more suitable than them for providing the physical consistency of the SPH method.

procedure could obviously be successfully invoked for correcting the divergence operator. We could define, indeed,

$$\begin{aligned} D_a^\gamma \{\mathbf{A}_b\} &\doteqdot \frac{\partial}{\partial \mathbf{r}_a} \cdot \mathbf{J}_a^\gamma \{A_b\} \\ &= \frac{1}{\gamma_a} \left(D_a \{\mathbf{A}_b\} - \mathbf{J}_a^\gamma \{\mathbf{A}_b\} D_a \{\mathbf{1}\} \right) \\ D_a^\gamma \{\mathbf{1}\} &= \mathbf{0} \end{aligned} \tag{5.158}$$

The calculation (5.125) could advantageously be used for writing other forms of corrected gradients, for example:

$$\mathbf{G}_a^{\gamma,k} \{A_b\} \doteqdot \frac{1}{\gamma_a} \left(\mathbf{G}_a^k \{A_b\} - J_a^\gamma \{A_b\} \mathbf{G}_a^k \{1\} \right) \tag{5.159}$$

verifying, of course, the same property ($\mathbf{G}_a^{\gamma,k} \{1\} = \mathbf{0}$). On the same basis, divergence operators $\tilde{D}_a^{\gamma,k}$ correcting the formulas (5.130) could be constructed. Just because of the presence of the $1/\gamma_a$ factor, however, they are deprived of every property of symmetry with respect to the particle labels, unlike (5.126). Furthermore, the renormalized operators $\tilde{D}_a^{\gamma,k}$ and $\mathbf{G}_a^{\gamma,k}$ are not skew-adjoint, contrary to $\tilde{D}_a^k$ and $\mathbf{G}_a^k$ (refer to the end of Section 5.2.4). This circumstance does not favour them, as will be explained in Section 5.3.3. A more suitable variant will be introduced in Section 6.3.2.

Let us now turn to the SPH interpolation consistency from a different point of view. The above arguments are based on the observation that the elementary interpolation is not even exact at the zeroth order when it is applied to the constant functions, which becomes clear, in particular, in (5.154), where it can be seen that the discrete gradient of a constant is not identically zero. As a consequence, without the kernel correction which we have just set out, the discrete estimation of the constants depends on the positions of particles, that is it is not translationally invariant. It is, of course, a challenge in terms of the mechanical invariance properties mentioned in Chapter 1 and satisfied by the continuous equations of Chapter 3. We will attempt to get a better insight into these properties in terms of the continuous and discrete interpolations (in the light of Vignjevic et al., 2006). By analogy with Section 1.4.3, the continuous interpolation shall preserve, indeed, by translation, the constant functions, that is by linearity, the unit function:

$$[1]_c (\mathbf{r} + \boldsymbol{\epsilon}, t) = [1]_c (\mathbf{r}, t) \tag{5.160}$$

for every vector $\boldsymbol{\epsilon}$. Coming back to the definition (5.2) of the $[\cdot]_c$ operator, that yields

$$\int_{\Omega_\mathbf{r}} w_h \left(|\mathbf{r} + \boldsymbol{\epsilon} - \mathbf{r}'| \right) d^n \mathbf{r}' = \int_{\Omega_\mathbf{r}} w_h \left(|\mathbf{r} - \mathbf{r}'| \right) d^n \mathbf{r}' \tag{5.161}$$

Through a first-order Taylor series expansion in $\boldsymbol{\epsilon}$ (as in eqn (1.130))[35], we get:

$$\forall \boldsymbol{\epsilon}, \quad \left[\int_{\Omega_\mathbf{0}} \nabla w_h (\tilde{r}) \, d^n \tilde{\mathbf{r}} \right] \cdot \boldsymbol{\epsilon} = 0 \tag{5.162}$$

[35] As stated in Chapter 1, the translational invariance through an infinitesimal vector immediately yields the invariance through arbitrary translations, using additivity.

hence

$$\int_{\Omega_0} \nabla w_h(\tilde{r})\, d^n\tilde{\mathbf{r}} = \mathbf{0} \tag{5.163}$$

This equation is identical to the first condition in (5.20), as verified by the kernel for the inner points of the domain (the kernel correction procedure (5.146) would allow us to extend that property to any point). Let us now look at the discrete interpolation operator in that respect. The condition $[1]_d(\mathbf{r}, t) = [1]_d(\mathbf{r} + \boldsymbol{\epsilon}, t)$ is written as

$$\sum_b V_b w_h(|\mathbf{r} + \boldsymbol{\epsilon} - \mathbf{r}_b|) = \sum_b V_b w_h(|\mathbf{r} - \mathbf{r}_b|) \tag{5.164}$$

Through a first-order expansion in $\boldsymbol{\epsilon}$, we get the discrete counterpart of (5.163):

$$\sum_b V_b \nabla w_h(|\mathbf{r} - \mathbf{r}_b|) = \mathbf{0} \tag{5.165}$$

Applying that condition to $\mathbf{r} = \mathbf{r}_a$, we get

$$\mathbf{G}_a\{1\} = \sum_b V_b w'_{ab} \mathbf{e}_{ab} = \mathbf{0} \tag{5.166}$$

Now, we have just said that this condition is a priori not satisfied $\left(\boldsymbol{\gamma}'_a = \mathbf{G}_a\{1\} \neq \mathbf{0}\right)$. Thus, the discrete interpolation is not translationally invariant, but the renormalized kernel (5.153) solves that problem through the relation (5.157).

Let us now examine the rotational invariance. By analogy with (1.140), we wish we would find the property

$$[1]_c(\mathbf{r} + \boldsymbol{\epsilon} \times \mathbf{r}, t) = [1]_c(\mathbf{r}, t) \tag{5.167}$$

that is

$$\int_{\Omega_\mathbf{r}} w_h\left(\left|\mathbf{r} + \boldsymbol{\epsilon} \times \mathbf{r} - \mathbf{r}'\right|\right) d^n\mathbf{r}' = \int_{\Omega_\mathbf{r}} w_h\left(\left|\mathbf{r} - \mathbf{r}'\right|\right) d^n\mathbf{r}' \tag{5.168}$$

After a first-order expansion and a rearrangement of the terms, we get the condition

$$\mathbf{r} \times \int_{\Omega_0} \nabla w_h(\tilde{r})\, d^n\tilde{\mathbf{r}} = \mathbf{0} \tag{5.169}$$

which is satisfied owing to (5.163) for the inner points. Applying the same condition to the discrete case is just as simple; thus, the equality $[1]_d(\mathbf{r} + \boldsymbol{\epsilon} \times \mathbf{r}, t) = [1]_d(\mathbf{r}, t)$ yields

$$\mathbf{r} \times \sum_b V_b \nabla w_h(|\mathbf{r} - \mathbf{r}_b|) = \mathbf{0} \tag{5.170}$$

which is a priori not better satisfied than (5.166), except with the renormalized kernel. The above calculations demonstrate that using a corrected kernel (i.e. renormalized operators) provides natural properties of spatial invariance upon the discrete interpolation. These operators then have physical instead of merely numerical virtues. We will come back to this matter in Section 6.3.2.

Lastly, let us turn to another natural condition which is linked to the notion of rotation, namely the nullity of strain for a rigid motion (refer to Chapter 1, Section 1.5.1).

As we have known from the beginning of Chapter 3, this condition can be reduced to the second line of the system (3.33) in Chapter 2, for a rigid velocity field, as given by equation (1.191) in Chapter 1. We would be pleased if this relation was accurately satisfied for the interpolated velocity fields, which yields, with the continuous interpolator:

$$[\mathbf{grad}\,(\mathbf{V} + \mathbf{\Omega} \times \mathbf{r}, t)]_c = \boldsymbol{\omega} \tag{5.171}$$

where the tensor $\boldsymbol{\omega}$ is defined by (3.30). The constant term $\mathbf{V}$ does not affect the preceding relation, owing to the property (5.163). By means of the first relation in (3.33), we ultimately come to the reduced expression

$$\boldsymbol{\omega} \int_{\Omega_{\mathbf{r}}} \mathbf{r}' \otimes \nabla w_h\left(\left|\mathbf{r} - \mathbf{r}'\right|\right) d^n\mathbf{r}' = \boldsymbol{\omega} \tag{5.172}$$

Deleting in that equality the tensor $\boldsymbol{\omega}$, whose components may be arbitrarily chosen, then combining with (5.163), we get the following condition:

$$\int_{\Omega_{\mathbf{0}}} \nabla w_h\,(\tilde{r}) \otimes \tilde{\mathbf{r}}\, d^n\tilde{\mathbf{r}} = -\mathbf{I}_n \tag{5.173}$$

This relation is satisfied for the points within the domain (second line in eqn (5.20)). Within the context of discrete interpolation, there is no need for an expansion of the (identical) steps showing that the identity $[\mathbf{grad}]_d\,(\mathbf{V} + \mathbf{\Omega} \times \mathbf{r}, t) = \mathbf{0}$ is reduced to

$$\sum_b V_b \mathbf{r}_{ab} \otimes w'_{ab}\mathbf{e}_{ab} = -\mathbf{I}_n \tag{5.174}$$

that is

$$\mathbf{G}_a\,\{\mathbf{r}_b\} = \mathbf{I}_n \tag{5.175}$$

Due to the nullity of the strain under pure rotation, the discrete gradient operator should then be accurate for the linear functions, and not merely for the constant functions. Taking (5.166) into account, the condition (5.175) may also be written as

$$\mathbf{G}_a\,\{\mathbf{A} + B\mathbf{r}_b\} = B\mathbf{I}_n \tag{5.176}$$

We then should go beyond the correction (5.156), which a priori does not satisfy this new condition. If, however, we decide to start from the operator $\tilde{\mathbf{G}}_a^0$, which at least is advantageous in that it satisfies $\tilde{\mathbf{G}}_a^0\,\{1\} = \mathbf{0}$ (translational invariance), we may contemplate correcting it by multiplying, in the formulas (5.131) and (5.136), the kernel gradient times a *renormalization matrix* $\mathbf{R}_a$:

$$\begin{aligned} \tilde{\mathbf{G}}_a^{R,0}\,\{A_b\} &\doteqdot -\sum_b V_b A_{ab} w'_{ab}\mathbf{R}_a\mathbf{e}_{ab} \\ \tilde{\mathbf{G}}_a^{R,0}\,\{\mathbf{A}_b\} &\doteqdot -\sum_b V_b \mathbf{A}_{ab} \otimes \left(w'_{ab}\mathbf{R}_a\mathbf{e}_{ab}\right) \end{aligned} \tag{5.177}$$

(where the superscript ^{R} refers to renormalized operators). The condition (5.175) applied to the renormalized operator (i.e. $\tilde{\mathbf{G}}_a^{R,0}\{\mathbf{r}_b\} = \mathbf{I}_n$) then yields

$$\begin{aligned}\mathbf{R}_a &= -\left[\sum_b V_b \mathbf{r}_{ab} \otimes \left(w'_{ab}\mathbf{e}_{ab}\right)\right]^{-1} \\ &= \left[\tilde{\mathbf{G}}_a^0\{\mathbf{r}_b\}\right]^{-1}\end{aligned} \tag{5.178}$$

This algorithm implies the calculation of the inverse of a $n \times n$ matrix for each particle. This technique may also be tailored to the corrected kernel (5.153) (refer, for instance, to Bonet and Lok, 1999). Moreover, it could be generalized to the case of all the operators (5.130), which would provide renormalized gradients $\tilde{\mathbf{G}}_a^{R,k}$. Along the lines of (5.177), we also can construct renormalized discrete divergence operators $D_a^{R,k}$ and $\mathbf{D}_a^{R,k}$ that would be the skew-adjoints of the $\tilde{\mathbf{G}}_a^{R,k}$ operators (Vila, 1999). For $k = 0$, a calculation analogous to (5.138) and (5.139) will give, for example

$$\begin{aligned}\left\langle\tilde{\mathbf{G}}_a^{R,0}\{A_b\}, \{\mathbf{B}_a\}\right\rangle &= -\rho_a \sum_{a,b} V_a V_b A_{ab} w'_{ab} \mathbf{B}_b^T \mathbf{R}_a \mathbf{e}_{ab} \\ &= -\frac{1}{2}\rho_a \sum_{a,b} V_a V_b A_{ab}\left(\mathbf{B}_b^T \mathbf{R}_a + \mathbf{B}_a^T \mathbf{R}_b\right) w'_{ab}\mathbf{e}_{ab} \\ &= -\rho_a \sum_{a,b} V_a V_b A_a \left(\mathbf{B}_b^T \mathbf{R}_a + \mathbf{B}_a^T \mathbf{R}_b\right) w'_{ab}\mathbf{e}_{ab} \\ &= -\left(\{A_a\}, D_a^{R,0}\{\mathbf{B}_b\}\right)\end{aligned} \tag{5.179}$$

with

$$D_a^{R,0}\{\mathbf{B}_b\} \doteqdot \sum_b V_b\left(\mathbf{B}_b^T \mathbf{R}_a + \mathbf{B}_a^T \mathbf{R}_b\right) \cdot w'_{ab}\mathbf{e}_{ab} \tag{5.180}$$

to be compared with (5.127).

As we have seen, there are two kinds of renormalization in SPH. The first consists in correcting the kernel near the boundaries, in order to improve the discrete interpolation in the vicinity of them (operators like $\mathbf{G}_a^{\gamma,k}$). The second consists in modifying the kernel gradient in order to get first-order interpolation consistency, and hence no velocity gradients for a rigid body motion (operators like $\tilde{\mathbf{G}}_a^{R,k}$).

5.3 Discrete forms of the fluid equations

We now address the question of writing the fluid dynamic equations in SPH language. From the considerations of the previous section, we derive several forms of the discrete continuity, momentum and energy equations for non-viscous, weakly compressible flows. We then show how the discrete momentum equations can be derived from a least action principle, using the skew-adjointness properties of the discrete operators. We eventually review the laws of discrete conservation in the SPH framework.

5.3.1 Discrete continuity equation

We are now going to utilize items from the previous sections to set up discrete forms of the Euler equations (3.129). To do this, we will implement the nearly incompressible formalism of Chapter 3 (Sections 3.4.2 and 3.4.3). Chapters 1 and 2 already gave a first insight into such discrete forms in the equation of motion as (1.240). For several reasons, however, the mere data from these two chapters are not sufficient for numerically solving these equations. In particular, Section 1.5.4 has highlighted the difficulty of modelling the dependence of the densities with respect to the positions. The next sections aim at clarifying the gaps by means of SPH formalism, trying to remain consistent, on the one hand, with those properties regarding the mechanics of the material point (especially the conservation laws), on the other hand with the continuous equations given in Chapter 3.

From the discrete differentiation method established in the previous paragraph, we can formulate the continuity equation (3.43) of a compressible or nearly incompressible fluid. We may recall that this equation is written as

$$\frac{d\rho}{dt} = -\rho \operatorname{div}\mathbf{u} \tag{5.181}$$

Using the approximation (5.127) (D_a^0 operator) for the right-hand side, we get

$$\forall a, \ \frac{d\rho_a}{dt} = -\rho_a \sum_b V_b \left(\mathbf{u}_a + \mathbf{u}_b\right) \cdot w'_{ab}\mathbf{e}_{ab} \tag{5.182}$$

One of the main properties of the SPH approach consists in inherently conserving the mass, if the particle masses are chosen to be constant, as we have.[36] Moreover, working in a compressible formalism means that the volume of each particle is a priori not conserved. Let us turn to the volume variation of a particle a, starting from the continuous equation (3.19) and the model (5.182) chosen for discretizing div**u**:

[36] Some authors, however, prefer working with non-constant masses (see e.g. Vila, 1999).

$$\begin{aligned} \frac{dV_a}{dt} &= V_a \left(\operatorname{div}\mathbf{u}\right)_a \\ &= \sum_b V_a V_b \left(\mathbf{u}_a + \mathbf{u}_b\right) \cdot w'_{ab}\mathbf{e}_{ab} \end{aligned} \tag{5.183}$$

This equation shows that the contribution of a particle b to that variation is written as

$$q_{V,ba} = V_a V_b \left(\mathbf{u}_a + \mathbf{u}_b\right) \cdot w'_{ab}\mathbf{e}_{ab} \tag{5.184}$$

We have called this quantity $q_{V,ba}$ because it represents a volume flux directed from b towards a, that is the volume transferred from b to a per unit time. Now, because of the kernel gradient antisymmetry (5.8) (or (5.121)), we find that the volume flux (5.121) is antisymmetric ($q_{V,ab} = -q_{V,ba}$), which means that the whole volume of matter transferred from b to a is gained by a; in other words, the whole volume of the continuous medium is conserved. Now, in this case, this interesting result is not physical since we are working in the context of a compressible fluid. With the above model, though the velocity field divergence in (5.183) is not necessarily zero, that is though the volume is not conserved locally, conversely it is conserved globally, which is at odds with the reality. The above discussion shows that we must adopt another form for the continuity equation, thus another operator for discretizing div**u**. If we use the formula (5.132) ($\tilde{D}_a^1$ operator) for modelling the velocity field divergence in (5.181), then we get

$$\forall a, \quad \frac{d\rho_a}{dt} = \sum_b m_b \mathbf{u}_{ab} \cdot w'_{ab} \mathbf{e}_{ab} \tag{5.185}$$

The volume flux is henceforth given by

$$q_{V,ba} = -\frac{1}{\rho_a^2} m_a m_b \mathbf{u}_{ab} \cdot w'_{ab} \mathbf{e}_{ab} \tag{5.186}$$

It no longer obeys any symmetry or antisymmetry condition, because of the term in ρ_a^2, so that the volume is no longer conserved. That discretization of the continuity equation is then a priori better than (5.182). Let us now consider a particle b moving towards a at a velocity $\mathbf{u}_b$ having a higher norm than that of $\mathbf{u}_a$, these two velocities being aligned (refer to Fig. 5.5 (a)). We then have

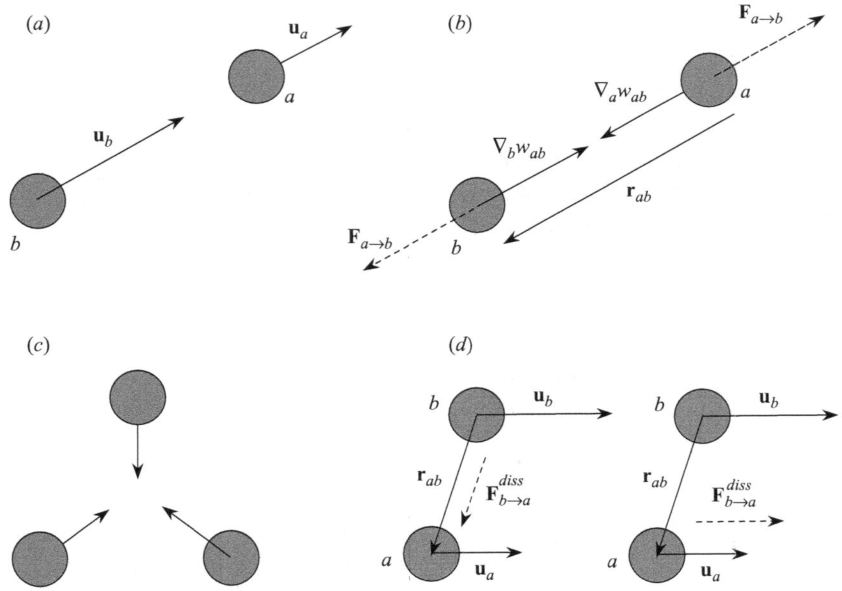

Fig. 5.5 A few illustrations of the processes in the SPH method: (a) two particles moving in the same direction; (b) conservation of linear momentum and angular momentum, based on forces collinear to the radius vector connecting two particles; (c) several particles attempting to converge onto one point; (d) two possible models of the viscous forces.

$\mathbf{u}_{ab} \cdot \mathbf{e}_{ab} > 0$; now, physical intuition here leads us to contemplate a volume flux being directed from b towards a, that is $q_{V,ba} > 0$. The derivative w'_{ab} then has to be negative, which, according to (5.122), leads to the general rule

$$w'_h(\tilde{r}) < 0 \tag{5.187}$$

that is if we return to the dimensionless kernel as defined by (5.36):

$$f'(q) < 0 \tag{5.188}$$

Thus, we get a further condition which the kernel should verify, namely its decrease versus distance. In the case where $|\mathbf{u}_b| < |\mathbf{u}_a|$, (5.186) results in a negative volume flux $q_{V,ba}$, that is a positive flux from a towards b ($q_{V,ab}$), true to intuition. Moreover, the form (5.184) provides a volume flux which is directed from a towards b irrespective of the sign of $\mathbf{u}_{ab} \cdot \mathbf{e}_{ab}$ (if b is moving towards a). This result, which is not physical, substantiates our choice.

We then must drop (5.182) and adopt the form (5.185) of the discrete continuity equation. This equation is the first equation in the 'standard' SPH model. It is used for globally assessing all the densities $\{\rho_a\}$ at each time step, knowing their positions $\{\mathbf{r}_a\}$ and velocities $\{\mathbf{u}_a\}$, by means of numerical schemes which will explicitly be disclosed later on (Section 5.4).

In the light of the above, we can see that choosing an SPH form for a given differential operator is not insignificant; we will realize it on several occasions. The considerations about the interparticular fluxes will be of the utmost importance in the following, because all of the other discretized equations will exhibit analogous symmetry or antisymmetry properties, resulting (when they are suitably chosen) in accurate and physical conservation laws. We have just explained that a $\tilde{D}_a^k$ type symmetric operator is required for discretizing the continuity equation, though our option (5.185) is not exclusive: any of the general formulations (5.130) would be possible instead of (5.132), which would yield

$$\forall a, \quad \frac{d\rho_a}{dt} = \frac{1}{\rho_a^{2k-1}} \sum_b V_b (\rho_a \rho_b)^k \mathbf{u}_{ab} \cdot w'_{ab} \mathbf{e}_{ab} \tag{5.189}$$

that is with $k = 0$:

$$\forall a, \quad \frac{d\rho_a}{dt} = \rho_a \sum_b V_b \mathbf{u}_{ab} \cdot w'_{ab} \mathbf{e}_{ab} \tag{5.190}$$

This equation may well be substituted for (5.185), which corresponds to $k = 1$.

The presently introduced method is not the only one for calculating the particle density. This is because we may simply return to the discrete interpolation of an arbitrary quantity (5.123) and apply it to the density raised to some power ($A = \rho^k$), which yields, with (5.79):

$$\rho_a^k \approx J_a \left\{ \rho_b^k \right\} = \sum_b V_b \rho_b^k w_{ab} \tag{5.191}$$

Dividing by ρ_a^{k-1}, we get a possible estimation of the density as

$$\forall a,\ \rho_a = \sum_b m_b \left(\frac{\rho_b}{\rho_a}\right)^{k-1} w_{ab} \tag{5.192}$$

With $k = 1$, we get, for example:

$$\forall a,\ \rho_a = \sum_b m_b w_{ab} \tag{5.193}$$

This approach then consists of calculating the densities $\{\rho_a\}$ knowing all the positions $\{\mathbf{r}_a\}$, that is the quantities $w_{ab} = w_h(r_{ab})$. It offers the advantage of not requiring any numerical resolution, unlike (5.185). However, it gives rise to stability problems, particularly as regards the free surfaces (Monaghan, 1994). Vila (1999), conversely, finds advantages in (5.193). It is noteworthy that these two approaches are consistent with each other, in some respects. To be sure, one only needs to derive (5.192) with respect to time and get, with (5.79):

$$\forall a,\ \frac{d\rho_a}{dt} = \rho_a \sum_b V_b \left(\frac{\rho_b}{\rho_a}\right)^k \frac{dw_{ab}}{dt} \tag{5.194}$$

Afterwards, the derivative of $w_{ab} = w_h(r_{ab})$ with respect to time is written using (5.117), (5.118) and (5.121) and yields:

$$\begin{aligned}\frac{dw_h(r_{ab})}{dt} &= \frac{\partial w_{ab}}{\partial \mathbf{r}_a}\cdot\frac{d\mathbf{r}_a}{dt} + \frac{\partial w_{ab}}{\partial \mathbf{r}_b}\cdot\frac{d\mathbf{r}_b}{dt} \\ &= w_h'(r_{ab})\,\mathbf{e}_{ab}\cdot\mathbf{u}_{ab}\end{aligned} \tag{5.195}$$

Combining the latter two equations, we encounter again (5.189). It should be pointed out that the presently demonstrated compatability between these two approaches is based on an integration of the position over time in (5.195). Now, from a numerical point of view, the integration over time is inevitably approximate,[37] which accounts for the different behaviours of the two formulations.

[37] In that regard, refer to Section 5.4.

It is interesting to address the discrete continuity equation through its relationship with the continuous continuity equation in its integral form (3.41), temporarily returning to the continuous interpolation (5.2) as discussed in Section 5.2.1. To that purpose, let us consider the interpolation $[A]_c(\mathbf{r}, t)$ of a field $A \doteqdot \rho B$, and let us write its Lagrangian derivative using the property (3.50) of Chapter 3:

$$\begin{aligned}\frac{d[A]_c(\mathbf{r},t)}{dt} &= \frac{d}{dt}\int_{\Omega_\mathbf{r}} \rho(\mathbf{r}')\,B(\mathbf{r}')\,w_h(|\mathbf{r}-\mathbf{r}'|)\,d^n\mathbf{r}' \\ &= \int_{\Omega_\mathbf{r}} \rho(\mathbf{r}')\,\frac{dB(\mathbf{r}')\,w_h(|\mathbf{r}-\mathbf{r}'|)}{dt}\,d^n\mathbf{r}'\end{aligned} \tag{5.196}$$

The case $A \equiv \rho$ corresponds to $B \equiv 1$. The Lagrangian derivative of $w_h\left(\left|\mathbf{r}-\mathbf{r}'\right|\right)$ is calculated as in (5.195) to yield

$$\begin{aligned}\frac{d\left[\rho\right]_c(\mathbf{r},t)}{dt} &= \int_{\Omega_\mathbf{r}} \rho\left(\mathbf{r}'\right)\frac{dw_h\left(\left|\mathbf{r}-\mathbf{r}'\right|\right)}{dt}d^n\mathbf{r}' \\ &= \int_{\Omega_\mathbf{r}} \rho\left(\mathbf{r}'\right)\nabla w_h\left(\left|\mathbf{r}-\mathbf{r}'\right|\right)\cdot\left[\mathbf{u}\left(\mathbf{r}\right)-\mathbf{u}\left(\mathbf{r}'\right)\right]d^n\mathbf{r}'\end{aligned} \tag{5.197}$$

For the time being, this calculation is perfectly exact. Considering the location $\mathbf{r}_a$ of a particle instead of $\mathbf{r}$, and using the discrete approximation (5.86) to approximate the integral, we have

$$\frac{d\left[\rho\right]_c(\mathbf{r}_a,t)}{dt} \approx \sum_b m_b \mathbf{u}_{ab}\cdot\nabla w_h\left(r_{ab}\right) \tag{5.198}$$

which is identical to (5.185), provided that $\left[\rho\right]_c(\mathbf{r}_a,t)$ is made identical to $\rho_a(t)$. Through that procedure, we find that the discrete continuity equation may be likened to a balance equation, those terms occurring under the discrete sum of the right-hand sides being possibly treated as discrete fluxes, as earlier explained. The same is true of all the equations to be provided in the following, which gives the SPH method some physical consistency (developments of this principle can be found in Vila, 1999, in the form of weak formulations for SPH).

This methodology can be matched to the nearly incompressible or compressible flow. The strictly incompressible flows will be dealt with in Section 6.2.5.

5.3.2 Discrete Euler equation

For treating the equation of motion according to the SPH formalism, some authors suggest, particularly within the context of elasticity, to go back to the generic equation of motion of the continua, that is the Cauchy equation (3.71), modelling the divergence term of the stress tensor by means of the $\mathbf{D}^k$ operator (eqn (5.126) extended to the second-order tensors as in (5.135)), which simply yields:

$$\forall a,\quad \frac{d\mathbf{u}_a}{dt} = \sum_b m_b \frac{\rho_b^{2k}\boldsymbol{\sigma}_a + \rho_a^{2k}\boldsymbol{\sigma}_b}{\left(\rho_a\rho_b\right)^{k+1}}\cdot w'_{ab}\mathbf{e}_{ab} + \mathbf{g} \tag{5.199}$$

The stress tensor $\boldsymbol{\sigma}_a$ attached to each particle is then modelled using the behaviour law of the medium being considered (refer to that notion in Chapter 3, particularly Section 3.4.1). This approach is advantageous because it is simple, but has demonstrated weaknesses, especially in terms of numerical stability. As regards fluid mechanics, since the knowledge of the behaviour law enables us to come to specific equations of motion, we prefer to directly discretize the latter.

In order to address all the possible forms of the discretized Euler or Navier–Stokes equations of motion, we will first consider a non-dissipative force context (the viscous forces will be discussed in Section 6.2.2). The Euler equation of motion (3.129) may be treated like the continuity equation dealt with in the previous section. To that purpose, we will employ the general form (5.126) of the discrete gradient ($\mathbf{G}_a^k$ operators), which leads to a general form of the discrete equation of motion:[38]

[38] In order to duly review all the particle forces, we will never use dynamic pressure in Part II.

$$\forall a, \quad \frac{d\mathbf{u}_a}{dt} = -\sum_b m_b \frac{\rho_b^{2k} p_a + \rho_a^{2k} p_b}{(\rho_a \rho_b)^{k+1}} w'_{ab} \mathbf{e}_{ab} + \mathbf{g} \tag{5.200}$$

which could immediately have been derived from (5.199) through the behaviour law (3.116) of a non-viscous fluid. The particular forms (5.127) ($k = 0$) and (5.128) ($k = 1$) of the discrete gradient give, respectively

$$\forall a, \quad \frac{d\mathbf{u}_a}{dt} = -\frac{1}{\rho_a} \sum_b V_b \left(p_a + p_b\right) w'_{ab} \mathbf{e}_{ab} + \mathbf{g} \tag{5.201}$$

and

$$\forall a, \quad \frac{d\mathbf{u}_a}{dt} = -\sum_b m_b \left(\frac{p_a}{\rho_a^2} + \frac{p_b}{\rho_b^2}\right) w'_{ab} \mathbf{e}_{ab} + \mathbf{g} \tag{5.202}$$

Let us now return to the general equation of motion (2.186) established in Chapter 1, without dissipative forces:

$$\begin{aligned} \forall a, \ m_a \frac{d\mathbf{u}_a}{dt} &= \mathbf{F}_a \\ &= \mathbf{F}_a^{int} + \mathbf{F}_a^{ext} \\ &= \sum_b \mathbf{F}_{b \to a}^{int} + \mathbf{F}_a^{ext} \end{aligned} \tag{5.203}$$

Since the external force being considered is gravity ($\mathbf{F}_a^{ext} = m_a \mathbf{g}$), we find that the right-hand side in (5.200) has a form analogous to (5.203), with

$$\mathbf{F}_{b \to a}^{int} = -m_a m_b \frac{\rho_b^{2k} p_a + \rho_a^{2k} p_b}{(\rho_a \rho_b)^{k+1}} w'_{ab} \mathbf{e}_{ab} = -\mathbf{F}_{a \to b}^{int} \tag{5.204}$$

(the antisymmetry results from (5.121)). The internal forces (pressure forces, in accordance with Section 1.5.4) then occur in an antisymmetric form duly ensuring Newton's third law, that is the conservation of the total momentum of an isolated system (refer to Fig. 5.5 (*b*)). As previously mentioned in the continuous formalism (Section 3.3.2), the forces consist of momentum fluxes. The remark we have just made is then analogous to the (erroneous) symmetry of the volume fluxes (5.184); this time, it definitely constitutes a physical reality which the general model (5.200) advantageously satisfies. We may note that with the general form (5.130) for the discrete gradient ($\tilde{\mathbf{G}}_a^k$ operators), this important property would not be verified any longer.[39]

[39] Let us also note the importance, within that context, of the assumed contant smoothing length h.

Equation (5.200) is used for constantly calculating the set $\{\mathbf{u}_a\}$ of velocities at each time, the quantities $\{\mathbf{r}_a\}$, $\{\mathbf{u}_a\}$, $\{p_a\}$ and $\{\rho_a\}$ being known a small instant before. The positions are then derived from the velocities by integrating equation (5.83). As in Chapter 1, the momentum $\mathbf{p}_a = m_a \mathbf{u}_a$ of a particle (eqn (1.53)) can be defined and the state parameters of a particle can be gathered in one vector $\mathbf{Y}_a \doteqdot \left(\mathbf{p}_a^T, \mathbf{r}_a^T, \rho_a\right)^T$ (in accordance with the definition (1.266)). We then obtain a system in the following form:

$$\forall a, \quad \frac{d\mathbf{Y}_a}{dt} = \dot{\mathbf{Y}}_a\left(\{\mathbf{r}_a\}, \{\mathbf{u}_a\}, \{p_a\}, \{\rho_a\}\right) \tag{5.205}$$

with

$$\dot{\mathbf{Y}}_a \doteqdot \begin{pmatrix} \mathbf{F}_a \\ \mathbf{u}_a \\ \dot{\rho}_a \end{pmatrix} \tag{5.206}$$

$\dot{\rho}_a$ standing for the quantity $d\rho_a/dt$, as given by Section 5.3.1.

Lastly, the question arises as to the calculation of pressure, which can be made in several ways. The simplest approach, for a weakly compressible and isentropic flow, consists in resorting to the barotropic state equation (2.113). The density fluctuations will then be given by the formula (3.170) of Chapter 3. As explained later on in Section 5.4.4, however, considering the true speed of sound c is numerically very burdensome when solving the discrete equations. Thus, a numerical sound velocity c_0 smaller than c will preferably be contemplated and the following explicit formula will be written:

$$\forall a, \quad p_a = \frac{\rho_0 c_0^2}{\gamma}\left[\left(\frac{\rho_a}{\rho_0}\right)^{\gamma} - 1\right] \tag{5.207}$$

This equation is attractive since it contains the physics as set out in Section 2.3.3. Moreover, from a strictly numerical point of view, it may accidentally yield negative pressure values if ever the density locally and momentarily takes a lower value than its reference value ρ_0. This phenomenon ready takes place if the standard values as provided by Table 3.2 in Chapter 3 (Section 3.4.1) are chosen for ρ_0. Negative pressures obviously have no physical meaning, as evidenced in Chapter 2 (Section 2.3.1), as well as in Section 3.4.1, on the basis of the definition of the stress tensor of a fluid. Adding a 'background pressure' is therefore often preferred, which amounts to adding a positive constant D to the expression between brackets in (5.207):

$$\begin{aligned} p_a &= \frac{\rho_0 c_0^2}{\gamma}\left[\left(\frac{\rho_a}{\rho_0}\right)^{\gamma} - 1 + D\right] \\ &= \frac{\rho_0 c_0^2}{\gamma}\left[\left(\frac{\rho_a}{\rho_0}\right)^{\gamma} + C\right] \end{aligned} \tag{5.208}$$

with

$$C \doteqdot D - 1 \geqslant -1 \tag{5.209}$$

This trick does not create any effect on the pressure forces, since we know that only the pressure deviations in space are significant (refer to Section 3.4.3).

It is noteworthy that the weakly compressible nature of this method can be interpreted in a simple way: when several particles tend to come closer to each other, the velocity field locally converges, and the Gauss theorem shows that div$\mathbf{u} < 0$. The continuity equation (5.181) then indicates that the density is slightly increasing. The ideas expressed in Section 5.3.1 show that it is still true in the discrete approach, and the state equation (5.207) leads to a local pressure rise, which is very large because of the value of c_0^2 and the superscript γ.

A high pressure gradient is then created, being directed towards the particle convergence point. The resulting pressure force serves as a repellent in the equation of motion in order to keep the particles away (refer to Fig. 5.5 (*c*)). Through this mechanism[40] slight density variations are ensured, which keep the interparticular distance close to its characteristic value δr, being defined in Section 5.2.3. According to (5.80), the volume of each particle is then preserved at first sight. The relative density deviations, according to (3.170), will now be given by

[40]This is exactly the physical mechanism taking place within a nearly-incompressible flow, as explained in Chapter 2 (Section 2.3.3).

$$\frac{\delta\rho}{\rho_0} \sim Ma_0^2 \max\left(St, 1, \frac{1}{Fr^2}\right) \qquad (5.210)$$

with a numerical Mach number, defined by analogy with (3.159):

$$Ma_0 \doteqdot \frac{U}{c_0} \qquad (5.211)$$

U being an order of magnitude of the velocity in the flow. If the latter is steady, and if inertia prevails among the gravity forces, $St = 0$ and $Fr \gg 1$ (refer to Section 3.4.3). It will therefore be possible to consider the flow as weakly compressible if Ma_0^2 is lower than a sufficiently small arbitrary critical value, which is usually set to 0.01. As a precaution, in order to satisfy that condition everywhere in the flow, the value of the maximum velocity $U_{\max}$ should once again be kept for U. Thus, the numerical sound velocity will be chosen as follows:

$$c_0 = 10U_{\max} \qquad (5.212)$$

This approach, which is recommended in particular by Monaghan (1992), has by and large proved its worth, although in some cases, particularly in the absence of gravity, Issa et al. (2004) have demonstrated that the factor 10 is sometimes insufficient. If gravity has major effects, (5.210) and (3.154) show that

$$c_0 = 10 \max\left(U_{\max}, \sqrt{gL}\right) \qquad (5.213)$$

should be substituted for (5.212), L being a characteristic vertical dimension of the flow being considered.

Pressure could also be calculated from a more general state equation analogous to (2.98), which requires the estimation of internal energy. Since we are referring to the context of the weakly compressible flows without viscous effects, we may utilize equation (3.117). With the $\tilde{D}_a^k$ operator, we get, for example

$$\forall a, \quad \frac{de_{int,a}}{dt} = \frac{p_a}{\rho_a^{2k+1}} \sum_b V_b \left(\rho_a \rho_b\right)^k \mathbf{u}_{ab} \cdot w'_{ab} \mathbf{e}_{ab} \qquad (5.214)$$

With the D_a^k operator, we find

$$\forall a, \quad \frac{de_{int,a}}{dt} = \frac{p_a}{\rho_a} \sum_b V_b \frac{\rho_b^{2k} \mathbf{u}_a + \rho_a^{2k} \mathbf{u}_b}{\left(\rho_a \rho_b\right)^k} \cdot w'_{ab} \mathbf{e}_{ab} \qquad (5.215)$$

which might supersede (5.214). Further forms of that equation, having more pleasant symmetry properties, could be provided. Let us resume the form (5.214), for example, which, for $k = 0$, corresponds to the following approximation:

$$-\left(\frac{p}{\rho}\mathrm{div}\mathbf{u}\right)_a \approx p_a \sum_b m_b \mathbf{u}_{ab} \cdot w'_{ab}\mathbf{e}_{ab} \tag{5.216}$$

Furthermore, the relation (5.137) also allows us to write

$$-\frac{1}{\rho_a}\left(p\mathrm{div}\mathbf{u}\right)_a \approx \frac{1}{\rho_a}\sum_b V_b p_b \mathbf{u}_{ab} \cdot w'_{ab}\mathbf{e}_{ab} \tag{5.217}$$

Considering the average of the latter two formulas, we get

$$\forall a, \quad \frac{de_{int,a}}{dt} = \frac{1}{2}\sum_b m_b \frac{p_a + p_b}{\rho_a \rho_b}\mathbf{u}_{ab} \cdot w'_{ab}\mathbf{e}_{ab} \tag{5.218}$$

All these equations can also be obtained, obviously, from (1.255) and (1.260), which yield

$$\frac{de_{int,a}}{dt} = \frac{p_a}{\rho_a^2}\frac{d\rho_a}{dt} \tag{5.219}$$

The form (5.189) of the discrete continuity equation, for instance, then once again provides (5.214).

We will consider in Section 6.2.5 of Chapter 6 another approach for the calculation of pressure, which is more suitable for the incompressible flows.

5.3.3 SPH through Lagrange

We have seen in the previous paragraph that the equations of motion as discretized through the SPH method enable us, provided that the appropriate discrete operators are chosen, to satisfy the law of conservation of the total momentum of an isolated system. Choosing antisymmetric operators leads to that result, providing momentum fluxes (in other words, forces) which are symmetrical with respect to the particle labels. It is an important result,[41] and we already catch a glimpse of the idea that the discrete equation (5.200) could, as in Chapter 1, be constructed directly from a least action principle,[42] without going through the continuous Euler equation. This would then automatically result in other conservation laws, in accordance with the foundations of mechanics as described in Chapter 1. This is what we intend to do in this section, generalizing the results of both Bonet and Lok (1999) and Oger et al. (2007). One may also refer to Bonet and Rodriguez-Paz (2005) for a Hamiltonian approach based on the arguments of Section 1.4.2.

[41] It is, of course, linked to the Galilean invariance of the discrete equations.

[42] This is because we have disregarded the dissipative forces in this chapter.

Let us first return to the Lagrangian of a particle system as written in (1.238). With the internal energy per unit mass $e_{int,b} \doteqdot E_{int,b}/m_b$ as defined in Section 1.5.4 and the potential (1.52) of the force of gravity, this gives

$$L = \sum_b \frac{1}{2}m_b u_b^2 - \sum_b m_b e_{int,b}\left(\{\mathbf{r}_a\}\right) - \sum_b m_b g z_b \tag{5.220}$$

which is a discrete approximation of the continuous Lagrangian as yielded by (3.300) and (3.341) for a perfect fluid flow. Here we have written the internal forces as a function of the positions instead of the densities, in accordance with the remark made at the end of Section 1.5.4. The corresponding Hamiltonian is given by

$$\begin{aligned} H &= \sum_b \frac{1}{2} m_b u_b^2 + \sum_b m_b e_{int,b}\left(\{\mathbf{r}_a\}\right) + \sum_b m_b g z_b \\ &= \sum_b \frac{1}{2} m_b u_b^2 + \sum_b E_{p,b}\left(\{\mathbf{r}_a\}\right) \end{aligned} \tag{5.221}$$

where $E_{p,b} = m_b e_{int,b} + E_{ext,b}$ represents the total potential. The discrete equation of motion of each particle will then be provided by the Lagrange equations (1.31), which are recalled below:

$$\forall a, \quad \frac{d}{dt}\frac{\partial L}{\partial \mathbf{u}_a} = \frac{\partial L}{\partial \mathbf{r}_a} \tag{5.222}$$

As in Chapter 1, the derivative of L with respect to $\mathbf{u}_a$ (the other parameters being set) is calculated from (5.220) and is nothing but the momentum of particle a (eqn (1.53)):

$$\frac{\partial L}{\partial \mathbf{u}_a} = m_a \mathbf{u}_a \tag{5.223}$$

Besides, (5.220) also yields the force as received by a:

$$\begin{aligned} \frac{\partial L}{\partial \mathbf{r}_a} &= -\sum_b m_b \frac{\partial e_{int,b}}{\partial \mathbf{r}_a} - \sum_b m_b g \frac{\partial z_b}{\partial \mathbf{r}_a} \\ &= -\sum_b m_b \left(\frac{\partial e_{int}}{\partial \rho}\right)_b \frac{\partial \rho_b}{\partial \mathbf{r}_a} - \sum_b m_b g \delta_{ab} \mathbf{e}_z \\ &= -\sum_b m_b \frac{p_b}{\rho_b^2}\frac{\partial \rho_b}{\partial \mathbf{r}_a} + m_a \mathbf{g} \end{aligned} \tag{5.224}$$

(we have used the relation (1.256)). There arises the derivative of the density of b with respect to the position of a. That makes sense, because we have explained in Section 1.5.4 that the mutual positions of the particles do determine the pressure forces. In order to estimate that quantity, we must, for the first time in this section, resort to the discrete SPH interpolation. We will write, using the interpolation formula (5.192), for an arbitrarily chosen real k, here being rearranged with changed indices:

$$\rho_b = \sum_c m_c \left(\frac{\rho_c}{\rho_b}\right)^{k-1} w_{bc} \tag{5.225}$$

As explained in Section 5.3.1, ρ_b thereby implicitly depends on the set $\{\mathbf{r}_c\}$ of positions, via the quantities w_{bc}, and now we can carry out the derivation with respect to $\mathbf{r}_a$, which will result in quantities of the $\nabla_a w_{ab}$ type. To that purpose, it should be pointed out that $\mathbf{r}_a$ may occur in two ways in (5.225), that is once when the dummy sub-script c equals a, as well as in each term of the sum if $b = a$, which yields:

$$\frac{\partial \rho_b}{\partial \mathbf{r}_a} = m_a \left(\frac{\rho_a}{\rho_b}\right)^{k-1} \nabla_a w_{ab} + \delta_{ab} \sum_c m_c \left(\frac{\rho_c}{\rho_b}\right)^{k-1} \nabla_a w_{ac} \tag{5.226}$$

Inserting (5.226) into (5.224) and taking advantage of (5.118), we get:

$$\begin{aligned} \frac{\partial L}{\partial \mathbf{r}_a} &= -\sum_b m_a m_b \frac{\rho_a^{k-1} p_b}{\rho_b^{k+1}} w'_{ab} \mathbf{e}_{ab} \\ &\quad - \sum_{b,c} m_b m_c \frac{p_b}{\rho_b^2} \left(\frac{\rho_c}{\rho_a}\right)^{k-1} \delta_{ab} w'_{ac} \mathbf{e}_{ac} + m_a \mathbf{g} \end{aligned} \tag{5.227}$$

The only subsisting term from the second discrete sum on b is the term corresponding to $b = a$, because of the Kronecker symbol δ_{ab}:

$$\begin{aligned} \sum_{b,c} m_b m_c \frac{p_b}{\rho_b^2} \left(\frac{\rho_c}{\rho_a}\right)^{k-1} \delta_{ab} w'_{ac} \mathbf{e}_{ac} &= \sum_c m_a m_c \frac{p_a}{\rho_a^2} \left(\frac{\rho_c}{\rho_a}\right)^{k-1} w'_{ac} \mathbf{e}_{ac} \\ &= \sum_b m_a m_b \frac{\rho_b^{k-1} p_a}{\rho_a^{k+1}} w'_{ab} \mathbf{e}_{ab} \end{aligned} \tag{5.228}$$

(we have made a change of dummy subscript $c \leftarrow b$ on the second line). Returning to (5.227), we ultimately get out of it, with the Lagrange equation (5.222) and dividing by m_a:

$$\forall a, \quad \frac{d\mathbf{u}_a}{dt} = -\sum_b m_b \frac{\rho_b^{2k} p_a + \rho_a^{2k} p_b}{(\rho_a \rho_b)^{k+1}} w'_{ab} \mathbf{e}_{ab} + \mathbf{g} \tag{5.229}$$

As expected, this equation is a true copy of (5.200).

It should be noted that this calculation is based on the interpolation (5.225). It then seems that an equation of motion (that is a value of k) was chosen because an interpolation method was adopted for calculating the density; in other words, it seems that the $\mathbf{G}_a^k$ operator for calculating the pressure gradient is compatible with the interpolation (5.225) of the quantities $\{\rho_b^k\}$. Now, we have shown in Section 5.3.1 that ρ may be calculated from a discrete form of the continuity equation; consequently, there is probably a way of deriving the equation of motion from the Lagrange equations through that procedure. To do this, we first return to the discretized continuity equation in its general form (5.189), with a slightly different arrangement to express the masses rather than the volumes:

$$\frac{d\rho_b}{dt} = \frac{1}{\rho_b^{2k-2}} \sum_c m_c (\rho_b \rho_c)^{k-1} \mathbf{u}_{bc} \cdot w'_{bc} \mathbf{e}_{bc} \tag{5.230}$$

Multiplying by dt and using (5.83), we then can write the differential $d\rho_b$:

$$\begin{aligned} d\rho_b &= \frac{1}{\rho_b^{2k-2}} \sum_c m_c \left(\rho_b \rho_c\right)^{k-1} \left(d\mathbf{r}_b - d\mathbf{r}_c\right) \cdot w'_{bc} \mathbf{e}_{bc} \\ &= \sum_a \frac{\partial \rho_b}{\partial \mathbf{r}_a} \cdot d\mathbf{r}_a \end{aligned} \tag{5.231}$$

(the first line of (5.231) could have been obtained as well by multiplying (5.194) times dt, then using (5.195)). The second line in (5.231) is nothing but the definition of a differential; through a term-wise identification of the two expressions, we find

$$\frac{\partial \rho_b}{\partial \mathbf{r}_a} = \frac{1}{\rho_b^{2k-2}} \sum_c m_c \left(\rho_b \rho_c\right)^{k-1} \left(\delta_{ab} - \delta_{ac}\right) w'_{bc} \mathbf{e}_{bc} \tag{5.232}$$

We now may enter (5.232) into the expression (5.224) of the force, which yields (dropping $m_a \mathbf{g}$ for the sake of clarity):

$$\begin{aligned} \frac{\partial L}{\partial \mathbf{r}_a} &= -\sum_{b,c} m_b m_c \frac{p_b}{\rho_b^{2k}} \left(\rho_b \rho_c\right)^{k-1} \left(\delta_{ab} - \delta_{ac}\right) w'_{bc} \mathbf{e}_{bc} \\ &= -\sum_c m_a m_c \frac{p_a}{\rho_a^{2k}} \left(\rho_a \rho_c\right)^{k-1} w'_{ac} \mathbf{e}_{ac} \\ &\quad + \sum_b m_b m_a \frac{p_b}{\rho_b^{2k}} \left(\rho_b \rho_a\right)^{k-1} w'_{ba} \mathbf{e}_{ba} \\ &= -\sum_b m_a m_b \frac{\rho_b^{2k} p_a + \rho_a^{2k} p_b}{\left(\rho_a \rho_b\right)^{k+1}} w'_{ab} \mathbf{e}_{ab} \end{aligned} \tag{5.233}$$

(we have utilized the properties of the Kronecker symbols to reduce the expression to a simple summation, then changed the dummy subscript on the last line; in addition, we have taken the property of symmetry (5.121)) into account. The final force is identical to that we have previously established; so, once again, the same discrete equation of motion is found.

We have desired above to provide detailed indicial calculations, because such manipulations of indices are essential to learn how to construct the discrete equations of the SPH method. We could, however, have employed a simpler procedure, based on the definition of the discrete operators. Thus, the discrete continuity equation (5.189) may also be written, noting that $d\mathbf{r}_c = \mathbf{u}_c dt$, as:

$$d\rho_b = -\rho_b \tilde{D}_b^k \left\{d\mathbf{r}_c\right\} \tag{5.234}$$

Furthermore, we can write the work[43] which the internal forces would exert if the system experienced a small shift as determined by the quantities $\{d\mathbf{r}_c\}$, on the basis of formulas (1.247) and (1.256):

$$\sum_a \mathbf{F}_a^{int} \cdot d\mathbf{r}_a = -\sum_{a,b} m_b \frac{p_b}{\rho_b^2} \frac{\partial \rho_b}{\partial \mathbf{r}_a} \cdot d\mathbf{r}_a \tag{5.235}$$

[43] The notion of virtual work lies behind this procedure.

$$= -\sum_b V_b \frac{p_b}{\rho_b} d\rho_b$$

$$= \sum_b V_b p_b \tilde{D}_b^k \{d\mathbf{r}_c\}$$

$$= \left(\{p_a\}, \tilde{D}_a^k \{d\mathbf{r}_b\}\right)$$

$$= -\left(\mathbf{G}_a^k \{p_b\}, \{d\mathbf{r}_a\}\right)$$

(we have used (5.231), (5.234), and afterwards the discrete scalar products (5.138) and the adjunction relation (5.141)). Since the above relation is valid for every set of infinitesimal shifts $\{d\mathbf{r}_c\}$, we come to the following conclusion:

$$\frac{1}{m_a}\mathbf{F}_a^{int} = -\frac{1}{\rho_a}\mathbf{G}_a^k \{p_b\} \tag{5.236}$$

which exactly constitutes the right-hand side in the discrete equation of motion (5.229) without the external forces. It is noteworthy that the calculation (5.235), taking (1.256) into account, could also be written as:

$$\begin{aligned} \frac{1}{\rho_a}\mathbf{G}_a^k \{p_b\} &= \sum_b m_b \frac{p_b}{\rho_b^2}\frac{\partial \rho_b}{\partial \mathbf{r}_a} \\ &= \sum_b m_b \frac{\partial e_{int,b}}{\partial \rho_b}\frac{\partial \rho_b}{\partial \mathbf{r}_a} \\ &= \frac{\partial e_{int,b}}{\partial \mathbf{r}_a} \end{aligned} \tag{5.237}$$

which gives the force (5.224) again.

A possible lesson from the above is that, from a variational point of view, each form of the equation of motion is consistent with a given form of the continuity equation. Generally speaking, we can state that (5.200) is compatible with (5.189). This reinforces the choices made for the discrete operators which are used for modelling the pressure gradient and the velocity divergence: let us say that the $\mathbf{G}_a^k$ and $\tilde{D}_a^k$ operators are compatible, which is because they are mutually skew-adjoint.[44] That means, in particular, that the equations (5.201) and (5.190) on the one hand, (5.202) and (5.185) on the other hand, are variationally compatible.[45] They are, a priori, to be kept apart from each other. Some numerical tests (Violeau, 2009b), however, suggest that utilizing inconsistent operators—in the meaning we have defined—does not substantially affect the results.[46]

the recommended general system of equations is then:

$$\begin{aligned} \forall a, \quad \frac{d\mathbf{u}_a}{dt} &= -\frac{1}{\rho_a}\mathbf{G}_a^k \{p_b\} + \mathbf{g} \\ \frac{d\rho_a}{dt} &= -\rho_a \tilde{D}_a^k \{\mathbf{u}_b\} \end{aligned} \tag{5.238}$$

and duly exhibits a formal conformity with the continuous equations (3.129). Considering the definition of the operators, this system is also written as

[44] Remember that we already had resorted to the adjoint nature of the continuous operators $-\mathbf{grad}$ and div in Chapter 3, Section 3.4.4 to demonstrate the Lagrange equations and the conservation laws in the continuous formalism (also refer to the four-dimensional versions (3.308) and (3.334)). Besides, these considerations obviously are not exclusively regarding the SPH method.

[45] Bonet and Lok (1999) state that the form (5.201) of the equation of motion is variationally compatible with the continuity equation, whereas the form (5.202) is compatible with the density interpolation equation. We have just explained that this result is erroneous, since all the forms (5.229) of the momentum equation can be associated with both density calculation methods, the main thing being the coherence of the superscript k.

[46] Bonet and Lok (1999) hold the opposite view. It seems, however, from numerical tests that the formulation based on an interpolation equation for density does provide poorer results that the continuity equation, regardless of the chosen superscripts k.

$$\forall a,\ \frac{d\mathbf{u}_a}{dt} = -\sum_b m_b \frac{\rho_b^{2k} p_a + \rho_a^{2k} p_b}{(\rho_a \rho_b)^{k+1}} w'_{ab}\mathbf{e}_{ab} + \mathbf{g}$$
$$\frac{d\rho_a}{dt} = \frac{1}{\rho_a^{2k-2}} \sum_b m_b (\rho_a \rho_b)^{k-1} \mathbf{u}_{ab} \cdot w'_{ab}\mathbf{e}_{ab} \quad (5.239)$$

that is for the $k = 0$ and 1 values, respectively:

$$\forall a,\ \frac{d\mathbf{u}_a}{dt} = -\frac{1}{\rho_a} \sum_b V_b (p_a + p_b)\, w'_{ab}\mathbf{e}_{ab} + \mathbf{g}$$
$$\frac{d\rho_a}{dt} = \rho_a \sum_b V_b \mathbf{u}_{ab} \cdot w'_{ab}\mathbf{e}_{ab} \quad (5.240)$$

and

$$\forall a,\ \frac{d\mathbf{u}_a}{dt} = -\sum_b m_b \left(\frac{p_a}{\rho_a^2} + \frac{p_b}{\rho_b^2}\right) w'_{ab}\mathbf{e}_{ab} + \mathbf{g}$$
$$\frac{d\rho_a}{dt} = \sum_b m_b \mathbf{u}_{ab} \cdot w'_{ab}\mathbf{e}_{ab} \quad (5.241)$$

Examplary applications of these systems of equations, together with the equation of state (5.207) and the kinematic equation (5.83), can be found in Chapters 7 and 8. These simulations take into account the arguments which will be discussed in Sections 5.4 (temporal schemes) and 6.2 (discrete viscosity).

The discrete equations can be corrected on the basis of the renormalization as set out at the end of Section 5.2.5, while remaining variationally compatible. The following variant of the equation (5.238), for instance, could be written as:

$$\forall a,\ \frac{d\mathbf{u}_a}{dt} = -\frac{1}{\rho_a} \tilde{\mathbf{G}}_a^{R,0} \{p_b\} + \mathbf{g}$$
$$\frac{d\rho_a}{dt} = -\rho_a D_a^{R,0} \{\mathbf{u}_b\} \quad (5.242)$$

where the discrete operators are defined by equations (5.177) and (5.180). The issue of the conservation of total momentum then arises, since we know that the symmetric operators $\tilde{\mathbf{G}}$ do not have the suitable properties, which is contrary to the antisymmetry property (5.204). To remedy this situation, Vila (1999) suggests writing the equation of motion as

$$\forall a,\ \frac{d\mathbf{u}_a}{dt} = -\frac{1}{\rho_a} \sum_b V_b (p_a \mathbf{R}_a + p_b \mathbf{R}_b)\, w'_{ab}\mathbf{e}_{ab} + \mathbf{g} \quad (5.243)$$

by analogy with (5.127). According to the principles which we have just established, (5.243) should be accompanied by an adjoint continuity equation. A calculation readily yields the following result:

$$\forall a,\ \frac{d\rho_a}{dt} = \rho_a \sum_b V_b \mathbf{u}_{ab}^T \mathbf{R}_a w'_{ab}\mathbf{e}_{ab} \quad (5.244)$$

This system can be assessed against equation (5.240). That procedure advantageously restores the conservation of momentum (antisymmetry of forces), but wrecks the rigorous nature of the renormalization.

5.3.4 Discrete conservation laws

As previously mentioned, the fact that the equations of motion result from a least action principle directly accounts for the conservation of the total momentum of an isolated system (i.e., in particular, in the absence of gravity), thus of the law of action-reaction (antisymmetry of the forces (5.204)):[47]

$$\sum_a \mathbf{F}_a = \mathbf{0} \tag{5.245}$$

where $\mathbf{F}_a = \mathbf{F}_a^{int}$. The calculation (5.233) recalls that the force is the derivative of the Lagrangian with respect to the position; it can then be understood that the discrete momentum conservation law is directly translationally related to the invariance of the discrete Lagrangian (5.220), as in the theory of Chapter 1. As we know (Sections 1.4.1 and 1.4.3), we then should naturally obtain the conservation of angular momentum on the one hand, of energy on the other hand, because the discrete Lagrangian is isotropic and time-invariant as well. We are going to deal with these conservation laws, still through an approach in which time is considered as a continuous variable.

[47] We may note that, due to eqn (5.199), this property would still hold true within the more general context of the discrete modelling of the non-fluid continuous media. It should also be pointed out that the discussions at the end of this section regard the conservativity in the absence of any temporal numerical scheme. The temporal discretization, as we will explain it in Section 5.4, may be contrary to these properties.

We will demonstrate that the angular momentum is actually conserved for an isolated system (no gravity), noting that the forces occurring in (5.239) are aligned with the vector $\mathbf{e}_{ab}$ (refer to Fig. 5.5 (*b*)):

$$\forall a, \quad \frac{d\mathbf{u}_a}{dt} = -\sum_b m_b B_{ab} w'_{ab} \mathbf{e}_{ab} = \frac{1}{m_a}\mathbf{F}_a \tag{5.246}$$

where B_{ab} is defined by

$$B_{ab} \doteqdot \frac{\rho_b^{2k} p_a + \rho_a^{2k} p_b}{(\rho_a \rho_b)^{k+1}} \tag{5.247}$$

and verifies

$$B_{ab} = B_{ba} \tag{5.248}$$

With the equality (1.148), we now have:

$$\begin{aligned} \frac{d\mathbf{M}}{dt} &= \sum_a \mathbf{r}_a \times \mathbf{F}_a \\ &= -\sum_{a,b} m_a m_b B_{ab} w'_{ab} \mathbf{r}_a \times \mathbf{e}_{ab} \end{aligned} \tag{5.249}$$

That sum can be split up into a couple of contributions, depending on whether a is lower of higher than b, which yields

$$\begin{aligned}\frac{d\mathbf{M}}{dt} &= -\sum_{a<b} m_a m_b B_{ab} w'_{ab} \mathbf{r}_a \times \mathbf{e}_{ab} \\ &\quad -\sum_{a>b} m_a m_b B_{ab} w'_{ab} \mathbf{r}_a \times \mathbf{e}_{ab} \\ &= -\sum_{a<b} m_a m_b B_{ab} w'_{ab} \mathbf{r}_a \times \mathbf{e}_{ab} - \sum_{a<b} m_b m_a B_{ba} w'_{ba} \mathbf{r}_b \times \mathbf{e}_{ba} \\ &= -\sum_{a<b} m_a m_b B_{ab} w'_{ab} \mathbf{r}_{ab} \times \mathbf{e}_{ab} \\ &= \mathbf{0}\end{aligned} \tag{5.250}$$

(the dummy sub-scripts were inverted on the second line, and the laws of symmetry (5.121) and (5.248)) were subsequently used. It can be observed that that conservation law is directly related to the fact that the forces are aligned[48] with $\mathbf{e}_{ab}$. The importance of the kernel symmetry in that property should be emphasized: the calculation (5.250) is based, indeed, on the relation (5.118), which stems from (5.33), that is ultimately from (5.31). Having chosen a kernel which only depends on the distance r_{ab} between particles (instead of the orientation of the vector $\mathbf{r}_{ab}$ which connects them) is then essential in this case. We find again the basic property of Section 1.4.3, because of which the conservation of angular momentum results from the rotational invariance: this law holds true at the SPH level since the kernel has no preferential direction. In other words, a particle equitably behaves with respect to the nearby particles, regardless of the orientation of the radius vector which separates them.

[48]Which could already have been noticed in Section 2.4.3 of Chapter 2, though that remark was made about the friction forces, which are no exception. We will come back to this point in Section 6.2.

It should be pointed out that these discrete conservation properties are directly geared to the fact that the chosen equations derive from a least action principle, that is initially, to the skew-adjoint nature of the $\mathbf{G}_a^k$ and $\tilde{D}_a^k$ operators, as explained in the preceding paragraph. Besides, that property nearly explicitly appears in the calculation (5.249) we have just made, which is based of the same set of dummy subscripts as (5.139). It could be shown that the $\tilde{\mathbf{G}}_a^k$ and D_a^k operators as defined by the formulas (5.132) and (5.126) are also mutually compatible, because they are adjoint (refer to the end of Section 5.2.3). Since they are contrary to the requisite principles of symmetry (particularly (5.204)), however, they do not lead to the proper conservation laws.

The conservation of energy with the Euler equation does not normally require any demonstration, since here the calculation (1.73) is identically applied.[49] It is interesting, however, to study this property in more detail in the SPH discrete formalism and to set it out in a different light. It is instructive, in particular, to look at how the law of conservation of energy is linked to the discrete variational principle which we have introduced in Section 5.3.3 for the Lagrangian spatial discretization, explicitly calculating the temporal derivative of energy for an isolated system. In this case, we may ignore the potential energy of the external forces, and we express the internal energy e_{int} with respect to the density:

[49]Here, thus, the conservation of energy is not linked to the spatial discretization scheme, but only to the fact that the Lagrange equations are satisfied; we will come back to this point within the context of a temporal scheme in Section 5.4.

$$
\begin{aligned}
E &= E_k + E_{int} \\
&= \sum_a \frac{1}{2} m_a \left|\mathbf{u}_a\right|^2 + \sum_a m_a e_{int,a}\left(\rho_a\right)
\end{aligned} \tag{5.251}
$$

The derivative of kinetic energy is provided by

$$
\begin{aligned}
\frac{dE_k}{dt} &= \sum_a m_a \mathbf{u}_a \cdot \frac{d\mathbf{u}_a}{dt} \\
&= \sum_a \mathbf{u}_a \cdot \mathbf{F}_a^{int}
\end{aligned} \tag{5.252}
$$

(refer to eqn (1.243)). Likewise, for the internal energy, we get

$$
\begin{aligned}
\frac{dE_{int}}{dt} &= \sum_a m_a \frac{\partial e_{int,a}}{\partial \rho_a} \frac{d\rho_a}{dt} \\
&= \sum_a m_a \left(\frac{\partial e_{int}}{\partial \rho}\right)_a \dot{\rho}_a
\end{aligned} \tag{5.253}
$$

Using the relation (1.256), we infer

$$
\frac{dE}{dt} = \sum_a \left(\mathbf{u}_a \cdot \mathbf{F}_a^{int} - \frac{m_a p_a}{\rho_a^2} \dot{\rho}_a\right) \tag{5.254}
$$

In order to calculate the term under the summation symbol, we return to the discrete models as given by (5.234) and (5.236). The total energy derivative is then formally written, resorting to the discrete scalar products as yielded by (5.138) and using (5.141):

$$
\begin{aligned}
\sum_a \left(\mathbf{u}_a \cdot \mathbf{F}_a^{int} - \frac{m_a p_a}{\rho_a^2} \dot{\rho}_a\right) &= \left\langle \{\mathbf{u}_a\}, \left\{\frac{\mathbf{F}_a^{int}}{V_a}\right\}\right\rangle - \left(\left\{\frac{p_a}{\rho_a}\right\}, \{\dot{\rho}_a\}\right) \\
&= \left\langle \{\mathbf{u}_a\}, \mathbf{G}_a^k \{p_b\}\right\rangle + \left(\{p_a\}, \tilde{D}_a^k \{\mathbf{u}_b\}\right) \\
&= -\left\langle \tilde{D}_a^k \{\mathbf{u}_b\}, \{p_a\}\right\rangle + \left(\{p_a\}, \tilde{D}_a^k \{\mathbf{u}_b\}\right) \\
&= 0
\end{aligned} \tag{5.255}
$$

As shown by our calculation, and in accordance with the observations made in that respect in Section 5.3.3, that property is closely linked to the choice of compatible (i.e. skew-adjoint) operators $\mathbf{G}^k$ and $\tilde{D}^k$ to discretize the equations of motion and continuity giving the numerical variation rates for velocity and density. It is little wonder, since the generic calculation (1.73) leading to the conservation of energy generally speaking is based on the continuous Lagrange equations. Furthermore, we explained by the end of Section 3.6.3 that the conservation of energy of a fluid was directly related to the adjoint nature of the continuous operators div and $-\mathbf{grad}$. The computation (5.255) is thus formally identical to (3.369).

The same property can be utilized for studying the changes in the total internal energy from (5.214), which is written, along the lines of (5.238), as:

$$\forall a, \quad \frac{de_{int,a}}{dt} = -\frac{p_a}{\rho_a}\tilde{D}_a^k\{\mathbf{u}_b\} \tag{5.256}$$

Since the mass of each particle is constant, the temporal derivative of the total internal energy is written, successively using equation (5.256), (5.141) and (5.236):

$$\begin{aligned} \frac{dE_{int}}{dt} &= \sum_a m_a \frac{de_{int,a}}{dt} \\ &= -\sum_a V_a p_a \tilde{D}_a^k\{\mathbf{u}_b\} \\ &= -\left(\{p_a\}, \tilde{D}_a^k\{\mathbf{u}_b\}\right) \\ &= \left\langle\{\mathbf{u}_a\}, \mathbf{G}_a^k\{p_b\}\right\rangle \\ &= -\sum_a \mathbf{u}_a \cdot \mathbf{F}_a^{int} \end{aligned} \tag{5.257}$$

This equation is identical to (1.245) (Chapter 1).

Let us also address the question of discrete energy conservation in another formalism. Section 1.5.4 in Chapter 1 presents the conservation of energy as a consequence of the Hamilton equations, which is expressed for a collection of particles. The main property ensuring that conservation law consists in the antisymmetric nature of the matrix $(\boldsymbol{\omega}_7)_{ab}$ as provided by the relation (1.272). Within the context of the SPH method, the uppermost left-hand four terms in that matrix obviously satisfy this property, which is a reflection of equations (5.83) and (5.203) where only the external force $\mathbf{F}_a^{ext}$ would be kept, these expressions reflecting the fact that the velocity is the derivative of the position and that the force is the derivative of the momentum. As regards the other terms in the matrix, they require an examination of the internal forces related to pressure and density. Now, due to the definitions (1.270), the discrete equations (5.239) can be written as

$$\begin{aligned} \dot{\mathbf{p}}_a &= -\sum_b \begin{bmatrix} \delta_{ab}\sum_c m_c \left(\frac{\rho_c}{\rho_a}\right)^{k-1} w'_{ac}\mathbf{e}_{ac} \\ +m_a\left(\frac{\rho_a}{\rho_b}\right)^{k-1} w'_{ab}\mathbf{e}_{ab} \end{bmatrix} \frac{\partial H}{\partial \rho_b} \\ \dot{\rho}_a &= \sum_b \begin{bmatrix} \delta_{ab}\sum_c m_c \left(\frac{\rho_c}{\rho_a}\right)^{k-1} w'_{ac}\mathbf{e}_{ac} \\ -m_b\left(\frac{\rho_b}{\rho_a}\right)^{k-1} w'_{ab}\mathbf{e}_{ab} \end{bmatrix} \cdot \frac{\partial H}{\partial \mathbf{p}_b} \end{aligned} \tag{5.258}$$

(where the points, as usual, denote a temporal derivative). The terms between brackets in the relations (5.258) are the coefficients being sought in the matrix $(\boldsymbol{\omega}_7)_{ab}$, corresponding to the $\partial\rho_b/\partial\mathbf{r}_a$ terms of the formula (1.272) in Chapter 1. One can readily understand that these coefficients satisfy the antisymmetry relations (1.274), which provides for the accurate conservation of the discrete energy.

The discrete conservation laws, as emphasized in Chapter 3, are directly linked to the related continuous laws, which can be found again in the SPH formalism by writing the integral of an arbitrary quantity A in the approximate form

$$\int_{\Omega_{\mathbf{r}}} \rho\left(\mathbf{r}', t\right) A\left(\mathbf{r}', t\right) d^n\mathbf{r}' \approx \sum_a m_a A_a \tag{5.259}$$

after the model of (5.86). Thus, deriving with respect to time, with the conservation of the individual masses of the particles, we find

$$\frac{d}{dt}\int_{\Omega_{\mathbf{r}}} \rho\left(\mathbf{r}', t\right) A\left(\mathbf{r}', t\right) d^n\mathbf{r}' \approx \sum_a m_a \frac{dA_a}{dt} \tag{5.260}$$

With $A \equiv 1$, that yields

$$\frac{dM}{dt} = \frac{d}{dt}\int_{\Omega_{\mathbf{r}}} \rho\left(\mathbf{r}', t\right) d^n\mathbf{r}' \approx 0 \tag{5.261}$$

where M is the total mass of the system, that is the sum of the m_a. The total mass conservation then immediately results from that of the individual masses. The approximation of this conservation law written in a continuous form is analogous to the error made in (5.259) (refer to Section 5.2.3). Likewise, putting[50] $A \equiv \mathbf{u}$ in (5.260) and using the equation of motion and the discrete conservation law (5.245), we get:

[50] (5.259) obviously remains valid for a vector $\mathbf{A}$.

$$\begin{aligned} \frac{d\mathbf{P}}{dt} &= \frac{d}{dt}\int_{\Omega_{\mathbf{r}}} \rho\left(\mathbf{r}', t\right) \mathbf{u}\left(\mathbf{r}', t\right) d^n\mathbf{r}' \\ &\approx \sum_a m_a \frac{d\mathbf{u}_a}{dt} \\ &= \sum_b \mathbf{F}_a = \mathbf{0} \end{aligned} \tag{5.262}$$

which may be compared to (3.68). The conservation of angular momentum would be found again with $A \equiv \rho\mathbf{r} \times \mathbf{u}$ and using (1.144), in a form analogous to (3.76). Lastly, putting $A \equiv e_{int}$, an approximation of a continuous balance of internal energy (refer to Section 3.3.3) can be established.

5.4 Temporal integration

In order to find solutions of the SPH discrete governing equations, we now propose various time integration schemes. This leads us to the idea of symplectic integrators, which possess nice properties of conservation in a discrete time framework. They are analysed from the simple pendulum case. We then show how it is possible to recover such properties, in the case of the SPH method, from discrete Lagrange equations. Finally, we investigate the numerical stability properties of the abovementionned schemes, in order to specify a first stability criterion.

5.4.1 Numerical integrators

The spatial discretization which is the subject of the SPH method should be accompanied, upon the solution of the resulting ordinary differential equations, by a temporal discretization based upon a time step δt. As a rule, by aggregating the parameters characterizing a particle in the state vector $\mathbf{Y}_a$ (refer to Section 5.3.2), the evolution of the latter at the time point $m+1$ is written as a function of its value at the previous instant, in the form of a discrete approximation of (5.205):

$$\forall a,\ \mathbf{Y}_a^{m+1} = \mathbf{Y}_a^m + \dot{\mathbf{Y}}_a \delta t \tag{5.263}$$

$\dot{\mathbf{Y}}_a$ being the rate of change of the parameters, as yielded by (5.206). The superscripts ^{m} refer to the iteration being considered (that is at the time $t^m = t_0 + mdt$). The question then arises as to the choice of the temporal scheme, as for every such problem.[51] To formulate the problem in more concrete terms, should $\dot{\mathbf{Y}}_a$ be chosen either at the time m or at the time $m+1$? A first idea consists of performing a first-order Taylor series expansion of every quantity A:

$$A^{m+1} = A^m + \left(\frac{dA}{dt}\right)^m \delta t + O\left(\delta t^2\right) \tag{5.264}$$

In this proposed approach, the variation rate[52] of A, that is dA/dt, is calculated at the time point m; the scheme is said to be *explicit*. That amounts to approximating the Lagrangian derivatives though a first-order scheme:[53]

$$\dot{A}^m \doteqdot \left(\frac{dA}{dt}\right)^m = \frac{A^{m+1} - A^m}{\delta t} + O\left(\delta t\right) \tag{5.265}$$

(in accordance with (3.184)). For those quantities with which we are concerned,[54] the model (5.264) gives

$$\begin{aligned} \forall a,\ \mathbf{u}_a^{m+1} &= \mathbf{u}_a^m + \frac{1}{m_a}\mathbf{F}_a^m \delta t \\ \mathbf{r}_a^{m+1} &= \mathbf{r}_a^m + \mathbf{u}_a^m \delta t \\ \rho_a^{m+1} &= \rho_a^m + \dot{\rho}_a^m \delta t \end{aligned} \tag{5.266}$$

where $\mathbf{F}_a^m$ is the total force (internal force + external force) received by the particle a at the time point m. $\mathbf{F}_a^m$ and $\dot{\rho}_a^m$ are yielded by the equations of the chosen model, for example (5.239) (or by any other model, possibly accompanied by additional forces, as discussed in Sections 6.2 and 6.3). We also may write (5.266) as follows:

$$\forall a,\ \mathbf{Y}_a^{m+1} = \mathbf{Y}_a^m + \dot{\mathbf{Y}}_a^m \delta t \tag{5.267}$$

There is nothing, however, to prevent us from approaching the derivatives differently from (5.265) and writing, for example:

$$\dot{A}^m = \frac{A^m - A^{m-1}}{\delta t} + O\left(\delta t\right) \tag{5.268}$$

[51] Most of the arguments in this section and the next two are far from being confined to the SPH method. We have preferred, however, to keep them in the second part of this book, because of their purely numerical nature.

[52] We may note it is a *Lagrangian* temporal variation which implicitly involves the advection (refer to Chapter 3, Section 3.2.1). With a Eulerian approach, the advection would explicitly appear and its discretization would give rise to the 'numerical diffusion' process which does not exist in the Lagrangian approaches.

[53] Note the obvious affinity to the expression (2.13).

[54] We may recall that pressure, in the approach being set out, does not obey any differential equation. If, for instance, the equation of state (5.207) or any other variant is chosen, then p is explicitly given as a function of ρ.

Applying this scheme to the derivatives of position and density, we would get

$$\begin{aligned}\forall a,\ \mathbf{u}_a^{m+1} &= \mathbf{u}_a^m + \frac{1}{m_a}\mathbf{F}_a^m \delta t \\ \mathbf{r}_a^{m+1} &= \mathbf{r}_a^m + \mathbf{u}_a^{m+1}\delta t \\ \rho_a^{m+1} &= \rho_a^m + \dot{\rho}_a^{m+1}\delta t\end{aligned} \tag{5.269}$$

With this new scheme, the velocities are then updated before the positions and the densities, not simultaneously, and so the sequence order of equation (5.269) is now important, the first equation having to be solved before the other two equations. The term $\dot{\rho}_a^{m+1}$ means that the temporal derivative of densities is here calculated from the positions[55] $\left\{\mathbf{r}_b^{m+1}\right\}$and the velocities $\left\{\mathbf{u}_b^{m+1}\right\}$ in the discrete continuity equation. We also could apply the approximation (5.268) to the velocity derivative, which would yield another scheme:

$$\begin{aligned}\forall a,\ \mathbf{u}_a^{m+1} &= \mathbf{u}_a^m + \frac{1}{m_a}\mathbf{F}_a^{m+1} \delta t \\ \mathbf{r}_a^{m+1} &= \mathbf{r}_a^m + \mathbf{u}_a^{m+1}\delta t \\ \rho_a^{m+1} &= \rho_a^m + \dot{\rho}_a^{m+1}\delta t\end{aligned} \tag{5.270}$$

or else

$$\forall a,\ \mathbf{Y}_a^{m+1} = \mathbf{Y}_a^m + \dot{\mathbf{Y}}_a^{m+1}\delta t \tag{5.271}$$

Thus, we now have three schemes; the first is fully explicit (5.266), the second is semi-implicit (5.269), the third is fully implicit (5.270). It should be pointed out that the solution of the system by means of the latter would require a specific treatment, since none of the three equations can be solved apart from the others. Non-linear equations would then have to be simultaneously solved, because $\mathbf{F}_a^{m+1}$ and $\mathbf{u}_a^{m+1}$ implicitly depend on both $\left\{\mathbf{u}_b^{m+1}\right\}$ and $\left\{\mathbf{r}_b^{m+1}\right\}$. If the time step is not too large, linearizing[56] the expressions (5.270) remains feasible, leading to the solution of a linear system, which makes the use of that kind of scheme possible, even though it is more complex than the other two schemes.[57] We will not provide details about the procedures that make it possible to address the kind of problem; we will however look at the respective properties of these three schemes in terms of relations to the physical reality.

Which is the best of these three schemes? And what can be said about the opportunities offered to the modelling specialist, particularly that which would consist in only implicitating the equation providing the positions, but not the equation providing the density? Part of the answer comes from the essential notion of conservativity. We have learnt in Sections 5.3.3 and 5.3.4 that the usual conservation laws are satisfied within a discrete context with the SPH method, subject to adequate operators being used for spatial discretization. Now, we have only established these laws within the context of a continuously elapsing time. As time is discretized, then the conservation laws should be dealt with in a different way, since time is crucial in the notion of conservation

[55] In the discrete continuity equation (second equation in the system (5.239) for instance), the positions play a role through the kernel.

[56] This procedure looks like the calculations which are contemplated further on to address the question of numerical stability (Section 5.4.4).

[57] To the best of our knowledge, few works have been published about the implicitation of the equations of the SPH method. Nevertheless, refer to Knapp (2000).

of a quantity. We will go deeper into this matter in Sections 5.4.2 and 5.4.3. We will initially show how a temporal numerical scheme can be constructed from the fundamental equations of mechanics which were discussed in Chapter 1.

Since the equations in the SPH model can be derived from a variational principle, it seems advisable to seek their formal solutions from those tools mentioned in Section 1.4.2, which are based upon the Hamilton equations. In order to simplify our calculations, however, and remain in line with the variational derivation of the discrete equations carried out in Section 1.4.2, we will write the forces as a function of the positions (not of the densities), which amounts to gathering the forces in the form $\mathbf{F}_a = \mathbf{F}_a^{int} + \mathbf{F}_a^{ext}$, or even gathering both the potential of the external forces and the internal energy of a given particle a under one potential $E_{p,a}(\{\mathbf{r}_b\})$ as defined by (1.237), as in equation (5.221). Through that procedure, our considerations can be confined to the equations with respect to the vector

$$\forall a,\ \mathbf{X}_a \doteqdot \begin{pmatrix} \mathbf{p}_a \\ \mathbf{r}_a \end{pmatrix} \tag{5.272}$$

(this definition is identical to eqn (1.99)), instead of the vector $\mathbf{Y}_a$ containing the density. In accordance with Section 1.4.2, (eqn (1.117)) we will write the solution of the Hamilton equations at time point t^{m+1} as a function of the system state at time point t^m as

$$\forall a,\ \mathbf{X}_a^{m+1} = \Psi_{\delta t}^{m,m+1} \mathbf{X}_a^m \tag{5.273}$$

where $\delta t \doteqdot t^{m+1} - t^m$ was noted and $\Psi_{\delta t}^{m,m+1}$ is an operator as yielded by

$$\Psi_{\delta t}^{m,m+1} \doteqdot \exp\left[\int_{t^m}^{t^{m+1}} D_H(t)\, dt\right] \tag{5.274}$$

(on the basis of eqn (1.118)). We may also recall that the operator D_H is yielded by (1.113) and the Poisson bracket of two quantities is given by (1.115); for the sake of convenience, these formulas are as follows:

$$\begin{aligned} D_H f &\doteqdot \{f, H\} \\ &= \sum_a \left(\frac{\partial f}{\partial \mathbf{r}_a} \cdot \frac{\partial H}{\partial \mathbf{p}_a} - \frac{\partial f}{\partial \mathbf{p}_a} \cdot \frac{\partial H}{\partial \mathbf{r}_a} \right) \end{aligned} \tag{5.275}$$

Here the Hamiltonian is yielded by (5.221), that is as the sum of two components, namely a kinetic one and a potential one:

$$H = H_{\mathbf{p}} + H_{\mathbf{r}} \tag{5.276}$$

with

$$\begin{aligned} H_{\mathbf{p}} &\doteqdot \sum_a \frac{p_a^2}{2m_a} \\ H_{\mathbf{r}} &\doteqdot \sum_a E_{p,a}(\{\mathbf{r}_b\}) \end{aligned} \tag{5.277}$$

The D_H operator, after the definition (5.275) is then written as:

$$D_H = D_{H_\mathbf{p}} + D_{H_\mathbf{r}} \tag{5.278}$$

with

$$\begin{aligned} D_{H_\mathbf{p}} f &\doteqdot \{f, H_\mathbf{p}\} \\ D_{H_\mathbf{r}} f &\doteqdot \{f, H_\mathbf{r}\} \end{aligned} \tag{5.279}$$

The two operators being introduced here can be written in more detail as:

$$\begin{aligned} \frac{\partial H_\mathbf{p}}{\partial \mathbf{X}_a} &= \begin{pmatrix} \mathbf{u}_a \\ \mathbf{0} \end{pmatrix} \\ \frac{\partial H_\mathbf{r}}{\partial \mathbf{X}_a} &= \begin{pmatrix} \mathbf{0} \\ -\mathbf{F}_a \end{pmatrix} \end{aligned} \tag{5.280}$$

We get, from the definitions, the following operators:

$$\begin{aligned} D_{H_\mathbf{p}} &= \sum_a \mathbf{u}_a \cdot \frac{\partial}{\partial \mathbf{r}_a} \\ D_{H_\mathbf{r}} &= \sum_a \mathbf{F}_a \cdot \frac{\partial}{\partial \mathbf{p}_a} \end{aligned} \tag{5.281}$$

It can be seen that they occur in a form analogous to the general formula (1.122). We can calculate their commutators $\left[D_{H_\mathbf{p}}, D_{H_\mathbf{r}}\right]$ (as defined by (1.121)), by analogy with the calculation (1.123)–(1.125). We initially write

$$\begin{aligned} D_{H_\mathbf{r}} D_{H_\mathbf{p}} f &= \sum_b \frac{\partial}{\partial \mathbf{p}_b} \left(\sum_a \frac{\partial f}{\partial \mathbf{r}_a} \cdot \mathbf{u}_a \right) \cdot \mathbf{F}_b \\ &= \sum_{a,b} \left[\frac{\partial^2 f}{\partial \mathbf{p}_b \partial \mathbf{r}_a} \cdot \mathbf{u}_a + \left(\frac{\partial f}{\partial \mathbf{r}_a} \right)^T \frac{\partial \mathbf{u}_a}{\partial \mathbf{p}_b} \right] \cdot \mathbf{F}_b \\ &= \sum_{a,b} \mathbf{F}_b^T \frac{\partial^2 f}{\partial \mathbf{p}_b \partial \mathbf{r}_a} \mathbf{u}_a + \sum_a \frac{\partial f}{\partial \mathbf{r}_a} \cdot \frac{1}{m_a} \mathbf{F}_a \end{aligned} \tag{5.282}$$

An analogous calculation yields

$$D_{H_\mathbf{p}} D_{H_\mathbf{r}} f = \sum_{a,b} \mathbf{F}_b^T \frac{\partial^2 f}{\partial \mathbf{p}_b \partial \mathbf{r}_a} \mathbf{u}_a + \sum_a \left(\frac{\partial f}{\partial \mathbf{p}_a} \right)^T \left(\frac{\partial \mathbf{F}}{\partial \mathbf{r}} \right)_a \mathbf{u}_a \tag{5.283}$$

As expected, the second-order derivatives are identical in (5.282) and (5.283), but the other terms persist without cancelling each other out, which means that the two operators a priori do not commute.

We are now going to utilize the decomposition (5.278) to try to construct an adequate temporal numerical scheme by separating the integrations of positions and velocities. More exactly, the velocity will be desirably integrated ignoring the effect of this process on the positions, and vice versa. The issue with the $\Psi_{\delta t}^{m,m+1}$ operator as yielded by (5.274) is, indeed, that it cannot be exactly integrated, because the velocities and the positions are linked and continuously change from one Hamilton equation to the other one. Now, from (5.274) and (5.278), we may write

$$\Psi_{\delta t}^{m,m+1} = \exp\left[\int_{t^m}^{t^{m+1}} D_{H_\mathrm{p}}(t)\,dt + \int_{t^m}^{t^{m+1}} D_{H_\mathrm{r}}(t)\,dt\right] \tag{5.284}$$

To make progress in this formal calculation, we might well be tempted to separate the two integrals by writing the exponential of a sum as the product of two exponentials. Unfortunately, the usual properties of the exponential calculation cannot be applied to non-commutative operators. To appreciate this, let us start from the definition (1.119) to write the exponential of the sum of two operators P and Q:

$$\begin{aligned} e^{P+Q} &= Id + P + Q + \frac{1}{2}(P+Q)^2 + O\left(\left\|(P+Q)^3\right\|\right) \\ &= Id + P + Q + \frac{1}{2}\left(P^2 + PQ + QP + Q^2\right) + O\left(\|P\|^3, \|Q\|^3\right) \end{aligned} \tag{5.285}$$

Moreover:

$$\begin{aligned} e^P e^Q &= \left[Id + P + \frac{1}{2}P^2 + O\left(\|P\|^3\right)\right] \\ &\quad \times \left[Id + Q + \frac{1}{2}Q^2 + O\left(\|Q\|^3\right)\right] \\ &= Id + P + Q + \frac{1}{2}\left(P^2 + 2PQ + Q^2\right) + O\left(\|P\|^3, \|Q\|^3\right) \end{aligned} \tag{5.286}$$

Combining (5.285) and (5.286), we find:

$$\begin{aligned} e^{P+Q} &= e^P e^Q - \frac{1}{2}[P, Q] + O\left(\|P\|^3, \|Q\|^3\right) \\ &= e^Q e^P + \frac{1}{2}[P, Q] + O\left(\|P\|^3, \|Q\|^3\right) \end{aligned} \tag{5.287}$$

In view of the above, the application sequence order of the operators is important. (5.287), in particular, provides the following relation:

$$\left[e^P, e^Q\right] = [P, Q] + O\left(\|P\|^3, \|Q\|^3\right) \tag{5.288}$$

Thus, since the commutator $\left[D_{H_\mathrm{p}}, D_{H_\mathrm{r}}\right]$ is non-zero, the selected sequence order of the operators affects the result of the calculation. However, a crucial remark is to be made: each of the D_{H_p} and D_{H_r} operators acts upon a single variable. D_{H_r} depends on the potential, that is on the forces and thereby acts upon the velocity of a particle. For its part, D_{H_p} depends on the kinetic energy, that is on the velocity, and acts upon the position. Thus, upon the integration of the position through the action of D_{H_p}, no force is applied to the concerned particle. Pursuant to the principle of inertia (refer to Section 1.2.1), its velocity then remains constant. Likewise, its position remains unchanged during the very short period of time in which its velocity is re-evaluated due to D_{H_r}. Thus, separating the operators allows us to assume that the parameters $\mathbf{u}_a$ and $\mathbf{F}_a$ are constant in the formulas (5.281), which amounts to considering that the operators $D_{H_\mathbf{p}}$ and $D_{H_\mathbf{r}}$ are constant over the time interval δt. The following operators can then be defined:

$$\Psi_{\mathbf{p},\delta t}^{m,m+1} \doteqdot \exp\left[\int_{t^m}^{t^{m+1}} D_{H_\mathbf{p}}(t)\,dt\right] = e^{\delta t D_{H_\mathbf{p}}(t^m)}$$
$$\Psi_{\mathbf{r},\delta t}^{m,m+1} \doteqdot \exp\left[\int_{t^m}^{t^{m+1}} D_{H_\mathbf{r}}(t)\,dt\right] = e^{\delta t D_{H_\mathbf{r}}(t^m)} \tag{5.289}$$

In view of the formulas (5.281), we then have:

$$\begin{aligned}\Psi_{\mathbf{p},\delta t}^{m,m+1}\mathbf{X}_a^m &= \exp\left[\delta t \sum_b \left(\mathbf{u}_b^m\right)^T \frac{\partial \mathbf{X}_b^m}{\partial \mathbf{r}_a^m}\right]\\ &= \exp\left[\delta t \sum_b \left(\mathbf{u}_b^m\right)^T \delta_{ab}\begin{pmatrix}\mathbf{0}_n & \mathbf{I}_n\end{pmatrix}\right]\\ &= \exp\left[\delta t \begin{pmatrix}\mathbf{0}_n \\ \mathbf{u}_a^m\end{pmatrix}\right]\end{aligned} \tag{5.290}$$

and likewise, after a brief calculation:

$$\Psi_{\mathbf{r},\delta t}^{m,m+1}\mathbf{X}_a^m = \exp\left[\delta t \begin{pmatrix}\mathbf{F}_a^m \\ \mathbf{0}_n\end{pmatrix}\right] \tag{5.291}$$

As we felt, the $\Psi_{\mathbf{p},\delta t}^{m,m+1}$ operator acts upon the position via the velocity $\mathbf{u}_a^m$, whereas $\Psi_{\mathbf{r},\delta t}^{m,m+1}$ acts upon the momentum (that is upon the velocity) through the force $\mathbf{F}_a^m$. It is noteworthy that the operators $D_{H_\mathbf{p}}$ and $D_{H_\mathbf{r}}$ are nilpotent, which can easily be seen from (5.281). Thus, their exponentials are simply $\mathbf{I}_n + D_{H_\mathbf{p}}$ and $\mathbf{I}_n + D_{H_\mathbf{r}}$, respectively. This ultimately yields:[58]

$$\Psi_{\mathbf{p},\delta t}^{m,m+1}\mathbf{X}_a^m = \begin{pmatrix}\mathbf{p}_a^m \\ \mathbf{r}_a^{m+1}\end{pmatrix}$$
$$\Psi_{\mathbf{r},\delta t}^{m,m+1}\mathbf{X}_a^m = \begin{pmatrix}\mathbf{p}_a^{m+1} \\ \mathbf{r}_a^{m}\end{pmatrix} \tag{5.292}$$

[58] For that reason, these operators are often referred to as 'Drift' ($D_{H_\mathbf{p}}$) and 'Kick' ($D_{H_\mathbf{r}}$) in scientific literature. The duality existing between position and momentum can be noticed, $D_{H_\mathbf{p}}$ acting upon $\mathbf{r}$ whereas $D_{H_\mathbf{r}}$ acts upon $\mathbf{p}$. We may recall that such duality is at the core of the symplectic nature of the Hamilton equation (Section 1.4.2).

From the above, it is evident that the commutator $\left[\Psi_{\mathbf{p},\delta t}^{m,m+1}, \Psi_{\mathbf{r},\delta t}^{m,m+1}\right]$ is of the order δt. From (5.284) and (5.287), we may decide to combine these two operators in the following sequence order:

$$\Psi_{\delta t}^{m,m+1} = \Psi_{\mathbf{p},\delta t}^{m,m+1}\Psi_{\mathbf{r},\delta t}^{m,m+1} + O\left(\delta t^2\right) \tag{5.293}$$

The definitions (5.292) can then be used for writing that scheme as (dropping the terms in δt^2):

$$\forall a, \ \mathbf{u}_a^{m+1} = \mathbf{u}_a^m + \frac{1}{m_a}\mathbf{F}_a^m \delta t$$
$$\mathbf{r}_a^{m+1} = \mathbf{r}_a^m + \mathbf{u}_a^{m+1}\delta t \tag{5.294}$$

We find again (except for the density) the semi-implicit scheme (5.269). This scheme is obviously an approximation, because of the ignored term in $O\left(\delta t^2\right)$, but it offers the advantage of explaining why the parameters are to be separated to be consecutively treated. We will provide a variational justification of that method in Section 5.4.2.

Applying the operators in the reverse order (that is considering the operator $\Psi_{\mathbf{r},\delta t}^{m,m+1}\Psi_{\mathbf{p},\delta t}^{m,m+1}$ rather than (5.293)) would, of course, have given another scheme in which the positions would be integrated before the velocities. Anyway, it is clear that the explicit (5.266) and fully implicit (5.270) schemes cannot result from such an approach, since they consider positions and velocities on an equal footing. It seems therefore that the semi-implicit scheme is better than the other two, because it directly stems from an application of the Hamilton equations. The next two Sections will substantiate that intuition.

We note that if our calculations had taken into account the density on the basis of the arguments in Section 1.5.4, our results would have been left unchanged, but we should have considered an additional operator $\Psi_{\rho,\delta t}^{m,m+1}$ acting upon the densities and stemming from that part of the Hamiltonian which corresponds to the internal energy, decomposing the Hamiltonian as follows, instead of (5.276):

$$H = H_{\mathbf{p}} + H_{\mathbf{r}} + H_\rho \tag{5.295}$$

H_ρ corresponding to the internal energy e_{int}. The whole semi-implicit scheme (5.269) would then amount to approximating the exact integrator $\Psi_{\delta t}^{m,m+1}$ through the combination $\Psi_{\rho,\delta t}^{m,m+1}\Psi_{\mathbf{p},\delta t}^{m,m+1}\Psi_{\mathbf{r},\delta t}^{m,m+1}$, acting upon the state vector $\mathbf{Y}_a^m$. In any case, since the density is a function of the particle positions, the result which we have just highlighted shows that the equation with respect to ρ_a has to be implicited, as in (5.269). A a rule, the state parameters are to be successively re-evaluated, in an arbitrary sequence order, provided that the values of the parameters which were updated by the preceding stages are used.

5.4.2 Symplectic integrators

In order to investigate in further detail the properties of the semi-implicit operator as defined in the previous section, we will consider the conservation of energy. We explained in Section 1.4.2 that a Hamiltonian (i.e. non-dissipative) system obeys equations of motion involving a linear operator $\Psi_{\delta t}$, which is found to be the exponential of an energy operator D_H. The integrator of the previous paragraph, however, is based on the compound operator $\Psi_{\delta t}^{m,m+1}$, which is in the form $\exp\left(\delta t D_{H_\mathbf{p}}\right)\exp\left(\delta t D_{H_\mathbf{p}}\right)$ (refer to eqns (5.289) and (5.293)), different from $\exp\left[\delta t\left(D_{H_\mathbf{p}} + D_{H_\mathbf{r}}\right)\right]$. It would be interesting to try to write $\Psi_{\delta t}^{m,m+1}$ in the form of an exponential, since that would mean that the numerical scheme stems from a Hamiltonian, which would then be exactly conserved. It should first be pointed out that (5.285) can be utilized for writing

$$e^{P+Q+\frac{1}{2}[P,Q]} = e^P e^{Q+\frac{1}{2}[P,Q]} - \frac{1}{2}\left[P, Q + \frac{1}{2}[P, Q]\right] + O\left(\|P\|^3, \|Q\|^3\right) = e^P e^{Q+\frac{1}{2}[P,Q]} - \frac{1}{2}[P, Q] + O\left(\|P\|^3, \|Q\|^3\right) \tag{5.296}$$

(noting that $[P, [P, Q]]$ is a third-order term). Likewise, using the formula (1.119), we find:

$$e^{Q+\frac{1}{2}[P,Q]} = e^Q e^{\frac{1}{2}[P,Q]} - \frac{1}{2}\left[Q, \frac{1}{2}[P, Q]\right] + O\left(\|P\|^3, \|Q\|^3\right) = e^Q + \frac{1}{2}[P, Q] + O\left(\|P\|^3, \|Q\|^3\right) \tag{5.297}$$

Combining the last two results, we come to the following approximation:

$$e^{P+Q+\frac{1}{2}[P,Q]} = e^P e^Q + O\left(\|P\|^3, \|Q\|^3\right) \tag{5.298}$$

Thus, the product $e^P e^Q$ can be approximated by an exponential to within the second order. Likewise, through an iterative procedure, the Baker–Campbell–Hausdorff formula, which was stated for the first time by Poincaré in 1899, would be obtained:

$$e^P e^Q = e^R \quad R \doteqdot P + Q + \frac{1}{2}[P, Q] + \frac{1}{12}[P - Q, [P, Q]] + \dots \tag{5.299}$$

(for a comprehensive statement of the residual terms and a demonstration of this formula, please refer to Dynkin, 1947). From (5.298), only keeping the terms below second order, we may therefore affirm that the $\exp\left(\delta t D_{H_\mathbf{p}}\right) \exp\left(\delta t D_{H_\mathbf{p}}\right)$ operator is the $\exp\left(\delta t D_{H_{num}}\right)$ exponential of an operator as given by

$$\delta t D_{H_{num}} = \delta t D_{H_\mathbf{p}} + \delta t D_{H_\mathbf{r}} + \frac{\delta t^2}{2}\left[D_{H_\mathbf{p}}, D_{H_\mathbf{r}}\right] + O\left(\delta t^3\right) \tag{5.300}$$

The definition (5.275) then shows that this operator corresponds to the Poisson bracket with a numerical Hamiltonian, which is denoted here as H_{num}, such that, for every function f, we get the relation

$$\{f, H_{num}\} = \left\{f, H_\mathbf{p}\right\} + \{f, H_\mathbf{r}\} + \frac{\delta t}{2}\left[D_{H_\mathbf{p}}, D_{H_\mathbf{r}}\right] f + O\left(\delta t^2\right) \tag{5.301}$$

Besides, the Jacobi identity (1.128) may be advantageously utilized for writing

$$\left[D_{H_\mathbf{p}}, D_{H_\mathbf{r}}\right] f = \left\{f, \left\{H_\mathbf{r}, H_\mathbf{p}\right\}\right\} \tag{5.302}$$

Relating this result to (5.301), we find

$$H_{num} = H_{\mathbf{p}} + H_{\mathbf{r}} + \frac{\delta t}{2}\left\{H_{\mathbf{r}}, H_{\mathbf{p}}\right\} + O\left(\delta t^2\right) \tag{5.303}$$

A major remark is to be made right now. All the terms in the expansion in the Baker–Campbell–Hausdorff formula (5.299) are similar to commutator brackets, and so the Jacobi identity makes it possible to exactly write the $D_{H_{num}} f$ operator as $\{f, H_{num}\}$, which shows that the numerical Hamiltonian is exactly conserved:

$$H_{num}^{m+1} = H_{num}^{m} \tag{5.304}$$

The calculation which we have just made only includes its first terms, but it is ensured that the semi-implicit numerical integrator (5.269) corresponds to an exactly conserved Hamiltonian. Let us now calculate the Poisson bracket of $H_{\mathbf{r}}$ and $H_{\mathbf{p}}$ from the definitions (5.277):

$$\begin{aligned}\left\{H_{\mathbf{r}}, H_{\mathbf{p}}\right\} &= \sum_a \left(\frac{\partial H_{\mathbf{r}}}{\partial \mathbf{r}_a} \cdot \frac{\partial H_{\mathbf{p}}}{\partial \mathbf{p}_a} - \frac{\partial H_{\mathbf{r}}}{\partial \mathbf{p}_a} \cdot \frac{\partial H_{\mathbf{p}}}{\partial \mathbf{r}_a}\right) \\ &= \sum_a \frac{\partial H_{\mathbf{r}}}{\partial \mathbf{r}_a} \cdot \frac{\partial H_{\mathbf{p}}}{\partial \mathbf{p}_a} \\ &= -\sum_a \mathbf{F}_a \cdot \mathbf{u}_a\end{aligned} \tag{5.305}$$

Thus, the numerical Hamiltonian associated with the semi-implicit scheme is written as

$$\begin{aligned}H_{num} &= H - \frac{\delta t}{2} \sum_a \mathbf{F}_a \cdot \mathbf{u}_a + O\left(\delta t^2\right) \\ &= H + O\left(\delta t\right)\end{aligned} \tag{5.306}$$

and then contains an error of order δt with respect to the theoretical Hamiltonian H. The above result is of the utmost importance for the integration of such discrete systems as the SPH method equations: with such an integrator, a given quantity H_{num} is conserved over time, approximating the energy with a time-bounded error, which can arbitrarily be minimized through a suitably chosen time step (if both forces and velocities are bounded). This result provides an estimation, in terms of energy, of the error made upon the (5.293).

This property of conservation can be approached in two other ways. First, we note that the semi-implicit scheme (5.294) can be inverted to give the initial state vector $\mathbf{X}_a^m$ as a function of the final vector $\mathbf{X}_a^{m+1}$:

$$\begin{aligned}\forall a,\ \mathbf{u}_a^m &= \mathbf{u}_a^{m+1} - \frac{1}{m_a}\mathbf{F}_a^m \delta t \\ \mathbf{r}_a^m &= \mathbf{r}_a^{m+1} - \mathbf{u}_a^{m+1}\delta t\end{aligned} \tag{5.307}$$

The achieved result is formally identical, to within the time sign,[59] provided that m and $m+1$ are swapped. Such a 'backwards' scheme (i.e. working back in time and among the temporal indices) would have to be implemented

[59] The sign would be unchanged if it had been considered that, looking back in time, the velocity sign had to be modified.

applying the second equation before the first, unlike the initial scheme. This amounts to writing

$$\Psi_{\mathbf{p},\delta t}^{m,m+1}\Psi_{\mathbf{r},\delta t}^{m,m+1} = \Psi_{\mathbf{r},-\delta t}^{m+1,m}\Psi_{\mathbf{p},-\delta t}^{m+1,m} \tag{5.308}$$

Thus, the scheme is time-reversible, which is in line with its property of conservation of an energy, pursuant to Chapter 1 (Section 1.4.1). It should be pointed out, however, that this reversibility is achieved at the cost of the operator swapping, that is provided that the operations are made in the reverse order.

Moreover, the implicit scheme (5.270) can obviously be rearranged as:

$$\begin{aligned} \forall a,\ \mathbf{u}_a^m &= \mathbf{u}_a^{m+1} - \frac{1}{m_a}\mathbf{F}_a^{m+1}\delta t \\ \mathbf{r}_a^m &= \mathbf{r}_a^{m+1} - \mathbf{u}_a^{m+1}\delta t \end{aligned} \tag{5.309}$$

which is similar to the explicit scheme (5.266), the temporal indices and the time sign being inverted. Thus, the fully implicit scheme may be considered as a reverse explicit scheme. Thus, neither of them is time-reversible, which characterizes their non-conservativity.

We explained in Section 2.2.2 that the conservation of energy is accompanied by the conservation of volumes in the phase space. Using the scheme (5.294), we may write, by analogy with (2.14), the Jacobian matrix of the transformation[60] $\left(\mathbf{p}_b^m, \mathbf{r}_b^m\right) \rightarrow \left(\mathbf{p}_a^{m+1}, \mathbf{r}_a^{m+1}\right)$ (i.e. $\mathbf{X}_b^m \rightarrow \mathbf{X}_a^{m+1}$), on the model of (2.16):

$$\begin{aligned} \mathbf{J}_{ab}^{m,m+1} &\doteq \begin{pmatrix} \dfrac{\partial \mathbf{p}_a^{m+1}}{\partial \mathbf{p}_b^m} & \dfrac{\partial \mathbf{p}_a^{m+1}}{\partial \mathbf{r}_b^m} \\ \dfrac{\partial \mathbf{r}_a^{m+1}}{\partial \mathbf{p}_b^m} & \dfrac{\partial \mathbf{r}_a^{m+1}}{\partial \mathbf{r}_b^m} \end{pmatrix} \\ &= \begin{pmatrix} \delta_{ab}\mathbf{I}_n & \dfrac{\partial \mathbf{F}_a^m}{\partial \mathbf{r}_b^m}\delta t \\ \mathbf{0} & \delta_{ab}\mathbf{I}_n \end{pmatrix} \end{aligned} \tag{5.310}$$

which verifies

$$\det \mathbf{J}_{ab}^{m,m+1} = \frac{\partial\left(\mathbf{X}_a^{m+1}\right)}{\partial\left(\mathbf{X}_b^m\right)} = \delta_{ab} \tag{5.311}$$

We may recall that the theoretical calculation (2.15) (Section 2.2.2) leading to the preservation of the volumes in the phase space for a molecular system was based on the arbitrary smallness of the infinitesimal time step dt, unlike the numerical time step δt. It is then essential for the term $\partial \mathbf{r}_a^{m+1}/\partial \mathbf{p}_b^m$ to be zero in the calculation of the Jacobian matrix $\mathbf{J}_{ab}^{m,m+1}$, unlike the time continuous case in Chapter 2. With the explicit and fully implicit schemes, this property would no longer hold true, because the positions at time $m+1$ would depend on the momentum at time m, i.e. $\partial \mathbf{r}_a^{m+1}/\partial \mathbf{p}_b^m \neq \mathbf{0}$ (in particular, refer to Marsden and West, 2001).

[60] This is not the matrix of the $\Psi_{\mathbf{p},\delta t}^{m,m+1}\Psi_{\mathbf{r},\delta t}^{m,m+1}$ operator, due to the quasilinear nature of the Hamilton equations (refer to Section 1.4.2).

We would come to the same conclusion by describing the system through the same vector including the density. The Jacobian matrix of the transformation $\left(\mathbf{p}_b^m, \mathbf{r}_b^m, \rho_b^m\right) \to \left(\mathbf{p}_a^{m+1}, \mathbf{r}_a^{m+1}, \rho_a^{m+1}\right)$ (i.e. $\mathbf{Y}_a^m \to \mathbf{Y}_a^{m+1}$) is then written as

$$\overline{\mathbf{J}}_{ab}^{m,m+1} \doteq \begin{pmatrix} \dfrac{\partial \mathbf{p}_a^{m+1}}{\partial \mathbf{p}_b^m} & \dfrac{\partial \mathbf{p}_a^{m+1}}{\partial \mathbf{r}_b^m} & \dfrac{\partial \mathbf{p}_a^{m+1}}{\partial \rho_b^m} \\ \dfrac{\partial \mathbf{r}_a^{m+1}}{\partial \mathbf{p}_b^m} & \dfrac{\partial \mathbf{r}_a^{m+1}}{\partial \mathbf{r}_b^m} & \dfrac{\partial \mathbf{r}_a^{m+1}}{\partial \rho_b^m} \\ \left(\dfrac{\partial \rho_a^{m+1}}{\partial \mathbf{p}_b^m}\right)^T & \left(\dfrac{\partial \rho_a^{m+1}}{\partial \mathbf{r}_b^m}\right)^T & \dfrac{\partial \rho_a^{m+1}}{\partial \rho_b^m} \end{pmatrix} = \begin{pmatrix} \delta_{ab}\left[\mathbf{I}_n\right] & \left[\dfrac{\partial \left(\mathbf{F}_a^{ext}\right)^m}{\partial \mathbf{r}_b^m}\right]\delta t & \left[\dfrac{\partial \left(\mathbf{F}_a^{int}\right)^m}{\partial \rho_b^m}\right]\delta t \\ [\mathbf{0}] & \delta_{ab}\left[\mathbf{I}_n\right] & \mathbf{0} \\ \mathbf{0}^T & \mathbf{0}^T & \delta_{ab} \end{pmatrix} \tag{5.312}$$

It verifies $\det \overline{\mathbf{J}}_{ab}^{m,m+1} = \delta_{ab}$ due to the nullity of the term $\partial \rho_a^{m+1} / \partial \mathbf{p}_b^m$, which would be false if the equation giving the changes of ρ_a was not written in an implicit form. The determinant, however, would also equal 1 if the term $\partial \rho_a^{m+1} / \partial \mathbf{r}_b^m$ was not zero. Thus, we could just calculate $\dot{\rho}_a^{m+1}$ from the $\left\{\mathbf{u}_a^{m+1}\right\}$ and the $\left\{\mathbf{r}_a^m\right\}$, that is only implicate the updated velocities in the discrete continuity equation.

It can be observed that, as in the continuous time case (Section 2.2.2), it is the phase space volume associated with the parameters of each particle which is conserved, so that the transformation is symplectic. Because of its conservative nature with respect to both energy and volumes of the phase space, the semi-implicit scheme is then sometimes referred to as symplectic, because it preserves the Hamiltonian structure of the equations of mechanics within the context of a discretized time. In order to clarify that notion, we may recall, indeed, that the concept of successive operators which are applied to the state vector of each particle amounts to periodically exerting a punctual force, which is tantamount to writing an approximation of the Hamiltonian (5.221) as

$$H_{\delta t}\left(\{\mathbf{p}_b\}, \{\mathbf{r}_b\}\right) \doteq \sum_b \frac{|\mathbf{p}_b|^2}{2m_b} + \sum_b E_{p,b}(\{\mathbf{r}_a\})\delta t \sum_m \delta\left(t - m\delta t\right) \tag{5.313}$$

where δ is the Dirac distribution.[61] The discrete equations of the symplectic scheme could easily be derived by applying the Hamilton equations to (5.313). With the explicit scheme (5.266), for example, that property would no longer be true and the preservation of the phase volumes would not be complied with. The energy would therefore not be conserved, which is consistent with the irreversible nature of that scheme. The same conclusions can be applied to the fully implicit scheme (5.270).

Thus, the symplectic scheme is recommended,[62] because it ensures that the energy never departs from its theoretical value by more than a small bounded

[61] Refer to Appendix B for the definition of that distribution. δ is here homogeneous to a frequency.

[62] By some happy coincidence, it is also the simplest one to be implemented, since it consists of successively updating quantities, which fits a sequential algorithm. That is why many developers (including the author of this book) have implemented it without initially knowing its advantages.

amount, whereas an explicit scheme would predict an energy liable to arbitrarily drift from its initial value after a large number of iterations, regardless of the time step value. This is an important result, because the time step cannot be arbitrarily chosen, as we will explain in Section 5.4.4.

The results we have just established are still valid in the very general case of some mechanical system or other. In order to highlight it, let us consider the simple pendulum case which we have already mentioned on several occasions (in particular in Section 1.4.4), which offers the advantage of becoming independent of any spatial discretization. Here the explicit scheme would be written as

$$\begin{aligned} \dot{\theta}^{m+1} &= \dot{\theta}^m + \ddot{\theta}^m \delta t \\ \theta^{m+1} &= \theta^m + \dot{\theta}^m \delta t \end{aligned} \tag{5.314}$$

With the equation of motion (1.153), the dimensionless state vector (1.174) at the time point $m+1$ could be expressed as a function of its value at the previous time point:

$$\begin{aligned} \mathbf{X}^{+m+1} &= \mathbf{\Psi}^+_{\delta t,\exp} \mathbf{X}^{+m} \\ \mathbf{\Psi}^+_{\delta t,\exp} &\doteq \begin{pmatrix} 1 & -\delta t^+ \\ \delta t^+ & 1 \end{pmatrix} \end{aligned} \tag{5.315}$$

where $\delta t^+ \doteq \omega_0 \delta t$ is the dimensionless time, as in (1.173). It can be observed that the formula (5.315) is nothing but a first order expansion in δt^+ of the continuous equation (1.175) involving the matrix $\mathbf{\Psi}^+$. The energy, according to formula (1.176), evolves in accordance with the following law:

$$\begin{aligned} H^{m+1} &= \frac{mg\ell}{2} \left| \mathbf{X}^{+m+1} \right|^2 \\ &= \frac{mg\ell}{2} \left(\mathbf{X}^{+m} \right)^T \left(\mathbf{\Psi}^+_{\delta t,\exp} \right)^T \mathbf{\Psi}^+_{\delta t,\exp} \mathbf{X}^{+m} \end{aligned} \tag{5.316}$$

Unlike the matrix $\mathbf{\Psi}^+$, however, $\mathbf{\Psi}^+_{\delta t,\exp}$ verifies

$$\left(\mathbf{\Psi}^+_{\delta t,\exp} \right)^T \mathbf{\Psi}^+_{\delta t,\exp} = \left(1 + \delta t^{+2} \right) \mathbf{I}_2 \tag{5.317}$$

to be compared with (1.177). Thus, the system energy increases at each time step, being multiplied times the factor $1 + \delta t^{+2}$. The volume in the phase space then changes too, as evidenced by the determinant in $\mathbf{\Psi}^+_{\delta t,\exp}$, which does not equal one:

$$\det \mathbf{\Psi}^+_{\delta t,\exp} = 1 + \delta t^{+2} \tag{5.318}$$

a formula to be compared with (2.12). A fully implicit scheme would yield, instead of (5.314):

$$\begin{aligned} \dot{\theta}^{m+1} &= \dot{\theta}^m + \ddot{\theta}^{m+1} \delta t \\ \theta^{m+1} &= \theta^m + \dot{\theta}^{m+1} \delta t \end{aligned} \tag{5.319}$$

that is

$$\begin{aligned}\mathbf{X}^{+m+1} &= \boldsymbol{\Psi}^{+}_{\delta t,\mathrm{imp}}\mathbf{X}^{+m}\\ \boldsymbol{\Psi}^{+}_{\delta t,\mathrm{imp}} &\doteq \begin{pmatrix} 1 & \delta t^{+} \\ -\delta t^{+} & 1 \end{pmatrix}^{-1}\end{aligned} \tag{5.320}$$

We can see that the implicit matrix is related to the explicit matrix by

$$\boldsymbol{\Psi}^{+}_{\delta t,\mathrm{imp}} = \left(\boldsymbol{\Psi}^{+}_{-\delta t,\exp}\right)^{-1} \tag{5.321}$$

which substantiates the fact that the implicit and fully explicit schemes are mutually inverse. The relation (5.317) is then substituted for by

$$\begin{aligned}\left(\boldsymbol{\Psi}^{+}_{\delta t,\mathrm{imp}}\right)^{T}\boldsymbol{\Psi}^{+}_{\delta t,\mathrm{imp}} &= \left(\left(\boldsymbol{\Psi}^{+}_{\delta t,\mathrm{imp}}\right)^{-1}\left[\left(\boldsymbol{\Psi}^{+}_{\delta t,\mathrm{imp}}\right)^{-1}\right]^{T}\right)^{-1}\\ &= \left[\boldsymbol{\Psi}^{+}_{-\delta t,\exp}\left(\boldsymbol{\Psi}^{+}_{-\delta t,\exp}\right)^{T}\right]^{-1}\\ &= \frac{1}{1+\delta t^{+2}}\mathbf{I}_2\end{aligned} \tag{5.322}$$

and (5.318) is substituted for by

$$\det \boldsymbol{\Psi}^{+}_{\delta t,\mathrm{imp}} = \frac{1}{1+\delta t^{+2}}$$

Unsurprisingly, the phase space volume decreases, as well as the energy. Thus, an explicit scheme is excitating, whereas a fully implicit scheme is dissipative, which once again reflects the fact that they may be deemed symmetric to each other through a time reversal.

The proper equilibrium is then provided by a symplectic scheme which, for instance, would here take the following form:

$$\begin{aligned}\dot{\theta}^{m+1} &= \dot{\theta}^{m} + \ddot{\theta}^{m}\delta t\\ \theta^{m+1} &= \theta^{m} + \dot{\theta}^{m+1}\delta t\end{aligned} \tag{5.323}$$

that is

$$\begin{aligned}\mathbf{X}^{+m+1} &= \boldsymbol{\Psi}^{+}_{\delta t,\mathrm{symp}}\mathbf{X}^{+m}\\ \boldsymbol{\Psi}^{+}_{\delta t,\mathrm{symp}} &\doteq \begin{pmatrix} 1 & -\delta t^{+} \\ \delta t^{+} & 1-\delta t^{+2} \end{pmatrix}\end{aligned} \tag{5.324}$$

This time, the matrix $\boldsymbol{\Psi}^{+}_{\delta t,\mathrm{symp}}$ verifies

$$\det \boldsymbol{\Psi}^{+}_{\delta t,\mathrm{symp}} = 1 \tag{5.325}$$

The phase space volumes are then preserved. On the basis of Section 2.2.2, it should then be expected that the energy would be conserved, which would correspond to an orthogonal matrix $\boldsymbol{\Psi}^{+}_{\delta t,\mathrm{symp}}$ (as with eqn (1.177) in the case of a continuous time), but it is an ill-founded intuition. We get, instead

$$\left(\boldsymbol{\Psi}^{+}_{\delta t,\mathrm{symp}}\right)^{-1} = \begin{pmatrix} 1-\delta t^{+2} & \delta t^{+} \\ -\delta t^{+} & 1 \end{pmatrix} \neq \left(\boldsymbol{\Psi}^{+}_{\delta t,\mathrm{symp}}\right)^{T} \tag{5.326}$$

The reason for this inequality is that, upon the time reversal, the operators are applied in the reverse order, as previously explained, which amounts to first integrating the position θ, and subsequently the velocity $\dot{\theta}$. This can easily be observed, since the transformation as given by the $\boldsymbol{\Psi}^{+}_{\delta t,\mathrm{symp}}$ matrix is decomposed as follows:

$$\boldsymbol{\Psi}^{+}_{\delta t,\mathrm{symp}} = \begin{pmatrix} 1 & 0 \\ \delta t^{+} & 1 \end{pmatrix} \begin{pmatrix} 1 & -\delta t^{+} \\ 0 & 1 \end{pmatrix} \tag{5.327}$$

which is identical to the decomposition $\Psi^{m,m+1}_{\mathbf{p},\delta t}\Psi^{m,m+1}_{\mathbf{r},\delta t}$ of eqn (5.293) for a particle. The system is then time reversible, but the conserved quantity corresponding to that invariance is just an approximation of the energy, as previously explained. We may introduce the matrix

$$\mathbf{A} \doteqdot \begin{pmatrix} 1 & \dfrac{\delta t^{+}}{2} \\ \dfrac{\delta t^{+}}{2} & 1 \end{pmatrix} \tag{5.328}$$

and define a numerical energy

$$\begin{aligned} H_{num}\left(p^{+}, q^{+}\right) &= \frac{mg\ell}{2}\left(\mathbf{X}^{+}\right)^{T}\mathbf{A}\mathbf{X}^{+} \\ &= \frac{mg\ell}{2}\left(p^{+2} + q^{+2} + p^{+}q^{+}\delta t^{+}\right) \\ &= H + \dot{\theta}\theta\delta t \end{aligned} \tag{5.329}$$

As in (5.306), H_{num} differs from H by an error of about δt. By analogy with (5.316), that quantity evolves in accordance with the law

$$H^{m+1}_{num} = \frac{mg\ell}{2}\left(\mathbf{X}^{+m}\right)^{T}\left(\boldsymbol{\Psi}^{+}_{\delta t,\mathrm{symp}}\right)^{T}\mathbf{A}\left(\boldsymbol{\Psi}^{+}_{\delta t,\mathrm{symp}}\right)\mathbf{X}^{+m} \tag{5.330}$$

The following property is easily verified:

$$\left(\boldsymbol{\Psi}^{+}_{\delta t,\mathrm{symp}}\right)^{T}\mathbf{A}\boldsymbol{\Psi}^{+}_{\delta t,\mathrm{symp}} = \mathbf{A} \tag{5.331}$$

and shows that the numerical energy is conserved. This is a specific case for which the numerical energy is exactly written to the first order in δt, the higher order terms being identically zero (because the system is linear).

Let us now return to the case of a collection of particles. Other symplectic schemes can be constructed by differently decomposing the operator D_H, with the general form

$$D_H = \sum_{i=1}^{K}\left(C_{pi}D_{H_{\mathbf{p}}} + C_{ri}D_{H_{\mathbf{r}}}\right) \tag{5.332}$$

instead of (5.278), with the condition

$$\sum_{i=1}^{K} C_{pi} = \sum_{i=1}^{K} C_{qi} = 1 \tag{5.333}$$

These relations make it possible to satisfy the equality

$$H = \sum_{i=1}^{K} \left(C_{pi} H_{\mathbf{p}} + C_{ri} H_{\mathbf{r}} \right) \tag{5.334}$$

which supersedes (5.276). Each operator $C_{pi} D_{H_{\mathbf{p}}}$ and $C_{ri} D_{H_{\mathbf{r}}}$ then obeys relations analogous to (5.289), and their exponentials[63] $\Psi_{\mathbf{p},C_{pi}\delta t} = \exp\left(C_{pi}\delta t D_{H_{\mathbf{p}}}\right)$ and $\Psi_{\mathbf{r},C_{ri}\delta t} = \exp\left(C_{ri}\delta t D_{H_{\mathbf{r}}}\right)$ yield

[63] Here we delete the superscripts, ^{m} etc., for the sake of clarity in the notations.

$$\begin{aligned} \Psi_{\mathbf{p},C_{pi}\delta t} \mathbf{X}_a(t) &= \begin{pmatrix} \mathbf{p}_a(t) \\ \mathbf{r}_a\left(t + C_{pi}\delta t\right) \end{pmatrix} \\ \Psi_{\mathbf{r},C_{ri}\delta t} \mathbf{X}_a(t) &= \begin{pmatrix} \mathbf{p}_a\left(t + C_{ri}\delta t\right) \\ \mathbf{r}_a(t) \end{pmatrix} \end{aligned} \tag{5.335}$$

to be compared with (5.292). Thus, we get a family of K-ordered schemes in which the time step is fragmented into K sub-iterations. The operator (5.293) is replaced by

$$\Psi_{\delta t} = \Psi_{\mathbf{p},C_{pK}\delta t} \Psi_{\mathbf{r},C_{rK}\delta t} \ldots \Psi_{\mathbf{p},C_{p1}\delta t} \Psi_{\mathbf{r},C_{r1}\delta t} + O\left(\delta t^{K+1}\right) \tag{5.336}$$

With $K = 1$ and $C_{p1} = C_{r1} = 1$, we are back to the previously discussed first-order symplectic scheme. With the option $K = 2$, $C_{p1} = C_{p2} = 1/2$, $C_{r1} = 0$ and $C_{r2} = 1$, we find

$$\Psi_{\delta t} = \Psi_{\mathbf{p},\delta t/2} \Psi_{\mathbf{r},\delta t} \Psi_{\mathbf{p},\delta t/2} + O\left(\delta t^3\right) \tag{5.337}$$

The relations (5.335) then allow us to write this scheme in the following expanded form:

$$\begin{aligned} \forall a,\ \mathbf{r}_a^{m+1/2} &= \mathbf{r}_a^m + \mathbf{u}_a^m \frac{\delta t}{2} \\ \mathbf{u}_a^{m+1} &= \mathbf{u}_a^m + \frac{1}{m_a} \mathbf{F}_a^{m+1/2} \delta t \\ \mathbf{r}_a^{m+1} &= \mathbf{r}_a^{m+1/2} + \mathbf{u}_a^{m+1} \frac{\delta t}{2} \end{aligned} \tag{5.338}$$

with a second-order accuracy in δt. The equations in (5.338) should be applied in their present sequence order. We have provided with superscript $^{m+1/2}$ all that corresponds to time $t_m + \delta t/2$. This so-called Störmer–Verlet scheme can then be seen as a predictor-corrector scheme (i.e. of second order), the positions being initially estimated and subsequently corrected as a result of the re-evaluation of the velocities (Leimkuhler and Reich, 2005). $\mathbf{F}_a^{m+1/2}$ should then be understood as the force received by a as calculated from the predicted positions, that is $\mathbf{r}_a^{m+1/2}$. It could readily be demonstrated that this scheme stems from the discrete Hamiltonian

$$H_{\delta t}\left(\{\mathbf{p}_b\},\{\mathbf{r}_b\}\right) \doteqdot \sum_b \frac{|\mathbf{p}_b|^2}{2m_b} + \sum_b E_{p,b}(\{\mathbf{r}_a\})\delta t \sum_m \delta\left[t - \left(m + \frac{1}{2}\right)\delta t\right] \tag{5.339}$$

in lieu of (5.313). This scheme is advantageous in that the error in the prediction of the energy is on the order of δt^2, instead of δt with the first-order scheme (eqn (5.306)). On the other hand, it is more computationally time-consuming. Please refer, for example, to Monaghan (2005) as regards its use within the context of the SPH method.

As a rule, it should be remembered that using a symplectic scheme, even of a low order, is always advantageous over a non-symplectic scheme, even of a high order. In other words, the symplectic (conservative) nature should be given prevalence over the order of accuracy. A fourth-order (non-symplectic) Runge-Kutta scheme will, for instance, give an error which is very weak during the first iterations but is liable to arbitrarily increase afterwards. It is a crucial matter for SPH applications in which the time step is often quite small (refer to Section 5.4.4), which requires many iterations.

5.4.3 Discrete Lagrange equations

The Hamiltonian nature of the symplectic scheme which we have introduced in the previous two sections suggests that we could advantageously utilize the formalism of the Lagrange equations to give it a rigorous variational nature, as we did in Chapter 1, later on for the continua in Chapter 3 (Section 3.6.3), and lastly within the framework of the SPH method provided with a continuous time (Section 5.3.3). We will write that the system's Lagrangian, as in Section 5.3.3, equals

$$L\left(\{\mathbf{r}_b\},\{\mathbf{u}_b\}\right) = \sum_b \frac{1}{2} m_b |\mathbf{u}_b|^2 - \sum_b E_{p,b}\left(\{\mathbf{r}_a\}\right) \tag{5.340}$$

(as in Section 5.4.1, we restrict the internal forces to a function of the positions alone). In order to apply a least action principle to this Lagrangian, we should integrate it over time for defining the action of the system, which implies that a quadrature will be chosen for both terms in (5.340). If we initially decide to approximate the velocities on the basis of the scheme (5.269), then we get:

$$\left(\frac{d\mathbf{r}_a}{dt}\right)^m = \mathbf{u}_a^m + O\left(\delta t\right) \tag{5.341}$$

with

$$\mathbf{u}_a^m = \frac{\mathbf{r}_a^m - \mathbf{r}_a^{m-1}}{\delta t} \tag{5.342}$$

and the Lagrangian at the time point m becomes

$$L\left(\left\{\mathbf{r}_b^m\right\}_b, \left\{\mathbf{r}_b^{m-1}\right\}_b\right) = \frac{1}{\delta t^2}\sum_b \frac{1}{2} m_b \left|\mathbf{r}_b^m - \mathbf{r}_b^{m-1}\right|^2 - \sum_b \left[E_{p,b}\left(\{\mathbf{r}_a\}\right)\right]^m \tag{5.343}$$

The last index in the notation $\left\{\mathbf{r}_b^m\right\}_b$ means that the curly braces $\{\cdot\}$ stand for the whole set of positions for all the subscripts b, the superscript m being set. Note that in the case of the scheme (5.342), the velocities are no longer parameters which are independent of the positions, and the quantities $\mathbf{u}_b^m$ are just auxiliary notations the definition of which is incidentally chosen, because we could have defined the velocities in an another way, as set out later. Besides, we can see that the contribution of each particle to the Lagrangian is identical; hence, the action associated with (5.343) is written as

$$\tilde{S}\left(\left\{\mathbf{r}_b^m\right\}_{b,m}\right) = \sum_{b,m} L_b\left(\mathbf{r}_b^m, \mathbf{r}_b^{m-1}\right)\delta t \tag{5.344}$$

where, this time, the curly braces in the notation $\left\{\mathbf{r}_b^m\right\}_{b,m}$ involve both indices b and m, and with

$$\forall b,\ L_b\left(\mathbf{r}_b^m, \mathbf{r}_b^{m-1}\right) \doteq \frac{1}{2\delta t^2} m_b \left|\mathbf{r}_b^m - \mathbf{r}_b^{m-1}\right|^2 - \left[E_{p,b}\left(\{\mathbf{r}_a\}\right)\right]^m \tag{5.345}$$

Equation (5.344) is a discrete spatio-temporal form of (3.325). After a summation in time and space, the parameters now comprise the whole set $\left\{\mathbf{r}_b^m\right\}_{b,m}$ of positions of all the particles at all the points in time, just as the generalized coordinates $\mathbf{q}$ in Section 3.6 happened to be fields varying in both time and space.

We will subsequently proceed as in Section 3.6, writing how the action varies due to an infinitesimal variation $\delta\mathbf{r}_b^m$ of its independent parameters. To that purpose, we necessarily enter the notations $\partial L_b/\partial\mathbf{x}$ and $\partial L_b/\partial\mathbf{y}$ to denote the derivatives of L_b with respect to the current $\left(\mathbf{r}_b^m\right)$ and earlier $\left(\mathbf{r}_b^{m-1}\right)$ positions of a particle b, respectively, which leads to:

$$\begin{aligned}\delta\tilde{S} &= \sum_{b,m}\left[\begin{array}{c}\dfrac{\partial L_b}{\partial\mathbf{x}}\left(\mathbf{r}_b^m, \mathbf{r}_b^{m-1}\right)\cdot\delta\mathbf{r}_b^m \\ +\dfrac{\partial L_b}{\partial\mathbf{y}}\left(\mathbf{r}_b^m, \mathbf{r}_b^{m-1}\right)\cdot\delta\mathbf{r}_b^{m-1}\end{array}\right]\delta t \\ &= \sum_{b,m}\left[\frac{\partial L_b}{\partial\mathbf{x}}\left(\mathbf{r}_b^m, \mathbf{r}_b^{m-1}\right) + \frac{\partial L_b}{\partial\mathbf{y}}\left(\mathbf{r}_b^{m+1}, \mathbf{r}_b^m\right)\right]\cdot\delta\mathbf{r}_b^m\,\delta t\end{aligned} \tag{5.346}$$

(we have offset the temporal index m in the second term of the right-hand side in order to unite the two sums).[64] Writing that this variation should be stationary for every chosen set of the $\delta\mathbf{r}_b^m$, we find

$$\forall(a,m),\ \frac{\partial L_a}{\partial\mathbf{x}}\left(\mathbf{r}_a^m, \mathbf{r}_a^{m-1}\right) + \frac{\partial L_a}{\partial\mathbf{y}}\left(\mathbf{r}_a^{m+1}, \mathbf{r}_a^m\right) = \mathbf{0} \tag{5.347}$$

which constitutes a spatio-temporal discrete form of (3.310), which may be called *discrete Lagrange equations* (Quinlan and Tremaine, 1990).

[64] This is only possible because the assumed variation of the positions $\delta\mathbf{r}_a^0$ and $\delta\mathbf{r}_a^N$ is null at the time limits (i.e. $m = 0$ and $m = N$, N being the number of iterations). This is because the initial and final conditions prescribe the values of the $\left\{\mathbf{r}_a^0\right\}$ and $\left\{\mathbf{r}_a^N\right\}$, as within the context of the variational approach for the continuous media (refer to Chapter 3, Section 3.6.2, eqn (3.330)).

With the Lagrangian (5.345), we find

$$\begin{aligned}\frac{\partial L_a}{\partial \mathbf{x}}\left(\mathbf{r}_a^m, \mathbf{r}_a^{m-1}\right) &= \frac{1}{\delta t^2} m_a \left(\mathbf{r}_a^m - \mathbf{r}_a^{m-1}\right) - \left(\frac{\partial E_{p,a}}{\partial \mathbf{r}_a}\right)^m \\ \frac{\partial L_a}{\partial \mathbf{y}}\left(\mathbf{r}_a^{m+1}, \mathbf{r}_a^m\right) &= -\frac{1}{\delta t^2} m_a \left(\mathbf{r}_a^{m+1} - \mathbf{r}_a^m\right)\end{aligned} \tag{5.348}$$

or else, taking (5.342) and the definition of the forces into account:

$$\begin{aligned}\frac{\partial L_a}{\partial \mathbf{x}}\left(\mathbf{r}_a^m, \mathbf{r}_a^{m-1}\right) &= \frac{1}{\delta t} m_a \mathbf{u}_a^m + \mathbf{F}_a^m \\ \frac{\partial L_a}{\partial \mathbf{y}}\left(\mathbf{r}_a^{m+1}, \mathbf{r}_a^m\right) &= -\frac{1}{\delta t} m_a \mathbf{u}_a^{m+1}\end{aligned} \tag{5.349}$$

The discrete Lagrange equations (5.347) are then written as

$$\forall (a, m), \; m_a \left(\mathbf{u}_a^m - \mathbf{u}_a^{m-1}\right) + \mathbf{F}_a^m \delta t = 0 \tag{5.350}$$

in accordance with (5.269). We find that the temporal discretization of the velocities is consistent, from a variational point of view, with that of the positions. Thus, a property of the symplectic scheme is that it results from a discrete variational principle (Marsden and West, 2001).

The explicit scheme does not have that property. We can observe, indeed, that the expansion (5.266) corresponds to the following approximation for the velocities:

$$\mathbf{u}_a^m = \frac{\mathbf{r}_a^{m+1} - \mathbf{r}_a^m}{\delta t} \tag{5.351}$$

Writing the action as (5.344) and applying the discrete Lagrange equations (5.347) would then provide the following equation of motion:

$$\forall (a, m), \; m_a \left(\mathbf{u}_a^{m-1} - \mathbf{u}_a^m\right) + \mathbf{F}_a^m \delta t = 0 \tag{5.352}$$

which is not consistent with (5.266). On the contrary, (5.352) corresponds to an implicit scheme for the velocities, which corresponds to another type of symplectic scheme, stemming from the time integration operator $\Psi_{\mathbf{r},\delta t}^{m,m+1} \Psi_{\mathbf{p},\delta t}^{m,m+1}$, in accordance with the notation in Section 5.4.1.

Once again, the above discussion advocates for the semi-implicit (symplectic) scheme, and we are going to strengthen that assertion by looking deeper into the notion of conservation within the context of a discrete time. This is because the Lagrange equations take us directly back to the properties of conservation of the systems whose evolution is based on a variational principle. We mentioned, especially in Section 5.3.3, the properties of conservation of those schemes which have been selected for the spatial discretization of the forces within the frame of the SPH method. Bringing in a temporal scheme may interfere with these properties, as happened with the energy in Section 5.4.2. The reason for this is as follows: we have already emphasized, at the end of Section 1.4.3, that the laws of conservation of momentum and angular momentum may be written in two different yet equivalent (continuous and discrete) forms. Now, with a discretized time, only such laws as (5.249), involving continuous temporal derivatives, can be implemented. Thus, although the sum of the moments of the forces exerted on an isolated system is zero, the conservation of the angular momentum of that system is not necessarily implied if time is discretized. In other words, the continuous conservation does not necessarily imply the discrete conservation; it depends on the temporal scheme.

Since the symplectic scheme (5.269) originates from a variational principle, it should be able, pursuant to Section 1.4, to exactly verify the required conservation laws,[65] as discussed in the following. We will first examine what the total discrete momentum of an isolated system becomes between two consecutive time points. By analogy with Section 1.4.3, it is first noteworthy that translating the position vectors $\{\mathbf{r}_b^m\}_b$ of a constant infinitesimal vector $\boldsymbol{\epsilon}$ leaves the discrete action (5.344) unchanged; thus, we may write

[65] By analogy with Chapter 1 (Section 1.4), a discrete Noether theorem could be formally established.

$$\forall \boldsymbol{\epsilon}, \quad \sum_{b,m} L_b\left(\mathbf{r}_b^m, \mathbf{r}_b^{m-1}\right) = \sum_{b,m} L_b\left(\mathbf{r}_b^m + \boldsymbol{\epsilon}, \mathbf{r}_b^{m-1} + \boldsymbol{\epsilon}\right) \tag{5.353}$$

to be compared with (1.129). A first-order Taylor series expansion in $\boldsymbol{\epsilon}$ yields

$$\forall \boldsymbol{\epsilon}, \quad \sum_b \left[\frac{\partial L_b}{\partial \mathbf{x}}\left(\mathbf{r}_b^m, \mathbf{r}_b^{m-1}\right) + \frac{\partial L_b}{\partial \mathbf{y}}\left(\mathbf{r}_b^m, \mathbf{r}_b^{m-1}\right)\right] \cdot \boldsymbol{\epsilon} = 0 \tag{5.354}$$

Deleting the vector $\boldsymbol{\epsilon}$ and using the discrete Lagrange equations (5.347), we get

$$\sum_a \frac{\partial L_a}{\partial \mathbf{y}}\left(\mathbf{r}_a^{m+1}, \mathbf{r}_a^m\right) = \sum_a \frac{\partial L_a}{\partial \mathbf{y}}\left(\mathbf{r}_a^m, \mathbf{r}_a^{m-1}\right) \tag{5.355}$$

This formula is a conservation law. It illustrates, indeed, the constant nature, from one time point m to the time point $m+1$, of the quantity

$$\begin{aligned} \mathbf{P}^m &\doteq -\delta t \sum_a \frac{\partial L_a}{\partial \mathbf{y}}\left(\mathbf{r}_a^m, \mathbf{r}_a^{m-1}\right) \\ &= \sum_a \frac{1}{\delta t} m_a \left(\mathbf{r}_a^m - \mathbf{r}_a^{m-1}\right) \\ &= \sum_a m_a \mathbf{u}_a^m \end{aligned} \tag{5.356}$$

which coincides with the system's total momentum. We may note that the definition (5.356) suggests that the role of the momentum of a particle is played by the quantity

$$\mathbf{p}_a^m = -\delta t \frac{\partial L_a}{\partial \mathbf{y}}\left(\mathbf{r}_a^m, \mathbf{r}_a^{m-1}\right) \tag{5.357}$$

Let us check the conservation of that quantity with the symplectic scheme (5.269):

$$\begin{aligned}\mathbf{P}^{m+1} &= \sum_a m_a \mathbf{u}_a^{m+1} \\ &= \sum_a m_a \mathbf{u}_a^m + \left(\sum_a \mathbf{F}_a^m\right)\delta t \\ &= \mathbf{P}^m\end{aligned} \tag{5.358}$$

(we have used the relation (5.245), which stipulates that the sum of the forces in such a system is exactly zero with an adequate spatial discretization). Continuous and discrete conservations here go hand in hand. Thus, we can see that the total momentum is exactly conserved from a discrete point of view, provided that a variational (symplectic) temporal scheme is chosen.[66]

Likewise, the discrete conservation of the total angular momentum results from the invariance of the action under an infinitesimal rotation, as in Section 1.4.3. That formally leads to the conservation of the quantity

$$\begin{aligned}\mathbf{M}^m &\doteqdot -\delta t \sum_a \frac{\partial L_a}{\partial \mathbf{y}}\left(\mathbf{r}_a^m, \mathbf{r}_a^{m-1}\right) \times \mathbf{r}_a^{m-1} \\ &= \sum_a \mathbf{p}_a^m \times \mathbf{r}_a^{m-1}\end{aligned} \tag{5.359}$$

where (5.357) has been used. By means of the vector product properties and the definition (5.342), that quantity is readily reduced to the standard expression

$$\begin{aligned}\mathbf{M}^m &= \sum_a \frac{1}{\delta t} m_a \left(\mathbf{r}_a^m - \mathbf{r}_a^{m-1}\right) \times \mathbf{r}_a^{m-1} \\ &= \sum_a \frac{1}{\delta t} m_a \left(\mathbf{r}_a^m - \mathbf{r}_a^{m-1}\right) \times \mathbf{r}_a^m \\ &= \sum_a m_a \mathbf{u}_a^m \times \mathbf{r}_a^m\end{aligned} \tag{5.360}$$

Once again, let us check that rule in the case of the symplectic scheme (5.269):

$$\begin{aligned}\mathbf{M}^{m+1} &= \sum_a m_a \mathbf{r}_a^{m+1} \times \mathbf{u}_a^{m+1} \\ &= \sum_a m_a \mathbf{r}_a^m \times \mathbf{u}_a^{m+1} + \left(\sum_a m_a \mathbf{u}_a^{m+1} \times \mathbf{u}_a^{m+1}\right)\delta t \\ &= \sum_a m_a \mathbf{r}_a^m \times \mathbf{u}_a^m + \left(\sum_a \mathbf{r}_a^m \times \mathbf{F}_a^m\right)\delta t \\ &= \mathbf{M}^m\end{aligned} \tag{5.361}$$

It should be pointed out that, to simplify this calculation, we have used the property $\mathbf{A} \times \mathbf{A} = \mathbf{0}$, as well as the calculation (5.249) and (5.250) which, it should be remembered, is only valid for forces aligned with the radius vectors connecting the particles, corresponding to the discrete conservation of the angular momentum. Moreover, the explicit scheme (5.266) would give

[66] This property would hold true with the explicit scheme (5.266). This is because the momentum consists of a linear quantity with respect to the dynamic variables of the problem, unlike the angular momentum, as explained just below.

$$\begin{aligned}\mathbf{M}^{m+1} &= \sum_a m_a \mathbf{r}_a^{m+1} \times \mathbf{u}_a^{m+1} \\ &= \sum_a m_a \mathbf{r}_a^{m} \times \mathbf{u}_a^{m+1} + \left(\sum_a m_a \mathbf{u}_a^{m} \times \mathbf{u}_a^{m+1}\right)\delta t \\ &= \sum_a m_a \mathbf{r}_a^{m} \times \mathbf{u}_a^{m} + \left(\sum_a \mathbf{r}_a^{m} \times \mathbf{F}_a^{m}\right)\delta t + \left(\sum_a \mathbf{u}_a^{m} \times \mathbf{F}_a^{m}\right)\delta t^2 \\ &= \mathbf{M}^m + O\left(\delta t^2\right)\end{aligned} \tag{5.362}$$

This time, we can see that the angular momentum is only conserved to the first order, which reveals a variationally non-compatible scheme. With the Störmer–Verlet scheme (5.338), whose symplectic nature was shown earlier, the angular momentum behaves as follows:

$$\begin{aligned}\mathbf{M}^{m+1} &= \sum_a m_a \mathbf{r}_a^{m+1} \times \mathbf{u}_a^{m+1} \\ &= \sum_a m_a \mathbf{r}_a^{m+1/2} \times \mathbf{u}_a^{m+1} + \frac{1}{2}\left(\sum_a m_a \mathbf{u}_a^{m+1} \times \mathbf{u}_a^{m+1}\right)\delta t \\ &= \sum_a m_a \mathbf{r}_a^{m+1/2} \times \mathbf{u}_a^{m} + \left(\sum_a \mathbf{r}_a^{m+1/2} \times \mathbf{F}_a^{m+1/2}\right)\delta t \\ &= \sum_a m_a \mathbf{r}_a^{m} \times \mathbf{u}_a^{m} + \frac{1}{2}\left(\sum_a m_a \mathbf{u}_a^{m} \times \mathbf{u}_a^{m}\right)\delta t \\ &= \mathbf{M}^m\end{aligned} \tag{5.363}$$

It is noteworthy that the exact conservation of the total linear momentum then brings about the conservation of the total angular momentum in all the coordinate systems, according to the relation (1.151) in Chapter 1.

The question of energy is more complex, because it is physically linked to the invariance of the action (or Lagrangian) with respect to any time translation, which does not explicitly appear. Besides, we learnt in Section 5.4.2 that we may only expect an approximate conservation of the actual energy, with an error of δt or so for a first-order scheme over time. We must then perform a first-order Taylor series expansion of the discrete Lagrangian (5.345):

$$L^{m+1} = L^m + \sum_b \frac{dL_b}{dt}\left(\mathbf{r}_b^m, \mathbf{r}_b^{m-1}\right)\delta t + O\left(\delta t^2\right) \tag{5.364}$$

The temporal derivative of the Lagrangian L_b is written, taking the definition (5.341) and the discrete Lagrange equations (5.347) into account, as:

$$\begin{aligned}\frac{dL_b}{dt}\left(\mathbf{r}_b^m, \mathbf{r}_b^{m-1}\right) &= \frac{\partial L_b}{\partial \mathbf{x}}\left(\mathbf{r}_b^m, \mathbf{r}_b^{m-1}\right)\cdot\left(\frac{d\mathbf{r}_b}{dt}\right)^m \\ &+ \frac{\partial L_b}{\partial \mathbf{y}}\left(\mathbf{r}_b^m, \mathbf{r}_b^{m-1}\right)\cdot\left(\frac{d\mathbf{r}_b}{dt}\right)^{m-1}\end{aligned} \tag{5.365}$$

$$= \frac{\partial L_b}{\partial \mathbf{x}}\left(\mathbf{r}_b^m, \mathbf{r}_b^{m-1}\right) \cdot \mathbf{u}_b^m + \frac{\partial L_b}{\partial \mathbf{y}}\left(\mathbf{r}_b^m, \mathbf{r}_b^{m-1}\right) \cdot \mathbf{u}_b^{m-1}$$
$$+ O(\delta t)$$
$$= -\frac{\partial L_b}{\partial \mathbf{y}}\left(\mathbf{r}_b^{m+1}, \mathbf{r}_b^m\right) \cdot \mathbf{u}_b^m + \frac{\partial L_b}{\partial \mathbf{y}}\left(\mathbf{r}_b^m, \mathbf{r}_b^{m-1}\right) \cdot \mathbf{u}_b^{m-1}$$
$$+ O(\delta t)$$

As expected, (5.365) then only leads to an approximate conservation, which is written as:

$$E_{num}^{m+1} = E_{num}^m + O\left(\delta t^2\right) \tag{5.366}$$

with a numerical energy as defined by

$$\begin{aligned} E_{num}^m &\doteqdot -\delta t \sum_b \frac{\partial L_b}{\partial \mathbf{y}}\left(\mathbf{r}_b^m, \mathbf{r}_b^{m-1}\right) \cdot \mathbf{u}_b^{m-1} - L^m \\ &= \sum_a \mathbf{p}_a^m \cdot \mathbf{u}_a^{m-1} - L^m \end{aligned} \tag{5.367}$$

(using (5.357)). This definition may also be compared to (1.69). Returning to the definition of the Lagrangian and using the relation (5.342), we find

$$\begin{aligned} E_{num}^m &= \sum_a \left\{ \begin{array}{l} \dfrac{1}{\delta t} m_a \left(\mathbf{r}_a^m - \mathbf{r}_a^{m-1}\right) \cdot \mathbf{u}_a^{m-1} \\ -\dfrac{1}{2} m_a \left|\mathbf{u}_a^m\right|^2 + \left[E_{p,a}(\{\mathbf{r}_b\})\right]^m \end{array} \right\} \\ &= \sum_a \left\{ \frac{1}{2} m_a \left|\mathbf{u}_a^m\right|^2 - m_a \mathbf{u}_a^m \cdot \left(\mathbf{u}_a^m - \mathbf{u}_a^{m-1}\right) + \left[E_{p,a}(\{\mathbf{r}_b\})\right]^m \right\} \end{aligned} \tag{5.368}$$

Lastly, the symplectic scheme (5.269) makes it possible to relate $\mathbf{u}_a^m$ to $\mathbf{u}_a^{m-1}$ and write

$$\begin{aligned} E_{num}^m &= E^m - \frac{\delta t}{2} \sum_a \mathbf{F}_a^{m-1} \cdot \mathbf{u}_a^m \\ &= E^m - \frac{\delta t}{2} \sum_a \mathbf{F}_a^m \cdot \mathbf{u}_a^m + O\left(\delta t^2\right) \end{aligned} \tag{5.369}$$

because $\mathbf{F}_a^m - \mathbf{F}_a^{m-1}$ is of the order of δt. This formula may be compared with (5.306) before coming to the conclusion that the numerical energy E_{num} is actually equivalent to the numerical Hamiltonian H_{num}. Unlike the discrete momentum and angular momentum, it is only equivalent to the true (discrete) energy with a (bounded) error. It should be pointed out that here we have not demonstrated the exact conservation of anything. To that purpose, one should return to the discussion in Section 5.4.2. This calculation, however, shows that, within the context of the variational integrators, the discrete Lagrange equations and the temporal invariance lead to the approximate conservation of the energy, just as the spatial invariance has led us to the discrete conservation of linear and angular momentum.

5.4.4 Numerical stability

Let us now turn to those conditions which the above briefly discussed numerical schemes should satisfy upon their implementation, in order to ensure the numerical stability of the solutions in every circumstance. It has been well-known since the work by Courant, Friedrichs and Lewy (1928, refer to the English translation published in 1967) that solving a system of partial derivative equations through a discrete approach should satisfy several conditions regarding the time step δt, in order to prevent the propagation of numerical waves which are subject to an undetermined amplification making the solution diverge. Within the frame of the SPH method, these instabilities may occur in various forms ranging from the aggregation of particles in pairs in static phase to the occurrence of an exponential disturbance ('blowing up'). Although these phenomena are purely theoretical, they may be addressed like the acoustic waves discussed in Section 3.4.3 of Chapter 3, by linearizing the equations around a reference situation. Here we generalize the results achieved by Balsara (1995) and Swegle et al. (1995) to the case of an arbitrary dimension. One may also refer to Belytschko and Xiao (2002) and Sigalotti et al. (2009).

First of all, we write a system providing the temporal variation rates of the system state variables in the phase space, according to the general rule (5.263), without prejudging, in a first step, the explicit, symplectic or implicit natures of the time integration scheme. As usual, we add a dot on top of those quantities which are derived with respect to time. We keep the general models as provided by equations (5.83) and (5.239). Using the state equation (5.208) to delete pressure in the equation of motion, we can write this system as

$$\begin{aligned}
\dot{\mathbf{u}}_a &= \frac{1}{m_a}\mathbf{F}_a = -\sum_b m_b \left(\frac{\rho_b^{k-1} p_a}{\rho_a^{k+1}} + \frac{\rho_a^{k-1} p_b}{\rho_b^{k+1}} \right) w'_{ab}\mathbf{e}_{ab} \\
&= -\frac{\rho_0 c_0^2}{\gamma} \sum_b m_b \left[\frac{\rho_b^{k-1}}{\rho_a^{k+1}} \left(\frac{\rho_a^\gamma}{\rho_0^\gamma} + C \right) + \frac{\rho_a^{k-1}}{\rho_b^{k+1}} \left(\frac{\rho_b^\gamma}{\rho_0^\gamma} + C \right) \right] w'_{ab}\mathbf{e}_{ab} \\
\dot{\mathbf{r}}_a &= \mathbf{u}_a \\
\dot{\rho}_a &= \sum_b m_b \left(\frac{\rho_b}{\rho_a} \right)^{k-1} \mathbf{u}_{ab} \cdot w'_{ab}\mathbf{e}_{ab}
\end{aligned} \tag{5.370}$$

We will consider in this section the wave-like solutions of these equations, given a small-amplitude wave, which allows the linearization[67] of equation (5.370). To that purpose, we seek the effect of small variations $\delta\mathbf{r}_a$, $\delta\mathbf{u}_a$ or $\delta\rho_a$ of the state variables of an arbitrary particle a upon the variation rates $\dot{\mathbf{u}}_a$ $\dot{\mathbf{r}}_a$ and $\dot{\rho}_a$. We will write, for example, that $\dot{\mathbf{u}}_a$ varies by the amount $\delta_{\mathbf{r}}\dot{\mathbf{u}}_a$ as a result of a variation $\delta\mathbf{r}_a$ of the positions. Let us initially consider the quantity $\dot{\mathbf{u}}_a$, which does not depend on the velocities. The first equation in (5.370) shows that the position of an arbitrary particle $\mathbf{r}_c$ occurs once in the sum of the right-hand side (for $b = c$), as well as in each of the forces if $c = a$ (this reasoning is analogous to that we employed in Section 5.3.3 and will be applied throughout this section). Using the notations $\mathbf{W}_{ab}$ as defined by (5.144), we find:

[67] The procedure being used in this case, which is known as *Von Neumann stability analysis*, is formally analogous to the theoretical arguments which prevail in the study of the stability of the solutions of the continuous equations for fluids. We briefly mentioned these works at the beginning of Section 4.2.2.

$$\delta_{\mathbf{r}}\dot{\mathbf{u}}_a = -\sum_c \left[\begin{array}{c} m_c \left(\frac{\rho_c^{k-1} p_a}{\rho_a^{k+1}} + \frac{\rho_a^{k-1} p_c}{\rho_c^{k+1}} \right) \mathbf{W}_{ac} \\ -\delta_{ac} \sum_b m_b \left(\frac{\rho_b^{k-1} p_a}{\rho_a^{k+1}} + \frac{\rho_a^{k-1} p_b}{\rho_b^{k+1}} \right) \mathbf{W}_{ab} \end{array} \right] \delta \mathbf{r}_c$$

$$= \sum_b m_b \left(\frac{\rho_b^{k-1} p_a}{\rho_a^{k+1}} + \frac{\rho_a^{k-1} p_b}{\rho_b^{k+1}} \right) \mathbf{W}_{ab} \left(\delta \mathbf{r}_a - \delta \mathbf{r}_b \right) \tag{5.371}$$

Through an analogous procedure, we find $\delta_\rho \dot{\mathbf{u}}_a$, keeping in mind that the density also affects the force via the pressures:

$$\delta_\rho \dot{\mathbf{u}}_a = -\frac{\rho_0 c_0^2}{\gamma} \sum_b m_b \left[\begin{array}{c} (k-1) \frac{\rho_a^{k-2}}{\rho_b^{k+1}} \left(\frac{\rho_b^\gamma}{\rho_0^\gamma} + C \right) \delta \rho_a \\ +(k-1) \frac{\rho_b^{k-2}}{\rho_a^{k+1}} \left(\frac{\rho_a^\gamma}{\rho_0^\gamma} + C \right) \delta \rho_b \\ +(\gamma - k - 1) \frac{\rho_b^{k-1} \rho_a^{\gamma-k-2}}{\rho_0^\gamma} \delta \rho_a \\ +(\gamma - k - 1) \frac{\rho_a^{k-1} \rho_b^{\gamma-k-2}}{\rho_0^\gamma} \delta \rho_b \\ -(k+1) C \left(\frac{\rho_b^{k-1}}{\rho_a^{k+2}} \delta \rho_a + \frac{\rho_a^{k-1}}{\rho_b^{k+2}} \delta \rho_b \right) \end{array} \right] w'_{ab} \mathbf{e}_{ab} \tag{5.372}$$

To simplify our calculations, we will henceforth consider a numerical wave superimposed on a homogeneous initial state[68] featured by the reference density ρ_0 together with a pressure p_0. Equations (5.371) and (5.372) are then simplified and yield:

$$\delta_{\mathbf{r}}\dot{\mathbf{u}}_a = \frac{2p_0}{\rho_0^2} \sum_b m_b \mathbf{W}_{ab} \left(\delta \mathbf{r}_a - \delta \mathbf{r}_b \right) \tag{5.373}$$

$$\delta_\rho \dot{\mathbf{u}}_a = -\frac{(\gamma - 2D)\, p_0}{D \rho_0^3} \sum_b m_b w'_{ab} \mathbf{e}_{ab} \left(\delta \rho_a + \delta \rho_b \right)$$

with

$$p_0 \doteqdot \frac{D \rho_0 c_0^2}{\gamma} \tag{5.374}$$

D being defined by (5.209). We are going to treat the system (5.373) by approximating each of the discrete sums through an integral, invoking the approximation made in Section 5.2.3, this time in the opposite direction.[69] Thus, it is the particle with a dummy subscript b which should be likened to the continuous variable of spatial continuous integration, namely $\mathbf{r} \equiv \mathbf{r}_b$. The integration volume element $d^n\mathbf{r}$ is then treated as equivalent to the volume m_b/ρ_0, whereas the continuous integration is performed over the volume Ω_a encompassing the points interacting with particle a. Integration, however, can

[68] One may imagine that this state is associated with a spatial distribution of those particles corresponding to a regular Cartesian grid.

[69] This approximation is subject to the accuracy provided by the relevant section. It should be pointed out that it is all the better since the integrands vary slowly in space, that is in this case for large wavelengths as compared with the smoothing length, as we will see later on. In a way, that comes down to considering the stability of the chosen system of equations within the context of the continuous SPH approximation.

be performed over the whole space, it being understood that, in fact, the presence of the kernel restricts the integrations to Ω_a (refer to Section 5.2.1):

$$\begin{aligned}\delta_{\mathbf{r}}\dot{\mathbf{u}}_a &= \frac{2p_0}{\rho_0^2}\left[\left(\sum_b m_b \mathbf{W}_{ab}\right)\delta\mathbf{r}_a - \sum_b m_b \mathbf{W}_{ab}\delta\mathbf{r}_b\right] \\ &\approx -\frac{2p_0}{\rho_0}\left(\int_\Omega \frac{\partial \nabla w_h}{\partial \mathbf{r}}\left(|\mathbf{r}-\mathbf{r}_a|\right) d^n\mathbf{r}\right)\delta\mathbf{r}_a \\ &+ \frac{2p_0}{\rho_0}\int_\Omega \frac{\partial \nabla w_h}{\partial \mathbf{r}}\left(|\mathbf{r}-\mathbf{r}_a|\right)\delta\mathbf{r}\, d^n\mathbf{r} \\ &= \frac{2p_0}{\rho_0}\int_\Omega \nabla\nabla w_h\left(|\mathbf{r}-\mathbf{r}_a|\right)\delta\mathbf{r} d^n\mathbf{r}\end{aligned} \tag{5.375}$$

(the first integral is identically zero, provided that the particle a is located far off the boundary, as in Section 5.2.1).[70] It is noteworthy that the gradient of the kernel gradient appears in (5.375), whereas it was discretized by the quantities $\mathbf{W}_{ab}$ so far. Since equations (5.373) are linear with respect to the unknowns $\delta\mathbf{r}$ and $\delta\rho$, we may consider that each of the Fourier modes in a solution will obey the same equations, which allows us to only consider one of these modes, that is for the fluctuation of the positions:

[70] We ignore all effects of boundaries in the present study.

$$\delta\mathbf{r} = \mathbf{R}(t)\exp(-i\mathbf{K}\cdot\mathbf{r}) \tag{5.376}$$

where $\mathbf{R}(t)$ is an unknown vector function and $\mathbf{K}$ is a wave vector.[71] The approximation (5.375) then yields, putting $\tilde{\mathbf{r}} \doteqdot \mathbf{r} - \mathbf{r}_a$:

[71] The above assumption that the integrands should vary slowly then implies that K is not too large. Morris (1996a) makes a more accurate discrete analysis which, however, is substantially restricted to one space dimension.

$$\begin{aligned}\delta_{\mathbf{r}}\dot{\mathbf{u}}_a &\approx \frac{2p_0}{\rho_0}\left[\int_\Omega \nabla\nabla w_h\left(|\mathbf{r}-\mathbf{r}_a|\right)\exp(-i\mathbf{K}\cdot\mathbf{r})\, d^n\mathbf{r}\right]\mathbf{R}(t) \\ &= \frac{2p_0}{\rho_0}\left[\int_\Omega \nabla\nabla w_h(\tilde{r})\exp\left(-i\mathbf{K}\cdot\tilde{\mathbf{r}}\right) d^n\tilde{\mathbf{r}}\right]\mathbf{R}(t)\exp(-i\mathbf{K}\cdot\mathbf{r}_a)\end{aligned} \tag{5.377}$$

(the gradient may equally well be taken with respect to either $\mathbf{r}$ or $\tilde{\mathbf{r}}$). We can see that the Fourier transform of the kernel double gradient appears.[72] Using equation (B.41) in Appendix B, we derive from it:

[72] The role of the kernel Fourier transform in the stability of the SPH method was highlighted, in particular, by Sweggle et al. (1995), as well as by Morris (1996a). One may also refer to Monaghan (2005).

$$\begin{aligned}\delta_{\mathbf{r}}\dot{\mathbf{u}}_a &\approx \frac{2p_0}{\rho_0}\widehat{\nabla\nabla w_h}(\mathbf{K})\,\mathbf{R}(t)\exp(-i\mathbf{K}\cdot\mathbf{r}_a) \\ &= -\frac{2p_0}{\rho_0}\widehat{w_h}(K)\,(\mathbf{K}\otimes\mathbf{K})\,\mathbf{R}(t)\exp(-i\mathbf{K}\cdot\mathbf{r}_a)\end{aligned} \tag{5.378}$$

We treat the second equation of (5.373) in the same way, seeking a numerical fluctuation of density in the following form:

$$\delta\rho = R(t)\exp(-i\mathbf{K}\cdot\mathbf{r}) \tag{5.379}$$

where $R(t)$ is an unknown time function,[73] which gives, using (B.40):

$$\begin{aligned}\delta_\rho \dot{\mathbf{u}}_a &\approx \frac{(\gamma - 2D)\, p_0}{D\rho_0^2} \int_\Omega \nabla w_h(|\mathbf{r} - \mathbf{r}_a|)\, \delta\rho d^n\mathbf{r} \\ &= \frac{(\gamma - 2D)\, p_0}{D\rho_0^2} R(t)\, \widehat{\nabla w_h}(\mathbf{K}) \exp(-i\mathbf{K}\cdot\mathbf{r}_a) \\ &= \frac{i(\gamma - 2D)\, p_0}{D\rho_0^2} R(t)\, \widehat{w_h}(K)\, \mathbf{K} \exp(-i\mathbf{K}\cdot\mathbf{r}_a)\end{aligned} \tag{5.380}$$

[73]The assumed low amplitude disturbance is then reduced to $R(t) \ll \rho_0$. The same remark can be applied to the other involved quantities.

Adding (5.378) and (5.380), we can lastly write the variation of $\dot{\mathbf{u}}_a$ resulting from the wave being considered, namely $\delta\dot{\mathbf{u}}_a = \delta_{\mathbf{r}}\dot{\mathbf{u}}_a + \delta_\rho\dot{\mathbf{u}}_a$:

$$\begin{aligned}\delta\dot{\mathbf{u}}_a &\approx \frac{p_0}{\rho_0} \widehat{w_h}(K) \\ &\times \left[\frac{i(\gamma - 2D)}{D\rho_0} R(t)\, \mathbf{K} - 2(\mathbf{K}\otimes\mathbf{K})\, \mathbf{R}(t)\right] \exp(-i\mathbf{K}\cdot\mathbf{r}_a)\end{aligned} \tag{5.381}$$

(Monaghan, 2005, sets out a one-dimensional version of that equation). Let us now turn to the second equation of the system (5.370). Proceeding likewise, we find the quite simple result:

$$\delta\dot{\mathbf{r}}_a = \mathbf{U}(t) \exp(-i\mathbf{K}\cdot\mathbf{r}_a) \tag{5.382}$$

where $\mathbf{U}(t)$ is an unknown vector which depends on time, defining the fluctuation of velocity by

$$\delta\mathbf{u} = \mathbf{U}(t) \exp(-i\mathbf{K}\cdot\mathbf{r}) \tag{5.383}$$

Lastly, the third equation in (5.370) quickly yields, to calculate $\delta\dot{\rho}_a = \delta_{\mathbf{r}}\dot{\rho}_a + \delta_{\mathbf{u}}\dot{\rho}_a + \delta_\rho\dot{\rho}_a$, the following discrete variations:

$$\begin{aligned}\delta_{\mathbf{r}}\dot{\rho}_a &= \sum_b m_b \left(\frac{\rho_b}{\rho_a}\right)^{k-1} \mathbf{u}_{ab}^T \mathbf{W}_{ab}(\delta\mathbf{r}_b - \delta\mathbf{r}_a) \\ \delta_{\mathbf{u}}\dot{\rho}_a &= \sum_b m_b \left(\frac{\rho_b}{\rho_a}\right)^{k-1} w'_{ab}\mathbf{e}_{ab}\cdot(\delta\mathbf{u}_a - \delta\mathbf{u}_b) \\ \delta_\rho\dot{\rho}_a &= (k-1)\sum_b m_b \left(\frac{\rho_b}{\rho_a}\right)^{k-1} \mathbf{u}_{ab}\cdot w'_{ab}\mathbf{e}_{ab}\left(\frac{\delta\rho_b}{\rho_b} - \frac{\delta\rho_a}{\rho_a}\right)\end{aligned} \tag{5.384}$$

Starting once again from a homogeneous initial state, only the second term subsists. As previously, we can approximate it, of course, through an integral:

$$\delta\dot{\rho}_a \approx \rho_0\left[-\int_\Omega \nabla w_h\left(|\mathbf{r}-\mathbf{r}_a|\right)d^n\mathbf{r}\right]\cdot\delta\mathbf{u}_a \qquad (5.385)$$
$$+\rho_0\int_\Omega \nabla w\left(|\mathbf{r}-\mathbf{r}_a|\right)\cdot\delta\mathbf{u}\,d^n\mathbf{r}$$
$$=\rho_0\left[\int_\Omega \nabla w_h\left(\tilde{r}\right)\exp\left(-i\mathbf{K}\cdot\tilde{\mathbf{r}}\right)d^n\tilde{\mathbf{r}}\right]\cdot\mathbf{U}\left(t\right)\exp\left(-i\mathbf{K}\cdot\mathbf{r}_a\right)$$
$$=i\rho_0\widehat{w_h}\left(K\right)\mathbf{K}\cdot\mathbf{U}\left(t\right)\exp\left(-i\mathbf{K}\cdot\mathbf{r}_a\right)$$

Gathering all the calculations which have been made, that is (5.381), (5.382) and (5.385), we get the following system, which provides the rate of variation (continuous over time) of the fluctuations occurring when the numerical wave being sought is passing by:

$$\forall a,\ \delta\dot{\mathbf{Y}}_a=\begin{pmatrix}\dfrac{p_0}{\rho_0}\widehat{w_h}\left(K\right)\begin{bmatrix}\dfrac{i\left(\gamma-2D\right)}{D\rho_0}R\left(t\right)\mathbf{K}\\-2\left(\mathbf{K}\otimes\mathbf{K}\right)\mathbf{R}\left(t\right)\end{bmatrix}\\\mathbf{U}\left(t\right)\\ \\i\rho_0\widehat{w_h}\left(K\right)\mathbf{K}\cdot\mathbf{U}\left(t\right)\end{pmatrix}\exp\left(-i\mathbf{K}\cdot\mathbf{r}_a\right)\qquad(5.386)$$

where, as previously explained, the state vector $\mathbf{Y}_a$ is defined by $\mathbf{Y}_a \doteqdot \left(\mathbf{p}_a^T, \mathbf{r}_a^T, \rho_a\right)^T$ (eqn (1.266)).

We are going to seek discrete solutions over time, which requires us to choose a temporal scheme. With the explicit scheme (5.266) as introduced in Section 5.4, we will have $\delta\mathbf{Y}_a^{m+1} = \delta\mathbf{Y}_a^m + \delta\dot{\mathbf{Y}}_a^m\delta t$, that is

$$\forall a,\ \delta\mathbf{u}_a^{m+1}=\delta\mathbf{u}_a^m+\delta\dot{\mathbf{u}}_a^m\delta t \qquad (5.387)$$
$$\delta\mathbf{r}_a^{m+1}=\delta\mathbf{r}_a^m+\delta\dot{\mathbf{r}}_a^m\delta t$$
$$\delta\rho_a^{m+1}=\delta\rho_a^m+\delta\dot{\rho}_a^m\delta t$$

Introducing this scheme into the linear system (5.386) leads us to

$$\mathbf{U}\left(t^{m+1}\right)=\mathbf{U}\left(t^m\right)+\frac{p_0}{\rho_0}\widehat{w_h}\left(K\right)\begin{bmatrix}\dfrac{i\left(\gamma-2D\right)}{D\rho_0}R\left(t^m\right)\mathbf{K}\\-2\left(\mathbf{K}\otimes\mathbf{K}\right)\mathbf{R}\left(t^m\right)\end{bmatrix}\delta t$$
$$\mathbf{R}\left(t^{m+1}\right)=\mathbf{R}\left(t^m\right)+\mathbf{U}\left(t^m\right)\delta t \qquad (5.388)$$
$$R\left(t^{m+1}\right)=R\left(t^m\right)+i\rho_0\widehat{w_h}\left(K\right)\mathbf{K}\cdot\mathbf{U}\left(t^m\right)\delta t$$

We are now seeking the amplitudes as harmonic functions of time, that is

$$\mathbf{U}\left(t\right)=\mathbf{U}_0\exp\left(i\omega t\right)$$
$$\mathbf{R}\left(t\right)=\mathbf{R}_0\exp\left(i\omega t\right) \qquad (5.389)$$
$$R\left(t\right)=R_0\exp\left(i\omega t\right)$$

where ω is a frequency. This is tantamount to prescribing the numerical fluctuations of the state vector components in the form of a progressive wave. As regards density, for instance, (5.379) and (5.389) yield

$$\delta\rho = R_0 \exp\left[i\left(\omega t - \mathbf{K}\cdot\mathbf{r}\right)\right] \tag{5.390}$$

which is identical to (3.164) (Chapter 3, Section 3.4.3). Putting

$$\chi \doteqdot \exp\left(i\omega\delta t\right) \tag{5.391}$$

we ultimately come to the following system:

$$\begin{aligned} \mathbf{U}_0\chi &= \mathbf{U}_0 + \frac{p_0}{\rho_0}\widehat{w_h}(K)\left[\frac{i(\gamma-2D)}{D\rho_0}R_0\mathbf{K} - 2(\mathbf{K}\otimes\mathbf{K})\mathbf{R}_0\right]\delta t \\ \mathbf{R}_0\chi &= \mathbf{R}_0 + \mathbf{U}_0\delta t \\ R_0\chi &= R_0 + i\rho_0\widehat{w_h}(K)\,\mathbf{K}\cdot\mathbf{U}_0\delta t \end{aligned} \tag{5.392}$$

whose unknowns are the constants R_0, $\mathbf{R}_0$ and $\mathbf{U}_0$. The existence of a solution requires that the determinant in that system be zero, which provides a characteristic equation whose unknown is χ. The definitions (5.389) then show that the wave amplitude varies over time[74] in χ^m, and so a necessary and sufficient condition for stability[75] reads as

$$|\chi| \leqslant 1 \tag{5.393}$$

If the calculation of the system determinant is preferably to be avoided, the following relation[76] can be derived from (5.392):

$$(\mathbf{K}\otimes\mathbf{K})\,\mathbf{R}_0 = -\frac{\gamma(\chi-1)^2}{c_0^2\delta t^2 A_1\widehat{w_h}(K)}\mathbf{R}_0 \tag{5.394}$$

with

$$\begin{aligned} A_1 &\doteqdot (\gamma-2D)\,\widehat{w_h}(K) + 2D \\ &= \gamma\widehat{w_h}(K) + 2D\left(1-\widehat{w_h}(K)\right) > 0 \end{aligned} \tag{5.395}$$

With the Gaussian kernel (5.45), whose Fourier transform (5.77) is positive and smaller than 1, A_1 is obviously positive. The same applies to the Wendland kernel (5.57). We henceforth will assume that these conditions are satisfied.

The relation (5.394) presents the scalar quantity in the right-hand side as an eigenvalue of the tensor $\mathbf{T} \doteqdot \mathbf{K}\otimes\mathbf{K}$ associated with the eigenvector $\mathbf{R}_0$. Now, $\mathbf{T}$ has two eigenvalues. The first one is 0, an $(n-1)$-ordered value whose eigenspace is the hyperplane constituting the kernel of the linear operator being associated with $\mathbf{T}$, which corresponds to $\chi = 1$, satisfying the condition (5.393). The second eigenvalue is K^2 (where $K \doteqdot |\mathbf{K}|$ is the wavenumber), associated with the one-dimensional eigenspace directed by the vector $\mathbf{K}$, and which yields

$$\chi = 1 \pm i\frac{Co}{Ma_0}K^+\sqrt{\frac{A_1\widehat{w_h}(K)}{\gamma}} \tag{5.396}$$

[74] Under this hypothesis, a growing numerical disturbance eventually takes values which are inconsistent with the assumed low amplitude values. This means that those equations having governed the establishment of the stability criterion give birth to solutions which antagonize the hypotheses, as is often the case with the hyperbolic systems of differential equations. The following criterion will nonetheless remain valid as a necessary condition for stability.

[75] Another approach consists of considering the frequency as a complex number whose imaginary part determines the increase or the decrease of the amplitude. The above condition is then written as $\mathrm{Im}(\omega) \geqslant 0$. This kind of condition is typical in the stability analysis, including within the context of the continuous formalism (refer to the remarks made at the end of Chapter 4 (Section 4.2.2) about the Orr–Sommerfeld equation. As regards the SPH method, Morris (1996a,b) carries out that analysis, albeit with the approximation $\omega\delta t \ll 1$, which makes it possible to linearize $\chi \approx 1 + i\omega\delta t$ (see further on).

[76] To that purpose, we use (A.44). The result below is in accordance with Morris (1996a,b), although the latter has restricted his arguments to a one-dimensional problem.

where K^+ and Ma_0 are defined by (5.73) and (5.211), whereas Co is called the *Courant number*, based on the flow velocity U scale

$$Co \doteqdot \frac{U\delta t}{h} \tag{5.397}$$

(the particular case $D = 0$ gives $\chi = 1 \pm iCoK^+\widehat{w_h}(K)/Ma_0$). Regardless of the value of K, at least one of these roots infringes the condition (5.393), and so the scheme is unconditionally unstable. In the following, we will only deal with the case in which $D = 0$ (which corresponds to a zero background pressure p_0[77]). This hypothesis will simplify our calculations without substantially altering our results. In this case, $A_1 = \gamma\widehat{w_h}(K)$ and equation (5.394) is reduced to

[77] One should remember, under this hypothesis, that such equations as (5.375) include an undetermination, without any change in our results.

$$(\mathbf{K} \otimes \mathbf{K})\,\mathbf{R}_0 = -\frac{(\chi - 1)^2 K^2}{A_2}\mathbf{R}_0 \tag{5.398}$$

with

$$A_2 \doteqdot \left[\frac{Co}{Ma_0}K^+\widehat{w_h}(K)\right]^2 > 0 \tag{5.399}$$

The result which we have just achieved is based on a number of options for the model being considered, particularly on the use of a discrete continuity equation for calculating the densities. We will come to a similar result by considering the interpolation equation (5.192) instead of the continuity equation. Its linearization yields, indeed

$$\begin{aligned} \delta_{\mathbf{r}}\rho_a &= \sum_b m_b \left(\frac{\rho_b}{\rho_a}\right)^{k-1} \nabla_b w_{ab} \cdot (\delta\mathbf{r}_a - \delta\mathbf{r}_b) \\ \delta_\rho \rho_a &= (k-1)\left(\frac{\rho_b}{\rho_a}\right)^{k-1} \sum_b m_b w_{ab}\left(\frac{\delta\rho_b}{\rho_b} - \frac{\delta\rho_a}{\rho_a}\right) \end{aligned} \tag{5.400}$$

Starting from a homogeneous initial state, we get

$$\delta\rho_a = \sum_b m_b \nabla_b w_{ab} \cdot (\delta\mathbf{r}_a - \delta\mathbf{r}_b) + \frac{k-1}{\rho_0}\sum_b m_b w_{ab}(\delta\rho_b - \delta\rho_a) \tag{5.401}$$

That expression can be approximated through

$$\begin{aligned} \delta\rho_a \approx {} & \{(k-1)[\widehat{w_h}(K) - 1]\,R(t) - i\rho_0\widehat{w_h}(K)\,\mathbf{K}\cdot\mathbf{R}(t)\} \\ & \times \exp(-i\mathbf{K}\cdot\mathbf{r}_a) \end{aligned} \tag{5.402}$$

Since there is no numerical scheme for that equation, its introduction into the linearized system will not involve χ:

$$\begin{aligned} \mathbf{U}_0\chi &= \mathbf{U}_0 + i\frac{c_0^2 R_0}{\rho_0}\widehat{w_h}(K)\,\mathbf{K}\delta t \\ \mathbf{R}_0\chi &= \mathbf{R}_0 + \mathbf{U}_0\delta t \\ [k - (k-1)\,\widehat{w_h}(K)]\,R_0 &= -i\rho_0\widehat{w_h}(K)\,\mathbf{K}\cdot\mathbf{R}_0 \end{aligned} \tag{5.403}$$

(we may recall that here and henceforth, $D = 0$). This system will provide us with the following eigenvalue equation:

$$(\mathbf{K} \otimes \mathbf{K})\,\mathbf{R}_0 = \frac{\left[k - (k-1)\,\widehat{w_h}(K)\right](\chi - 1)^2 K^2}{A_2}\mathbf{R}_0 \tag{5.404}$$

whose solutions which differ from one are

$$\chi = 1 \pm \frac{CoK^+\widehat{w_h}(K)}{Ma_0\sqrt{k - (k-1)\,\widehat{w_h}(K)}} \tag{5.405}$$

(the quantity under the radical is positive). Since one of these roots is higher than 1, the system is unconditionally unstable again.

Let us return to the case which is based on the continuity equation with $D = 0$. Considering the effect of the symplectic scheme[78] (5.269) consists in substituting

$$\begin{aligned}
\mathbf{U}_0\chi &= \mathbf{U}_0 + i\frac{c_0^2 R_0}{\rho_0}\widehat{w_h}(K)\,\mathbf{K}\delta t \\
\mathbf{R}_0\chi &= \mathbf{R}_0 + \mathbf{U}_0\chi\,\delta t \\
R_0\chi &= R_0 + i\rho_0\widehat{w_h}(K)\,\mathbf{K}\cdot\mathbf{U}_0\chi\,\delta t
\end{aligned} \tag{5.406}$$

[78] We may recall that this scheme, which was discussed in Sections 5.4.2, consists in treating both positions and densities implicitely. One can easily understand that the implicitation of the position does not affect the solutions in the system being dealt with. This is because $\mathbf{R}_0$ is only involved in the second equation, the form of which is then irrelevant. Thus, the implicitation of the continuity equation is crucial for establishing the results below.

for the system (5.392).

As compared with (5.392), the only difference (apart from D being zero) is the presence of χ to the right of the second and third equations. This system is reduced to

$$(\mathbf{K} \otimes \mathbf{K})\,\mathbf{U}_0 = -\frac{(\chi - 1)^2 K^2}{A_2\chi}\mathbf{U}_0 \tag{5.407}$$

Comparing that equation with (5.398), the new feature is the factor χ in the denominator, which gives as a characteristic equation (disregarding the solution $\chi = 1$):

$$\chi^2 - (2 - A_2)\,\chi + 1 = 0 \tag{5.408}$$

One can easily understand that the solutions to (5.408) may only satisfy the condition (5.393) if

$$0 \leqslant A_2 \leqslant 4 \tag{5.409}$$

The solutions for χ are then complex conjugate, and (5.408) shows that their norm equals 1. Given the definition (5.399) of A_2 and its positiveness, the condition (5.409) is reduced to

$$\widehat{w_h}(K)\,K^+ \leqslant \frac{2Ma_0}{Co} \tag{5.410}$$

and should be satisfied for all the wavenumbers. With the Gaussian kernel, the left-hand side in (5.410) is written as $K^+e^{-K^{+2}/4}$ and has a maximum which equals $\sqrt{2/e}$ for $K^+ = \sqrt{2}$. Thus, there is a sufficient stability condition, namely:

$$\frac{Co}{Ma_0} \leqslant \sqrt{2e} \tag{5.411}$$

Taking the definitions (5.397) and (5.211) of Co and Ma_0 into account, this condition is also written as

$$\delta t \leqslant \sqrt{2e}\frac{h}{c_0} \approx 2.33\frac{h}{c_0} \tag{5.412}$$

The system is then conditionally stable, associated with the Courant–Friedrich–Levy (CFL) condition (5.412).

Here we should make a remark about the result we have just come to. The abundant literature dealing with the specific problems of numerical stability in the SPH method discusses the role played by the sign of the normal forces σ_{ii}, that is the diagonal part of the stress tensor (refer to Sections 3.3.2 and 3.4.1). Most of the authors consider that a stability criterion is provided by the necessary condition $w_h''\sigma_{ii} \leqslant 0$, where w_h'' is the second derivative of the kernel (refer, for instance, to Swegle et al., 1995). Now, for a weakly compressible fluid, since two particles are spaced at least δr apart to a first approximation, the ratio $q = r_{ab}/h$ at least equals $\delta r/h$, that is it ranges from 0.5 to 1 (refer to eqn (5.81)). For these values, Fig. 5.2 shows that we are close to the kernel inflection point, corresponding to a sign change of w_h'', generating uncertainties as regards the sign of $w_h''\sigma_{ii}$. The resulting instabilities may become stronger if the pressure is poorly estimated (Swegle et al., 1995). This result, however, is established for a one-dimensional system, and also considers that a given particle is only linked to a pair of nearby particles on either side of it. The condition (5.410) is a generalization thereof for an arbitrary spatial dimension and an arbitrary number of nearby particles.

Thus, we can see that the stability of the SPH method, such as we have set it out, depends on the time scheme being chosen, the symplectic scheme being recommended. When added to the discussion of Section 5.4.2, that conclusion advocates even more for that scheme. Lastly, it is noteworthy that our calculations are based on the hypothesis of a kernel whose Fourier transform is positive and decreasing, which favours the Wendland kernel (5.57) rather than the B-splines (refer to Section 5.2.2). The numerical experience (see Robinson, 2009) substantiates that opinion, even though the numerical wavenumbers can hardly reach the values at which the Fourier transforms of the B-splines oscillate. Besides, it is interesting to study what the condition (5.412) becomes with the Wendland kernel, even though Section 5.2.2 only gives its Fourier transform in a dimension $n = 1$. The maximum of the $\widehat{w_h}(K)\, K^+ = \alpha_{w,1}\widehat{f}\left(K^+\right) K^+$ function (refer to eqn (5.71)) equals approximately 1.13, which only slightly changes the result (5.412), the constant 2.33 becoming 1.78. Figure 5.6 illustrates the shape of the $\widehat{w_h}(K)\, K^+$ function for both selected kernels of this section.

As regards the explicit scheme, it does not necessarily become worthless, because our analysis is based on a number of approximations and assumptions. In particular, the numerical interpolation tends to stabilize the systems, and so each of these schemes may be considered practically stable, provided that such a kind of condition as (5.412) is adjoined to it, that is

$$\delta t = C_{\delta t,1}\frac{h}{c_0} \tag{5.413}$$

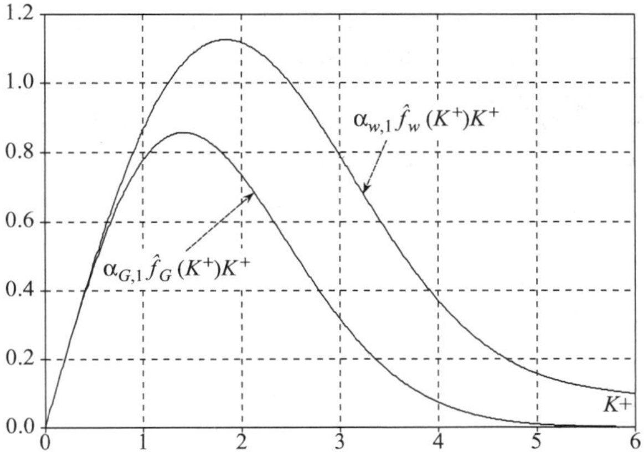

Fig. 5.6 Graphs of the $\alpha_{w,1}\widehat{f}\left(K^+\right)K^+$ function for the Gaussian kernel and the Wendland kernel (dimension n=1).

We may recall that the analysis we have made in this section is only valid for high wavenumbers. It is obvious, however, that the development of a numerical wave is only possible if the latter's wavelength is at least twice as large as the size δr of a particle, which corresponds to a maximal value of K^+ as given by

$$K^+_{\max} = \frac{\pi h}{\delta r} \tag{5.414}$$

Because of the usual values chosen for the $h/\delta r$ ratio (eqn (5.81)), $K^+_{\max}$ then approximately equals 5 and the approximation being performed is probably found to be defective. Practically, the condition (5.413) should then be associated with a coefficient which is lower than the value as yielded by (5.412). Starting from numerical tests, Monaghan (1992) recommends $C_{\delta t,1}$ of at most 0.4. It is well-known, however (for instance, refer to Randles and Libersky, 1996) that making a scheme partly implicit inevitably increases the stability and especially makes it possible to increase the constant $C_{\delta t,1}$ occurring in (5.413), possibly up to such values as 2 or so. The fact that the sound velocity affects the time step is excessively restricting: the usually high value of c_0 (refer to Section 2.3.3), indeed, leads to a very small time step. This is why, immediately upon Section 5.3.2, we have recommended that the numerical sound velocity be lower than the physical sound velocity (equation (5.213)).

In the case of low frequency numerical waves, the approximation $\omega\delta t \ll 1$ can be made, which yields $\chi \approx 1 + i\omega\delta t$. The formula (5.407) then makes it possible to write a dispersion relation as

$$\omega = \pm c_0 \widehat{w_h}(K)\, K \tag{5.415}$$

By analogy with equation (3.167) in Chapter 3, a numerical propagation velocity can be defined for a disturbance with a given wavenumber:

$$c_K \doteqdot \frac{\omega}{K} \tag{5.416}$$

which yields

$$c_K^2 = \left[c_0 \widehat{w_h}(K)\right]^2 \tag{5.417}$$

The propagation velocity of the numerical waves is then close to the numerical sound velocity c_0 as soon as $\widehat{w_h}(K)$ is close to 1. With the Gaussian kernel, (5.77) shows that it happens for small values of K^+.

To complete this paragraph, let us mention the case of the fully implicit scheme (5.270) as discussed in Section 5.4. It would easily be demonstrated that the characteristic equation leads to such solutions as

$$\chi = \frac{1}{1 \pm i A_2} \tag{5.418}$$

whose norm is given by

$$|\chi|^2 = \frac{1}{1 + \left[\dfrac{Co}{Ma_0} K^+ \widehat{w_h}(K) \right]^2} < 1 \tag{5.419}$$

This scheme is then unconditionally stable, which is quite advantageous since the time step can be arbitrarily chosen, provided that it is sufficiently shorter than the characteristic time of the phenomenon being modelled. It should be recalled, however, that this scheme (apart from its technical challenges) is not conservative, as explained in Section 5.4.2. We have learnt through the simple pendulum case that this kind of scheme is artificially dissipative, which is intuitively consistent with the property of stability which we have just highlighted.

Lastly, a remark should be made about the selected options for the spatial discretization of the differential operators. Since Section 5.3.1, we have repeatedly laid stress on the virtue of the symmetric shapes by emphasizing their conservative nature. Morris (1996a,b), however, has demonstrated that antisymmetric, that is non-conservative, shapes, on the contrary, have better properties of stability.

We have previously explained the importance of the Wendland kernel in the control of the numerical stability in SPH. Among the other solutions for alleviating the issue of instabilities, Monaghan (1999) suggests adding an artificial pressure, which induces a beneficial modification of the dispersion relation, whereas Rabczuk et al. (2004) introduces deformable particles. We also must mention the idea expressed by Dyka et al. (1997), who argue in favour of such an approach as the 'decolocalization', a method which consists of considering two collections of 'dual' particles, one of them used for calculating the velocities and the other for calculating the pressures.[79]

More information about this approach for numerical stability is given by Violeau and Leroy (2014).

[79] This is a conventional methodology for the methods involving meshes.

Advanced hydraulics with SPH

6

6.1 Introduction

The development of the SPH method was naturally matched with a willingness to broaden the scope of the modelling of the physical phenomena. Now, the tools in Chapter 5 have been constructed within a non-dissipative, that is diffusionless, especially viscosity-free, framework. The concept of viscosity for SPH, however, soon appears within a context of stabilized solutions of the discrete equations, along the lines of the available techniques which have already been tested for the methods involving a grid. Yet a formalism can be established for representing in a discrete form those physical viscous forces at work in the Navier–Stokes equations. More generally, the issue of the discrete treatment of the second-order operators has been dealt with in many publications (for instance, refer to Monaghan, 1992) before coming to a number of mutually related formulations taking the form of the dissipative forces as discussed in Chapter 2. It can then be found, indeed, that in addition to their purely physical nature, the discrete viscous forces as formalized in the SPH language have a stabilizing effect from a strictly numerical point of view.

These tools are attractive since they also allow the treatment of equations with respect to scalars and energy, taking the diffusion processes into account. The solution of Poisson's equation also happens to be possible in this framework, paving the way for strictly incompressible schemes which were developed as far back as the early 1990s. The incompressible approaches for SPH quickly proved their advantages due to shorter computational times and more regular modelled fields, in particular for pressure prediction.

The desire for representing increasingly complex flows and processes was not limited to these considerations among the modelling specialists. Among the major issues which they had to address when the SPH method emerged in hydraulics was the treatment of boundary conditions. Solid wall modelling was the subject matter of lots of publications implementing quite different techniques with varying results (Monaghan, 1992, Marongiu, 2007; Ferrand et al., 2011). It seems that the variational nature of SPH has been a possibly significant part of the solution from the early 2000s (Kulasegaram et al., 2004). The treatment of the interfaces between two fluids in surface tension conditions was abundantly dealt with too (for example, refer to Hu and Adams, 2006), as well as the modelling of bluff or floating bodies (Monaghan et al., 2003). As regards the specific treatment of the open boundaries, it is largely an ... open problem. The scope of all these questions covers, of course, the coupling of various phenomena (interactions between fluids and structures) as well as the coupling of the SPH method with other numerical approaches (finite volumes, finite elements). These presently emerging problems are being actively and promisingly tackled.

It will be recalled that the importance of the turbulent phenomena was highlighted in Chapter 4. Now, the question of turbulence modelling with SPH has only recently been addressed, because of the strictly numerical problems which the scientists were faced with, generating instabilities and inaccuracies which obscured the fine scale of the details required by the modelling of intricate physical phenomena. Yet, nowadays, due to the increasingly numerous attempts regarding this topic, it is about to become a classic (Violeau and Issa, 2007a). The two-equation methods, in particular (Chapter 4), can be treated on the basis of the previously discussed discrete scalar equations. Yet, the future of turbulence with SPH probably lies in the large eddy simulation (LES), because of its Lagrangian nature. Through the development of the intensive calculation (refer to Chapter 8), it will probably become a standard approach within a few years.

In this chapter, we will initially deal with the discrete modelling of the second-order differential operators, which will make it possible to extend the discrete Euler equations (Chapter 5) to formulations of the discrete Navier–Stokes equation, then to write SPH forms of the scalar equations, particularly for energy, starting from the continuous models as discussed in Chapter 3. Through another link to Chapter 3, we will establish a discrete incompressible scheme. We then will turn to such questions as surface tension modelling, then the treatment of solid walls (particularly on the basis of variational approaches), in connection with Chapters 1 and 3. Next, we will discuss the case of the fluid-driven solid bodies. Lastly, we will review the main methods for simulating the turbulent flows with SPH, still referring to Chapter 4, as well as to the end of Chapter 2.

6.2 Discrete viscosity and diffusion

We start this chapter by stating a discrete form of second-order operators in the SPH language. This tool is then used to derive discrete forms of the viscous forces in the Navier–Stokes equations, and we study the effects of these forces on numerical stability. An equation of energy with diffusion is then written in a discrete form. Finally, we show how it is possible to treat truly incompressible flows with SPH by solving a discrete Poisson equation for pressure.

6.2.1 Second-order SPH operators

In the preceding chapter, we have purposefully ignored the second-order operators, which are necessary for writing the equations of motion of a fluid, for example the Laplacian operator. This work could be started using a method analogous to that described in Section 5.2.3 for discretizing the first-order operators. For instance, applying the Laplacian to the first equation in (5.89), we find

$$\left(\nabla^2 A\right)_a \approx \sum_b V_b A_b \nabla^2 w_h\left(r_{ab}\right) \tag{6.1}$$

where $\nabla^2 w_h$ is the kernel Laplacian. This approach, however, involves the second derivatives of w_h, which induces a great sensitivity to the particle disorder[1] (refer to Monaghan, 1992). Monaghan (2005) additionally records a couple of drawbacks to that method: first, it disrupts the discrete conservativity of the quantity A, because it cannot ensure the antisymmetry of the flux $A_b \nabla^2 w_h\left(r_{ab}\right)$ (refer to Chapter 3, Section 3.3.1). Furthermore, this flux has no determined sign, insofar as the kernel second derivatives are liable to change sign. Now, as explained in Section 3.3.1, the flux of a quantity is necessarily directed towards its decreasing values. Gonzales et al. (2009) also demonstrate that this kind of approach is numerically questionable. Although some authors utilize the method (refer, for example, to Takeda et al., 1994), another procedure shall preferably be adopted. Another idea consists in noting that the continuous relation $\nabla^2 = \operatorname{div}(\mathbf{grad})$, when transcribed into the SPH discrete formalism, can be written as

$$\begin{aligned}\left(\nabla^2 A\right)_a &\approx \tilde{D}_a^0\left\{\mathbf{G}_b^0\left\{A_c\right\}\right\} \\ &= -\sum_{b,c} V_b V_c\left[\left(A_a + A_c\right) w'_{ac}\mathbf{e}_{ac} - \left(A_b + A_c\right) w'_{bc}\mathbf{e}_{bc}\right] \cdot w'_{ab}\mathbf{e}_{ab}\end{aligned} \tag{6.2}$$

or any other combination of the first-order operators as defined in Section 5.2.3. Although some authors utilize this procedure (for example, refer to Bonet and Rodriguez-Paz, 2005), it appears, however, that this formulation is based upon a computationally time-consuming two-fold sum.

Thus, in this Section, we will seek another method for writing a discrete approximation of the second-order operators. To provide an idea of the kind of result to be achieved, let us initially consider the general case of the vector $[\mathbf{div}\left(J_{\mathbf{A}}\mathbf{grad}\,\mathbf{A}\right)]_a$, where $J_{\mathbf{A}}$ is the (dynamic) diffusion coefficient for the vector field $\mathbf{A}$. Using the operator D^k (eqn (5.126)) for estimating the divergence, we get:

$$\begin{aligned}&\mathbf{D}_a^k\left\{\left(J_{\mathbf{A}}\mathbf{grad}\,\mathbf{A}\right)_b\right\} \\ &= \sum_b V_b \frac{\rho_b^{2k} J_{A,a}\left(\mathbf{grad}\,\mathbf{A}\right)_a + \rho_a^{2k} J_{A,b}\left(\mathbf{grad}\,\mathbf{A}\right)_b}{\left(\rho_a\rho_b\right)^k} \cdot w'_{ab}\mathbf{e}_{ab}\end{aligned} \tag{6.3}$$

As explained above, the gradients appearing in (6.3) could now be estimated by means of the $\mathbf{G}^k$ or $\tilde{\mathbf{G}}^k$ operator, but this would result in a two-fold summation over dummy particle subscripts, making the algorithm heavier. In order to settle that question, we may perform a Taylor series expansion of $\mathbf{A}_b$ about the point $\mathbf{r}_a$ by writing

$$\mathbf{A}_b = \mathbf{A}_a - \left(\mathbf{grad}\,\mathbf{A}\right)_a \cdot \mathbf{r}_{ab} + \mathbf{O}\left(r_{ab}^2\right) \tag{6.4}$$

(the 'minus' sign comes from the definition $\mathbf{r}_{ab} = \mathbf{r}_a - \mathbf{r}_b$). Hence we get

$$\left(\mathbf{grad}\,\mathbf{A}\right)_a \cdot \mathbf{e}_{ab} \approx \frac{\mathbf{A}_{ab}}{r_{ab}} \tag{6.5}$$

[1] We could further explore this point through such an analysis of stability as that in Section 5.4.4. The discrete viscous forces will, indeed, be studied this way in Section 6.2.3, but starting from the form provided later on in this Section.

with, as already mentioned, $\mathbf{A}_{ab} \doteqdot \mathbf{A}_a - \mathbf{A}_b$. Swapping the subscripts in (6.5), we get to

$$\begin{aligned}(\mathbf{grad}\ \mathbf{A})_b \cdot \mathbf{e}_{ab} &= -(\mathbf{grad}\ \mathbf{A})_b \cdot \mathbf{e}_{ba} \\ &\approx -\frac{\mathbf{A}_{ba}}{r_{ba}} = \frac{\mathbf{A}_{ab}}{r_{ab}}\end{aligned} \tag{6.6}$$

Introducing (6.5) and (6.6) into (6.3), we come to a new family of SPH operators, which in this case are second-order operators:

$$\begin{aligned}\mathbf{L}_a^k\left[\{J_{\mathbf{A},b}\},\{\mathbf{A}_b\}\right] &\doteqdot \sum_b V_b \frac{\rho_b^{2k} J_{\mathbf{A},a} + \rho_a^{2k} J_{\mathbf{A},b}}{(\rho_a \rho_b)^k} \frac{\mathbf{A}_{ab}}{r_{ab}} w'_{ab} \\ &\approx [\mathbf{div}\,(J_{\mathbf{A}}\mathbf{grad}\ \mathbf{A})]_a\end{aligned} \tag{6.7}$$

For $k = 0$, we find in particular

$$\mathbf{L}_a^0\left[\{J_{\mathbf{A},b}\},\{\mathbf{A}_b\}\right] \doteqdot 2\sum_b V_b \frac{\overline{J}_{\mathbf{A},ab}\mathbf{A}_{ab}}{r_{ab}} w'_{ab} \tag{6.8}$$

with, for an arbitrary field B:

$$\overline{B}_{ab} \doteqdot \frac{B_a + B_b}{2} \tag{6.9}$$

The same method could be applied to a scalar, and we would get

$$L_a^0\left[\{J_{A,b}\},\{A_b\}\right] \doteqdot 2\sum_b V_b \frac{\overline{J}_{A,ab} A_{ab}}{r_{ab}} w'_{ab} \approx [\mathrm{div}\,(J_A \mathbf{grad}\ A)]_a \tag{6.10}$$

Thus, we have got a discrete form of a second-order operator[2] only involving first-order partial derivatives of the kernel. This method, however, can only be laboriously extended to any second-order operators, the general form of which, as applied to a scalar A of a diffusion coefficient J_A, is written as

$$\mathbf{L} \doteqdot \mathbf{grad}\left[J_A\,(\mathbf{grad}\ A)\right] = \frac{\partial}{\partial x_i}\left(J_A \frac{\partial A}{\partial x_j}\right)\mathbf{e}_i \otimes \mathbf{e}_j \tag{6.11}$$

We are then going to steer our quest towards a close approach, giving rise, however, to a single family of discrete operators. In accordance with the approach adopted by Monaghan (2005) and Español and Revenga (2003),[3] we are going to show that the ij–th component of $\mathbf{L}_a$ can be approximated as follows:

$$\begin{aligned}&\left[\frac{\partial}{\partial x_i}\left(J_A \frac{\partial A}{\partial x_j}\right)\right]_a \\ &= \sum_b V_b \frac{\overline{J}_{A,ab} A_{ab}}{r_{ab}} \left[(n+2)\, e_{ab,i} e_{ab,j} - \delta_{ij}\right] w'_{ab} + O\left(h^2\right)\end{aligned} \tag{6.12}$$

where $e_{ab,i}$ is the i–th component of the vector $\mathbf{e}_{ab}$ (here n is still the geometric dimension of the problem), and, as previously:

$$A_{ab} \doteqdot A_a - A_b \tag{6.13}$$

[2] Note that the latter vanishes for any constant field, i.e. $L_a^0[\{J_b\},\{1\}] = 0$.

[3] This reference concerns the Lagrangian numerical method dissipative particle dynamics (DPD), formally close to SPH, although it is rather used in the area of molecular dynamics. Some variants exist of the formulation proposed herein, with similar properties (see *e.g.* Schwaiger, 2007; 2008).

In a tensorial form, the latter result may also be written as:

$$\mathbf{L}_a = \sum_b V_b \frac{\overline{J}_{A,ab} A_{ab}}{r_{ab}} \left[(n+2)\,\mathbf{e}_{ab} \otimes \mathbf{e}_{ab} - \mathbf{I}_n\right] w'_h(r_{ab}) + O\left(h^2\right) \quad (6.14)$$

(also refer to Yildiz et al., 2009 and Violeau, 2009b). To demonstrate this result, it should first be observed that the tensor $\mathbf{L}_a$ is inherently symmetric. Next, an integral approximation of the right-hand side in (6.14) can be written, according to the elementary interpolation (5.84):

$$\mathbf{J} \doteq \int_{\Omega_0} \frac{\left(\overline{J}_{A,ab} A_{ab}\right)(\tilde{\mathbf{r}})}{\tilde{r}} \left[(n+2)\frac{\tilde{\mathbf{r}} \otimes \tilde{\mathbf{r}}}{\tilde{r}^2} - \mathbf{I}_n\right] w'_h(\tilde{r})\, d^n\tilde{\mathbf{r}} \quad (6.15)$$

With the elements of Section 5.2.3, we know that $\mathbf{J}$ approximates the discrete sum (6.14). It then remains to be shown that $\mathbf{J}$ is a good approximation of $\mathbf{L}_a$, which is defined as the value of $\mathbf{L}$ at the point being occupied by the particle a. First, a Taylor expansion of $J_A A$ near the particle a, which is analogous to (6.4) but to the second-order, yields:

$$\begin{aligned} J_{A,b} A_b = {} & J_{A,a} A_a - J_{A,a} (\mathbf{grad}\, A)_a \cdot \mathbf{r}_{ab} - A_a (\mathbf{grad}\, J_A)_a \cdot \mathbf{r}_{ab} \\ & + \frac{1}{2}\mathbf{L}_a : (\mathbf{r}_{ab} \otimes \mathbf{r}_{ab}) + \frac{1}{2}\left[(\mathbf{grad}\, J_A)_a \cdot \mathbf{r}_{ab}\right]\left[(\mathbf{grad}\, A)_a \cdot \mathbf{r}_{ab}\right] \\ & + O\left(r_{ab}^3\right) \end{aligned} \quad (6.16)$$

We now invoke the expansions of J_A and A which are identical to (6.5):

$$\begin{aligned} (\mathbf{grad}\, A)_a \cdot \mathbf{r}_{ab} &= A_{ab} + O\left(r_{ab}^2\right) \\ (\mathbf{grad}\, J_A)_a \cdot \mathbf{r}_{ab} &= J_{A,ab} + O\left(r_{ab}^2\right) \end{aligned} \quad (6.17)$$

Combining (6.16) and (6.17), we find

$$\overline{J}_{A,ab} A_{ab} \approx (J_A \mathbf{grad}\, A)_a \cdot \mathbf{r}_{ab} - \frac{1}{2}\mathbf{L}_a : (\mathbf{r}_{ab} \otimes \mathbf{r}_{ab}) + O\left(r_{ab}^3\right) \quad (6.18)$$

Afterwards, the integral (6.15) can be written as

$$\mathbf{J} = \left[\mathbf{J}_1 (J_A \mathbf{grad}\, A)_a + \frac{1}{2}\mathbf{J}_2 : \mathbf{L}_a + \mathbf{J}_3\right] \quad (6.19)$$

where the integrals $\mathbf{J}_1$, $\mathbf{J}_2$ and $\mathbf{J}_3$ are third-, fourth- and second-order tensors, respectively, which are defined by

$$\begin{aligned} \mathbf{J}_1 &\doteq \int_{\Omega_0} \frac{1}{\tilde{r}}\left[(n+2)\frac{1}{\tilde{r}^2}\left(\tilde{\mathbf{r}} \otimes \tilde{\mathbf{r}} \otimes \tilde{\mathbf{r}}\right) - \mathbf{I}_n \otimes \tilde{\mathbf{r}}\right] w'_h(\tilde{r})\, d^n\tilde{\mathbf{r}} \\ \mathbf{J}_2 &\doteq -\int_{\Omega_0} \frac{1}{\tilde{r}}\left[(n+2)\frac{1}{\tilde{r}^2}\left(\tilde{\mathbf{r}} \otimes \tilde{\mathbf{r}} \otimes \tilde{\mathbf{r}} \otimes \tilde{\mathbf{r}}\right) - \mathbf{I}_n \otimes \tilde{\mathbf{r}} \otimes \tilde{\mathbf{r}}\right] w'_h(\tilde{r})\, d^n\tilde{\mathbf{r}} \\ \mathbf{J}_3 &\doteq \int_{\Omega_0} \frac{1}{\tilde{r}} O\left(\tilde{r}^3\right)\left[(n+2)\frac{1}{\tilde{r}^2}\tilde{\mathbf{r}} \otimes \tilde{\mathbf{r}} - \mathbf{I}_n\right] w'_h(\tilde{r})\, d^n\tilde{\mathbf{r}} \end{aligned} \quad (6.20)$$

The property (5.34) shows that $\mathbf{J}_1$ is equal to zero. However, the term $O\left(\tilde{r}^3\right)$ appearing in $\mathbf{J}_3$ does not allow us to state that this integral is equal to zero. The least we can say is that it is an $\mathbf{O}\left(h^2\right)$, such that (6.19) yields

$$\mathbf{J} = \frac{1}{2}\mathbf{J}_2 : \mathbf{L}_a + \mathbf{O}\left(h^2\right) \tag{6.21}$$

To calculate $\mathbf{J}_2$, we write $d^n\tilde{\mathbf{r}} = \tilde{r}^{n-1}d\tilde{r}d\omega$ (as in Section 5.2.2), which makes it possible to separate the variables:

$$\mathbf{J}_2 = -\int_0^{h_t} w_h'\left(\tilde{r}\right)\tilde{r}^n d\tilde{r}\begin{bmatrix}(n+2)\int_{\Sigma_n}\mathbf{e}\otimes\mathbf{e}\otimes\mathbf{e}\otimes\mathbf{e}\,d\omega\\ -\mathbf{I}_n\otimes\int_{\Sigma_n}\mathbf{e}\otimes\mathbf{e}\,d\omega\end{bmatrix} \tag{6.22}$$

where $\mathbf{e}$ is the local unit vector $\tilde{\mathbf{r}}/\tilde{r}$, running over the unit $n-$sphere Σ_n. The first integral in (6.22) is given by the normalization of the kernel (5.38):

$$\begin{aligned}\int_0^{h_t} w_h'(r)\,r^n dr &= \left[w_h(r)\,r^n\right]_0^{h_t} - n\int_0^{h_t} w_h(r)\,r^{n-1}dr\\ &= -n\alpha_{w,n}\int_0^{R_f} f(q)\,q^{n-1}dq\\ &= -\frac{n}{S_n}\end{aligned} \tag{6.23}$$

The last two integrals in (6.22) are calculated from symmetry considerations, as we did for the calculation of the second-order moment of the kernel in Section 5.2.1. We will first calculate the second one: formula (5.63), with $F\left(\tilde{r}\right) = 1/\tilde{r}^2$, may also be applied to the sphere Σ_n, which immediately yields

$$\begin{aligned}\int_{\Sigma_n}\mathbf{e}\otimes\mathbf{e}\,d\omega &= \frac{1}{n}\left(\int_{\Sigma_n}d\omega\right)\mathbf{I}_n\\ &= \frac{S_n}{n}\mathbf{I}_n\end{aligned} \tag{6.24}$$

Through a reasoning analogous to that which enabled us to establish (5.63), we now can calculate the integral of $\mathbf{e}\otimes\mathbf{e}\otimes\mathbf{e}\otimes\mathbf{e}$, which consists of a fourth-order tensor. To that purpose, we will denote as e_i those components of $\mathbf{e}$ (not to be confused with the vectors in the $\mathbf{e}_i$, which are written in bold). Thus, the only non-zero components in the tensor being considered are the integrals of such quantities as e_i^4 or $e_i^2e_j^2$ with $i \neq j$. In addition, we may note that a rotation by a $\pi/4$ angle about the origin within the plane (i, j) leaves the sphere unchanged, which yields

$$\int_{\Sigma_n} e_i^4 d\omega = \int_{\Sigma_n}\left(\frac{e_i + e_j}{\sqrt{2}}\right)^4 d\omega \tag{6.25}$$

Expanding the last integral and using symmetry and isotropy arguments, we find

$$\int_{\Sigma_n} e_i^4 d\omega = 3\int_{\Sigma_n} e_i^2 e_j^2 d\omega \tag{6.26}$$

Besides, we may also write

$$\int_{\Sigma_n}\left(\sum_i e_i^2\right)^2 d\omega = \int_{\Sigma_n}\left(\sum_i e_i^4 + \sum_{i\neq j} e_i^2 e_j^2\right) d\omega \tag{6.27}$$

which yields

$$S_n = n\int_{\Sigma_n} e_i^4 d\omega + n\,(n-1)\int_{\Sigma_n} e_i^2 e_j^2 d\omega \tag{6.28}$$

Combining with (6.26), we get

$$\int_{\Sigma_n} e_i^4 d\omega = \frac{3S_n}{n\,(n+2)} \tag{6.29}$$

The results which are found can be gathered for writing the integral being sought in the following simple and smart form:

$$\int_{\Sigma_n} \mathbf{e}\otimes\mathbf{e}\otimes\mathbf{e}\otimes\mathbf{e}\, d\omega = \frac{S_n}{n\,(n+2)}\left(\delta_{ij}\delta_{kl}+\delta_{ik}\delta_{jl}+\delta_{il}\delta_{jk}\right)\mathbf{e}_i\otimes\mathbf{e}_j\otimes\mathbf{e}_k\otimes\mathbf{e}_l \tag{6.30}$$

(the terms in e_i^4 are implicitly included in the other ones). Ultimately, (6.22), (6.24) and (6.30) yield

$$\mathbf{J}_2 = \left(\delta_{ik}\delta_{jl}+\delta_{il}\delta_{jk}\right)\mathbf{e}_i\otimes\mathbf{e}_j\otimes\mathbf{e}_k\otimes\mathbf{e}_l \tag{6.31}$$

Returning to (6.21) and using the symmetry of $\mathbf{L}_a$, we get the expected result:

$$\begin{aligned}\frac{1}{2}\mathbf{J}_2 : \mathbf{L}_a &= \frac{1}{2}\left(\delta_{ik}\delta_{jl}+\delta_{il}\delta_{jk}\right)\left[\frac{\partial}{\partial x_k}\left(J_A\frac{\partial A}{\partial x_l}\right)\right]_a \mathbf{e}_i\otimes\mathbf{e}_j \\ &= \left[\frac{\partial}{\partial x_i}\left(J_A\frac{\partial A}{\partial x_j}\right)\right]_a \mathbf{e}_i\otimes\mathbf{e}_j = \mathbf{L}_a\end{aligned} \tag{6.32}$$

which demonstrates (6.12).

From that formula, we now can derive discrete approximations of numerous second-order operators which are useful in fluid mechanics. For example, we may apply (6.12) to an arbitrary component A_l of a vector, then multiply times the unit vector corresponding to the axis m and get:

$$\begin{aligned}&\left[\frac{\partial}{\partial x_i}\left(J_A\frac{\partial A_l}{\partial x_j}\right)\right]_a \mathbf{e}_m \\ &\approx \sum_b V_b \frac{\overline{J}_{A,ab}A_{ab,l}}{r_{ab}}\left[(n+2)\,e_{ab,i}e_{ab,j}-\delta_{ij}\right]w'_{ab}\mathbf{e}_m\end{aligned} \tag{6.33}$$

We then will get various second-order discrete operators by contracting pairs of suitably chosen indices from (6.33). Let us first prescribe $l = m$ and $i = j$; we find

$$[\mathbf{div}\,(J_A\mathbf{grad}\ \mathbf{A})]_a \approx 2\sum_b V_b\frac{\overline{J}_{A,ab}\mathbf{A}_{ab}}{r_{ab}}w'_{ab} \tag{6.34}$$

which is identical to (6.8). Returning to (6.33) and this time putting $l = i$ and $m = j$, we get

$$\left[\mathbf{div}\left(J_A\,(\mathbf{grad}\ \mathbf{A})^T\right)\right]_a \approx \sum_b V_b \frac{\overline{J}_{A,ab}}{r_{ab}} \left[(n+2)\left(\mathbf{A}_{ab}\cdot\mathbf{e}_{ab}\right)\mathbf{e}_{ab} - \mathbf{A}_{ab}\right] w'_{ab} \tag{6.35}$$

Then, letting $l = j$ and $m = i$:

$$\left[\mathbf{grad}\,(J_A \mathrm{div}\mathbf{A})\right]_a \approx \sum_b V_b \frac{\overline{J}_{A,ab}}{r_{ab}} \left[(n+2)\left(\mathbf{A}_{ab}\cdot\mathbf{e}_{ab}\right)\mathbf{e}_{ab} - \mathbf{A}_{ab}\right] w'_{ab} \tag{6.36}$$

(it should be pointed out that the latter two obtained discrete forms are identical, although the corresponding continuous operators are only identical if J_A is constant). If the diffusion is a tensor $\mathbf{J_A}$ (as in eqn (4.251) in Chapter 4), the above formulas remain valid, *mutatis mutandis*, for instance:

$$\left[\mathbf{div}\,(\mathbf{J}_A \mathbf{grad}\ \mathbf{A})\right]_a \approx 2\sum_b V_b \frac{\overline{\mathbf{J}}_{A,ab}\mathbf{A}_{ab}}{r_{ab}} w'_{ab} \tag{6.37}$$

The case of a scalar is obtained by letting $i = j$ in (6.12). Since the vector $e_{ab,i}$ is normalized, we find:

$$\left[\mathrm{div}\,(J_B \mathbf{grad}\ A)\right]_a \approx 2\sum_b V_b \frac{\overline{J}_{A,ab}A_{ab}}{r_{ab}} w'_{ab} \tag{6.38}$$

which corresponds to the $L_a^0\left[\{J_{A,b}\},\{A_b\}\right]$ operator as defined by (6.10). We find that with this model, the flux of A between two particles has the sign of $A_{ab}w'_{ab}$, and the decrease of the kernel as given by (5.187) allows us to ensure that it is exerted from the particle having the highest value of A to the other, as expected.

A last remark should be made about the presence of the interparticle distances r_{ab} at the denominators of most of the second-order operators we have disclosed. In order to prevent any singularity (division by zero), the common practice consists of adding a small quantity to it, that is for instance, carrying out the following substitution in (6.38):

$$\frac{w'_{ab}}{r_{ab}} \longleftarrow \frac{w'_{ab}r_{ab}}{r_{ab}^2 + \eta^2} \tag{6.39}$$

where η is a distance of about $0.1h$ (Monaghan, 1992). For the sake of simplicity in this book, we will systematically omit that correction, without, however, questioning its relevance.

6.2.2 Discrete Navier–Stokes equations

We are now going to utilize the discussions of the preceding Section for modelling the viscous forces in the discrete equations of motion. In that respect, it is not possible to refer to equation (5.199) by means of Stokes' behaviour law (3.107), because the second derivatives would arise, and we explained in Section 6.2.1 that it is not advisable; on the contrary, we will directly consider the Navier–Stokes viscous terms and use the second-order discrete operators as disclosed in the preceding Section. Equation (3.120) shows that they are written in a continuous form $(\mathbf{div}\ \boldsymbol{\tau})/\rho$, with

$$\mathbf{div}\ \boldsymbol{\tau} = \mathbf{div}\left[\mu\left(\mathbf{grad}\ \mathbf{u}\right)\right] + \mathbf{div}\left[\mu\left(\mathbf{grad}\ \mathbf{u}\right)^T\right] + \mathbf{grad}\left(\lambda \mathrm{div}\mathbf{u}\right) \tag{6.40}$$

By means of the discrete approximations (6.34), (6.35) and (6.36) we can approximate (6.40) as

$$(\mathbf{div}\ \boldsymbol{\tau})_a \approx \sum_b \frac{V_b}{r_{ab}} \begin{bmatrix} (n+2)\left(\overline{\mu}_{ab} + \overline{\lambda}_{ab}\right)\left(\mathbf{u}_{ab} \cdot \mathbf{e}_{ab}\right)\mathbf{e}_{ab} \\ + \left(\overline{\mu}_{ab} - \overline{\lambda}_{ab}\right)\mathbf{u}_{ab} \end{bmatrix} w'_{ab} \tag{6.41}$$

$\overline{\mu}_{ab}$ and $\overline{\lambda}_{ab}$ being defined as in (6.9), whereas $\mathbf{u}_{ab} \doteqdot \mathbf{u}_a - \mathbf{u}_b$, according to (1.192) and (6.13).

Let us go more deeply into a nearly incompressible flow, that is with the condition $\mathrm{div}\mathbf{u} = 0$. This amounts to putting $\lambda = 0$, by virtue of (6.40). The preceding model is then simplified and gives

$$(\mathbf{div}\ \boldsymbol{\tau})_a \approx \sum_b V_b \frac{\overline{\mu}_{ab}}{r_{ab}} \left[(n+2)\left(\mathbf{u}_{ab} \cdot \mathbf{e}_{ab}\right)\mathbf{e}_{ab} + \mathbf{u}_{ab}\right] w'_{ab} \tag{6.42}$$

If we additionally assume that the viscosity μ is constant in space, we may also write

$$\begin{aligned}(\mathbf{div}\ \boldsymbol{\tau})_a &= \mu\left[\mathbf{div}\left(\mathbf{grad}\ \mathbf{u}\right)\right]_a \\ &\approx 2\mu \sum_b V_b \frac{\mathbf{u}_{ab}}{r_{ab}} w'_{ab}\end{aligned} \tag{6.43}$$

(Morris et al., 1997). Lastly, still in the nearly incompressible case with a constant viscosity, we may invoke the nullity of the second term in (6.40) (refer to eqn (3.107)) to write an ultimate model of the viscous forces:

$$\begin{aligned}(\mathbf{div}\ \boldsymbol{\tau})_a &= \mu\left[\mathbf{div}\left(\mathbf{grad}\ \mathbf{u} + 2\left(\mathbf{grad}\ \mathbf{u}\right)^T\right)\right]_a \\ &\approx 2(n+2)\mu \sum_b V_b \frac{\mathbf{u}_{ab} \cdot \mathbf{e}_{ab}}{r_{ab}} w'_{ab}\mathbf{e}_{ab}\end{aligned} \tag{6.44}$$

(Monaghan, 1992).

We then have four formulas,[4] the last three being restricted to the nearly incompressible flows and the last two to the constant viscosities. They are discrete forms of the dissipative forces $\mathbf{F}_a^{diss}$ as introduced in Chapter 2 (Section 2.4.3). For instance, with (6.44), we write

$$\mathbf{F}_a^{diss} = 2(n+2)\mu \sum_b V_a V_b \frac{\mathbf{u}_{ab} \cdot \mathbf{e}_{ab}}{r_{ab}} w'_{ab}\mathbf{e}_{ab} \tag{6.45}$$

As in Section 2.4.3, we note that all of these four forms are combinations of friction forces $\mathbf{F}_{b \to a}^{diss}$ between pairs of particles, each one linearly depending

[4] We may note that, when μ is constant, (6.42) is the average of (6.43) and (6.44). It is also worth mentioning the idea of some authors leading to slightly different viscosity coefficients. Cleary (1996), for instance, proposes considering $\mu_a \mu_b / \overline{\mu}_{ab}$. This author also discusses the dimensionless coefficient occurring on the left of the discrete sum (6.44), suggesting the 2×4 and 2×4.9633 values on the basis of numerical tests, which remains consistent with the $2(n+2)$ coefficient set out.

on the velocities, in accordance with the general formula (2.182). Thus, the equation of motion (5.203) now takes the following more complete shape:

$$\forall a,\ m_a \frac{d\mathbf{u}_a}{dt} = \sum_b \left(\mathbf{F}^{int}_{b\to a} + \mathbf{F}^{diss}_{b\to a} \right) + \mathbf{F}^{ext}_a \tag{6.46}$$

in accordance with (2.186). The discrete friction forces correspond to kinetic matrices (refer to Chapter 2) as provided by

$$\boldsymbol{\alpha}_{ab} = -\frac{V_a V_b}{r_{ab}} w'_{ab} \left[(n+2) \left(\overline{\mu}_{ab} + \overline{\lambda}_{ab} \right) \mathbf{e}_{ab} \otimes \mathbf{e}_{ab} + \left(\overline{\mu}_{ab} - \overline{\lambda}_{ab} \right) \mathbf{I}_n \right] \tag{6.47}$$

repectively, for (6.41), then

$$\boldsymbol{\alpha}_{ab} = -V_a V_b \frac{\overline{\mu}_{ab}}{r_{ab}} w'_{ab} \left[(n+2)\, \mathbf{e}_{ab} \otimes \mathbf{e}_{ab} + \mathbf{I}_n \right] \tag{6.48}$$

for (6.42),

$$\boldsymbol{\alpha}_{ab} = -2\mu \frac{V_a V_b}{r_{ab}} w'_{ab} \mathbf{I}_n \tag{6.49}$$

for (6.43), lastly

$$\boldsymbol{\alpha}_{ab} = -2\,(n+2)\,\mu \frac{V_a V_b}{r_{ab}} w'_{ab} \mathbf{e}_{ab} \otimes \mathbf{e}_{ab} \tag{6.50}$$

for (6.44). In the formulas we have just established, the dependence of the $\boldsymbol{\alpha}_{ab}$ with respect to the distances r_{ab} between particles is achieved through the gradient of the kernel $w'_{ab} \doteqdot w'_h(r_{ab})$. The Onsager's symmetry conditions (2.181) are satisfied by these four models, which implies in particular $\mathbf{F}^{diss}_{b\to a} = -\mathbf{F}^{diss}_{a\to b}$, as in (5.204) for the pressure forces. This result ensures the conservation of the total momentum of an isolated system, which means that equation (5.245) is still valid with $\mathbf{F}_a = \mathbf{F}^{int}_a + \mathbf{F}^{diss}_a$. Conversely, the definite positiveness of the kinetic matrices (refer to Section 2.4.3) is not satisfied yet. This is because the general form of the energy dissipation (2.185) here leads to

$$F = -\frac{1}{2} \sum_{a,b} \frac{V_a V_b}{r_{ab}} \left[\begin{array}{c} (n+2) \left(\overline{\mu}_{ab} + \overline{\lambda}_{ab} \right) \left(\mathbf{e}_{ab} \cdot \mathbf{u}_{ab} \right)^2 \\ + \left(\overline{\mu}_{ab} - \overline{\lambda}_{ab} \right) u^2_{ab} \end{array} \right] w'_{ab} \tag{6.51}$$

This is a discrete form of the integral occurring in the continuous balance (3.114). The sign of that quantity is a priori unknown, because of the presence of the $\overline{\mu}_{ab} - \overline{\lambda}_{ab}$ factor. Its non-positiveness reveals a flaw in the model (6.14) which we have adopted for estimating the second-order derivatives. As regards the weakly compressible flows, however, the specific forms (6.48) to (6.50) respectively give

$$F = -\frac{1}{2} \sum_{a,b} V_a V_b \frac{\overline{\mu}_{ab}}{r_{ab}} \left[(n+2) \left(\mathbf{e}_{ab} \cdot \mathbf{u}_{ab} \right)^2 + u^2_{ab} \right] w'_{ab} \tag{6.52}$$

for (6.48),

$$F = -\mu \sum_{a,b} \frac{V_a V_b}{r_{ab}} u^2_{ab} w'_{ab} \tag{6.53}$$

for (6.49), lastly

$$F = -(n+2)\,\mu \sum_{a,b} \frac{V_a V_b}{r_{ab}} (\mathbf{e}_{ab} \cdot \mathbf{u}_{ab})^2\, w'_{ab} \tag{6.54}$$

for (6.50). Their positiveness is ensured, because $w'_{ab} < 0$ for $b \neq a$, by virtue of (5.187).[5]

The three proposed viscous formulas still have different properties when they are assessed against the discussion of Section 2.4.3 (refer to Fig. 5.5 (*d*)). The choice (6.43), indeed, is advantageous in that it reaches friction forces which are collinear to the velocity difference $\mathbf{u}_{ab}$, in accordance with what is suggested by the theory of incompressible viscous fluids (eqn (3.107)). The discussion in Section 2.4.3, however, shows that the conservation of the total angular momentum of an isolated system, as well as the nullity of the shear stresses for a rigid motion, requires the condition $\boldsymbol{\alpha}_{ab} = \gamma_{ab}\mathbf{r}_{ab} \otimes \mathbf{r}_{ab}$ (eqn (2.204)), which is only verified by (6.50) (i.e. (6.44)) with

$$\gamma_{ab} = -2\,(n+2)\,\mu \frac{V_a V_b}{r_{ab}^3} w'_{ab} > 0 \tag{6.55}$$

which then is the seemingly best proposed form.[6] In this case, the friction forces are aligned with the radius vector $\mathbf{e}_{ab}$, which provides for the conservation of the angular momentum. In other words, the calculations (5.246) to (5.250) remain valid with $\mathbf{F}_a = \mathbf{F}_a^{int} + \mathbf{F}_a^{diss}$. The formula (6.43), which does not verify that property, is nevertheless often used in the literature dealing with SPH and also works well. However, it seems to be interesting to give prevalence to the option (6.44), which ensures exact conservation properties. The dissipation function (6.54) then corresponds to (2.197). Each particle is the locus of a dissipation as given by

$$F_a = -(n+2)\,\mu \sum_{b} \frac{V_a V_b}{r_{ab}} (\mathbf{e}_{ab} \cdot \mathbf{u}_{ab})^2\, w'_{ab} \tag{6.56}$$

in accordance with (2.189). Due to the positiveness of the quantities F_a, the total macroscopic kinetic energy decreases in favour of the internal energy (refer to Chapter 2).

The discrete equation of motion can now take different forms, according to the model selected for the pressure forces and the viscous forces. Combining equations (5.201) and (6.43), it has, for example, the following shape:

$$\forall a,\ \frac{d\mathbf{u}_a}{dt} = -\frac{1}{\rho_a} \sum_b V_b \left[(p_a + p_b)\,\mathbf{e}_{ab} - 2\mu \frac{\mathbf{u}_{ab}}{r_{ab}} \right] w'_{ab} + \mathbf{g} \tag{6.57}$$

whereas (5.202) and (6.44) yield

$$\forall a,\ \frac{d\mathbf{u}_a}{dt} = -\sum_b m_b \left[\frac{p_a}{\rho_a^2} + \frac{p_b}{\rho_b^2} - 2\,(n+2)\,\mu \frac{\mathbf{u}_{ab} \cdot \mathbf{e}_{ab}}{\rho_a \rho_b r_{ab}} \right] w'_{ab}\mathbf{e}_{ab} + \mathbf{g} \tag{6.58}$$

In the case where the viscosity is not constant, for example for a mixture of several fluids with various molecular viscosities, the model (6.42) shall a priori (still in the nearly incompressible hypothesis) be resorted to for the

[5] It is a consequence of the sign of the viscous momentum flux, as noted at the end of Section 6.2.1.

[6] Our numerical tests, however, demonstrate that all these presented forms give mutually equivalent results, as regards the specific case discussed in Section 7.2.1. Gonzales et al. (2009) propose a comparison of several viscosity models for the SPH method, with similar conclusions.

viscous forces, even though the conservation of the angular momentum is dropped:

$$\forall a, \quad \frac{d\mathbf{u}_a}{dt} = -\sum_b m_b \left[\begin{array}{c} \left(\frac{p_a}{\rho_a^2} + \frac{p_b}{\rho_b^2} \right) \mathbf{e}_{ab} \\ -\frac{\mu_a + \mu_b}{2\rho_a \rho_b r_{ab}} \left[(n+2) \left(\mathbf{u}_{ab} \cdot \mathbf{e}_{ab} \right) \mathbf{e}_{ab} + \mathbf{u}_{ab} \right] \end{array} \right] w'_{ab} + \mathbf{g} \tag{6.59}$$

These equations are substituted for (5.200) for the calculation of the velocities when viscous forces are present. They are duly invariant under Galilean transformation, because of their construction. When provided with a discrete continuity equation (Section 5.3.1) and an equation of state (refer to the end of Section 5.3.2), they make up complete system a few applications of which are disclosed in Chapter 7 and 8.

The temporal schemes as contemplated in Section 5.4.1 are a priori unchanged after the viscous forces are added, and the latter can be either explicitly or implicitly treated in the time discretized equation of motion. No such variational argument as in the discussion of Section 5.4.3 makes it possible to decide between the two treatments, since these forces do not conserve the macroscopic kinetic energy. Although they can preferably be explicitly treated because this is easy, attempts have been made to implicit them, which requires the solution of a linear system, but offers numerical advantages for the treatment of flows with very low Reynolds numbers (for example, refer to Cueille, 2005).

Keeping in mind the importance of the choice of the temporal scheme for the conservation of the energy in the conservative case (Section 5.4.2), its effect upon the accuracy of the estimation of the dissipated energy is interesting. With a scheme looking like the symplectic scheme (5.269), the equation for updating the velocity of each particle now reads as

$$\forall a, \ \mathbf{u}_a^{m+1} = \mathbf{u}_a^m + \frac{1}{m_a} \left(\mathbf{F}_a^{int} + \mathbf{F}_a^{diss} + \mathbf{F}_a^{ext} \right)^m \delta t \tag{6.60}$$

We must turn to the variation of macroscopic kinetic energy upon an iteration, only due to the dissipative forces. The variation rate of $E_{k,a}$ is written, only keeping $\mathbf{F}_a^{diss}$ in (6.60), as:

$$\begin{aligned} \frac{E_{k,a}^{m+1} - E_{k,a}^m}{\delta t} &= \frac{1}{2\delta t} m_a \left(\left| \mathbf{u}_a^{m+1} \right|^2 - \left| \mathbf{u}_a^m \right|^2 \right) \\ &= \frac{1}{2\delta t} m_a \left(\mathbf{u}_a^{m+1} - \mathbf{u}_a^m \right) \cdot \left(\mathbf{u}_a^{m+1} + \mathbf{u}_a^m \right) \\ &= \left(\mathbf{F}_a^{diss} \right)^m \cdot \frac{1}{2} \left(\mathbf{u}_a^{m+1} + \mathbf{u}_a^m \right) \end{aligned} \tag{6.61}$$

It is, in accordance with Chapter 2 (eqn (2.193)), the strength of the dissipative forces, which is calculated on the basis of the mean velocity between the two points in time being considered. Because of the discrete nature of time, one may not state that this quantity coincides with the dissipation rate F_a as given by (2.191) (i.e. here (6.56)). Taking (6.60) again, only on the basis of the dissipative forces, the difference is written as:

$$\begin{aligned}
F_a^{num} - F_a &= \left(\mathbf{F}_a^{diss}\right)^m \cdot \frac{1}{2}\left(\mathbf{u}_a^{m+1} + \mathbf{u}_a^m\right) - \left(\mathbf{F}_a^{diss} \cdot \mathbf{u}_a\right)^m \\
&= \left(\mathbf{F}_a^{diss}\right)^m \cdot \frac{1}{2}\left(\mathbf{u}_a^{m+1} - \mathbf{u}_a^m\right) \\
&= \frac{1}{2m_a}\left(\left|\mathbf{F}_a^{diss}\right|^2\right)^m \delta t
\end{aligned} \tag{6.62}$$

Thus, it is a first-order error in time. Besides, it is positive, and so the model is liable to dissipate too much energy. The dissipation error can then accumulate as the successive iterations occur. Unlike the conservative forces, for which it was explained in Section 5.4.2 that they make a time bounded error as regards energy (eqn (5.306)), the dissipative forces dissipate an amount of energy which a priori can indefinitely deviate from the theoretical predictions. A scheme making it possible to bound the dissipation error still has to be constructed.

A further remark should be made. Historically, the viscosity phenomenon was introduced for strictly numerical reasons into the discrete equations of the SPH method (for instance, refer to Monaghan, 1992), on the basis of the experience gained in the field of the grid-based numerical methods. That approach is supported by a search for numerical stability of the Euler equations and leads to an artificial viscous force, which is based on a viscosity depending on sound velocity c_0. These artifical forces, however, do take a form analogous to the above disclosed formula (6.44). There is no obvious reason for dropping the 'physical' description of viscosity which we have provided, since it does not add complexity, while effectively ensuring a numerical stabilization of the system, as will be explained in the next section.

6.2.3 Stabilization by viscosity

In this section, we will discuss the effect of viscosity upon the numerical stability properties of the SPH model. To do this, we take again the arguments of Section 5.4.4, adding thereto the viscous forces as calculated in accordance with the model (6.43) for a constant viscosity. Thus, in the calculation of the variation rates of the state parameters of a particle, as given by (5.370) for the Euler equations, a further term is added to $\dot{\mathbf{u}}_a$, namely

$$\dot{\mathbf{u}}_a^{diss} = 2\mu \sum_b m_b \frac{\mathbf{u}_{ab}}{\rho_a \rho_b r_{ab}} w'_{ab} \tag{6.63}$$

This term will not affect the quantities $\delta_{\mathbf{r}}\dot{\mathbf{u}}_a$ and $\delta_\rho \dot{\mathbf{u}}_a$ arising in (5.381), because the quantity $\mathbf{u}_{ab} = \mathbf{u}_a - \mathbf{u}_b$ will cancel the corresponding variations, since our calculations of stability were made assuming that the initial state is homogeneous. Moreover, the forces now depend on the velocities, and a term written as $\delta_{\mathbf{u}}\dot{\mathbf{u}}_a$ will therefore be added to (5.381):

$$\delta_{\mathbf{u}}\dot{\mathbf{u}}_a = \frac{2\nu}{\rho_0}\sum_b m_b \frac{w'_{ab}}{r_{ab}}\left(\delta\mathbf{u}_a - \delta\mathbf{u}_b\right) \tag{6.64}$$

We will treat that term, as in Section 5.4.4, approximating it by means of integrals:

$$\begin{aligned}\delta_{\mathbf{u}}\dot{\mathbf{u}}_a &\approx 2\nu\left[\left(\int_\Omega \frac{w'_h(\tilde{r})}{\tilde{r}} d^n\tilde{\mathbf{r}}\right)\delta\mathbf{u}_a - \int_\Omega \frac{w'_h(\tilde{r})}{\tilde{r}}\delta\mathbf{u}\, d^n\tilde{\mathbf{r}}\right]\\ &= 2\nu\left[\int_\Omega \frac{w'_h(\tilde{r})}{\tilde{r}} d^n\,\tilde{\mathbf{r}} - \int_\Omega \frac{w'_h(\tilde{r})}{\tilde{r}}\exp\left(-i\mathbf{K}\cdot\tilde{\mathbf{r}}\right) d^n\tilde{\mathbf{r}}\right]\\ &\quad\times\mathbf{U}(t)\exp\left(-i\mathbf{K}\cdot\mathbf{r}_a\right)\end{aligned} \tag{6.65}$$

For treating the presently arising integrals, we may note that

$$\begin{aligned}&\frac{d}{d\mathbf{K}}\int_\Omega \frac{w'_h(\tilde{r})}{\tilde{r}}\exp\left(-i\mathbf{K}\cdot\tilde{\mathbf{r}}\right) d^n\tilde{\mathbf{r}}\\ &= -i\int_\Omega \nabla w_h(\tilde{r})\exp\left(-i\mathbf{K}\cdot\tilde{\mathbf{r}}\right) d^n\tilde{\mathbf{r}}\\ &= \widehat{w_h}(K)\,\mathbf{K}\end{aligned} \tag{6.66}$$

In the case of the Gaussian kernel (5.45), whose Fourier transform is given by (5.77), we then will get[7]

$$\begin{aligned}\int_\Omega \frac{w'_h(\tilde{r})}{\tilde{r}}\exp\left(-i\mathbf{K}\cdot\tilde{\mathbf{r}}\right) d^n\tilde{\mathbf{r}} &= -\frac{2}{h^2}\exp\left(-\frac{h^2K^2}{4}\right)\\ &= -\frac{2}{h^2}\widehat{w_h}(K)\end{aligned} \tag{6.67}$$

[7] In the general case of some kernel or other, the above calculation leads to the antiderivative of the kernel's Fourier transform.

In particular, for $\mathbf{K} = \mathbf{0}$ we get

$$\int_\Omega \frac{w'_h(\tilde{r})}{\tilde{r}} d^n\tilde{\mathbf{r}} = -\frac{2}{h^2}\widehat{w_h}(0) \tag{6.68}$$

Combining (6.65), (6.67) and (6.68), we can ultimately write

$$\delta_{\mathbf{u}}\dot{\mathbf{u}}_a \approx \frac{4\nu}{h^2}\left[\widehat{w_h}(K) - \widehat{w_h}(0)\right]\mathbf{U}(t)\exp\left(-i\mathbf{K}\cdot\mathbf{r}_a\right) \tag{6.69}$$

The total numerical fluctuation of velocity is now written as $\delta\dot{\mathbf{u}}_a = \delta_{\mathbf{r}}\dot{\mathbf{u}}_a + \delta_{\mathbf{u}}\dot{\mathbf{u}}_a + \delta_\rho\dot{\mathbf{u}}_a$. With the explicit weakly compressible scheme, the linearized system (5.406) is then corrected and gives

$$\begin{aligned}\mathbf{U}_0\chi &= \mathbf{U}_0 + i\frac{c_0^2 R_0}{\rho_0}\widehat{w_h}(K)\,\mathbf{K}\delta t\\ &\quad+\frac{4\nu}{h^2}\left[\widehat{w_h}(K) - \widehat{w_h}(0)\right]\mathbf{U}_0\delta t\\ R_0\chi &= R_0 + i\rho_0\widehat{w_h}(K)\,\mathbf{K}\cdot\mathbf{U}_0\delta t\end{aligned} \tag{6.70}$$

(the second line in the system (5.406) is useless here). The eigenvalue equation is now written, with the notation (5.391), as:

$$(\mathbf{K} \otimes \mathbf{K})\,\mathbf{U}_0 = -\frac{(\chi - 1)(\chi - 1 + A_5)\,K^2}{A_2}\mathbf{U}_0 \tag{6.71}$$

where A_2 is given by (5.399), whereas

$$A_5 \doteq \frac{4Co}{Ma_0 Re_0}\left[\widehat{w_h}(0) - \widehat{w_h}(K)\right] > 0 \tag{6.72}$$

We may recall that Ma_0 and Co are given by (5.211) and (5.397), whereas Re_0 is a numerical Reynolds number as defined by

$$Re_0 \doteq \frac{c_0 h}{\nu} \tag{6.73}$$

(please refer to Section 3.4.3 in Chapter 3 as regards the meaning of the Reynolds number). The positiveness of A_5 is provided by the decrease of the kernel's Fourier transform, a condition which is satisfied in the case of the Gaussian kernel (refer to Section 5.2.2). Equation (6.71) duly gives (5.398) again in the absence of viscous forces, that is when $A_5 = 0$.

Something new is appearing as compared with the calculations in Section 5.4.4: the zero eigenvalue of the tensor $\mathbf{T} \doteq \mathbf{K} \otimes \mathbf{K}$ provides a necessary condition of stability, together with the following solution:

$$\chi = 1 - A_5 \tag{6.74}$$

Owing to the positiveness of A_5, the stability condition (5.393) is then reduced to $A_5 \leqslant 2$, that is

$$\delta t \leqslant \frac{h^2}{2\nu\left[\widehat{w_h}(0) - \widehat{w_h}(K)\right]} \tag{6.75}$$

which holds true for every value of wavenumber K. Since the Fourier transform $\widehat{w_h}(K)$ is assumed to be decreasing,[8] we also may write

$$\delta t \leqslant \frac{h^2}{2\widehat{w_h}(0)\,\nu} = \frac{1}{2}\frac{h^2}{\nu} \tag{6.76}$$

[8] Remind that the wavenumber K is numerically strictly limited by $\pi/\delta r$, because the particle spacing δr provides a minorant of the half-wavelength of the numerical fluctuations which are liable to occur. That would induce substantially no change in the result being found.

Let us now turn to the eigenvalue K^2 of the operator $\mathbf{T}$. The corresponding characteristic equation is written as

$$\chi^2 - (2 - A_5)\,\chi + 1 - A_5 + A_2 = 0 \tag{6.77}$$

An analysis would show that the discriminant of this trinomial can still take either negative or positive values, according to the wavenumber, regardless of the viscosity value. We will successively discuss these two possibilities. When this discriminant is negative, the complex conjugate roots have a module as given by

$$|\chi|^2 = 1 - A_5 + A_2 \tag{6.78}$$

The stability condition is then written as

$$A_5 - 2 \leqslant A_2 \leqslant A_5 \tag{6.79}$$

As we have just explained that the previous condition implies $A_5 \leqslant 2$, the positiveness of A_2 automatically ensures the left-hand condition in (6.79), whereas the right-hand condition is also written as

$$\delta t \leqslant \frac{1}{A_3}\frac{\nu}{c_0^2} \tag{6.80}$$

where A_3 is given by

$$A_3 \doteq \frac{\left[K^+\widehat{w_h}\,(K)\right]^2}{4\left[\widehat{w_h}\,(0) - \widehat{w_h}\,(K)\right]} > 0 \tag{6.81}$$

With a Gaussian kernel, A_3 is a decreasing function of $K^+ \doteq hK$, and $A_3\left(K^+ = 0\right) = 1$, hence $A_3 \leqslant 1$. Thus, the condition (6.80) can be written as

$$\delta t \leqslant \frac{\nu}{c_0^2} \tag{6.82}$$

Let us now consider the case in which the discriminant of (6.77) is positive. The (real) roots are then written as:

$$\chi = 1 - \frac{A_5}{2}\left(1 \pm \sqrt{1 - \frac{4A_2}{A_5^2}}\right) \tag{6.83}$$

and the stability condition comes back to $A_5 \leqslant 2$, a condition which was previously investigated.

With the viscous forces, the explicit scheme then becomes conditionally stable, and we have come to a couple of additional conditions with respect to the time step, (6.76) and (6.82), which are added to (5.413). It should be pointed out, however, that the condition (6.82) is quite binding because of the low viscosity and the high sound velocity. Yet, this condition is automatically satisfied with (6.76) provided that

$$\nu \geqslant \frac{c_0 h}{\sqrt{2}} \tag{6.84}$$

With the fully explicit scheme, a sufficient viscosity shall necessarily be prescribed for the sake of numerical stability. Since the lower bound which we have just determined depends on numerical parameters, (5.212) shall preferably be invoked for returning (6.84) to

$$\frac{h}{L} \leqslant \frac{\sqrt{2}}{10}\frac{\nu}{UL} \approx \frac{0.14}{Re} \tag{6.85}$$

where L is a flow-specific length scale. This condition is only achieved at the cost of a sufficiently high viscosity, which then does not necessarily have the physical value of the molecular viscosity and therefore plays the role of a numerical viscosity. In view of this, the condition (6.82) prevails over the CFL condition (5.413). In practice, this condition is too binding; the explicit scheme can then be made conditionally stable by the viscosity, at too high a cost.

Let us consider the question of the stability of the discrete viscous system by means of the symplectic scheme (5.269). It can easily be explained that the relation (6.77) becomes

$$\chi^2 - (2 - A_2 - A_5)\,\chi + 1 - A_5 = 0 \tag{6.86}$$

As previously, the discriminant Δ in this trinomial has no determined sign. When it is negative, it can readily be demonstrated that the stability condition (6.76) is unchanged, whereas the condition (6.82) disappears. Now, for Δ to be negative, the following should be satisfied

$$\frac{Co}{Ma_0} \leqslant \frac{\widehat{w_h}(K)\,\dfrac{K^+}{2} - \dfrac{1}{Re_0}\left[\widehat{w_h}(0) - \widehat{w_h}(K)\right]}{\left[\widehat{w_h}(K)\,\dfrac{K^+}{2}\right]^2} \tag{6.87}$$

where K^+, it should be remembered, is defined by (5.73). Surprisingly, this condition is more binding than (5.410) (an analysis would provide a similar result in the case where Δ is positive). Thus, from this perspective, it looks as though the viscous forces do not play any part in the stability of the numerical system. Yet, a closer analysis demonstrates the opposite. To do this, let us consider the case of the low frequency waves as discussed at the end of Section 5.4.4. Starting from (6.77) and linearizing into $\omega\delta t$, a dispersion relation which is different from (5.415) can be found:

$$(\omega\delta t)^2 - iA_5\omega\delta t - A_2 = 0 \tag{6.88}$$

Its roots are as follows:

$$\omega = \frac{2\nu}{h^2}\left[\widehat{w_h}(0) - \widehat{w_h}(K)\right] \times \left\{\pm\sqrt{\left[\frac{Re_0 K^+\widehat{w_h}(K)}{2\left[\widehat{w_h}(0) - \widehat{w_h}(K)\right]}\right]^2 - 1} + i\right\} \tag{6.89}$$

to be compared with the relation (3.166) in Chapter 3. Compared with the viscosity-free case (5.415), we now can see that an imaginary part appears in the expression of ω. In view of (5.389), the wave amplitudes are then reduced by the following factor:

$$\exp\left\{-2\left[\widehat{w_h}(0) - \widehat{w_h}(K)\right]\frac{\nu t}{h^2}\right\} \tag{6.90}$$

to be compared with (3.168). Thus, as in the continuous case, here the viscosity has a damping effect on the numerical waves, limiting the risk of instability (the same result would be found by calculating the numerical celerity of the waves and comparing it with (5.417)).

Yet, as previously explained, the condition (6.87) prescribes a time step δt which is shorter than in the absence of viscous forces. One can readily understand, however, that the term which depends on ν is negligible in it. Thus, in practice, we may consider that the CFL condition (5.413) is unchanged in the presence of viscosity.[9] Regarding the condition (6.76), it is essential. As in

[9] It is noteworthy that the right-hand side in (6.87) may theoretically take negative values. This artifact would be solved by treating the viscous forces with an implicit time integration scheme.

the non-viscous case, it is only valid, strictly speaking, for high wavenumbers. It can be generalized as

$$\delta t \leqslant C_{\delta t,2}\frac{h^2}{\nu} \tag{6.91}$$

Morris et al. (1997) suggest $C_{\delta t,2} = 0.125$. The ration between the CFL condition and the viscous condition is written as

$$\frac{C_{\delta t,2}}{C_{\delta t,1}} Re_0 = \frac{10C_{\delta t,2}}{C_{\delta t,1}} Re\frac{h}{L} \tag{6.92}$$

In view of the usually involved high Reynolds numbers and the spatial discretization compatible with the practical simulations, the CFL condition often remains the most binding one. However, the viscous condition (6.91) should be kept for modelling the low Reynolds number flows[10] (Morris et al., 1997). Based on (5.413) and (6.91), the following can then be written:

[10] As explained later on in Section 6.4.1, the same care should be exercised for modelling the turbulent flows.

$$\delta t = \min\left(C_{\delta t,1}\frac{h}{c_0}, C_{\delta t,2}\frac{h^2}{\nu}\right) \tag{6.93}$$

The reader can find a more rigourous approach in Violeau and Leroy (2014).

6.2.4 Discrete scalar and energy balances

In order to model the transport-diffusion equation of a scalar, we use the discrete model (6.38) for describing the diffusion term of a scalar B associated with a dynamic molecular diffusion coefficient J_B (refer to Section 3.3.1). With the definition (6.9), the transport equation (3.60) then takes the following discrete form:

$$\forall a, \quad \frac{dB_a}{dt} = S_{B,a} + \sum_b m_b \frac{\left(J_{B,a} + J_{B,b}\right) B_{ab}}{\rho_a \rho_b r_{ab}} w'_{ab} \tag{6.94}$$

where $S_{B,a}$ is the source term associated with the particle a, and

$$B_{ab} \doteqdot B_a - B_b \tag{6.95}$$

in accordance with (6.13). The term under the summation symbol in (6.94) represents the scalar flux between two particles. It can be observed that it is directed from a to b if $B_a > B_b$, because $w'_{ab} < 0$ (refer to the end of Section 6.2.1).

In the absence of a source term, the following relation can easily be derived from (6.94):

$$\begin{aligned}\frac{d}{dt}\sum_a m_a B_a &= \sum_{a,b} V_a V_b \frac{\left(J_{B,a} + J_{B,b}\right) B_{ab}}{r_{ab}} w'_{ab} \\ &= 0\end{aligned} \tag{6.96}$$

(the latter equality is obtained through the process which was previously implemented in such circumstances, that is using the symmetry of the summed terms). Equation (6.96) is an exact discrete form of (3.52), that is a total scalar conservation law (the total scalar, it should be remembered, is reckoned per unit mass according to our conventions).

Let us now return to the discrete modelling of internal energy, which was previously discussed at the end of Section 5.3.2. We provided there three discrete modellings (eqns (5.214), (5.215) and (5.218)) of its continuous transport equation (3.117), which is set out in Chapter 3. We know from the above that eqn (5.215) is a priori preferred, because the internal energy flux between two particles is then antisymmetric. Yet, in Section 5.3.4, we have explained that (5.214) (i.e. (5.256)) is variationally compatible with the system (5.238), which is suitable for the exact conservation of the total momentum. A choice then has to be made; here we will choose (5.218) because of its smartness.

In the presence of viscous forces, Section 3.4.1 made it possible to write the most general form (3.119) of the internal energy discrete equation, containing a source term originating from the molecular dissipation $\widetilde{f} = \nu s^2$ (eqn (3.111)). It is well known that such a quantity corresponds to the dissipation of macroscopic kinetic energy resulting from the viscous friction. In the SPH discrete formalism, the energy dissipation rate F as calculated in Section 6.2.2 corresponds to that quantity and is likened to the volume integral of the quantity[11] $\widetilde{f}$ of Section 3.4.1. That result had already been mentioned in Section 3.3.3, in which $\widetilde{f}$ (eqn (3.91)) had been compared with the discrete dissipation rate F_a of Chapter 2 (Section 2.5, equation (2.189)). In principle, the quantity F_a then has to be added to the right-hand side in equation (5.218) if one wants to predict the internal energy of the particles. With the model (6.56), this yields

$$\begin{aligned}\forall a, \quad \frac{de_{int,a}}{dt} &= \frac{1}{2} \sum_b m_b \frac{p_a + p_b}{\rho_a \rho_b} \mathbf{u}_{ab} \cdot w'_{ab} \mathbf{e}_{ab} \\ &\quad - (n+2)\,\mu \sum_b \frac{V_a V_b}{r_{ab}} \left(\mathbf{e}_{ab} \cdot \mathbf{u}_{ab}\right)^2 w'_{ab}\end{aligned} \tag{6.97}$$

Care ensures the conservation of the total (kinetic + internal) energy of the fluid in the presence of viscous forces. This principle may also be found by writing a discrete equation for the entropy of the particles, as in Chapter 2 (Section 2.5.1), provided that the antisymmetry of the coefficients $(\omega_8)_{ab}$ involved in the system (2.219) is satisfied, as we have established the antisymmetry of the coefficients $(\omega_7)_{ab}$ by the end of Section 5.3.4. Due to the positiveness of the dissipation rates F_a, the increase of internal energy, and hence of entropy, is provided.

It may be useful to model the evolution of the total particle energy. We will adopt the operator $\tilde{D}^1_a$ for approximating the divergence, in order to satisfy the conservation of total energy. A discrete approximation of the second equation in the system (3.119) can then readily be constructed:

$$\forall a, \quad \frac{de_a}{dt} = -\rho_a \sum_b m_b \left(\frac{p_a \mathbf{u}_a + 2\mu_a \mathbf{s}_a \mathbf{u}_a}{\rho_a^2} + \frac{p_b \mathbf{u}_b + 2\mu_b \mathbf{s}_b \mathbf{u}_b}{\rho_b^2} \right) \cdot w'_{ab} \mathbf{e}_{ab} \tag{6.98}$$

where $\mathbf{s}_a$ is the rate of strain tensor attached to a particle. It is defined as in (3.15):

[11] Care shall obviously be taken not to confuse the dissipation rate f with the dimensionless kernel $f(q)$.

$$\mathbf{s}_a = \frac{1}{2}\left[(\mathbf{grad}\ \mathbf{u})_a + (\mathbf{grad}\ \mathbf{u})_a^T\right] \tag{6.99}$$

In order to calculate it, one may resort to the model (5.135) for estimating the components of the velocity gradient tensor:

$$(\mathbf{grad}\ \mathbf{u})_a \approx \mathbf{G}_a^0\{\mathbf{u}_b\} = \sum_b V_b\left(\mathbf{u}_a + \mathbf{u}_b\right) \otimes w'_{ab}\mathbf{e}_{ab} \tag{6.100}$$

or else, component after component (the subscripts $_i$ and $_j$ denoting the vector components):

$$\left(\frac{\partial u_i}{\partial x_j}\right)_a \approx \sum_b V_b\left(u_{i,a} + u_{i,b}\right) w'_{ab} e_{j,ab} \tag{6.101}$$

We note that the latter three equations can also be used for calculating the scalar deformation rate s^2 on the basis of the definition (3.26), which would allow a different computation of the discrete energy dissipation, through (3.111). That approach which, besides, is more complex than the previously described one, would however be disadvantageous because it does not provide the exact value of the kinetic energy which is dissipated under the action of the friction forces.

6.2.5 Discrete incompressible scheme

The weakly compressible scheme which we introduced at the end of Section 5.3.2 is advantageously simple and clear. It does come within the physical scope of the first Part of this book and is the usual basis for the SPH modelling specialists. Yet, the analysis of stability which was proposed in Section 5.4.4 discloses a need to resort to a very short time step. Moreover, the numerical experience highlights practical drawbacks, especially the occurrence of artificial cavities near the solid edge irregularities (for example, refer to Issa et al., 2004). That shortcoming can be partly remedied by means of a background pressure (refer to the end of Section 5.3.2), or alternatively by increasing the sound speed, which is detrimental since that further reduces the time step, according to equation (5.413). The simulations based on the weakly compressible approach also highlight large pressure fluctuations caused by the great sensitivity of the state equation. Modelling specialists are familiar with that phenomenon which is not specific to the SPH method, the grid methods being affected by that numerical artifact too. Is is well known that a solution consists of decolocalizing the quantites, that is estimating the pressure and the velocity at different places in each computational mesh (for instance, pressure along the edges and velocity at the centre of the mesh cells).[12]

The Riemann solvers are an alternative to that problem for treating incompressible flows, which is well known within the context of the grid methods. Since the works of Vila (1999) and other authors, there is an ample literature about that matter within the frame of the SPH method, which we can only briefly talk of here. The idea consists of calculating both densities and pressures using the exact formulations resulting from the solution of a Riemann problem between each pair of particles. In particular, this approach was associated with a conservative arbitrary Lagrangian–Eulerian (ALE) scheme[13]

[12] That idea, applied to the SPH scope, was mentioned by the end of Section 5.4.4 (refer to Dyka et al., 1997).

[13] It consists in considering that the particles move at a velocity which is not necessarily their Lagrangian velocity but, instead, an arbitrary velocity, which amounts to rewriting the equations of motion starting from balances analogous to (3.37).

by Marongiu (2007), with outstanding results as regards the prediction of the pressure at fixed or moving walls. In some versions of this approach (in particular, refer to Vila, 1999), the discrete equations are written in such a way that the mass is not exactly conserved, which is consistent with a non-fully Lagrangian approach. These numerical schemes are usually accompanied by viscous stabilizing terms, formally taking the same shape as the discrete viscous forces being dealt with in Section 6.2.2. Yet, the relative complexity of this kind of approach should be emphasized, since a profitable implementation of the Riemann solvers implies numerical precautions.

A short time step nevertheless remains in all the variants and amendments we have just mentioned, which compels us to seek another solution for the calculation of pressure, which would fit the case of strictly incompressible flows,[14] that is with $\forall a,\ \rho_a = \rho_0$. It is therefore interesting to consider pressure as a Lagrange multiplier (refer to Chapter 1, Section 1.2.4) associated with the incompressibility constraint for a fluid (Chapter 4, Section 3.6.3). Equation (3.355), indeed, gives the contribution of pressure to the total Lagrangian of a fluid in the incompressible formalism. With the discrete approximation (5.84), we can write

$$L_{incomp} = -\sum_b V_b p_b \ln \frac{\rho_b}{\rho_0} \tag{6.102}$$

We may recall that using a Lagrangian implies that the densities ρ_b be kept virtually variable before varying L_{incomp}. To the first order, however, the V_b may be considered as constant for calculating the variation of L_{incomp}. The corresponding force is then written by deriving with respect to the position of an arbitrary particle, using the first line of the calculation (5.237):

$$\begin{aligned} \frac{\partial L_{incomp}}{\partial \mathbf{r}_a} &= -\sum_b V_b p_b \frac{1}{\rho_b} \frac{\partial \rho_b}{\partial \mathbf{r}_a} \\ &= -\frac{1}{\rho_a} \mathbf{G}_a^k \{p_b\} \end{aligned} \tag{6.103}$$

We note that this force is identical to (5.236). The incompressibility condition then induces the pressure forces within the discrete context of the SPH method as in the context of continua, with the adequate discrete gradient operator.[15]

For that reason, we may invoke Section 3.4.4 and try to match the continuous predictor-corrector scheme to the discrete SPH formalism (Cummins and Rudman, 1999). We may recall that the scheme in Section 3.4.4 is primarily based on the adjoint nature of the div and $-$**grad** operators (eqn (3.198)). Now, we pointed out at the end of Section 5.2.3 that the discrete operators $\tilde{D}_a^k$ and $\mathbf{G}_a^k$ satisfy the same relation (eqn (5.141)), indicating that the projection method as defined in the continuous case may be exactly transcribed within the discrete context.[16] In other words, an arbitrary velocity field $\{\mathbf{u}_a\}$ can be projected onto a field $\left\{\mathbf{u}_a^d\right\}$ which exactly verifies $\tilde{D}_a^k \left\{\mathbf{u}_a^d\right\} = 0$. Owing to that property, we can state the predictor-corrector scheme (3.205) can be matched to the discrete case, taking a discrete form of Poisson's equation (3.206) into account for calculating the pressure giving rise to the internal forces during the corrective step:

[14] With all the reservations we have formulated in Section 3.4.2 about that concept.

[15] We would get a similar result by prescribing the condition $J_a \left\{\rho_b^k\right\} = \rho_0^k$.

[16] The adjoint nature of the discrete operators plays a nearly obvious role in the calculation (6.103), which could have been made along the lines of (5.235).

$$
\begin{aligned}
\forall a,\ \mathbf{u}_a^{m*} &= \mathbf{u}_a^m + \left[\frac{1}{m_a}\left(\mathbf{F}_a^{diss}\right)^m + \mathbf{g}\right]\delta t \\
\left(\nabla^2 p\right)_a^{m+1} &= \frac{\rho_0}{\delta t}\left(\operatorname{div}\mathbf{u}^{m*}\right)_a \\
\mathbf{u}_a^{m+1} &= \mathbf{u}_a^{m*} + \frac{1}{m_a}\left(\mathbf{F}_a^{int}\right)^{m+1}\delta t \\
\mathbf{r}_a^{m+1} &= \mathbf{r}_a^m + \mathbf{u}_a^{m+1}\delta t
\end{aligned}
\tag{6.104}
$$

A remark is to be made right now about the time points labelled as ^{m*}. Some authors prefer to use an intermediate time point $m+1/2$ for the predictive step, as we previously did for the Störmer–Verlet scheme (5.338) in Section 5.4.2. That notation was then justified by the fact that this scheme is based on a discrete Lagrangian (5.313) analogous to (5.339) but explicitly giving rise to a term in $\delta t/2$. In other words, within the scope of the Störmer–Verlet scheme, the time $m+1/2$ has a true meaning, the acceleration being discretized to second order instead of the simple formula (5.265). On the other hand, that notation would be inappropriate here, because one might be led to wrongly believe that the 'time point' $t^{m+1/2}$ has a physical meaning. Instead, the 'time point' t^{m*} should be considered a practical notation, the prediction step being nothing but an attempt to be subsequently corrected.[17]

[17] It is not ruled out, however, that a symplectic nature may be imparted to the incompressible scheme (6.104).

To discretize Poisson's equation which is the second line in the system (6.104), we must take into account the fact that the continuous Laplacian occurring in the Poisson's equation is defined by $\nabla^2 \doteqdot \operatorname{div}(\mathbf{grad})$, the latter two operators being skew-adjoint. Thus, in order to make a proper projection, the discrete Laplacian should be modelled by the compound operator $\tilde{D}_a^k\left\{\mathbf{G}_b^k\right\}$, which corresponds to equation (6.2). As outlined above, this kind of formulation sometimes happens to be so costly that practical applications cannot be contemplated. Thus, the exactness of the projection has to be sacrificed using, for modelling the discrete Laplacian, one of the second-order operators constructed in Section 6.2.1, in this case $\left(\nabla^2 p\right)_a \approx L_a^0\left[\{1\},\{p_b\}\right]$, where L_a^0 is given by (6.10). Keeping the $\tilde{D}_a^0$ operator to approximate the divergence (eqn (5.131)), that yields:

$$
\forall a,\ \sum_b V_b \frac{p_{ab}^{m+1}}{r_{ab}^m} w_{ab}'^m = -\frac{\rho_0}{2\delta t}\sum_c V_c \mathbf{u}_{ac}^{m*}\cdot\left(w_{ac}'\mathbf{e}_{ac}\right)^m \tag{6.105}
$$

with, it should be remembered, $p_{ab} \doteqdot p_a - p_b$. Equation (6.105) is in the form of a linear system whose unknowns are the pressures $\{p_b\}$:

$$
\forall a,\ \{M_{ab}\}\{p_b\}^{m+1} = \{N_a\} \tag{6.106}
$$

where the coefficients $\{M_{ab}\}$ make up a tensor[18] $N_p \times N_p$ and the $\{N_a\}$ make up a vector with N_p components, these being defined by

$$
\begin{aligned}
M_{ab} &\doteqdot V_b\left(\delta_{ab}-1\right)\frac{w_{ab}'^m}{r_{ab}^m} + \delta_{ab}\sum_c V_c \frac{w_{ac}'^m}{r_{ac}^m} \\
N_a &\doteqdot -\frac{\rho_0}{2\delta t}\sum_c V_c \mathbf{u}_{ac}^{m*}\cdot\left(w_{ac}'\mathbf{e}_{ac}\right)^m
\end{aligned}
\tag{6.107}
$$

[18] Here, of course, M_{ab} does not take the form of the difference of two terms. It is a matrix whose subscripts actually are a and b, unlike the matrices $\boldsymbol{\alpha}_{ab}$ as defined in Section 2.4.3, for example. Here we prefer to resort to indicial rather than tensorial notations so as not to confuse them with such vectors of the ordinary space as velocity.

Thus, the projection achieved through this approach is only approximated.[19] This system should be solved through a suitable numerical method which still has some degree of approximation. Since the matrix $\{A_{ab}\}$ has none of the standard qualities which the numericians can expect within this context (it is not very sparse and has no symmetry property), one shall preferably utilize such a method as Bi-CGSTAB (Van der Vorst, 1992).

[19]This is because $L_a^0 \neq \tilde{D}_a^0\{\mathbf{G}_b^0\}$.

The discrete forms of the forces appearing in the equations of the scheme (6.104) are simplified, as compared with the weakly compressible case, due to the fact that the density is now constant, which especially makes it possible to substitute $\rho_0 V_a$ for m_a. In particular, the pressure forces (6.103) no longer depend on the selected superscript k:

$$\begin{aligned}
\mathbf{F}_a^{int} &= -\sum_b V_a V_b \left(p_a + p_a\right) w'_{ac}\mathbf{e}_{ac} \\
\mathbf{F}_a^{diss} &= 2\left(n+2\right)\mu \sum_b V_a V_b \left(\mathbf{u}_{ab}\cdot\mathbf{e}_{ab}\right) w'_{ab}\mathbf{e}_{ab}
\end{aligned} \tag{6.108}$$

(here we have adopted (6.44) for the description of the viscous forces).

We have not yet discussed the treatment of the boundary conditions for the SPH method, since it will be dealt with in a part of a subsequent Section 6.3. Suffice it to say that the wall pressure condition (3.207) should be transcribed into the SPH language, which will be made in Section 6.3.2. In addition, a brief comment is to be made as regards the free surfaces, about which we have learnt in Section 3.4.2 that the pressure, which as a first approximation may be assumed to be zero (eqn (3.134)), must be prescribed. With the nearly incompressible scheme, that condition is established in a natural way, nothing has to be done to prescribe it. This is due to the state law (5.207), which sets a positive pressure when $\rho_a \geqslant \rho_0$ and a negative pressure when $\rho_a < \rho_0$. Thus, when a particle moves away from a free surface where it laid thus far, its density decreases according to the process described by the end of Section 5.3.2 and its pressure falls. It is then subject to a return force which tends to bring it back[20] to the surface, and so it comes closer to other particles, and so both its density and pressure rise. A new equilibrium is attained when the particle pressure is nearly zero. Conversely, with the incompressible algorithm being introduced in this section, the pressure at the surface should be prescribed. To that purpose, it is necessary to determine which particles are lying on the surface, which, it should be acknowledged, raises a problem of ambiguity. Nevertheless, we are going to provide a couple of suitable criteria. The first consists in noting that the property as formulated in the second line of (2.106), namely div$\mathbf{r} = n$, is only approximatively true for the surface particles, which are affected by a lack of nearby particles,[21] like the particles lying near a wall (Fig. 5.4). Thus, it is possible to calculate an approximation of that order of magnitude based on the discrete operator $\tilde{D}_a^0$ as given by (5.131), and decide that a particle is lying on the surface if that estimation is below some threshold (the following value is suggested by Lee et al., 2008):

[20]Unless another force counteracts it, especially inertia.

[21]Unless the renormalization (Section 5.2.5) is used.

$$-\sum_c V_c \mathbf{r}_{ac}^m \cdot \left(w'_{ac}\mathbf{e}_{ac}\right)^m < 0.75n \tag{6.109}$$

Another approach, which is proposed by Doring (2005), is based on the use of the renormalization matrix $\mathbf{R}_a$ as given by (5.178). This paper shows that a particle a may be considered as lying on a free surface if the smallest eigenvalue of $\mathbf{R}_a$ is lower than a threshold value which typically equals 0.3. There are more efficient variants of this algorithm (for instance, refer to Marrone et al., 2010).

To complete this section, we will explore the stability properties of that scheme on the basis of those considerations as introduced in Section 5.4.4, confining ourselves to the case without viscous forces. Contrary to what was done for the explicit and semi-explicit schemes, here we have to differentiate the time points at which the various quantities are estimated immediately upon the construction of the variation rates, because the initial equations (6.104) involve quantities which are estimated at different time points, and so the state variables for each particle are $\left(\mathbf{u}_a^{m*}, \mathbf{u}_a^{m+1}, \mathbf{r}_a^{m+1}, p_a^{m+1}\right)$ instead of $(\mathbf{u}_a, \mathbf{r}_a, \rho_a)$. Tu put it differently, since we are dealing with a predictor-corrector scheme, we must start from the system (6.104) with time-discretized equations (without writing any such variation rate as $\dot{\mathbf{r}}_a$, etc.). We use again (6.104) (without dissipative forces) to write the numerical fluctuations of the state parameters:

$$\begin{aligned} \delta \mathbf{u}_a^{m*} &= \delta \mathbf{u}_a^m \\ \delta \left(\nabla^2 p\right)_a^{m+1} &= \frac{\rho_0}{\delta t} \delta \left(\operatorname{div} \mathbf{u}^{m*}\right)_a \\ \delta \mathbf{u}_a^{m+1} &= \delta \mathbf{u}_a^{m*} + \frac{1}{m_a} \delta \left(\mathbf{F}_a^{int}\right)^{m+1} \delta t \\ \delta \mathbf{r}_a^{m+1} &= \delta \mathbf{r}_a^m + \delta \mathbf{u}_a^{m+1} \delta t \end{aligned} \qquad (6.110)$$

Compared with the nearly incompressible case of Section 5.4.4, a prediction equation for velocity has appeared and the continuity equation has disappeared and has been replaced by the Poisson's equation for pressure. Since the latter does not give the pressure variation rate over time, however, it shall initially conveniently be simplified in order to determine a relation between the pressure fluctuations δp and the velocity fluctuations $\delta \mathbf{u}$, which will allow us to delete pressure from the linearized equations, as we did through the equation of state at the beginning of Section 5.4.4. We will adequately seek δp in the following form:

$$\delta p = P(t) \exp\left(-i \mathbf{K} \cdot \mathbf{r}\right) \qquad (6.111)$$

The left-hand side in the discrete Poisson's equation (6.105) depends on both pressures and positions. However, because of the quantities $p_{ab}^{m+1} = p_a^{m+1} - p_b^{m+1}$, the share of the variation rate resulting from the fluctuations of the positions cancels itself out when seeking a solution starting from a homogeneous initial state, as in Section 5.4.4. It is therefore sufficient to evaluate the variation rate caused by the pressure variations, using, as previously, an integral approximation:

$$
\begin{aligned}
&\delta_p \left(\sum_b m_b \frac{p_{ab}^{m+1}}{r_{ab}^m} w_{ab}'^m \right) \\
&= \sum_b m_b \frac{w_{ab}'^m}{r_{ab}^m} \left(\delta p_a^{m+1} - \delta p_b^{m+1} \right) \\
&\approx \rho_0 \left[\left(\int_\Omega \frac{w'(\tilde{r})}{\tilde{r}} d^n \tilde{\mathbf{r}} \right) \delta p_a^{m+1} - \int_\Omega \frac{w'(\tilde{r})}{\tilde{r}} \delta p^{m+1} d^n \tilde{\mathbf{r}} \right] \\
&= \rho_0 \left[\int_\Omega \frac{w'(\tilde{r})}{\tilde{r}} d^n\, \tilde{\mathbf{r}} - \int_\Omega \frac{w'(\tilde{r})}{\tilde{r}} \exp\left(-i\mathbf{K}\cdot\tilde{\mathbf{r}}\right) d^n \tilde{\mathbf{r}} \right] \\
&\times P\left(t^{m+1}\right) \exp\left(-i\mathbf{K}\cdot\mathbf{r}_a\right)
\end{aligned}
\tag{6.112}
$$

Invoking (6.67) and (6.68), we come to

$$
\begin{aligned}
&\delta_p \left(\sum_b m_b \frac{p_{ab}^{m+1}}{r_{ab}^m} w_{ab}'^m \right) \\
&\approx \rho_0 \frac{2}{h^2} \left[\widehat{w_h}(K) - \widehat{w_h}(0) \right] P\left(t^{m+1}\right) \exp\left(-i\mathbf{K}\cdot\mathbf{r}_a\right)
\end{aligned}
\tag{6.113}
$$

As to the right-hand side in Poisson's equation, for the same reasons as above, only considering its variation induced by the velocity fluctuations is enough:

$$
\begin{aligned}
&\delta_{\mathbf{u}} \left(\sum_c m_c \mathbf{u}_{ac}^{m*} \cdot w_{ac}'^m \mathbf{e}_{ac}^m \right) \\
&= \sum_c m_c w_{ac}'^m \mathbf{e}_{ac}^m \cdot \left(\delta\mathbf{u}_a^{m*} - \delta\mathbf{u}_b^{m*} \right) \\
&\approx \rho_0 \begin{bmatrix} -\left(\int_\Omega \nabla w_h(\tilde{r})\, d^n \tilde{\mathbf{r}} \right) \cdot \delta\mathbf{u}_a^{m*} \\ + \int_\Omega \nabla w_h(\tilde{r}) \cdot \delta\mathbf{u}^{m*} d^n \tilde{\mathbf{r}} \end{bmatrix} \\
&= \rho_0 \widehat{\nabla w_h}(K) \cdot \mathbf{U}\left(t^{m*}\right) \exp\left(-i\mathbf{K}\cdot\mathbf{r}_a\right) \\
&= i\rho_0 \widehat{w_h}(K)\, \mathbf{K} \cdot \mathbf{U}\left(t^{m*}\right) \exp\left(-i\mathbf{K}\cdot\mathbf{r}_a\right)
\end{aligned}
\tag{6.114}
$$

We may note that the term $\mathbf{U}(t^{m*})$ has no practical physical sense, taking into account the remark which was made above as regards the 'time point' t^{m*}. We will subsequently see that this term disappears from the linearized equations. Equalling (6.112) and (6.114), we get the simplified form of Poisson's equation:

$$
\left[\widehat{w_h}(0) - \widehat{w_h}(K) \right] P\left(t^{m+1}\right) = \frac{i\rho_0 h^2}{4\delta t} \widehat{w_h}(K)\, \mathbf{K} \cdot \mathbf{U}\left(t^{m*}\right) \tag{6.115}
$$

relating the pressure wave to the velocity wave. The dynamic equations in the system (6.110) are still to be linearized. We will simply provide details about the equation giving $\delta\mathbf{u}_a^{m+1}$, the other two equations being elementary. The term $\delta_{\mathbf{r}} \left(\mathbf{F}_a^{int}\right)^{m+1}$ is unchanged as compared with the calculation in Section 5.4.4 leading to (5.378), whereas the pressure fluctuation effect is written as

$$
\begin{aligned}
&\frac{1}{m_a}\delta_p\left(\mathbf{F}_a^{int}\right)^{m+1}\\
&= -\frac{1}{\rho_0^2}\sum_b m_b\left(\delta p_a^{m+1}+\delta p_b^{m+1}\right)\left(w'_{ab}\mathbf{e}_{ab}\right)^m\\
&\approx -\frac{1}{\rho_0}\begin{bmatrix} -\delta p_a^{m+1}\int_\Omega \nabla w_h\left(\tilde{r}\right)d^n\tilde{\mathbf{r}}\\ -\int_\Omega \delta p\left(t^{m+1}\right)\nabla w_h\left(\tilde{r}\right)d^n\tilde{\mathbf{r}}\end{bmatrix}\\
&= \frac{P\left(t^{m+1}\right)}{\rho_0}\int_\Omega \exp\left(-i\mathbf{K}\cdot\mathbf{r}\right)\nabla w_h\left(\tilde{r}\right)d^n\tilde{\mathbf{r}}\\
&= \frac{iP\left(t^{m+1}\right)}{\rho_0}\widehat{w_h}\left(K\right)\mathbf{K}\exp\left(-i\mathbf{K}\cdot\mathbf{r}_a\right)
\end{aligned}
\tag{6.116}
$$

Combining (5.378), (6.115) and (6.116), we find:

$$
\begin{aligned}
&\frac{1}{m_a}\delta\left(\mathbf{F}_a^{int}\right)^{m+1}\\
&\approx \frac{\widehat{w_h}\left(K\right)}{\rho_0}\left[iP\left(t^{m+1}\right)\mathbf{K}-2p_0\left(\mathbf{K}\otimes\mathbf{K}\right)\mathbf{R}\left(t^m\right)\right]\\
&\times\exp\left(-i\mathbf{K}\cdot\mathbf{r}_a\right)\\
&= -\widehat{w_h}\left(K\right)\left(\mathbf{K}\otimes\mathbf{K}\right)\begin{Bmatrix}\dfrac{h^2\widehat{w_h}\left(K\right)}{4\left[\widehat{w_h}\left(0\right)-\widehat{w_h}\left(K\right)\right]\delta t}\mathbf{U}\left(t^{m*}\right)\\ +\dfrac{2p_0}{\rho_0}\mathbf{R}\left(t^m\right)\end{Bmatrix}\\
&\times\exp\left(-i\mathbf{K}\cdot\mathbf{r}_a\right)
\end{aligned}
\tag{6.117}
$$

The pressure then has disappeared from the linearized system, as we wanted. With the above, (6.110) is rewritten as

$$
\begin{aligned}
&\mathbf{U}\left(t^{m*}\right)=\mathbf{U}\left(t^m\right)\\
&\mathbf{U}\left(t^{m+1}\right)=\mathbf{U}\left(t^{m*}\right)-\widehat{w_h}\left(K\right)\left(\mathbf{K}\otimes\mathbf{K}\right)\\
&\qquad\times\begin{Bmatrix}\dfrac{h^2\widehat{w_h}\left(K\right)}{4\left[\widehat{w_h}\left(0\right)-\widehat{w_h}\left(K\right)\right]}\mathbf{U}\left(t^{m*}\right)\\ +\dfrac{2p_0}{\rho_0}\mathbf{R}\left(t^m\right)\delta t\end{Bmatrix}\\
&\mathbf{R}\left(t^{m+1}\right)=\mathbf{R}\left(t^m\right)+\mathbf{U}\left(t^{m+1}\right)\delta t
\end{aligned}
\tag{6.118}
$$

The first equation in the system (6.118) enables to delete $\mathbf{U}\left(t^{m*}\right)$, which yields, with the notations (5.389) and (5.391), the following eigenvalue equation:

$$
\left(\mathbf{K}\otimes\mathbf{K}\right)\mathbf{U}_0=-\frac{\left(\chi-1\right)^2K^2}{2A_4\chi-A_3}\mathbf{U}_0
\tag{6.119}
$$

A_3 being given by (6.81), whereas we have defined here

$$A_4 \doteq \frac{A_3}{2} + \left(Ma_0 Co K^+\right)^2 \widehat{w_h}\left(K\right) > 0 \qquad (6.120)$$

The quantities K^+, Ma_0 and Co are yielded by (5.73), (5.211) and (5.397), whereas a reference velocity U has been defined:

$$U \doteq \sqrt{\frac{p_0}{\rho_0}} \qquad (6.121)$$

U has the same order of magnitude as the main velocity within the flow, as suggested by equation (3.157) in Chapter 3. The characteristic equation resulting from (6.119) is written (once again ignoring the solution $\chi = 1$) as:

$$\chi^2 - 2\left(1 - A_4\right)\chi + 1 - A_3 = 0 \qquad (6.122)$$

A careful analysis would enable us to show that there are still such wavenumbers that the discriminant in that system is positive, some others making it negative. Both cases should then be studied. If it is negative, the solutions are complex conjugate, and equation (6.122) shows that their norms are given by

$$|\chi|^2 = 1 - A_3 \qquad (6.123)$$

Since A_3 is positive, the stability condition (5.393) is satisfied in all circumstances. If, on the contrary, the discriminant in (6.122) is positive, then the solutions are real and can be written as

$$\chi = 1 - A_4\left(1 \pm \sqrt{1 - \frac{2A_4 - A_3}{A_4^2}}\right) \qquad (6.124)$$

In this case, the stability condition is reduced to

$$A_4 + \sqrt{A_4^2 - 2A_4 + A_3} \leqslant 2 \qquad (6.125)$$

It can easily be demonstrated that this is equivalent to $A_3 \leqslant 4$ (we have just seen that this condition is always satisfied) and $A_4 \leqslant 2$, that is coming back to the notations K^+ and Co:

$$8Co^2 \leqslant \frac{16}{K^{+2}\widehat{w_h}\left(K\right)} - \frac{\widehat{w_h}\left(K\right)}{\widehat{w_h}\left(0\right) - \widehat{w_h}\left(K\right)} \qquad (6.126)$$

With the Gaussian kernel, this is also written as:

$$8Co^2 \leqslant \frac{4}{K^+ e^{-K^+}} - \frac{e^{-K^+}}{1 - e^{-K^+}} \qquad (6.127)$$

The function of the right-hand side in (6.127) is positive and has a minimum[22] of 10.25 for $K^+ = 0.9056$, and so the stability condition is ultimately written as:

$$\delta t \leqslant C_{\delta t,3}\frac{h}{U} \qquad (6.128)$$

[22] This minimum is nearly insensitive to the second term in the right-hand side of (6.127), which results from the variation $\delta_p \dot{\mathbf{u}}_a^{m+1}$. This is an evidence of the excessively stable nature of the system with respect to the pressure variations.

with $C_{\delta t,3} = 1.13$. The incompressible scheme which we have presented is then conditionally stable and associated with a CFL type condition, just like the nearly incompressible scheme being discussed in Section 5.4.4. On the other hand, the time step upper bound is now based on the fluid velocity instead of the sound speed as in (5.413), which enables a significant increase of the time step. Considering the definition (5.212) of c_0, equation (5.413) shows, indeed, that the incompressible scheme allows a time step which is at least ten times larger than the weakly compressible scheme. Although the solution to the linear system (6.106) implies additional calculations, the incompressible scheme is then much faster for simulating a phenomenon extending over a given physical time. If the viscous forces are taken into account, it can readily be demonstrated that the condition (6.91) adds to the condition (6.128) and provides such a mixed condition type as (6.93). More information about the numerical stability of the SPH incompressible algorithm can be found in Violeau and Leroy (2015).

There are several variants of the just set forth incompressible approach (for example, refer to Shao and Lo, 2003; Hu and Adams, 2008). In all of them, the pressure behaves as the quantity enabling us to make the velocity field incompressible, as is the case with the continuous media (Section 3.4.4). Ellero et al. (2007) utilize this principle to introduce a similar approach in which the pressure explicitly appears, in the discrete equations, as the Lagrange multiplier associated with the incompressibility condition, which we have demonstrated by means of the calculation (6.103). These authors' work is also based upon a Hamiltonian expression of the thermodynamic equations which is analogous to what we explained in Section 2.5.1. The main difference lies in the assumed constant density, ans so only entropy plays a role as a thermodynamic state parameter.

The incompressible algorithms are suitable for largely solving the problem of the numerical fluctuations of pression, but, on the other hand, impart a very ordered motion to all the particles, which may be troublesome in the absence of gravity. Particles initially arranged on a Cartesian mesh and moving about, for instance, as determined by a Taylor–Green vortex network, tend to keep an arrangement exhibiting constrictions alternating with voids, due to the strain of the medium (refer to Fig. 6.1 (*a*)). Xu et al. (2009) suggest solving that problem by introducing an artificial random disorder reallotting their positions in a more homogeneous way, as in Fig. 6.1 (*b*) (we will discuss this idea again in Section 6.5.3).

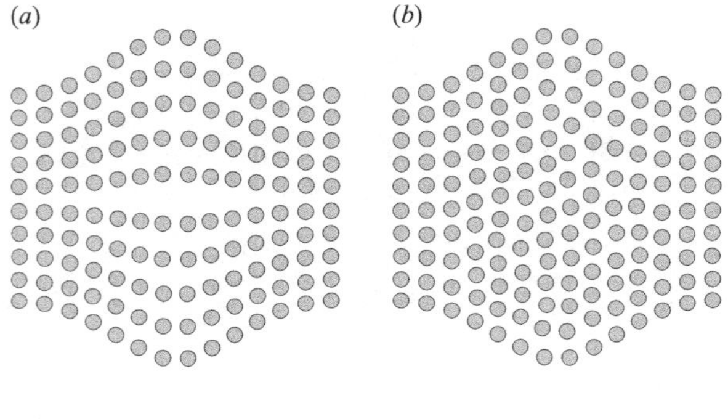

Fig. 6.1 Discrete incompressible approach: too orderly evolution of the particle positions (*a*) and random correction (*b*).

Applications of equations (6.104) to (6.109) are dealt with in Sections 7.2.2, 7.3.3, 7.4.1 and 7.4.2.

6.3 Boundary treatment

The boundary condition question, which we have put aside so far, is one of the most awkward issues when setting up an SPH algorithm. In this section, we will deal with the conditions at the interfaces between two fluids, introducing the discrete modelling of the surface tension. We will subsequently turn to the modelling of the conditions at a solid wall, which will make it possible to deal with the solid bodies moving about in a fluid. We ignore the open boundary question (inflow or outflow), which has been so far rather unfrequently dealt with in the SPH-related literature. As regards the latter point, however, one may refer to Lastiwka et al. (2008).

6.3.1 Discrete surface tension

The SPH method is attractive, among others, because one may disregard the free surface condition, at least with the weakly compressible scheme (nevertheless, Section 6.2.5 mentions that a pressure has to be prescribed for it with the incompressible scheme). If, however, the scales being considered are of same order of magnitude as (or smaller than) the length L_β as defined in (3.226), the surface tension phenomena have to be taken into account. That approach is also relevant for modelling multiphase flows, which is readily allowed by the SPH method ascribing the characteristics of a given phase to each particle. In the following, we will consider the case of a multiphase flow.

In this case, we must add a term into the right-hand side of the equation of motion (first line in the system (5.239), for instance), representing a discrete form of these forces, which we now have to determine. Some authors (particularly Hu and Adams, 2006) propose resorting to a simple form based on the equation of motion (5.199), after defining a surface tensor for each particle through the continuous formalism formula (3.218). However, we are going to set forth a slightly different approach, inspired by Monaghan (1998). One may also advantageously refer to Grenier et al. (2009), Rogers et al. (2009), Hu et al. (2009) and Adami et al. (2010).

We will initially use the discrete operators for approximating the continuous form of these forces. In the first place, we write the following relation, which is applicable to every pair of vectors **A** and **B** (refer to Appendix A, eqn (A.69)):

$$(\mathrm{div}\mathbf{B})\,\mathbf{A} = \mathbf{div}\,(\mathbf{A}\otimes\mathbf{B}) - (\mathbf{grad}\;\mathbf{A})\,\mathbf{B} \tag{6.129}$$

Thus, the mass force as yielded by (3.224) (Chapter 3) makes it possible to write the surface tension force as received by a particle a as follows:

$$\mathbf{F}_a^{ST} = \beta V_a\left[(\mathbf{grad}\;\mathbf{grad}\;C)\,\mathbf{n}^I - \mathbf{div}\left(\mathbf{grad}\;C\otimes\mathbf{n}^I\right)\right]_a \tag{6.130}$$

β being the surface tension coefficient between the two fluids, and $\mathbf{n}^I$ being the unit vector locally normal to the interface (Section 3.4.5). It is now necessary to discretize the characteristic function C. Let us consider the case of two fluids,[23] and let us note as C the function equalling one within the fluid 1, equalling zero within the fluid 2. We use the discrete gradient operator as defined by (5.123) to write[24]

[23] We will subsequently turn to the case of an arbitrary number of fluids.

[24] Note that this argument would not be valid with the $\tilde{\mathbf{G}}_a^k$ operator because of the property (5.133) nor with a renormalized operator (Section 5.2.5).

$$(\mathbf{grad}\ C)_a \approx \mathbf{G}_a\{1\} = \sum_{b\in\Omega_1} V_b w'_{ab}\mathbf{e}_{ab} \tag{6.131}$$

where Ω_1 denotes all the particles belonging to the fluid 1. Thus, taking (3.216) into account, a vector normal to the interface can be defined for each particle:

$$\mathbf{n}_a^I = \frac{\sum_{c\in\Omega_1} V_c w'_{ac}\mathbf{e}_{ac}}{\left|\sum_{c\in\Omega_1} V_c w'_{ac}\mathbf{e}_{ac}\right|} \tag{6.132}$$

It should be pointed out that, according to the approach of the surface tension mass forces introduced in Section 3.4.5, this vector becomes zero for particles being sufficiently distant from the interface, just as is **grad** C. The kernel here plays the role of the distribution δ_Σ (eqn (3.220)), as everywhere in the SPH formalism (refer to Section 5.2.1). We now can discretize the first term in (6.130) through a term-wise derivation:

$$(\mathbf{grad\ grad}\ C)_a \approx \sum_{b\in\Omega_1} V_b \frac{\partial w'_{ab}\mathbf{e}_{ab}}{\partial \mathbf{r}_a} \tag{6.133}$$

We are now trying to estimate the derivative occurring in the right-hand side of (6.130), which is a second-order tensor. To that purpose, we utilize the quantities as defined by the formula (5.144), which are recalled below:

$$\begin{aligned}\mathbf{W}_{ab} &\doteq \frac{\partial w'_{ab}\mathbf{e}_{ab}}{\partial \mathbf{r}_b} \\ &= \left(\frac{w'_{ab}}{r_{ab}} - w''_{ab}\right)\mathbf{e}_{ab}\otimes\mathbf{e}_{ab} - \frac{w'_{ab}}{r_{ab}}\mathbf{I}_n\end{aligned} \tag{6.134}$$

Since $w'_{cd}\mathbf{e}_{cd}$ only depends on $\mathbf{r}_a$ if one of the two subscripts c and d equals a, we get the more general relation:

$$\frac{\partial w'_{cd}\mathbf{e}_{cd}}{\partial \mathbf{r}_a} = (\delta_{ad} - \delta_{ac})\,\mathbf{W}_{cd} \tag{6.135}$$

Inserting the latter relation into (6.133), after making $c = a$ and $d = b$, we get:

$$\left[(\mathbf{grad\ grad}\ C)\,\mathbf{n}^I\right]_a \approx \sum_{b\in\Omega_1} V_b(\delta_{ab} - 1)\,\mathbf{W}_{ab}\mathbf{n}_a^I \tag{6.136}$$

The following approximation can easily be shown using the same procedure:

$$\left[\mathbf{div}\left(\mathbf{grad}\ C\otimes\mathbf{n}^I\right)\right]_a \approx \sum_{b\in\Omega_1} V_b(\delta_{ab} - 1)\,\mathbf{W}_{ab}\mathbf{n}_b^I \tag{6.137}$$

Combining (6.136) and (6.137), the discretized surface tension force (6.130) becomes

$$\mathbf{F}_a^{ST} = -\beta \sum_{b \in \Omega_1} V_a V_b \mathbf{W}_{ab} \mathbf{n}_{ab}^I \tag{6.138}$$

with

$$\mathbf{n}_{ab}^I \doteq \mathbf{n}_a^I - \mathbf{n}_b^I = -\mathbf{n}_{ba}^I \tag{6.139}$$

(to write (6.138), we have taken into account the fact that $\mathbf{n}_{ab}^I$ is zero when $b = a$, which makes it possible to drop the Kronecker symbol δ_{ab}). Taking (6.134) into account, the relation being found may also be written as

$$\mathbf{F}_a^{ST} = -\beta \sum_{b \in \Omega_1} V_a V_b \left[\left(w''_{ab} - \frac{w'_{ab}}{r_{ab}} \right) \left(\mathbf{e}_{ab} \cdot \mathbf{n}_{ab}^I \right) \mathbf{e}_{ab} + \frac{w'_{ab}}{r_{ab}} \mathbf{n}_{ab}^I \right] \tag{6.140}$$

It can be seen that this force involves the kernel's second derivative, which is natural since the continuous form (6.130) depends on the interface curvature, which is a second-order quantity. As a consequence of the remarks made in Section 6.2.1 about the hazardous nature of that option, it would probably be advisable to provide a variant of it that would be based on second-order operators analogous to those introduced in the same section (that is not what we will do here).[25]

In addition, if there is an arbitrary number of fluids, a force similar to (6.140) should exist in each particle a, the discrete sum extends over all those particles belonging to the same fluid as a, with appropriate coefficients β_{ab} (refer, for example, to Grenier et al., 2009).

We are going to explain how the discrete surface tension forces can be derived from a least action principle, through the procedure as developed in Section 5.3.3 for the other forces. This is possible since these forces are conservative, as explained in Chapter 3. In accordance with the generic interpolation (5.84), let us initially write a discrete form of the tension potential (3.222):[26]

$$E_p^{ST} \approx \beta \sum_b V_b \left| \mathbf{grad}\, C \right|_b \tag{6.141}$$

Coming back to the definition (3.216) of $\mathbf{n}^I$, and taking the approximation (6.132) into account, we get a discrete form of the potential involvoing the kernel, which can then be added (with a 'minus' sign) to the Lagrangian (5.220) to write:

$$\begin{aligned} L = & \sum_b \frac{1}{2} m_b u_b^2 - \sum_b m_b e_{int,b} - \sum_b m_b g z_b \\ & - \beta \sum_b V_b \left| \sum_c V_c w'_{bc} \mathbf{e}_{bc} \right| \end{aligned} \tag{6.142}$$

[25] The formula (6.33) proposed in Section 6.2.1 would, however, be insufficient here because, if it was applied to the operator (6.130), it would yield zero.

[26] In the following, for the sake of clarity, we will purposefully not specify that the discrete sums regard particles as belonging to the same fluid.

The related discrete force exerted on a will be written by deriving, as usual, with respect to the components of the relevant particle:

$$\begin{aligned}\mathbf{F}_a^{ST} &= -\beta \frac{\partial}{\partial \mathbf{r}_a} \sum_b V_b \left| \sum_c V_c w'_{bc} \mathbf{e}_{bc} \right| \\ &= -\beta \frac{\partial}{\partial \mathbf{r}_a} \sum_b V_b \frac{\partial}{\partial \mathbf{r}_a} \left| \sum_c V_c w'_{bc} \mathbf{e}_{bc} \right|\end{aligned} \tag{6.143}$$

In order to derive the norm occurring in that expression, we write $|\mathbf{B}|\,(\mathbf{grad}\,|\mathbf{B}|) = (\mathbf{grad}\,\mathbf{B})\,\mathbf{B}$, which yields:

$$\begin{aligned}&\frac{\partial}{\partial \mathbf{r}_a} \left| \sum_c V_c w'_{bc} \mathbf{e}_{bc} \right| \\ &= \frac{1}{\left| \sum_c V_c w'_{bc} \mathbf{e}_{bc} \right|} \sum_d V_d \frac{\partial w'_{bd} \mathbf{e}_{bd}}{\partial \mathbf{r}_a} \sum_c V_c w'_{bc} \mathbf{e}_{bc} \\ &= \sum_d V_d \frac{\partial w'_{bd} \mathbf{e}_{bd}}{\partial \mathbf{r}_a} \mathbf{n}_b^I\end{aligned} \tag{6.144}$$

To calculate the derivative occurring in this result, we make $c = b$ in (6.135), which yields

$$\mathbf{F}_a^{ST} = -\beta \sum_b \sum_d V_b V_d \left(\delta_{ab} - \delta_{ad} \right) \mathbf{W}_{bd} \mathbf{n}_b^I \tag{6.145}$$

Because of the term $\delta_{ab} - \delta_{ad}$, two sums emerge from the above expression, one of them corresponding to the cases $b = a$ and $d \neq a$ on the one hand, $b \neq a$ and $d = a$ on the other hand. Hence:

$$\mathbf{F}_a^{ST} = -\beta \sum_{d \neq a} V_a V_d \mathbf{W}_{ad} \mathbf{n}_a^I + \beta \sum_{b \neq a} V_b V_a \mathbf{W}_{ba} \mathbf{n}_b^I \tag{6.146}$$

Lastly, we make a change of dummy subscript in the first sum of (6.146), then we use the symmetry property of $\mathbf{W}_{ab}$ as given by (5.144), and ultimately write

$$\mathbf{F}_a^{ST} = -\beta \sum_b V_a V_b \mathbf{W}_{ab} \mathbf{n}_{ab}^I \tag{6.147}$$

(as previously, the case $b = a$ has no effect on the sum because of the vector $\mathbf{n}_{ab}^I$). Unsurprisingly, the form (6.138) derived from the discrete equations can be found again.

This force may be split up into individual contributions, as with the pressure and viscosity forces:

$$\begin{aligned}\mathbf{F}^{ST}_{b\to a} &= -\beta V_a V_b \mathbf{W}_{ab}\mathbf{n}^I_{ab} \\ &= -\beta V_a V_b \left[\left(w''_{ab} - \frac{w'_{ab}}{r_{ab}}\right)\left(\mathbf{e}_{ab}\cdot\mathbf{n}^I_{ab}\right)\mathbf{e}_{ab} + \frac{w'_{ab}}{r_{ab}}\mathbf{n}^I_{ab}\right] \\ &= -\mathbf{F}^{ST}_{a\to b}\end{aligned} \tag{6.148}$$

The fact that these forces derive from a discrete least action principle has major effects, of course. In particular, the usual conservation properties are satisfied. The relation (6.148) induces, among others, the conservation of linear momentum, whereas the conservation of angular momentum appears through a calculation analogous to (5.250), on the basis of the relation $\mathbf{r}_{ab}\times\mathbf{e}_{ab} = \mathbf{0}$:

$$\begin{aligned}\frac{d\mathbf{M}^{ST}}{dt} &= \sum_{a<b}\mathbf{r}_{ab}\times\mathbf{F}^{ST}_{b\to a} \\ &= -\beta\sum_{a<b} V_a V_b \frac{w'_{ab}}{r_{ab}}\mathbf{r}_{ab}\times\mathbf{n}^I_{ab} \\ &= -\beta\sum_{a,b} V_a V_b w'_{ab}\mathbf{e}_{ab}\times\mathbf{n}^I_a\end{aligned} \tag{6.149}$$

(we have utilized the opposite procedure to that used in the calculation (5.250) and only considered the quantities $\mathbf{n}^I_a$, ranging over all the pairs of dummy subscripts, and used the definition $\mathbf{e}_{ab}$). Turning back to the definition (6.132), this equation is also written as:

$$\frac{d\mathbf{M}^{ST}}{dt} = -\beta\sum_a \frac{V_a}{\left|\sum_d V_d w'_{ad}\mathbf{e}_{ad}\right|}\sum_{b,c} V_b V_c w'_{ab} w'_{ac}\mathbf{e}_{ab}\times\mathbf{e}_{ac} \tag{6.150}$$

But since $\mathbf{e}_{ab}\times\mathbf{e}_{ac} = -\mathbf{e}_{ac}\times\mathbf{e}_{ab}$, the two-fold sum is zero ranging over both dummy subscripts b and c.

As regards energy, we already explained in Section 1.4.1 that its conservation is natural for a force deriving from a potential, as shown by the general literal calculation (1.73). The forces $\mathbf{F}^{ST}_a$ are then to be included among the conservative forces in the discrete equation of motion, when relevant, to yield

$$\forall a,\ m_a\frac{d\mathbf{u}_a}{dt} = \sum_b\left(\mathbf{F}^{int}_{b\to a} + \mathbf{F}^{ST}_{b\to a} + \mathbf{F}^{diss}_{b\to a}\right) + \mathbf{F}^{ext}_a \tag{6.151}$$

Compared with the calculation we have just made, here we have added the viscous fources, which enables a comparison with (6.46).

From the calculations of numerical stability as disclosed in Sections 5.4.4, 6.2.3 and 6.2.5, a necessary condition for stability with respect to the discrete surface tension forces could be established; a simple dimensional reasoning (refer to Section 3.4.3) easily shows that it is written as

$$\delta t \leqslant C_{\delta t,4}\sqrt{\frac{\rho_0 h^3}{\beta}} \tag{6.152}$$

with $C_{\delta t,4} = 0.10$ (Adami et al., 2010). That condition adds to (6.93). We may note that the condition (6.152) is also written in the equivalent form $Co \leqslant C_{\delta t,4}\sqrt{We_0}$, where $We_0 \doteqdot \rho_0 U^2 h/\beta$ is a numerical Weber number, according to the definition (3.225), whereas the Courant number Co is yielded by (5.397).

Other forces of molecular origin, such as the Van der Waals forces, can be modelled on the basis of similar approaches (refer to Tartakofsky et al., 2009).

6.3.2 Solid wall modelling

Applying the principles of the SPH method to the real flow modelling nearly always implies[27] that the presence of solid walls is taken into account. The latter can be modelled in several ways: we will disclose some of the most frequent and consistent.

Among the available models in the specialized literature, the wall itself is nearly always modelled by means of particles representing the solid material, arranged in one layer and lying at a substantially constant distance apart, corresponding to the diameter δr of one particle[28] as given at the beginning of Section 5.2.3. We will call them *edge particles* or *wall particles*. When a fluid particle comes closer to a wall, the Fig. 5.4 shows that a void space gradually arises within its range of influence. This leads to a pressure force whose component perpendicular to the wall is insufficient to force the fluid particle to remain within the fluid domain: the thus-modelled wall is not tight.

This problem can be overcome in various ways. The dummy particle method, in particular, encompasses many forms, and we are going to explain its simplest version, which has proved its validity in many implementation cases. The idea consists of 'filling' the possible voids with particles lying on the other side of the wall. A first, simple approach is based on the prior

[27]Except in astrophysics, for example. Besides, that discipline has given rise to the SPH method, maximizing the benefits of its fully Lagrangian nature (for instance, refer to Springel et al., 2001).

[28]Some authors prefer to reduce that distance, describing the wall by means of a finer discretization than the fluid.

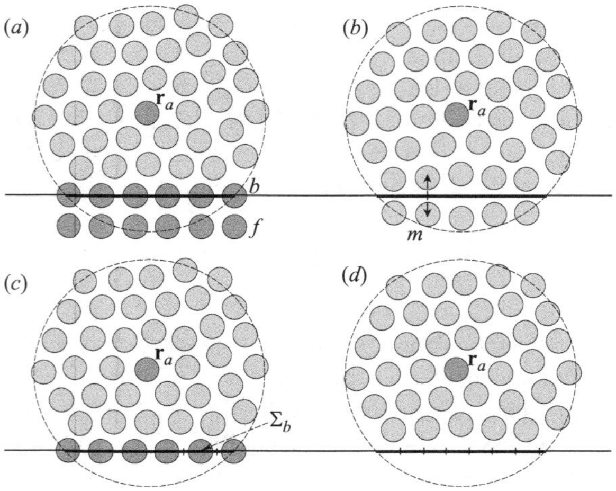

Fig. 6.2 Four approaches for modelling walls with SPH: dummy particles (a), mirror particles (b), boundary particles (c) and mesh (d).

establishment of a network of evenly spaced stationary particles, arranged in several layers beyond the edge particles (refer to Fig. 6.2 (*a*)). The number of layers theoretically depends on the kernel radius, but experience shows that two or three layers are enough (a single one is illustrated in Fig. 6.2 (*a*)). Their action is as follows: these dummy particles are considered 'dead', insofar as their related quantities are constant over time. Their mere presence, and particularly their having a non-zero density (equalling, for instance, the reference value ρ_0 of the equation of state (5.207)) confers on them an action in the discrete sum occurring in the continuity equation of the system (5.239). Thus, they affect the density of the fluid particles lying near a wall. Within the context of the weakly compressible model, the continuity equation shows that the contribution of a dummy particle f to the volume of a fluid particle a equals

$$q_{V,fa} = -m_a m_f \frac{\rho_f^{k-1}}{\rho_a^{k+1}} \mathbf{u}_{af} \cdot w'_{af} \mathbf{e}_{af} \tag{6.153}$$

according to (5.186), which corresponds to the option $k = 1$. Putting $\rho_f = \rho_0$ and $\mathbf{u}_f = \mathbf{0}$ (which is always possible when the position is determined within the wall's specific coordinate system if the wall is mobile), we find

$$q_{V,fa} = -m_a m_f \frac{\rho_0^{k-1}}{\rho_a^{k+1}} \mathbf{u}_a \cdot w'_{af} \mathbf{e}_{af} \tag{6.154}$$

Since the derivative w'_{af} is negative (refer to Section 5.3.1), this contribution has the same sign as $\mathbf{u}_a \cdot \mathbf{e}_{af}$, that is negative if the fluid particle moves towards the wall. In such a case, the volume of a tends to decrease insignificantly; since its mass is constant, its density imperceptibly increases, as well as its pressure by virtue of the equation of state (5.207). The dummy particles then behave like natural 'repellents', playing the role of the wall due to their prescribed position. It should right now be specified that, as suggested by our explanation, this method enables us to easily model moving walls the motion of which is chosen by the modeller (e.g. a wave paddle, as discussed later on in Section 8.2). The case of bodies whose motion is established by the fluid motion and interacts with the latter will be dealt with in Section 6.3.3.

Within the frame of the incompressible algorithm (Section 6.2.5), the above argument is no longer valid. A condition should therefore be prescribed for pressure. It may be expected that the right-hand side in the condition (3.207) $\mathbf{u}^{m*} \cdot \mathbf{n}_{wall}$ is almost zero, which amounts to prescribing

$$\left(\frac{\partial p}{\partial \mathbf{n}_{wall}} \right)_b^{m+1} = 0 \tag{6.155}$$

for every edge particle b. An approximate but fairly simple means consists in considering the edge particle pressures as unknowns in the linear system to be solved (6.106). That pressure is subsequently duplicated for the immediately underlying dummy particles. This approach, however, implies that each dummy particle is skilfully linked to an edge particle, which raises issues, particularly near the corners.

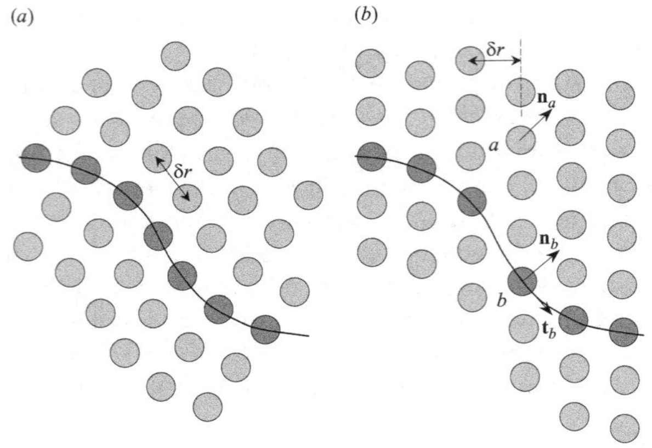

Fig. 6.3 Two possible layouts for arranging the edge and dummy particles on a curved wall. Configuration (*b*) is recommended.

Regardless of the chosen algorithm (either weakly compressible or incompressible), one must beware, in this approach, of the arrangement of the dummy particles. Numerical experience does show that, in the absence of any force of gravity, the fluid behaviour near curved walls (as is the case in the example of Section 7.2.1) generates void pockets, if the edge and dummy particles are not carefully arranged. In particular, upon the construction of the curved walls, two options, as described in Fig. 6.3, are a priori offered. Graph (*a*) illustrates (two-dimensionally) a wall along which the particle are evenly spaced, whereas (*b*) shows them more distant apart along the wall, while exhibiting a constant spacing along the x and z axes. Numerical experience teaches that the latter option is suitable. Tests have demonstrated, indeed, that the arrangement illustrated in the graph (*a*) generates the above mentioned voids. Furthermore, the right-hand construction is obviously much more easily achievable. It may seem odd that the density of edge particles per unit length is variable, which seemingly creates a 'hole' at the place where the wall inflection point is located; the presence of dummy particles with a constant density *per unit volume*, however, corrects that apparent gap. The dummy particle technique is used in all the application cases as disclosed in Chapters 7 and 8.

There are many variants to that technique, among which are in particular the *mirror particle* type approaches, in which these particles are analogous to the above described dummy particles but are reconstructed upon each iteration by symmetry across the walls. The procedure consists in adding—to each fluid particle c which is close enough to a wall—a mirror particle m provided with parameters which are symmetric to those of c (Fig. 6.2 (*b*)) The influence of these particles then results in symmetric terms in the discrete dynamic equations. One of the advantages in that procedure is that a no-slip condition ($\mathbf{u} = \mathbf{0}$) can more easily be achieved for the wall by prescribing $\mathbf{u}_m = -\mathbf{u}_c$.

The above described methods, as well as all their variants, are advantageously simple, although in a somewhat rough way, but at the cost of extra computational time and, above all, additional memory space, particularly in three dimensions. Furthermore, they can hardly be applied to the very deformed walls or to the description of small-sized solid bodies along a space direction (membranes, beams). We will disclose an alternative in which only edge particles are taken into account. It consists of getting rid of every dummy particle by providing the edge particles with carefully selected repulsive forces which are added to the equation of motion (Fig. 6.2 (*c*)). There are several kinds of them, out of which the most frequently used in fluid mechanics is based on the Lennard–Jones molecular model which was briefly set out in Section 1.3.2, but with different power coefficients. In line with the formula (1.46), we may, indeed, propose a force taking the following form, for every pair of particles a (fluid) and b (edge):

$$\mathbf{F}_{b\to a}^{LJ} = -\frac{\partial E_p^{LJ}}{\partial \mathbf{r}_a} = \frac{E_0}{r_{ab}}\left[\left(\frac{r_0}{r_{ab}}\right)^{P_2} - \left(\frac{r_0}{r_{ab}}\right)^{P_1}\right]\mathbf{e}_{ab} \tag{6.156}$$

with, for example, $P_1 = 4$ and $P_2 = 2$ (Monaghan, 1994). The reference potential E_0 has the dimension of a mass multiplied by the square of a velocity. For those flows which involve gravity, it should be set as follows:

$$E_0 = CmgH \tag{6.157}$$

where m is the mean mass of a particle, g denotes the acceleration of gravity, H is a characteristic flow vertical length and C denotes a dimensionless constant ranging from 5 to 10. At to r_0, here it represents the range of influence of the wall beyond which the force as imparted by (6.156) vanishes; it is usually slightly shorter than the initial mean interparticular distance δr. It can be seen in (6.156) that r_{ab}/r_0 is theoretically greater than one, which makes it possible to ensure that the wall force is duly directed towards the fluid particle. There are in the literature other models than that we have just disclosed, but they do not offer any further significant advantage (for instance, refer to Monaghan, 1994). The model (6.156) is compared with that of the dummy particles in Section 7.3.1, the latter being preferred.

In spite of its simplicity, the above approach provides for wall sealing, but lacks physical rigour and sometimes gives rise to a questionable numerical behaviour near the walls. Moreover, introducing a wall force will a priori break down the so far preserved strict conservation of linear momentum. Here we disclose a more rigorous method as developed by Kulasegaram et al. (2004), of which we take up the major points (also refer to Bonet and Rodriguez-Paz, 2005). The idea, which was discussed at the beginning of Section 5.2.5, consists of noting that the too low number of nearby particles which a particle a meets near a wall can be corrected through the corrected continuous interpolation (5.146). In particular, we note that the interpolation (5.193) of the density, in this formalism, is written as:

$$\forall a,\ \rho_a = \frac{1}{\gamma_a}\sum_b m_b w_{ab} \tag{6.158}$$

where γ_a is defined through a discrete (5.150) or continuous (5.151) approximation. Using that procedure, we have renormalized the density. In order to examine its implications, we will get into the context of the weakly compressible algorithm. The calculation (5.194) and (5.195) having led to the continuity equation now has to be revised. In order to illustrate this point, let us here consider the option in which the superscript of the discrete operators is $k = 1$. Multipying (6.158) by γ_a then deriving with respect to time, we find an additional term, due to the fact that γ_a depends on $\mathbf{r}_a$, that is implicitly on time:

$$\forall a, \ \gamma_a \frac{d\rho_a}{dt} + \rho_a \frac{d\gamma_a}{dt} = \sum_b m_b \mathbf{u}_{ab} \cdot w'_{ab} \mathbf{e}_{ab} \tag{6.159}$$

The new term may also be written as

$$\begin{aligned} \frac{d\gamma_a(\mathbf{r}_a)}{dt} &= \frac{\partial \gamma_a}{\partial \mathbf{r}_a} \cdot \frac{d\mathbf{r}_a}{dt} \\ &= \boldsymbol{\gamma}'_a \cdot \mathbf{u}_a \end{aligned} \tag{6.160}$$

where $\boldsymbol{\gamma}'_a$ is the gradient of γ_a, as given by (5.154) or (5.155), the latter showing that it is directed towards particle a. Equation (6.159) yields

$$\forall a, \ \frac{d\rho_a}{dt} = \frac{1}{\gamma_a} \sum_b m_b \mathbf{u}_{ab} \cdot w'_{ab} \mathbf{e}_{ab} - \frac{\rho_a \boldsymbol{\gamma}'_a \cdot \mathbf{u}_a}{\gamma_a} \tag{6.161}$$

which is substituted for the continuity equation (5.185). In addition to the corrective term γ_a appearing in the denominator (equalling one for a particle located far off any wall, if it is defined in the form of a continuous integral), we find that an additive term has appeared in the right-hand side. In order to provide an interpretation of it, let us consider the case of a planar wall being directed by the constant vector $\mathbf{n}_{wall}$: $\boldsymbol{\gamma}'_a$ is then collinear with the latter. If particle a moves parallel to the wall $\left(\boldsymbol{\gamma}'_a \cdot \mathbf{u}_a = 0\right)$, the contribution of this term then equals zero, whereas it will tend to increase the density of a particle moving towards the wall $\left(\boldsymbol{\gamma}'_a \cdot \mathbf{u}_a < 0\right)$.

As explained in Section 5.3.3, modifying the continuity equation implies that the equation of motion is revised by virtue of the variational principle. Thus, in the calculation (5.224) which led us to the discrete equation of motion (5.229), the force deriving from the internal energy should be modified. We may recall, indeed, that this calculation highlighted the pressure forces having the following shape (we restrict ourselves to the option $k = 1$, but generalizing to an arbitrary superscript k would be easy):

$$-\frac{\partial E_{int}}{\partial \mathbf{r}_a} = -\sum_b m_b \frac{p_b}{\rho_b^2} \frac{\partial \rho_b}{\partial \mathbf{r}_a} \tag{6.162}$$

With the corrected approximation (6.158), the calculation (5.226) gives, as in the continuity equation, a corrective factor γ_a in the denominator, as well as an additive term:

$$\frac{\partial \rho_b}{\partial \mathbf{r}_a} = \frac{m_a}{\gamma_b} \nabla_a w_{ab} + \frac{\delta_{ab}}{\gamma_b} \sum_c m_c \nabla_a w_{ac} - \frac{\delta_{ab}}{\gamma_b^2} \sum_c m_c w_{bc} \boldsymbol{\gamma}'_a \tag{6.163}$$

Introducing that development into (6.162), we get, after some algebra:

$$-\frac{\partial E_{int}}{\partial \mathbf{r}_a} = -\sum_b m_a m_b \left(\frac{p_a}{\gamma_a \rho_a^2} + \frac{p_b}{\gamma_b \rho_b^2} \right) w'_{ab} \mathbf{e}_{ab} + m_a \frac{p_a}{\gamma_a^2 \rho_a^2} \sum_b m_b w_{ab} \boldsymbol{\gamma}'_a \tag{6.164}$$

Thus, the conservative forces are corrected by the factor $1/\gamma_a$, as compared with (5.202). In addition, it can be seen that a contact force $\mathbf{F}_a^{cont}$ appears due to the presence of the wall and acts somewhat like the previously set out Lennard–Jones force $\mathbf{F}_{b\to a}^{LJ}$, but it is constructed on a more reliable physical base. Taking equation (6.158) into account, we may ultimately write:

$$\begin{aligned} \mathbf{F}_{b\to a}^{int} &= -m_a m_b \left(\frac{p_a}{\gamma_a \rho_a^2} + \frac{p_b}{\gamma_b \rho_b^2} \right) w'_{ab} \mathbf{e}_{ab} \\ \mathbf{F}_a^{cont} &= \frac{V_a p_a}{\gamma_a} \boldsymbol{\gamma}'_a \end{aligned} \tag{6.165}$$

Thus, the equation of motion now has the following shape:

$$\forall a,\ m_a \frac{d\mathbf{u}_a}{dt} = \sum_b \left(\mathbf{F}_{b\to a}^{int} + \mathbf{F}_{b\to a}^{ST} + \mathbf{F}_{b\to a}^{diss} \right) + \mathbf{F}_a^{cont} + \mathbf{F}_a^{ext} \tag{6.166}$$

(here we have added both dissipative and surface tension forces). For illustrating the contact forces, let us resume the example of the planar wall oriented by $\mathbf{n}_{wall}$: $\mathbf{F}_a^{cont}$ is directed from the wall towards the fluid particle in all circumstances, as expected (if the pressure is duly positive).

One can realize the opportunity of the expression of $\mathbf{F}_a^{cont}$ by resuming the continuous interpolation (5.15) of a gradient (for instance, refer to Marongiu, 2007). As regards pressure, we get:

$$[\mathbf{grad}\ p]_c (\mathbf{r}_a, t) = \oint_{\partial\Omega_\mathbf{r}} p(\mathbf{r}', t)\, w_h(|\mathbf{r}' - \mathbf{r}_a|)\, \mathbf{n}(\mathbf{r}')\, d^{n-1}\Gamma(\mathbf{r}') + \int_{\Omega_\mathbf{r}} p(\mathbf{r}', t) \frac{\partial w_h(|\mathbf{r}' - \mathbf{r}_a|)}{\partial \mathbf{r}_a} d^n \mathbf{r}' \tag{6.167}$$

The second term in the right-hand side is the continuous approximation of the pressure gradient as it naturally appears in the calculations leading to the discrete equations. The first term is a boundary integral which is only zero when the fluid particle is located far off the walls, as explained in Section 5.2.1. Otherwise, this term is reduced to:[29]

$$\int_{\Sigma_a} p(\mathbf{r}', t)\, w_h(|\mathbf{r}' - \mathbf{r}_a|)\, \mathbf{n}(\mathbf{r}')\, d^{n-1}\Gamma(\mathbf{r}') \tag{6.168}$$

Σ_a being defined in Fig. 5.4. The above expression may be compared with the contact force $\mathbf{F}_a^{cont}$ reduced to the particle unit volume, taking (5.155) and (6.165) into account:

[29] With the boundary integral, it is possible to show that the continuous interpolation is accurate up to the first order in h even near a wall, contrary to the original case (5.12).

$$\frac{\mathbf{F}_a^{cont}}{V_a} \approx \frac{p_a}{\gamma_a} \int_{\Sigma_a} w_h\left(\left|\mathbf{r}' - \mathbf{r}_a\right|\right) \mathbf{n}\left(\mathbf{r}'\right) d^{n-1}\Gamma\left(\mathbf{r}'\right) \tag{6.169}$$

(Monaghan and Kajtar, 2009 get an analogous formula from heuristic considerations). Thus, the contact force provided by the variational method corresponds to a boundary edge term resulting from the integration by parts of the pressure gradient (also refer to De Leffe et al., 2009). A suitable treatment of the friction forces at the wall could be established according to the same principle[30] (Ferrand et al., 2010, 2012).

[30]As a rule, all the quantities would have to be treated in an analogous way to provide for the flow consistency at the wall.

We have constructed the equations of motion in the presence of contact forces, resorting to the variational compatibility of both continuity equation and equation of motion. This can be more accurately formalized by introducing, from the definitions (6.161) and (6.164), modified discrete gradient and divergence operators:

$$\begin{aligned} \mathbf{G}_a^{\Gamma,1}\{A_b\} &\doteqdot \rho_a \sum_b m_b \left(\frac{A_a}{\gamma_a \rho_a^2} + \frac{A_b}{\gamma_b \rho_b^2}\right) w'_{ab}\mathbf{e}_{ab} - \frac{A_a}{\gamma_a}\boldsymbol{\gamma}'_a \\ \tilde{D}_a^{\Gamma,1}\{\mathbf{A}_b\} &\doteqdot -\frac{1}{\rho_a \gamma_a} \sum_b m_b \mathbf{A}_{ab} \cdot w'_{ab}\mathbf{e}_{ab} + \frac{\boldsymbol{\gamma}'_a \cdot \mathbf{A}_a}{\gamma_a} \end{aligned} \tag{6.170}$$

which is also written as:

$$\begin{aligned} \mathbf{G}_a^{\Gamma,1}\{A_b\} &= \mathbf{G}_a^1\left\{\frac{A_b}{\gamma_b}\right\} - \frac{A_a}{\gamma_a}\boldsymbol{\gamma}'_a \\ \tilde{D}_a^{\Gamma,1}\{\mathbf{A}_b\} &= \frac{1}{\gamma_a}\tilde{D}_a^1\{\mathbf{A}_b\} + \frac{\boldsymbol{\gamma}'_a \cdot \mathbf{A}_a}{\gamma_a} \end{aligned} \tag{6.171}$$

where $\mathbf{G}_a^1$ and $\tilde{D}_a^1$ are defined by (5.128) and (5.132). Taking the definition (5.154) into account, this is also written as

$$\begin{aligned} \mathbf{G}_a^{\Gamma,1}\{A_b\} &= \mathbf{G}_a^1\left\{\frac{A_b}{\gamma_b}\right\} - \frac{A_a}{\gamma_a}\mathbf{G}_a\{1\} \\ \tilde{D}_a^{\Gamma,1}\{\mathbf{A}_b\} &= \frac{1}{\gamma_a}\left(\tilde{D}_a^1\{\mathbf{A}_b\} + \mathbf{G}_a\{1\} \cdot \mathbf{A}_a\right) \end{aligned} \tag{6.172}$$

which may be compared to (5.156) and (5.158). If the identity operator is denoted as Id, and if an operator Γ is defined such that $\Gamma\{A_b\} \doteqdot \{A_b/\gamma_b\}$, then both above relations are also written in a condensed form

$$\begin{aligned} \mathbf{G}_a^{\Gamma,1} &= \left(\mathbf{G}_a^1 - \mathbf{G}_a\{1\}\, Id\right) \circ \Gamma \\ \tilde{D}_a^{\Gamma,1} &= \Gamma \circ \left(\tilde{D}_a^1 + \mathbf{G}_a\{1\} \cdot Id\right) \end{aligned} \tag{6.173}$$

The following scalar product (in the sense of eqn (5.138)) can easily be calculated:

$$\begin{aligned}
&\left\langle \mathbf{G}_a^{\Gamma,1}\{A_b\}, \{\mathbf{B}_a\}\right\rangle \\
&= \left\langle \mathbf{G}_a^{1}\left\{\frac{A_b}{\gamma_b}\right\}, \{\mathbf{B}_a\}\right\rangle - \left\langle \left\{\frac{A_a}{\gamma_a}\boldsymbol{\gamma}'_a\right\}, \{\mathbf{B}_a\}\right\rangle \\
&= -\left(\left\{\frac{A_a}{\gamma_a}\right\}, \tilde{D}_a^{1}\{\mathbf{B}_b\}\right) - \sum_a V_a \frac{A_a}{\gamma_a}\boldsymbol{\gamma}'_a\cdot\mathbf{B}_a \qquad (6.174)\\
&= -\left(\{A_a\}, \frac{1}{\gamma_a}\tilde{D}_a^{1}\{\mathbf{B}_b\}\right) - \left(\{A_a\}, \left\{\frac{\boldsymbol{\gamma}'_a\cdot\mathbf{B}_a}{\gamma_a}\right\}\right) \\
&= -\left(\{A_a\}, \tilde{D}_a^{\Gamma,1}\{\mathbf{B}_b\}\right)
\end{aligned}$$

(we have used (5.141)). Thus, the operators (6.171) are skew-adjoint, that is variationally compatible, in the sense of Section 5.3.3 (Ferrand et al., 2010). The variational principle will then naturally jointly reveal them. The modified equations which we have got through that procedure could as well be based on the calculation (5.235), *mutatis mutandis*. These modified operators are therefore a more suitable renormalization than the operators (5.156) and (5.158) as defined in Section 5.2.5. Continuous versions of these operators might readily be suggested, as we did in Section 5.2.1, and a similar adjunction property could be highlighted, such as equation (5.30) (see also Mayrhofer et al., 2013).

The appropriateness of chosen operators is also based on the calculation leading to formula (6.169), that is on the Gauss theorem, which enables us to reduce a volume integral to a boundary integral. A discrete formulation of that theorem can be written through the approximation (5.84):

$$\int_\Omega \mathbf{grad}\, A(\mathbf{r}')\, d^n\mathbf{r}' \approx \sum_a V_a \mathbf{G}_a^{\Gamma,1}\{A_b\} \qquad (6.175)$$

Based on the definition (6.171), we are going to calculate a quantity which looks like the right-hand side of this relation:

$$\begin{aligned}
\sum_a V_a \mathbf{G}_a^{\Gamma,1}\{A_b\} &= \sum_a V_a \mathbf{G}_a^{1}\left\{\frac{A_b}{\gamma_b}\right\} - \sum_a V_a \frac{A_a}{\gamma_a}\boldsymbol{\gamma}'_a \\
&= \sum_{a,b} m_a m_b \left(\frac{A_a}{\gamma_a\rho_a^2} + \frac{A_b}{\gamma_b\rho_b^2}\right) w'_{ab}\mathbf{e}_{ab} - \sum_a V_a \frac{A_a}{\gamma_a}\boldsymbol{\gamma}'_a
\end{aligned} \qquad (6.176)$$

The first discrete sum in (6.176) is zero due to the antisymmetry. We get

$$\sum_a V_a \mathbf{G}_a^{\Gamma,1}\{A_b\} = -\sum_a V_a \frac{A_a}{\gamma_a}\boldsymbol{\gamma}'_a \qquad (6.177)$$

which is a discrete form of the Gauss theorem, the sum in the right-hand side being possibly approximated through a boundary integral by means of (5.155). This calculation shows, on the other hand, that the operators $\mathbf{G}_a^1$ and $\tilde{D}_a^1$ have a zero discrete integral (i.e. without any boundary term), due to their antisymmetry.

In spite of these qualities, the operators (6.170) have the drawback of not ensuring the property (5.176) of the renormalized operators. De Leffe et al. (2009) propose a way to address this, but at the cost of dropping the conservation properties which are specific to the variationally compatible operators. Nevertheless, the numerical experience promotes the latter approach (Ferrand et al., 2010).

Numerical simulation shows that the model of contact forces $\mathbf{F}_a^{cont}$ works well[31] provided that γ_a and $\boldsymbol{\gamma}'_a$ are defined through their continuous versions (5.151) and (5.155) (Ferrand et al., 2010). In such a case, their calculation for a planar wall and at a given spherical kernel is theoretically possible, but leads to complex formulae. Kulasegaram et al. (2004) advocate for an approximation based on a tabulation of the integral, if the third-order kernel as given by (5.48) is adopted:

$$\begin{aligned} \gamma_a &= 1 + (0.0625 - 0.0531\epsilon)(\epsilon - 2)^3 \\ \boldsymbol{\gamma}'_a &= \frac{1}{h}(0.2937 - 0.2124\epsilon)(\epsilon - 2)^2\,\mathbf{n}_a \end{aligned} \tag{6.178}$$

where $\epsilon \doteqdot y/h$, h being the smoothing length and y the distance to the nearest wall, whereas $\mathbf{n}_a$ is the unit vector connecting the particle a to the wall through the shortest path, directed towards a (refer to Fig. 6.3). Furthermore, in the case of arbitrary-shaped walls, the presently disclosed method proves to be costly and complex, or even impossible. In addition, it can be observed that the approximation (6.178) may provide values of γ_a which are greater than 1. However, there are variants (Feldman and Bonet, 2007; Ferrand et al., 2010; Mayrhofer et al., 2014).

[31] Ferrand et al. (2010), however, suggest a continuity equation which is different from (6.161), based on a connection to the discrete interpolation (5.193), starting from the calculation (5.195).

Another more precise possibility consists in evaluating the quantities γ_a and $\boldsymbol{\gamma}'_a$ from a Eulerian description, the solid edges being provided with a mesh, as in Fig. 6.2 (*d*) (Marongiu, 2007). The areas of the mesh elements are then accurately known, as well as their normal vectors. This method is also advantageous in that it better describes the geometry of the walls when they are complex, and properly fits into the context of an arbitrary Eulerian–Lagrangian (ALE) approach for the treatment of the fluid-solid interactions, as outlined at the beginning of Section 6.2.5. When such a description of the walls is not avalaible, it is preferable to keep to the discrete formulations (5.150) and (5.154). Lastly, Ferrand et al. (2010, 2012) suggest calculating γ_a by writing a Lagrangian transport equation:

$$\forall a, \ \frac{d\gamma_a}{dt} = \boldsymbol{\gamma}'_a \cdot \mathbf{u}_a \tag{6.179}$$

This equation, which is analogous to (1.268), can be integrated by means of a numerical scheme which is inspired from Section 5.4. This method makes it possible to reduce the problem to the calculation of $\boldsymbol{\gamma}'_a$. To that purpose, formula (5.155) suggests a discrete approximation, provided that the wall surface element[32] attached to the edge particle b (Fig. 6.2 (*c*)) is denoted as Σ_b:

$$\boldsymbol{\gamma}'_a \approx \sum_b \Sigma_b w_{ab} \mathbf{n}_b \tag{6.180}$$

where the sum covers all the edge particles b, $\mathbf{n}_b$ being the unit vector normal to the wall at the point where the particle b is located (Fig. 6.3). With that

[32] This surface element can be approximated through the volume-to-radius ratio, that is $\Sigma_p = V_p/\delta r$ (δr being, it should be remembered, the particle diameter).

form, the contact force (6.165) can be split up into contributions due to each edge particle:

$$\mathbf{F}_{b\to a}^{cont} = \frac{p_a}{\gamma_a} V_a \Sigma_b w_{ab} \mathbf{n}_b \tag{6.181}$$

That formulation explicitly highlights the fact that the contact force $\mathbf{F}_a^{cont}$ as given by the second line in (6.165) is normal to the wall, as well as its proportionality to the contact surface area Σ_b (refer to Chapter 1, Section 1.5.4, equation (1.253)). Moreover, it can be seen it is a repulsive force, in accordance with the desired effect. As a conclusion to the above discussion, as compared with the above introduced dummy particle model, the model which we have just discussed is advantageous in that it is based on a variational principle and then reveals further terms having a physical meaning. In particular, these forces satisfy the conventional conservation laws. Moreover, their scope of validity can be extended to the incompressible algorithm (see Leroy et al., 2014).

To complete this section, let us explain how an alternative method could be constructed using a different variational principle, based on the observance of the wall boundary condition (3.144). Within the context of a variational method for a continuous fluid, we have stated in Section 3.6.3 that the wall boundary condition may be prescribed on the basis of a Lagrange multiplier, the latter being the pressure if the condition is written according to the formula (3.360). We may recall that, in that formula, $\mathbf{r}$ and $\mathbf{r}'$ stand for two arbitrary points in the fluid and the wall, respectively. Using the elements of Sections 5.2.1 and 5.2.3, we can write a discrete approximation of that condition. To do this, it is necessary to approximate the Dirac distribution $\delta\left(\left|\mathbf{r}-\mathbf{r}'\right|\right)$ through the kernel $w_h\left(r_{cb}\right) = w_{cb}$, $\mathbf{r}_c$ and $\mathbf{r}_b$ standing for $\mathbf{r}$ and $\mathbf{r}'$. Considering that the surface element $d\Gamma\left(\mathbf{r}'\right)$ can be approximated through Σ_b, we get:

$$L_{wall} \approx \sum_{b,c} V_c \Sigma_b p_c w_{cb} \mathbf{r}_{cb} \cdot \mathbf{n}_b \tag{6.182}$$

where the subscripts b and c range over the edge and fluid particles, respectively. The quantity (6.182) is the wall contribution to the system's Lagrangian and should then be added to the Lagrangian (5.220), generating a new force. For applying the Lagrange equations (1.31), L_{wall} has to be derived with respect to the parameters regarding a fluid particle (the edge particles having a prescribed motion within that context).[33] We find that $\partial L_{wall}/\partial \mathbf{u}_a = \mathbf{0}$. To calculate $\partial L_{wall}/\partial \mathbf{r}_a$, we may assume a rigid wall, and so the areas Σ_b may be considered as constant. It is theoretically necessary to take the variation of $V_c = m_c/\rho_c$ with $\mathbf{r}_a$, which is based on the relation (6.162), into account. For the sake of clarity, however, we will assume that the volume remains constant. We get

[33] The case of moving solid bodies will be dealt with in the next section.

$$\frac{\partial L_{wall}}{\partial \mathbf{r}_a} = p_a V_a \sum_b \Sigma_b \left(w_{ab}\mathbf{n}_b + w'_{ab}\mathbf{r}_{ab} \cdot \mathbf{n}_b\right) \tag{6.183}$$

Thus, the force exerted by b or a is written, using the relation (A.44) of Appendix A, as:

$$\mathbf{F}_{p\to a}^{cont} = p_a V_a \Sigma_b \left(w_{ab}\mathbf{I}_n + w'_{ab} r_{ab} \mathbf{e}_{ab} \otimes \mathbf{e}_{ab}\right) \mathbf{n}_b \tag{6.184}$$

with, as usually, $\mathbf{r}_{ab} = r_{ab}\mathbf{e}_{ab}$ (eqn (1.58)). Compared with (6.181), it can be found that the factor $1/\gamma_a$ has disappeared, as in the approach leading to (6.168). Besides, w_{ab} is replaced by $w_{ab}\mathbf{I}_n + w'_{ab} r_{ab}\mathbf{e}_{ab} \otimes \mathbf{e}_{ab}$. The force (6.184) then has a component which is tangential to the wall if a is not exactly over b. Furthermore, the normal component of the force may be attractive if the particle a is sufficiently far off the wall. For a particle which is close to the wall, conversely, and if $\mathbf{r}_{ab}$ is collinear with $\mathbf{n}_b$, the normal component of the force is, as a first approximation, proportional to $w_{ab} > 0$, that is repulsive, as expected.[34]

[34]This requires that the kernel derivative tends to a finite constant for small values of its argument, as is the case with the examples in Section 5.2.2.

6.3.3 Mobile rigid bodies

Let us now examine how the motion of a rigid body under the action of the forces which a fluid exerts on it can be modelled. We have known since Chapter 1 (Section 1.5.1) that the motion of a solid is determined by the total force and the total moment it receives. The discussion in Section 3.4.3 has taught us that the force can be estimated from a drag coefficient, but the latter is usually unknown, except in simple cases, from experimental data. Even so, this model is only applicable to a steady motion in a presumably 'infinite' space. As regards the moment of forces, it might be subjected to an analogous theoretical treatment, with the same restrictions. In view of the study of any motion of a body under the action of a mixture of fluids possibly exhibiting a free surface, and in non-steady conditions, we must resort to numerical simulation. Here we disclose a possible approach through the SPH method (for instance, refer to Monaghan et al., 2003; Violeau et al., 2007).

First of all, let us consider the case of a single rigid body, hereafter called the 'solid'. As for the fluid, one might imagine that it is discretized by a collection Ω_S of particles, along the lines of Section 1.5.1. This approach may seem complicated for modelling a rigid object, consequently having few degrees of freedom. The underlying idea is that those particles making up the solid are considered as edge particles which are possibly accompanied by dummy particles, the solid then playing the role of a wall moving according to the fluid load. We can then truly speak of a fluid–structure interaction. It should be specified that all the wall treatment methods set out in the previous section can be used for implementing this approach.

Within this context, the mass and inertia tensor of the solid are defined using the formulas (1.180) and (1.203), which are gathered below:

$$\begin{aligned} M &= \sum_{a\in\Omega_S} m_a \\ \mathbf{J} &= \sum_{a\in\Omega_S} m_a \left[|\mathbf{r}_a - \mathbf{R}|^2 \mathbf{I}_3 - (\mathbf{r}_a - \mathbf{R}) \otimes (\mathbf{r}_a - \mathbf{R}) \right] \end{aligned} \tag{6.185}$$

where $\mathbf{R}$ is the centre of inertia of the solid (we have used the definition (1.183)). We may recall that $\mathbf{J}$ is constant and invertible, according to Section 1.5.1. There are cases in which the solid has a simple shape, for which M and $\mathbf{J}$ can be calculated on a theoretical basis from continuous integrals among which (6.185) are approximations. In the general case, however, the discrete forms (6.185) are necessary. In the following, upon each iteration, **R** will have to be calculated according to (1.182):

$$\mathbf{R} = \frac{1}{M} \sum_{a \in \Omega_S} m_a \mathbf{r}_a \tag{6.186}$$

In the above equations, the masses of the particles[35] making up the solid are calculated as for the fluid particles (refer to the beginning of Section 5.2.3). Since the solid density is usually not identical to that of the fluid, it is all the same with the particle masses.

[35] In the case of a two-dimensional simulation, it should be remembered that they are masses per unit length. Hence, it is all the same with M and $\mathbf{J}$.

Here, the equations of motion (1.232), providing the velocity **V** of the centre of inertia and the rotation vector $\mathbf{\Omega}$ of the solid, are applied to the case in which the forces result from the action of both pressure ($\mathbf{F}_a^{int}$) and friction ($\mathbf{F}_a^{diss}$) as exerted by each of the surrounding fluid particles, on the basis of (1.214) and (1.217). As to the gravity force moment, it is yielded by (1.231):

$$\begin{aligned} M\frac{d\mathbf{V}}{dt} &= \sum_{a \in \Omega_S} \left(\mathbf{F}_a^{int} + \mathbf{F}_a^{diss}\right) + M\mathbf{g} \\ \mathbf{J}\frac{d\mathbf{\Omega}}{dt} &= \sum_{a \in \Omega_S} (\mathbf{r}_a - \mathbf{R}) \times \left(\mathbf{F}_a^{int} + \mathbf{F}_a^{diss}\right) \end{aligned} \tag{6.187}$$

Note that in the second line of (6.187), the gravity contribution has been cancelled out thanks to (6.186). For writing the fluid forces $\mathbf{F}_a^{int}$ and $\mathbf{F}_a^{diss}$, we will make the choice as given by equation (5.204) with $k = 1$ (which corresponds to (5.202)) and by eqn (6.45)). The equations of motion (6.187) then become similar to (6.58):

$$\begin{aligned} M\frac{d\mathbf{V}}{dt} &= -\sum_{a \in \Omega_S, b \in \Omega_F} m_a m_b B'_{ab} w'_{ab} \mathbf{e}_{ab} + M\mathbf{g} \\ \mathbf{J}\frac{d\mathbf{\Omega}}{dt} &= -\sum_{a \in \Omega_S, b \in \Omega_F} m_a m_b B'_{ab} w'_{ab} \left(\mathbf{r}_a - \mathbf{R}\right) \times \mathbf{e}_{ab} + M\mathbf{R} \times \mathbf{g} \end{aligned} \tag{6.188}$$

where Ω_F denotes the set of fluid particles, whereas the quantities

$$B'_{ab} \doteqdot \frac{p_a}{\rho_a^2} + \frac{p_b}{\rho_b^2} - 2\left(n+2\right)\mu \frac{\mathbf{u}_{ab} \cdot \mathbf{e}_{ab}}{\rho_a \rho_b r_{ab}} \tag{6.189}$$

have been defined by analogy with (5.246) and (5.247).

The two-fold sum occurring in the second line of the system (6.188) is tricky, because a and b do not belong to the same sets. It is therefore impossible to make the manipulations of the calculation (5.250), which would have consequently cancelled the total moment.

The system (6.188) should be accompanied by the kinematic equations in order to become identical in form to (1.219), where **F** and $\mathbf{F}_\Omega$ denote the total force and the total moment of the forces, as given by the right-hand sides of the equations in (6.188). A discrete form over time can then be produced along the lines of (5.266), where each of the pairs (**R**, **V**) and ($\boldsymbol{\theta}$, $\mathbf{\Omega}$) is treated separately. Due to the discussions in Section 5.4.1, the symplectic scheme (Section 5.4.2), is recommended and then gives

$$\begin{aligned}\mathbf{V}^{m+1} &= \mathbf{V}^m + \frac{1}{M}\mathbf{F}^m\delta t\\ \mathbf{R}^{m+1} &= \mathbf{R}^m + \mathbf{V}^{m+1}\delta t\\ \boldsymbol{\Omega}^{m+1} &= \boldsymbol{\Omega}^m + \mathbf{J}^{-1}\mathbf{F}^m_\Omega\delta t\\ \boldsymbol{\theta}^{m+1} &= \boldsymbol{\theta}^m + \boldsymbol{\Omega}^{m+1}\delta t\end{aligned} \tag{6.190}$$

The particles making up the solid are still to be moved. As a first approach, it may be written that their velocity is governed by the second line of equation (1.191), whereas their position may be incremented using the second line in the particle symplectic scheme (5.269):

$$\begin{aligned}\forall a, \quad \mathbf{u}_a^{m+1} &= \mathbf{V}^{m+1} + \boldsymbol{\Omega}^{m+1} \times \left(\mathbf{r}_a^m - \mathbf{R}^m\right)\\ \mathbf{r}_a^{m+1} &= \mathbf{r}_a^m + \mathbf{u}_a^{m+1}\delta t\end{aligned} \tag{6.191}$$

We may note, however, that this system does not define a really rigid motion, because the calculation (1.194) made in Chapter 1 does not exactly persists. (6.191) yields, indeed

$$\mathbf{u}_{ab}^{m+1} = \boldsymbol{\Omega}^{m+1} \times \mathbf{r}_{ab}^m \tag{6.192}$$

As such, we get $\mathbf{u}_{ab}^{m+1} \cdot \mathbf{r}_{ab}^m = 0$, which looks like (1.194). Here, however, that does not involve $r_{ab}^{m+1} = r_{ab}^m$, because of the temporal integration scheme. The distance between two arbitrary particles does vary, in accordance with the following law:

$$\begin{aligned}\left(r_{ab}^{m+1}\right)^2 &= \left(\mathbf{r}_{ab}^m + \mathbf{u}_{ab}^{m+1}\delta t\right)^2\\ &= \left(r_{ab}^m\right)^2 + 2\mathbf{r}_{ab}^m \cdot \mathbf{u}_{ab}^{m+1}\delta t + \left(u_{ab}^{m+1}\delta t\right)^2\\ &= \left(r_{ab}^m\right)^2 + 2\mathbf{r}_{ab}^m \cdot \left(\boldsymbol{\Omega}^{m+1} \times \mathbf{r}_{ab}^m\right)\delta t + \left(\boldsymbol{\Omega}^{m+1} \times \mathbf{r}_{ab}^m\delta t\right)^2\\ &= \left(r_{ab}^m\right)^2 + \left(\boldsymbol{\Omega}^{m+1} \times \mathbf{r}_{ab}^m\delta t\right)^2\end{aligned} \tag{6.193}$$

(the central term on the third line is zero by virtue of the vector product properties). A second-order error in δt can be found in the case of a rotational motion, as in the calculation (5.362). Substituting $\mathbf{r}_a^{m+1}$for $\mathbf{r}_a^m$ in the scheme (6.191) may seem to be a solution to that problem,[36] since the discrete motion would then verify $\mathbf{u}_{ab}^{m+1} \cdot \mathbf{r}_{ab}^{m+1} = 0$, which is now identical to (1.194). Nevertheless, a calculation similar to (6.193) would show that a further error is made. The condition $r_{ab} = cst$ prevails, indeed, over (1.194) because the latter only involves the former within the context of a continuous time. This phenomenon is analogous to the non-compliance to the conservation with the non-symplectic temporal schemes (refer to Section 5.4.3).

[36]That would require, however, solving the new system, which would illustrate the positions in an implicit form.

It should be pointed out that the relative error is very small. This is because the norm of the vector $\boldsymbol{\Omega}$ has the same order of magnitude as U/L, where U is a flow velocity scale and L is a spatial scale characterizing the solid. With the definitions (5.212) and (5.413), we then get

$$\begin{aligned}\frac{\left(r_{ab}^{m+1}\right)^2-\left(r_{ab}^{m}\right)^2}{\left(r_{ab}^{m}\right)^2} &= \left(\frac{\boldsymbol{\Omega}^{m+1}\times \mathbf{r}_{ab}^{m}\delta t}{r_{ab}^{m}}\right)^2 \\ &\sim (\Omega\delta t)^2 \\ &\sim \left(\frac{h\Omega}{c_0}\right)^2 \\ &\sim \left(\frac{h}{10L}\right)^2\end{aligned} \tag{6.194}$$

which is typically on the order of 10^{-5}. The error, however, will produce significant effects after 10^5 iterations. Such a scheme as the Störmer–Verlet scheme may be utilized to reduce that error (refer to Section 5.4.2). The error upon each iteration is then in δt^3 (Kajtar and Monaghan, 2008). A more rigorous solution consists in applying the fourth line of the system (6.190) (left unused so far), in order to update the components of the vector $\boldsymbol{\theta}$. The solid can then be reconstructed from its initial configuration in the new coordinate system with axes $\left(\theta_1^{m+1},\theta_2^{m+1},\theta_3^{m+1}\right)$. To that purpose, it is necessary to resume the discussion conducted in Section 1.5.1, just before equation (1.190), that is to subject the solid, in its inertial frame of reference, to a rotation by an angle $\left|\boldsymbol{\theta}^{m+1}-\boldsymbol{\theta}^{m}\right| = \left|\boldsymbol{\Omega}^{m+1}\right|\delta t$ about the axis directed by $\boldsymbol{\Omega}^{m+1}$. This process strictly requires the use of a transformation the matrix of which is provided by (1.136). In the limit of a very short time step, the approximation (1.138) then gives

$$\mathbf{r}_a'^{m+1} \approx \mathbf{r}_a'^{m} + \boldsymbol{\Omega}^{m+1}\times \mathbf{r}_a'^{m}\delta t \tag{6.195}$$

which corresponds to a discrete form of (1.190) over time. Equation (6.195) could be readily found again by means of the scheme (6.191). Thus, the latter is, indeed, nothing but an approximation of an exact motion of a rigid body.

If the particles making up the solid are considered as edge and dummy particles, in the sense as introduced in the previous section, they will impart to the fluid particles forces which will be symmetric to those they receive. Thus, the conservation of the total linear and angular momenta in the system consisting of the fluid and the solid is ensured. Each of these two subsystems is not isolated, of course, and then does not obey these conservation laws, since linear and angular momentum fluxes exist between them, as given by equation (6.187). There are further methods for treating the effect of the solid on the liquid while satisfying these conservation laws (for instance, refer to Kajtar and Monaghan, 2008).

In all cases, the normal to the wall $\mathbf{n}_b$ for each edge particle b should follow the rotational motion of the solid. Equation (3.146) then yields

$$\frac{d\mathbf{n}_b}{dt} = \boldsymbol{\Omega}\times\mathbf{n}_b \tag{6.196}$$

That equation may be integrated into a scheme identical to that of the solid motion (6.191):

$$\forall b,\quad \mathbf{n}_b^{m+1} = \mathbf{n}_b^{m} + \boldsymbol{\Omega}^{m+1}\times\mathbf{n}_b^{m} \tag{6.197}$$

The presently disclosed model (eqns (6.188) to (6.191)) is illustrated by a practical application in Section 8.4.2.

The case of several solids is treated likewise, with an additional difficulty, since the inter-solid contacts have to be handled. To that purpose, the methods for calculating the wall forces as mentioned in Section 6.3.2 can be matched, but there are dedicated approaches. In that respect, methods of the 'contact algorithms' type, which too are based on a variational approach, are worth mentioning. One may refer, for example, to Campbell et al. (2000). Kajtar and Monaghan (2008) treat yet another problem of the same kind, namely several solid bodies interconnected by joints.

6.4 Discrete turbulence: basic models

To start with turbulence modelling within the SPH method, we begin by writting the Reynolds-averaged momentum equation in a discrete form, then we analyse SPH forms of the first-order turbulence closure models: mixing length, one-equation and two-equation models, up to explicit algebraic models. Eventually, the $k-\varepsilon$ model is studied from a numerical stability viewpoint.

6.4.1 Discrete Reynolds equations

Turbulence can be modelled in many ways through SPH. In the next sections, we will use the discussions of Chapter 4 to disclose some of them, keeping the near-incompressibility hypothesis in mind.

Wishing to begin with the simplest models, we may introduce for each particle a statistical averaging operator identical to that in Section 4.3.1, and work in Reynolds averages. Before writing the basic equations of this approach, we must question whether the notion of mean Lagrangian derivative as defined by (4.40) is actually justified. What is meant, indeed, by the motion of a particle 'following its mean velocity $\overline{\mathbf{u}}_a$'? This is a statistical notion which makes some sense in a continuous formalism, but which could be very puzzling within the context of a discrete Lagrangian method. To clarify this point, let us consider the behaviour of a particle as a macroscopic fluid mass over time: if it is roughly likened to an $n-$sphere[37] of diameter δr, Fig. 6.4 (*b*) shows that the turbulent structures of the same order of magnitude as δr, by exerting a space variable Lagrangian motion, will take part in the deformation of that sphere. After a given period of time on the order of the time scale characterizing the space scales, that is $\left(\delta r^2/\varepsilon\right)^{1/3}$ according to equation (4.120), the particle will be too strained to be possibly likened to a sphere. After an even longer lapse of time, it will even be spread over a large part of the fluid domain,[38] preventing us considering any longer a discrete model representing the flow.

In order to get through that issue, one has to imagine that the particles are regularly reconstructed (theoretically at each time step), in the same way as in Fig. 6.4 (*c*). In this formalism, the particles then regularly exchange mass (while keeping constant individual masses), as well as linear momentum and energy. This interparticular transfer of information is illustrated by fluxes, which exactly happen to be the turbulent fluxes as discussed in Section 4.3.2.

[37] Most of the graphic representations illustrating the SPH method utilize this convienient trick. It should be recalled, however, that it is a matter of intellectual comfort, since, from a mechanical point of view, a particle is likened to a material point. Since this point carries a finite mass, however, it represents a volume of fluid which spreads through space, but in an indefinite way.

[38] In the same way as the dispersal of information for a strange attractor, according to the notions introduced in Section 4.2.2. This is a manifestation of the mixing process which we extensively discussed in Chapter 4.

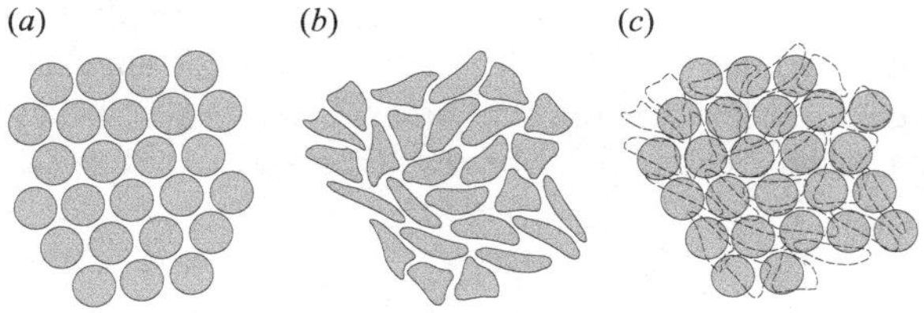

Fig. 6.4 Deformation ((*a*) and (*b*)) and reconstruction (*c*) of the volumes representing the particles, under the effect of stretching.

The notion of particle eddy viscosity then comes into its own as a model enabling us to redistribute the mean quantities for modelling the exchanges having given rise to scales which are not solved by the discretization being chosen.[39] This is the context in which the Lagrangian derivatives 'by following the mean motion' D/Dt should be understood. This is also true for all the types of derivatives contemplated throughout Chapter 4 (i.e. $\tilde{D}/Dt$ and $\breve{D}/Dt$), which we will use here too. It shall be kept in mind, however, that in practice these notations stand for the same thing as d/dt for the establishment of the numerical schemes[40] in Section 5.4. It should be mentioned, however, that this does not need any particular procedure from the programmer.

Thus, mean quantities $\overline{\mathbf{u}}_a$, $\overline{p}_a$, and so on, as well as an eddy viscosity $\nu_{T,a}$ will be considered for each particle a. In addition, an operator of a Lagrangian derivative 'by following the mean motion', identical to (4.40), is defined:

$$\frac{DA_a}{Dt} \doteqdot \frac{\partial A_a}{\partial t} + (\mathbf{grad}A)_a \cdot \overline{\mathbf{u}}_a \tag{6.198}$$

The numerical schemes remain unchanged as compared with the discussion in Section 5.4. The equations of continuity (4.39) and motion (4.82) may also be discretized like the equations with respect to the instantaneous fields, for example with such a system as (5.241), provided with viscous forces analogous to those in Section 6.2. We must remember, however, that the eddy viscosity is not constant in space (nor in time), which forces us to resort to the model (6.59) for treating the viscous term, hence losing the conservation of angular momentum. The system of discrete equations governing the evolution of both mean velocity and density of particles then takes the following form:

$$\forall a,\ \frac{D\rho_a}{Dt} = \sum_b m_b \overline{\mathbf{u}}_{ab} \cdot w'_{ab}\mathbf{e}_{ab} \tag{6.199}$$

for the continuity, and

$$\forall a,\ \frac{D\overline{\mathbf{u}}_a}{Dt} \tag{6.200}$$

$$= -\sum_b m_b \left[\begin{array}{c} \left(\dfrac{\overline{p}_a}{\rho_a^2} + \dfrac{\overline{p}_b}{\rho_b^2}\right)\mathbf{e}_{ab} \\ -\dfrac{\mu_{m,a}+\mu_{m,b}}{2\rho_a\rho_b r_{ab}}\left[(n+2)\left(\overline{\mathbf{u}}_{ab}\cdot\mathbf{e}_{ab}\right)\mathbf{e}_{ab} + \overline{\mathbf{u}}_{ab}\right] \end{array} \right] w'_{ab} + \mathbf{g}$$

for the momentum. The dynamic eddy viscosity of each particle is defined according to the formulas (4.83) and (4.74):

$$\mu_{m,a} \doteqdot \rho_a\left(\nu_a + \nu_{T,a}\right) \tag{6.201}$$

[39] It should be pointed out that what has just been stated is analogous to the molecular viscosity process which, from a Boltzmann statistics point of view, represents the exchanges of mass, momentum and energy between mesoscopic particles in the continuous medium theory (refer to Chapter 3, Section 3.5.2). We may recall, however, that it is a very unsophisticated model within the context of turbulence.

[40] The arguments in the next sections are obviously not linked to the retained option for the treatment of (incompressible or nearly incompressible) pressure.

It is always possible to get back to a form identical to (6.58) for the viscous term, but with a variable viscosity. The conservation of angular momentum is then recovered, but the resulting accuracy is no longer strictly consistent with the arguments in Section 6.2. The resulting equation of motion is then as follows:

$$\forall a, \quad \frac{D\overline{\mathbf{u}}_a}{Dt} = -\sum_b m_b \begin{bmatrix} \dfrac{\overline{p}_a}{\rho_a^2} + \dfrac{\overline{p}_b}{\rho_b^2} \\ -(n+2)\left(\mu_{m,a} + \mu_{m,b}\right) \dfrac{\overline{\mathbf{u}}_{ab} \cdot \mathbf{e}_{ab}}{\rho_a \rho_b r_{ab}} \end{bmatrix} w'_{ab} \mathbf{e}_{ab} + \mathbf{g} \tag{6.202}$$

Lastly, with a viscosity of the (6.57) type, we will get

$$\forall a, \quad \frac{d\mathbf{u}_a}{dt} = -\frac{1}{\rho_a} \sum_b V_b \left[\left(\overline{p}_a + \overline{p}_b\right) \mathbf{e}_{ab} - \left(\mu_{m,a} + \mu_{m,b}\right) \frac{\overline{\mathbf{u}}_{ab}}{r_{ab}} \right] w'_{ab} + \mathbf{g} \tag{6.203}$$

sacrificing both rigour and conservation of total momentum in favour of an even greater simplicity.

Since the pressure is averaged here, it will be suitably calculated using mean quantities. With the incompressible algorithm as described in Section 6.2.5, the mean velocity field resulting from the prediction step (first line in eqn (6.104)) should, indeed, be used for calculating the right-hand side of the discrete Poisson's equation (second line in (6.104)). Thus, the scheme (6.104) is formally unchanged, all the quantities being considered as averages, and the dissipative forces taking into account the eddy viscosity according to the just described model. In the weakly compressible case, the equation of state (5.207) is left unchanged, the mean pressure being related to the mean density, which is the same thing as ρ by virtue of the comments made in Section 4.3.2:

$$\overline{p}_a = \frac{\rho_0 c_0^2}{\gamma} \left[\left(\frac{\rho_a}{\rho_0} \right)^\gamma + 1 \right] \tag{6.204}$$

Let us now turn to the closure models for eddy viscosity. It is well-known that the simplest model is based on the notion of mixing length (Section 4.4.4), which varies in both space and time and then probably varies from one particle to the other. We will write, in accordance with (4.140):

$$\nu_{T,a} = L_{m,a}^2 S_a \tag{6.205}$$

with, according to equation (4.98), a specific mean rate-of-strain for each particle being defined by

$$S_a = \sqrt{2\mathbf{S}_a : \mathbf{S}_a} \tag{6.206}$$

where $\mathbf{S}_a$ is the mean rate-of-strain tensor of particle a, being defined by equations identical to (4.61) and (4.72):

$$\mathbf{S}_a = \frac{1}{2} \left[(\mathbf{grad}\,\overline{\mathbf{u}})_a + (\mathbf{grad}\,\overline{\mathbf{u}})_a^T \right] \tag{6.207}$$

For explicitly calculating S_a, a simple method consists of estimating the tensor $\mathbf{S}_a$, for example on the basis of an approximation analogous to (6.100). Equation (6.206) can subsequently be directly used (Violeau et al., 2001). In the next section, however, we will explain that another method is sometimes preferably used.

Velocity is subject to a wall boundary condition as given by formula (4.189) in Chapter 4, that is for each edge particle b:

$$\overline{\mathbf{u}}_b = u_{*,b}\left(\frac{1}{\kappa}\ln\frac{\delta}{k_{s,b}} + C_*\right)\mathbf{t}_b \tag{6.208}$$

where $u_{*,b}$, $k_{s,b}$ and $\mathbf{t}_b$ denote the (algebraic) shear velocity, the wall roughness and the wall tangential vector, at the point occupied by the particle b (Fig. 6.3), respectively. As to $u_{*,b}$, it is calculated by the formula (4.190). The estimation of velocity $\overline{\mathbf{u}}\left(z = \delta'\right)$ at the distance δ' from the wall can be made by means of a discrete interpolation analogous to (5.89).

Lastly, it should be said that the viscous condition (6.91) as regards the time step is rearranged as follows in a turbulent regime:

$$\delta t \leqslant C_{\delta t,2}\frac{h^2}{\max\limits_a \nu_{T,a}} \tag{6.209}$$

With the model (6.205), and estimating the strain rate by the order of magnitude $S_a \sim U_{\max}/L$ (where L is a characteristic mixing length for the flow), we can write the ratio of the CFL condition (5.413) to the viscous condition as

$$\frac{C_{\delta t,2}}{C_{\delta t,1}}\frac{c_0 h}{\max\limits_a\left(L_{m,a}^2 S_a\right)} \sim \frac{10C_{\delta t,2}}{C_{\delta t,1}}\left(\frac{L}{L_m}\right)^2\frac{h}{L} \tag{6.210}$$

As compared with the applicable relation (6.92) in a laminar regime, we can see that this ratio does not depend on the Reynolds number, in accordance with the principles established in Chapter 4. With the estimation $L_m/L \sim 1/10$ (refer to Section 4.4.4) and a reasonable spatial discretization, the viscous condition now prevails. This result is the consequence of the fact that the eddy viscosity is significantly greater than the molecular viscosity, as is well-known.

The mixing length model is a so-called 'zero equation' turbulence model, that is calculating the eddy viscosity without solving any additional differential equation. Other models from the same category have been tested with the SPH method. In particular, refer to López et al. (2009a), who propose a model in which the eddy viscosity is based on the vorticity of the mean flow (4.229). This kind of model, which was not mentioned in Chapter 4, however, produces good results in the specific case disclosed by these authors (hydraulic jump).

6.4.2 One- and two-equation discrete models

The models with either one or two equations as set out in Section 4.4.4 can readily be implemented within the frame of the SPH method. Equations (6.199), (6.200) and (6.204) persist, the eddy viscosity being calculated otherwise. Thus, the equation governing the evolution of the turbulent kinetic energy (4.145) can be discretized along the lines of the generic equation (6.94) and yields:

$$\forall a, \quad \frac{Dk_a}{Dt} = P_a - \varepsilon_a + \sum_b m_b \frac{\left(\mu_{k,a} + \mu_{k,b}\right) k_{ab}}{\rho_a \rho_b r_{ab}} w'_{ab} \tag{6.211}$$

with, according to (4.144) and (6.95):

$$\begin{aligned} \mu_{k,a} &\doteq \rho_a \left(\nu_a + \frac{\nu_{T,a}}{\sigma_k} \right) \\ k_{ab} &\doteq k_a - k_b \end{aligned} \tag{6.212}$$

(Violeau, 2004; Violeau and Issa, 2007a). The production of energy is constructed along the lines of (4.97):

$$P_a = \nu_{T,a} S_a^2 \tag{6.213}$$

unless one prefers the semi-linear model (4.158) which, with (4.117), is written as

$$P_a = \min\left(1, \sqrt{C_\mu} \frac{k_a}{\varepsilon_a} S_a\right) \sqrt{C_\mu} k_a S_a \tag{6.214}$$

Both the eddy viscosity and dissipation rate of each particle can be estimated by the models (4.147) and (4.146):

$$\begin{aligned} \nu_{T,a} &= C_\mu^{1/4} k_a^{1/2} L_{m,a} \\ \varepsilon_a &= C_\mu^{3/4} \frac{k_a^{3/2}}{L_{m,a}} \end{aligned} \tag{6.215}$$

When it is desired to ignore such data as the mixing lengths $L_{m,a}$, which are empirical or even unknown, one may adopt the $k - \varepsilon$ model, discretizing the dissipation equation (4.148) in the same form as (6.211) (Violeau and Issa, 2007a):

$$\forall a, \quad \frac{D\varepsilon_a}{Dt} = \frac{\varepsilon_a}{k_a} \left(C_{\varepsilon 1} P_a - C_{\varepsilon 2} \varepsilon_a\right) + \sum_b m_b \frac{\left(\mu_{\varepsilon,a} + \mu_{\varepsilon,b}\right) \varepsilon_{ab}}{\rho_a \rho_b r_{ab}} w'_{ab} \tag{6.216}$$

with, according to (4.149) and (6.95):

$$\begin{aligned} \mu_{\varepsilon,a} &\doteq \rho_a \left(\nu_a + \frac{\nu_{T,a}}{\sigma_\varepsilon} \right) \\ \varepsilon_{ab} &\doteq \varepsilon_a - \varepsilon_b \end{aligned} \tag{6.217}$$

In this case, we may recall that the eddy viscosity is given by (4.137):

$$\nu_{T,a} = C_\mu \frac{k_a^{3/2}}{\varepsilon_a} \tag{6.218}$$

The definition (6.201) is then used for solving the equation of motion (6.200). The model's constants are summarized in Table 4.1, Section 4.4.4 (Chapter 4). The boundary conditions are provided by the formula (4.189). They are applicable to each edge particle b (refer to Section 6.3.2):

$$k_b = \frac{u_{*,b}^2}{\sqrt{C_\mu}}$$
$$\varepsilon_b = \frac{\left|u_{*,b}\right|^3}{\kappa\delta} \tag{6.219}$$

The model proposed in Section 6.3.2 based on the γ_a correction, however, could be used to prescribe turbulent boundary conditions in a better way (Ferrand et al., 2011, 2012; Leroy et al., 2014).

As regards the velocity boundary condition at a wall, it is unchanged as compared with (6.208). As to the treatment of the free surfaces, the condition (4.191) with respect to k can hardly be matched to a discrete context. It is frequently preferred to not prescribe any condition. On the other hand, the condition (4.192) with respect to ε is easily applicable provided that those particles lying at the surface can be identified (refer to Section 6.2.5), and that the water depth can be defined without ambiguity.

A remark should be made here about the calculation of the strain rate, which is necessary for calculating the rate of production of turbulent kinetic energy P_a. We may recall that, in the continuous theory of Section 4.4.3, the kinetic energy lost by the mean velocity field is fully recovered by the turbulent eddies. On the basis of that assumption, we have got the global dissipation equation (4.100) which gives the energy lost in the whole fluid. Using the elementary discrete integration (5.84) and the definition (6.201), we can approximate that global dissipation rate through the following discrete sum:

$$F \approx \sum_a V_a \mu_{m,a} S_a^2 \tag{6.220}$$

which means that a mean field kinetic energy dissipation as given by

$$\begin{aligned} F_a &\approx V_a \mu_{m,a} S_a^2 \\ &\approx m_a P_a \end{aligned} \tag{6.221}$$

takes place in each particle in accordance with (4.99). Now, the discussion in Section 6.2.2 provides us with a discrete form of F_a. With the presently adopted model for the viscous forces, Equation (6.56) can be generalized to a mean field which is subject to an eddy viscosity:

$$F_a = -\frac{1}{4}\sum_b V_a V_b \frac{\mu_{m,a} + \mu_{m,b}}{r_{ab}} \left[(n+2)\left(\mathbf{e}_{ab} \cdot \bar{\mathbf{u}}_{ab}\right)^2 + \bar{u}_{ab}^2\right] w'_{ab} \tag{6.222}$$

Identifying (6.220) and (6.222), we ultimately find the following approximation:

$$S_a^2 \approx -\frac{1}{4\mu_{m,a}}\sum_b V_b \frac{\mu_{m,a} + \mu_{m,b}}{r_{ab}} \left[(n+2)\left(\mathbf{e}_{ab} \cdot \bar{\mathbf{u}}_{ab}\right)^2 + \bar{u}_{ab}^2\right] w'_{ab} \tag{6.223}$$

This formula gives the value which the mean rate-of-scale S_a of each particle should have in order for the dissipation of energy as modelled by the eddy viscosity (i.e. $V_a \mu_{m,a} S_a^2$) to be compatible with the adopted model for discretizing the viscous forces.[41] It is a priori advisable to use (6.223) for

[41] In Section 6.2.4, we already have taken the same precaution for internal energy, which recovers the whole macroscopic kinetic energy being lost through viscous friction. Here, however, it should be pointed out that this approach only works, as it is disclosed, with the quadratic model (6.213) for the production.

calculating S_a, instead of the method being suggested in the previous section and given by equations (6.100) and (6.206). The presently described method may also be used within the context of the mixing length model and, besides, advantageously resorts to a single discrete sum instead of the n^2 (= 4 ou 9) sums which are required for calculating the components of the $\mathbf{S}_a$ tensor with (6.207). In addition, we may note that (6.223) always defines a real S_a, thanks to the positiveness of F (refer to Section 6.2.2). Lastly, for the sake of simplicity, the model (6.54) may be utilized for calculating F_a, on the basis of a constant eddy viscosity, even though it is merely an approximation of reality. The following, less calculation-intensive model would then be achieved:

$$S_a^2 = -\frac{1}{2}\sum_b \frac{V_b}{r_{ab}} \left[(n+2)\left(\mathbf{e}_{ab} \cdot \overline{\mathbf{u}}_{ab}\right)^2 + \overline{u}_{ab}^2 \right] w'_{ab} \tag{6.224}$$

There is convincing evidence that the other standard two-equation models, such as the model $k - \omega$ (refer to Section 4.4.4) are suitable for a discrete approach equivalent to what has just been explained. Thus, a discrete form of equation (4.153) is yielded by

$$\forall a, \quad \frac{D\omega_a}{Dt} = C_{\omega 1} \frac{\omega_a}{k_a} P_a - C_{\omega 2} \omega_a^2 + \sum_b m_b \frac{\left(\mu_{\omega,a} + \mu_{\omega,b}\right) \omega_{ab}}{\rho_a \rho_b r_{ab}} w'_{ab} \tag{6.225}$$

(Issa et al., 2009a) with obvious notations, referring to the formulas (4.151) and (4.154):

$$\begin{aligned} \mu_{\omega,a} &\doteqdot \rho_a \left(\nu_a + \frac{\nu_{T,a}}{\sigma_\omega} \right) \\ \omega_{ab} &\doteqdot \omega_a - \omega_b \end{aligned} \tag{6.226}$$

The algebraic explicit model for the Reynolds stresses (non-linear $k - \varepsilon$ model), which was set out in Section 4.5.3, may also fit the SPH method. We may recall it is restricted to a class of flows which are invariant in one direction. It shall therefore only be used in $n = 2$ dimensions. This model amounts to contemplating a viscosity tensor as given by the formula (4.253). In order to match it to the SPH formalism, we may note that, in two dimensions, the second relation in equation (4.227) may be rearranged as

$$\begin{aligned} \mathbf{\Omega} &= \Omega_{xz} \begin{pmatrix} 0 & 1 \\ -1 & 0 \end{pmatrix} \\ &= -\tilde{\Omega} \boldsymbol{\omega}_2 \end{aligned} \tag{6.227}$$

where the unitary symplectic matrix $\boldsymbol{\omega}_2$ is given by (1.93), whereas $\tilde{\Omega}$ is simply given by

$$\tilde{\Omega} \doteqdot \Omega_{xz} \tag{6.228}$$

Using (4.142), (4.253) may then be rearranged, for each particle a:

$$\begin{aligned}\mu_{T,a} &= \rho_a C_\mu \frac{k_a^2}{\varepsilon_a}\left(\mathbf{I}_2 - 2B_2\frac{k_a\tilde{\Omega}_a}{\varepsilon_a}\boldsymbol{\omega}_2\right)\\ &= \mu_{m,a}\left(\mathbf{I}_2 - 2B_2\tilde{\Omega}_a^*\boldsymbol{\omega}_2\right)\end{aligned} \tag{6.229}$$

with,[42] in the same way as (4.236):

$$\tilde{\Omega}_a^* = \frac{k_a\tilde{\Omega}_a}{\varepsilon_a} \tag{6.230}$$

[42] We may notice that $\mu_{T,a}$ has been likened here to $\mu_{m,a}$, ignoring the effects of the molecular viscosity, in accordance with Chapter 4.

The equation of motion (6.203) is now written (Violeau and Issa, 2007a; 2007b) as:

$$\forall a,\ \frac{D\mathbf{u}_a}{Dt} = -\frac{1}{\rho_a}\sum_b V_b\left[\begin{array}{c}\left(\overline{p}_a+\overline{p}_b\right)\mathbf{e}_{ab} - \dfrac{\mu_{m,a}+\mu_{m,b}}{r_{ab}}\overline{\mathbf{u}}_{ab}\\ +2B_2\dfrac{\mu_{m,a}\tilde{\Omega}_a^*+\mu_{m,b}\tilde{\Omega}_b^*}{r_{ab}}\boldsymbol{\omega}_2\overline{\mathbf{u}}_{ab}\end{array}\right]w'_{ab}+\mathbf{g} \tag{6.231}$$

It may seem strange that (6.203), rather than (6.200) or (6.202), was utilized here for the modelling of the viscous forces, since Section 6.2 has taught us that this form has flaws. That option, however, has a major advantage which can be highlighted by writing the kinetic matrix corresponding to the just introduced discrete viscous forces (refer to Chapter 2, Section 2.4.3):

$$\boldsymbol{\alpha}_{ab} = -\frac{V_aV_b}{r_{ab}}w'_{ab}\left[\left(\mu_{m,a}+\mu_{m,b}\right)\mathbf{I}_2 - 2B_2\left(\mu_{m,a}\tilde{\Omega}_a^*+\mu_{m,b}\tilde{\Omega}_b^*\right)\boldsymbol{\omega}_2\right] \tag{6.232}$$

(to be compared with (6.49)). These coefficients partly satisfy the Onsager symmetry principle (refer to Chapter 2, Section 2.4.3, eqn (2.181)), since $\boldsymbol{\alpha}_{ab} = \boldsymbol{\alpha}_{ba}$. On the other hand, they do not satisfy the condition $\boldsymbol{\alpha}_{ab}^T = \boldsymbol{\alpha}_{ab}$. Furthermore, they are not written in the requisite form for the conservation of angular momentum, that is $\boldsymbol{\alpha}_{ab} = \gamma_{ab}\mathbf{r}_{ab}\otimes\mathbf{r}_{ab}$ (eqn (2.204)).[43] However, the associated dissipation of mean kinetic energy, as given by formula (2.185), is written in a simple form:

[43] From a theoretical point of view, we could palliate that drawback by integrating the anti-symmetric part of $\boldsymbol{\alpha}_{ab}$ (i.e. the non-linear viscous term in $\boldsymbol{\omega}_2$) into the Hamiltonian matrix $\boldsymbol{\Omega}_{2N}$ in eqn (2.207). This idea is consistent with the fact that this term does not dissipate energy, as evidenced by the following equation.

$$\begin{aligned}F &= -\frac{1}{2}\sum_{a,b}\frac{V_aV_b}{r_{ab}}\\ &\times\left[\begin{array}{c}\left(\mu_{m,a}+\mu_{m,b}\right)\overline{\mathbf{u}}_{ab}\cdot\overline{\mathbf{u}}_{ab}\\ -2B_2\left(\mu_{m,a}\tilde{\Omega}_a^*+\mu_{m,b}\tilde{\Omega}_b^*\right)\overline{\mathbf{u}}_{ab}\cdot\left(\boldsymbol{\omega}_2\overline{\mathbf{u}}_{ab}\right)\end{array}\right]w'_{ab}\\ &= -\frac{1}{2}\sum_{a,b}\frac{V_aV_b}{r_{ab}}\left(\mu_{m,a}+\mu_{m,b}\right)\overline{u}_{ab}^2w'_{ab}\end{aligned} \tag{6.233}$$

This result is due to the fact that $\boldsymbol{\omega}_2\overline{\mathbf{u}}_{ab}$ is orthogonal to $\overline{\mathbf{u}}_{ab}$. Equation (6.233), to be compared with (6.53), provides an always positive dissipation (we may recall that the derivative of the kernel w'_{ab} is negative), which is tantamount to asserting the definite positiveness of the kinetic matrices $\boldsymbol{\alpha}_{ab}$

as provided by (6.232). This result is important in itself, because it shows that the chosen model (6.203) for the viscous forces satisfies the principle of entropy increase (refer to Chapter 2), whereas the models (6.200) and (6.202) would not satisfy that property with the non-linear eddy viscosity model we have just described. With (6.202), for example, a term proportional to $(\mathbf{e}_{ab} \cdot \overline{\mathbf{u}}_{ab})(\boldsymbol{\omega}_2\mathbf{e}_{ab} \cdot \overline{\mathbf{u}}_{ab})$, of undetermined sign, would subsist in (6.233). Furthermore, this option shows that the term in $\boldsymbol{\omega}_2$ in the model (6.229) plays no part in the dissipation of energy, which is consistent with the calculations made in Section 4.5.3 within the context of the continuous fluids (refer to the calculation (4.254)).

The discrete algebraic explicit model we have just disclosed is unsurprinsingly accompanied by equations (6.201), (6.211), (6.212), (6.216), (6.217) and (6.218) of the discrete model $k - \varepsilon$. We may recall, however, that the production can no longer be calculated by means of the basic models (6.213) or (6.214). Section 4.5.3, indeed, provides the more precise model (4.256)–(4.257), which here should be used as is for calculating P^*. The production is subsequently calculated by means of the definition (4.242), that is

$$P_a = P_a^* \varepsilon_a \tag{6.234}$$

Equation (4.257) requires the values of the mean rate-of-strain S_a^* and mean scalar vorticity Ω_a^* of each particle, which are related to S_a and Ω_a via equation (4.236). The model (6.224) may no longer be kept for calculating S_a, since (6.213) is invalidated; it is then necessary to return to the calculation method as defined in the previous section (eqns (6.206) and (6.207)). Likewise, we have

$$\Omega_{xz,a} = \frac{1}{2}\left[\left(\frac{\partial \overline{u}_x}{\partial z}\right)_a - \left(\frac{\partial \overline{u}_z}{\partial x}\right)_a\right] \tag{6.235}$$

which is calculated by means of (6.101). That yields Ω_a thanks to the definition (4.229) (we may recall that we are in two dimensions here):

$$\Omega_a = 2\left|\Omega_{xz,a}\right| \tag{6.236}$$

$\tilde{\Omega}_a^*$ is subsequently calculated through (6.230), with $\tilde{\Omega}_a = \Omega_{xz,a}$, to be used in equation (6.231).

Applications of the discrete turbulence models which have been disclosed so far can be found throughout Chapters 7 and 8 (in particular, refer to Section 7.2.3, 7.3.2, 7.4.3, 8.3 and 8.4.2). Compared with the mixing length model as set out in the previous paragraph, the discrete standard $k - \varepsilon$ model provides better results in terms of velocity profiles, as explained by De Padova et al. (2010) about the case of a hydraulic jump.

6.4.3 Stability of the two-equation turbulent models

The numerical solution of the differential equations (6.211) and (6.216) requires that a numerical scheme be selected. As with the solution of the equations of motion (refer to Section 5.4), a first-order scheme, either explicit or implicit, may be chosen. Denoting the diffusion terms as D_k and D_ε, the first one gives

$$\forall a,\ k_a^{m+1} = k_a^m + \left[\left(P_a + D_{k,a}\right)^m - \varepsilon_a^m\right]\delta t$$
$$\varepsilon_a^{m+1} = \varepsilon_a^m + \frac{\varepsilon_a^m}{k_a^m}\left[\left(C_{\varepsilon 1}P_a + D_{\varepsilon,a}\right)^m - C_{\varepsilon 2}\varepsilon_a^m\right]\delta t \tag{6.237}$$

Here, however, the choice of a possible partial implicitation is not guided by such conservation reasons as in Section 5.4.2, but by the fact that the quantities k and ε should always be positive. Now, the dissipation terms within these equations may give rise, in particular if the time step is insufficiently small, to numerical values which are either locally and/or temporarily negative. In order to prevent that artifact, a scheme provided with partly implicit dissipative terms may be selected. The following will then be written:

$$\forall a,\ k_a^{m+1} = k_a^m + \left[\left(P_a + D_{k,a}\right)^m - \frac{\varepsilon_a^m}{k_a^m}k_a^{m+1}\right]\delta t$$
$$\varepsilon_a^{m+1} = \varepsilon_a^m + \left[\frac{\varepsilon_a^m}{k_a^m}\left(C_{\varepsilon 1}P_a + D_{\varepsilon,a}\right)^m - C_{\varepsilon 2}\frac{\varepsilon_a^m}{k_a^m}\varepsilon_a^{m+1}\right]\delta t \tag{6.238}$$

we have expressed in that formula, as a factor of the explicit terms k_a^{m+1} and ε_a^{m+1}, the characteristic time of the large eddies at time m, that is $\tau_a^m \doteqdot k_a^m/\varepsilon_a^m$ according to equation (4.142). That system may then be rearranged in an explicit form to yield

$$\forall a,\ k_a^{m+1} = \frac{k_a^m + \left(P_a + D_{k,a}\right)^m \delta t}{1 + \frac{\delta t}{\tau_a^m}}$$
$$\varepsilon_a^{m+1} = \frac{\varepsilon_a^m + \left(C_{\varepsilon 1}P_a + D_{\varepsilon,a}\right)^m \frac{\delta t}{\tau_a^m}}{1 + C_{\varepsilon 2}\frac{\delta t}{\tau_a^m}} \tag{6.239}$$

Except for the diffusion terms, whose signs are a priori unknown, the right-hand sides in that system only comprise positive terms, unlike (6.237). Thus, the dissipation terms cannot generate artificial negative values, which ensures the positiveness of the quantities being sought for a turbulence in equilibrium. These terms, however, duly take part in the decrease of k and ε, because the denominators in (6.239) are strictly greater than 1. Besides, a first-order expansion of (6.239) into δt would yield (6.237) again.

We will briefly discuss the numerical stability[44] of the $k-\varepsilon$ model on the basis of the method dealt with in Sections 5.4.4 and 6.2.3, starting from the explicit scheme (6.238). A thorough study of that stability would prove rather difficult, because it would involve simultaneous treatments of the equations of motion, the continuity equation and both turbulent model equations. In order to make our work easier, we will then assume, as in Sections 5.4.4 and 6.2.3, that the possible numerical instability evolves from a homogeneous situation, which corresponds to constant initial values k_0, ε_0 and $\nu_{T,0}$, allowing us to uncouple the equations with respect to k_0 and ε_0 from the others have which been already treated. Besides, this hypothesis leads to zero velocity gradients, hence to a zero production of turbulent kinetic energy. This working environment is obviously favourable to the stability of the solutions, and consequently

[44] To tell the truth, the following is not related to the chosen numerical method for treating that system, except for the diffusion terms, which nevertheless will disappear in this analysis.

we only will get a necessary condition for stability. Under these hypotheses, the variation rates of both state parameters k and ε are written, by virtue of (6.211) and (6.216), as:

$$\dot{k}_a = -\varepsilon_a + \sum_b m_b \frac{\left(\mu_{k,a} + \mu_{k,b}\right) k_{ab}}{\rho_a \rho_b r_{ab}} w'_{ab}$$
$$\dot{\varepsilon}_a = -C_{\varepsilon 2} \frac{\varepsilon_a^2}{k_a} + \sum_b m_b \frac{\left(\mu_{\varepsilon,a} + \mu_{\varepsilon,b}\right) \varepsilon_{ab}}{\rho_a \rho_b r_{ab}} w'_{ab} \tag{6.240}$$

Next, we ignore the molecular viscosity in favour of the eddy viscosity, which, according to (6.212) and (6.217), is tantamount to likening μ_k to $\rho_0 \nu_{T,0}/\sigma_k$ and μ_ε to $\rho_0 \nu_{T,0}/\sigma_\varepsilon$. The effect of a parameter fluctuation upon the variation rate of k is then revealed by the following fluctuations of energy:

$$\delta_k \dot{k}_a = \frac{2\nu_{T,0}}{\sigma_k \rho_0} \sum_b m_b \frac{w'_{ab}}{r_{ab}} \left(\delta k_a - \delta k_b\right)$$
$$\delta_\varepsilon \dot{k}_a = -\delta \varepsilon_a \tag{6.241}$$

(the fluctuations of position do not offer any variation of $\dot{k}_a$, if one starts from a homogeneous initial state, because of the term $k_{ab} \doteqdot k_a - k_b$). As usual, we will seek the numerical wave components in the generic form

$$\delta k = k\,(t) \exp\left(-i\mathbf{K} \cdot \mathbf{r}\right)$$
$$\delta \varepsilon = E\,(t) \exp\left(-i\mathbf{K} \cdot \mathbf{r}\right) \tag{6.242}$$

$\mathbf{K}$ being a wave vector. After introducing (6.242) into (6.241), then approximating the discrete sum through an integral as in Section 5.4.4 (for instance, refer to equation (5.375)), we get the following relations:

$$\delta_k \dot{k}_a \approx \frac{4\nu_{T,0}}{\sigma_k h^2} \left[\widehat{w_h}\,(K) - \widehat{w_h}\,(0)\right] k\,(t) \exp\left(-i\mathbf{K} \cdot \mathbf{r}_a\right)$$
$$\delta_\varepsilon \dot{k}_a = -E\,(t) \exp\left(-i\mathbf{K} \cdot \mathbf{r}_a\right) \tag{6.243}$$

The total fluctuation of turbulent energy is yielded by the sum $\delta \dot{k}_a = \delta_k \dot{k}_a + \delta_\varepsilon \dot{k}_a$. Through a similar process for the equation with respect to ε, we ultimately get the following linearized system:

$$\delta \dot{k}_a \approx -\left[\frac{A_{5,T}}{\sigma_k \delta t} k\,(t) + E\,(t)\right] \exp\left(-i\mathbf{K} \cdot \mathbf{r}_a\right)$$
$$\delta \dot{\varepsilon}_a \approx -\left[\left(\frac{A_{5,T}}{\sigma_\varepsilon \delta t} + \frac{2C_{\varepsilon 2}}{\tau_0}\right) E\,(t) - \frac{C_{\varepsilon 2}}{\tau_0^2} k\,(t)\right] \exp\left(-i\mathbf{K} \cdot \mathbf{r}_a\right) \tag{6.244}$$

where[45]

$$A_{5,T} \doteqdot \frac{4Co}{Ma_0 Re_0} \left[\widehat{w_h}\,(0) - \widehat{w_h}\,(K)\right] > 0$$
$$\tau_0 \doteqdot \frac{k_0}{\varepsilon_0} \tag{6.245}$$

[45] Here we shall take care not to confuse τ_0 with the time scale of the small eddies, a notation which was used in Chapter 3. Here the subscript $_0$ refers to the initial state, but it is the characteristic time τ of the large eddies, in accordance with (4.142).

We have defined $A_{5,T}$ by analogy with (6.72), with

$$\overline{Re_0} \doteqdot \frac{c_0 h}{\nu_{T,0}} \tag{6.246}$$

in accordance with the notations (6.73) and (4.201). We are now seeking the temporal variations in the following form:

$$\begin{aligned} k\,(t) &= K_0 \exp(i\omega t) \\ E\,(t) &= E_0 \exp(i\omega t) \end{aligned} \tag{6.247}$$

ω being a frequency associated to the numerical wave being sought. With the explicit numerical scheme[46] (6.237), the system (6.244) becomes

[46]In that respect, the semi-explicit scheme (6.238) would lead to similar results.

$$\begin{aligned} K_0\chi &= \left(1 - \frac{A_{5,T}}{\sigma_k}\right) K_0 - E_0\delta t \\ E_0\chi &= \left(1 - \frac{A_{5,T}}{\sigma_\varepsilon} - 2C_{\varepsilon 2}\frac{\delta t}{\tau_0}\right) E_0 + \frac{C_{\varepsilon 2}}{\tau_0^2} K_0 \delta t \end{aligned} \tag{6.248}$$

where χ is defined, as in Section 5.4.4, by (5.391), that is $\chi \doteqdot \exp(i\omega\delta t)$. Since this system is simpler than those disclosed in Sections 5.4.4 and 6.2.3, we can easily show how the characteristic equation results from its homogeneous linear nature. We may, indeed, rearrange (6.248) as

$$\begin{pmatrix} 1 - \dfrac{A_{5,T}}{\sigma_k} - \chi & -\delta t \\ C_{\varepsilon 2}\dfrac{\delta t}{\tau_0^2} & 1 - \dfrac{A_{5,T}}{\sigma_\varepsilon} - 2C_{\varepsilon 2}\dfrac{\delta t}{\tau_0} - \chi \end{pmatrix} \begin{pmatrix} K_0 \\ E_0 \end{pmatrix} = 0 \tag{6.249}$$

A solution for (K_0, E_0) clearly requires the nullity of the determinant in that matrix, that is

$$\begin{aligned} &\chi^2 - \left[2 - A_{5,T}\left(\frac{1}{\sigma_k} + \frac{1}{\sigma_\varepsilon}\right) - 2A_6\right]\chi \\ &+ \left(1 - \frac{A_{5,T}}{\sigma_k}\right)\left(1 - \frac{A_{5,T}}{\sigma_\varepsilon} - 2A_6\right) + \frac{A_6^2}{C_{\varepsilon 2}} = 0 \end{aligned} \tag{6.250}$$

with

$$A_6 \doteqdot C_{\varepsilon 2}\frac{\delta t}{\tau_0} > 0 \tag{6.251}$$

It is obvious, from the experience in Section 6.2.3, that the diffusion terms will introduce, for the time step, a condition analogous to the first condition in (6.209). We consequently will merely deal with the case in which the diffusion is ignored,[47] which amounts to putting $A_{5,T} = 0$:

[47]Thus, as previously stated, the following discussion becomes independent of the spatial discretization method.

$$\chi^2 - 2(1 - A_6)\chi + 1 - 2A_6 + \frac{A_6^2}{C_{\varepsilon 2}} = 0 \tag{6.252}$$

The discriminant is positive and the (real) roots are given by

$$\chi = 1 - A_6\left(1 \pm \sqrt{1 - \frac{1}{C_{\varepsilon 2}}}\right) \tag{6.253}$$

The stability condition (5.393) then prescribes

$$A_6 \leqslant \frac{2}{1 + \sqrt{1 - \frac{1}{C_{\varepsilon 2}}}} \tag{6.254}$$

The time step is ultimately conditioned by

$$\delta t \leqslant \frac{2}{C_{\varepsilon 2} + \sqrt{C_{\varepsilon 2}^2 - C_{\varepsilon 2}}}\tau_0 \approx 0.62\tau_0 \tag{6.255}$$

Back to the notation τ for the characteristic time of the large eddies in arbitrary conditions, we may then assert that the time step should satisfy

$$\delta t \leqslant C_{\delta t,5}\tau \tag{6.256}$$

It seems natural for δt to be incremented by the characteristic time of the large turbulent eddies. Unlike all the previously established numerical conditions, this is a relation with a physical rather than numerical meaning, because it does not depend on the spatial discretization nor on the speed of sound. We may compare (6.255) with the CFL condition (5.413). The characteristic time τ may, indeed, be estimated from (4.117), (4.146) and (4.187), which yields:

$$\frac{c_0\tau}{h} \sim \frac{10U_{\max}L_m}{\sqrt{C_\mu}u_* h} \sim 500\frac{L_m}{h} \tag{6.257}$$

Thus, the CFL condition prevails. We may recall (Section 6.4.1) that it is in turn governed by the turbulent viscous condition (6.209).

6.5 Discrete turbulence: advanced models

We continue our overview of SPH turbulence models with advanced turbulence modelling. It turns out that the LES and α approaches are the most natural techniques to deal with turbulent flows with SPH. In particular, it is shown how the latter is connected to the mirror systems theory presented at the end of Chapter 2. We finally introduce a stochastic approach based on a Langevin equation, from the considerations of Chapters 2 and 4.

6.5.1 Discrete LES

The SPH formalism perfectly fits the large eddy simulation (LES), which was dealt with in Section 4.6.1 within the context of the continuous fluids (Gotoh et al., 2004; Issa et al., 2005; Dalrymple and Rogers, 2006). A first indication for such an affinity can be found in the definition of the LES filtering operator (4.258). With a filter obeying the condition (4.260), it occurs in the same form as the SPH continuous interpolation (5.2), being provided with the condition (5.31):

$$\begin{aligned}\tilde{A}(\mathbf{r},t) &= \int_{\Omega} A\left(\mathbf{r}',t\right) G_{\Delta}\left(\left|\mathbf{r}-\mathbf{r}'\right|\right) d^3\mathbf{r}' \\ [A]_c(\mathbf{r},t) &= \int_{\Omega} A\left(\mathbf{r}',t\right) w_h\left(\left|\mathbf{r}-\mathbf{r}'\right|\right) d^n\mathbf{r}'\end{aligned} \tag{6.258}$$

These two formulas suggest that the kernel may be likened to a filter whose characteristic size Δ is given by the smoothing length h. This idea is confirmed by the kernel properties. First, its Fourier transform is most often decreasing (with the Wendland kernel (5.57)), like that of the gate filter whose transform is given by (4.264). Besides, the SPH continuous integration commutes with the partial derivative operators (compare (4.268) with (5.17)). Lastly, the kernel support is usually compact. The properties (4.267) and (5.14), stemming from the assumed symmetry of the filter and the kernel, may be parallelled too. It therefore seems appropriate to define a discrete approximation of the filtered quantities $\tilde{A}$ by means of the continuous approximation of $[A]_c(\mathbf{r},t)$, as given by the operator J_a (first line in eqn (5.123)):

$$\tilde{A}_a \approx J_a\{A_b\} \tag{6.259}$$

(Issa, 2005). The discrete equations (6.199) and (6.202) which we have constructed in Section 6.4.1 for treating the Reynolds equations may be transcribed, *mutatis mutandis*, for writing the discrete forms of the continuous equations (4.271) and (4.279):

$$\begin{aligned}\forall a,\quad \frac{\tilde{D}\rho_a}{Dt} &= \sum_b m_b \tilde{\mathbf{u}}_{ab}\cdot w'_{ab}\mathbf{e}_{ab} \\ \frac{\tilde{D}\tilde{\mathbf{u}}_a}{Dt} &= -\sum_b m_b \left[\begin{array}{c}\dfrac{\tilde{p}_a}{\rho_a^2}+\dfrac{\tilde{p}_b}{\rho_b^2} \\ -(n+2)\left(\mu_{M,a}+\mu_{M,b}\right)\dfrac{\tilde{\mathbf{u}}_{ab}\cdot\mathbf{e}_{ab}}{\rho_a\rho_b r_{ab}}\end{array}\right] w'_{ab}\mathbf{e}_{ab} \\ &+\mathbf{g}\end{aligned} \tag{6.260}$$

The equivalent viscosity of each particle, which is denoted as $\mu_{M,a}$, is defined as in (4.280), in reference to the notation (6.201):

$$\mu_{M,a} \doteqdot \rho_a\left(\nu_a+\nu_{S,a}\right) \tag{6.261}$$

where $\nu_{S,a}$ is the *sub-particle viscosity*[48] of the particle a. It is linked to the filter size, namely h here, by the formulas (4.277) and (4.281) (Violeau and Issa, 2007a):

$$\nu_{S,a} = \left(c_S h\right)^2 \tilde{s}_a \tag{6.262}$$

where the Smagorinsky constant c_S is given by (4.291). As to the filtered rate-of-strain, it can be calculated in several ways, as explained in Section 6.4.1 and 6.4.2. The formulation (6.224), for example, now yields

[48]Within the context of the continuous fluids, we had called it 'subgrid viscosity', in reference to the Eulerian numerical modelling, in Chapter 4.

$$\tilde{s}_a^2 = -\frac{1}{2}\sum_b \frac{V_b}{r_{ab}}\left[(n+2)\left(\mathbf{e}_{ab}\cdot\tilde{\mathbf{u}}_{ab}\right)^2 + \tilde{u}_{ab}^2\right] w'_{ab} \tag{6.263}$$

The pressure is calculated, by analogy with (6.204), through the usual equation of state:

$$\tilde{p}_a = \frac{\rho_0 c_0^2}{\gamma}\left[\left(\frac{\rho_a}{\rho_0}\right)^\gamma + C\right] \tag{6.264}$$

The incompressible model disclosed in Section 6.2.5 does not raise more difficulties in this formalism.

This model is advantageously simple, because the equations are formally unchanged as compared with the mixing length model, as was previously pointed out in Chapter 4 within the context of the continuous fluid turbulence. However, the requirements of this method are to be remembered now: the calculation should be three-dimensional and the spatial discretization $\Delta = h$ should satisfy the condition (4.275). Since the size of the large eddies is approximately one tenth of the spacial scale of the flow L (refer to Chapter 4), here we get

$$\frac{h}{L} \ll \frac{1}{10C_\mu^{3/4}} \approx 0.54 \tag{6.265}$$

This condition is usually rather easily verified. Lastly, we may recall the question of the initial conditions which were discussed at the end of Section 4.6.1; LES, indeed, requires a plausible initial turbulence state to be workable. The SPH method is then quite helpful: the Lagrangian nature of that method is sufficient for generating small numerical instabilities giving rise to turbulence in all circumstances, provided that the viscosity is not too large (refer to Section 6.2.3), which is the case with the LES approach (Issa et al., 2005). Applications of SPH LES can be found in Sections 7.2.3 and 7.3.1.

The above circumstance also argues in favour of the DNS as an approach for turbulence modelling by means of the SPH method, with all the restrictions we expressed in Chapter 4 (in particular, refer to the end of Section 4.4.6). Among the (few) attempts made to that purpose so far, it is worth mentioning Robinson (2009). This work is restricted to two-dimensional tubulence, which was not dealt with in Chapter 4. Though turbulence is definitely a mainly three-dimensional phenomenon, a two-dimensional approach is possible for studying such very flat flows as the atmosphere or the sea currents on the continental shelves. Developments analogous to those in Section 4.4.3 then highlight a different behaviour of the energy spectra, with respect to the three-dimensional case. Robinson's work shows that the SPH method can predict a satisfactory statistical and spectral behaviour within that context, although it leads to an overestimation of dissipation at the smallest scales. The transition towards turbulence (refer to Chapter 4, Section 4.2.2) was also superficially investigated through the SPH method (in particular, refer to (Jiang et al., 2006)).

6.5.2 α–XSPH model

Because of its Lagrangian nature, the SPH method is prone to disorder, as mentioned in the preceding section. In order to prevent the erratic motions of the particles from making them unfortunately prone to disorder, sometimes inducing cross-trajectories, some authors utilize the so-called *XSPH* variant (according to Monaghan, 2002). The idea consists of moving the particles no longer with the velocity $\mathbf{u}_a$, but with a modified velocity denoted as $\breve{\mathbf{u}}_a$, and defined by

$$\breve{\mathbf{u}}_a \doteq \mathbf{u}_a - \epsilon \sum_b m_b \frac{\mathbf{u}_{ab}}{\overline{\rho}_{ab}} w_{ab} \tag{6.266}$$

where ϵ is a dimensionless, positive parameter nearly equalling one, whereas $\overline{\rho}_{ab}$ $(= \overline{\rho}_{ba})$ is defined[49] as the harmonic average of the densities of particles a and b:

$$\overline{\rho}_{ab} \doteq \frac{2}{\dfrac{1}{\rho_a} + \dfrac{1}{\rho_b}} = \overline{\rho}_{ba} \tag{6.267}$$

[49]We may note that the definition of $\overline{\rho}_{ab}$ is different from the formula (6.9), contrary to what is suggested by the notation.

Thus, in this section, instead of (5.83), we have:

$$\forall a, \ \frac{d\mathbf{r}_a}{dr} = \breve{\mathbf{u}}_a \tag{6.268}$$

We may first notice that the conservations of total linear and angular momenta in an isolated system are not altered by that change. The usual properties of symmetry may be easily used, indeed, for noting that the former quantity is left unchanged by the application of the operator (6.266):

$$\begin{aligned}\breve{\mathbf{P}} &\doteq \sum_a m_a \breve{\mathbf{u}}_a \\ &= \sum_a m_a \mathbf{u}_a - \epsilon \sum_{a,b} m_a m_b \frac{\mathbf{u}_{ab}}{\overline{\rho}_{ab}} w_{ab} \\ &= \sum_a m_a \mathbf{u}_a = \mathbf{P}\end{aligned} \tag{6.269}$$

(the sum over a and b vanishes by antisymmetry). As to the angular momentum, it is therefore modified, but its conservation persists:

$$\begin{aligned}\frac{d\mathbf{M}}{dt} &= \sum_a m_a \left(\frac{d\mathbf{r}_a}{dt} \times \mathbf{u}_a + \mathbf{r}_a \times \frac{d\mathbf{u}_a}{dt} \right) \\ &= \sum_a m_a \left(\breve{\mathbf{u}}_a \times \mathbf{u}_a + \mathbf{r}_a \times \frac{d\mathbf{u}_a}{dt} \right) \\ &= -\epsilon \sum_{a,b} m_a m_b \frac{\mathbf{u}_{ab} \times \mathbf{u}_a}{\overline{\rho}_{ab}} w_{ab} + \sum_a m_a \mathbf{r}_a \times \frac{d\mathbf{u}_a}{dt} \\ &= -\epsilon \sum_{a<b} m_a m_b \frac{\mathbf{u}_{ab} \times \mathbf{u}_{ab}}{\overline{\rho}_{ab}} w_{ab} + \sum_a m_a \mathbf{r}_a \times \frac{d\mathbf{u}_a}{dt}\end{aligned} \tag{6.270}$$

The first sum in that result is zero, because each of the $\mathbf{u}_{ab} \times \mathbf{u}_{ab}$ terms is zero too; the second one is the right-hand side in (5.249), which leads to the conservation of $\mathbf{M}$.

The idea behind this modified velocity field consists of smoothing the velocity in order to stabilize the motion, as can be seen by writing a continuous approximation of the discrete terms in (6.266) as

$$\begin{aligned}\sum_b m_b \frac{\mathbf{u}_a}{\overline{\rho}_{ab}} w_{ab} &= \frac{1}{2}\int_\Omega \left(1+\frac{\rho\left(\mathbf{r}'\right)}{\rho_a}\right)\mathbf{u}_a w_h\left(\left|\mathbf{r}'-\mathbf{r}_a\right|\right) d^n\mathbf{r}' \\ &\quad + O\left(h^2\right) \\ \sum_b m_b \frac{\mathbf{u}_b}{\overline{\rho}_{ab}} w_{ab} &= \frac{1}{2}\int_\Omega \left(1+\frac{\rho\left(\mathbf{r}'\right)}{\rho_a}\right)\mathbf{u}\left(\mathbf{r}'\right) w_h\left(\left|\mathbf{r}'-\mathbf{r}_a\right|\right) d^n\mathbf{r}' \\ &\quad + O\left(h^2\right)\end{aligned} \tag{6.271}$$

where $\mathbf{r}'$ stands for $\mathbf{r}_b$ in the continuous formalism. It is noteworthy that the terms in $O\left(h^2\right)$ appearing in these approximations are of the same order of magnitude, because $\mathbf{u}\left(\mathbf{r}'\right)$ is close to $\mathbf{u}_a$. A calculation analogous to those in Section 5.2.1 would show that their difference is in $O\left(h^3\right)$. Writing the difference of the two lines in (6.271), we get

$$\breve{\mathbf{u}}_a = \mathbf{u}_a + \frac{\epsilon}{2\rho_a}\int_\Omega \left(\rho\left(\mathbf{r}'\right)+\rho_a\right)\left[\mathbf{u}\left(\mathbf{r}'\right)-\mathbf{u}_a\right] w_h\left(\left|\mathbf{r}'-\mathbf{r}_a\right|\right) d^n\mathbf{r}' + O\left(h^3\right) \tag{6.272}$$

In order to simplify the following discussion, we will provisionally assume that the density is constant, which will not change our conclusions. Through the kernel normalization condition (5.5), we can write:

$$\breve{\mathbf{u}}_a \approx (1-\epsilon)\,\mathbf{u}_a + \epsilon\int_\Omega \mathbf{u}\left(\mathbf{r}'\right) w_h\left(\left|\mathbf{r}'-\mathbf{r}_a\right|\right) d^n\mathbf{r}' \tag{6.273}$$

That formulation expresses $\breve{\mathbf{u}}_a$ as a barycentre and suggests that a value greater than 1 of ϵ would be meaningless (we will come back to that). Let us now consider the spatial Fourier transform in (6.273), noting that the transform of a convolution equals the product of the transforms (refer to Appendix B):

$$\begin{aligned}\widehat{\breve{\mathbf{u}}_a}(K) &\approx (1-\epsilon)\,\widehat{\mathbf{u}_a}(K) + \epsilon\widehat{\mathbf{u}_a}(K)\,\widehat{w_h}(K) \\ &= \widehat{\mathbf{u}_a}(K)\left\{1-\epsilon\left[1-\widehat{w_h}(K)\right]\right\}\end{aligned} \tag{6.274}$$

With the Gaussian kernel, whose Fourier transform (5.77) is less than 1, we then get $\widehat{\breve{\mathbf{u}}_a}(K) < \widehat{\mathbf{u}_a}(K)$, and the modified field then contains less energy than the original field. But (6.274) primarily teaches us that this damping is even more efficient since one considers large wavenumbers, because $\widehat{w_h}(K)$ is decreasing. Thus, the operator (6.266) smoothes the velocity field in the same way as the operator (6.259) does.

To get a deeper insight into this mechanism, let us return to the expression (6.272), and let us proceed in a fairly similar way as in the discussion of Section 5.2.1, writing a second-order Taylor series expansion of $\mathbf{u}\left(\mathbf{r}'\right)$:

$$\mathbf{u}\left(\mathbf{r}'\right) = \mathbf{u}_a + (\mathbf{grad}\,\mathbf{u})_a \cdot \left(\mathbf{r}' - \mathbf{r}_a\right) \\ +\frac{1}{2}\left(\mathbf{r}' - \mathbf{r}_a\right)^T (\mathbf{grad}\,\mathbf{grad}\,\mathbf{u})_a \left(\mathbf{r}' - \mathbf{r}_a\right) + O\left(\left|\mathbf{r}' - \mathbf{r}_a\right|^3\right) \tag{6.275}$$

Inserting (6.275) into (6.272), we find

$$\breve{\mathbf{u}}_a = \mathbf{u}_a + \epsilon\left[(\mathbf{grad}\,\mathbf{u})_a \mathbf{J}_1 + \frac{1}{2}(\mathbf{grad}\,\mathbf{grad}\,\mathbf{u})_a : \mathbf{J}_2\right] + O\left(h^3\right) \tag{6.276}$$

where $\mathbf{J}_1$ and $\mathbf{J}_2$ are two integrals as defined by

$$\mathbf{J}_1 \doteqdot \int_\Omega \left(\mathbf{r}' - \mathbf{r}_a\right) w_h\left(\left|\mathbf{r}' - \mathbf{r}_a\right|\right) d^n\mathbf{r}' = \mathbf{0} \\ \mathbf{J}_2 \doteqdot \int_\Omega \left(\mathbf{r}' - \mathbf{r}_a\right) \otimes \left(\mathbf{r}' - \mathbf{r}_a\right) w_h\left(\left|\mathbf{r}' - \mathbf{r}_a\right|\right) d^n\mathbf{r}' \tag{6.277}$$

The former is zero by virtue of the kernel properties being dealt with in Section 5.2.2. As to the latter, we learn from equation (5.64) that it equals

$$\mathbf{J}_2 = \frac{C_{w,2}}{n} h^2 \mathbf{I}_n \tag{6.278}$$

where $C_{w,2}$, as given by (5.65), is positive. This integral has the dimension of the square of a distance and the order of magnitude of its components is h^2. Hence, returning to (6.276), we get:

$$\breve{\mathbf{u}}_a = \mathbf{u}_a + \alpha^2\left(\nabla^2\mathbf{u}\right)_a + O\left(h^3\right) \tag{6.279}$$

where the length α is defined by

$$\alpha \doteqdot \sqrt{\frac{\epsilon C_{w,2}}{2n}} h \sim h \tag{6.280}$$

Before carrying on with our investigations, we may note that we have got a form of the discrete Laplacian which differs from those we have established in Section 6.2.1:

$$\left(\nabla^2\mathbf{u}\right)_a \approx -\frac{2n}{C_{w,2}h^2} \sum_b m_b \frac{\mathbf{u}_{ab}}{\overline{\rho}_{ab}} w_{ab} \tag{6.281}$$

We may, for instance, compare it with (6.43) and find that they look like each other. Their proximity is ensured because

$$\frac{\mathbf{u}_{ab}}{r_{ab}} w'_{ab} \sim \frac{\mathbf{u}_{ab}}{h} \frac{-2w_{ab}}{h} \tag{6.282}$$

(we may recall that the kernel has a negative derivative). The form (6.281), however, would correspond to kinetic matrices (refer to Chapter 2, Section 2.4.3) as provided by

$$\boldsymbol{\alpha}_{ab} = \frac{2n}{C_{w,2}h^2} m_a m_b \frac{w_{ab}}{\overline{\rho}_{ab}} \mathbf{I}_n \tag{6.283}$$

These quantities look like (6.49) and verify the Onsager symmetry conditions (2.181), but do not verify the required conditions for properly describing the frictions, as formulated in Section 2.4.3.

Let us now return to the discrete form of $\breve{\mathbf{u}}_a$ and show that the just introduced formalism may be relevant for the formal approach of the mirror Lagrangian systems, as introduced in Section 2.5.2 of Chapter 2. (6.266) may, indeed, also be rearranged as

$$\breve{\mathbf{u}}_a = \sum_b m_b A_{ab} \mathbf{u}_b \tag{6.284}$$

where the A_{ab} quantities have been defined by:[50]

$$A_{ab} \doteqdot \frac{1}{m_a}\left[1 - \epsilon\left(\sum_c \frac{m_c w_{ac}}{\overline{\rho}_{ac}}\right)\right]\delta_{ab} + \epsilon \frac{w_{ab}}{\overline{\rho}_{ab}} \tag{6.285}$$

[50]Once again, they should not be confused with the difference of two terms. That notation refers to the notations A_{ij} in Section 2.5.2.

The velocities $\breve{\mathbf{u}}_a$ here play the part of the mirror velocities denoted as $\mathbf{u}_a^*$ in Section 2.5.2. The A_{ab} make up a symmetric tensor, a and b here being tensor sub-scripts. We are going to explain that it has, up to the fourth-order in h, an inverse tensor. To that purpose, we note that the definition (6.266) allows us to write

$$\breve{\mathbf{u}}_{ab} = \mathbf{u}_{ab} - \epsilon \sum_c m_c \left(\frac{\mathbf{u}_{ac}}{\overline{\rho}_{ac}} w_{ac} - \frac{\mathbf{u}_{bc}}{\overline{\rho}_{bc}} w_{bc}\right) \tag{6.286}$$

We then calculate the following expression:

$$\breve{\mathbf{u}}_a + \epsilon \sum_b m_b \frac{\breve{\mathbf{u}}_{ab}}{\overline{\rho}_{ab}} w_{ab} = \mathbf{u}_a - \epsilon^2 \sum_b \frac{m_b}{\overline{\rho}_{ab}} w_{ab} \sum_c m_c \left(\frac{\mathbf{u}_{ac}}{\overline{\rho}_{ac}} w_{ac} - \frac{\mathbf{u}_{bc}}{\overline{\rho}_{bc}} w_{bc}\right) \tag{6.287}$$

Through the calculation leading to (6.279) we know that each of the discrete sums in the right-hand side of (6.287), multiplied by ϵ, is an operator equal to $\alpha^2\nabla^2 + O\left(h^3\right)$. Since α has the same order of magnitude as h, it becomes obvious that

$$\begin{aligned}
&\epsilon^2 \sum_b \frac{m_b}{\overline{\rho}_{ab}} w_{ab} \sum_c m_c \left(\frac{\mathbf{u}_{ac}}{\overline{\rho}_{ac}} w_{ac} - \frac{\mathbf{u}_{bc}}{\overline{\rho}_{bc}} w_{bc}\right) \\
&= \alpha^4 \left(\nabla^4 \mathbf{u}\right)_a + O\left(h^5\right) \\
&= O\left(h^4\right)
\end{aligned} \tag{6.288}$$

The equality (6.287) is then reduced to

$$\mathbf{u}_a = \breve{\mathbf{u}}_a + \epsilon \sum_b m_b \frac{\breve{\mathbf{u}}_{ab}}{\overline{\rho}_{ab}} w_{ab} + O\left(h^4\right) \tag{6.289}$$

Thus, we have managed to invert (to the fourth order in h) the definition (6.266), which shows us that the inverse of the tensor A_{ab}, having components B_{ab}, is yielded by

$$B_{ab} \doteqdot \frac{1}{m_a}\left[1 + \epsilon\left(\sum_c \frac{m_c w_{ac}}{\overline{\rho}_{ac}}\right)\right]\delta_{ab} - \epsilon \frac{w_{ab}}{\overline{\rho}_{ab}} + O\left(h^4\right) \tag{6.290}$$

This formula is identical to (6.285), provided that $-\epsilon$ is substituted for ϵ. The above discussion shows that inverting the velocity fields $\mathbf{u}_a$ and $\breve{\mathbf{u}}_a$ is tantamount to changing the sign of ϵ, or even formally going back in time,[51] which changes the sign of the velocities. This is consistent with the basic idea of the mirror systems as discussed in Section 2.5.2, which aims at modelling a dissipation of energy by highlighting an energy flux between the basic system and the mirror system. Hence, without making any calculation, we can invert (6.279) and write

$$\mathbf{u}_a = \breve{\mathbf{u}}_a - \alpha^2 \left(\nabla^2 \mathbf{u}\right)_a + O\left(h^3\right) \tag{6.291}$$

[51] The strictly formal nature of this concept is reflected by the fact that α^2 then becomes negative. The irreversible nature of the dissipative systems can be found again, and it can be seen that the positiveness of ϵ is relevant.

The XSPH model then takes the form of a discrete approximation of the α model of turbulence (Section 4.6.2), the formula we have just determined being the analogous of (4.298). That idea is further supported by the fact that, for reasonable values of h, the length α satisfies the condition (4.326), as evidenced by the relation (6.280). We may then consider the velocities $\breve{\mathbf{u}}_a$ as regularized velocities, and we will denote as $\breve{A}$ all the quantities A being calculated from that velocity field. Thus, it is unsurprisingly a turbulence model, because of the quadradic aspect of the stresses in a turbulent flow (Chapter 4). We may recall that the quadratic nature of the friction forces is a specific feature of the mirror systems in Section 2.5.2.

Due to the above arguments, we have to seek the equations of motions from the Lagrange equations. To do this, by analogy with (2.259) and (4.295), we introduce the following Lagrangian:

$$\begin{aligned} L &= \sum_a \frac{1}{2} m_a \mathbf{u}_a \cdot \breve{\mathbf{u}}_a \\ &= \sum_{a,b} \frac{1}{2} m_a m_b A_{ab} \mathbf{u}_a \cdot \mathbf{u}_b \\ &= \sum_{a,b} \frac{1}{2} B_{ab} \breve{\mathbf{u}}_a \cdot \breve{\mathbf{u}}_b \end{aligned} \tag{6.292}$$

(we will purposefully neglect both internal energy and external potential, which have no effect on our arguments in this sections; if need be, they will be added at the very end of our calculations). According to equation (2.260) in Section 2.5.2, the second of these formulas corresponds to the formalism of Section 1.4.1 with $\mathbf{\Lambda}_{ab} = m_a m_b A_{ab} \mathbf{I}_n$. This brings us back to the standard SPH approach (i.e. $A_{ab} = \delta_{ab}/m_a$) when $\epsilon = 0$, according to (6.285).

Using the definition of $\breve{\mathbf{u}}_a$ to calculate the Lagrangian (here being limited to the kinetic energy), we get

$$\begin{aligned} L &= \sum_a \frac{1}{2} m_a \mathbf{u}_a \cdot \mathbf{u}_a - \frac{\epsilon}{2} \sum_{a,b} m_a m_b \frac{\mathbf{u}_{ab}}{\overline{\rho}_{ab}} w_{ab} \cdot \mathbf{u}_a \\ &= \sum_a \frac{1}{2} m_a u_a^2 - \frac{\epsilon}{4} \sum_{a,b} m_a m_b \frac{u_{ab}^2}{\overline{\rho}_{ab}} w_{ab} \end{aligned} \tag{6.293}$$

and, *mutatis mutandis*:

$$L = \sum_a \frac{1}{2} m_a \breve{u}_a^2 + \frac{\epsilon}{4} \sum_{a,b} m_a m_b \frac{\breve{u}_{ab}^2}{\overline{\rho}_{ab}} w_{ab} \tag{6.294}$$

(where the velocities in standard characters refer to norms, as usual). Equation (6.294) indicates that the kinetic energy is positive definite, in connection with Section 4.6.2, and constitutes a discrete approximation of (4.312), as can be seen from the previously introduced procedure (eqn (6.272).

The discussion in Section 2.5.2 enables us to immediately assert, from (6.292):

$$\begin{aligned} \mathbf{p}_a &= \frac{\partial L}{\partial \mathbf{u}_a} = m_a \breve{\mathbf{u}}_a \\ \breve{\mathbf{p}}_a &= \frac{\partial L}{\partial \breve{\mathbf{u}}_a} = m_a \mathbf{u}_a \end{aligned} \tag{6.295}$$

The second equation in (6.295) will provide the equation of motion, applying the Lagrange equations. In that respect, we may note that the velocity $\breve{\mathbf{u}}_a$ is associated with the position $\mathbf{r}_a$, by virtue of (6.268), and so the Lagrange equation takes the following form:

$$\frac{\breve{D}}{Dt} \frac{\partial L}{\partial \breve{\mathbf{u}}_a} = \frac{\partial L}{\partial \mathbf{r}_a} \tag{6.296}$$

The strictly formal resort to the notation $\breve{D}/Dt$ is justified by the above remark about the close relationship between that formalism and the model α in Section 4.6.2. The general calculation (2.252) in Section 2.5.2 could be used for calculating the force received by the particle, but here it is simpler to start again from the definition (6.293) of the Lagrangian:

$$\frac{\partial L}{\partial \mathbf{r}_a} = -\frac{\epsilon}{4} \sum_{b,c} m_b m_c u_{bc}^2 \frac{\partial}{\partial \mathbf{r}_a} \left(\frac{w_{bc}}{\rho_b} \right) \tag{6.297}$$

(the symmetry of the expression under the summation sign makes it possible to substitute ρ_b for $\overline{\rho}_{bc}$). Once again, we will consider that the density does not vary, which is tantamount to only keeping the influence of the positions on the kernel value in (6.297) (Monaghan, 2002 provides a more complete calculation). We find:

$$\begin{aligned} \frac{\partial w_{bc}}{\partial \mathbf{r}_a} &= \delta_{ab} \frac{\partial w_{ac}}{\partial \mathbf{r}_a} + \delta_{ac} \frac{\partial w_{ab}}{\partial \mathbf{r}_a} \\ &= -\delta_{ab} w'_{ac} \mathbf{e}_{ac} - \delta_{ac} w'_{ab} \mathbf{e}_{ab} \end{aligned} \tag{6.298}$$

Then, inserting (6.298) into (6.297) and acting upon the sub-scripts, we get

$$\frac{\partial L}{\partial \mathbf{r}_a} = \frac{\epsilon}{4} \sum_{b,c} m_b m_c u_{bc}^2 \frac{1}{\rho_b} \left(\delta_{ab} w'_{ac} \mathbf{e}_{ac} + \delta_{ac} w'_{ab} \mathbf{e}_{ab} \right) \tag{6.299}$$

$$= \frac{\epsilon}{4}\sum_c m_a m_c u_{ac}^2 \frac{1}{\rho_a} w'_{ac} \mathbf{e}_{ac} + \frac{\epsilon}{4}\sum_b m_b m_a u_{ba}^2 \frac{1}{\rho_b} w'_{ab} \mathbf{e}_{ab}$$

$$= \frac{\epsilon}{2}\sum_b m_a m_b \frac{u_{ab}^2}{\overline{\rho}_{ab}} w'_{ab} \mathbf{e}_{ab}$$

Lastly, with (6.295) and (6.299) the Lagrange equations (6.296) yield:

$$\frac{\breve{D}\mathbf{u}_a}{Dt} = -\frac{\epsilon}{2}\sum_b m_b \frac{u_{ab}^2}{\overline{\rho}_{ab}} w'_{ab} \mathbf{e}_{ab} \tag{6.300}$$

It should be kept in mind that we have ignored the internal energy which, as usual, would give rise to the same pressure forces as in the previous sections.

It can be found that the force which we have determined in (6.300) depends on the velocities and is therefore a viscous force, analogous to the $-\mathbf{div}\ \boldsymbol{\tau}_\alpha$ term of equation (4.304) in the continuous α model theory, which gives the XSPH model the nature of a discrete LES type approach, with a numerical mixing length having the order of magnitude of h, according to (6.280), which is still consistent with the results of Section 6.5.1. The smoothing length h should then satisfy, within the frame of this model, the condition (4.325).

As suggested above, it appears that the viscous force (6.300) is no longer linear in $\mathbf{u}_{ab}$ as in Section 6.2, but is quadratic, as in Section 2.5.2. Thus, using the above approach, we can find again the quadratic aspect of the turbulent viscous forces, which was already emphasized in Section 4.4.4. That force then has the general shape (2.179) as determined in Section 2.4.3, but with kinetic matrices which now depend on the velocities:

$$\boldsymbol{\alpha}_{ab} = \frac{\epsilon}{2} m_a m_b \frac{w'_{ab}}{\overline{\rho}_{ab}} \left(\mathbf{e}_{ab} \otimes \mathbf{u}_{ab}\right) \tag{6.301}$$

This form satisfies the symmetry condition $\boldsymbol{\alpha}_{ab} = \boldsymbol{\alpha}_{ba}$, as well as the condition (2.196), which substantiates the conservation of angular momentum, but not (2.204), and so the viscous forces subsist in the presence of a rigid body motion. Besides, the condition $\boldsymbol{\alpha}_{ab}^T = \boldsymbol{\alpha}_{ab}$ is disregarded. [52]

Comparing (6.301) with (6.50), an estimation of the sub-particle viscosity $\breve{\nu}_a$ can be provided in the XSPH formalism:

$$\breve{\nu}_a \sim \frac{\epsilon r_{ab} u_{ab}}{4\,(n+2)} \sim \frac{\epsilon h u_{ab}}{4\,(n+2)} \tag{6.302}$$

Let us interpret that result. Noting that $\breve{\mathbf{u}}_a \sim \mathbf{u}_a$, the regularized velocity gradient (or rate-of-strain of the regularized velocity) can be estimated by

$$\breve{s}_a \sim \frac{u_{ab}}{h} \tag{6.303}$$

and so (6.302) is rearranged as

$$\breve{\nu}_a \sim (c_S h)^2\, \breve{s}_a \tag{6.304}$$

[52] In that respect, a close relationship can be found between this approach and the discrete algebraic explicit model as described in Section 6.4.2.

(in accordance with the continuous model (4.309)) provided that we put

$$c_S = \sqrt{\frac{\epsilon}{4\,(n+2)}} \tag{6.305}$$

The XSPH model may then be considered as a form of LES along with a Smagorinsky closure model, in accordance with (4.281). This result substantiates the implicit link existing between SPH and LES, as previously mentioned in Section 6.5.1.

Let us now turn to energy. Section 2.5.2 states that it is defined by (2.262) and is constant. The kinetic energy lost by the mirror system (i.e. the regularized velocity field) can be calculated through (6.300):

$$\begin{aligned} \frac{d\breve{E}}{dt} &\doteq \frac{d}{dt}\sum_a \frac{1}{2} m_a \breve{u}_a^2 \\ &= \sum_a m_a \breve{\mathbf{u}}_a \cdot \frac{d\breve{\mathbf{u}}_a}{dt} \\ &= -\frac{\epsilon}{4}\sum_{a,b} m_a m_b \frac{u_{ab}^2}{\overline{\rho}_{ab}} \left(\breve{\mathbf{u}}_{ab}\cdot\mathbf{e}_{ab}\right) w'_{ab} \end{aligned} \tag{6.306}$$

(we have carried out a symmetrization of the subscripts on the last line). Formula (6.306) is analogous to (2.263) (without potential energy), and the energy flux released by each particle is derived from the formula as:

$$\breve{P}_a = \frac{\epsilon}{4}\sum_b m_b \frac{u_{ab}^2 \left(\breve{\mathbf{u}}_{ab}\cdot\mathbf{e}_{ab}\right)}{\overline{\rho}_{ab}} w'_{ab} \tag{6.307}$$

(the sum now only regarding b). This is the local production of turbulent kinetic energy, as predicted by the XSPH model. As expected, it is a cubic function of the velocities, in accordance with Section 2.5.2 and Chapter 4. We can provide its order of magnitude based on the above arguments:[53]

$$\breve{P}_a \sim \frac{\epsilon h^2 \breve{s}_a^3}{4} \tag{6.308}$$

[53]The following formula confirms that the energy dissipation rate varies like $u^2\breve{u}$, in accordance with the continuous case within the frame of the α model.

Besides, the velocity fluctuations, defined as the velocity deviation from the regularized velocity, are determined by

$$\begin{aligned} \mathbf{u}'_a &\doteq \mathbf{u}_a - \breve{\mathbf{u}}_a \\ &= \epsilon \sum_b m_b \frac{\mathbf{u}_{ab}}{\overline{\rho}_{ab}} w_{ab} \\ &\sim \epsilon h \breve{s}_a \end{aligned} \tag{6.309}$$

(the ultimate estimation made in (6.309) corresponds to each of the components of this vector). The turbulent kinetic energy (i.e. included in the regularized velocity fluctuations) can then be estimated by calculating the half of the squared norm of $\mathbf{u}'_a$, according to (4.76):

$$\breve{k}_a \sim \frac{3}{2}\left(\epsilon h \breve{s}_a\right)^2 \tag{6.310}$$

Estimating the turbulent energy dissipation by $\breve{\varepsilon}_a \sim \breve{P}_a$ (as in Section 4.4.2, eqn (4.105)), we lastly can provide an order of magnitude of the dimensionless rate-of-strain of the regularized velocity field:

$$\breve{s}_a^* \doteqdot \frac{\breve{k}_a \breve{s}_a}{\breve{\varepsilon}_a} \sim 6\epsilon \tag{6.311}$$

If the value found in Section 4.4.6 is adopted as the order of magnitude of $\breve{s}_a^*$ for a logarithmic area (eqn (4.205)), we can derive from it a recommended value for the dimensionless parameter ϵ involved in the definition of the regularized velocity (6.266):

$$\epsilon \sim \frac{1}{6\sqrt{C_\mu}} \approx 0.55 \tag{6.312}$$

Most of the authors utilizing the XSPH regularization have adopted the values ranging from 0.5 to 2/3, without providing further explanations. We have just shown that it is the order of magnitude allowing us to ensure an energy dissipation of the regularized field which is compatible with the commonly accepted theory in a near-wall turbulent area. Returning to the estimation of the Smagorinsky coefficient (6.305), we derive from it its value:

$$c_S = \sqrt{\frac{\epsilon}{4\,(n+2)}} \sim \frac{1}{\sqrt{24\,(n+2)}C_\mu^{1/4}} \approx 0.17 \tag{6.313}$$

(we refer, of course, to a dimension $n = 3$, as is normal for a LES type approach). Interestingly, this is the value proposed by Lilly, according to the discussion in Section 4.6.1 (eqn (4.291)).

Thus, the XSPH regularization, provided with a value of the same order of magnitude as (6.312) for the parameter ϵ, constitutes a discrete version of the turbulence model α, as pointed out by Monaghan (2002). In the light of the current analysis, it also behaves like a Lagrangian large eddy simulation, provided with a sub-particle viscosity in line with the orders of magnitude as prescribed by the theoretical investigations of Chapter 4. Interestingly, this model handles the turbulence without the assistance of any further equation, the only trick being the regularization (6.266).

Let us now return to the procedure through which we could get the equation of motion (6.300); we may note that its derivation may be extended to the pressure and gravity forces, provided that both internal energy and gravity potential are deleted in the Lagrangian (6.292). Returning to the case of a slightly variable density, the resulting forces will be left unchanged as compared with the already discussed discrete systems, but we have learnt in Section 5.3.3 that the pressure force discretization is closely related to that of the continuity equation. From the calculation extending from (5.231) to (5.233), the velocity appearing in the latter is by definition obviously $d\mathbf{r}_a/dt$, that is here $\breve{\mathbf{u}}_a$. This result is consistent with the continuous case (4.301), in which the regularized velocity $\breve{\mathbf{u}}$ plays a role in the continuity equation. Thus, taking (6.266) into account and if the system (5.241) is chosen for the discretization of the pressure term, the whole XSPH is written as:

$$\forall a,\ \frac{\breve{D}\mathbf{u}_a}{Dt} = -\sum_b m_b\left(\frac{p_a}{\rho_a^2}+\frac{p_b}{\rho_b^2}+\frac{\epsilon}{2}\frac{u_{ab}^2}{\overline{\rho}_{ab}}\right)w'_{ab}\mathbf{e}_{ab}+\mathbf{g}$$
$$\breve{\mathbf{u}}_a = \mathbf{u}_a - \epsilon\sum_b m_b\frac{\mathbf{u}_{ab}}{\overline{\rho}_{ab}}w_{ab}$$
$$\frac{d\mathbf{r}_a}{dr} = \breve{\mathbf{u}}_a \tag{6.314}$$
$$\frac{\breve{D}\rho_a}{Dt} = \sum_b m_b\breve{\mathbf{u}}_{ab}\cdot w'_{ab}\mathbf{e}_{ab}$$

It should be remembered that the formulation $\breve{D}/Dt$ only has a formal role, and that it may be considered as a mere Lagrangian derivative for the numerical resolution. Thus, the preceding equations may be subject to one of the temporal schemes as disclosed in Section 5.4.

As can be seen, the viscous term takes a much simpler form than before, paradoxical as it may seem for a turbulence model. One cannot help but emphasize the outstanding smartness of this model, since it is variationally compatible with the discussion in Section 4.6.2 within a continuous context. That model was only seldom numerically substantiated (refer, however, to Monaghan, 2011), but the success of the continuous α model suggests that its translation into SPH formalism is expected to yield good results.

It is arguable that our calculations were partly based on a presumably constant density, whereas we came back, for establishing the system (6.314), to an approach governed by a variable density continuity equation; it is justifiable, however, to consider that approach as suitable within the frame of weakly compressible flows.

It is noteworthy that the continuity equation might be written in a form analogous to (5.238), with a discrete divergence operator which is different from $\tilde{D}_a^k$. It could then be argued that the discrete gradient operator chosen for modelling the pressure gradient is not skew-adjoint to it. In spite of this shortcoming, our choice is satisfactory. It should be kept in mind, indeed, that the α model does not satisfy the conservation of energy in the ordinary sense (refer to Section 4.6.2). It should also be pointed out that there is no obstacle to associating this model with the incompressible algorithm as described in Section 6.2.5.

As regards the stability of the XSPH model, it can be studied through a procedure analogous to what was done in Sections 5.4.4 and 6.2.3. Section 5.4.4 taught us that the discrete formulation of the kinematic equation $d\mathbf{r}_a/dr = \breve{\mathbf{u}}_a$ does not play any part when one considers a continuity equation accompanied by an equation of state without background pressure. As to the viscous term in the equation of motion, Section 6.2.3 states that it only acts via the variation term $\delta_{\mathbf{u}}\dot{\mathbf{u}}_a$, which here is linear in $\mathbf{u}_{ab}$, and hence zero if one starts from a uniform initial state. Thus, the stability analysis of the system (6.314) is unchanged as compared to what was written in Section 5.4.4, the only difference being that the continuity equation is now based upon the regularized field $\breve{\mathbf{u}}_a$. Starting from (6.266), we will write a slight variation of this field as

$$\delta\breve{\mathbf{u}}_a = \delta\mathbf{u}_a - \frac{\epsilon}{\rho_0}\sum_b m_b w_{ab}\left(\delta\mathbf{u}_a - \delta\mathbf{u}_b\right) \tag{6.315}$$

Approximating that discrete equation through integrals is no more difficult than in Section 5.4.4, by seeking the numerical velocity fluctuation $\delta\mathbf{u}$ in the form (5.383), then using the kernel normalization condition (5.5):

$$\begin{aligned}
\delta\breve{\mathbf{u}}_a &\approx \delta\mathbf{u}_a - \epsilon\left[\left(\int_\Omega w_h\left(\tilde{r}\right)d^n\tilde{\mathbf{r}}\right)\delta\mathbf{u}_a - \int_\Omega w_h\left(\tilde{r}\right)\delta\mathbf{u}\, d^n\tilde{\mathbf{r}}\right]\\
&= \left[1 - \epsilon\int_\Omega w_h\left(\tilde{r}\right)d^n\tilde{\mathbf{r}} + \epsilon\int_\Omega w_h\left(\tilde{r}\right)\exp\left(-i\mathbf{K}\cdot\tilde{\mathbf{r}}\right)d^n\tilde{\mathbf{r}}\right] \\
&\quad\times\mathbf{U}\left(t\right)\exp\left(-i\mathbf{K}\cdot\mathbf{r}_a\right)\\
&= \left[1 - \epsilon\left(1 - \widehat{w_h}\left(K\right)\right)\right]\mathbf{U}\left(t\right)\exp\left(-i\mathbf{K}\cdot\mathbf{r}_a\right)
\end{aligned} \tag{6.316}$$

Thus, the system (5.406), whose last equation is derived from the continuity equation, is then modified and yields

$$\begin{aligned}
\mathbf{U}_0\chi &= \mathbf{U}_0 + i\frac{c_0^2 R_0}{\rho_0}\widehat{w_h}\left(K\right)\mathbf{K}\delta t\\
R_0\chi &= R_0 + i\rho_0\widehat{w_h}\left(K\right)\left[1 - \epsilon\left(1 - \widehat{w_h}\left(K\right)\right)\right]\mathbf{K}\cdot\mathbf{U}_0\chi\,\delta t
\end{aligned} \tag{6.317}$$

(we may recall that the second equation in (5.406) has no effect on the result). Equations (5.407) and (5.408) are then unchanged, the A_2 coefficient as defined in (5.399) being substituted by

$$\begin{aligned}
A_{2,T} &= \left[\frac{Co}{Ma_0}K^+\widehat{w_h}\left(K\right)\right]^2\left[1 - \epsilon\left(1 - \widehat{w_h}\left(K\right)\right)\right]\\
&= A_2\left[1 - \epsilon\left(1 - \widehat{w_h}\left(K\right)\right)\right] > 0
\end{aligned} \tag{6.318}$$

with the usual definitions (5.73), (5.211) and (5.397) for K^+, Ma_0 and Co. Thus, the condition (5.409) remains, but now regards $A_{2,T}$, that is $A_{2,T} \leqslant 4$, and leads to a CFL type condition analogous to (5.410). We may note, however, that with a Gaussian kernel (i.e. smaller than 1) and if $\epsilon \leqslant 1$ (a previously mentioned intuitive idea), the definition (6.318) gives $A_{2,T} \leqslant A_2$. The condition $A_{2,T} \leqslant 4$ is then more readily satisfied, so much more as the wavenumber is high. In particular, the maximum of $A_{2,T}$ is lower than that of A_2. Thus, the XSPH model tends to stabilize the numerical system, even better since ϵ is great, which is true to intuition. For $\epsilon = 1$, it can easily be demonstrated that the coefficient $\sqrt{2e} \approx 2.33$ appearing in the CFL condition (5.410) is substituted for by $\sqrt{3e} \approx 2.86$, whereas this coefficient would be close to 2.6 for the above recommended value $\epsilon = 0.55$.

6.5.3 Discrete stochastic models

We will conclude this overview of the turbulent SPH models by mentioning the stochastic models, on the basis of the discussions in Section 4.6.3. This is because the Lagrangian nature of the SPH approach incites us to represent the turbulent mixing by means of a random force, that is utilizing a Langevin type equation (Welton and Pope, 1997; Violeau et al., 2001). Equation (4.359), for instance, can be applied to each fluid particle, provided that the discrete time step δt is substituted for the infinitesimal time dt, which yields

$$\forall a, \quad \frac{d\mathbf{u}_a}{dt} = \left(-\frac{1}{\rho}\,\mathbf{grad}\,\overline{p}^* - \frac{C_R}{2}\frac{\varepsilon}{k}\mathbf{u}' + C_P\boldsymbol{\Gamma}\mathbf{u}'\right)_a + \sqrt{\frac{C_{0,a}\varepsilon_a}{\delta t}}\boldsymbol{\xi}_a \qquad (6.319)$$

where $C_{0,a}$ is defined by (4.358) and depends on the particle a via the $P_a^* \doteq P_a/\varepsilon_a$ ratio of the kinetic energy production to its dissipation rate. In equation (6.319), $\boldsymbol{\xi}_a$ denotes a random vector ascribed to particle a and changing upon each iteration. It should satisfy the last three relations of the system (4.330), that is to say it should be statistically independent of $\mathbf{u}_a$ and be a normalized centred random vector of independent components:

$$\forall a, \quad \begin{aligned} \overline{\boldsymbol{\xi}_a} &= \mathbf{0} \\ \overline{\boldsymbol{\xi}_a \otimes \boldsymbol{\xi}_a} &= \mathbf{I}_n \end{aligned} \qquad (6.320)$$

In dimension $n = 2$, such a random vector can be got from two real random variables $\zeta_{1,a}$ and $\zeta_{2,a}$, which are evenly distributed over the segment [0, 1]:

$$\forall a, \; \boldsymbol{\xi}_a = \sqrt{-2\ln\zeta_{1,a}}\begin{pmatrix}\cos\left(2\pi\zeta_{2,a}\right)\\ \sin\left(2\pi\zeta_{2,a}\right)\end{pmatrix} \qquad (6.321)$$

The components of the $\boldsymbol{\xi}_a$ vector are then independent standard normal variables (Carter, 1994), thus satisfying (6.320).

With, as usual, $\overline{p}_a^* \doteq \overline{p}_a + \rho g z_a$ and $\mathbf{u}_a' = \mathbf{u}_a - \overline{\mathbf{u}}_a$, taking into account equation (4.358) giving $C_{0,a}$, and resorting to the model (5.241) for discretizing the pressure gradient, equation (6.319) is also written as

$$\begin{aligned} \forall a, \quad \frac{d\mathbf{u}_a}{dt} &= -\sum_b m_b\left(\frac{\overline{p}_a}{\rho_a^2} + \frac{\overline{p}_b}{\rho_b^2}\right) w'_{ab}\mathbf{e}_{ab} \\ &\quad + \left(-\frac{C_R}{2}\frac{\varepsilon_a}{k_a}\mathbf{I}_n + C_P\boldsymbol{\Gamma}_a\right)(\mathbf{u}_a - \overline{\mathbf{u}}_a) \\ &\quad + \sqrt{\frac{2\left[C_P P_a + (C_R - 1)\,\varepsilon_a\right]}{3\delta t}}\boldsymbol{\xi}_a + \mathbf{g} \end{aligned} \qquad (6.322)$$

Thus, each particle is provided with a velocity gradient $\boldsymbol{\Gamma}_a$ (as defined by (4.61)), in addition to the other variables which were previously defined in Section 6.4.2 (we may recall that the constants C_P and C_R are given by the Table 4.2 in Chapter 4). It is a discrete form of the generalized Langevin model, the quantities k_a, ε_a, P_a and $\boldsymbol{\Gamma}_a$ being possibly calculated by means of the equations (6.211), (6.216), (6.214)–(6.224) and a formula analogous to (6.100):

$$\boldsymbol{\Gamma}_a \approx \mathbf{G}_a^0\{\overline{\mathbf{u}}_b\} = \sum_b V_b\left(\overline{\mathbf{u}}_a + \overline{\mathbf{u}}_b\right) \otimes w'_{ab}\mathbf{e}_{ab} \tag{6.323}$$

When it is preferred to not use the discrete model $k - \varepsilon$ for calculating these quantities, the following reasoning may be applied. Considering the production-dissipation equilibrium (4.105) and the definition (4.97) of P, we will write

$$P_a = \varepsilon_a = \nu_{T,a} S_a^2 \tag{6.324}$$

S_a being calculated as in Section 6.4.2 (eqn (6.224)). As to k, it may also be related to ε via equation (6.215), on the basis of an *ad hoc* estimation of the mixing length.

As specified in Section 4.6.3, the question of the estimation of the mean velocity and pressure has to be addressed. One possible approach consists in utilizing the J_a operator (first line in eqn (5.123)) to write

$$\begin{aligned} \overline{p}_a &\approx J_a\{A_b\} \\ \overline{\mathbf{u}}_a &\approx J_a\{\mathbf{u}_b\} \end{aligned} \tag{6.325}$$

Comparing this formula with (6.259), however, shows that this approximation is conspicuously not fully relevant. We could as well resort to the equations of an averaged model in the discrete Reynolds sense (Section 6.4.1 and 6.4.2), definitely with an extra computational time. As indicated at the end of Section 4.6.3, most of the benefit of the stochastic approach would then be lost. Thus, it seems one has to be satisfied with the approximation (6.325).

The case of the simplified Langevin model (eqns (4.360) and (4.362)) is reduced to the following discrete form:

$$\begin{aligned} \forall a,\ \frac{d\mathbf{u}_a}{dt} &= -\sum_b m_b\left(\frac{\overline{p}_a}{\rho_a^2} + \frac{\overline{p}_b}{\rho_b^2}\right) w'_{ab}\mathbf{e}_{ab} \\ &\quad -\frac{C_R}{2}\frac{\varepsilon_a}{k_a}\left(\mathbf{u}_a - \overline{\mathbf{u}}_a\right) \\ &\quad +\sqrt{\frac{2\left(C_R - 1\right)\varepsilon_a}{3\delta t}}\boldsymbol{\xi}_a + \mathbf{g} \end{aligned} \tag{6.326}$$

Whether the model (6.322) or (6.326) is chosen, it can be found that a stochastic force appears, and (6.166) may therefore be substituted by

$$\forall a,\ m_a\frac{d\mathbf{u}_a}{dt} = \sum_b \mathbf{F}_{b\to a}^{int} + \mathbf{F}_a^{diss} + \mathbf{F}_a^{stoc} + \mathbf{F}_a^{ext} \tag{6.327}$$

according to equation (2.280) in Section 2.5.3 (here we have ignored the surface tension and wall forces). The stochastic force $\mathbf{F}_a^{stoc}$ models the mixing effect exerted by the turbulent eddies, and thus is added to the dissipative forces $\mathbf{F}_a^{diss}$, which are here reduced to the second line of (6.326). This equation, together with a discrete continuity equation (refer to Section 5.3.1), can be integrated by means of a temporal scheme (refer to Section 5.4.1). As stated in Section 4.6.3, the time step δt should satisfy the conditions as given by equation (4.329) for the stochastic model to have a physical meaning.

If, however, that condition is not satisfied, in particular if the time step is larger than or of the same order as[54] τ_a, it is also possible to utilize the model (6.326) provided that an exact integration is made, on the basis of the formula (2.305), which here can be applied with $\mathbf{u}_a^0 \equiv \mathbf{u}_a'^m$ and $\mathbf{u}_a(t) \equiv \mathbf{u}_a'^{m+1}$, as well as $T_R = 2\tau_a^m / C_R$ (refer to Section 4.6.3, eqn (4.328) as well as the comments after eqn (4.361)):

[54] Taking into account the conditions prescribed for the time step for the stability of the SPH numerical model (refer to Section 5.4.4), it is unlikely that this case will occur. The stability of the $k-\varepsilon$ model even requires the inverse condition (Section 6.4.3, Equation (6.247)).

$$\begin{aligned} \mathbf{u}_a'^{m+1} &= \mathbf{u}_a'^m \exp\left(-\frac{C_R \delta t}{2\tau_a^m}\right) \\ &+ \frac{2\tau_a^m}{C_R m_a}\left[1-\exp\left(-\frac{C_R \delta t}{2\tau_a^m}\right)\right]\left(\mathbf{F}_a^{int}+\mathbf{F}_a^{ext}\right)^m \\ &+ \sqrt{\frac{C_0 k_a^m}{C_R}\left[1-\exp\left(-\frac{C_R \delta t}{\tau_a^m}\right)\right]}\boldsymbol{\xi}_a^m \end{aligned} \tag{6.328}$$

Within the limit of the very small time steps, that is $\delta t \ll \tau_a^m$, this model provides the following approximation

$$\begin{aligned} \forall a,\ \mathbf{u}_a'^{m+1} &= \mathbf{u}_a'^m + \frac{1}{m_a}\left(\mathbf{F}_a^{int}+\mathbf{F}_a^{ext}\right)^m \delta t \\ &- \frac{C_R}{2\tau_a^m}\mathbf{u}_a'^m \delta t + \sqrt{C_0 \varepsilon_a^m \delta t}\,\boldsymbol{\xi}_a^m \end{aligned} \tag{6.329}$$

which is an explicit discretization of (6.327), and complies with (2.307).

It will probably have been noticed that the equation of motion (6.326) does not present the dissipative and stochastic forces in a symmetric form, and therefore the action-reaction principle is not perfectly satisfied, contrary to what has been done so far. Thus, the presently disclosed model only ensures the conservation of momentum of an isolated system on an average. The latter result is directly due to the fact that the Langevin models present a fluctuating velocity signal the average of which obeys the Reynolds equation of motion.

The stability of the presently described model would still have to be studied, in the same way as in Section 5.4.4. This study, however, could only be undertaken on the basis of the mean behaviour of the system. Practically, this kind of model leads to a comparatively disorderly motion of the particles, readily giving rise to numerical instabilities caused by the resulting major pressure fluctuations. This problem could be remedied by joining the simplified Langevin equation to the incompressible scheme as set out in Section 6.2.5. Furthermore, that would advantageously disorder the particles to a sufficient extent to prevent the problem illustrated in Fig. 6.1 (*b*). Thus, the stochastic forces would induce a rearrangement of the particles in space over time, whereas Poisson's equation (6.105) for the treatment of pressure would restore the stability of the system, each of these two processes correcting the errors in the other. To the best of our knowledge, this kind of model still has to be devised.

A stochastic turbulence application for SPH is disclosed in Section 7.2.3. Besides, this kind of model is suitable for simulating a wide variety of diffusive processes, as is the case with the fluids in the porous media. Such an approach was adopted by Tartakofsky et al. (2008) for simulating this mechanism through the SPH method. The thermal fluctuations are modelled by means of an analogous technique by Litvinov et al. (2008) and Vasquez-Quesada et al. (2009).

7 SPH method validation

7.1 Introduction

Developing software necessarily implies a validation step, in order to test the conceptual foundations of the adopted numerical method (namely, for the purpose of this book, the SPH method), as well as to check both implementation and programming for quality. To do this, a battery of test cases is usually set up in the form of simple, sometimes very schematic, applications suitable for successively testing the software functions on the basis of theoretical solutions or experimental data. In fluid mechanics, the flows for which there are mathematical solutions of the equations are rather rare, as explained throughout the Chapters 3 and 4, and poorly reflect the everyday reality, since they are most often laminar. Using a model based upon another numerical method is then attractive, allowing validation on the basis of a proper matching of the achievement of the model being tested to those of a previously evaluated model. We will sometimes utilize this method, considering reference models based on techniques involving grids. Nevertheless, we will focus our attention on empirical observations for a really satisfactory validation of the modelled physical processes.

The test cases disclosed in this chapter have been chosen through experience gained in the development of the validated software, namely SPARTACUS (*Smoothed Particle Hydrodynamics for AcCurate flow Simulation*), which has been developed since 1999 at EDF R&D's LNHE department (Laboratoire National d'Hydraulique et Environnement), first in Fortran77, then in Fortran90. It is based on the SPH method, implementing some of the equations as presented in Chapters 5 and 6, with the following options:

- the systems of equations (5.240) and (5.241) are indiscriminately used for momentum and continuity;
- the viscous forces are indiscriminately modelled by means of the models (6.57) and (6.58);
- the equation of state (5.207) is utilized in the frame of the weakly compressible approach and the (6.106)–(6.108) system is employed in the case of the incompressible algorithm;
- no surface tension effect is considered;
- the symplectic scheme (5.269) for integration over time is adopted;
- no renormalization is considered;

- the solid wall effect is taken into account by the dummy particle method disclosed in Section 6.3.2. The Lennard–Jones repulsive forces (6.156) are also considered, as well as the conservative approach as provided by the equations (6.161), (6.165) and (6.166), in only one instance;
- the mobile rigid bodies are modelled through the technique disclosed in Section 6.3.3;
- turbulence is addressed by means of the mixing length model (eqns (6.199) to (6.207)), the two-equation models (6.211) to (6.218), or even the LES approach (6.259) to (6.264). The non-linear $k - \varepsilon$ model (6.231) and the stochastic model (6.326) are only seldom tested too.

Care should be exercised with the equations as specified throughout the last two chapters, in particular as regards the choice of the sound speed, which is governed by the conditions (5.212) and (5.213). The time step is set by the CFL conditions (5.413) and (6.128), depending on the chosen algorithm for the calculation of pressure.

We need to say a few words about the basic precautions to be taken when programming a model which is based on the SPH method. Throughout Chapters 5 and 6, we have highlighted the symmetry (or antisymmetry) of those terms appearing in the specific discrete sums of the SPH method. That symmetry characterizes fluxes, namely mass, momentum (force) and energy (power) fluxes and provides discrete conservation properties (in particular, refer to the remark just after eqn (5.204), as well as Section 5.3.4). In addition, it has an algorithmic advantage, since the fluxes can be calculated only once for each pair of particles, which saves computation time and memory space.

As mentioned in Section 5.2.3, each particle interacts with a finite number of nearby particles, provided that the selected kernel has compact support. This choice is highly recommended, because it leads to an algorithm the size of which is proportional to N_p (number of particles) instead of N_p^2. For each particle a, the particles b contribution to the evolution of the quantities related to a lie, as previously explained, within a $n-$ball of radius h_t. A procedure has to be devised for seeking the particles lying near each particle a, and should be implemented upon each iteration, since they are mobile due to the Lagrangian nature of the SPH method. Several methods are available to achieve this (in particular, refer to Dominguez et al., 2010). One of the simplest methods, which is rather economical, consists of splitting the computational domain into $n-$cubes of side length h_t, which can be made once for all ar the beginning of the calculation. Upon each iteration and for each particle a, all those particles b which are located either within the cube containing a or within the adjacent $3^n - 1$ cubes can then be reviewed, only retaining those verifying[1] the condition $r_{ab} \doteqdot |\mathbf{r}_a - \mathbf{r}_b| \leqslant h_t$.

As previously explained, the geometry of some flows should be considered as definitely three-dimensional ($n = 3$), whereas other ones may be regarded, in terms of simulation, as two-dimensional ($n = 2$); such is the case with the open channel being dealt with in Section 4.4.5. In this case, the modelling requires much less computation time and memory space. The case of the channel in Section 4.4.5 also highlights the periodical nature of some flows.[2] Given these conditions, two periodic, that is virtually communicating boundaries

[1] From an algorithmic point of view, considering the criterion $r_{ab}^2 \leqslant h_t^2$ is more efficient, since no superfluous calculation of a square root has to be made.

[2] The mentioned channel has a stronger property, namely the invariance under longitudinal translation, which actually makes it a one-dimensional flow, since the variables only depend of the coordinate which is perpendicular to the bed.

should be considered. It should then be ensured that a particle moving out through a periodic boundary immediately reappears in the domain across the other boundary. But there is more: it should also be borne in mind that the particles lying at a distance h_t from each of these boundaries are liable to interact with those lying within the same distance from the other boundary, through the above described process.

This chapter will deal with flows exhibiting a growing complexity, starting with steady cases, that is governed by pressure and viscosity; it will then become possible to validate the spatial discretization schemes of the associated terms. We then will discuss flows with a quickly deformed free surface, which will validate the SPH suitability for easily modelling them, a property which primarily results from its Lagrangian nature. Thus, it will be possible to assess the suitability of that method for predicting the forces exerted on a solid, which will lastly advantageously be used for treating flows involving moving and stationary bodies. We will review the functions we have just mentioned, particularly to show the contribution of the incompressible scheme and the turbulence models. For those flows involving interfaces, the spatial scales of the flows are presumably large enough for the Weber number (3.225) to be high enough, so that the surface tension forces are, at first glance, negligible (refer to Section 3.4.5).

There are several approaches for graphically illustrating the results of a Lagrangian code. The visualization of the particles, that is of their positions and possibly their paths, is unquestionably advantageous since it remains true to the spirit of the SPH method. However, it is hardly suitable for the visualization of the continuous fields, albeit at the basis of the theory of fields, as explained in Chapter 3. As such, a post-processing of the particle data will sometimes be preferably carried out to display the data in a more readable form, which additionally facilitates the comparison with the grid methods. The results which are discussed later utilize the discrete interpolation (5.86), which associates a discretely interpolated continuous field, being denoted as $[A]_d(\mathbf{r}, t)$, to each point in space $\mathbf{r}$, each point in time t and each discrete collection of quantities $\{A_b\}$. This procedure may be applied to the points

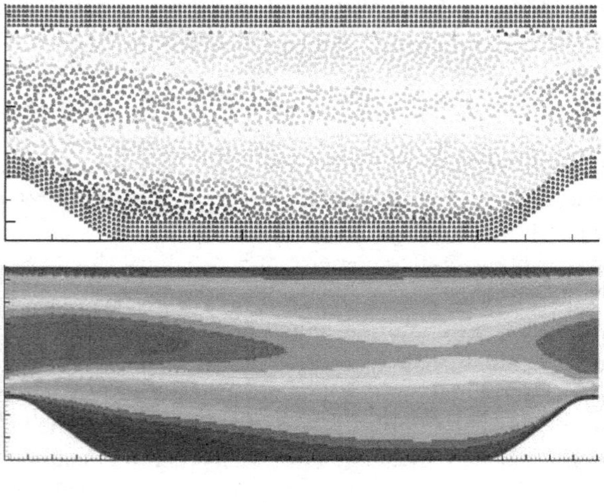

Fig. 7.1 Application of discrete interpolation (5.86) to a configuration of particles. It is the velocity field of the test case in Section 7.2.1. The dummy particles (refer to Section 6.3.2) are involved in the discrete sum but are not represented in the interpolated field.

corresponding to the positions $\mathbf{r}_a$ of the particles, which is equivalent to the discrete operator $J_a\{A_b\}$ in Section 5.2.4 (first line in the system (5.119)). It is nevertheless often more convenient to apply $[A]_d(\mathbf{r}, t)$ to all the points $\mathbf{r}$ of a grid, which provides the result appearing in Fig. 7.1. We will almost systematically use this procedure.

7.2 Steady flows

Steady flows can be used to assess the suitability of a model for properly predicting the viscous phenomena, which play a major part in it as regards the spatial distribution of the velocities. The wall boundary conditions and the numerical stability of the method are the issues at stake within that context.

7.2.1 Periodic hill

It is interesting to assess the suitability of the SPH method for simulating gravity-less confined flows. With this proposed test case, we also attempt to predict a recirculation area in a configuration having a periodic geometry. Through this case, the reprentativeness of the boundary conditions can then be assessed both at the solid walls and at the periodic boundaries. A laminar flow regime is set in order to avoid the difficulties raised by the definition of turbulent boundary conditions for a spatially variable flow (refer to Section 4.4.6). In particular, this makes it possible to validate the results by comparing them with those achieved through a finite volume type approach. The following is in accordance with Issa et al. (2004), as well as Violeau and Issa (2007a).

We consider an infinite length channel which is invariant along the transverse direction y, periodic along axis x, and featured by the periodic occurrence of hills at the lower wall. This case can then be treated two-dimensionally and, because of the periodicity of the configuration, a single hill may be contemplated. The geometry of the latter is defined by its ordinate z, as given by the following six functions (refer to Uribe and Laurence, 2002):

$$z(x) = \begin{cases} a_{00} + a_{02}x^2 + a_{03}x^3 & \text{if } \quad 0 \leqslant x < 9 \\ a_{10} + a_{11}x + a_{12}x^2 + a_{13}x^3 & \text{if } \quad 9 \leqslant x < 14 \\ a_{20} + a_{21}x + a_{22}x^2 + a_{23}x^3 & \text{if } 14 \leqslant x < 20 \\ a_{30} + a_{31}x + a_{32}x^2 + a_{33}x^3 & \text{if } 20 \leqslant x < 30 \\ a_{40} + a_{41}x + a_{42}x^2 + a_{43}x^3 & \text{if } 30 \leqslant x < 40 \\ a_{50} + a_{51}x + a_{52}x^2 + a_{53}x^3 & \text{if } 40 \leqslant x < 54 \end{cases} \tag{7.1}$$

where x and z stand for the horizontal and vertical distances, respectively, being reckoned from the origin as indicated by the halftone shaded disc in Fig. 7.2. If these distances are measured in millimeters, the coefficients $\{a_{ij}\}$ are given by Table 7.1. The lengths corresponding to Fig. 7.2 then are $L = 252$ mm, $l = 54$ mm, $h_H = 28$ mm and $h_I = 56.98$ mm.

The fluid being considered is water ($\rho_0 = 1,000$ kg/m^3 and $\nu = 10^{-6}$ m^{2}s^{-1}, as per Table 3.2). Steady flow conditions are set, featured by a Reynolds number $Re \doteq U h_H/\nu$. The latter is constructed from the height of the hill and the averaged velocity U, which is defined by a formula identical to (4.193)

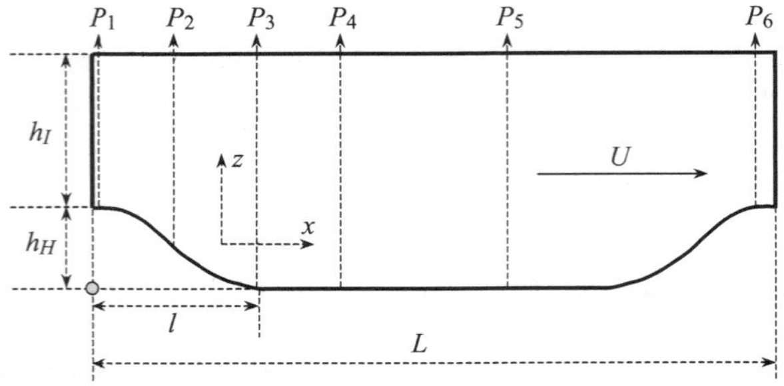

Fig. 7.2 Periodic hill geometry, after Issa et al. (2004). The spatial origin is illustrated by the halftone shaded disc. The solid edges are illustrated by thick lines, the boundaries provided with periodicity conditions are illustrated by thicker lines.

Table 7.1 Periodic hill case. Coefficients defining the geometry as illustrated in Fig. 7.2 (after Uribe and Laurence, 2002).

a_{ij}	$j = 0$	$j = 1$	$j = 2$	$j = 3$
$i = 0$	28.000	0	6.775×10^{-3}	-2.125×10^{-3}
$i = 1$	25.074	9.755×10^{-1}	-1.016×10^{-1}	1.889×10^{-3}
$i = 2$	25.796	8.207×10^{-1}	-9.055×10^{-2}	1.627×10^{-3}
$i = 3$	40.464	-1	1.946×10^{-2}	-2.070×10^{-4}
$i = 4$	17.925	8.744×10^{-1}	-5.567×10^{-2}	6.277×10^{-4}
$i = 5$	56.390	-1	1.645×10^{-2}	2.675×10^{-5}

through an integration over an arbitrary vertical section. A laminar flow regime is considered, characterized by a Reynolds number $Re = 50$, which corresponds to $U = 1.786 \times 10^{-3}$ m/s. The flow is determined by a mass-specific horizontal force F^{ext} which is added into the right-hand side of the Navier–Stokes equation of motion and which acts as a driving pressure gradient (refer to Section 4.4.5). F^{ext} is then to be regarded as one of the external forces and here is substituted for $\left|\mathbf{F}_a^{ext}\right| / m_a$ in equation (5.203). Since the velocity profile is not known in advance, there is no theoretical link between F^{ext} and the value of U to be desirably prescribed. This shortcoming can be palliated by matching F^{ext} upon each iteration ^{m} according to the law

$$F^{ext,m+1} = \frac{2\left(U^m - U\right) - \left(U^{m-1} - U\right)}{2\delta t} \tag{7.2}$$

where U^m denotes the averaged velocity as constructed from those velocities calculated upon the corresponding iteration.[3] The initial value $F^{ext,0}$ may be taken as zero.

[3]The above-mentioned process should, in principle, be analyzed through a numerical stability study, on the same grounds as Section 5.4.4.

At the initial time, the particles lie a distance $\delta r = 1$ mm apart on a regular Cartesian grid and are not moving about at any velocity. Their density is set to the reference value ρ_0 and their pressure is zero. Here a fourth-order, B-spline type kernel (eqn (5.51)) is considered, for a ratio $h/\delta r = 1.2$. The walls are modelled by edge particles and three layers of dummy particles, according to the procedure described in Section 6.3.2. Using these data, the system gathers 19,548 fluid particles and 2,008 particles for wall modelling purposes.

The weakly compressible scheme based on an equation of state (refer to Section 5.3.2) is adopted. The numerical speed of sound c_0 is equal to 10 times the maximum flow velocity $U_{\max}$, in accordance with the recommendations in Section 5.3.2 (eqn (5.212), in the absence of gravity actions). Only knowing the average velocity, it is estimated that $U_{\max}$ is about twice as much, which remains consistent with the applicable law (4.196) in a rectilinear channel. We then choose $c_0 = 0.03$ m/s. Hence, the simulation time step, which here is set by the CFL condition in equation (5.413), equals $\delta t = 1.6 \times 10^{-2}$ s. The number of iterations is set to 1.7×10^5, and so a particle can be transported over a distance 20 times longer than the channel length. This option is sufficient for establishing a steady regime; the simulated physical time is then 2.720 s.

Since no theoretical or experimental data is available for this case, the results achieved with SPH are compared with those stemming from the Eulerian calculation model CODE_SATURNE, a software which was developed and validated at EDF R&D from a finite volume method (Uribe and Laurence, 2002). Figure 7.3 shows that the axial and vertical velocity fields are quite similar to those obtained with CODE_SATURNE. Furthermore, as shown in Fig. 7.4, the recirculation area is properly modelled by SPH. These conclusions, however, should be mitigated by specifying that the results in Fig. 7.3 were obtained through a temporal averaging of the results, once the system convergence was established (Issa, 2005), as allowed by the steady nature

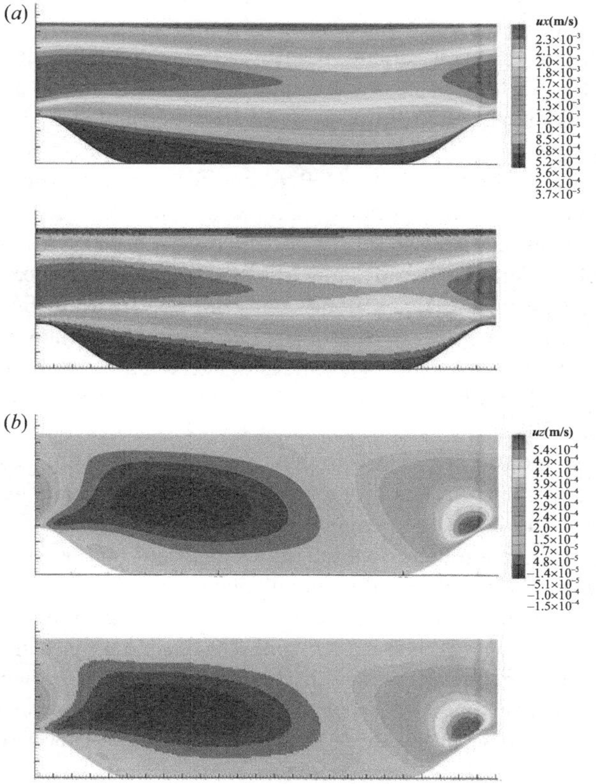

Fig. 7.3 Periodic hill case. Spatial distributions of the horizontal (*a*) and vertical (*b*) velocities on completion of convergence. Comparison between the finite volume (top) and SPH (bottom) methods after Violeau and Issa (2007a, with permission from Wiley & Sons).

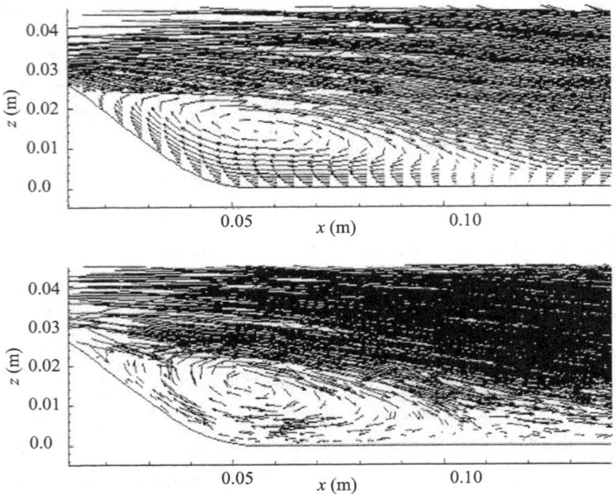

Fig. 7.4 Periodical hill case. Comparison between the finite volume (top) and SPH (bottom) methods. Velocity vectors at the recirculation, after Violeau and Issa (2007a, with permission from Wiley & Sons).

of the flow. Without this procedure, the results are not as smooth as those regarding the finite volumes (particularly in terms of vertical velocities). These irregularities, which will not be discussed here, can partly be ascribed to the Lagrangian nature of the SPH method and remain quantitatively permissible. There are several means which are more attractive than a temporal average to get free of them, for example the incompressible algorithm as disclosed in Section 6.2.5 (not applied in this case).

In order to provide that validation with a more quantitative nature, we will now consider the horizontal velocity profiles along the cross sections $\{P_i\}$ $(i = 1, \ldots, 6)$ illustrated in Fig. 7.2, having respective abscissae as given by $x(P_i)/h_H = \alpha_i$, with $\{\alpha_i\} = (0.05; 1; 2; 3; 5; 8)$. Figure 7.5 shows that the profiles achieved with CODE_SATURNE and SPH are very similar. It should be pointed out that the SPH profiles in Fig. 7.5 result from a three-dimensional modelling (Moulinec et al. 2008). As compared with the just disclosed two-dimensional modelling, the advantage is that the number of particles interacting with a given particle is significantly greater (refer to Section 5.2.3). The profiles, which are achieved through the discrete interpolation (5.86), then become much smoother. The two-dimensional modelling, however, provides outstanding results with respect to the velocity profiles, albeit less regular (refer to Issa et al., 2004).

In addition, the profiles related to the P_2, P_3 and P_4 sections clearly reveal the presence of the recirculation area. The latter can be characterized by the abscissa x_S of its detachment point, as well as the abscissa x_R of its reattachment point, though these two quantities can only be observed in a rather imprecise way. The resulting values are displayed in Table 7.2. It can be found that the place of the detachment point is perfectly determined through SPH. The difference as regards the reattachment point is greater and may be related to the Lagrangian nature of the model. This is because the particle motion only allows a statistical estimation of that point. It is noteworthy, however, that there is no reason to decide in favour of either of the results of the two models as long as experimental data are not available.

Table 7.2 Periodic hill case. Abscissae of the detachment (x_S) and reattachment (x_R) points as predicted through numerical calculation.

Method	SPH	Finite volume
x_S (mm)	14.6	14.7
x_R (mm)	115	120

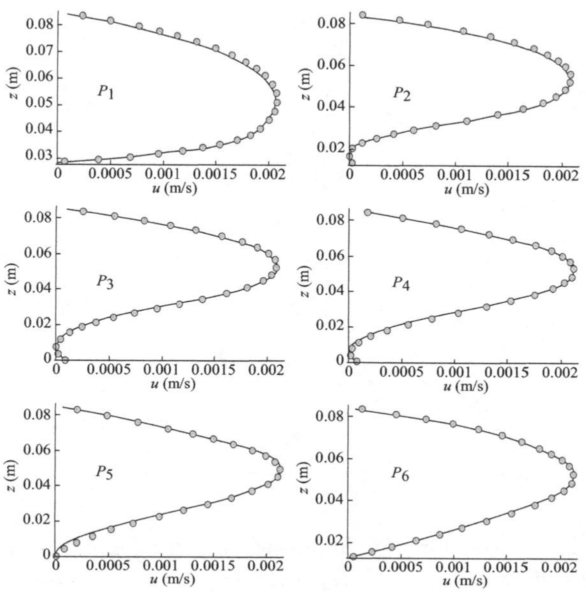

Fig. 7.5 Periodic hill case. Comparison between the finite volume (solid lines) and SPH (symbols) methods. Axial velocity profiles for each of the cross sections as shown in Fig. 7.2, after Moulinec et al. (2008).

This test case demonstrates the SPH method's ability to simulate incompressible non-gravity flows featured by the presence of a recirculation area in a periodic regime. This is because the velocity profiles are very similar to those stemming from a well tried and tested Eulerian model. Moreover, the boundary conditions as prescribed at the walls and the periodic boundaries are satisfactory.

This case was tested for sensitivity of the results to various parameters (Violeau, 2009b). The velocity profiles are found to be hardly sensitive to the discrete SPH operators selected for modelling the equations of continuity and motion (systems (5.240) and (5.241)). Variationally inconsistent operators (refer to Section 5.3.3) do not impair the simulation in this particular case. Lastly, all three discrete models for the viscous forces in a weakly compressible regime (eqns (6.42), (6.43) and (6.44)) give quite similar results. On the contrary, the interpolation equation (5.192) for the density is insufficient in this case for predicting well-founded results. All those we have set out were obtained through the discrete continuity equation (5.194), the value of the superscript k having no effect on the results.

The case of the three-dimensional periodic hill has also been addressed in a turbulent flow regime ($Re = 10,595$) by Issa (2005), on the basis of the LES as disclosed in Section 4.6.1 and 6.5.1. A further similar, laminar case in the form of a periodic backward-facing step can be found in Issa et al. (2004).

7.2.2 Lid-driven cavity flow

Here we propose a relevant case for testing the SPH ability to replicate the driven flow of a fluid in a closed cavity, induced by the motion of one of the walls and in the absence of gravity. Due to the viscous forces, the fluid is

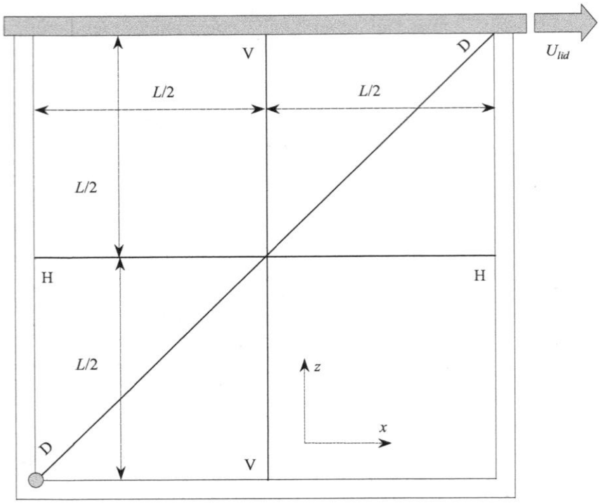

Fig. 7.6 Lid-driven cavity flow case. Description of the geometry after Lee et al. (2008), and positions of the horizontal (HH), vertical (VV) and diagonal (DD) monitoring sections. The spatial origin is illustrated by the halftone shaded disc.

then set into a rotational motion in the cavity, getting a more or less complex structure according to the problem parameters. This case is fully described in Rogers (2006). The wall condition modelling is important in this case, more specifically the modelling of the viscous term in the Navier–Stokes equation, which accounts for the diffusion of the momentum within the flow, hence, ultimately, for the flow itself. Here we follow the ideas of Lee et al. (2008).

The square cavity of side L illustrated in Fig. 7.6 is considered, the flow being presumably two-dimensional. The top wall is moving rightward with a constant velocity U_{lid}. The fluid (water, i.e. $\rho_0 = 1,000$ kg/m^3 and $\nu = 10^{-6}$ m^{2}s^{-1}) will then start rotating substantially clockwise. To be honest, the numerical values of these parameters are not critical, and it is the same with L and U_{lid}, since the validations which are proposed below are dimensionless. The analysis in Section 3.4.3 shows, indeed, that the characteristics of this flow are only based on the value of the Reynolds number $Re \doteqdot U_{lid}L/\nu$. Here we consider the values $Re = 400$ and $Re = 1,000$, for the measurements and simulations that are available in the scientific literature. Thus, they are laminar flows. These values of Re are governed by the choice[4] of U_{lid}.

At the initial time, the particles are spaced a distance $\delta r = L/N_d$ apart on a regular Cartesian grid and their velocity is zero. It is contemplated to study the sensitivity of the results to the value of the integer N_d, to which the values 40, 70, 100 and 160 are assigned. The inital density of the particles equals the reference value ρ_0 and their pressure is zero. A fourth-order, B-spline type kernel (eqn (5.51)) is being considered, for a ratio $h/\delta r = 1.3$. The walls are modelled by edge particles and three layers of dummy particles, according to the procedure described in Section 6.3.2.

At first, the weakly compressible sheme based on an equation of state (refer to Section 5.3.2) is utilized. The numerical speed of sound c_0 is determined by eqn (5.212), as expected in the absence of gravity. $U_{\max}$ is not known in advance, but it seems obvious that it should equal U_{lid}. Initial tests, however, highlight a void pocket at the centre of the cavity, which can be reduced provided that a factor 100 is substituted for the factor 10 in (5.212). We then

[4] On the contrary, U_{lid} and L might be set to one, the value of Re being determined through the selected viscosity ν. We may recall, indeed, that the factor $1/Re$ plays the part of viscosity within a dimensionless context (refer to Section 3.4.3).

ultimately choose $c_0 = 100U_{lid}$. Thus, the simulation time step, which is here set by equation (5.413), is yielded by

$$\delta t = 5.2 \times 10^{-3} \frac{L}{N_d U_{lid}} \tag{7.3}$$

The required number of iterations for establishing a steady regime is then independent of the L and U_{lid} parameters, as well as the discretization N_d, of course. On the other hand, this number depends on Re, as long as the latter is not very high (refer to Section 3.4.3). The numerical experience shows that 40,000 iterations are enough for the presently adopted values of Re.

We also intend to test the incompressible scheme as described in Section 6.2.5. The time step is then based on the fluid velocity, as it is known (condition (6.128)), which allows 100 times less iterations, with the option we have adopted for c_0.

Both lengths and velocities will be non-dimensionalized as follows, in accordance with (3.151):

$$\begin{aligned} \forall i,\ x_i^+ &\doteqdot \frac{x}{L} \\ u_i^+ &\doteqdot \frac{u_i}{U_{lid}} \end{aligned} \tag{7.4}$$

Figure 7.7 illustrates the shape of the velocity field near the upper left and lower right corners for $Re = 1,000$. The fluid trend to rotate clockwise can

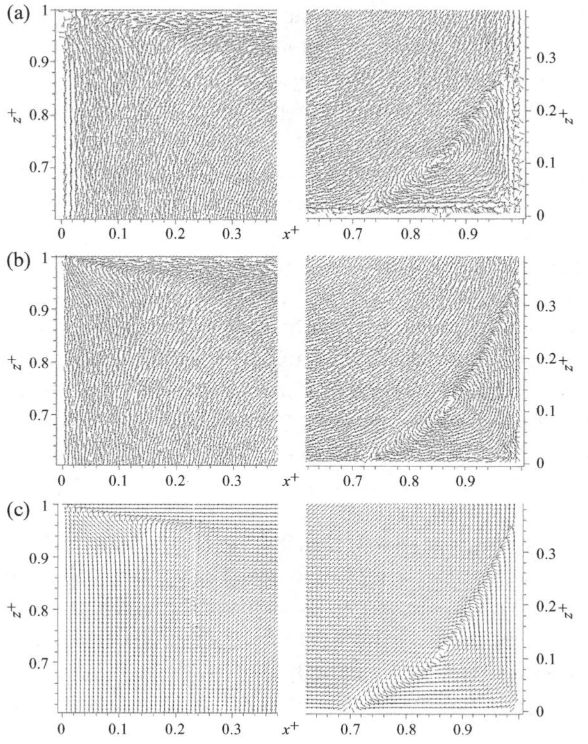

Fig. 7.7 Lid-driven cavity flow case, $Re = 1,000$. Spatial distribution of the velocity fields in the upper left (left) and lower right (right) corners. Predictions of the nearly-incompressible (*a*) and incompressible (*b*) SPH methods, and finite volumes (*c*). After Lee et al. (2008, with permission from Elsevier).

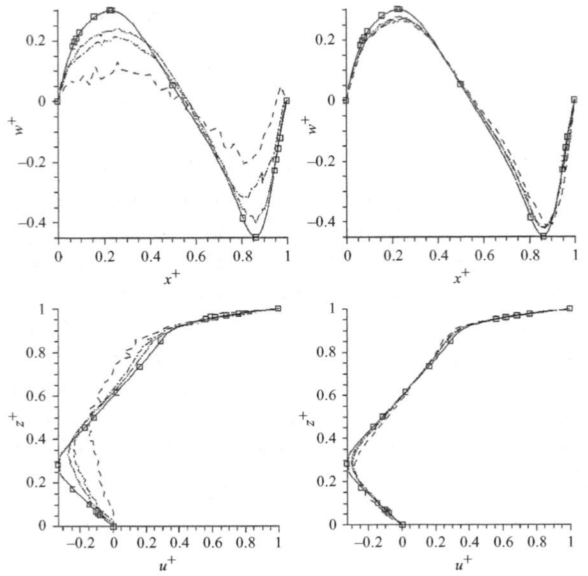

Fig. 7.8 Lid-driven cavity flow case, $Re = 400$. Profiles of dimensionless velocity u_i^+ along the sections in Fig. 7.6. Vertical velocity along HH (top) and horizontal velocity along VV (bottom). Comparison of the nearly-incompressible (left) and incompressible (right) SPH methods for several discretizations, that is several values of N_d: 40, 70 and 100 (dotted lines), with the finite volumes (STAR_CD model, solid lines) and the observations made by Ghia et al. (1982, symbols). After Lee et al. (2008, with permission from Elsevier).

be seen, with a counter-clockwise recirculation near the lower right corner. This flow configuration is confirmed by Chern et al. (2005). There the weakly compressible and incompressible schemes of the SPH method are compared with the predictions made by the commercially available software STAR_CD, which is based on a finite volume technique, with a discretization corresponding to $N_d = 128$. It can be found that the general shape of the flow is properly replicated by both SPH schemes, the incompressible algorithm exhibiting, however, a smoother and higher-grade velocity field, particularly as regards the recirculation. Figure 7.8 quantitatively confirms that qualitative analysis by illustrating profiles of dimensionless velocity u_i^+ for $Re = 400$. It is the horizontal distribution of vertical velocity ($i \equiv z$) along axis HH, as well as the vertical distribution of horizontal velocity ($i \equiv x$) along axis VV, both of them being illustrated in Fig. 7.6. The spatial origin is shown by the halftone shaded disc in the same figure. It can be seen that those profiles achieved by the weakly compressible scheme are excessively fluctuating, whereas those provided by the incompressible method are significantly smoother, close to the finite volume profiles and the measurements proposed by Ghia et al. (1982). Figure 7.8 also suggests the convergence of the SPH method, the profiles becoming similar to those predicted by STAR_CD as N_d increases. $N_d = 100$ provides profiles which are satisfactory in that respect.

Figure 7.9 illustrates the spatial distribution of pressure with both schemes in the SPH method, for $Re = 1,000$. Pressure is very poorly predicted by the weakly compressible scheme, as can be seen, exhibiting great spatial variations which make interpretation of the figure barely feasible, whereas the incompressible scheme provides incomparably more regular profiles. Quantitatively, a dimensionless pressure may attractively be defined along the lines of eqn (3.151):

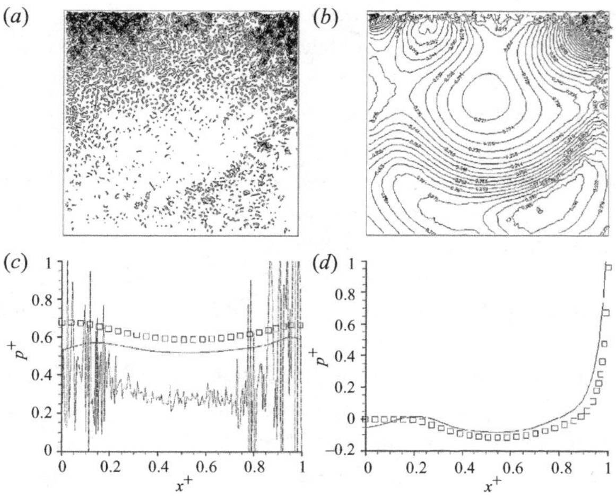

Fig. 7.9 Lid-driven cavity flow case, $Re = 1,000$. Pressure distribution within the cavity with nearly-incompressible (*a*) and incompressible (*b*) SPH. Profiles of dimensionless pressure p^+ along the sections HH (*c*) and DD (*d*) in Fig. 7.6: comparison of the nearly-incompressible (dotted lines) and incompressible (solid lines) SPH methods for $N_d = 160$, with the finite volumes (symbols). After Lee et al. (2008, with permission from Elsevier).

$$p^+ \doteqdot \frac{p}{\frac{1}{2}\rho_0 U_{lid}^2} \tag{7.5}$$

(we may note, however, the presence of a $1/2$ factor with respect to (3.151)). Figure 7.9 illustrates the spatial distributions of p^+ along two profiles, namely HH and DD, a diagonal profile displayed in Fig. 7.6. For $N_d = 160$, the dimensionless scheme is very close to the predictions through the finite volumes.

7.2.3 Infinite open channel

Let us now turn to a simple generic case, namely the infinite channel, whose theory is set out in Chapter 4, Section 4.4.5. In spite of its seeming simplicity, it will make it possible, for the first time in this chapter, to test the validity of the turbulence models we have described in it, as well as of their discrete forms as disclosed in Chapter 6. Thus, this case is crucial for testing the efficiency of the modelling of the Reynolds equation viscous term having a spatially-varying eddy viscosity. For the first time, we also consider the influence of gravity. Here we follow the ideas of Issa (2005) and Violeau and Issa (2007a).

The channel of Fig. 4.7 is considered, the flow being presumably two-dimensional, with a water depth $H = 0.4$ m. Since the flow is invariant along x, the channel length does not matter, the flow being periodical along x. The fluid with a reference density $\rho_0 = 1,000$ kg/m^3 flows at the average velocity U under the action of a driving force (refer to Section 7.2.1). A laminar flow, featured by $U = 0.25$ m/s and $\nu = 10^{-2}$ m^{2}s^{-1}, corresponding to a Reynolds number $Re \doteqdot UH/\nu$ of 10 is first considered, then a turbulent regime ($U = 0.753$ m/s and $\nu = 10^{-6}$ m^{2}s^{-1}, that is $Re = 3.012 \times 10^5$). In the turbulent case, an equivalent bottom roughness $k_s = 1$ cm (refer to Section 4.4.5) is defined. The shear velocity, as provided by (4.194), then equals $u_* = 0.0435$ m/s. The shear Reynolds number, as defined by (4.182), equals $Re_* = 435$: this is a rough turbulent regime.

At the initial time, the particles are spaced by a distance $\delta r = 1$ cm apart on a regular Cartesian grid, and their velocity obeys a linear profile between the bottom and the surface. Their density equals the reference value ρ_0 and their pressure is zero. Here a fourth order, B-spline type kernel (eqn (5.51)) is considered, for a ratio $h/\delta r = 1.5$. The walls are modelled by edge particles and three layers of dummy particles, according to the procedure as described in Section 6.3.2.

The weakly compressible scheme based on an equation of state (refer to Section 5.3.2) is adopted. The numerical sound velocity c_0 is determined by eqn (5.213), as it should be in the presence of gravity. As a precaution, as in the preceding section, we further increase the factor 10 in the law (5.213), and here put $c_0 = 30U_{\max}$. In the laminar case, the maximum velocity is derived from the law (4.196); thus, we choose $c_0 = 11$ m/s. The simulation time step, which here is set by the CFL condition of Equation (5.413), then equals $\delta t = 5.45 \times 10^{-4}$ s. In the turbulent case, the maximum velocity, being derived from the theoretical profile (4.183), equals $U_{\max} = 0.858$ m/s, which yields $c_0 = 26$ m/s and $\delta t = 2.31 \times 10^{-4}$ s. In both cases, the results are treated after a steady regime has been achieved, which occurs after about 10^5 iterations.

For modelling the turbulence, we contemplate the discrete models $k - L_m$ (with the profile (4.175) for the mixing length) and $k - \varepsilon$ (§ 6.4.2). The condition (4.192) for the particles lying at the surface is prescribed.

Figure 7.10 illustrates the pressure and longitudinal velocity profiles achieved in a laminar regime, as compared with the theoretical profiles (4.166) and (4.171). It can be found that both of them are fairly well replicated by the model. In a turbulent regime, Fig. 7.11 illustrates the profiles of the turbulent kinetic energy k, its dissipation rate ε, of the eddy viscosity ν_T and of the mean longitudinal velocity $\overline{u}$ as a function of $z^+ \doteqdot z/H$. The first three above quantities are adimensionalized and denoted as k^+, ε^+, ν_T^+, in accordance with the formulas (4.188). As can be seen, ε and $\overline{u}$ perfectly match the semi-analytical formulas (4.183) and (4.188), whereas the profile of k as predicted by the model $k - L_m$ is very close to the values as calculated by Warner et al. (2005). In addition, the figure indicates the empirical profile as proposed by Nezu and Nakagawa (1993):

$$k^+ = C_k e^{-2z^+} \tag{7.6}$$

with $C_k = 4.78$. As to ν_T, the model $k - \varepsilon$ provides predictions which are rather close to the experiments conducted by Nezu and Nakagawa (1993). It should be pointed out that the numerical prediction of that quantity is a delicate

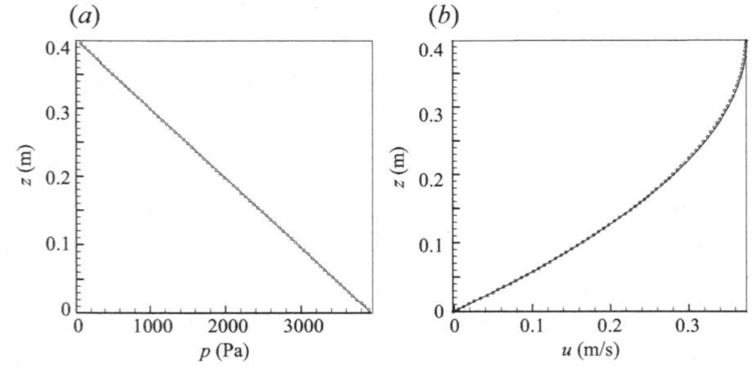

Fig. 7.10 Infinite channel case, laminar regime. Pressure (*a*) and velocity (*b*) profiles between the bottom and the surface. The SPH results (symbols) are compared with the theoretical profiles (4.166) and (4.171) (solid lines).

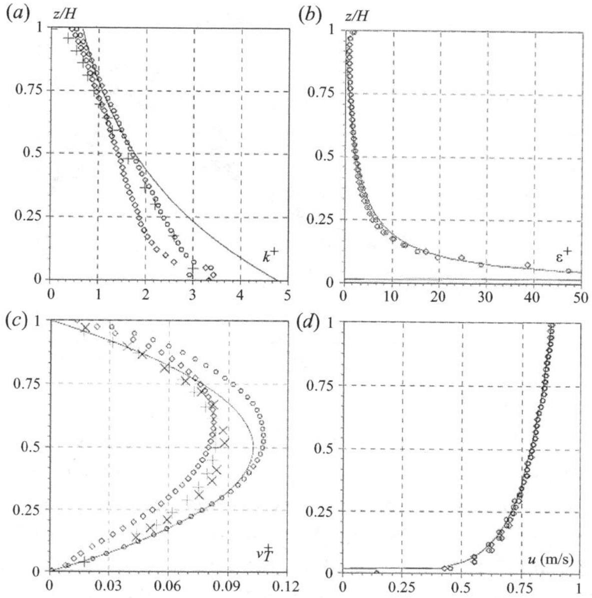

Fig. 7.11 Infinite channel case, turbulent regime. Profiles of dimensionless quantities k^+ (*a*), ε^+ (*b*), ν_T^+ (*c*) and $\overline{u}$ (*d*) betwen the bottom and the surface. The SPH results with the $k - L_m$ (o) and $k - \varepsilon$ (◇) models are compared with the formulas (4.183), 4.188) and (7.6) (solid lines) and with experimental (×, after Nezu and Nakagawa, 1993) and numerical (+, after Warner et al., 2005) data. This figure is taken from Violeau and Issa (2007a, with permission from Wiley & Sons).

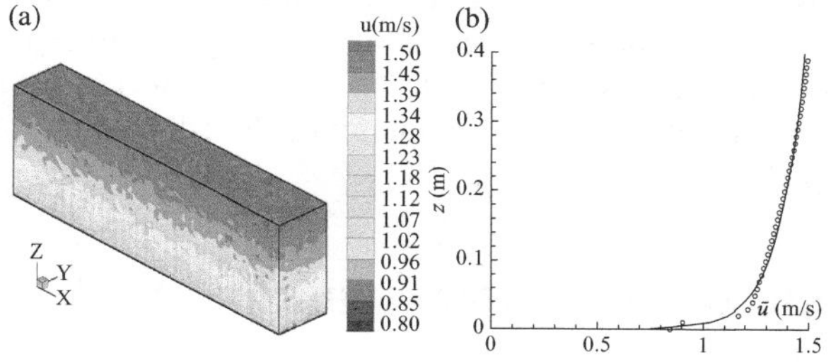

Fig. 7.12 Infinite channel case, turbulent regime. SPH calculations with a LES type model for the turbulent closure. Spatial distribution of the filtered longitudinal velocities (*a*) and vertical distribution of their Reynolds average (*b*). The solid line represents the logarithmic law (4.183). After Violeau and Issa (2007a, with permission from Wiley).

task, and that there is no precise 'standard' profile in as simple a case as that adopted here. As regards the measurements, they are naturally rather difficult to carry out. The present profiles are in accordance with the DNS available in the literature (for instance, refer to Moser et al., 1998). The $k - \omega$ model (which is not disclosed here) provides results which are similar to the $k - \varepsilon$ model. Violeau and Issa (2007a) explain that, in this case, the non-linear model $k - \varepsilon$ (refer to Section 6.4.2) relatively trivially improves the predictions.

The LES approach (refer to Section 6.5.1) was applied to this case (as expected in three dimensions) by Issa (2005), with a 0.2 m wide and 1.2 m long channel, for a depth-averaged velocity $U = 1.345$ m/s (i.e. $Re = 5.38 \times 10^5$). The method provides good results in terms of mean longitudinal velocities $\overline{u}$ (refer to Fig. 7.12). To this purpose, the fields are time-averaged, considering that this is a steady flow. Besides, the figure also provides an instantaneous view of the filtered longitudinal velocity field $\tilde{u}$, that is taking the large eddy fluctuations into account (refer to Section 4.6.1).

Lastly, an analogous test case was the subject of a stochastic modelling of the turbulence, on the basis of the simplified Langevin model as disclosed in

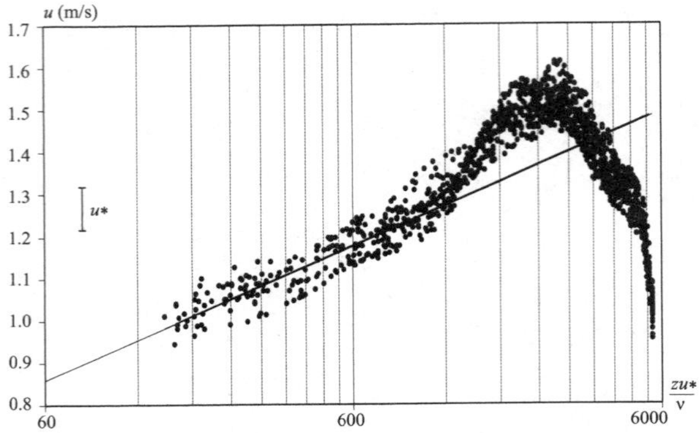

Fig. 7.13 Infinite closed channel case, turbulent regime. SPH calculations with a stochastic model for the turbulent closure. Distribution of the stochastic longitudinal velocities as a function of the distance to the lower wall, the latter being represented on a logarithmic scale. The solid line represents the logarithmic law (4.183). After Violeau et al. (2001).

Section 6.5.3 (refer to Violeau et al., 2001). This time, it is a closed channel, a wall being substituted for the free surface, and accordingly gravity may be disregarded. It is shown (for instance, refer to Pope, 2000) that the theoretical arguments in Section 4.4.5 are unchanged in some vicinity of each of the two walls. Figure 7.13 shows the longitudinal velocity field resulting from equation (6.326), as a function of the distance to the lower wall, which is adimensionalized on the basis of the shear velocity and the molecular viscosity. It can be found that the rough logarithmic profile (4.183) is properly replicated on an average, the fluctuations of longitudinal velocity u'_a duly having the order of magnitude of u_* (refer to Chapter 4). The closed channel case was also simulated with good results by means of the mixing length model (Section 6.4.1) (Violeau et al., 2001).

7.3 Collapse of a water column

The collapse of a water column is a class of simplified flows that is particularly propitious for assessing one of the specificities of the SPH method, namely its suitability for modelling greatly strained flows which are determined by both gravity and inertia, along with a comparatively strong impact onto a wall. Many quantitative and qualitative validations are possible.

7.3.1 Collapse onto a dry bottom

The presently discussed case is suitable for testing the SPH ability to predict the behaviour of a rapidly varying free surface turbulent flow, in the presence of solid walls under the action of gravity. This case consists of a rectangular fluid column collapsing onto a two-dimensional basin, which may be likened to a schematic dam failure. The wall condition modelling faces a serious challenge, since the break wave rapidly propagates towards the opposite vertical wall closing the basin. In addition, this case shows that complex free surfaces can be treated, since, upon the return of the wave, a wave-breaking takes place and is followed by an intense turbulent mixing. Here we follow the ideas of Issa (2005), as well as Violeau and Issa (2007a).

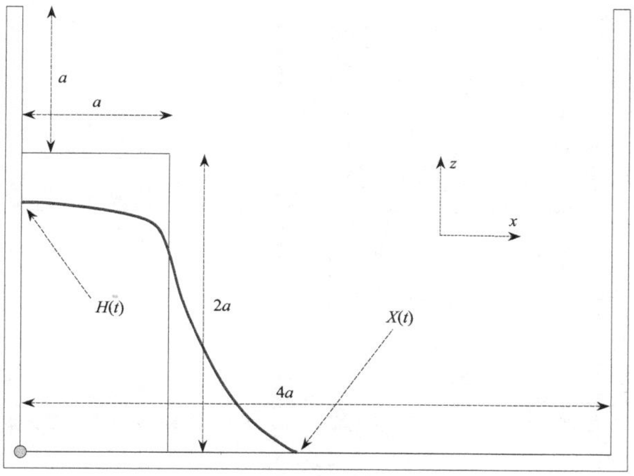

Fig. 7.14 Case of the collapse of a water column onto a dry bottom, after Koshizuka and Oka (1996). The initial configuration is shown by a thin line and a subsequent configuration by a thick line. X and H are also defined, the spatial origin being illustrated by the halftone shaded disc.

Figure 7.14 illustrates the presently adopted geometry, in order to refer to the experiment conducted by Koshizuka and Oka (1996). It is considered that the flow is two-dimensional, the origin of the coordinates lying in the lower left corner of the fluid (halftone shaded disc in the figure). Here the length a equals 1 m, which is not really important since the subsequently proposed validations are dimensionless. The fluid is water ($\rho_0 = 1,000$ kg/m^3 and $\nu = 10^{-6}$ m^{2}s^{-1}).

At the initial time, the particles are spaced by a distance $\delta r = 1$ cm apart on a regular Cartesian grid, and are not moving about at any velocity. Their density equals the reference density ρ_0 and their pressure is zero. Here a fourth-order, B-spline type kernel (eqn (5.51)) is considered, for a ratio $h/\delta r = 1.5$. The walls are modelled by edge particles and three layers of dummy particles, according to the procedure described in Section 6.3.2. With these data, the system gathers 20,000 fluid particles and 5,045 particles for wall modelling.

The weakly compressible scheme based on a equation of state (refer to Section 5.3.2) is adopted. The numerical speed of sound c_0 is determined by equation (5.213), as it should be in the presence of gravity. U_{max} is not known in advance, but it is close to 12 m/s, as evidenced by a preliminary numerical experience. We then choose $c_0 = 120$ m/s. The condition with respect to the water depth in (5.213) does not alter that value, unless the depth exceeds 15 m upon the uprush following the impact against the wall, but this is most unlikely to happen. In addition, it should be pointed out that this condition is based on the assumption of a hydrostatic pressure, which probably proves invalid upon that runup phase due to the significant vertical inertia of the fluid during that stage. Thus, the simulation time step, here being set by equation (5.413), is 5×10^{-5} s. The number of iterations is set to 5×10^4, which corresponds to a simulated physical duration of 2.5 s.

Here the flow is turbulent, but the inertia, gravity and pressure forces initially prevail over the Reynolds stresses.[5] Utilizing a turbulent model would therefore be useless, at least during the first stages of the flow. On the other hand, immediately upon the impact of the fluid against the right-hand wall of the reservoir, the high deformations generate an intense turbulent kinetic

[5] As stated in the introduction to Chapter 4, a comprehensive theory still has to be developed for generally assessing the extent of turbulence within a flow, particularly when it has a free surface. Here we can intuitively justify our assertions, which would be substantiated by a calculation of orders of magnitude through a method analogous to that in Section 3.4.3.

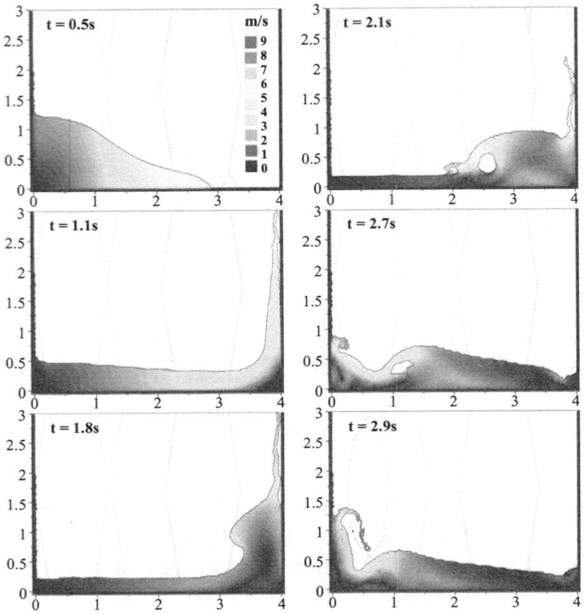

Fig. 7.15 Case of the collapse of a water column on a dry bed. From right to left and bottom to top: changes in the flow upon the impact against the wall, then upon the wave-breaking (the shading illustrates the velocity norm). After Violeau and Issa (2007a, with permission from Wiley & Sons).

energy (refer to Chapter 4) which diffuses the momentum and exerts a strong impact on the shape of the free surface. Without any experience for accurately validating that effect, we consider for now that sticking to a turbulence-free model is satisfactory. This approximation is corroborated by simulations with turbulence models which do not alter the first flow phases.

Figure 7.15 illustrates the evolution of the resulting flow, as well as the spatial distribution of the velocity norm at various times. In order to perform the validation of the initial instants, we will use the maximum abscissa X of the fluid particles, as well as the maximum water depth H along the left-hand wall (refer to Fig. 7.14). The evolution of these two quantities over time t is considered. Distance a and gravity g are obviously the parameters involved in this process. Thus, for *each* of the X and H quantities, we have $n = 4$ parameters (X or H, g, t, a) out of which $m = 2$ are independent, and the considerations in Section 3.4.3 allow to state that we can construct $n - m = 2$ independent dimensionless numbers. Hence we get

$$\begin{aligned} X^+ &= \phi_X\left(t^+\right) \\ H^+ &= \phi_H\left(t^+\right) \end{aligned} \tag{7.7}$$

where ϕ_X and ϕ_H are unknown functions, and with the conventional definitions

$$\begin{aligned} t^+ &\doteq t\sqrt{\frac{2g}{a}} \\ X^+ &\doteq \frac{X}{a} \\ H^+ &\doteq \frac{H}{2a} \end{aligned} \tag{7.8}$$

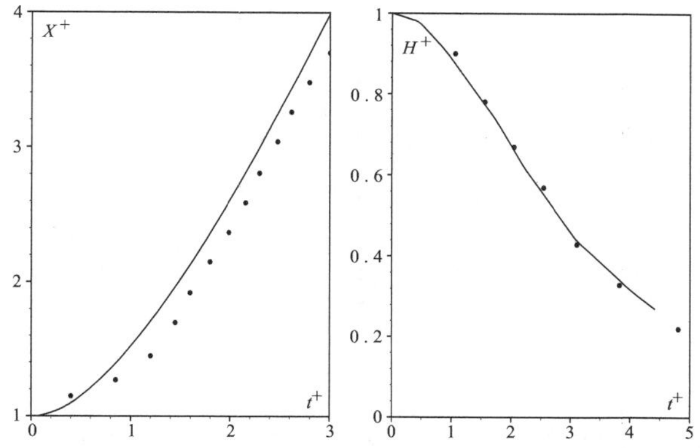

Fig. 7.16 Case of the collapse of a water column on a dry bed. Evolutions of X^+ and H^+ as a function of time t^+ (eqn (7.8)): comparison of the SPH results (solid lines) with the measurements made by Koshizuka and Oka (1996) (symbols). After Violeau and Issa (2007a).

Figure 7.16 displays the resulting graphs, being compared with the experimental results as achieved by Koshizuka and Oka (1996). It can be found that the results are satisfactory, in spite of a slight offset along X^+, which is likely to be caused, especially during the early flow, by the solid boundary condition near the front, where the detailed experimental observations reveal the occurrence of an initial breaking phenomenon which can barely be modelled (Stansby et al., 1998). The observations made by Koshizuka and Oka (1996) show, indeed, at the very beginning of the column collapse, an acceleration of the front which is greater than that suggested by our simulations, which acceleration might result from this kind of behaviour.

We may recall that the simulations set out so far are based on a 'dummy particles' type approach for the treatment of the walls. It is interesting to study the behaviour of the wall force model (6.156) disclosed in Section 6.3.2. The coefficient C occurring in the formula (6.157) is then set to 7.5 for $H \doteqdot 2a$, which yields $E_0 = 15\ \mathrm{kg.m^2s^{-2}}$. The total number of particles is then reduced from 25,045 to 21,001. Figure 7.17 shows that the flow shape is globally identical, but with significant local differences near the edges. It can be seen that the dummy particle method is more stable, although it has the shortcoming of letting fluid particles adhere to the wall and stay on it when the fluid itself has flown away.

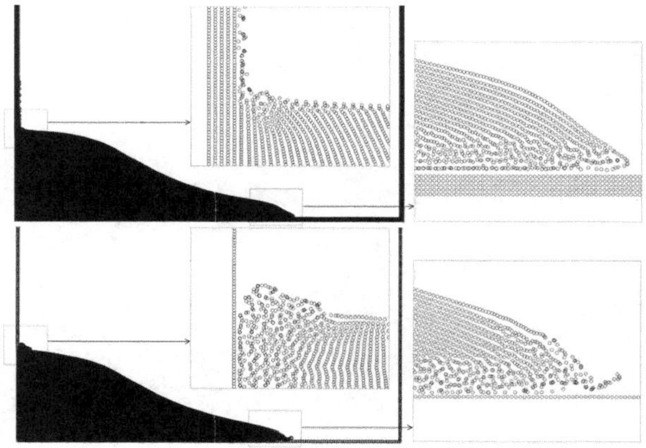

Fig. 7.17 Case of the collapse of a water column on a dry bed. Positions of the particles for t = 0.5 s, on top with the dummy particle model, on bottom with the Lennard–Jones wall forces.

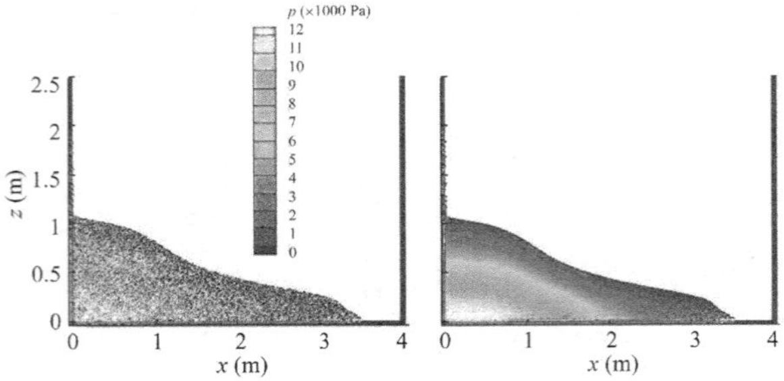

Fig. 7.18 Case of the collapse of a water column on a dry bed. Spatial distribution of the pressure at $t = 0.6$ s, without density smoothing (left) and with the smoothing (7.9) (right).

Figure 7.18 illustrates the spatial distribution of the pressure at $t = 0.6$ s. It exhibits great spatial variabilities resulting from the use of the weakly compressible scheme. As in the case described in Section 7.2.2, the incompressible scheme disclosed in Section 6.2.5 here provides for a much better prediction of pressure, as demonstrated by Lee et al. (2008). Another way of abating these fluctuations, which is less computationally costly and more in line with the spirit of the SPH method, consists of utilizing a gravity smoothing analogous to the XSPH model (6.266):

$$\forall a, \ \rho_a \leftarrow \breve{\rho}_a \doteq \rho_a + \epsilon_\rho \sum_b m_b \frac{\rho_{ab}}{\overline{\rho}_{ab}} w_{ab} \tag{7.9}$$

where $\overline{\rho}_{ab}$ is defined by the general formula (6.9),[6] whereas $\rho_{ab} \doteq \rho_a - \rho_b$. The parameter ϵ_ρ is set to 0.05. Figure 7.18 then shows markedly improved results, but without achieving the same performance as the incompressible scheme.

[6]Care should be exercised not to confound this with the definition (6.267).

The same case may be subjected to a three-dimensional simulation, which facilitates the comparison with an experience. Issa (2005) carried out this work with an LES type model (Section 6.5.1) for the turbulent closure. Figure 7.19 shows that the qualitative prediction of the open surface is satisfactory (also refer to Issa et al., 2005). The effect of the selected turbulence model over a water column collapse will be discussed in the next section.

Fig. 7.19 Case of the collapse of a water column on a dry bed. Three-dimensional SPH simulation with a LES type model, compared with an experience (after Issa, 2005).

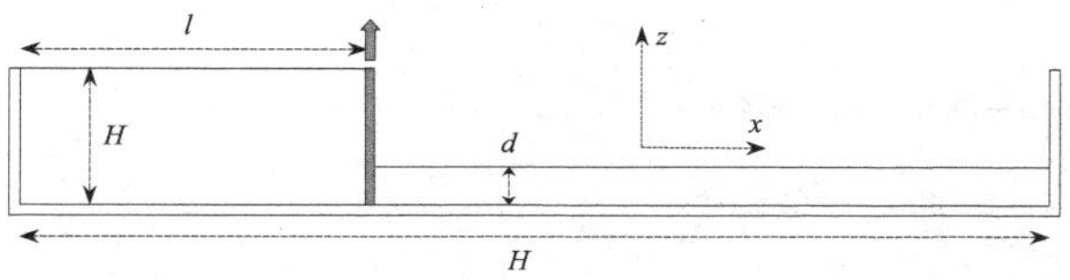

Fig. 7.20 Case of the collapse of a water column on a wet bed. Geometry and initial configuration (after Janosi et al., 2004).

7.3.2 Collapse onto a wet bottom

Let us now turn to a case which is similar to that in the preceding section, but with a slightly different geometry, and particularly a reservoir which is filled with water up to a given height before the column collapses. The break wave which propagates in it then breaks, which may be compared with the experience conducted by Janosi et al. (2004). Here we follow the ideas of Violeau and Issa (2007b) and (2007c).

Figure 7.20 illustrates the once again two-dimensional geometry being adopted here. We may note that the experience was obviously conducted using a gate separating the two parts of the reservoir, being lifted at the rate of 1.5 m/s from the very beginning. This gate shall preferably be taken into account in the model, because an instantaneous collapse would result in slightly different results. The selected distances are $L = 1.18$ m, $l = 0.38$ m and $H = 0.15$ m. As to the depth of the right-hand part of the reservoir, two values are adopted: $d = 18$ mm and $d = 38$ mm. The fluid is water ($\rho_0 = 1,000$ kg/m^3 and $\nu = 10^{-6}$ m^{2}s^{-1}).

At the initial time, the particles are spaced by a distance $\delta r = 1$ mm apart on a regular Cartesian grid and do not move about at any velocity. Their density equals the reference value ρ_0 and their pressure is zero. Here a fourth-order, B-spline type kernel (eqn (5.51)) is considered, for a ratio $h/\delta r = 1.5$. The walls are modelled by edge particles and three layers of dummy particles, according to the procedure described in Section 6.3.2. With these data, the system gathers 71,310 fluid particles and 8,450 particles for wall modelling. From the very beginning, the particles making up the gate are moved vertically at the required velocity.

The weakly compressible scheme based on a equation of state (refer to Section 5.3.2) is adopted. The numerical speed of sound c_0 is determined by equation (5.213), as it should be in the presence of gravity. Preliminary tests show that U_{max} is about 2.4 m/s, hence we choose $c_0 = 24$ m/s. The condition with respect to the water depth in (5.213) does not alter this value.[7] Thus, the simulation time step, which here is set by the condition in equation (5.413), is 2.5×10^{-5} s. The number of iterations is set to 2.4×10^4, which corresponds to a simulated physical period of time of 0.6 s, which is enough for the wave to propagate to the right-hand wall.

[7] According to eqn (3.170), in case of rapidly varying flow an additional condition should control the time step, which is based on the Strouhal number. In the present case and later on, however, the latter condition has no effect.

As in Section 7.3.1, this flow is turbulent, but inertia, gravity and pressure prevail over it. The wave breaking, however, gives rise to local strains which generate turbulent kinetic energy (refer to Section 4.4.2), liable to affect the shape of the breaking wave. Accordlingly, turbulence is taken into account by means of those models described in Section 6.4. Here the one-equation model $k - L_m$ and the non-linear model $k - \varepsilon$ disclosed in Section 6.4.2 are adopted. Since we do not know the spatio-temporal distribution of the mixing

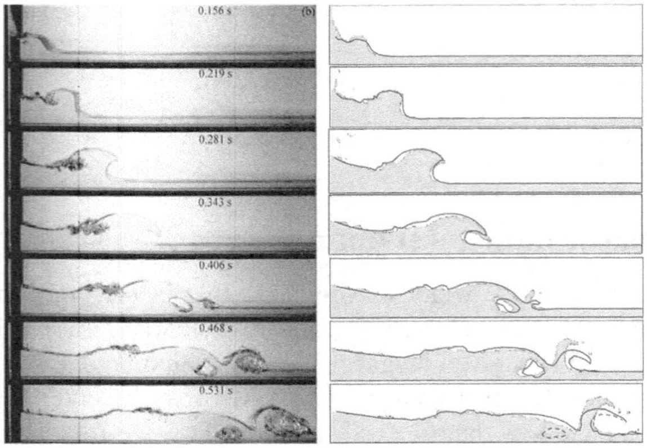

Fig. 7.21 Case of the collapse of a water column on a wet bed. Comparison of the numerical models (Violeau and Issa, 2007b) with the experience conducted by Janosi et al. (2004, with permission from Springer). Case $d = 18$ mm. The thick curved lines reproduce the experimental free surface.

length L_m in such a complex case, we make it equal to the particle size, which comes down to a large eddy simulation (refer to Section 6.5.1). In order to assess the extent of the spatial variability of the turbulent phenomena, we also contemplate a model in which the eddy viscosity is constant. Starting from the discussion in Chapter 4 (eqn (4.140)), a literal estimation of it may be provided as

$$\nu_T = L_m^2 S \sim \frac{1}{100} U_{\max} d \tag{7.10}$$

(assuming $L_m \sim d/10$, as recommended in Chapter 4). This estimation amounts to stating that the Reynolds number of the mean flow $\overline{Re}$, as defined by (4.84), is in the region of 100, which is consistent with the order of magnitude as provided in Section 4.4.6. Numerically, we get $\nu_T \sim 7 \times 10^{-4}$ $m^2 s^{-1}$, a value which we will prescribe for the simulation under a constant viscosity.

Janosi et al. (2004) provide fairly accurate photographs of the experience being replicated here; they are suitable for a validation as regards the shape of the unsteady free surface. Figures 7.21 and 7.22 allow us to make a comparison of the various models with the experience for both selected values of d, and show that the general shape of the flow is properly replicated. The shape of the experimental free surface is roughly replicated in solid lines on the graphs plotted through the simulation, with as much accuracy as allowed by the experience on the one hand (actually sensitive to slight three-dimensional effects), and through the quality of the photographs on the other hand. We may note that all the authors having worked on this test case have found a temporal offset of about 0.043 s between the times displayed on the photographs and those to be used for setting the results in the best possible conditions, which may be caused by an ambiguous choice of the initial time (chosen either just before or after the beginning of the gate motion). Accordingly, we will adopt that corrective action, which amounts to choosing graphic printout instants occurring 0.043 s prior to the measurement instants (the latter, however, will be adopted as reference time instants, because they are mentioned on the experimental photographs).

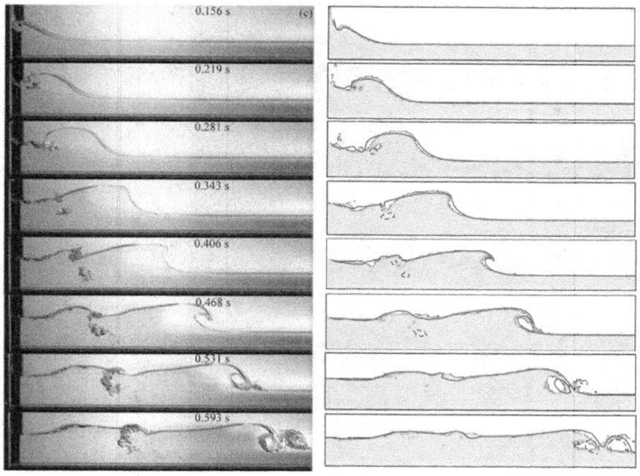

Fig. 7.22 Case of the collapse of a water column on a wet bed. Comparison of the numerical results (Violeau and Issa, 2007b) with the experience conducted by Janosi et al. (2004, with permission from Springer). Case $d = 38$ mm. The thick curved lines reproduce the experimental free surface.

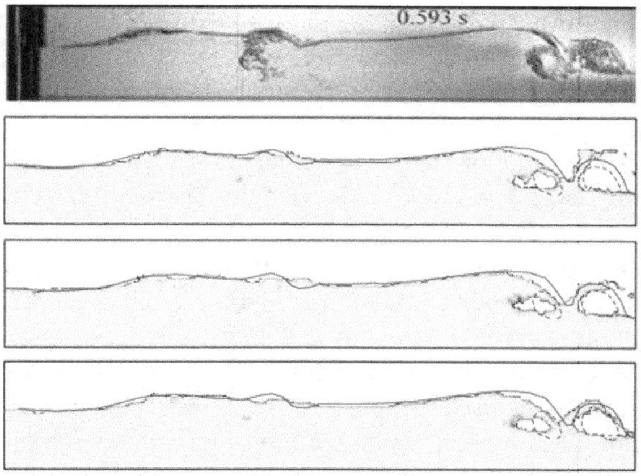

Fig. 7.23 Case of the collapse of a water column on a wet bed, with $d = 38$ mm. Comparison of three turbulence models, after Violeau and Issa (2007b). From top to bottom: constant viscosity, $k - L_m$ model, non-linear $k - \varepsilon$ model. The continuous curved lines reproduce the experimental free surface, after Janosi et al. (2004, with permission from Springer).

Figure 7.23, corresponding to $d = 38$ mm immediately after breaking, reveals the effect of the turbulence model. The constant viscosity model does not represent the free surface shape as accurately as the 'true' turbulent models. Similar results are highlighted with $d = 18$ mm (Violeau and Issa, 2007b, 2007c). Further simulations, which are not presented here, do not reveal any improvement with a standard model $k - \varepsilon$ as compared with the $k - L_m$ model. As to the mixing length model (refer to Section 6.4.1), in this case its results are poorer than those of the constant viscosity model, which confirms that it is not suitable for simulating quickly varying flows, as we stated in Section 4.4.4. Issa et al. (2009a) give an account of an attempt with the $k - \omega$ model yielding results which are qualitatively analogous to those of the $k - \varepsilon$ model.

Whichever turbulence model is adopted, the breaking is not perfectly replicated and the respective roles of the modelled physics and the numerical options being chosen cannot be brought out. The imperfection of the physical approach being contemplated here should be emphasized, indeed: on the one hand, the presence of air carried along by water (refer to Fig. 7.23) may

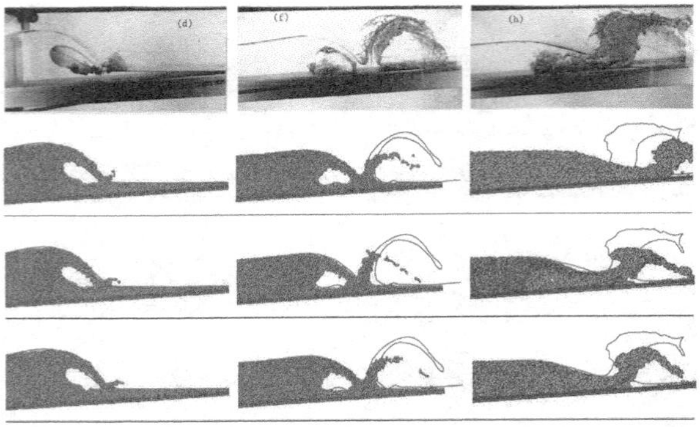

Fig. 7.24 Breaking of a single wave in a sloping plane. Comparison of three turbulence models, after Issa and Violeau (2009). From top to bottom: constant viscosity, mixing length model, $k - L_m$ model. The thick lines reproduce the experimental free surface, after Li and Raichlen (2003).

significantly change the shape of the free surface; on the other hand, the relative smallness of the flow scale (in the region of several centimetres) suggests that surface tension may play a significant role (refer to Section 3.4.5). For all that, it is hard to estimate the importance of a finer physical modelling. On the other hand, simulations have highlighted the importance of the numerical options, particularly the contribution of the renormalization (refer to Section 5.2.5), for finely modelling large variations of the velocity and the free surface shape, in particular the surface wave propagation (for example, refer to Guilcher et al., 2007). Discretization is obviously a major point in the class of the numerical parameters as well. Further simulations (which are not presented here) show that with a double particle size ($\delta r = 2$ mm), the free surface prediction is markedly less precise. Lee et al. (2008) also applied the incompressible algorithm as introduced in Section 6.2.5 to that flow, with similar results.

The breaking phenomenon which we have highlighted can more generally be modelled with the SPH method. A single wave propagating in a gently sloping plane, for instance, is illustrated in Fig. 7.24 (refer to Issa et al., 2009b; Issa and Violeau, 2009). The comparison with the experience conducted by Li and Raichlen (2003) corroborates the analysis we have just made as regards the role of the turbulence model in the prediction of that process. Once again, however, the choice of numerical options may prove to be quite as important. Khayyer et al. (2008) achieve good results without any turbulence model but with a renormalization method. Xu et al. (2010), using an analogous method, manage to perfectly predict the counter-breaking illustrated in Fig. 7.24 (c).

7.3.3 Collapse onto a motionless obstacle

For the first time in this chapter, we will discuss a three-dimensional case, in line with the collapse onto a dry bottom of Section 7.3.1. We will consider a parallelepipedic fluid block collapsing onto a three-dimensional basin, provided with a parallelepipedic obstacle lying on the bottom. The break wave will encounter the obstacle and cause a significant splash-up on it, before being reflected onto the opposite wall and moving back and forth in the basin.

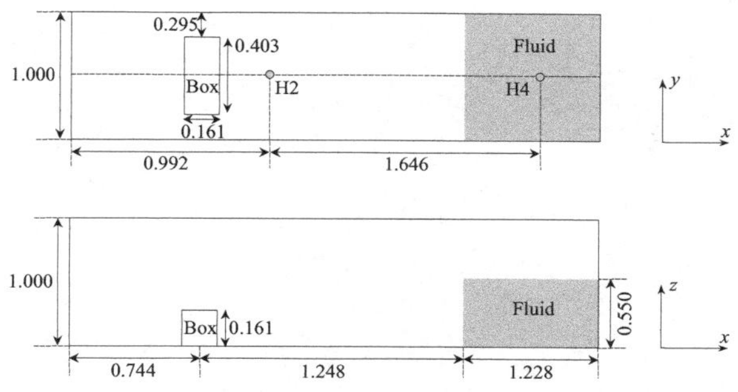

Fig. 7.25 Case of the collapse of a three-dimensional water column onto an obstacle. Description of the geometry, after Kleefsman et al. (2005). The distances are indicated in metres.

A detailed definition of this case is given by Issa (2006). Here we follow the ideas of Lee et al. (2010).

Figure 7.25 illustrates the presently adopted geometry, in order to refer to the experience conducted by Kleefsman et al. (2005). The fluid is water ($\rho_0 = 1,000$ kg/m^3 and $\nu = 10^{-6}$ m^{2}s^{-1}). At the initial time, the particles are spaced by a distance $\delta r = 1.8333$ cm apart on a regular Cartesian grid, which corresponds to 30 particles over the initial water depth of 0.55 m. They do not move at any velocity and their density equals the reference value ρ_0, whereas their pressure is zero. The walls are modelled by edge particles and three layers of dummy particles, according to the procedure described in Section 6.3.2. The system then gathers 108,540 fluid particles.

The weakly compressible scheme based on an equation of state (refer to Section 5.3.2) is initially utilized. In this case, a fourth-order, B-spline type kernel (eqn (5.51)) is considered, for a ratio $h/\delta r = 1.5$. With these data, 151,734 particles are required for modelling the walls. It can be found that this number has the same order of magnitude as the number of fluid particles, which highlights one of the major defects of the three-dimensional dummy particle technique. Since the total number of particles is comparatively large (260,274), a parallel algorithm will preferably be used for solving the equations (refer to Section 8.1). The numerical sound velocity c_0 is determined by equation (5.213), as it should be in the presence of gravity. U_{max} is not known in advance, but a first numerical experience shows that it is almost 6 m/s. Hence we choose $c_0 = 60$ m/s. The condition with respect to the water depth in (5.213) does not alter that value. Then, the simulation time step, which here is set by equation (5.413), is 1.83×10^{-4} s.

The incompressible scheme, the efficiency of which is known from the confined flow case in Section 7.2.2, is considered here too.[8] The reference velocity for the calculation of the time step is then U_{max}, as it is known (condition (6.128)), which would yield $\delta t = 1.83 \times 10^{-3}$ s. However, it is deliberately decided to reduce it to 10^{-3} s, in order to be in line with the numerical model of Kleefsman et al. (2005). The latter is based on a volume of fluids (VOF) technique, on the basis of slightly less than 1.22 million cells.

Here the flow is turbulent, but the inertia, gravity and pressure forces prevail over the Reynolds stresses in the first instants, as in Section 7.3.1. Though this argument is defeated, in particular, upon the impact against the obstacle, it is decided, for the sake of simplicity, not to model the turbulence here.

[8] In this case, the number of particles discretizing the walls is slightly lower than in the nearly-incompressible model, for reasons related to the model parallelism (refer to Lee et al., 2010).

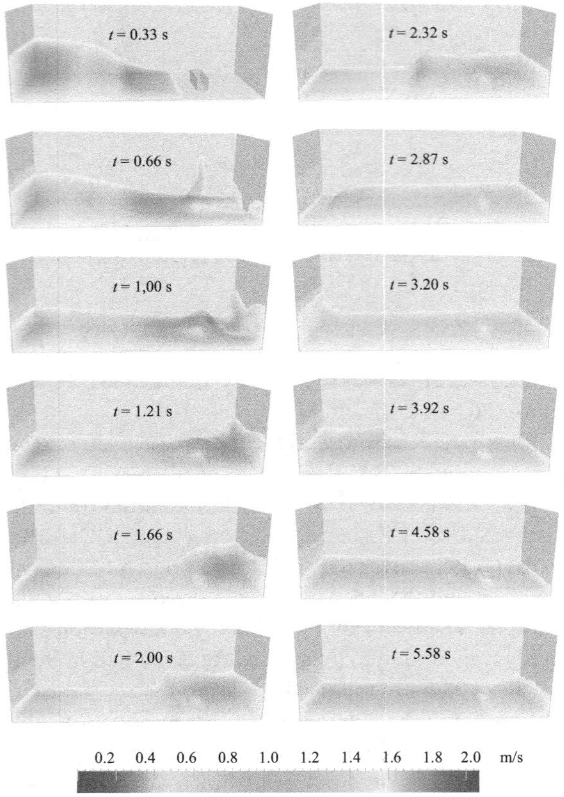

Fig. 7.26 Case of the collapse of a three-dimensional water column onto an obstacle. Images of the flow as simulated with the near-incompressible algorithm of the SPH method. The shading represents the velocity module, in m/s.

Figure 7.26 shows a few images achieved with the weakly compressible algorithm. The wave propagation can be seen, the impact against the obstacle then the wave return to the left-hand wall, exhibiting the expected to-and-fro motion.[9] Figure 7.27 displays some instantaneous views obtained with both algorithms, being compared to the pictures in the experiment conducted by Kleefsman et al. (2005). One can see that the free surface is substantially more rugged with the incompressible scheme. Now, the experimental observations confirm the existence of breaking generating an intense waterdrop splashing, but it is smoother than predicted by the incompressible scheme, more like the results of the weakly compressible scheme. Though this comparison is very qualitative, its seems therefore that the reality lies between the two algorithms as regards the shape of the surface. The best behaviour of the incompressible algorithm in some places may partly result from a better prediction of the pressure by Poisson's equation. On the other hand, the excessive particle splashing seems to be caused by a defect in the technique for the detection of free surface particles by means of the criterion (6.109).

[9] It is a sloshing phenomenon, for which the SPH method proves to be quite suitable (for instance, refer to Colagrossi et al., 2010).

In addition, it can be seen in Fig. 7.27 that the SPH prediction is slightly lagging behind the experience, upon the after-impact wave return. This finding is confirmed by Fig. 7.28, which illustrates the temporal evolution of the water depth at points H2 and H4 whose positions are shown in Fig. 7.25. As in

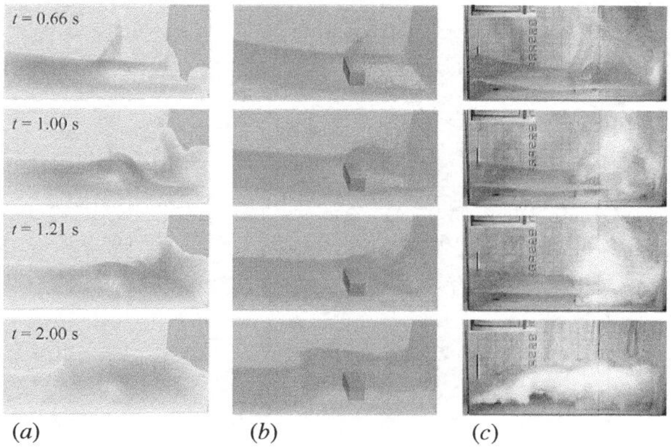

Fig. 7.27 Case of the collapse of a three-dimensional water column onto an obstacle. Images of the flow as simulated with SPH, nearly-incompressible (*a*) and incompressible (*b*) algorithms. Comparison with the experience conducted by Kleefsman et al. (2005, with permission from Elsevier) (*c*).

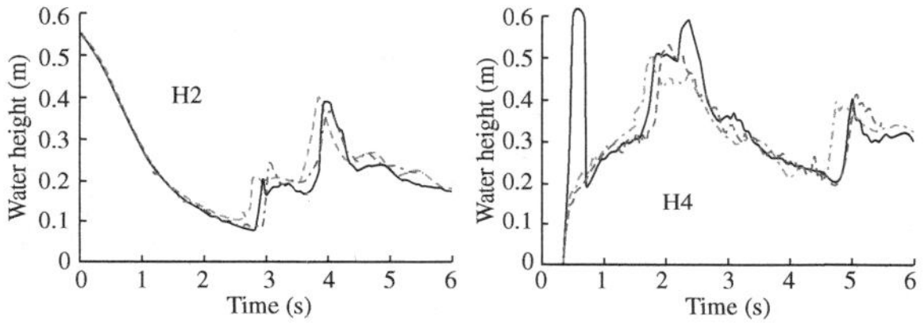

Fig. 7.28 Case of the collapse of a three-dimensional water column onto an obstacle. Temporal evolution of the water depth at points H2 and H4 (refer to Fig. 7.25) with SPH, weakly-incompressible (dashed-dotted line) and incompressible (solid line) algorithms. Comparison with the experience conducted by Kleefsman et al. (2005, dashed line). After Lee et al. (2010, with permission from Taylor and Francis).

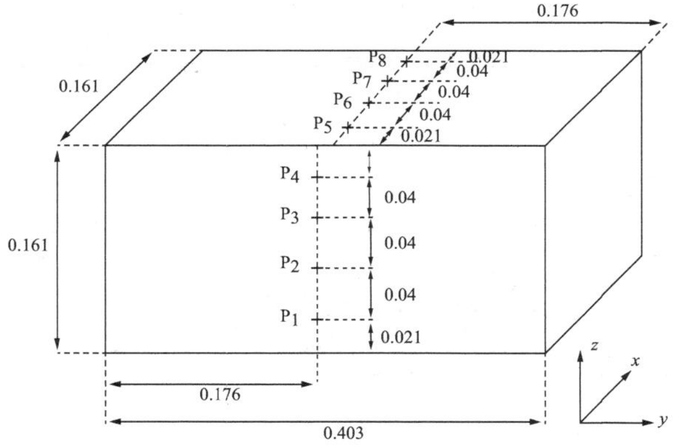

Fig. 7.29 Case of the collapse of a three-dimensional water column onto an obstacle. Locations of the pressure measurement points for the graphs in Fig. 7.30, after Kleefsman et al. (2005, with permission from Elsevier). Distances are indicated in metres.

the preceding section, the reasons for this drawback can be found either in the physical domain (turbulence, for instance) or in the numerical domain (insufficient discretization, in particular).

The H4 profile in Fig. 7.28 highlights a slightly less good representativity of the free surface with the incompressible algorithm, in particular with an initial peak that does not correspond to the observations. On the other hand, Fig. 7.30 shows that the pressure is much better predicted with the weakly compressible scheme. This figure regards the temporal evolution of pressure at points P1, P3, P5 and P7, which lie on the obstacle (their positions are shown in Fig. 7.29). Thus, as in Section 7.2.2, it appears that the prediction of pressure is signicantly better with the incompressible scheme, even in the presence of a free surface. However, an abnormal pressure peak can be noticed in the first instance at

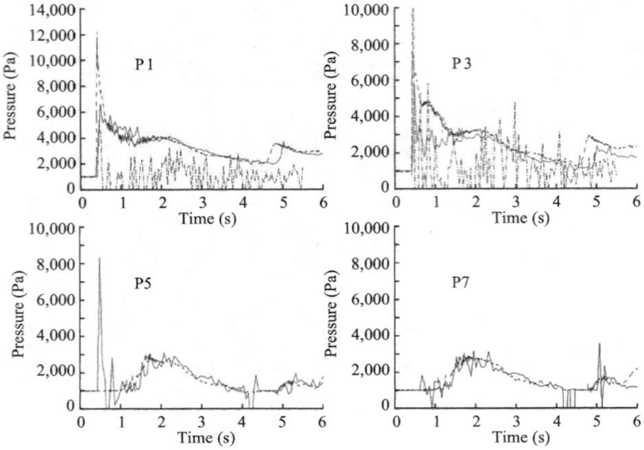

Fig. 7.30 Case of the collapse of a three-dimensional water column onto an obstacle. Temporal evolution of pressure at points P1, P3, P5 and P7 (refer to Fig. 7.29) with SPH, weakly-incompressible (dashed-dotted line) and incompressible (solid line) algorithms. Comparison with the experience conducted by Kleefsman et al. (2005, dashed line). After Lee et al. (2010, with permission from Taylor and Francis).

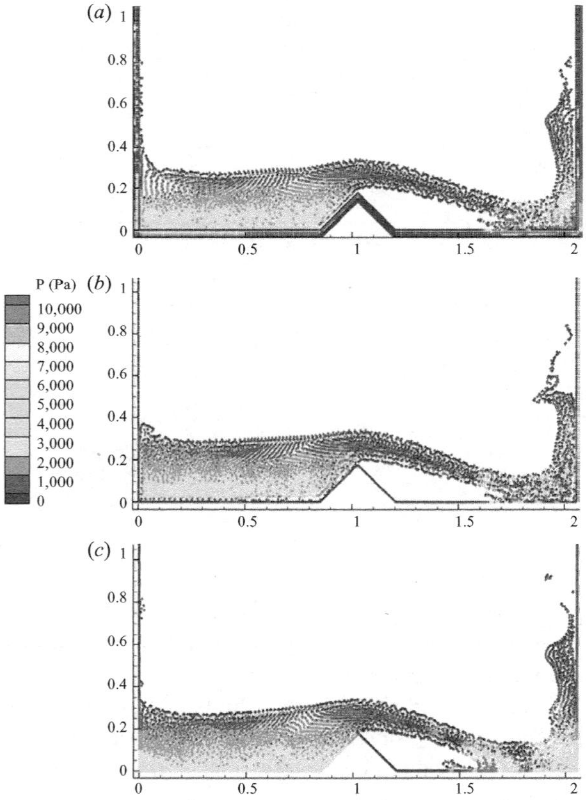

Fig. 7.31 Case of the collapse of a two-dimensional water column onto a wedge. Instantaneous views of the pressure as predicted by SPH with the dummy particles (*a*), Lennard–Jones repulsive forces (*b*), lastly the contact forces (*c*) methods (after Ferrand et al., 2010).

point P5. Lee et al. (2010), however, explain that this peak diminishes as the time step is shortened.

There are many similar cases in the scientific literature dealing with SPH. For instance, Ferrand et al. (2010), deal with a triangular obstacle within a two-dimensional basin which is identical to that in Section 7.3.1. These authors investigate, within that context, the respective efficiencies of all three methods as described in Section 6.3.2 for wall treatment, namely the dummy particle method (Fig. 7.31 (*a*)), the Lennard–Jones repulsive force method (eqn (6.156), Fig. 7.31 (*b*)) and lastly the contact force approach (eqns (6.161), (6.165) and (6.166), Fig. 7.31 (*c*)).[10] Although no validation is available in this case, the results, which speak for themselves, support the third approach, since the first two methods give rise to large pressure fluctuations near the walls, as well as small local void pockets, particularly in the case of the Lennard–Jones forces (see also Ferrand et al., 2012).

[10] In actual fact, it is a variant of this model, which we discussed just after eqn (6.177) in Chapter 6.

7.4 Immersed bodies

The SPH suitability for efficiently treating the fluid-structure interactions should be tested. The flows around motionless rigid bodies are an initial step before turning to the motion of a body motion induced by a fluid. The prediction of viscous forces and pressure, which was already discussed in previous sections of this chapter, is once again at the core of that issue, this time within the context of a slowly varying regime.

7.4.1 Motionless body in a channel

Here we will try to replicate the flow around a motionless body in the middle of a pipe. Depending on the conditions, the flow may either be steady or oscillate downstream of a body. This case is suitable for determining the SPH method's ability to properly predict the distribution of the velocity and pressure fields in this kind of configuration, even when they are unsteady. Here, as in Sections 7.2.1 and 7.2.2, the effect of gravity is ignored. We primarily adopt the views of Lee et al. (2007a, 2008).

The two-dimensional channel of Fig. 7.32, provided with a square obstacle at its centre, is considered. It is a periodic flow along axis x, which makes the treatment of the boundary conditions easier, avoiding the treatment of inflow and outflow boundaries, which are not easy to handle through SPH (refer to the introduction to Section 6.3). A steady flow is prescribed on the basis of the method described in Section 7.2.1, corresponding to an average velocity denoted as U_0. The values of the side d and velocity U_0 are irrelevant, since the quantities can be non-dimensionalized on the basis of the water characteristics ($\rho_0 = 1,000$ kg/m^3 and $\nu = 10^{-6}$ m^{2}s^{-1}). The discussion in Section 3.4.3 then shows that the flow is fully governed by the value of the Reynolds number $Re \doteqdot U_0 d/\nu$. Setting the value of U_0, the values $Re = 20$ and $Re = 100$ providing laminar flows are considered. For this kind of configuration, experience suggests that the regime remains steady for the first of these values, and that it has an axis of symmetry corresponding to the symmetry of the geometry. On

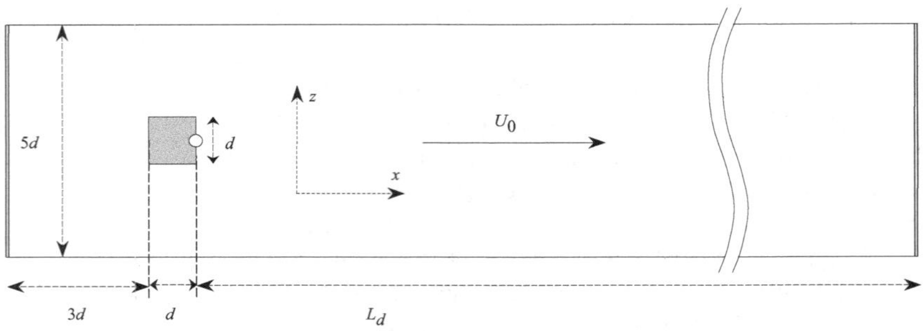

Fig. 7.32 Motionless solid body in a channel. Description of the geometry, after Lee et al. (2008). The periodicity boundaries are indicated by a double line.

the other hand, the second value of Re leads to the development of a vortex street with a temporal period T (refer to Section 4.2.2). The following relation can be highlighted through the dimensional analysis:

$$St \doteqdot \frac{d}{U_0 T} = \phi_{adim}(Re) \tag{7.11}$$

where the Strouhal number St has been defined according to (3.155).

The channel length depends on the selected distance L_d (refer to Fig. 7.32), which should be large enough to prevent the wake downwards of the body from significantly disturbing the channel inlet due to a periodicity effect. Thus, for $Re = 20$, $L_d = 28d$ is chosen, whereas $L_d = 98d$ is necessary when $Re = 100$, taking the presence of the vortex wake into account. The spatial origin, located aft of the solid, is illustrated by the white disc in Fig. 7.32.

At the initial time, the particles are separated by a distance $\delta r = d/20$ apart on a regular Cartesian grid and are not moving about at any velocity. Their density takes the reference value ρ_0 and their pressure is zero. Here a fourth-order, B-spline type kernel (eqn (5.51)) is considered, for a ratio $h/\delta r = 1.3$. The walls are modelled by edge particles and three layers of dummy particles, according to the procedure described in Section 6.3.2. With these data, the system gathers 63,018 fluid particles and 6,730 particles for wall modelling when $Re = 20$. With $Re = 100$, these numbers are increased up to 201,618 and 20,730, respectively.

The weakly compressible scheme based on an equation of state (refer to Section 5.3.2) is initially adopted. The numerical speed of sound c_0 is determined by equation (5.212), in the absence of gravity. $U_{\max}$ obviously has the same order of magnitude as U_0. For reasons previously set out in Section 7.2.2, however, we choose $c_0 = 100U_0$. Thus, the simulation time step, here being set by the CFL condition in equation (5.413), equals

$$\delta t = 2.6 \times 10^{-4} \frac{d}{U_0} \tag{7.12}$$

The required number of iterations for establishing a steady or periodic regime is then independent of the d and U_0 parameters. On the other hand, this number depends on Re, the latter not being very high. Numerical experience

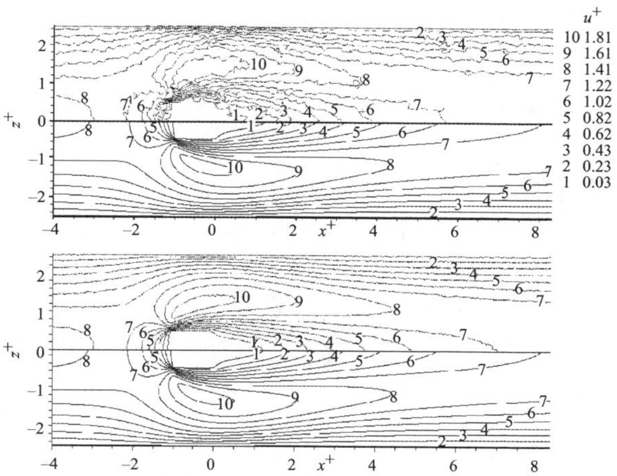

Fig. 7.33 Motionless solid body in a channel, $Re = 20$. Isocontour lines of the dimensionless longitudinal velocity u_x^+ obtained with SPH, Nearly-incompressible (top) and incompressible (bottom) schemes. The lower half of each graph exhibits a simulation in finite volumes. After Lee et al. (2008, with permission from Elsevier).

shows that 5×10^4 iterations are required for the presently adopted values of Re.

We also intend to test the incompressible scheme as described in Section 6.2.5. The time step is then based on the fluid velocity (condition (6.128)), which allows 100 times fewer iterations, with the choice we have made for c_0.

Time, lengths and velocities will be non-dimensionalized in the same way as (7.4):

$$\begin{aligned} t^+ &\doteq \frac{U_0 t}{d} \\ \forall i,\ x_i^+ &\doteq \frac{x}{d} \\ \forall i,\ u_i^+ &\doteq \frac{u_i}{U_0} \end{aligned} \tag{7.13}$$

Figure 7.33 displays the isovalue lines of dimensionless longitudinal velocity u_x^+ for $Re = 20$. The weakly incompressible and incompressible schemes of the SPH method are compared with the predictions made by the STAR_CD commercially available software, which is based on a finite volume technique, with a very fine discretization (approximately 1 million cells). As expected, the flow is steady and symmetric to the solid body axis. Both SPH schemes agree fairly well with the finite volumes; as in Section 7.2.2, the incompressible scheme proves to be more precise. Figure 7.34 illustrates the development of the vortex street for $Re = 100$, with the same notations. Once again, SPH is rather effective with the incompressible scheme, whereas the weakly compressible approach is not suitable for faithfully replicating the calculations in finite volumes, which were carried out on a grid of about 3 million cells. With the weakly incompressible approach, the vortices unfortunately tend to remain too much collapsed. Comparing the pressure distributions (not illustrated here) leads to conclusions which are similar to those in Section 7.2.2 (refer to Lee et al., 2008).

By analogy with equation (7.5), the dimensionless pressure is defined by

$$p^+ \doteq \frac{p}{\frac{1}{2}\rho_0 U_0^2} \tag{7.14}$$

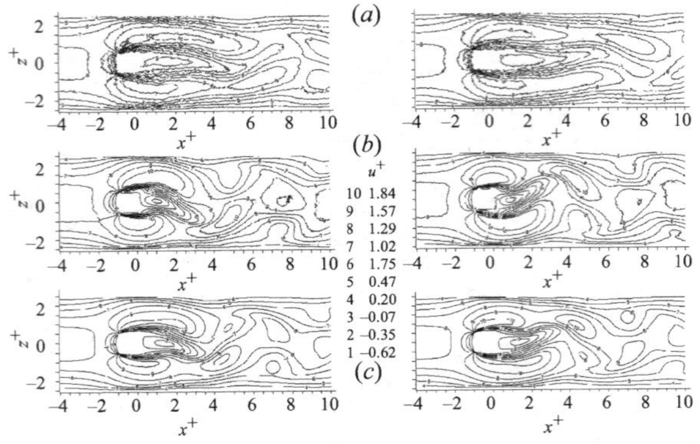

Fig. 7.34 Motionless solid body in a channel, $Re = 100$. Isocontour lines of the dimensionless longitudinal velocity u_x^+ at $t^+ = 0.25$ (left) and $t^+ = 0.75$ (right). Results achieved with SPH, nearly-incompressible (*a*) and incompressible (*b*) schemes, and a simulation in finite volumes (*c*). After Lee et al. (2008, with permission from Elsevier).

Table 7.3 Motionless solid body in a channel. Values of p_{av}^+, adimensionalized pressure as averaged at the square obstacle boundary, and of the Strouhal number St. WCSPH = Weakly compressible SPH; ISPH = incompressible SPH.

Method	WCSPH	ISPH	Finite volume
p_{av}^+ $(Re = 20)$	14.56	3.48	5.53
p_{av}^+ $(Re = 100)$	1.09	6.30	3.23
St $(Re = 100)$	0.19	0.23	0.25

The value of p^+ as averaged at the square body boundary is denoted as p_{av}^+; its values are given by Table 7.3, for the various simulations being disclosed here. The incompressible algorithm is not as precise as expected, while providing values of the same order of magnitude as the finite volumes, whereas the weakly compressible scheme yields quite inaccurate values. Table 7.3 also displays the predicted values of the Strouhal number for $Re = 100$, which once again support the incompressible scheme.

7.4.2 Mobile body within an enclosure

Since the preceding section has validated the ability to replicate the flow around a motionless body, it is interesting to consider the forced motion of a solid within an enclosure containing a fluid initially at rest. A thorough description of this case can be found in Colagrossi (2006). Compared to the preceding case, this test is suitable for determining the SPH method's ability to handle mobile solids, although the motion is not controlled by the fluid, but instead by the programmer. Besides, the flow is definitely unsteady, the body accelerates from the state of rest. As in the preceding section, here the effect of gravity is ignored. We follow the views of Lee (2007) and Lee et al. (2007b).

We consider the two-dimensional closed enclosure as illustrated in Fig. 7.35, internally provided with a square body. The latter gradually accelerates in order to get an asymptotic velocity denoted as U_0. If both velocity and acceleration

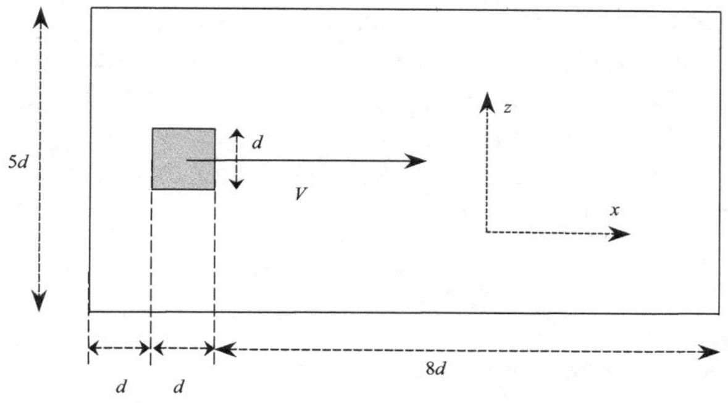

Fig. 7.35 Case of the mobile solid within an enclosure. Description of the geometry, after Colagrossi (2006).

are denoted as V and γ, then they can be non-dimensionalized, as well as the time, as follows:

$$\begin{aligned} t^+ &\doteqdot \frac{U_0 t}{d} \\ V^+\left(t^+\right) &\doteqdot \frac{V(t)}{U_0} \\ \gamma^+\left(t^+\right) &\doteqdot \frac{d\gamma(t)}{U_0^2} \end{aligned} \tag{7.15}$$

The graphs of $V^+(t)$ and $\gamma^+\left(t^+\right)$ are then shown in Fig. 7.36. The values of the solid side d and the velocity U_0 are of little importance, since the values will be non-dimensionalized on the basis of the water characteristics ($\rho_0 = 1,000$ kg/m^3 and $\nu = 10^{-6}$ m^{2}s^{-1}). The arguments in Section 3.4.3 then show that the flow fully depends on the Reynolds number value $Re \doteqdot U_0 d/\nu$, as in Sections 7.2.2 and 7.4.1. Acting on U_0, we consider the values $Re = 50$, 100 and 150, providing laminar flows.

At the initial point in time, the particles are seperated by a distance $\delta r = d/60$ apart on a regular Cartesian grid, and are not moving about at any velocity. Their density takes the reference value ρ_0 and their pressure is zero. As previously, a fourth-order, B-spline type kernel is considered (eqn (5.51)), for a ratio $h/\delta r = 1.3$. The walls are modelled by edge particles and three layers of dummy particles, according to the procedure described in Section 6.3.2. With these data, the system gathers 175,380 fluid particles and 10,200 particles for wall modelling.

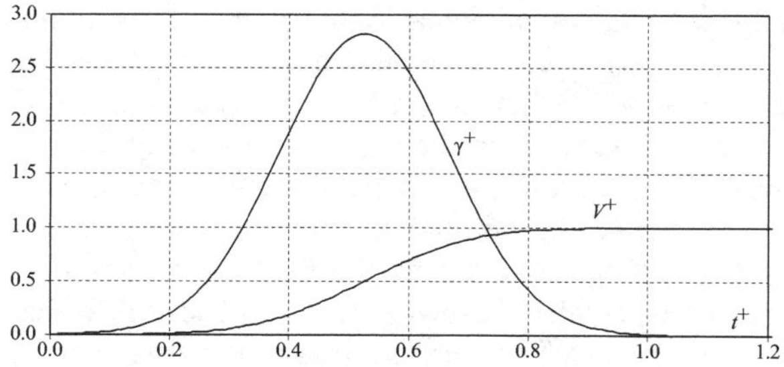

Fig. 7.36 Case of the mobile solid within an enclosure. Temporal evolution of the dimensionless velocity V^+ and the dimensionless acceleration γ^+ of the solid (eqn (7.17)), after Colagrossi (2006).

The weakly compressible scheme based on an equation of state (refer to Section 5.3.2) is initially adopted. The numerical sound speed c_0 is determined by equation (5.212), in the absence of gravity. U_{max} obviously has the order of magnitude of U_0. For reasons previously set out in Section 7.2.2, we choose $c_0 = 30U_0$. Thus, the simulation time step, which here is set by the CFL condition of equation (5.413), is yielded by

$$\delta t = 2.89 \times 10^{-4} \frac{d}{U_0} \tag{7.16}$$

We also intend to test the incompressible scheme as described in Section 6.2.5. The time step is then based on the fluid velocity (condition (6.128)), which allows 30 times fewer iterations, with the choice we have made for c_0.

As in Section 7.2.2, we non-dimensionalize the time, the coordinates, the velocity and the pressure of the fluid as follows:

$$\begin{aligned} t^+ &\doteqdot \frac{U_0 t}{d} \\ \forall i,\ x_i^+ &\doteqdot \frac{x}{d} \\ \forall i,\ u_i^+ &\doteqdot \frac{u_i}{U_0} \\ p^+ &\doteqdot \frac{p}{\rho_0 U_0^2} \end{aligned} \tag{7.17}$$

(we may notice, however, the absence of the $1/2$ factor as compared with (7.5), which is equivalent to the definition (3.151)). For $Re = 100$, the distributions of the norm ot both dimensionless velocity $u^+ \doteqdot |\mathbf{u}^+|$ and dimensionless pressure p^+ are illustrated in Fig. 7.37 at $t^+ = 5$ and in Fig. 7.38 at $t^+ = 8$. The results achieved with the two SPH schemes are compared with the predictions made by a finite difference method on a 180,000 cell grid (Colicchio et al., 2006). These images require the same comments as those of Sections 7.2.2 and 7.4.1: SPH yields results analogous to the finite differences with respect to velocity, with a slight advantage for the incompressible scheme. Conversely, only the latter is suitable for replicating the pressure fields, the weakly incompressible scheme giving rise to very wide numerical fluctuations. With that scheme, in addition, two small void pockets appear near the right-hand corners of the solid, as can be seen in Figs 7.37(a) and 7.38(a). These artificial cavities, however, disappear through an appropriately selected numerical sound speed, that is $c_0 = 100U_0$ instead of $30U_0$, as in Section 7.2.2. However, the time step is consequently reduced, which increases the number of iterations.

Knowing this, we will focus on the incompressible scheme results. Figure 7.39 illustrates the fields of dimensionless velocity and pressure being compared to the finite differences, for the other two values of Re. Usually a very good consistency between the two numerical methods can be seen, in spite of the significant differences in the prediction of pressure at $t^+ = 8$. The norm of the total pressure forces exerted by the fluid on the solid per width will advantageously be calculated in the cross-wise direction to the figures, the force here being denoted F^P. Along the lines of the drag coefficient as given by (3.175), the pressure coefficient is then defined by

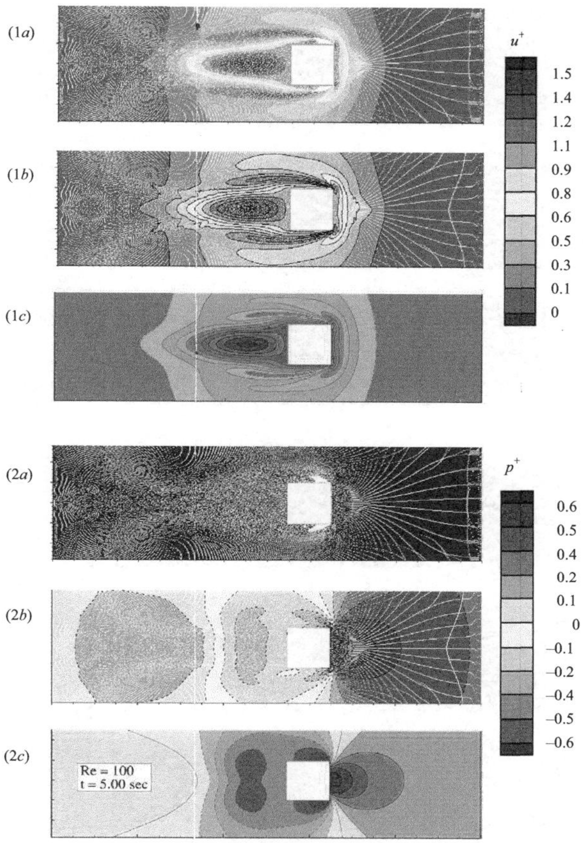

Fig. 7.37 Case of the mobile solid within an enclosure, $Re = 100$. Norm of the fluid dimensionless velocity u^+ and dimensionless pressure p^+ at $t^+ = 5$. Weakly compressible (*a*) and incompressible (*b*) SPH simulations, after Lee (2007). Comparison with a finite difference scheme (*c*), after Colicchio et al. (2006).

$$C_P\left(t^+\right) \doteqdot \frac{2F^P\left(t^+\right)}{\rho_0 U_0^2 d} \tag{7.18}$$

Figure 7.40 illustrates the evolution of C_P over time, for $Re = 150$. The incompressible algorithm in the SPH method is fairly consistent with the finite differences, in spite of several fluctuations which adversely affect SPH. The examination of C_P for the other values of Re leads to similar conclusions (Lee, 2007).

The SPH method proves quite suitable for the problems related to the motion of a body within a fluid, whether its motion is forced or not, on the basis of the equations in Section 6.3.3. Within that context, the behaviour of the material forming the solid may also be modelled on the basis of the discretized Cauchy equation (5.199) (for example, refer to Oger et al., 2005). We will deal with the free motion of an undeformable body under the action of a surrounding flow in the next section.

7.4.3 Sphere in a turbulent flow

The preceding two sections have confirmed SPH's suitability for modelling the flow around a fixed or mobile body, whose motion is forced. We now will

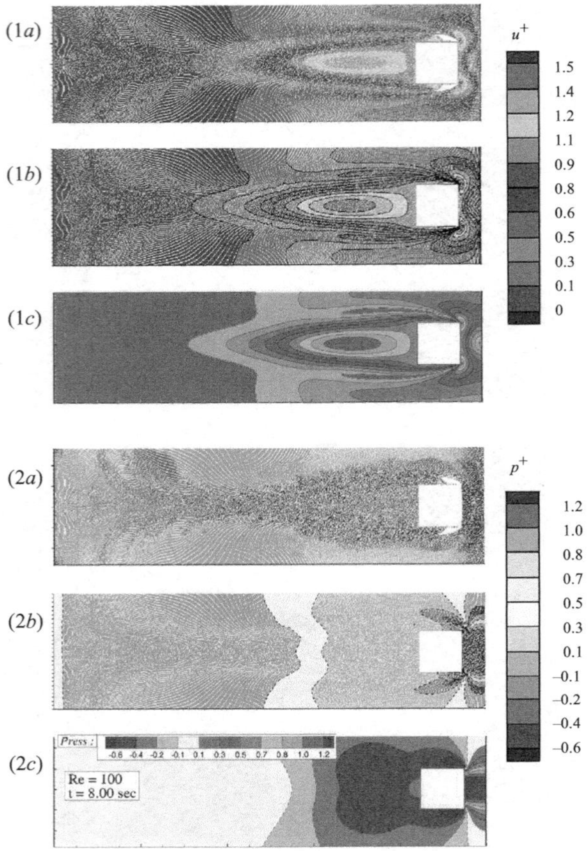

Fig. 7.38 Case of the mobile solid within an enclosure, $Re = 100$. Norm of the fluid dimensionless velocity u^+ and dimensionless pressure p^+ at $t^+ = 8$. Weakly incompressible (*a*) and incompressible (*b*) SPH simulations, after Lee (2007). Comparison with a finite difference scheme (*c*), after Colicchio et al. (2006).

investigate the possibility of letting the body move about under the action of the forces exerted on it by the fluid, on the basis of the arguments in Section 6.3.3. It is therefore a fluid/structure coupling, though the stresses within the structure are not calculated. We will focus on the motion of a solid under the action of a current and gravity in an open channel analogous to that in Section 7.2.3.

We consider the geometry of Fig. 7.41, showing a sphere lying in a plane under the action of gravity. As a matter of fact, it is rather an infinite cylinder, because the flow is presumably two-dimensional, that is invariant in the crosswise direction. It is assumed that the plane is the bottom of an open channel analogous to that in Section 7.2.3, with a flow caused by a steady stream having a superficial velocity U through the method as described in Section 7.2.1, that is based on a driving force F^{ext} (eqn (7.2)). The flow is periodic along the horizontal axis x.

As in many cases which we have tested in this chapter, the values of the diameter d of the sphere, the channel length L, the water depth H and the velocity U are of little importance, because the quantities may be adimensionalized. However, since the number of parameters is comparatively high here, we choose the following fixed values: $d = 0.01$ m, $L = 0.1$ m and

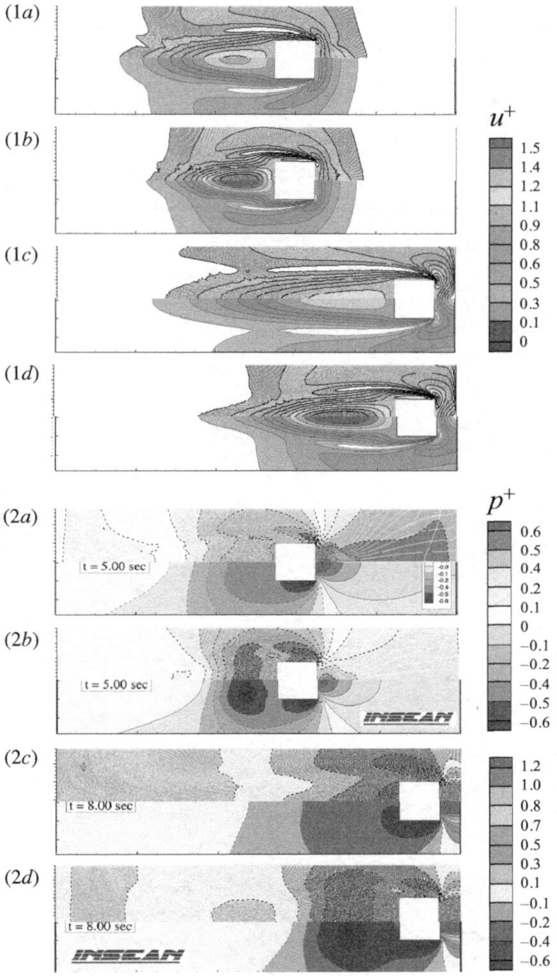

Fig. 7.39 Case of the mobile solid within an enclosure. Norm of the fluid dimensionless velocity u^+ (1) and dimensionless pressure p^+ (2) for $Re = 50$ (a and c) and 150 (b and d), at $t^+ = 5$ (a and b) and $t^+ = 8$ (c and d). On each graph, the upper half represents the results of the incompressible SPH scheme (after Lee, 2007), the lower half represents the results of the finite differences, after Colicchio et al. (2006).

$H = 0.06$ m. In order for the sphere to be considered isolated, the selected value of L would theoretically have to be much greater than d, as in Section 7.4.1. With these selected values, it should be considered that it is a periodic network of spheres. The selected fluid is water ($\rho_0 = 1,000$ kg/m^3 and $\nu = 10^{-6}$ m^{2}s^{-1}). Three values of the average velocities are considered, namely $U = 0.59$ m/s, 1.18 m/s and 1.77 m/s. The Reynolds number $Re \doteqdot UH/\nu$ is then higher than 35,400 (refer to Table 7.4), which provides a turbulent flow in all circumstances.

At the initial point in time, the particles are separated by a distance $\delta r = d/20 = 0.5$ mm apart on a regular Cartesian grid and are not moving about at any velocity. The fluid particle density equals the reference density ρ_0 and their pressure is zero. The particles making up the sphere are provided with a constant density $\rho_s = 2,600$ kg/m^3. The fourth-order, B-spline type kernel (eqn (5.51)) is kept, for a ratio $h/\delta r = 1.5$. The walls are modelled by edge particles and four layers of dummy particles, according to the procedure as

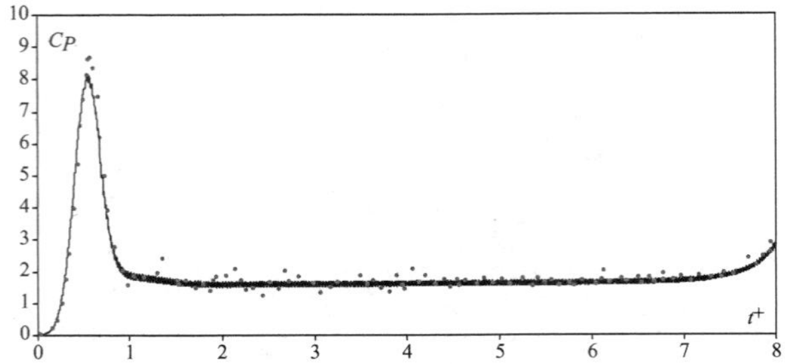

Fig. 7.40 Case of the mobile solid within an enclosure, $Re = 150$. Evolution of the pressure coefficient C_P over time (eqn (7.18)). Comparison of the incompressible SPH (symbols, after Lee, 2007), and finite difference schemes (solid line, after Colagrossi, 2006).

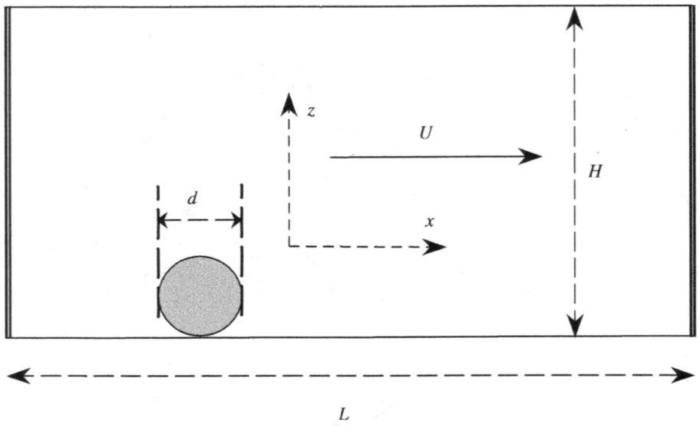

Fig. 7.41 Case of the sphere in a turbulent flow. Description of the geometry. The periodic boundaries are indicated by double lines.

Table 7.4 Case of the sphere in a turbulent flow. Values of the parameters related to the motion of the sphere as driven by the current.

U (m/s)	Re	u_* (m/s)	Re_d	T^+
0.59	35400	0.0566	566	0
1.18	70800	0.1131	1131	0.482
1.77	106200	0.1697	1697	2.34

described in Section 6.3.2. With these data, the system gathers 23,822 fluid particles, 1,000 particles for wall modelling and 178 for the sphere.

The weakly compressible scheme based on an equation of state (refer to Section 5.3.2) is adopted. The numerical sound speed c_0 is determined by equation (5.213), since gravity is present. U_{max} obviously has the same order of magnitude as U, which yields $c_0 = 10U = 25$ m/s. The simulation time step, which here is set by the CFL condition in equation (5.413), then equals 1.2×10^{-4} s.

Here the turbulence modelling is provided by the $k - L_m$ model as described in Section 6.4.2. Since the formula (4.175) is only valid for a steady flow in an infinite channel, the mixing length L_m is set to the particle size, which is equivalent to a large eddy simulation approach (refer to Section 4.6.1).

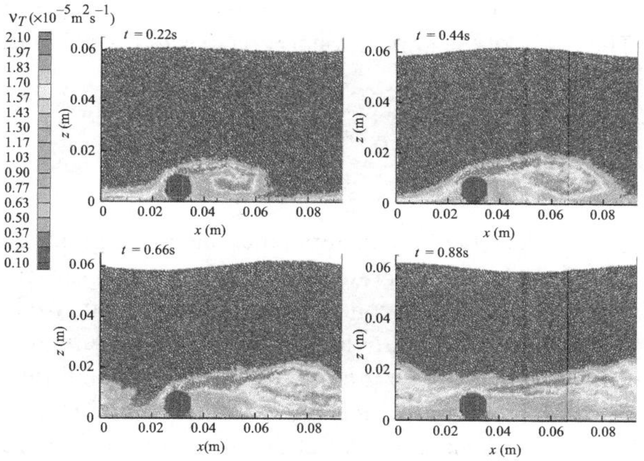

Fig. 7.42 Case of the sphere in a turbulent flow. Evolution of the eddy viscosity when the sphere is kept motionless, for $U = 0.59$ m/s.

Figure 7.42 illustrates the development of the turbulent wake downstream of the sphere when it is motionless, as well as the spatial distribution of the eddy viscosity, for $U = 0.59$ m/s. The estimate (7.10) yields $\nu_T \sim 6 \times 10^{-5}$ $\mathrm{m^2s^{-1}}$, in accordance with the range of obtained values. Figure 7.43 illustrates the distribution of the turbulent kinetic energy k when the sphere can freely move about, for $U = 1.77$ m/s. Equation (4.147) and the estimate (7.10) yield, with $L_m = \delta r = d/20$ and $S \sim U/d$:

$$\begin{aligned} k &= \frac{(L_m S)^2}{\sqrt{C_\mu}} \\ &\sim \frac{1}{400\sqrt{C_\mu}} U^2 \approx 8.3 \times 10^{-3} U^2 \end{aligned} \tag{7.19}$$

We get the order of magnitude $k \sim 5 \times 10^{-3}$ $\mathrm{m^2s^{-2}}$, in accordance with the numerical values.

In order to quantatively validate these simulations, we will establish a result of the sediment transport theory. The motion of a ball in a plane is a simplified model, indeed, of that of a sand grain on a river bed, in a current-induced bedload motion. In order to provide an estimate of the risk of entrainment

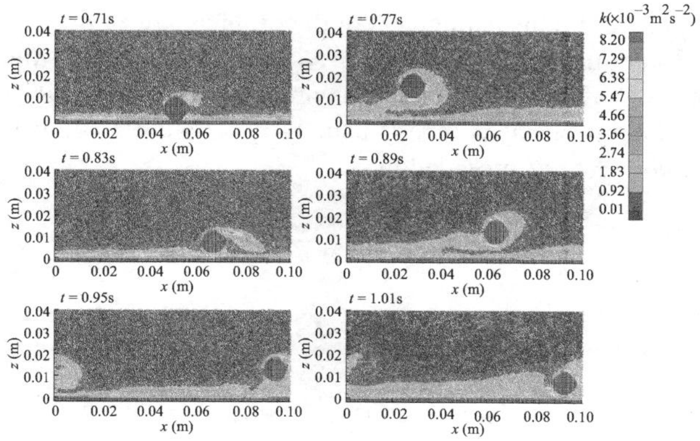

Fig. 7.43 Case of the sphere in a turbulent flow. Evolution of the sphere position when it can freely move about, and distribution of the turbulent kinetic energy, for $U = 1.77$ m/s.

of a grain having a diameter d, a balance of the forces it receives is initially established. The result of the gravitational force (second line in (1.57)) and the buoyancy (3.137) is written as

$$\begin{aligned} \mathbf{F}^g + \mathbf{F}^A &= M\mathbf{g} - \rho V \mathbf{g} \\ &= (\rho_s - \rho)\, V \mathbf{g} \end{aligned} \tag{7.20}$$

where $\rho_s > \rho$ is the density of the material making up the grain, and $V \propto d^3$ is its volume. This force keeps the grain on the bed, whereas the friction exerted by the fluid on its surface, on the contrary, tends to entrain it. We know that the bed shear stress τ_b exerted by a turbulent flow equals ρu_*^2 (refer to eqn (4.168)). The friction force exerted on the grain is found through multiplication by its apparent area $S \propto d^2$. The coefficients implicitly included in the just described laws of proportionality depend on the grain shape and the bed roughness, as well as on the way the grain is lying on the bed. In addition, we should mention the definite presence of other grains in its immediate vicinity, which makes the calculation of these coefficients absolutely impossible. Thus, the drag-to-gravity ratio (corrected for buoyancy) will be written as:

$$\begin{aligned} \frac{|\mathbf{F}^D|}{|\mathbf{F}^g + \mathbf{F}^A|} &= \frac{S \rho u_*^2}{(\rho_s - \rho)\, V g} \\ &\propto \frac{u_*^2}{\xi g d} \end{aligned} \tag{7.21}$$

with

$$\xi \doteq \frac{\rho_s}{\rho} - 1 > 0 \tag{7.22}$$

The ratio (7.21) denotes a dimensionless bed shear stress. It is sometimes also known as the *Shields parameter*, and is denoted here as τ^+; it characterizes the possibility of grain entrainment. The latter will occur if that ratio exceeds some critical threshold τ_c^+:

$$\tau^+ \doteq \frac{u_*^2}{\xi g d} \geqslant \tau_c^+ \tag{7.23}$$

The critical stress can be estimated through the dimensional analysis (refer to Section 3.4.3). Here the parameters of the problem are ρ, ρ_s, ν, g, d and u_*, that is $n = 7$ parameters out of which $m = 3$ are independent (three physical units). Thus, the Vaschy–Buckingham theorem states that there are $n - m = 4$ independent dimensionless numbers, namely τ_c^+, ξ and τ^+, as well as a *grain Reynolds number* Re_d as provided by

$$Re_d \doteq \frac{u_* d}{\nu} \tag{7.24}$$

Hence we may write

$$\begin{aligned} \tau_c^+ &= \phi_{dim}\left(\rho, \rho_s, \nu, g, d, u_*\right) \\ &= \phi_{adim}\left(Re_d, \xi, \tau^+\right) \end{aligned} \tag{7.25}$$

The condition (7.23) can then be rearranged in such a way as to delete the parameter τ^+ from the function ϕ_{adim} yielding τ_c^+. Besides, in most natural applications in sedimentology, the parameter ξ takes values which are very close to 2.6, which allows us to consider it as constant. Thus, we ultimately write

$$\tau_c^+ = \phi_{adim}\,(Re_d) \tag{7.26}$$

We have known since Section 3.4.3 that the effect of the Reynolds number fades away when the latter is high enough. In the case of a sediment grain, however, the relative smallness of the diameter d leads to moderate values of Re_d, which does not provide the independence of τ_c^+ with respect to Re_d. With the presently adopted diameter, however, the observations (for instance, refer to Van Rijn, 1984) show that such is the case given the values of Re_d (refer to Table 7.4). The experimental value of the critical stress then equals $\tau_c^+ \approx 0.055$. The deviation from the critical threshold (or *Shields parameter*) is then introduced by

$$T^+ \doteq \max\left(0; \frac{\tau^+}{\tau_c^+} - 1\right) \tag{7.27}$$

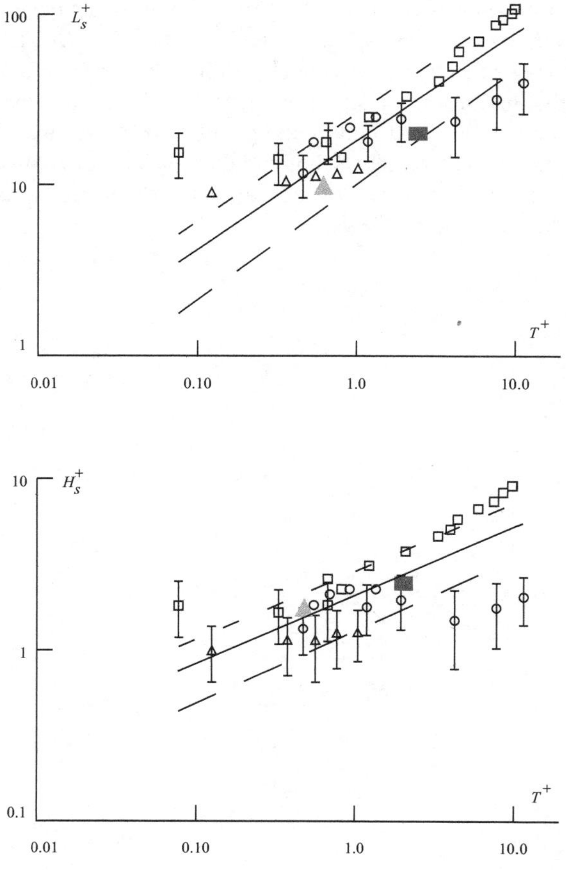

Fig. 7.44 Case of the sphere in a turbulent flow. Saltation parameters L_s^+ and H_s^+ as a function of T^+. SPH simulations for $U = 1.18$ m/s (triangles) and $U = 1.77$ m/s (squares). The other symbols represent experimental data, after Lee et al. (2000, with permission from ASCE).

Thus, the grain is then transported by the current when T^+ is strictly positive, that is when the condition (7.23) is satisfied. In this case, it describes a curved path before falling back onto the bed, and so on. This phenomenon is known as *saltation* and can be characterized by the length L_s and the height H_s of the path after each further leap. The same reasoning as previously immediately gives

$$L_s^+ \doteqdot \frac{L_s}{d} = \phi_{adim}\left(Re_d, \xi, \tau^+\right) \tag{7.28}$$

Equation (7.27) shows that τ^+ is a function of T^+ and τ_c^+, or else of T^+ and Re_d, because of (7.26). Furthermore, ξ may be assumed to be constant, as previously, and the grain Reynolds number is possibly high enough. Following the same reasoning about $H_s^+ \doteqdot H_s/d$, we get

$$\begin{aligned} L_s^+ &= \phi_L\left(T^+\right) \\ H_s^+ &= \phi_H\left(T^+\right) \end{aligned} \tag{7.29}$$

The functions ϕ_L and ϕ_H are provided experimentally (for instance, refer to Lee et al., 2000).

The presently set out results of the SPH calculation can be validated on the basis of the just mentioned theoretical elements. The shear velocity u_* is calculated by means of equation (4.170), substituting the driving force F^{ext} for $g \sin\theta$ (refer to eqn (4.165)), which yields $u_* = \sqrt{F^{ext} H}$. The corresponding values of Re_d and T^+ are provided by Table 7.4. For $U = 0.59$ m/s, T^+ is zero, that is the dimensionless stress is lower than the critical value: the grain remains motionless, as confirmed by the simulation. As regards the other two values of the superficial velocity, saltation takes place; Fig. 7.44 shows that the values of L_s^+ and H_s^+ as found through the numerical calculation are within the range of the experimental laws.

SPH applied to hydraulic works

8

8.1 Introduction

The practical applications of weakly compressible fluid mechanics, in particular the design of fluvial or coastal works, or even the investigation of flows in closed pipes and enclosures, were long exclusively restricted to small-scale models. The latter have been being used for quite a long time, either for studying a specific phenomenon or for establishing semi-empirical dimensioning formulas, or else for assessing the efficiency of a given layout in an actual facility. With the advent of scientific calculation, lots of flows could be addressed through numerical simulation, especially in the field of often confined industrial flows. Grid methods, which fit quite well this kind of simulation, also found a range of applications with the free surface flows provided that the latter remain both simple and not very mobile (as is the case, in particular, with the propagation of long waves). The behaviour of some free surface flows, however, has features which are hardly compatible with the principle of the Eulerian methods. Just think of a breaking wave, the sudden fall of a solid into a fluid initially at rest, or even a dam break wave.

The volume of fluid (VOF) technique, however, could provide suitable solutions for treating such flows through the grid methods, at the cost of susbstantially greater complexity. This is the context within which the Lagrangian methods will probably be applied in the most effective way, since they can easily handle heavily and quickly strained free surfaces. In particular, although the SPH method was developed for addressing astrophysical problems, it was already being used in first hydraulic applications rather close to reality by the end of the 1990s (refer, notably, to Monaghan and Kos, 1999). Nowadays, the case studies are mushrooming, increasingly focusing on hydraulics (for instance, refer to Campbell and Vignjevic, 2009; Marongiu et al., 2010; Crespo et al., 2010).

Since the practical applications in hydraulics are nowadays often three-dimensional, issues concerning computational time and memory space handling are raised, clearly to the disadvantage of SPH as compared with the Eulerian methods. One reason for this is the very small time step, primarily with the weakly compressible algorithm (refer to Section 5.4.4), but also and above all the large number of particles interacting with a given particle (refer to Section 5.2.3). This shortcoming becomes prohibitive where simulations involving several million particles are contemplated, which has become a frequent case in three dimensions. In order to cope with this problem, it is essential, during the developement of SPH-based software, to have a specific architecture in mind, for example harnessing the capabilities of the supercomputers through a massive parallelism (Moulinec et al., 2007; Moulinec et al., 2008; Issa et al., 2008; Maruzewski et al., 2010). A currently very popular alternative, which emerged in and around 2006, is based on the harnessing of the computational capabilities of graphics cards (*Graphic Processor Units*, or GPUs; for example, refer to Hérault et al., 2010; Crespo et al., 2010).

Applied scientific computing also requires the processing of input and output data. The first (pre-processing) consists in getting means for digitizing topographic or industrial data in order to discretize them in the form of a collection of fluid particles surrounded by solid edges which are formed as specified in Section 6.3.2, in both two and three dimensions. This kind of tool requires specific developments in order to meet the quality requirements as regards the modern numerical simulations (Oger et al., 2009; Lee et al., 2009; Mayrhofer et al., 2010 may be referred to). As to the processing of the output data (post-processing), it is based on software designed to display many particles provided with different quantities, for possibly deriving continuous fields therefrom through an interpolation. This should be achievable with visual effects ensuring a proper understanding of the results, even in three dimensions, and within periods of time consistent with the industrial exploitation of the results (for instance, refer to Jang et al., 2008; Oger et al., 2009; Marrone et al., 2010; Crespo et al., 2010).

In view of the satisfactory results of the schematic cases as disclosed in Chapter 7, we are going to set out four applications of the SPH method which nearly keep up with the industry requirements with respect to the possibilities and quality of the results. Thus, we will successively deal with the dimensioning of coastal works under the wave action, the functioning of a fish pass, the behaviour of an oil spill contained by a floating boom, and eventually the flow over the spillway of a large dam. In each case, we will first establish some theoretical results leading to either estimations of the required quantities or semi-empirical formulas. The implementation of the SPH method will be tested through these results.

8.2 Wave action upon waterworks

We consider here the behaviour of waves in the vicinity of coastal waterworks. We will see how dimensional analysis leads to semi-empirical formulae to estimate the overtopping rate of waves over a dyke. We then reproduce this phenomenon with the SPH method, before modelling the setup effect by wave breaking on a schematic coral reef.

8.2.1 Coastal work design elements

The design of the coastal works (breakwaters, groins, etc.) which are built for protecting a seashore against erosion or submersion by the sea is now more topical than ever. A thorough knowledge of the dynamics of wave interaction

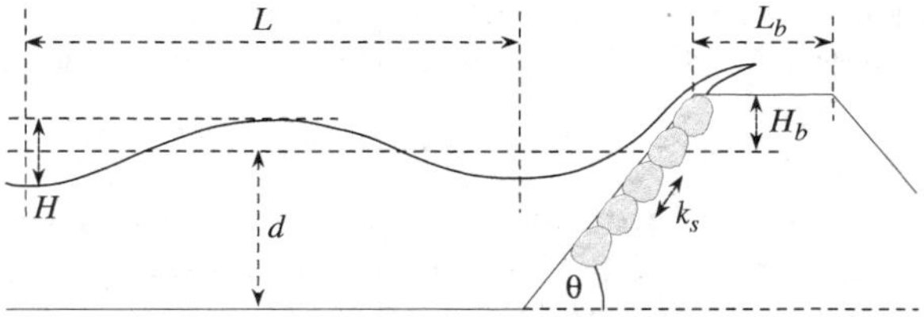

Fig. 8.1 Study of the monochromatic wave overtopping discharge rate acting upon a trapezoidal breakwater: notations.

with such a structure is obviously crucial for this dimensioning. Now, the consequences of the hurricanes, tsunamis and heavy storms having hit various regions on Earth in the recent years have highlighted further advances required in the control of these interactions. The presently popular design techniques for dykes first rely on two characteristic phenomena, namely the *runup*, which is the vertical extent of a wave uprush along the facing of a breakwater, and the *overtopping discharge rate*, that is the volume of water flowing over the work per unit time, due to the wave action. These poorly controlled phenomena are always evaluated with some inaccuracy, and their use as dimensioning factors too often leads to very compelling conservatisms, or—what is worse—to an underdimensioning of the works concerned. Numerical simulation may then prove useful in assisting the designers of coastal civil engineering works by providing further data in addition to those supplied by the scale models.

Here we do not intend to set out a wave theory (refer to Dean and Dalrymple, 1984). We will initially focus on some elementary theoretical and empirical results about the overtopping phenomenon. The overtopping discharge rate per unit length (reckoned in m^2/s) is denoted as q; this is the volume of water flowing over the work per unit time and per unit length of the structure being considered. It clearly depends on the physical parameters of the fluid (ρ and ν) as well as on the gravity g, on the geometrical characteristics of the work and ultimately on the characteristics of the waves acting upon it. If we consider the case of a simple breakwater with a trapezoidal cross-section, as in Fig. 8.1, which is acted upon by a monochromatic wave train, the list of these parameters is comparatively short. The geometry, indeed, fully depends on the slope $\tan\theta$, on the crest elevation above the mean sea level H_b, lastly on the berm width L_b. There are such other parameters as the size of the components (rock-fills, gravels, vegetation, etc.) making up the facing, which influence the wave runup through the friction exerted on the fluid. Referring to Section 4.4.5, the roughness caused by these components will suitably be denoted as k_s. The elevation d as determined by the mean sea level above the presumably flat bottom should be taken into account as well. The monochromatic wave theories show that they are fully characterized by two parameters[1], namely their period T and their amplitude (from crest to trough) H. It is sometimes more convenient, however, to consider the wavelength of the waves, related to the period and the water depth by an implicit relationship as given by the linear wave theory (Dean and Dalrymple, 1984):

$$L = \frac{gT^2}{2\pi} \tanh \frac{2\pi d}{L} \tag{8.1}$$

[1] This outcome is natural, since every oscillator is featured by a frequency (inverse of its period) and an energy. It is shown that the energy of small waves is proportional to the square of the wave amplitude. For random waves, which are closer to reality, the notions of significant height and mean period are to be introduced (Dean and Dalrymple, 1984).

Thus, the pairs (H, T) and (H, L) may be chosen interchangeably for characterizing the wave train. We then must seek a relation of the same kind as

$$q = \phi_{dim}\left(\rho, \nu, g, d, H, T, H_b, L_b, \tan\theta, k_s\right) \quad (8.2)$$

where ϕ_{dim} is an undetermined dimensional function. It is noteworthy that we have not introduced any velocity scale for the orbital wave motion, the latter's order of magnitude being naturally determined by the ratio[2] $U \equiv H/T$. A Reynolds number can then be constructed on the basis of that estimate:

[2] More exactly, we have $U_{\max} = \pi H/T$ in the linear wave theory.

$$Re \doteqdot \frac{H^2}{T\nu} \quad (8.3)$$

For ordinary values of H and T, Re largely exceeds $Re_{c,1} = 3,000$ (refer to Section 4.2.2), and often reaches values of about 10^6. Hence, this is a highly turbulent flow, and so the Reynolds number influence is insignificant, as shown by Section 3.4.3 and Chapter 4. That means that the molecular viscosity ν may be ignored in equation (8.2). Let us now refer to the arguments in Section 3.4.3. We have $n = 10$ parameters out of which $m = 3$ are dimensionally independent, which makes it possible to construct $n - m = 7$ dimensionless numbers and rearrange (8.2) as follows:

$$q^+ = \phi_{adim}\left(Fr, d^+, H_b^+, L_b^+, k_s^+, \tan\theta\right) \quad (8.4)$$

where the following has been defined

$$\begin{aligned} q^+ &\doteqdot \frac{q}{gHT} & H_b^+ &\doteqdot \frac{H_b}{H} \\ Fr &\doteqdot \frac{1}{T}\sqrt{\frac{H}{g}} & L_b^+ &\doteqdot \frac{L_b}{H} \\ d^+ &\doteqdot \frac{d}{H} & k_s^+ &\doteqdot \frac{k_s}{H_b} \end{aligned} \quad (8.5)$$

It will have been noticed that the density is not taken into account in these definitions. This is because here ρ is the single quantity which depends on the mass unit, and so it cannot be used for constructing a dimensionless number.[3] The pair of integers (n, m) could then have been reduced from $(10, 3)$ down to $(9, 2)$, with the same result.

[3] Going back to the discussion in Section 3.4.3, it can be found that changing the unit of measurement of mass would change the value of ρ without affecting the other parameters, which is another way of understanding that ρ plays no part here.

With these notations, the relation (8.1) becomes

$$2\pi Fr^2 L^+ = \tanh \frac{2\pi d^+}{L^+} \quad (8.6)$$

with

$$L^+ \doteqdot \frac{L}{H} \quad (8.7)$$

Formally, (8.6) can be inverted for writing

$$L^+ = L^+\left(Fr, d^+\right) \quad (8.8)$$

There are many empirical formulas of the (8.4) type in scientific literature; they are established as a result of small-scale model test campaigns. Three such formulas are provided below, beginning with that of Bradbury and Allsop (1988), as amended by Aminti and Franco (1988):

$$q^+ = \frac{C_1^{BA}}{Fr\left(Fr \cdot H_b^{+2}\right)^{C_2^{BA}}} \tag{8.9}$$

where C_1^{BA} and C_2^{BA} are two constants (refer to Table 8.1). This yields, from a dimensional point of view:

$$q = C_1^{BA} gHT \left(\frac{T\sqrt{gH^3}}{H_b^2}\right)^{C_2^{BA}} \tag{8.10}$$

That formula has the benefit of simplicity, but the drawback of not revealing any dependence of the d^+, L_b^+ and k_s^+ parameters. Ahrens and Heimbaugh (1988) partly correct that deficiency with a dependence on d^+:

$$q^+ = C_1^{AH} \exp\left[-C_2^{AH} \frac{H_b^+}{L^+\left(Fr, h^+\right)^{C_3^{AH}}} - C_4^{AH} Fr\right] \tag{8.11}$$

hence

$$q = C_1^{AH}\sqrt{gH^3} \exp\left[-C_2^{AH} \frac{H_b}{H}\left(\frac{H}{L}\right)^{C_3^{AH}} - \frac{C_4^{AH}}{T}\sqrt{\frac{H}{g}}\right] \tag{8.12}$$

d^+ occurs here via L^+ (or L), as given by equation (8.8) (or (8.1)). Lastly, Hebsgaard et al. (1998) propose a more exhaustive formulation exhibiting the breakwater slope and berm width, as well as the friction exerted by the facing:

$$q^+ = C_1^{HSJ}\left(k_s^+\right) \ln \frac{1}{2\pi Fr^2} \exp\left[-C_2^{HSJ}\left(k_s^+\right) \frac{H_b^+ + C_3^{HSJ} L_b^+}{(\tan\theta)^{C_4^{HSJ}}}\right] \tag{8.13}$$

or else

$$q = C_1^{HSJ}\sqrt{gH^3} \ln \frac{gT^2}{2\pi H} \exp\left[-C_2^{HSJ} \frac{H_b + C_3^{HSJ} L_b}{H(\tan\theta)^{C_4^{HSJ}}}\right] \tag{8.14}$$

The facing roughness acts through the variable coefficients $C_1^{HSJ}\left(k_s^+\right)$ and $C_2^{HSJ}\left(k_s^+\right)$. Two values of either of them are given by the authors; they correspond, on the one thand to a smooth facing $\left(k_s^+ = 0\right)$, on the other hand to a rubble-mounded facing $\left(k_s^+ \sim 1\right)$. Though the influence of roughness is taken into account in a rather shallow way here, the formula offers the advantage of depending on it, unlike the first two formulas, which are calibrated through tests conducted on rubble-mounded breakwaters. However, since the latter face type is the only one in the considered applications, here we only keep those values of C_1^{HSJ} and C_2^{HSJ} which correspond to $k_s^+ \sim 1$.

The coefficients appearing in the above formulas are given by Table 8.1.

These empirical formulas, of course, yield similar orders of magnitude. However, it would be naive to grant them a high predictive power, taking into account the complex processes governing the overtopping phenomenon,

Table 8.1 Calculation of the overtopping discharge rate as caused by monochromatic waves on the same breakwater as in Fig. 8.1. Values of the coefficients in the formulas developed by Bradbury and Allsop (1988, eqn (8.9)), Ahrens and Heimbaugh (1988, eqn (8.11)) and Hebsgaard et al. (1998, eqn (8.13)).

Formula	C_1	C_2	C_3	C_4
Bradbury and Allsop (1988)	2.27×10^{-5}	2.68	–	–
Ahrens and Heimbaugh (1988)	308	10.732	1/3	16.616
Hebsgaard et al. (1998)	300	4.92	0.175	0.3

inducing in particular large uncertainties in the experimental measurements. Figure 8.5 in the next section illustrates the quality of predictions to be reasonably expected from these equations.

Lastly, it should be pointed out that the mean elevation z_{surf} of the sea surface level plays a crucial part in the use of the above formulas, because it essentially acts through the level H_b of the breakwater crest and is reckoned with respect to said level, that is $H_b = z_{surf} - z_b$, z_b being the breakwater crest elevation in a arbitrary coordinate system. Now, of course, z_{surf} depends on various external circumstances, particularly on the tide and the atmospheric effects. It is noteworthy that the waves themselves influence it, for the following reason: as they propagate towards shallower bottoms, they tend to become stiffer and they ultimately break, in a way similar to that described in Section 7.3.2. In the process, they lose energy, a part of which is converted back into mean level-related potential energy, that is $\rho g z_{surf}$ (refer to eqn (3.138)). More specifically, this mechanism may be investigated by associating to the orbital motion of the breaking waves a velocity denoted as $\mathbf{u}^w$, which comes on top of the velocity possibly induced by the ambient current. It then appears appropriate to average the physical quantities over the wave period T for an arbitrary physical quantity A:

$$\{A\}_w \doteqdot \frac{1}{T}\int_0^T A dt \tag{8.15}$$

Considerations similar to those in Section 4.3.3 then show that a new stress tensor becomes established within the fluid; it is known as the *radiation stress tensor*, being denoted as $\mathbf{W}$ and given by:[4]

$$\mathbf{W} \doteqdot \{\mathbf{u}^w \otimes \mathbf{u}^w\}_w \tag{8.16}$$

(by analogy with the Cauchy and Reynolds stresses, provided by eqns (3.250) and (4.46), respectively). Thus, the breaking phenomenon induces increased shear stresses, which are offset at equilibrium by a pressure gradient, and therefore by a mean water level gradient. Thus, the mean value of the water level z_{surf} rises slightly, in the breaking zone, by a quantity which is called *setup* and is denoted here as Δz_{surf}. A theoretical description and experimental analyses of that process can be found in Dean and Dalrymple (1984).

In the next pages of this section, we will attempt to replicate the overtopping and setup phenomena by means of the SPH method. The above formulas will then be used for validating the numerical model for the prediction of overtopping. The setup phenomenon will be validated on the basis of experimental measurements and numerical simulations based on other methods.

[4] $\mathbf{W}$ should strictly be defined as a momentum flux (refer to Section 3.3.2). Effects of pressure variation should then be added to the presently given definition.

8.2.2 Numerical replication of overtopping

The SPH method is commonly used for investigating the wave interactions with either fixed or floating coastal works (refer, for example, to Shao et al., 2006; Dalrymple and Rogers, 2006). Its Lagrangian nature does lend itself to the modelling of quick and highly strained motions of a free surface, as discussed in Chapter 7 (also refer to Colagrossi et al., 2010). The overtopping dynamics over a coastal defense structure was studied on that basis, in particular, by Shao et al. (2006) and Didier and Neves (2009). We will initially focus on the overtopping phenomenon, following the ideas of Violeau and Issa (2005), Lee et al. (2006) and Issa et al. (2007).

The experiment conducted by Stansby and Feng (2004) will now be considered. The wave agitation channel in Fig. 8.2 is discretized in dimension $n = 2$, in the same way as in the test cases of Chapter 7 (in particular, refer to Section 7.3.2), and provided with a paddle (mobile wall) moving according to the following law:

$$X(t) = X_{\max} \sin \frac{2\pi t}{T} \tag{8.17}$$

This motion generates monochromatic waves having a period T and an amplitude H, as yielded by

$$H = X_{\max} \frac{2\pi d}{L} \tag{8.18}$$

d being the mean water depth and L the wavelength associated with d and T, as given by (8.1) (Dean and Dalrymple, 1984). It is contemplated to vary H and T, with such values as[5] $H \sim 0.15$ m and $T \sim 3$ s. The channel has a flat bed over a length L_1, before getting a gentle slope of $1/20^{th}$ over the length L_2, then coming against a trapezoidal breakwater. It is initially filled over a depth d as measured above the bed elevation in its flat section. The distances mentioned in Fig. 8.2 are $d = 0.36$ m, $L_1 = 1.90$ m and $L_2 = 5.60$ m, whereas the parameters characterizing the breakwater geometry (which are defined as in Fig. 8.1) equal $H_b = 0.04$ m, $L_b = 0.10$ m and $\tan\theta = 1/2$. Since the breakwater is comparatively smooth in the original experiment, a low roughness is selected: $k_s = 1$ mm.

A water particle size $\delta r = 1$ cm is considered, which gives a total of about 26,000 fluid particles and 6,600 particles for the walls, and $h/\delta r = 1.2$ is set. Here the weakly compressible model is adopted for the prediction of pressure. Since the wave orbital velocity has the order of magnitude $\pi H/T$ (refer to the preceding section), a sound speed $c_0 = 10 \max\left(\pi H/T, \sqrt{gd}\right)$ is prescribed (eqn (5.213)). The abovementioned orders of magnitude of H and T, as well as the value of d, then recommend $c_0 = 19$ m/s. Since the velocities may significantly increase upon the breaking and the overtopping, however, it was decided

[5] Let us remember that the initial experience is a small-scale model. Assuming that its spatial scale is $1/25^{th}$, $H = 0.1$ m corresponds to full-scale waves of 2.5 m. The Froude similarity (refer to Section 8.5.1) then prescribes a time scale of $1/5^{th}$. Hence, $T = 2$ s corresponds to a full-scale period of 10 s.

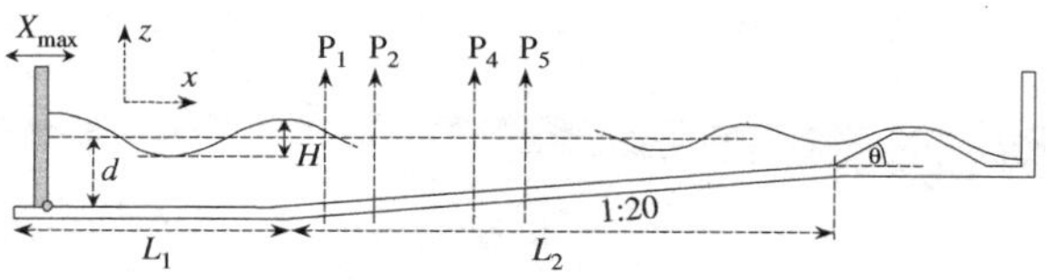

Fig. 8.2 Geometry of the experiment conducted by Stansby and Feng (2004).

to bring that value up to $c_0 = 30$ m/s, which gives a time step $\delta t = 1.6 \times 10^{-4}$ s. No turbulence model is taken into account, in order to avoid dissipating too much energy for keeping the wave train in its propagation. This approach is in line with the existing wave modelling theories, which recommend considering the flow as that of a perfect fluid (refer to Chapter 3). On the other hand, as explained in Section 7.3.2, turbulence may play a role upon the breaking, which will therefore probably be poorly replicated by this model.

Figure 8.3 provides several instantaneous views of the flow as replicated through the SPH method for $H = 0.108$ m and $T = 2.39$ s (i.e. $L = 4.301$ m and $X_{\max} = 0.103$ m). On these a wave can be seen propagating, breaking and being partly reflected from the breakwater, whereas some amount of water overflows it in each period. In order to get a quantitative estimate of the degree of accuracy of the numerical predictions, the evolution of the free surface elevation z_{surf} over time is initially considered at four points along the channel $\{P_i\}$ $(i = 1, 2, 4, 5)$ whose positions are indicated in Fig. 8.2. Their respective abscissae are provided by $x(P_i) = X_i$, with $\{X_i\} = (2.90$ m; 3.47 m; 4.47 m; 497 m), as reckoned from the spatial origin illustrated by the halftone shaded disk in the figure.

For validation purposes, it is interesting to compare the SPH results with those of another model, solving the Boussinesq equations[6] by means of a finite difference type method, what will henceforth be referred to as BFD (*Boussinesq finite difference*). Figure 8.4 illustrates the temporal evolution of z_{surf} at the points being considered; on these the results of both SPH and BFD methods are compared with the measurements made by Stansby and Feng (2004). The comparison is clearly in favour of SPH, which almost always

[6]Establishing and discussing these equations would lead beyond the scope of this book. Refer, for instance, to Nwogu (1993).

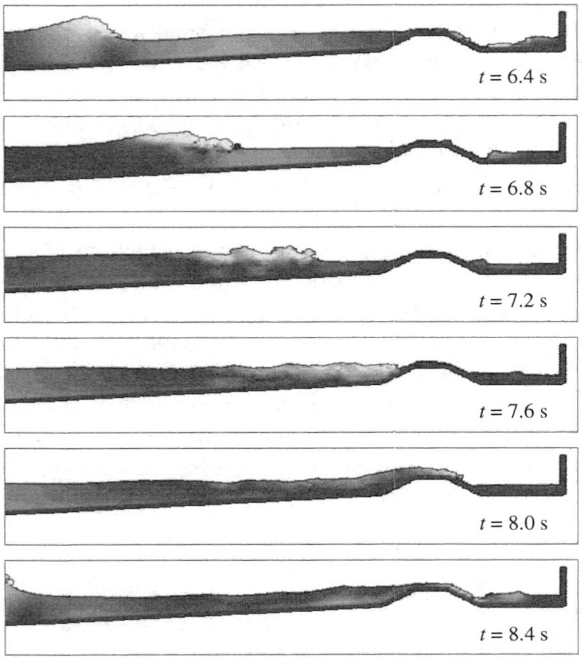

Fig. 8.3 Numerical replication, by means of the SPH method, of the experiment conducted by Stansby and Feng (2004).

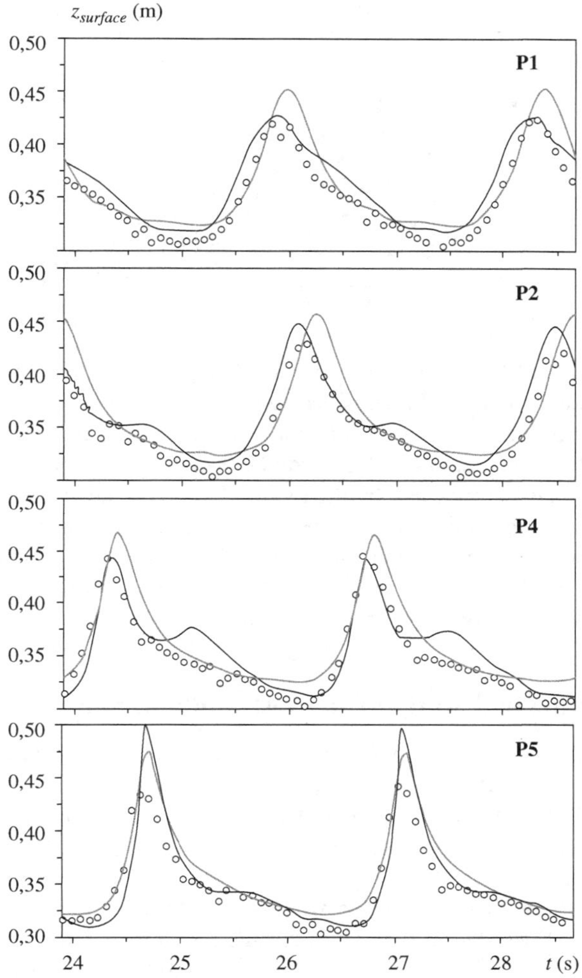

Fig. 8.4 Temporal evolution of the free surface elevation at the points $\{P_i\}$ occurring in Fig. 8.2. Comparison of the experiment conducted by Stansby and Feng (2004, thick lines) with the numerical predictions of the SPH (symbols) and BFD (thin lines) methods. After Lee et al. (2006) and Issa et al. (2009a, with permission from World Scientific).

results in a better prediction of the free surface profiles, except for the main peak exhibited by the point P_5. Besides, it can be noticed that the secondary peak highlighted by the measurement at points P_2 and P_4 is not replicated by either method. As we suggested it in Section 7.3.2, this SPH deficiency could largely be remedied though either a renormalization procedure (Section 5.2.5) or an incompressible model (Section 6.2.5) for the prediction of pressure. In addition, increasing the discretization significantly improves the results in this case, at the cost of a more time-consuming calculation. A test with $\delta r = 0.5$ cm (not set out here) proved being more suitable for predicting the peaks of z_{surf}.

A reckoning of the number of particles overflowing the breakwater crest makes it possible to estimate the overtopping discharge rate as predicted by the simulation, it being known that their individual volume (i.e. their area, in dimension 2) is provided by the formula (5.80). Varying H in the 0.1 m–0.2 m range and T in the 1.2 s–4.0 s range, the values of q plotted in Fig. 8.5 can be determined. Comparing the predicted discharge rates with

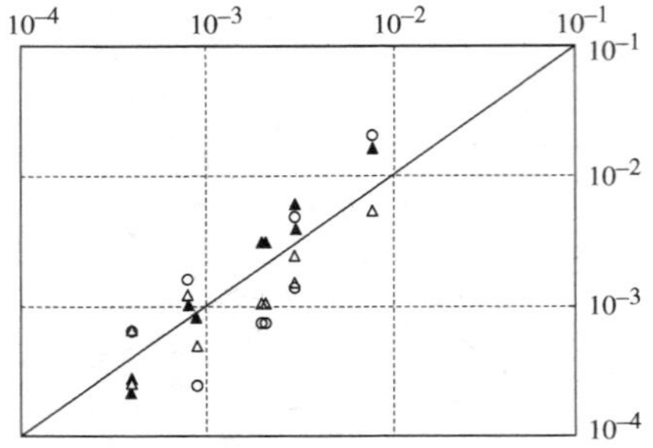

Fig. 8.5 Numerical replication of the experiment conducted by Stansby and Feng (2004) using the SPH method. Comparisons of the calculated overtopping discharge rates (horizontal axis) with the semi-empirical formulas obtained by Bradbury and Allsop (1988, eqn (8.9), o), Ahrens and Heimbaugh (1988, eqn (8.11), △) and Hebsgaard et al. (1998, eqn (8.13), ▲). After Issa et al. (2007).

the semi-empirical formulas of Section 8.2.1 (eqns (8.10), (8.12) and (8.14)) provides a satisfactory match, considering the reasonably expectable accuracy of the prediction of so complex a phenomenon as well as the disparity in the empirical predictions. In particular, it can be found that the increase of q with T and H is properly replicated by the model. However, the above conclusions should be mitigated, noting that Fig. 8.5 has logarithmic axes. Furthermore, the roughness of the breakwater facing would have to be consistent with the presence of rockfills for better predicting the fluid behaviour when overflowing an actual breakwater (refer to Section 8.2.1).

8.2.3 Numerical study of the setup phenomenon

We will now turn to the experiment conducted by Gourlay (1996), which comprises the flow of a wave train over a board schematically representing a coral reef. Through this study, we will address the issue of the wave breaking and the setup which is described by the end of Section 8.2.1. Here we follow the ideas of Lee et al. (2006) and Issa et al. (2007).

The adopted geometry, which is slightly simplified as compared with the experiment, is illustrated in Fig. 8.6, once again in dimension $n = 2$. The values $L_1 = L_4 = 1$ m, $L_2 = 7$ m, $L_3 = 8$ m, $d_1 = 0.08$ m, $d_2 = 0.4$ m, $d = 0.5$ m are adopted. Waves having a height of $H_0 \sim 0.15$ m and a period $T \sim 1$ s are generated by the method as set out in the preceding chapter. Particles having a size $\delta r = 1$ cm are considered, which gives a total amount of approximately 40,000 fluid particles and 9,100 particles for the walls, and $h/\delta r = 1.3$ is set. Here the weakly compressible model is utilized for the prediction of pressure. Since the wave orbital velocity is about $U_{\max} = \pi H_0/T$ (refer to Section 8.2.1), a speed of sound of $c_0 = 10 \max\left(U_{\max}, \sqrt{gd}\right)$ (eqn (5.213)) is prescribed. The above mentioned orders of magnitude of H_0 and T, as well as the value of d, then recommend $c_0 = 20$ m/s. The velocities may significantly

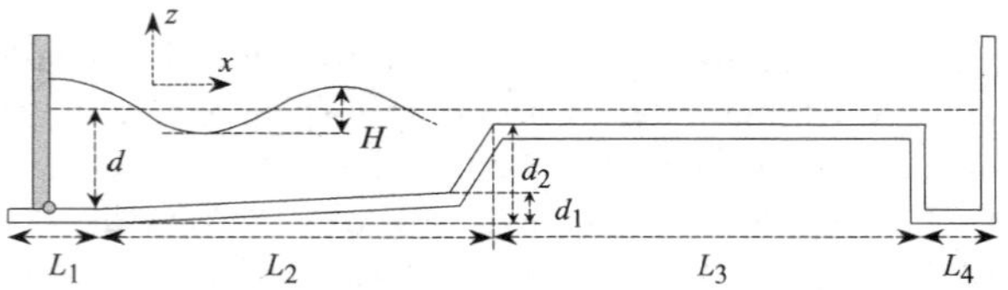

Fig. 8.6 Simplified geometry of the experiment conducted by Gourlay (1996).

increase during the breaking and the flow over the board, but that does not affect the choice of the value of c_0, because $\sqrt{gd}$ here substantially prevails over $U_{\max}$. The time step is then $\delta t = 2.6 \times 10^{-4}$ s. No turbulence model is taken into account, although that may affect the amount of macroscopic kinetic energy which is lost upon the breaking.

Figure 8.7 illustrates an example of the shape of the wave near the point where it reaches the board for $H_0 = 0.200$ m and $T = 1.48$ s. It can be observed that the waves do break there, which makes them lose a large amount of energy, inducing the setup phenomenon Δz_{surf}. The latter can be calculated by considering the elevation of the particles lying at the surface, for a given abscissa, which also makes it possible to determine the wave height H, which decreases from its incident value H_0, as Δz_{surf} increases. Figures 8.8 and 8.9 show, for $H_0 = 0.148$ m and $T = 1.10$ s, the distribution of H and Δz_{surf} as a function of the abscissa x, the origin being indicated by the halftone shaded disc in Fig. 8.6. Once again, we have decided to compare the results of the SPH method with the BFD method which was discussed in the preceding section, here proposing two variants to that approach. We also provide the numerical results obtained by Massel and Gourlay (2000). The comparisons are slightly in favour of the BFD model for H, whereas SPH substantially better predicts Δz_{surf}.

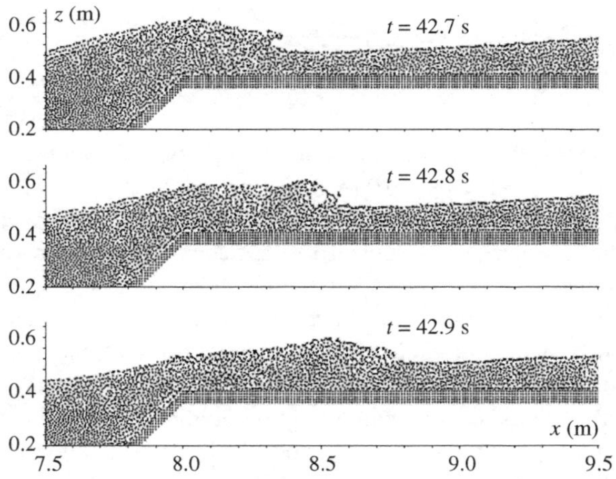

Fig. 8.7 Numerical replication of the experiment conducted by Gourlay (1996) with SPH, for $H_0 = 0.200$ m and $T = 1.48$ s: zooming in on the point where the wave reaches the board (after Lee et al., 2006) and Issa et al. (2009a, with permission from World Scientific).

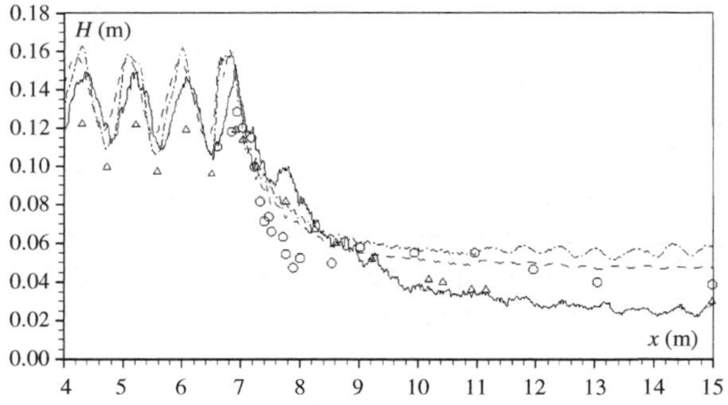

Fig. 8.8 Experiment conducted by Gourlay (1996) for $H_0 = 0.148$ m and $T = 1.10$ s: variation of the wave height H along with x. Comparison of the measurements made by Gourlay (o) with the numerical simulations: SPH (solid line), BFD models (dashed line), Massel and Gourlay (2000, Δ). After Issa et al. (2007) and Issa et al. (2009a, with permission from World Scientific).

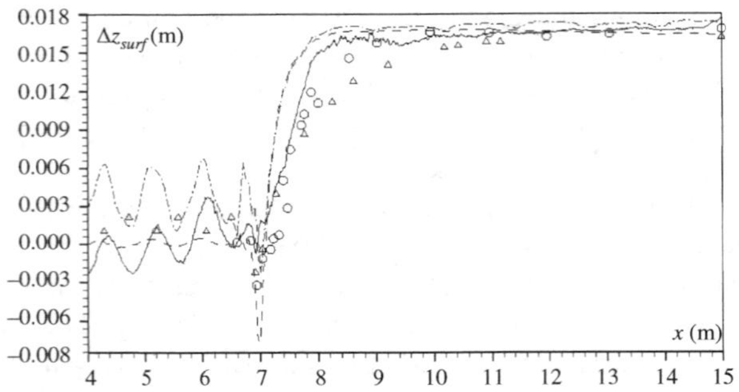

Fig. 8.9 Experiment conducted by Gourlay (1996) for $H_0 = 0.148$ m and $T = 1.10$ s: setup variation Δz_{surf} with x. Comparison of the measurements made by Gourlay (o) with the numerical simulations: SPH (solid line), BFD models (dashed lines), Massel and Gourlay (2000, Δ). After Issa et al. (2007) and Issa et al. (2009a, with permission from World Scientific).

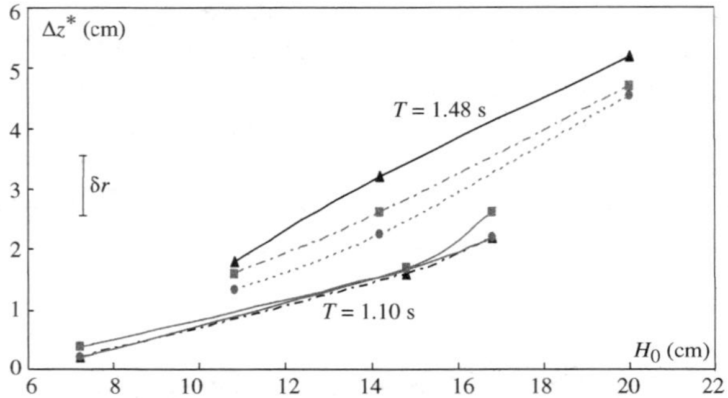

Fig. 8.10 Experiment conducted by Gourlay (1996): value of the asymptotic setup Δz^* for various pairs (H_0, T). Comparison of the measurements made by Gourlay (squares) with the numerical simulations: SPH (triangles) and the BFD models (circles). After Lee et al. (2006).

Not surprisingly, the setup reaches an asymptotic value above the board, being denoted as Δz^*. Figure 8.10 presents the values of Δz^* for several sets of pairs (H_0, T). The SPH predictions are quite satisfactory, especially as the position of the free surface is subject to an inaccuracy of the same order of magnitude as the size δr of a particle which is indicated in the figure.

Both the brief studies which we have just disclosed substantiate the effectiveness of SPH in the prediction of the interactions of the coastal waves and the fixed coastal structures, as confirmed by many papers, sometimes involving floating bodies, and leading to good predictions of the loads applied to the structures. As explained by Crespo et al. (2010), this approach may be extended to the three-dimensional simulation, provided that the available code is efficient as regards the computational time management (refer to Section 8.1).

8.3 Fish pass

The method of integral balances and the Kolmorov analysis make it possible to establish certain rules and orders of magnitude in order to proceed to the design of a waterwork such as a fish pass. After briefly presenting this analysis, we propose an SPH-based simulation. We also compare various turbulence models.

8.3.1 Theoretical calculation of a pool fish pass

Constructing a dam disrupts the biological continuity of a stream, changing its ecosystem and morphology. Failing specific measures, fish can no longer migrate as easily and therefore the biological cycles are upset. Thus, the dams are obstacles which fish must be able to get over in either direction for a successful life cycle. In particular, swimming upstream in spite of the obstacles they encounter on their way is crucial for the successful spawning of some species, which often takes place in very distant places upriver. The fish passes are designed to allow that journey, in all circumstances and for all the species concerned and therefore play a vital ecological role. Their dimensioning is based on the estimate of the spatial distribution of the current, which hinders the fish displacement, as well as on the characterization of the turbulent quantities. Whereas the smaller eddies do not affect the fish migration, the larger ones are confusing structures, indeed, and a medium-sized eddy, that is of the same order of magnitude as the animal size, will hamper the swimming motions. This is how the spectral nature of the turbulence, which involves exceedingly variable scales (refer to Chapter 4), is detrimental to the efficiency of these works.

Let us focus on a fairly common type of fish pass, which consists of pools communicating through vertical slots and provided with a resting area each (refer to Fig. 8.11). When there are enough pools, the flow may be considered

Fig. 8.11 Vertical slot fish pass with successive pools, on the Claies de Vire dam, in France (photo by A. Richard, Onema).

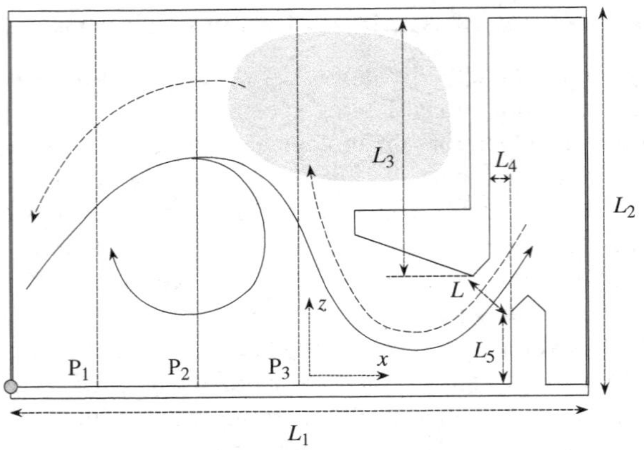

Fig. 8.12 Geometry of a pass pool (after Tarrade et al., 2008). The halftone shaded area indicates the resting area. The solid curves illustrate the current shape, whereas the dashed curves correspond to the approximate fish trajectories. The periodic boundaries are shown by thick lines.

as periodic, as here. Thus, we may focus on a single basin whose geometry is illustrated by a top view in Fig. 8.12. The presently adopted dimensions are $L_1 = 3$ m, $L_2 = 2$ m, $L_3 = 1.388$ m and $L_4 = 0.132$ m, which corresponds to $L = 0.300$ m. The walls are rather smooth, with an equivalent roughness $k_s = 0.54$ mm. The experiments (among others, refer to Tarrade et al., 2008) show that the flow is nearly two-dimensional, that is the mean velocity (in the Reynolds understanding) are essentially horizontal or, instead, parallel to the bottom, which is sloping with a constant slope being denoted as $I = \tan\theta$ and being order 0.1. That slope induces a very turbulent but steady flow (on average) and the mean velocity field exhibits a waviness together with lateral recirculations the largest of which is schematically illustrated in Fig. 8.12. The measuments made by Tarrade et al. (2008) reveal a spectrum which is consistent with the Kolmogorov law (4.119).

The balances achieved in Chapter 3 may advantageously be utilized for establishing some theoretical results which, in some respect, are not related to the exact geometry of the pools. To that purpose, we will turn back to equation (3.72), matching it to the context of Reynolds-averaged fields, and under a constant density. We know (refer to eqn (4.52)) that the Cauchy stress tensor then comes along with the Reynolds stress tensor, which yields

$$\int_{\Omega} \frac{\partial \overline{\mathbf{u}}}{\partial t} d\Omega = \oint_{\partial\Omega} \left(\frac{1}{\rho} \overline{\sigma} - \mathbf{R} - \overline{\mathbf{u}} \otimes \overline{\mathbf{u}} \right) \mathbf{n} d\Gamma + V\mathbf{g} \tag{8.19}$$

where V is the volume enclosed within the domain Ω, which here is extended to the whole inside of a pool, whereas the boundary $\partial\Omega$ of the latter corresponds to the vertically extruded edges of Fig. 8.12 (direction at right angles to that figure), to which the free surface and the bottom should be added. We will denote as $\partial\Omega_u$ the portion of $\partial\Omega$ representing the walls, including the bottom, and as $\partial\Omega_p$ the free surface (we borrow these notations from Section 3.4.4). Lastly, $\partial\Omega_Q$ stands for the boundary portion corresponding to the periodic boundaries, and so $\partial\Omega = \partial\Omega_u \cup \partial\Omega_p \cup \partial\Omega_Q$.

Since the flow is steady, the first integral among those in (8.19) is identically zero. Since the wall velocity $\mathbf{u}_{wall}$ is zero, the wall boundary conditions (third line in (4.159)) yields $(\overline{\mathbf{u}} \otimes \overline{\mathbf{u}})\,\mathbf{n} = (\overline{\mathbf{u}} \cdot \mathbf{n})\,\overline{\mathbf{u}} = \mathbf{0}$ on $\partial\Omega_u$. At a free surface (denoted as $\partial\Omega_p$), the steady nature of the flow and the condition (3.142) ensure $\overline{\mathbf{u}} \cdot \mathbf{n} = 0$, which makes it possible to reach the same conclusion. Lastly, the integral of $(\overline{\mathbf{u}} \otimes \overline{\mathbf{u}})\,\mathbf{n}$ on $\partial\Omega_Q$ is zero through periodicity, that is it has the same value with an opposite sign at both periodic boundaries.[7] Thus, the integral of $(\overline{\mathbf{u}} \otimes \overline{\mathbf{u}})\,\mathbf{n}$ over $\partial\Omega$ is identically zero. Hence we get:

$$\oint_{\partial\Omega} \left(\frac{1}{\rho} \overline{\sigma} - \mathbf{R} \right) \mathbf{n} d\Gamma + V \mathbf{g} = \mathbf{0} \tag{8.20}$$

The force $\mathbf{F}$ exerted by the fluid onto the pool can be derived from:

$$\begin{aligned} \mathbf{F} &= -\int_{\partial\Omega_u} (\overline{\sigma} - \rho\mathbf{R})\,\mathbf{n} d\Gamma \\ &= \int_{\partial\Omega_p \cup \partial\Omega_Q} (\overline{\sigma} - \rho\mathbf{R})\,\mathbf{n} d\Gamma - V\mathbf{g} \end{aligned} \tag{8.21}$$

Let us now invoke the models (4.71) and (4.73) to express the stress tensors. We include the gravity in the pressure by means of the dynamic pressure $\overline{p}^* \doteqdot \overline{p} + \rho g z$ (definition (3.138)), and we delete the air pressure $\overline{p}_{atm}$, which acts on the other side of the edges being contemplated. With the approximations (4.80) and (4.202), we find[8]

$$\begin{aligned} \mathbf{F} &= \int_{\partial\Omega_p \cup \partial\Omega_Q} \left[-\left(\overline{p}^* - \overline{p}_{atm} + \frac{2}{3}\rho k \right) \mathbf{n} + 2\rho\,(\nu + \nu_T)\,\mathbf{S} \cdot \mathbf{n} \right] d\Gamma \\ &\approx \int_{\partial\Omega_p \cup \partial\Omega_Q} \left[-\left(\overline{p}^* - \overline{p}_{atm} \right) \mathbf{n} + 2\rho\nu_T \mathbf{S} \cdot \mathbf{n} \right] d\Gamma \end{aligned} \tag{8.22}$$

The shear stress $\overline{\tau} = 2\rho\nu_T \mathbf{S} \cdot \mathbf{n}$ is zero on the free surface when there is no wind (second line in the system (4.159)). Furthermore, since the velocity field is nearly two-dimensional, this is a simple shear flow, as in Section 4.4.5 (save that there is no translational invariance). We may recall what this means: the velocity is substantially horizontal and varies substantially vertically. We therefore can state that the shear stress is negligible at the periodic boundaries. Moreover, we may infer that the pressure is distributed under a hydrostatic distribution (eqn (4.166)). If we now focus on the horizontal component $F^D = \mathbf{F} \cdot \mathbf{e_x}$ of the force, the x axis being oriented in the direction of the overall flow, the above arguments can be used to reduce the integration to the periodic boundaries $\partial\Omega_Q$ only keeping the pressure forces:[9]

$$\begin{aligned} F^D &\approx -\int_{\partial\Omega_Q} \left(\overline{p}^* - \overline{p}_{atm} \right) (\mathbf{n} \cdot \mathbf{e}_x)\, d\Gamma \\ &\approx -\rho g \cos\theta \int_{\partial\Omega_Q} z_{surf}\,(\mathbf{n} \cdot \mathbf{e}_x)\, d\Gamma \end{aligned} \tag{8.23}$$

The experiments show that the free surface elevation z_{surf} varies in space, but we will assume that it is relatively constant at each of the boundaries making up $\partial\Omega_Q$. The integral of z_{surf} at each of these boundaries is then constant and equals $dL_2 z_{surf}$, where d denotes the corresponding depth and L_2 the pool width. Since the $\mathbf{n}$ vectors are opposite and verify $\mathbf{n} \cdot \mathbf{e}_x = \pm 1/\cos\theta$, the integral (8.23) ultimately yields:

[7] This outcome is natural, since these boundaries are used solely for mental convenience. In fact, they are part of the inside of the flow.

[8] The $\nu \ll \nu_T$ assumption is justified by the very high value of the Reynolds number, which will be substantiated later on. On the other hand, with the presently considered slope, which is relatively steep, it seems that the relation (4.202) shows that the influence of the turbulent kinetic energy nonetheless amounts to about 20% of that of the pressure. We will clarify this point later.

[9] This does not mean, however, that the turbulent shear loads do not play any part in the establishment of the force being sought. The following shows that it is balanced by the bottom-slope induced pressure gradient, as in the infinite channel case discussed in Chapter 4 (refer to the first line in eqn (4.165)).

$$\begin{aligned} F^D &\approx \rho g d L_2 \delta z_{surf} \\ &= \rho g d L_1 L_2 I \end{aligned} \tag{8.24}$$

L_1 denoting the pool length, and $\delta z_{surf} = L_1 I$ being the differential elevation of the free surface from one pool to another.

The arguments in Section 3.4.3 also make it possible to write the force F^D on the basis of a drag coefficient which, for instance, is constructed on the area $L_1 L_2$ of the pool bottom (refer to eqn (3.175)):

$$F^D = \frac{1}{2}\rho L_1 L_2 C_D U^2 \tag{8.25}$$

where U is an order of magnitude of the mean velocity. Making both resulting expressions identical, we ultimately get

$$U = \sqrt{\frac{2}{C_D} g d I} \tag{8.26}$$

Thus, the velocity being established within the pool is proportional to the square root of the bottom slope, as corroborated by the measurements made by Tarrade et al. (2008) with an approximation of about 10%. For a set flow rate $Q = dL_2 U$, we may also derive the characteristic water depth:

$$d = \left(\frac{C_D Q^2}{2 g L_2^2 I} \right)^{1/3} \tag{8.27}$$

The validity of that law can be directly confirmed through measurements. For $Q = 0.736$ m^3/s, indeed, Tarrade et al. (2008) record $d = 1.12$ m for $I = 0.10$, which is yielded again by (8.27) provided that $C_D = 20.4$. For $I = 0.05$, (8.27) then yields $d = 1.41$ m, which is fairly consistent with the measurements ($d = 1.56$ m). It should be pointed out that the experimental water depths are averaged over the whole pool. For some values of the slope, Tarrade et al. (2008) also highlight an eddying detachment of the mean velocity field at the outlet of the slot, which makes the flow unsteady. This kind of flow is similar to the unstable flows discussed in Section 4.2.2, although the oscillations concern the Reynolds-averaged velocities. Thus, the deviations between theoretical calculations and measurements are due to the approximations we have made, in particular the steady nature of the flow.

The same conclusion could be drawn about the turbulent kinetic energy k, by generalizing (3.55) to the case of a mean field. As explained in Chapter 4, here it would give nothing but an equilibrium between the integrals of the production P and the dissipation ε, and would be no use for us to estimate k. Yet, the estimate (7.19) can be matched to this case. Our estimates of the turbulent quantities, however, will preferably be based upon the slot width L instead of L_2, which yields $L_m \sim L/10$ and $S \sim U'/L$, U' being the velocity across the slot. This is, indeed, the area having the largest velocity gradient, which causes most of the turbulent energy to be generated (Section 4.4.2).

The conservation of flow rate (refer to Chapter 3) also yields $LU' \approx L_2U$, the water depth hardly varying in space. Lasty, we can state

$$\begin{aligned} k &= \frac{(L_m S)^2}{\sqrt{C_\mu}} \\ &\sim 0.033 \left(\frac{L_2}{L} U\right)^2 \end{aligned} \tag{8.28}$$

The abovementioned flow rate corresponds to $U = 0.329$ m/s for $I = 0.10$ and $U = 0.236$ m/s for $I = 0.05$, which yields $k \sim 0.08$ m^{2}s^2 to 0.16 m^{2}s^2. The velocity values we have just found lead to a Reynolds number $Re \doteq UL_2/\nu$ which is higher than 4.7×10^5, which validates the hypothesis of a highly turbulent flow. The Froude number $Fr \doteq U/\sqrt{gd}$ then ranges from 0.060 to 0.10. It should then be pointed out that the estimates (4.202) and (4.203), which were made along an infinite channel in Section 4.4.5, are no longer strictly valid. Here we find, owing to (8.28):

$$\begin{aligned} \frac{\frac{2}{3}\rho k}{\overline{p} - \overline{p}_{atm}} &\sim \frac{1}{150\sqrt{C_\mu}} \left(\frac{L_2}{L}\right)^2 \frac{U^2}{gd} \\ &\approx 0.022 \left(\frac{L_2}{L}\right)^2 Fr^2 \end{aligned} \tag{8.29}$$

The factor $(L_2/L)^2$ makes all the difference. Yet, this result confirms that the turbulent kinetic energy is relatively small when compared to the pressure (refer to margin note 8 p. 525).

Conversely, when it is desired to more precisely characterize the size $L_t \sim L_m$ of the large turbulent eddies from their energies, it will be possible to invert (8.28):

$$\begin{aligned} L_t &= C_\mu^{1/4} \frac{k^{1/2}}{S} \\ &\sim 0.55 \frac{L^2 k^{1/2}}{L_2 U} \end{aligned} \tag{8.30}$$

As can be seen, the arguments in the theoretical chapters can be used for determining both laws and orders of magnitude in practical cases, from the dimensional analysis, as well as from the equations and the general reviews we have made. Let us now set out simulations of flow in the fish pass through the SPH method; they will be partly validated by means of those formulas we have just established, as well as with measurements and simulations provided by another numerical method.

8.3.2 Simulation of a pool fish pass

Several papers point to the suitability of the SPH method to simulate the flow in a fish pass of the same kind as that which we have just described (in particular, refer to Marivela et al., 2009, for a three-dimensional simulation). Here we follow Violeau et al. (2008), once again using the geometry of Fig. 8.12 being treated in dimension $n = 2$. Thus, this flow is seen from above, that is to say the free surface is not modelled in it, and no effect of gravity is taken into account. The adopted conditions correspond to the slope $I = 0.10$, for a mean depth $d = 1.12$ m.

A steady flow is induced within the pool by means of the method as described in Section 7.2.1. Water particles having a size $\delta r = 1$ cm are considered, which provide a total of about 60,000 and $h/\delta r$ is set to 1.5. In accordance with the preceding section, the selected flow rate is $Q = 0.736$ m^3/s. Thus, the flow velocity across the slot having a width L (refer to Fig. 8.12) is $U = 2.2$ m/s, a value which we prescribe in equation (7.2). Here the weakly compressible model will be applied for pressure prediction, with $U_{\max} = U$ as an order of magnitude of the maximum velocity. Since we do not take the effect of gravity into account, however, we decide to increase the factor 10 of equation (5.213) up to 30 (refer to Sections 7.2.2 and 7.2.3), which yields $c_0 = 60$ m/s. We then get $\delta t = 10^{-4}$ s as a time step.

Turbulence here plays a major part, as in all the confined steady flows (although the actual flow here exhibits a free surface). We decide to test the following four models: the one-equation $k - L_m$ model and the two-equation models $k - \varepsilon$, $k - \omega$ and non-linear $k - \varepsilon$ (Section 6.4.2). Since we do not know the spatio-temporal distribution of the mixing length L_m in so complex a case, it is made equal to the particle size, which looks like a large eddy simulation (refer to Section 6.5.1). In addition, we will try to prescribe a constant eddy viscosity, as we did in Section 7.3.2. An estimate of its value may stem from a discussion analogous to that which resulted in equation (7.10), estimating $L_m \sim L_2/10$ in it (as in the previous section). This yields $\nu_T \sim UL_2/100 = 0.044$ m^{2}s^{-1}. The turbulence models will highlight an eddy viscosity which, although it is variable, nonetheless has the same order of magnitude, albeit slightly less high. We ultimately choose $\nu_T = 0.03$ m^{2}s^{-1} for the constant viscosity model.

Figure 8.13 illustrates the distributions of the velocity magnitude after convergence. Three out of the four selected turbulence models are illustrated (the $k - \omega$ model is ignored here), as well as the constant viscosity model, and compared with the measurements made by Tarrade et al. (2008). Small void pockets can be found near the slot, which means that the incompressible algorithm would be more relevant here (refer to Chapter 7, particularly Sections 7.2.2 and 7.4.1).

It can be found that the $k - L_m$ model provides a velocity field which is more variable in space than expected from the measurements, whereas the other models are more precise, especially the two-equation models. A closer analysis of the simulations shows that the $k - L_m$ model is not suitable here to get a steady flow, exhibiting an eddying detachment downstream of the slot, at odds with the measurements for these values of the flow rate and the bottom slope. This may result from a too small mixing length value, the eddying detachment resembling large turbulent structures. In this case, the mixing length model (refer to Section 5.4.3) leads to similar conclusions (which are not presented here). This result is in line with the results of Fig. 8.14, illustrating the spatial distribution of the turbulent kinetic energy. It can be observed that the $k - L_m$ model does not generate enough eddying energy, underesti-

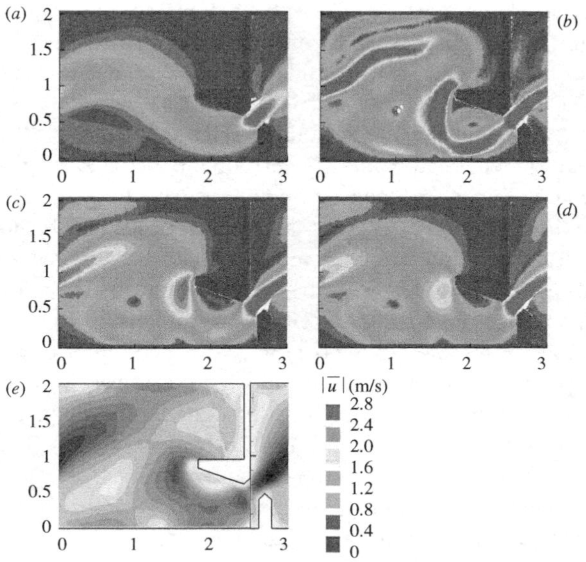

Fig. 8.13 Pool of a fish pass, for $I = 0.10$ and $Q = 0.736\ \mathrm{m^3/s}$. Distribution of the mean SPH-simulated velocity module for the constant viscosity (*a*), $k - L_m$ (*b*), $k - \varepsilon$ (*c*) and non-linear $k - \varepsilon$ (*d*) models. Comparison with the measurement made by Tarrade et al. (2008) (*e*).

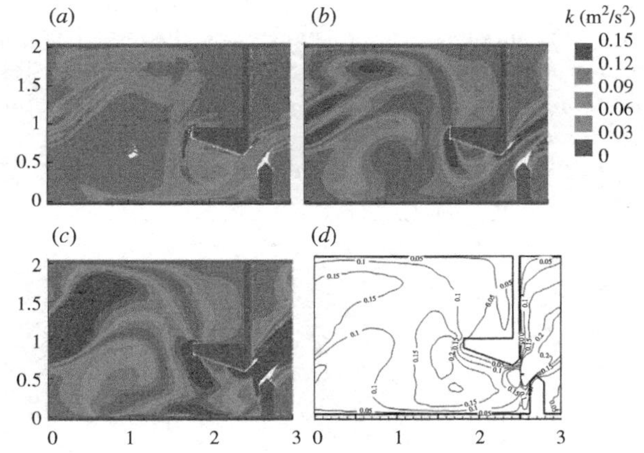

Fig. 8.14 Pool of a fish pass, for $I = 0.10$ and $Q = 0,736\ \mathrm{m^3/s}$. Distribution of the SPH-simulated turbulent kinetic energy for the $k - L_m$ (*a*), $k - \varepsilon$ (*b*) and non-linear $k - \varepsilon$ (*c*) models. Comparison with the measurements made by Tarrade et al. (2008) (*d*).

mating the eddy viscosity (refer to the first line in eqn (6.215)). The velocity diffusion is consequently reduced, leading to an unstable velocity field.

Whether it is for the prediction of the velocity modulus $|\overline{\mathbf{u}}|$ or the turbulent kinetic energy k, Figs 8.13 and 8.14 corroborate the $k - \varepsilon$ and non-linear $k - \varepsilon$ models. Whereas the former slightly better predicts $|\overline{\mathbf{u}}|$, the latter is more efficient for predicting k. The order of magnitude of k is $0.1\ \mathrm{m^2s^2}$, in line with the estimate made in the preceding paragraph (eqn (8.28)). In the resting area indicated in Fig. 8.12, k ranges from $0.05\ \mathrm{m^2s^2}$ to $0.1\ \mathrm{m^2s^2}$. The law (8.30) then gives the characteristic size of the large eddies: $L_t \sim 0.017$ m to 0.024 m. As expected, eddies of about $L/10$, that is a few centimetres, can be found again. They will probably only slightly disturb the fish species which are liable to use the pass.

Three cross-sections $\{P_i\}$ $(i = 1,\ 2,\ 3)$, shown in Fig. 8.12, have been selected for a quantitative validation of the quantities obtained. They are located at the abscissae $x(P_i) = X_i$, with $\{X_i\} = (0.5 \text{ m}; 1.0 \text{ m}; 1.5 \text{ m})$, versus the origin point represented by the halftone shaded disc in the same figure. For further enhancing the validation, we consider the results of the TELEMAC-2D software, which was developed and validated in EDF R&D with respect to a finite element technique (Hervouet, 2007). This software solves the two-dimensional Saint-Venant equations (or *non-linear shallow water equations*),[10] which are relevant in this case due to the nearly horizontal nature of the flow. Figure 8.15 shows the profiles of $|\bar{\mathbf{u}}|$ and k, along each of the three profiles, for all the turbulent models which are selected here with the SPH method, being compared with the measurement made by Tarrade et al. (2008) and the results of TELEMAC-2D.

The two-equation turbulent models in the SPH method fairly well replicate the velocity profiles, except in the vicinity of the upper wall, whereas the constant viscosity model gives slightly less good results. In addition, the $k-\varepsilon$ model was provided with the Yap correction (4.150), with substantially improved predictions. The Saint-Venant equations give results in line with those of SPH, albeit slightly less good. This is probably because they naturally induce a numerical diffusion phenomenon which does not exist in the Lagrangian methods. The turbulent kinetic energy profiles are markedly less well predicted in the upper half of the flow, where the kinetic energy is overestimated by the models. The constant viscosity model is not suitable, of course, for a proper quantitative prediction of that quantity. Tests (which are not dealt with here) conducted using the CODE_SATURNE finite volume

[10] We do not intend to present these equations here (for example, refer to Hervouet, 2007). It should be noted, however, that they result from an integration of the Reynolds equations (4.51) over each water column, leading to a system of two-dimensional equations. They can be derived, to some extent, from a least action principle similar to that in Section 3.6 (Rodriguez-Paz and Bonet, 2005). The model which is disclosed here takes the turbulent effects into account through a $k-\varepsilon$ model which is integrated over the water column (Rodi, 2000). Lastly, the two-dimensional Saint-Venant equations can be numerically treated on the basis of the SPH method (in particular, refer to De Padova et al., 2010). A one-dimensional version also exists for the flows along channels, which would immediately result in the equilibrium (8.26), and whose Boussinesq equations as discussed in Section 8.2.2 are a generalization.

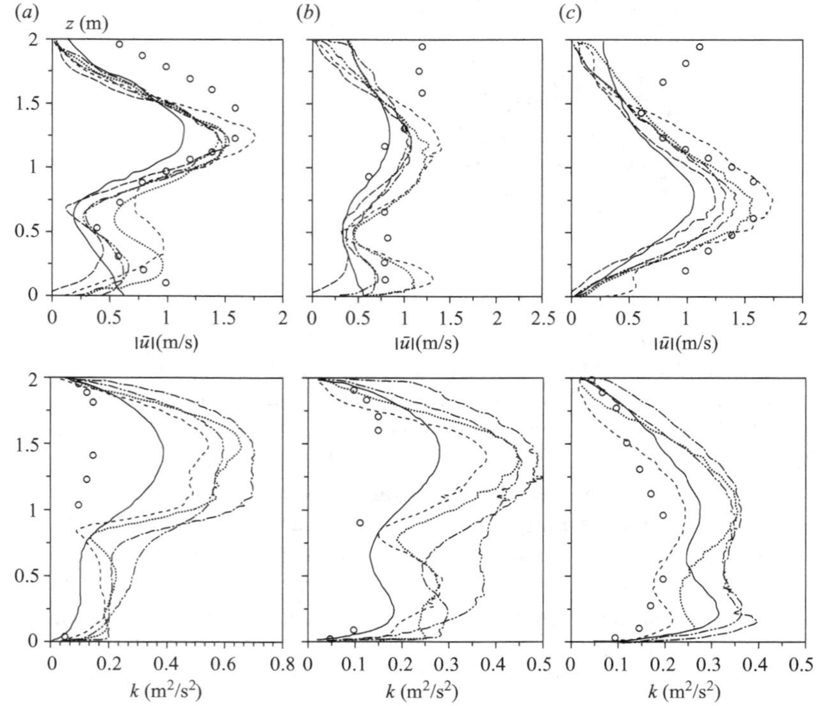

Fig. 8.15 Pool of a fish pass, for $I = 0.10$ and $Q = 0.736$ m^3/s. Distribution of the velocity and SPH-simulated turbulent kinetic energy module along P_1 (*a*), P_2 (*b*) and P_3 (*c*) profiles. Constant viscosity (large dash), $k-\varepsilon$ (dotted), $k-\omega$ (dash dotted), $k-\varepsilon$ non linear (solid lines), $k-\varepsilon$ with Yap correction (dash-dot-dotted) models. The small dash represent the results of the TELEMAC-2D modelling system (finite element Saint-Venant equations) and the symbols represent the measurements made by Tarrade et al. (2008).

software (refer to Section 7.2.1) suggest that a good prediction of k here implies the use of a second-order model (Section 4.5.2) or even an LES (Section 4.6.1). Yet, the Yap correction significantly improves the predictions of the $k - \varepsilon$ model about the P_3 profile. The CODE_SATURNE software has been the subject of a simulation based upon the LES model as described in Section 4.6.1. Figure 8.16 shows the shape of the norm of the filtered velocity field $|\tilde{\mathbf{u}}|$, the latter being compared with the mean velocity fields as provided by the SPH method and the TELEMAC-2D software.

As can be seen, the turbulence mechanisms are so complex that it is hard to accurately predict the desired quantities on actual geometries, even when they remain relatively simple. Fortunately, even a rough prediction of the eddy viscosity has relatively little impact on the velocity field, which makes it possible to have confidence in the numerical simulation results in that respect. On the other hand, if it becomes necessary to quantify either the kinetic energy or the eddy size for a practical application, the user must be aware of this model reliability.

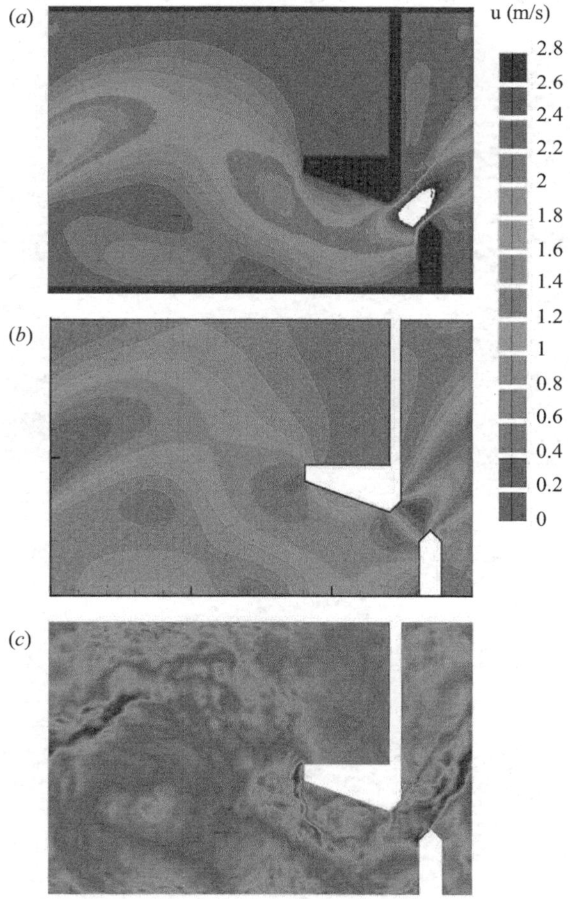

Fig. 8.16 Pool of a fish pass, for $I = 0.10$ and $Q = 0.736$ m^3/s. Distribution of the module of velocity as simulated by means of the SPH (*a*, Reynolds-averaged velocity), finite element (*b*, depth-averaged velocity) and finite volume (*c*, filtered velocity as meant by LES) methods. After Violeau et al. (2008).

8.4 Floating oil spill containment boom

We first study the stability of an oil spill floating over a water column in the presence of a current, in the vicinity of an oil containment boom. We establish a criterion for oil entrainment under the skirt of the boom. An SPH simulation is then proposed to confirm the latter analysis.

8.4.1 Oil spill instability criterion

When an oil tanker wrecks, one of the most commonly used antipollution techniques consists in containing the floating oil spill on the water surface by means of a floating boom before pumping the hydrocarbon. This kind of boom, which is illustrated in Fig. 8.17, consists of an inflated tube lying over a retaining skirt, and provided with mooring members. Under the action of current, however, the hydrocarbon may be pulled under the skirt (entrainment), whereas the occurrence of waves may contribute to force it over the top of the tube (submersion), as shown in Fig. 8.18. There are further oil spill escape patterns (refer to Violeau et al., 2007). In order to optimize the real-time deployment of these booms, it is of interest to attempt to quantify the risk

Fig. 8.17 View of a floating oil spill containment boom (photo courtesy of EDF).

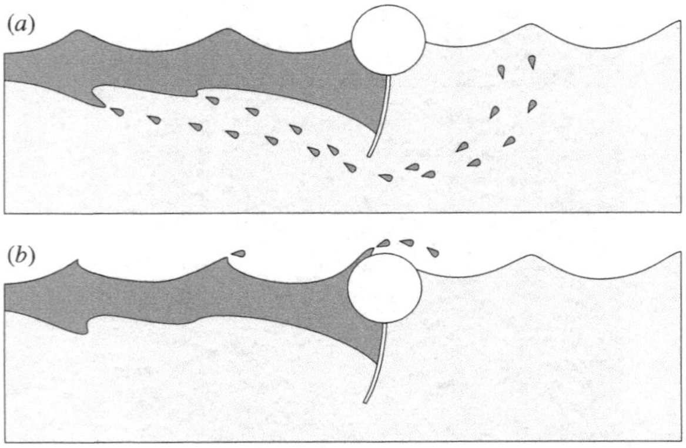

Fig. 8.18 Floating oil spill containment boom: escape caused by entrainment (*a*) and submersion (*b*). After Violeau et al. (2007).

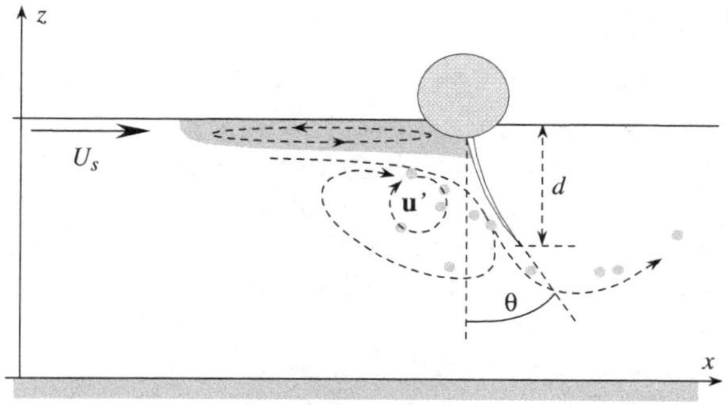

Fig. 8.19 Floating oil spill containment boom: notations.

of escape as a function of the external conditions, that is according to the wave parameters, the current velocity, the type of hydrocarbon and the boom dimensions.

If one initially only considers the current effect under a steady regime, it can be stated that there is a critical velocity beyond which the hydrocarbon is no longer contained; the order of magnitude of that velocity can be established through theoretical reasonings. Here we refer to the notations in Fig. 8.19, U_s denoting the velocity at the surface. The experiment (Violeau et al., 2007) shows that the volume of hydrocarbon is enclosed in a surface layer the thickness of which varies slowly, then is nearly constant. Entrainment takes place, at the interface, in a relatively restricted area which is located somewhat upstream of the boom. Assuming that the boom is anchored[11] by its mooring members, we may invoke the dimensional analysis as set out in Section 3.4.3 in order to estimate the critical entrainment velocity $U_{s,c}$ as a function of the parameters of that problem. In the first place, we will consider the water density and molecular viscosity, that is ρ_1 and ν_1, whereas the hydrocarbon is affected by $\rho_2 < \rho_1$ and ν_2. Afterwards, we will consider the gravity g, which plays a crucial role in the oil buoyancy process. The tube diameter will presumably not be involved in the oil escape process. On the other hand, the effect of the mean water depth H will be taken into account because the entrainment mechanism is governed by the turbulent eddies, whose size depends on H. That mechanism is induced, near the water–oil interface, by a conflict between the turbulent agitation, which tends to drive the oil droplets into water through a rotational motion, and the gravity $\mathbf{F}^g$ corrected for buoyancy $\mathbf{F}^A$ which, on the contrary, acts as a stabilizer, driving the hydrocarbon back to the surface. The force $\mathbf{F}^g + \mathbf{F}^A$, whose calculation is given by (7.20), is still valid here. Thus, by analogy with (7.22), we can define:

$$\xi \doteqdot \frac{\rho_1}{\rho_2} - 1 > 0 \tag{8.31}$$

The effects of gravity appear in the equations through the $\xi\rho_2$ factor, which make it possible to ignore ρ_1. Furthermore, it may be considered that once an oil droplet has been 'sucked' far enough below the interface, it will follow the mean water velocity field and flow across the skirt, whose depth d (i.e. the boom *draught*) consequently affects the process. We ultimately may write,

[11] This assumption can be made in the absence of waves, although a gentle vertical motion is possible. We will contemplate it in the next section within the frame of the numerical simulation. For a more rigorous approach, it would also be necessary to consider both mobility and deformation of the skirt, which is made of a composite fabric. Anyway, the angle θ made by the skirt with the vertical (Fig. 8.19) is not to be taken into account among the parameters, because it depends on the current velocity and the boom stiffness.

with a good approximation:

$$U_{s,c} = \phi_{dim}\left(\xi\rho_2, \nu_1, \nu_2, g, H, d\right) \tag{8.32}$$

Because of the oil droplet size, it would probably be necessary, for better accuracy, to take into account the surface tension phenomenon (Section 3.4.5), which we ignore here for the sake of simplicity. With $n = 7$ parameters out of which $m = 3$ are independent, we then have $n - m = 4$ dimensionless numbers, and the preceding relation is reduced to

$$Fr_c = \phi_{adim}\left(\frac{d}{H}, Re_1, Re_2\right) \tag{8.33}$$

where

$$\begin{aligned} Fr_c &\doteq \frac{U_{s,c}}{\sqrt{\xi g d}} \\ Re_1 &\doteq \frac{U_{s,c}d}{\nu_1} \\ Re_2 &\doteq \frac{U_{s,c}d}{\nu_2} \end{aligned} \tag{8.34}$$

When the flow is very turbulent, as is always the case in actual fact, the Reynolds number effect is no longer felt. We can therefore state that oil entrainment under the boom takes place for a critical Froude number Fr_c which now merely depends on d/H:

$$Fr_c = \phi_{adim}\left(\frac{d}{H}\right) \tag{8.35}$$

In order to search for an approximation of Fr_c, the mechanism we have just described should be quantified. A stability criterion of the oil spill will be written by comparing the power P of the (destabilizing) turbulence with the power P^{gA} of the gravity force $\mathbf{F}^g + \mathbf{F}^A = \xi\rho_2 V\mathbf{g}$, during the motion carrying a droplet from the top to the bottom of a large turbulent eddy having a size L_t and a velocity $\mathbf{u}'$ (Fig. 8.19), as reckoned per unit mass:

$$P < -P^{gA} = -\frac{1}{L_t}\int_C \xi\mathbf{g}\cdot\mathbf{u}'d\ell \tag{8.36}$$

Thus, the oil droplet traverses a half-circle (assuming that the eddy is perfectly circular). If the fluctuating velocity $\mathbf{u}'$ is constant in it, we get

$$\begin{aligned} P^{gA} &= -\xi\left|\mathbf{u}'\right|\mathbf{g}\cdot\int_{-\pi/2}^{\pi/2}\mathbf{e}_\theta d\theta \\ &= -2\xi g u' \end{aligned} \tag{8.37}$$

$\left|\mathbf{u}'\right| \approx u'$ can then be estimated by means of the formula (4.77), the turbulence presumably being isotropic.[12] Invoking the equation (4.97) yielding P, the criterion (8.36) is then rearranged as

$$\Theta < \Theta_c \doteq \sqrt{\frac{8}{3C_\mu}} \approx 5.44 \tag{8.38}$$

[12] This hypothesis is never true near an interface, since the gravity inevitably dampens the velocity fluctuations occurring at right angles to it (Violeau, 2009a).

where a dimensionless number Θ has been defined on the basis of the turbulent kinetic energy k and the mean rate-of-strain S in the place being considered, in some vicinity to the interface:

$$\Theta \doteqdot \frac{k^{1/2}S}{\xi g} \tag{8.39}$$

In order to estimate k, we utilize the first line in (7.19). The mixing length (i.e. the size of the large turbulent eddies) L_m is proportional, indeed, to the water depth H, what we will express as $L_m = \eta_\ell H$, defining the dimensionless factor η_ℓ, which is a priori unknown but has an order of magnitude of 0.1 (refer to Chapter 4). Hence we get

$$k = \frac{(\eta_\ell H S)^2}{\sqrt{C_\mu}} \tag{8.40}$$

We still have to characterize the rate of strain S, which is the most doubtful part of our calculation. Admittedly, the velocity gradient in the hydrocarbon layer can be approximated through $S \approx 2U_s/\delta$, δ being its thickness. The velocity gradient in the immediately underlying body of water, however, should be considered, because it will govern the generation of turbulent energy P accounting for the entrainment.[13] Now, the continuity of the stress $\overline{\boldsymbol{\tau}} = 2\rho\nu_T\mathbf{S}\cdot\mathbf{n}$ does not imply the continuity of the rate of strain, if only because the density abruptly changes across the interface. Since the velocity gradients in the water body depend in the first place on the draught d, one can merely provide the following estimate for the rate of strain:

[13] A momentum balance analogous to (8.19) would show that the quantity Θ decreases in the oil layer as the distance to the boom decreases. This kind of reasoning would then be insufficient to highlight the occurrence of such an entrainment as shown by the experiments.

$$S = \frac{\eta_s U_s}{d} \tag{8.41}$$

where η_s is a dimensionless factor of about one. In addition, the dimensional analysis indicates that it is constant, for a given boom geometry with a large Reynolds number. Combining (8.40) and (8.41), the stability criterion (8.38) becomes

$$Fr < Fr_c \tag{8.42}$$

where the Froude number has been defined by analogy with (8.34):

$$Fr \doteqdot \frac{U_s}{\sqrt{\xi g d}} \tag{8.43}$$

In the light of the foregoing, its critical value is provided by

$$\begin{aligned} Fr_c &\doteqdot \left(\frac{8\sqrt{C_\mu}}{3}\right)^{1/4} \frac{1}{\sqrt{\eta_\ell \eta_s}} \frac{d}{H} \\ &\approx \frac{0.946}{\sqrt{\eta_\ell \eta_s}} \sim 3\frac{d}{H} \end{aligned} \tag{8.44}$$

This result obeys the general formulation (8.35). The condition (8.42) gives the critical velocity for the appearance of entrainment as

$$U_{s,c} = \left(\frac{8\sqrt{C_\mu}}{3}\right)^{1/4} \frac{\sqrt{\xi g d^3}}{\sqrt{\eta_\ell}\eta_s H} \tag{8.45}$$

As an application, let us consider a relatively heavy oil, having a density $\rho_2 = 995$ kg/m^3, which corresponds to $\xi = 0.005$. for a boom with a draught $d = 0.75$ m, a water depth $H = 2.5$ m and arbitrarily setting $\eta_\ell = 0.1$ and $\eta_s = 1$, the formula (8.45) leads to a critical velocity $U_{s,c} = 0.17$ m/s. In the same conditions but for a light oil, ($\rho_2 = 850$ kg/m^3, i.e. $\xi = 0.176$), that yields $U_{s,c} = 1.02$ m/s. These orders of magnitude are consistent with the experimental observations, giving $U_{s,c} = 0.7$ m/s for a light oil (Violeau et al., 2007). The difference primarily lies in the choice of the dimensionless quantities η_ℓ and η_s, but it is also noteworthy that the heavy oil viscosity equals $\nu_2 = 3 \times 10^{-2}$ m^{2}s^{-1}. The Reynolds number associated to the oil layer, analogous to Re_2 (formulas (8.34)) but constructed over its thickness $\delta \approx 0.3$ m, then equals $Re'_2 \doteqdot U_{s,c}\delta/\nu_2 = 7$, and the flow is laminar in it. Thus, the critical Froude number (eqn (8.33)) is probably related to the latter: $Fr_c = \phi_{adim}\left(Re'_2\right)$ instead of being constant. On the other hand, with the light oil, the viscosity $\nu_2 = 3.32 \times 10^{-6}$ m^{2}s^{-1} provides for the turbulence of the flow within the oil spill.

8.4.2 Simulation of a floating boom

The multiphase flow modelling with SPH has been dealt with in lots of papers (in particular, refer to Monaghan and Kocharayan, 1995; Colagrossi and Landrini, 2003; Ataie-Ashtiani and Shobeyri, 2008; Valizadeh et al., 2008). On the other hand, the specific case of the behaviour of an oil spill near a floating boom has been tested on several occasions, but has seldom been the subject of publications. We are going to explain that those mechanisms which were highlighted in the previous section may be readily replicated according to the findings made by Buvat and Violeau (2006) and Violeau et al. (2007).

We will initially turn back to the data reported at the end of Section 8.4.1 about the draught and the characteristics of both hydrocarbons, and we will work in $n = 2$ dimensions within a vertical plane which is locally orthogonal to the boom. A steady flow is induced in a 16 m long periodic channel having a depth $H = 2.5$ m, using the method as described in Section 7.2.1. Particles with a size $\delta r = 2$ cm are considered, which gives a total of about 100,000 particles, and $h/\delta r = 1.2$ is set. The reference densities of the two fluids are set to $\rho_1 = 1,000$ kg/m^3 and $\rho_2 = 995$ kg/m^3, respectively. The selected flow rate corresponds to an average velocity U which may be related to the velocity U_s at the surface, assuming a rough logarithmic velocity profile (eqn (4.183)) sufficiently far upstream of the boom. The two values $U_s = 0.3$ m/s and $U_s = 0.7$ m/s are studied here. For these velocity values, the experiments (refer to Violeau et al., 2007) provide several values of the angle θ which can be seen in Fig. 8.19, according to the boom stiffness. Besides, it should be taken into account that this angle is variable according to the portion being considered

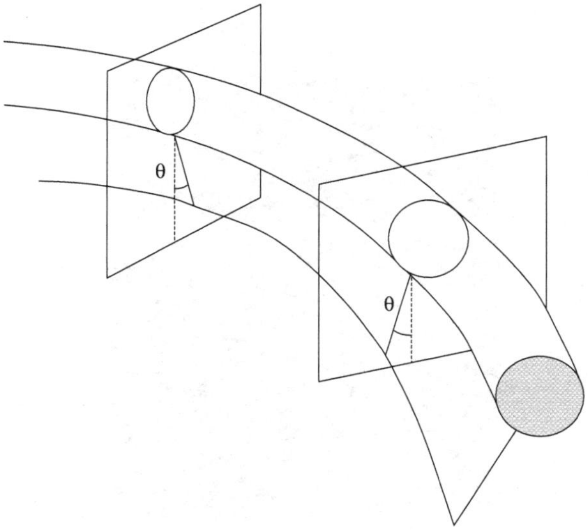

Fig. 8.20 Floating oil spill containment boom: variation of angle θ between the skirt and the vertical, according to the vertical section being considered.

along the boom (refer to Fig. 8.20), which prompts us to arbitrarily set θ. The value $\theta = +10°$ is examined here.

The weakly compressible model is adopted for pressure prediction, with $U_{max} = 2U$ as the order of magnitude of the maximum velocity, in order to take the increase of the velocity under the boom skirt into account. Given these conditions, the term $\sqrt{gH} = 5$ m/s prevails over U_{max} in equation (5.212). We decide, however, to raise the factor 10 in that equation up to 20 (refer to Sections 7.2.2 and 7.2.3), which yields $c_0 = 100$ m/s. We then have $\delta t = 1.2 \times 10^{-4}$ s as a time step.

Turbulence here plays a major role, as shown by the analysis in the preceding section, to represent the oil entrainment mechanism. Here we choose the one-equation model $k - L_m$ (Section 6.4.2), which may be considered a good trade-off between accuracy and rapidity, and which additionally provides a satisfactory estimate of the turbulent kinetic energy in this case, which is similar to that in Section 7.2.3. Since we do not know the spatial distribution of the mixing length L_m near the boom, it is put equal to the size of the particles, which looks like a large eddy simulation (refer to Section 6.5.1). However, in line with what was written at the end of Section 8.4.1, no turbulence model is considered within the heavy oil, which is too viscous. In accordance with the previously discussed theoretical elements, we ignore the surface tension effects. More precise simulations in that respect would require the use of the discrete model as disclosed in Section 6.3.1.

The boom is modelled using the floating-body technique as mentioned in Section 6.3.3. Its motion, however, is restricted to the vertical direction, in order to take the effect of buoyancy and the presence of mooring members into account. Here, this motion is actually of modest amplitude. The boom mass (per unit length in the transverse direction) is set to $M = 13$ kg/m. Yet, the particle densities are set to the water reference value ρ_0 to prevent errors from occurring upon the application of the SPH discrete interpolation operator, for the quantities depending on ρ.

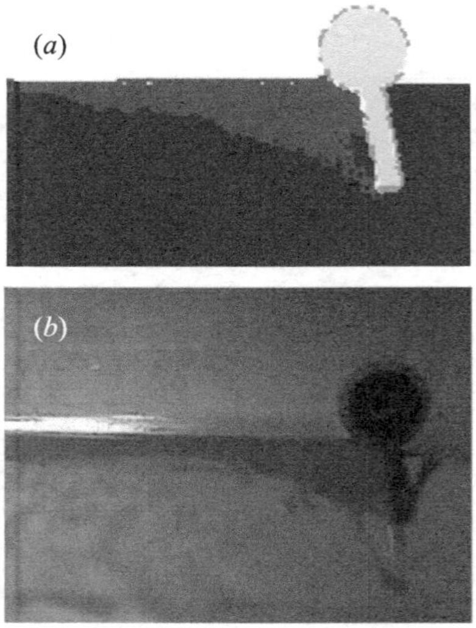

Fig. 8.21 Floating oil spill containment boom: origin area of an entrainment phenomenon for $U_s = 0.3$ m/s, in the case of a heavy oil. Comparison of the SPH simulation (*a*) with a laboratory experiment (*b*). After Violeau et al. (2007, with permission from Elsevier).

With the light oil, the simulations show that the oil spill remains stable for $U_s = 0.3$ m/s, whereas the entrainment takes place for $U_s = 0.7$ m/s, in accordance with the criterion as introduced in the preceding section. Likewise, the model predicts an entrainment-induced oil escape with both velocities in the heavy oil case. In the latter case, Fig. 8.21 provides a zoomed in view of the boom and the entrainment origin area, for $U_s = 0.3$ m/s. It can be seen that the interface shape is very similar to that resulting from an experiment on a small-scale model.[14]

[14]This model, which was prepared at the 1:15 scale at the La Rochelle University (France), is constructed on the basis of a Froude similarity, as expected for a flow in which the gravity prevails (refer to Section 3.4.3).

In the case of the light oil, Fig. 8.22 illustrates the spatial distribution of the temporal maximum $\Theta_{\max,t}$ of the parameter Θ as defined by formula (8.39). It can be found that when the velocity at the surface is greater, $\Theta_{\max,t}$ ($\mathbf{r}$) duly takes higher values, because the largest velocity gradients S additionally result in a high production of turbulent energy k. The maxima (in space) of $\Theta_{\max,t}$ occur near the interface and just under the skirt. Figure 8.23 illustrates the evolution of the spatial maximum $\Theta_{\max,\mathbf{r}}$ of Θ over time. For $U_s = 0.3$ m/s, it can be seen in it that $\Theta_{\max,\mathbf{r}}$ (t) remains below the critical value Θ_c as defined by (8.38). On the contrary, for $U_s = 0.3$ m/s, $\Theta_{\max,\mathbf{r}}$ (t) exceeds the threshold

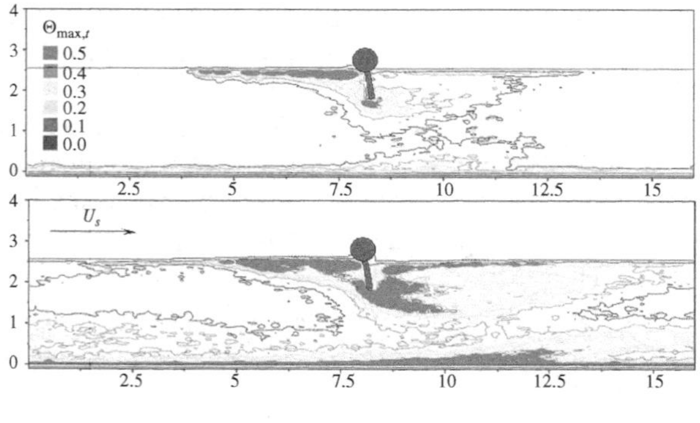

Fig. 8.22 Floating oil spill containment boom: SPH simulation in the case of a light oil. Spatial distribution of the temporal maximum $\Theta_{\max,t}$ ($\mathbf{r}$) of the parameter Θ (formula (8.39)) for $U = 0.3$ m/s (*a*) and $U = 0.7$ m/s (*b*). After Violeau et al. (2007, with permission from Elsevier).

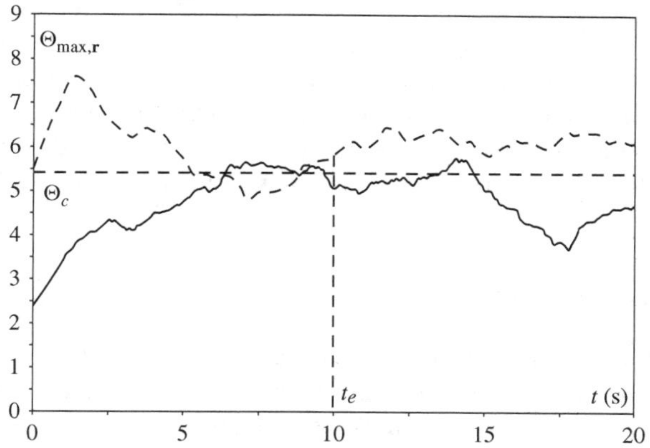

Fig. 8.23 Floating oil spill containment boom: SPH simulation in the case of a light oil. Temporal evolution of the spatial maximum $\Theta_{\mathbf{max,r}}(t)$ of the parameter Θ (formula (8.39)) for $U = 0.3$ m/s (solid line) and $U = 0.7$ m/s (dashed line). The time t_e indicates the occurrence of the oil entrainment in the second case, whereas the critical value Θ_c is provided by (8.38). After Violeau et al. (2007, with permission from Elsevier).

Θ_c. A first overshooting takes place in the very beginning of the simulation, when the velocity field is being established, generating high strains under the boom skirt, without any effect on oil spill stability. The second overshooting corresponds to the time t_e when the hydrocarbon begins being entrained, and the maximum $\Theta_{\mathbf{max,r}}(t_e)$ is located near the water-oil interface when this mechanism takes place (refer to Fig. 8.22). That analysis corroborates the validity of the stability criterion (8.38).

Other simulations of the same kind are discussed by Violeau et al. (2007), for various values of the skirt angle θ. In addition, a simulation is tested in that paper on the basis of a non-Newtonian hydrocarbon, that is the molecular viscosity of which depends on the rate of strain (refer to Section 3.4.1). According to the values of the surface current and θ, and as a function of the type of hydrocarbon being contemplated, other escape patterns are highlighted. Although these numerical modellings are not exhaustive, they are suitable for quantifying the possible oil escape from the boom in relatively general conditions.

The same work was conducted by Violeau et al. (2007) in the presence of waves, the boom and the slick being arranged in a wave channel which is

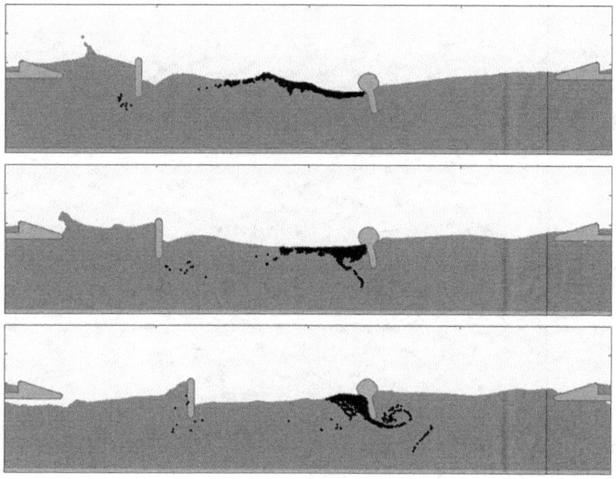

Fig. 8.24 Floating oil spill containment boom: behaviour in the presence of combined waves and current, simulated with the SPH method.

identical to that in Section 8.2.2. The vertical motion of the boom, as well as the rotation thereof in the vertical plane, are then induced by the wave orbital velocity distribution and should be damped for modelling the transverse stiffness of the boom. This can be easily achieved though an artificial restoring force device. Lastly, both current and waves can be combined by providing a paddle which partly covers the water column, as shown in Fig. 8.24.

8.5 Dam spillway

In the present section, dimensional analysis is first used to establish a rule giving the flow rate over a dam spillway as a function of the upstream head. We then propose a short SPH model reproducing this law in the case of a simple, Creager-type spillway. The Goulours dam spillway is then numerically investigated in three dimensions.

8.5.1 Dam release monitoring

During its service life, a large dam is subject to variable flow rates, some of which are so high that they require a water release in order to prevent an overtopping which might endanger the dam stability. That requirement is even more crucial since a high flow rate is associated with such a great water depth. Now, too high a water level at the dam generates a flood which is detrimental to the people living immediately upstream of it. Dams are therefore fitted with water release units which are known as spillways, which are dimensioned for discharging the surplus flow from the reservoir when that is necessary (refer to Fig. 8.25). Their shape and size are quite variable according to the structure configuration and the experienced flow rates. During the development of these works, a high dimensioning flow rate is established, indeed, on the basis of extreme statistical data, before conducting a hydraulic study. The latter is usually carried out on a scale model, although software has been used on a growing number of configurations since the 2000s.

Fig. 8.25 The spillway at the Grangent dam (France). Photo EDF.

The fact that the flowrate Q of water flowing over a work arranged on a channel or a river is a growing function of the associated water depth can be understood from the infinite channel case (without any water work) disclosed in Section 4.4.5. For a channel of width L, the formula (4.194) yields, indeed,

$$\begin{aligned} Q &= LHU \\ & Lu_* H \left(\frac{1}{\kappa} \ln \frac{H}{ek_s} + C_* \right) \end{aligned} \tag{8.46}$$

which does constitute a growing function of H. Apart from this simple theoretical result, intuition and experience confirm this behaviour. Besides, one could generalize, to some extent, some of the arguments expressed in Section 4.4.5 to the case of a channel provided with works having any geometry, as we did in the case of the fish pass in Section 8.3.1. For all that, the complexity of the involved spillway layout would not enable us to explicitly calculate any analytical relation between Q and H in the general case.

The dimensional analysis principles as established in Section 3.4.3, however, may advantageously be used to that purpose. The flow rate is sought as a function of the fluid characteristics (ρ and ν), the gravity g and the dimensions characterizing the dam and the spillway. To do this, a reference length denoted as H_d (which may be a characteristic distance of some element of the spillway, although this is not necessary) is selected, then one gets the maximum width L at the spillway, as well as the maximum water depth H as reckoned above the maximum elevation of the useful part of the spillway (refer to Fig. 8.26). The other lengths required for describing the spillway geometry are denoted as $\{L_i\}$ $(i = 1, \ldots k)$, as for the calculation of the force exerted onto a solid (refer to the end of Section 3.4.3). Thus, we will write

$$Q = \phi_{dim} \left(\rho, \nu, g, H, H_d, L, L_1, \ldots, L_k \right) \tag{8.47}$$

As in Section 8.2.1, no independent velocity scale is available here; instead, it is linked to the other parameters by the first line in (8.46), which remains

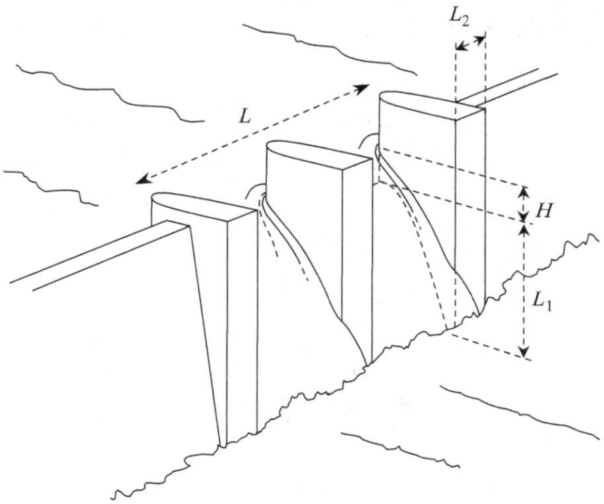

Fig. 8.26 Calculation of the spillway discharge law: notations.

general. A Reynolds number can then be formed:

$$Re \doteqdot \frac{Q}{LH\nu} \tag{8.48}$$

As is well known, its influence disappears if it is high enough. Furthermore, for the same reasons as in Section 8.2.1 the density cannot play any part, it only depends on the mass unit. Then there still are, in (8.47), $n = k + 5$ parameters out of which $m = 2$ are independent, which yields $m - n = k + 3$ dimensionless numbers. In the first place, the Froude number can be constructed as follows:

$$Fr \doteqdot \frac{Q}{L\sqrt{gH^3}} \tag{8.49}$$

The $k + 2$ remaining numbers are the length ratios H/H_d, L/H_d and $\{L_i/H_d\}$ $(i = 1, \ldots, k)$. Apart from the first one, which depends on the upstream water depth, all of them are parameters characterizing the spillway geometry. As in the case of the drag coefficient discussed in Section 3.4.3 (eqn (3.175)), we will write (8.47) in the following dimensionless form:

$$Q = L\sqrt{gH^3}C_Q\left(\frac{H}{H_d}, shape\right) \tag{8.50}$$

where C_Q is an undetermined function which is known as the works *discharge coefficient*, and only depends on its geometry, without taking the scale into account. It is determined through experience.

By way of example, here we describe one of the simplest possible, albeit rather frequent cases: the Creager weir. It consists of a locally lowered crest over a width L (refer to Fig. 8.27), exhibiting a very precise shape for the water nappe falling downstream to keep a pressure equal to the atmospheric pressure for a given value of the upstream energy. The upstream facing, extending between the points M_1 and M_2 in the figure, is defined by three arcs of circles having centres with coordinates $\{x_i\}$ and $\{z_i\}$ and radii $\{R_i\}$ $(i = 1, \ldots, k)$. Their values, as well as the arc boundaries, are given by the Table 8.2 (US Army Corps of Engineers, 1990). Downstream of the point M_2, the facing

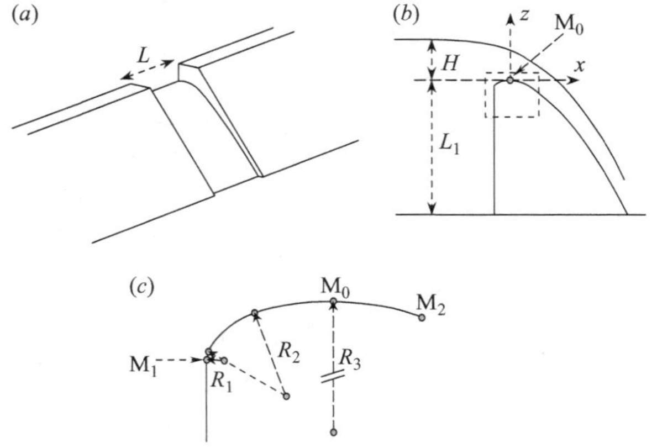

Fig. 8.27 Geometry of a Creager weir, after US Army Corps of Engineers (1990). Overall view (*a*), vertical profile in the indented area (*b*) and zooming in on the area surrounded by a dashed line (*c*).

Table 8.2 Geometry of a Creager weir: values of the abscissae x_i, ordinates z_i, radii R_i and validity domains of all three arcs of a circle defining the upstream face (Fig. 8.27), which are adimensionalized by the dimensioning load H_d. After US Army Corps of Engineers, (1990).

i	x_i/H_d	z_i/H_d	R_i/H_d	Limits
1	0	−0.5000	0.50	$-0.1750 \leqslant x/H_d \leqslant 0$
2	−0.1050	−0.2190	0.20	$-0.2760 \leqslant x/H_d < -0.1750$
3	−0.2418	−0.1360	0.04	$-0.2818 \leqslant x/H_d < -0.2760$

shape is defined by the following equation:

$$\frac{z}{H_d} = -\frac{1}{2}\left(\frac{x}{H_d}\right)^{\alpha} \tag{8.51}$$

with $\alpha = 1.85$, the origin being located on top of the weir (point M_0 in Fig. 8.27). Apart from H_d and the width L, here the single length parameter is the *height of weir* L_1 (refer to Fig. 8.27). The discharge coefficient then only depends on the ratios H/H_d and H/L_1; it is provided by the Brudenell's experimental formula (refer to Ven Te Chow, 1959):

$$C_Q\left(\frac{H}{H_d}\right) = C_1^B\left(\frac{H}{L_1}\right) \cdot \left(\frac{H}{H_d}\right)^{C_2^B} \tag{8.52}$$

where $C_2^B = 0.12$. As regards the function C_1^B, it may be considered as constant and equal to 0.700 for low values of H/L_1. This formula remains valid for $0.2 \leqslant H/H_d \leqslant 2$. A Creager weir may also be provided with piers (like the spillway in Fig. 8.26), which exert a contraction effect. They can be taken into account as follows:

$$C_Q\left(\frac{H}{H_d}, shape\right) = C_1^B \cdot \left(\frac{H}{H_d}\right)^{C_2^B}\left[1 - 2NK\frac{H}{L}\right] \tag{8.53}$$

where N is the number of piers, whereas $K \sim 0.02$ is a coefficient which depends on the layout of the piers (Ven Te Chow, 1959). Lots of other formulas can be found in the scientific literature dealing with the dimensioning of the hydraulic works, for conventional works with a great variety of forms.

The SPH method can easily be applied[15] to the simulation of a flow over a pierless Creager weir whose width L coincides with that of the channel. Thus, the flow may be considered as two-dimensional. We illustrate in Fig. 8.28 the variation of depth H versus the flow rate per unit length Q/L, as calculated using SPH for $H_d = 5$ m. We observe that the results suitably agree with the formulas (8.50) and (8.52), though the water depth is slightly underestimated by the simulation for the light flow rates. Similar simulations can be found in Agate and Guandalini (2010) and Mahmood et al. (2011). In addition, Shakibaeinia and Jin (2009) have disclosed an application to this case of the MPS Lagrangian method, which was briefly discussed in Section 5.1.

[15] However, input and output boundary conditions should be introduced, a matter which we mentioned in the introduction to Section 6.3.

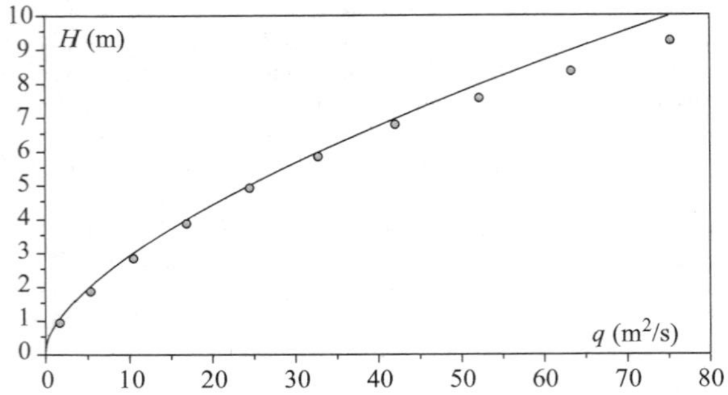

Fig. 8.28 Discharge law of a Creager weir for $H_d = 5$ m: SPH simulations (symbols) compared with the semi-empirical laws (8.50) and (8.52).

8.5.2 Simulation of the Goulours spillway

Numerical simulation of the large dams has already been the subject of several implementations of the SPH method, particularly for the break wave prediction (Cleary and Prakash, 2004; Žagar et al., 2009). The modelling of complex spillways was dealt with as well, in particular by López et al. (2009b). Here we explain an application to the operation of an existing spillway, in accordance with Lee et al. (2009, 2010).

We will consider the Goulours dam, which was built on the River Lauze (France) in 1946. Figure 8.29 illustrates the condition of that dam in 2006, which is 71 m wide on its top. Two spillways can be seen, one of them being a ski-jump spillway, whereas the other one, which is more recent, is a *piano key weir*, or PK-weir). We will deal with the former. In its upper part, it has the same geometry as the Creager weir which was described in the preceding section, over a width $L = 4$ m, between two training walls. The discharge carrier curve becomes inverted in its lower part, whereas it gets a narrower width, which is reduced down to 2.35 m at the outlet (refer to Fig. 8.30). There water follows a free fall path before hitting the river bed. Due to the discharge carrier length (15.5 m) and shape, the water jet does not wear the bottom away just at the foot of the dam, which otherwise could jeopardize its stability. This

Fig. 8.29 View of the Goulours dam during the building of the PK-weir spillway in 2006 (photo EDF).

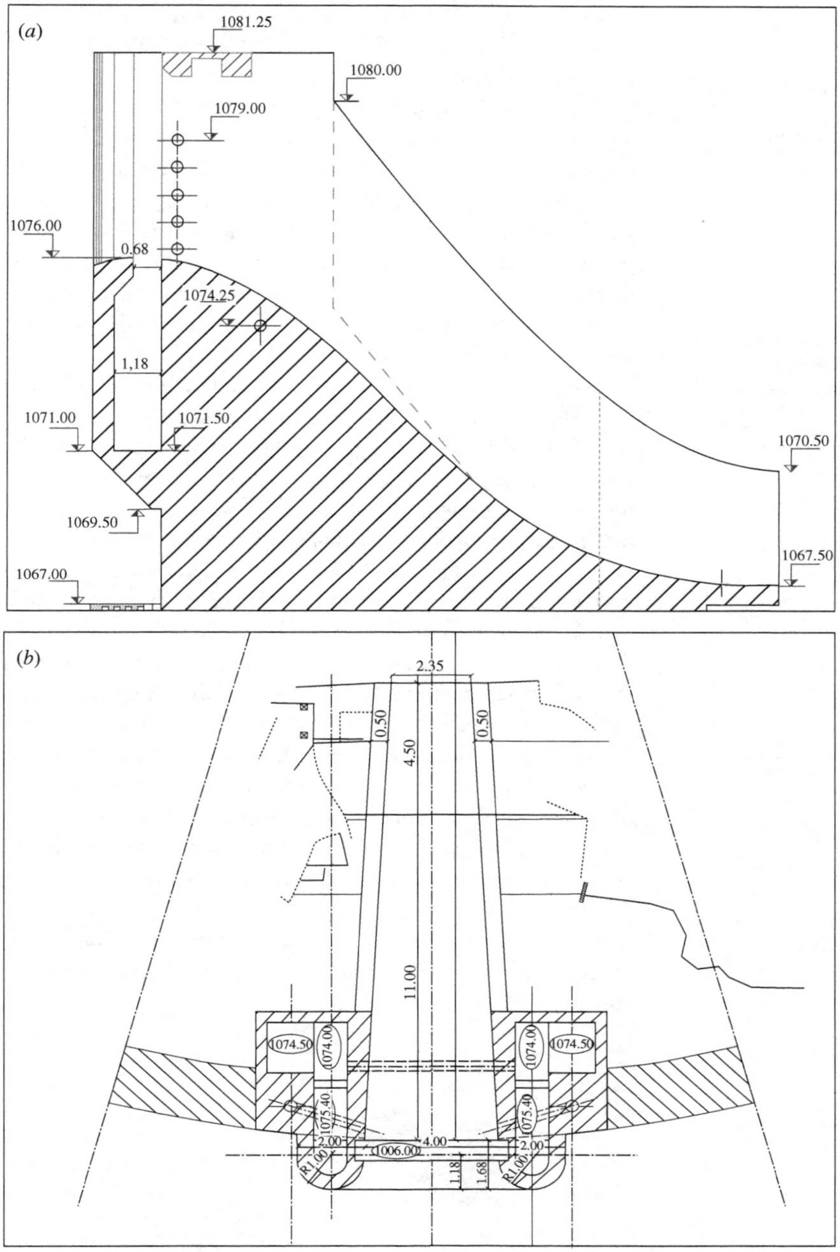

Fig. 8.30 Geometry of the ski-jump type spillway of the Goulours dam: side (*a*) and top (*b*) sketches (courtesy EDF).

spillway is dimensioned in accordance with a small-scale model which was built on a 1:20 scale at EDF R&D and is suitable for discharge of a flow at a dimensioned rate[16] $Q = 88\ \mathrm{m^3/s}$.

[16]The reassessment of that flow rate has required the construction of the second, above mentioned spillway in 2006.

Wishing to replicate a flow in as close to actual conditions as possible, the work is obviously conducted in $n = 3$ dimensions. The valley geometry is discretized from topographic data, then CAD data are utilized for representing the dam and the spillway (Fig. 8.31). A schematic reservoir is provided upstream of the latter and is filled with water at a sufficient initial elevation. For all these purposes, particles having a size $\delta r = 20$ cm are considered, which yields a total amount of about 1,292,807 particles, out of which 584,457 are dedicated to the wall modelling.[17] $h/\delta r = 1.5$ is set, whereas the fluid particles are given the water properties ($\rho_0 = 1,000\ \mathrm{kg/m^3}$, $\nu = 10^{-6}\ \mathrm{m^2s^{-1}}$).

[17]This is the main drawback of the dummy particle method as described in Section 6.3.2; it was already discussed in Section 7.3.3.

The weakly compressible SPH scheme will desirably be used. The reservoir, containing water initially at rest, will gradually empty out. Ignoring the energy transfers within the fluid during this process, it may be likened to a free fall motion. Integrating the equation of motion (1.56) then leads to a formula giving the order of magnitude of the velocity as a function of the water head H, that is $U_{\max} \sim \sqrt{2gH}$. With $H = 35$ m, that yields $U_{\max} \sim 26$ m/s. In order to take into account a possible fluid acceleration after the impact against the bed, the factor 10 in the formula (5.212) is raised too (refer to Sections 7.2.2 and 7.2.3), which ultimately yields $c_0 = 600$ m/s. Hence we get a time step $\delta t = 2 \times 10^{-4}$ s. Turbulence only plays a minor role in this flow, at least before the impact against the bed, where the strain is very large. However, since a prior simulation is desirably to be performed, this effect is ignored.

Figure 8.32 shows several instantaneous views of the water flow down the spillway, then of its fall onto the river bed. This simulation, which extends over 12 s of physical time, has required about 100 hrs of computational time over 1,024 nodes of an IBM BlueGene/L cluster. The flow is qualitatively consistent with the small-scale model results, as shown in Fig. 8.33, where the point of jet impact against the bed can be seen. A quantitative validation, however, would require that a steady regime be achieved, which implies that an upstream boundary condition be established for letting a constant flow rate appear in the reservoir.

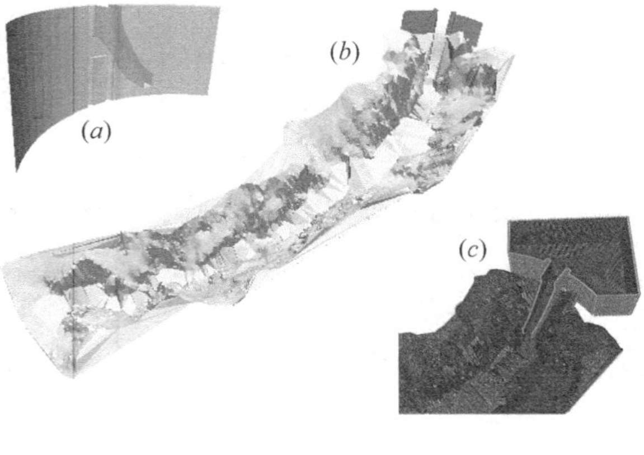

Fig. 8.31 Goulours dam: discretization elements for the dam and spillway geometry (*a*) and a part of the valley (*b*). Zooming in on the works and the reservoir, showing the particles which represent the walls (*c*).

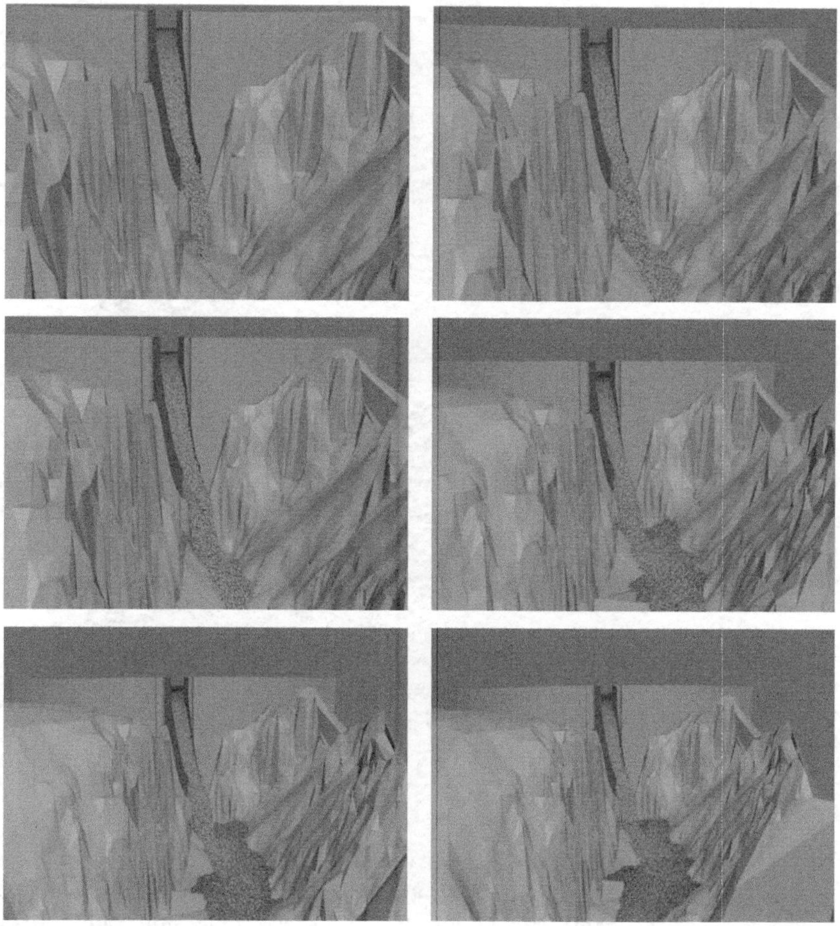

Fig. 8.32 Ski-jump spillway of the Goulours dam: SPH simulation of the reservoir emptying. Here the walls are represented by facets, although they are discretized by particles.

8.6 Conclusion: what future for SPH?

Through the simulations as set out in this chapter, confidence can logically be enhanced in the SPH method as a hydraulic flow prediction tool for subsequent applications in both industry and environment. High performance scientific computing, whether it is based on parallelism or the use of graphics cards (GPU), is suitable for circumventing one of the major drawbacks of the method. Thus, SPH, which is presently still innovative, is very likely to become a key method for free surface flow simulation. Its future looks alluring and probably lies in a greater use of its specificities, among which is the ability to predict phenomena involving a complex physics. At the time these lines are being written, that is in 2011, not all of its most recent fields of application can be mentioned here. They include the local modelling of sediment behaviour (Manenti et al., 2009), turbine dimensioning (Marongiu, 2007; Marongiu et al.,

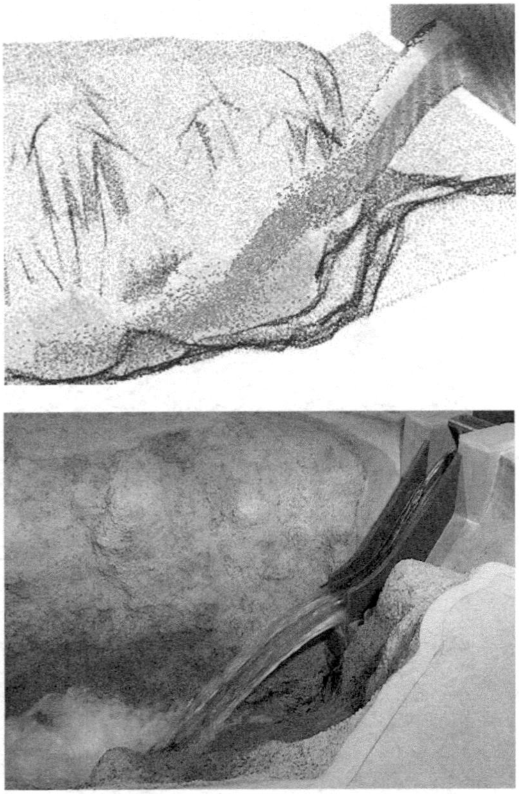

Fig. 8.33 Ski-jump spillway of the Goulours dam: point of jet impact. Comparison between the SPH simulation (*a*) and the small-scale model (*b*).

2009, 2010), fluid-structure interactions (Maurel, 2008; Lobovsky, 2008), but also such more 'exotic' issues such as cellular dynamics (Van Liedekerke et al., 2009). In addition, the two space dimension Saint-Venant equations, which were briefly discussed in Section 8.3.1 (refer to Hervouet, 2007), have recently been satisfactorily solved (Rodriguez-Paz and Bonet, 2005; De Padova et al., 2010), including in realistic cases (Vacondio et al., 2010).

However, it could be argued that the SPH method is rather unsuitable for spatial refinement (although very recent works considered this issue), which makes it difficult to implement in wide application areas for the prediction of fine-scale phenomena. As a partial answer to this question, one should mention one of the most topical areas, as induced by the coupling of SPH and grid methods. This approach gives rise to an ever increasing number of developments (Crespo et al., 2008; Narayanaswamy et al., 2010; Kassiotis et al., 2011), and is used in more and more numerous applications, particularly as regards fluid-structure coupling (Groenenboom and Cartwright, 2010; Fourey et al., 2010). These techniques create unprecedented opportunities and also contribute to promote SPH attractiveness for modelling specialists.

Tensorial formalism

A

In this appendix, we will disclose the basic elements of the tensorial calculation, which is often referred to in this book. The notion of tensor, which was introduced by Hamilton in 1846, gave rise to a sound initial differential theory in the late nineteenth century, particularly through the research conducted by Ricci and Levi-Civita (1900) in order to formalize crystallography. It actually took off under the impetus of Einstein, who used it to lay the foundations of the theory of general relativity as early as 1915. Here we do not intend to describe that theory in its general form, nor to present its rigorous mathematical grounds. Since we contemplate application to fluids, we work in a Cartesian orthonormal coordinate system, which avoids, in particular, the need to use covariant subscripts and contravariant superscripts, as well as the notion of metric tensor. For more detailed and rigorous explanations, one may refer, for instance, to either Arnold (1989) or Danielson (2003).

Let an orthonormed basis of vectors $B \doteqdot \{\mathbf{e}_i\}$ be in an n-dimensional Cartesian space. Hence we get, in particular:

$$\forall i, j \quad \mathbf{e}_i \cdot \mathbf{e}_j = \delta_{ij} \tag{A.1}$$

δ_{ij} representing the Kronecker symbol, equalling 1 if $i = j$, 0 if not. A point in the $\mathbf{x}$ space is then defined by its components $\{x_i\}$:

$$\mathbf{x} = \sum_{i=1}^{n} x_i \mathbf{e}_i \tag{A.2}$$

Due to the basis orthogonality, we may write

$$x_i = \mathbf{x} \cdot \mathbf{e}_i \tag{A.3}$$

where $(\cdot)$ denotes the ordinary scalar product. In the following, we will frequently use expressions in which dummy indices appearing twice in one monomial occur, as is the case with the preceding formula. In such a case, the *Einstein convention* (or repeated index convention) is applicable for omitting the summation sign without affecting the meaning of the expression, which significantly alleviates the notations. Thus, (A.2) may be rearranged as

$$\mathbf{x} = x_i \mathbf{e}_i \tag{A.4}$$

An index appearing twice in a monomial should then be considered as dummy, and covering the whole range of values from 1 to n.

Adopting now another orthonormal basis $\mathring{B} \doteqdot \{\mathring{\mathbf{e}}_i\}$, the matrix $\mathbf{M}$ is defined, the ij–th coefficient of which is the j–th component of $\mathring{\mathbf{e}}_i$ in B, hence, with (A.3):

$$\begin{aligned} \forall i, j \quad M_{ij} &\doteqdot \mathring{\mathbf{e}}_i \cdot \mathbf{e}_j \\ \forall i, \quad \mathring{\mathbf{e}}_i &= M_{ij}\mathbf{e}_j \end{aligned} \tag{A.5}$$

Taking (A.1) and (A.5) into account, the following relation may be written (with a summation sign over both indices k and l, being each repeated twice):

$$\begin{aligned} \mathring{\mathbf{e}}_i \cdot \mathring{\mathbf{e}}_j &= M_{ik}M_{jl}\mathbf{e}_k \cdot \mathbf{e}_l \\ &= M_{ik}M_{jl}\delta_{kl} \\ &= M_{ik}M_{jk} \end{aligned} \tag{A.6}$$

The equality (A.1), of course, still holds true for the vectors $\{\mathring{\mathbf{e}}_i\}$, which ultimately yields

$$M_{ik}M_{jk} = \delta_{ij} \tag{A.7}$$

That equality shows that the matrix $\mathbf{M}$ is the transpose of its inverse (i.e. it is *orthogonal*).

$$\mathbf{M} = \left(\mathbf{M}^T\right)^{-1} \tag{A.8}$$

Applying equation (A.3) to the basis $\mathring{B}$, we may now write the components $\{\mathring{x}_i\}$ of $\mathbf{x}$ in $\mathring{B}$ as:

$$\begin{aligned} \forall i, \quad \mathring{x}_i &= \mathbf{x} \cdot \mathring{\mathbf{e}}_i \\ &= x_j\mathbf{e}_j \cdot \mathring{\mathbf{e}}_i \\ &= M_{ij}x_j \end{aligned} \tag{A.9}$$

This relation is also written, in a vector form:

$$\mathring{\mathbf{x}} = \mathbf{M}\mathbf{x} \tag{A.10}$$

$\mathbf{M}$ is referred to as a *transformation matrix* from B towards $\mathring{B}$. Equation (A.8) yields the reciprocal relation:

$$\begin{aligned} \forall i, \quad x_i &= M_{ji}\mathring{x}_j \\ \mathbf{x} &= \mathbf{M}^T\mathring{\mathbf{x}} \end{aligned} \tag{A.11}$$

The formalism we have just described makes it possible to introduce the notion of *tensor*. By definition, a zeroth order tensor is a scalar field, that is a scalar function A which depends on the space coordinates:[1]

$$A = A(\mathbf{x}) \tag{A.12}$$

A first-order tensor is defined as a vector field, that is generalizing (A.2):

$$\begin{aligned} \mathbf{A} &= \mathbf{A}(\mathbf{x}) \\ &= \sum_{i=1}^{n} A_i(\mathbf{x})\,\mathbf{e}_i \end{aligned} \tag{A.13}$$

[1] This field, as well as all the tensors, may possibly depend on time. In order to alleviate the notations, we will systematically omit that dependence in this appendix. The dependence on space coordinates will sometimes be dropped as well.

In order to distinguish the higher than zeroth-order tensors, we write them in bold, according to an international convention. With the Einstein convention, and omitting the spatial dependence, we also may write

$$\mathbf{A} = A_i \mathbf{e}_i \tag{A.14}$$

This relation, however, is not sufficient to define a first-order tensor from a set of scalar quantities $\{A_i\}$. To that purpose, the tensor components must be transformed, upon the transition from the basis B to the basis $\mathring{B}$, through a law analogous to (A.10) (or (A.9)):

$$\begin{aligned} \mathring{\mathbf{A}} &= \mathbf{MA} \\ \forall i, \quad \mathring{A}_i &= M_{ij} A_j \end{aligned} \tag{A.15}$$

A second-order tensor is defined as a matrix field, that is by the data of $n \times n$ scalar fields $\{A_{ij}\}$:

$$\begin{aligned} \mathbf{A} &= \sum_{i,j=1}^{n} A_{ij}(\mathbf{x})\, \mathbf{e}_i \otimes \mathbf{e}_j \\ &= A_{ij} \mathbf{e}_i \otimes \mathbf{e}_j \end{aligned} \tag{A.16}$$

The symbol $\otimes$ denotes the *tensor product*. The expression $\mathbf{e}_i \otimes \mathbf{e}_j$ should be understood as the intersection of the $i-$th row and the $j-$th column of a matrix. For actually constituting a tensor, the quantities $\{A_{ij}\}$ should additionally be transformed according to the following law through a change of basis:

$$\forall i, j, \quad \mathring{A}_{ij} = M_{ik} M_{jl} A_{kl} \tag{A.17}$$

(with summation signs over k and l). Each of the subscripts of A_{ij} should then be subject to a transformation analogous to (A.15). This relation may also be arranged in the following matrix form:

$$\mathring{\mathbf{A}} = \mathbf{MAM}^T \tag{A.18}$$

A simple example of a second-order (constant) tensor is provided by the identity tensor:

$$\mathbf{I}_n \doteqdot \delta_{ij} \mathbf{e}_i \otimes \mathbf{e}_j \tag{A.19}$$

Since it is invariant under a change of basis, it logically obeys the transformation (A.17), as evidenced by the calculation (A.6).

As with a standard matrix, the *transpose* of a second-order tensor can be defined by

$$\mathbf{A}^T \doteqdot A_{ji} \mathbf{e}_i \otimes \mathbf{e}_j \tag{A.20}$$

Its *symmetric part* is given by

$$\mathbf{A}^S \doteqdot \frac{1}{2}\left(\mathbf{A} + \mathbf{A}^T\right) \tag{A.21}$$

whereas its *trace* is defined by

$$\text{tr}\,\mathbf{A} \doteqdot A_{ii} \tag{A.22}$$

Lastly, its *deviator* is defined as follows:

$$\mathbf{A}^D \doteqdot \mathbf{A} - \frac{1}{n}\,(\text{tr}\,\mathbf{A})\,\mathbf{I}_n \tag{A.23}$$

The above discussion may be generalized to define tensors of arbitrary orders. Thus, the notation $\mathbf{e}_i \otimes \mathbf{e}_j$ is generalized by introducing the tensor product of k basis vectors:

$$\mathbf{e}_{i_1} \otimes \cdots \otimes \mathbf{e}_{i_k} \tag{A.24}$$

A k-ordered tensor is then defined by the data of k scalar fields $\{A_{i_1 i_2 \dots i_k}\}$:

$$\mathbf{A} = A_{i_1 i_2 \dots i_k}\mathbf{e}_{i_1} \otimes \mathbf{e}_{i_2} \otimes \cdots \otimes \mathbf{e}_{i_k} \tag{A.25}$$

which should be transformed, upon a change of basis, according to the law

$$\forall i_1, \ldots, i_k, \quad \mathring{A}_{i_1 i_2 \dots i_k} = M_{i_1 j_1} M_{i_2 j_2} \cdots M_{i_k j_k} A_{j_1 j_2 \dots j_k} \tag{A.26}$$

with summations over all the dummy subscripts $j_1, \ldots, j_k$, which generalizes (A.17). It should therefore be pointed out that the data of arbitrary coefficients of any kind usually does not suffice for defining a tensor. The above condition is rather binding, indeed.

The tensors of given orders make up a vector space, which means that they can be added and multiplied by arbitrary constants. They can be subject to many other operations, in particular the *contraction of subscripts*. Contracting a k-ordered tensor over its $n-$th and $p-$th subscripts i_n and i_p amounts to putting $i_p = i_n$. The latter is then repeated twice, which assumes an implicit sum over i_n:

$$\begin{aligned} \mathbf{A}' = {} & A_{i_1 \dots i_{n-1} i_n i_{n+1} \dots i_{p-1} i_n i_{p+1} \dots i_k}\mathbf{e}_{i_1} \otimes \cdots \\ & \otimes \mathbf{e}_{i_{n-1}} \otimes \mathbf{e}_{i_{n+1}} \cdots \otimes \mathbf{e}_{i_{p-1}} \otimes \mathbf{e}_{i_{p+1}} \cdots \otimes \mathbf{e}_{i_k} \end{aligned} \tag{A.27}$$

This newly defined tensor $\mathbf{A}'$ now only comprises $k-2$ independent subscripts, which makes it a $(k-2)$-ordered tensor, with coefficients $A'_{i_1 \dots i_{n-1} i_{n+1} \dots i_{p-1} i_{p+1} \dots i_k} \doteqdot A_{i_1 \dots i_{n-1} i_n i_{n+1} \dots i_{p-1} i_n i_{p+1} \dots i_k}$. This is the reason why the vectors $\mathbf{e}_{i_n}$ and $\mathbf{e}_{i_p}$ have been deleted from the vector product, as can be seen. $\mathbf{A}'$ is then actually a tensor, as can be seen by contracting the subscripts i_n and i_p in the formula (A.26):

$$\begin{aligned} & \mathring{A}_{i_1 \dots i_{n-1} i_n i_{n+1} \dots i_{p-1} i_n i_{p+1} \dots i_k} \\ & = M_{i_1 j_1} \cdots M_{i_{n-1} j_{n-1}} M_{i_n j_n} M_{i_{n+1} j_{n+1}} \cdots \\ & M_{i_{p-1} j_{p-1}} M_{i_n j_p} M_{i_{p+1} j_{p+1}} \cdots \\ & M_{i_k j_k} A_{j_1 \dots j_{n-1} j_n j_{n+1} \dots j_{p-1} j_p j_{p+1} \dots j_k} \end{aligned} \tag{A.28}$$

The product $M_{i_n j_n} M_{i_n j_p}$ appears in that formulation; according to (A.7), its value is $\delta_{j_n j_p}$. The above formulation then yields

$$\begin{aligned}
&\mathring{A}'_{i_1 \dots i_{n-1} i_{n+1} \dots i_{p-1} i_{p+1} \dots i_k} \\
&= M_{i_1 j_1} \cdots M_{i_{n-1} j_{n-1}} M_{i_{n+1} j_{n+1}} \cdots \\
&M_{i_{p-1} j_{p-1}} M_{i_{p+1} j_{p+1}} \cdots \\
&M_{i_k j_k} A'_{j_1 \dots j_{n-1} j_{n+1} \dots j_{p-1} j_{p+1} \dots j_k}
\end{aligned} \tag{A.29}$$

which is exactly the condition (A.26) applied to $\mathbf{A}'$. As many pairs of subscripts as desired can obviously be contracted by deleting as many pairs of vectors from the basis. If p pairs of subscripts of a k-ordered tensor are contracted, the resulting tensor is of the $k - 2p$ order. The simple contraction of a second-order tensor can be performed in only one way and yields its trace (A.22) (zeroth-order tensor).

The *tensor product* (or *external product*) of two tensors of respective orders k and m is defined by a simple product of their coefficients:

$$\begin{aligned}
\mathbf{A} \otimes \mathbf{B} &= \left(A_{i_1 \dots i_k} \mathbf{e}_{i_1} \otimes \cdots \otimes \mathbf{e}_{i_k}\right) \\
&\otimes \left(B_{j_1 \dots j_m} \mathbf{e}_{j_1} \otimes \cdots \otimes \mathbf{e}_{j_m}\right) \\
&\doteqdot A_{i_1 \dots i_k} B_{j_1 \dots j_m} \mathbf{e}_{i_1} \otimes \cdots \otimes \mathbf{e}_{i_k} \otimes \mathbf{e}_{j_1} \otimes \cdots \otimes \mathbf{e}_{j_m}
\end{aligned} \tag{A.30}$$

This definition generalizes (A.24) and provides a $(k + m)$-ordered tensor. It will easily be demonstrated that it is actually a tensor, which means that the property (A.26) is satisfied. Thus, the tensor product of two vectors is the second-order tensor as defined by

$$\begin{aligned}
\mathbf{A} \otimes \mathbf{B} &= (A_i \mathbf{e}_i) \otimes \left(B_j \mathbf{e}_j\right) \\
&\doteqdot A_i B_j \mathbf{e}_i \otimes \mathbf{e}_j
\end{aligned} \tag{A.31}$$

In such a case, the relevant transformation law is trivially verified:

$$\begin{aligned}
\mathring{A}_i \mathring{B}_j &= (M_{ik} A_k)\left(M_{jl} B_l\right) \\
&= M_{ik} M_{jl} A_k B_l
\end{aligned} \tag{A.32}$$

The *internal product* (or simply *product*) of two k- and m-ordered tensors, respectively, is given by

$$\begin{aligned}
\mathbf{A} \cdot \mathbf{B} &= \left(A_{i_1 \dots i_{k-1} i_k} \mathbf{e}_{i_1} \otimes \cdots \otimes \mathbf{e}_{i_{k-1}} \otimes \mathbf{e}_{i_k}\right) \\
&\cdot \left(B_{j_1 j_2 \dots j_m} \mathbf{e}_{j_1} \otimes \mathbf{e}_{j_2} \otimes \cdots \otimes \mathbf{e}_{j_m}\right) \\
&\doteqdot A_{i_1 \dots i_{k-1} i_k} B_{i_k j_2 \dots j_m} \mathbf{e}_{i_1} \otimes \cdots \otimes \mathbf{e}_{i_{k-1}} \otimes \mathbf{e}_{j_2} \otimes \cdots \otimes \mathbf{e}_{j_m}
\end{aligned} \tag{A.33}$$

Thus, as with the external product, the coefficients have been multiplied, and the product was subsequently contracted over the last subscript of $\mathbf{A}$ and the first subscript of $\mathbf{B}$. As a combination of operations resulting in tensors, the internal product similarly yields a $(k + m - 2)$-ordered tensor. Thus, the internal product of two first-order tensors is their ordinary scalar product of two vectors:

$$\mathbf{A} \cdot \mathbf{B} \doteqdot A_i B_i \tag{A.34}$$

It can readily be understood, through the above elements, that it may also be written as

$$\mathbf{A}\cdot\mathbf{B} = \operatorname{tr}(\mathbf{A}\otimes\mathbf{B}) \tag{A.35}$$

The norm $|\mathbf{A}|$ of a first-order tensor is yielded by

$$\begin{aligned} |\mathbf{A}|^2 &\doteqdot A_i A_i \\ &= \mathbf{A}\cdot\mathbf{A} \\ &= \operatorname{tr}(\mathbf{A}\otimes\mathbf{A}) \end{aligned} \tag{A.36}$$

The internal product of a second-order tensor $\mathbf{A}$ and a first-order tensor $\mathbf{B}$ is the usual matrix-vector product:

$$\begin{aligned} \mathbf{A}\cdot\mathbf{B} &\doteqdot A_{ij} B_j \mathbf{e}_i \\ &= \mathbf{AB} \end{aligned} \tag{A.37}$$

(as indicated in the above formulation, the dot is sometimes omitted). Let us check, in this case, that it is actually a tensor:

$$\begin{aligned} \forall i,\quad \mathring{A}_{ij}\mathring{B}_j &= M_{ik}M_{jl}M_{jm}A_{kl}B_m \\ &= M_{ik}\delta_{lm}A_{kl}B_m \\ &= M_{ik}A_{kl}B_l \end{aligned} \tag{A.38}$$

(the property (A.7) was used).

The internal product is usually not commutative. Thus, the $\mathbf{B}\cdot\mathbf{A}$ product is written as

$$\begin{aligned} \mathbf{B}\cdot\mathbf{A} &\doteqdot B_i A_{ij} \mathbf{e}_j \\ &= \mathbf{B}^T\mathbf{A} \end{aligned} \tag{A.39}$$

Thus, once again, this is a matrix-vector product, the vector appearing to the left of the matrix. We may note that the dummy subscripts i and j can be permutated in the above formulation, which yields, through (A.37):

$$\begin{aligned} \mathbf{B}\cdot\mathbf{A} &= A_{ji} B_j \mathbf{e}_i \\ &= \mathbf{A}^T\mathbf{B} \end{aligned} \tag{A.40}$$

Let us provide a last example of an internal product, namely the product of two second-order tensors:

$$\begin{aligned} \mathbf{A}\cdot\mathbf{B} &\doteqdot A_{ik} B_{kj} \mathbf{e}_i \otimes \mathbf{e}_j \\ &= \mathbf{AB} \end{aligned} \tag{A.41}$$

Naturally, it consists of the ordinary product of two matrices, which is why the dot is omitted.

The external product may be repeated as many times as desired, for example, for second-, first- and first-order tersors $\mathbf{A}$, $\mathbf{B}$ and $\mathbf{C}$, respectively:

$$
\begin{aligned}
(\mathbf{A}\cdot\mathbf{B})\cdot\mathbf{C} &= A_{ij}B_jC_i \\
&= \mathbf{B}^T\mathbf{A}^T\mathbf{C}
\end{aligned}
\tag{A.42}
$$

Lastly, the following are two examples of a combination of the external product and the tensor product, first for a matrix **A** and a vector **B**:

$$
\begin{aligned}
\mathbf{A}\,(\mathbf{B}\otimes\mathbf{B}) &= A_{ik}B_kB_j\mathbf{e}_i\otimes\mathbf{e}_j \\
&= (\mathbf{AB})\otimes\mathbf{B}
\end{aligned}
\tag{A.43}
$$

then for two vectors:

$$
\begin{aligned}
(\mathbf{B}\otimes\mathbf{B})\,\mathbf{A} &= B_iB_jA_j\mathbf{e}_i \\
&= (\mathbf{A}\cdot\mathbf{B})\,\mathbf{B}
\end{aligned}
\tag{A.44}
$$

Let us turn now to the double product of two tensors. It is constructed like the product we have just defined, but contracting two pairs of subscripts, that is the last in **A** with the first in **B**, then the penultimate in **A** with the second in **B**:

$$
\begin{aligned}
\mathbf{A}:\mathbf{B} =\ & \left(A_{i_1\dots i_{k-2}i_{k-1}i_k}\mathbf{e}_{i_1}\otimes\cdots\otimes\mathbf{e}_{i_{k-2}}\otimes\mathbf{e}_{i_{k-1}}\otimes\mathbf{e}_{i_k}\right) \\
:\ & \left(B_{j_1j_2j_3\dots j_m}\mathbf{e}_{j_1}\otimes\mathbf{e}_{j_2}\otimes\mathbf{e}_{j_3}\otimes\cdots\otimes\mathbf{e}_{j_m}\right) \\
\doteqdot\ & A_{i_1\dots i_{k-2}i_{k-1}i_k}B_{i_ki_{k-1}j_3\dots j_m}\mathbf{e}_{i_1}\otimes\mathbf{e}_{i_2}\otimes\cdots \\
& \otimes\mathbf{e}_{i_{k-2}}\otimes\mathbf{e}_{j_3}\otimes\cdots\otimes\mathbf{e}_{j_m}
\end{aligned}
\tag{A.45}
$$

For the same reasons as previously, this always results in a $(k+m-4)$-ordered tensor. For instance, the double product of two second-order tensors is written as

$$
\begin{aligned}
\mathbf{A}:\mathbf{B} &\doteqdot A_{ij}B_{ji} \\
&= \mathrm{tr}\,(\mathbf{AB})
\end{aligned}
\tag{A.46}
$$

In particular, we get

$$
\mathrm{tr}\,\mathbf{A} = \mathbf{A}:\mathbf{I}_n \tag{A.47}
$$

Some formulations can often be rearranged by means of that notation. For instance, for a matrix **B** and two vectors **A** and **C**, we get

$$
\begin{aligned}
\mathbf{A}\cdot(\mathbf{B}\cdot\mathbf{C}) &= A_iB_{ij}C_j \\
&= \mathbf{B}:(\mathbf{A}\otimes\mathbf{C})
\end{aligned}
\tag{A.48}
$$

The notion of contracted product may be generalized, of course. By contracting over three pairs of subscripts, for instance, we may write

$$
\left(A_{ijk}\mathbf{e}_i\otimes\mathbf{e}_j\otimes\mathbf{e}_k\right)\vdots\left(B_{ijkl}\mathbf{e}_i\otimes\mathbf{e}_j\otimes\mathbf{e}_k\otimes\mathbf{e}_l\right)\doteqdot A_{ijk}B_{kjil}\mathbf{e}_l \tag{A.49}
$$

(with a sum over all four subscripts, according to the standard practice). A product of two, respectively k- and m-ordered tensors, which is contracted p times, is a $(k+m-2p)$-ordered tensor.

Since the tensors are in the form of fields, that is of spatially variable quantities, they can be subject to derivations. Thus, a *gradient* operator is defined by the following relation:

$$\begin{aligned} \mathbf{grad}\,\mathbf{A} &= \mathbf{grad}\left(A_{i_1 \dots i_k}\mathbf{e}_{i_1} \otimes \cdots \otimes \mathbf{e}_{i_k}\right) \\ &\doteqdot \frac{\partial A_{i_1 \dots i_k}}{\partial x_{i_{k+1}}}\mathbf{e}_{i_1} \otimes \cdots \otimes \mathbf{e}_{i_k} \otimes \mathbf{e}_{i_{k+1}} \end{aligned} \tag{A.50}$$

It apparently transforms a k-ordered tensor into a $(k+1)$-ordered tensor. In order to show that **grad A** actually is a tensor, we may note that the relation (A.11) yields

$$\forall i, j, \quad \frac{\partial x_i}{\partial \mathring{x}_j} = M_{ji} \tag{A.51}$$

Thus, the partial derivatives of the components of a tensor can be written in a rearranged form in the new basis by applying the chain derivation rule, the subscript j_{k+1} being dummy in the following formulation:

$$\begin{aligned} \frac{\partial \mathring{A}_{i_1 \dots i_k}}{\partial \mathring{x}_{i_{k+1}}} &= \frac{\partial \mathring{A}_{i_1 \dots i_k}}{\partial x_{j_{k+1}}}\frac{\partial x_{j_{k+1}}}{\partial \mathring{x}_{i_{k+1}}} \\ &= M_{i_1 j_1} M_{i_2 j_2} \dots M_{i_k j_k} M_{i_{k+1} j_{k+1}} \frac{\partial A_{j_1 j_2 \dots j_k}}{\partial x_{j_{k+1}}} \end{aligned} \tag{A.52}$$

(we have utilized (A.26) and (A.51)). The above discussion shows that the components in **grad A** are duly transformed according to the tensor rules. Let us provide several examples. The gradient of a zeroth-order tensor is a first-order tensor, corresponding to the usual gradient of a scalar:

$$\mathbf{grad}\, A \doteqdot \frac{\partial A}{\partial x_i}\mathbf{e}_i \tag{A.53}$$

The partial derivative of a quantity A with respect to a vector $\mathbf{B}$ can be defined by the following relation:

$$\frac{\partial A}{\partial \mathbf{B}} \doteqdot \frac{\partial A}{\partial B_i}\mathbf{e}_i \tag{A.54}$$

The gradient (A.53) is then written otherwise:

$$\mathbf{grad}\, A = \frac{\partial A}{\partial \mathbf{x}} \tag{A.55}$$

The gradient of a first-order tensor is a second-order gradient, that is the gradient of a vector field is a matrix field:

$$\mathbf{grad}\,\mathbf{A} = \mathbf{grad}\left(A_i \mathbf{e}_i\right) \doteqdot \frac{\partial A_i}{\partial x_j}\mathbf{e}_i \otimes \mathbf{e}_j \tag{A.56}$$

The above discussion reflects the fact that the gradient operator increases the order of a tensor by one unit. As regards the *divergence* operator, it reduces the order by the same extent:

$$\begin{aligned}\mathbf{div}\,\mathbf{A} &\doteqdot \mathbf{div}\left(A_{i_1\ldots i_k}\mathbf{e}_{i_1}\otimes\cdots\otimes\mathbf{e}_{i_k}\right)\\ &= \frac{\partial A_{i_1\ldots i_k}}{\partial x_{i_k}}\mathbf{e}_{i_1}\otimes\cdots\otimes\mathbf{e}_{i_{k-1}}\end{aligned} \tag{A.57}$$

It can easily be shown that the result is actually a tensor. To that purpose, the simplest way consists of observing the following property, which results from the definitions (A.45), (A.50) and (A.57):

$$\mathbf{div}\,\mathbf{A} = (\mathbf{grad}\,\mathbf{A}) : \mathbf{I}_n \tag{A.58}$$

In the case of a zeroth-order tensor, the relation (A.47) may also be used to write (A.58) in another form:

$$\mathbf{div}\,A = \mathrm{tr}\,(\mathbf{grad}\,A) \tag{A.59}$$

For instance, let us provide the divergence of a first-order tensor:

$$\mathrm{div}\mathbf{A} = \mathbf{div}\,(A_i\mathbf{e}_i) \doteqdot \frac{\partial A_i}{\partial x_i} \tag{A.60}$$

Then, for a second-order tensor:

$$\mathbf{div}\,\mathbf{A} = \mathbf{div}\left(A_{ij}\mathbf{e}_i\otimes\mathbf{e}_j\right) \doteqdot \frac{\partial A_{ij}}{\partial x_j}\mathbf{e}_i \tag{A.61}$$

A *nabla* operator, being denoted as ∇, is sometimes defined by

$$\nabla \doteqdot \frac{\partial}{\partial x_i}\mathbf{e}_i \tag{A.62}$$

The above definitions can then be used to write the gradient and divergence operators in a formal way, by means of the external and internal products:

$$\begin{aligned}\mathbf{grad}\,A &\doteqdot \nabla\otimes A\\ \mathrm{div}\mathbf{A} &\doteqdot \nabla\cdot\mathbf{A}\end{aligned} \tag{A.63}$$

The usual derivation rules are suitable for establishing various equations involving the differential operators and the various products which we have presented. For example, for a scalar A, we get, through the definition (A.19):

$$\begin{aligned}\mathbf{div}\,(A\mathbf{I}_n) &= \frac{\partial A\delta_{ij}}{\partial x_j}\mathbf{e}_i\\ &= \frac{\partial A}{\partial x_i}\mathbf{e}_i\\ &= \mathbf{grad}\,A\end{aligned} \tag{A.64}$$

For a vector **B**, we may write

$$\begin{aligned} \mathrm{div}\,(A\mathbf{B}) &= \frac{\partial AB_i}{\partial x_i} \\ &= \frac{\partial A}{\partial x_i}B_i + A\frac{\partial B_i}{\partial x_i} \\ &= (\mathbf{grad}\,A)\cdot\mathbf{B} + A\mathrm{div}\mathbf{B} \end{aligned} \tag{A.65}$$

and

$$\begin{aligned} \mathbf{grad}\,(A\mathbf{B}) &= \frac{\partial AB_i}{\partial x_j}\mathbf{e}_i\otimes\mathbf{e}_j \\ &= \frac{\partial A}{\partial x_j}B_i\mathbf{e}_i\otimes\mathbf{e}_j + A\frac{\partial B_i}{\partial x_j}\mathbf{e}_i\otimes\mathbf{e}_j \\ &= (\mathbf{grad}\,A)\otimes\mathbf{B} + A\mathbf{grad}\,\mathbf{B} \end{aligned} \tag{A.66}$$

We also get, if **B** is a matrix:

$$\begin{aligned} \mathbf{div}\,(A\mathbf{B}) &= \frac{\partial AB_{ij}}{\partial x_j}\mathbf{e}_i \\ &= \frac{\partial A}{\partial x_j}B_{ij}\mathbf{e}_i + A\frac{\partial B_{ij}}{\partial x_j}\mathbf{e}_i \\ &= \mathbf{B}\,(\mathbf{grad}\,A) + A\mathbf{div}\,\mathbf{B} \end{aligned} \tag{A.67}$$

Besides, if **A** is a matrix and **B** a vector:

$$\begin{aligned} \mathrm{div}\,(\mathbf{AB}) &= \frac{\partial A_{ik}B_k}{\partial x_i} \\ &= \frac{\partial A_{ik}}{\partial x_i}B_k + A_{ik}\frac{\partial B_k}{\partial x_i} \\ &= \left[\mathbf{div}\left(\mathbf{A}^T\right)\right]\mathbf{B} + \mathbf{A} : \mathbf{grad}\,\mathbf{B} \end{aligned} \tag{A.68}$$

Lastly, for two vectors **A** and **B**:

$$\begin{aligned} \mathbf{div}\,(\mathbf{A}\otimes\mathbf{B}) &= \frac{\partial A_iB_j}{\partial x_j} \\ &= \frac{\partial A_i}{\partial x_j}B_j + A_i\frac{\partial B_j}{\partial x_j} \\ &= (\mathbf{grad}\,\mathbf{A})\,\mathbf{B} + (\mathrm{div}\mathbf{B})\,\mathbf{A} \end{aligned} \tag{A.69}$$

The above operators can be combined. For instance:

$$\mathbf{grad}\,\mathbf{grad}\,\mathbf{A} = \frac{\partial^2 A_{i_1\,\ldots\,i_k}}{\partial x_{i_{k+1}}\partial x_{i_{k+2}}}\mathbf{e}_{i_1}\otimes\cdots\otimes\mathbf{e}_{i_{k+2}} \tag{A.70}$$

Combining the divergence and gradient olperators, we get the *Laplacian* of a tensor:

$$\begin{aligned}\nabla^2 \mathbf{A} &\doteqdot \mathbf{div}\,(\mathbf{grad}\,\mathbf{A}) \\ &= \frac{\partial^2 A_{i_1 \ldots i_k}}{\partial x_{i_{k+1}} \partial x_{i_{k+1}}} \mathbf{e}_{i_1} \otimes \cdots \otimes \mathbf{e}_{i_k}\end{aligned} \tag{A.71}$$

Thus, this is a tensor of the same order as **A**. The partial derivative commutation rules immediately induce the commutation of the Laplacian with the gradient and divergence operators. In the case of a zeroth-order tensor, equation (A.59) yields as well

$$\nabla^2 A = \mathrm{tr}\,(\mathbf{grad}\,\mathbf{grad}\,A) \tag{A.72}$$

To complete this appendix, we must mention Gauss' theorem,[2] which is frequently used throughout this book. Within a n-dimensional space, a $(n-1)$-dimensional, closed and differentiable variety $\partial\Omega$ is adopted. Thus, for $n = 2$, it is a regular closed curve; for $n = 3$, it is a surface having the same properties. In addition, $\partial\Omega$ is assumed to be orientable, and the field of unit vectors normal to $\partial\Omega$ and directed towards the outside of it is denoted as **n**. The part of the space included in $\partial\Omega$ is denoted as Ω. This theorem (which we will not demonstrate) then states that, for any tensor **A**:

[2] As stated by Arnold (1989), this theorem could be named Newton–Leibnitz–Gauss–Green–Ostrogradski–Stokes–Poincaré theorem. However, we will only refer to Gauss.

$$\int_\Omega \mathbf{div}\,\mathbf{A}\,d\Omega = \oint_{\partial\Omega} \mathbf{A}\cdot\mathbf{n}\,d\Gamma \tag{A.73}$$

One of the advantages of the tensor theory is that this result remains valid irrespective of the order of tensor **A**. The most popular version of this theorem can be found again when that order equals 1 (fields of vectors). If the field **A** is written as $\mathbf{A} = A\mathbf{I}_n$, then the property (A.64) makes it possible to derive therefrom

$$\int_\Omega \mathbf{grad}\,A\,d\Omega = \oint_{\partial\Omega} A\mathbf{n}\,d\Gamma \tag{A.74}$$

B Fourier transform

The goal of this appendix is to describe the main rules governing the calculation of the *Fourier transforms*. Studying the heat equation, J. Fourier provided the conceptual basis for this essential tool in physics, with the success we all know, by 1810. As in Appendix A, we do not intend to thoroughly demonstrate it, and certainly do not intend to lay the rigorous mathematical foundations of that theory. For further details, one may advantageously refer to Bracewell (2000).

In the first place, we consider a function $f(t)$ being defined over all the reals, having complex (possibly real) values, and integrable. In particular, f is assumed to be continuous and tending fowards 0 when $t \to \pm\infty$. Its Fourier transform $\widehat{f}$ is defined by

$$\widehat{f}(\omega) \doteq \int_{-\infty}^{+\infty} f(t)\, e^{-i\omega t} dt \tag{B.1}$$

That integral does exist, in the light of the hypotheses. If the variable t denotes the time, ω then has the dimension of an angular frequency,[1] and is considered as the conjugate variable of t. Likewise, the position in space and the wave vector may be associated, which is what we will do later on in an arbitrary dimension. Regardless of the variable being considered, its conjugate variable lies in the so-called *Fourier space*.

[1] The Fourier transform can be defined in several ways, depending on whether a factor 2π appears or not in the complex exponential. If so, ω still physically has the dimension of the inverse of a time, but represents a frequency rather than an angular frequency. Several values of the dimensionless coefficient—here being equal to one—placed before the integral are distinguished as well. By setting it equal to $1/\sqrt{2\pi}$, we get a formalism which is slightly smarter than here.

The properties of the Fourier transform are so numerous that fully reviewing them would be endless; we will therefore simply consider the most important ones. Let us initially observe the following relation:

$$\widehat{f}(0) = \int_{-\infty}^{+\infty} f(t)\, dt \tag{B.2}$$

The Fourier transform defines a continuous linear operator. For every constant a, the transform of the function $t \longrightarrow f(at)$ is yielded by the following calculation:

$$\begin{aligned}\widehat{f(at)}(\omega) &= \int_{-\infty}^{+\infty} f(at)\, e^{-i\omega t} dt \\ &= \frac{1}{a}\int_{-\infty}^{+\infty} f(t')\exp\left(-\frac{i\omega t'}{a}\right) dt' \\ &= \frac{1}{a}\widehat{f}\left(\frac{\omega}{a}\right)\end{aligned} \tag{B.3}$$

One can easily calculate $\widehat{f}(-\omega)$:

$$\widehat{f}(-\omega) = \int_{-\infty}^{+\infty} f(t)\, e^{i\omega t} dt \tag{B.4}$$

Taking the conjugate complex (denoted through a star *) of the resulting expression, we get

$$\widehat{f}(-\omega) = \left[\int_{-\infty}^{+\infty} f^*(t)\, e^{-i\omega t} dt\right]^* \tag{B.5}$$

Hence, if the function f has real values (i.e. $f = f^*$), we get

$$\widehat{f}(-\omega) = \widehat{f}^*(\omega) \tag{B.6}$$

that is its transform is even. Returning to the general case of a function with complex values, a mere change of variable is enough to write the transform of the complex conjugate function f^*:

$$\begin{aligned}\widehat{f^*}(\omega) &= \int_{-\infty}^{+\infty} f^*(t)\, e^{-i\omega t} dt \\ &= \left[\int_{-\infty}^{+\infty} f(t)\, e^{i\omega t} dt\right]^* \\ &= \left[\int_{-\infty}^{+\infty} f(-t)\, e^{-i\omega t} dt\right]^*\end{aligned} \tag{B.7}$$

If the function f presents a parity property, we get

$$\widehat{f^*}(\omega) = \pm \widehat{f}^*(\omega) \tag{B.8}$$

The sign $+$ occurs for an even function, whereas the sign $-$ is applicable to the odd functions. In particular, a real even function exhibits a Fourier transform verifying $\widehat{f} = \widehat{f}^*$, that is a real-valued function. On the other hand, a real odd function has a purely imaginary transform.

The convolution of two functions f and g is now defined by the following relation:

$$(f * g)(t) \doteqdot \int_{-\infty}^{+\infty} f(t')\, g(t - t')\, dt' = (g * f)(t) \tag{B.9}$$

(the symmetry is ensured by the change of variable $t'' \doteqdot t - t'$). The transform of the convolution of two functions is then written as

$$\begin{aligned}\widehat{(f * g)}(\omega) &= \int_{-\infty}^{+\infty}\int_{-\infty}^{+\infty} f(t')\, g(t - t')\, e^{-i\omega t} dt' dt \\ &= \int_{-\infty}^{+\infty} f(t') \left[\int_{-\infty}^{+\infty} g(t - t')\, e^{-i\omega t} dt\right] dt' \\ &= \int_{-\infty}^{+\infty} f(t')\, e^{-i\omega t'} \left[\int_{-\infty}^{+\infty} g(t'')\, e^{-i\omega t''} dt''\right] dt' \\ &= \widehat{f}(\omega)\, \widehat{g}(\omega)\end{aligned} \tag{B.10}$$

The Fourier transform has a fairly simple relationship with the notion of derivative. Thus, a mere integration by parts provides the following result:

$$\begin{aligned}\widehat{\frac{\partial f}{\partial t}}(\omega) &= \int_{-\infty}^{+\infty} \frac{\partial f}{\partial t} e^{-i\omega t} dt \\ &= \left[f(t)\, e^{-i\omega t} \right]_{-\infty}^{+\infty} + i\omega \int_{-\infty}^{+\infty} f(t)\, e^{-i\omega t} dt\end{aligned} \tag{B.11}$$

With the boundary conditions prescribed on f at infinity, that yields as well

$$\widehat{\frac{\partial f}{\partial t}}(\omega) = i\omega \widehat{f}(\omega) \tag{B.12}$$

A transform of an arbitrary k-ordered derivative can easily be derived from it through a mathematical induction:

$$\widehat{\frac{\partial^k f}{\partial t^k}}(\omega) = (i\omega)^k \widehat{f}(\omega) \tag{B.13}$$

The transform derivative is achieved by deriving the definition (B.1):

$$\begin{aligned}\frac{\partial \widehat{f}}{\partial \omega} &= -i \int_{-\infty}^{+\infty} f(t)\, t e^{-i\omega t} dt \\ &= -i \widehat{f(t)\, t}\end{aligned} \tag{B.14}$$

We now will calculate the transform of the following Gaussian function:

$$\phi(t) \doteqdot \exp\left(-\frac{t^2}{2}\right) \tag{B.15}$$

To that purpose, let us initially observe that ϕ satisfies the following ordinary differential equation:

$$\phi' + t\phi = 0 \tag{B.16}$$

Let us now consider the derivative of $\widehat{\phi}$. The equations (B.14), (B.16) and (B.12) successively yield

$$\begin{aligned}\frac{\partial \widehat{\phi}}{\partial \omega} &= -i\widehat{\phi(t)\, t} \\ &= i\widehat{\frac{\partial \phi}{\partial t}} \\ &= -\omega\widehat{\phi}\end{aligned} \tag{B.17}$$

which also yields $\widehat{\phi}' + \omega\widehat{\phi} = 0$. Thus, $\widehat{\phi}$ obeys the same differential equation (B.16) as ϕ by substituting ω for the variable t. Since it is a first-order homogeneous linear equation, its solutions as a whole make up a vector space of dimension 1. It can be inferred therefrom that $\widehat{\phi}$ is proportional to ϕ, their ratio being given by the respective values of the two functions in zero, for instance. Now, with (B.2), we get:

$$\widehat{\phi}\,(0) = \int_{-\infty}^{+\infty} \phi\,(t)\,dt = \sqrt{2\pi} \tag{B.18}$$

Lastly, we have

$$\widehat{\phi}\,(\omega) = \sqrt{2\pi}\phi\,(\omega) \tag{B.19}$$

More generally, the relation (B.3) gives the transform of an arbitrary Gaussian:

$$\widehat{\phi\,(at)}\,(\omega) = \frac{\sqrt{2\pi}}{a}\phi\left(\frac{\omega}{a}\right) \tag{B.20}$$

Due to this significant result, the inverse operator of the Fourier transform can be determined. To that purpose, let us consider the following quantity, which is defined for every non-zero ϵ:

$$\begin{aligned} f_\epsilon\,(t) &\doteq \frac{1}{2\pi}\int_{-\infty}^{+\infty} \widehat{f}\,(\omega)\exp\left[i\omega t - \frac{(\epsilon\omega)^2}{2}\right]d\omega \\ &= \frac{1}{2\pi}\int_{-\infty}^{+\infty} f\,(t')\int_{-\infty}^{+\infty} \exp\left[i\omega\left(t-t'\right)\right]\phi\,(\epsilon\omega)\,d\omega dt' \end{aligned} \tag{B.21}$$

In the second line of that definition, we can mentally invert the dummy variables $t - t'$ and ω, the time temporarily playing the role of a frequency, and conversely. The integral in $d\omega$ is then formally the Fourier transform of the function $\phi\,(\epsilon\omega)$, taken at the point $t - t'$. Taking (B.20) into account, that yields

$$\begin{aligned} \int_{-\infty}^{+\infty} \exp\left[i\omega\left(t-t'\right)\right]\phi\,(\epsilon\omega)\,d\omega &= \widehat{\phi\,(\epsilon\omega)}\left(t-t'\right) \\ &= \frac{\sqrt{2\pi}}{\epsilon}\phi\left(\frac{t'-t}{\epsilon}\right) \end{aligned} \tag{B.22}$$

A mere change of variable in (B.21) is then sufficient to write

$$\begin{aligned} f_\epsilon\,(t) &= \frac{1}{\sqrt{2\pi}\epsilon}\int_{-\infty}^{+\infty} f\,(t')\,\phi\left(\frac{t'-t}{\epsilon}\right)dt' \\ &= \frac{1}{\sqrt{2\pi}}\int_{-\infty}^{+\infty} f\,(\epsilon t''+t)\,\phi\,(t'')\,dt'' \end{aligned} \tag{B.23}$$

Equating (B.21) and (B.23), then making ϵ tend towards 0, we get

$$\begin{aligned} \lim_{\epsilon\to 0} f_\epsilon\,(t) &= \frac{1}{2\pi}\int_{-\infty}^{+\infty} \widehat{f}\,(\omega)\,e^{i\omega t}d\omega \\ &= \frac{1}{\sqrt{2\pi}}f\,(t)\int_{-\infty}^{+\infty} \phi\,(t'')\,dt'' \\ &= f\,(t) \end{aligned} \tag{B.24}$$

(we have ultimately used (B.18)). We have just shown that the Fourier transform is invertible and that its inverse is analogous to the initial definition (B.1), the sign in the complex exponential being changed and a normalizing coefficient appearing before the integral:[2]

[2]According to the preceding margin note, introducing a factor $1/\sqrt{2\pi}$ into the definition (B.1) would make it possible to get uniform arrangements of the transform and its inverse.

$$f(t) = \frac{1}{2\pi} \int_{-\infty}^{+\infty} \widehat{f}(\omega)\, e^{i\omega t}\, d\omega \tag{B.25}$$

This result may also be written as

$$f(t) = \frac{1}{2\pi} \widehat{\widehat{f}}(-t) \tag{B.26}$$

The inverse relation of (B.10) can then easily be shown, which means that the product transform is the convolution of the transforms.

Applying the inverse Fourier transform to the relation (B.12), the derivative of a function can then be related to its transform:

$$\frac{\partial f}{\partial t} = \frac{1}{2\pi} \int_{-\infty}^{+\infty} i\omega \widehat{f}(\omega)\, e^{i\omega t}\, dt \tag{B.27}$$

We are going to discuss a further major result which is known as the *Poisson summation formula*. To that purpose, let us initially consider a function having a period T. We try to decompose it into a *Fourier series*, that is write it as

$$F(t) = \sum_{k\in\mathbb{Z}} a_k e^{ik\omega_T t} \tag{B.28}$$

where each $\{a_k\}$ is a coefficient to be sought, whereas a conjugate frequency of T has been defined:

$$\omega_T \doteq \frac{2\pi}{T} \tag{B.29}$$

The analytical expression of each a_k is found by multiplying (B.28) by $e^{ik\omega_T t}$ then by integrating over an interval whose width is equal to the period. The order of integral and the discrete sum may be reversed according to the dominated convergence theorems:

$$\int_0^T F(t)\, e^{-im\omega_T t}\, dt = \sum_{k\in\mathbb{Z}} a_k \int_0^T e^{i(k-m)\omega_T t}\, dt \tag{B.30}$$

The integral of a complex exponential over that interval is zero, unless $k = m$, which yields

$$a_m = \frac{1}{T} \int_0^T F(t)\, e^{-im\omega_T t}\, dt \tag{B.31}$$

Due to that result, (B.28) may be rearranged as

$$F(t) = \frac{1}{T} \sum_{k\in\mathbb{Z}} \int_0^T F(t')\, e^{ik\omega_T (t-t')}\, dt' \tag{B.32}$$

For an integrable (not necessarily periodic) function f, let us now consider the function F as defined by

$$F(t) \doteqdot \sum_{m\in\mathbb{Z}} f(t+mT) \tag{B.33}$$

The sum converges, in accordance with the initial hypotheses about f. F is obviously periodic with a period T. Applying the formula (B.32) to it, we get

$$\sum_{m\in\mathbb{Z}} f(t+mT) = \frac{1}{T}\sum_{k\in\mathbb{Z}}\sum_{m\in\mathbb{Z}}\int_0^T f\left(t'+mT\right)e^{ik\omega_T(t-t')}dt' \tag{B.34}$$

In order to calculate the sum with respect to the relative integer m, we make a change of variable $t'' = t' + mT$ in each integral:

$$\begin{aligned}
&\sum_{m\in\mathbb{Z}}\int_0^T f\left(t'+mT\right)e^{ik\omega_T(t-t')}dt' \\
&= \sum_{m\in\mathbb{Z}}\int_{mT}^{(m+1)T} f\left(t''\right)e^{ik\omega_T(t-t'')}e^{-2\pi ikm}dt'' \\
&= e^{ik\omega_T t}\int_{-\infty}^{+\infty} f\left(t''\right)e^{-ik\omega_T t''}dt'' \\
&= e^{ik\omega_T t}\widehat{f}(k\omega_T)
\end{aligned} \tag{B.35}$$

(we have lumped the integrals and utilized the relation $e^{-2\pi ikm} = 1$). (B.34) is ultimately written as

$$\sum_{m\in\mathbb{Z}} f(t+mT) = \frac{1}{T}\sum_{k\in\mathbb{Z}} e^{ik\omega_T t}\widehat{f}(k\omega_T) \tag{B.36}$$

For $t = 0$, that yields the Poisson formula:

$$\sum_{m\in\mathbb{Z}} f(mT) = \frac{1}{T}\sum_{k\in\mathbb{Z}} \widehat{f}(k\omega_T) \tag{B.37}$$

The sum of the values of a function at the points in a regular network is then related to an analogous sum in the Fourier space.

The above notion can fairly easily be extended to those functions as defined over $\mathbb{R}^n$. It can then be found that the variables ($\mathbf{r}$, $\mathbf{K}$) are more relevant than (t, ω), referring to the notation $\mathbf{r}$ as used in physics for denoting the space variable, whereas $\mathbf{K}$ denotes a wave vector:

$$\widehat{f}(\mathbf{K}) \doteqdot \int_{\mathbb{R}^n} f(\mathbf{r})\,e^{-i\mathbf{K}\cdot\mathbf{r}}d^n\mathbf{r} \tag{B.38}$$

Properties which are similar to that of the one-dimensional transform then appear. Thus, the equations (B.6) (for those functions having real values) and (B.12) respectively become

$$\begin{aligned}
&\widehat{f}(-\mathbf{K}) = \widehat{f}^*(\mathbf{K}) \\
&\widehat{\frac{\partial f}{\partial x_j}} = iK_j\widehat{f}(\mathbf{K})
\end{aligned} \tag{B.39}$$

On the basis of the second of these two formulas, more general derivation rules can be established. Hence:

$$\widehat{(\mathbf{grad}\ f)}\,(\mathbf{K}) = i\mathbf{K}\widehat{f}\,(\mathbf{K}) \tag{B.40}$$

Likewise:

$$\widehat{(\mathbf{grad}\ \mathbf{grad}\ f)}\,(\mathbf{K}) = -\mathbf{K} \otimes \mathbf{K}\widehat{f}\,(\mathbf{K}) \tag{B.41}$$

(both gradient operator and tensor product being defined in Appendix A). Taking the trace of the above formula, we may invoke (A.36) and (A.72) and additionally write

$$\widehat{\left(\nabla^2 f\right)}\,(\mathbf{K}) = -K^2\widehat{f}\,(\mathbf{K}) \tag{B.42}$$

A rule which is identical to (B.10) is involved as well in the expression of the transform of a spatial convolution, the integral concerning $\mathbb{R}^n$.

It is easy to see that the transform of an isotropic (or radial) function, that is only depending on the norm $r \doteqdot |\mathbf{r}|$ of $\mathbf{r}$, has the same property, that is constitutes a function of $K \doteqdot |\mathbf{K}|$. One of the coordinate axes in the formula (B.38) may, indeed, be arbitrarily aligned with the wave vector $\mathbf{K}$, without affecting the value of $f\,(r)$ nor the integration volume element $d^n\mathbf{r}$. Hence, in this case:

$$\begin{aligned} f\,(\mathbf{r}) &= f\,(r \doteqdot |\mathbf{r}|) \\ \widehat{f}\,(\mathbf{K}) &= \widehat{f}\,(K \doteqdot |\mathbf{K}|) \end{aligned} \tag{B.43}$$

It should be pointed out that the above notation is deceptive, $\widehat{f}\,(K)$ meaning that the transform $\widehat{f}\,(K)$ of f (considered as a function of $\mathbf{r}$) is a function of K only. For all that, $\widehat{f}\,(K)$ is not the one-dimensional Fourier transform of f as seen as a function of r. However, $\widehat{f}\,(K)$ can be related to $f\,(r)$. In the two-dimensional case, for instance, the calculations are quite simple by integrating on the polar angle θ, then over r:

$$\widehat{f}\,(\mathbf{K}) = \int_0^{+\infty} f\,(r)\left(\int_0^{2\pi} e^{-iKr\cos\theta}\,d\theta\right) r\,dr \tag{B.44}$$

The function written between brackets is the zeroth-order Bessel function of the first kind, being denoted as J_0. Hence, we get

$$\widehat{f}\,(K) = 2\pi \int_0^{+\infty} r f\,(r)\, J_0\,(Kr)\, dr \tag{B.45}$$

It is no surprise that a Bessel function appears in the solution of a problem with respect to a radial function. This result may be generalized to the case of an arbitrary dimension n (Buhmann, 2003):

$$\widehat{f}\,(K) = (2\pi)^{n/2}\, K^{1-n/2} \int_0^{+\infty} r^{n/2} f\,(r)\, J_{n/2-1}\,(Kr)\, dr \tag{B.46}$$

the Bessel functions of the first kind being defined by

$$J_m\,(x) \doteqdot \frac{1}{2\pi}\int_{-\pi}^{\pi} e^{-i(m\theta - x\sin\theta)}\,d\theta \tag{B.47}$$

The formula (B.46) is similar to the one-dimensional Fourier transform (B.1), the Bessel functions being substituted for the complex exponential. This is what is known as the *Hankel transform* (see e.g. Gaskill, 1978).

The multidimensional transform of a Gaussian is a Gaussian too. This is because one may define, for every constant vector **a** having components $\{a_j\}$:

$$\phi_{\mathbf{a}}(\mathbf{r}) \doteqdot \phi\left(\sum_{j=1}^{n}(a_j x_j)^2\right) = \exp\left[-\frac{1}{2}\sum_{j=1}^{n}(a_j x_j)^2\right] \tag{B.48}$$

Hence we get, according to the definition (B.15),

$$\begin{aligned}\phi_{\mathbf{a}}(\mathbf{r})\,e^{-i\mathbf{K}\cdot\mathbf{r}} &= \exp\left\{-\sum_{j=1}^{n}\left[\frac{(a_j x_j)^2}{2} + iK_j x_j\right]\right\} \\ &= \prod_{j=1}^{n}\phi(a_j x_j)\,e^{-iK_j x_j}\end{aligned} \tag{B.49}$$

Integrating over $\mathbb{R}^n$, we find

$$\begin{aligned}\widehat{\phi_{\mathbf{a}}(\mathbf{r})}(K) &= \prod_{j=1}^{n}\int_{\mathbb{R}}\phi(a_j x_j)\,e^{-iK_j x_j}dx_j \\ &= \prod_{j=1}^{n}\widehat{\phi(a_j x_j)}(K_j) \\ &= (2\pi)^{n/2}\prod_{j=1}^{n}\frac{1}{a_j}\phi\left(\frac{K_j}{a_j}\right)\end{aligned} \tag{B.50}$$

In the specific case of a radial Gaussian, that is for a vector **a** whose components are equal to a constant a, that yields

$$\widehat{\phi(ar)}(K) = \left(\frac{\sqrt{2\pi}}{a}\right)^n\phi\left(\frac{K}{a}\right) \tag{B.51}$$

to be compared with the one-dimensional case (B.20). The transform is then radial, in accordance with (B.43).

The calculation (B.21)–(B.24) can easily be generalized from equation (B.51). The multidimensional inverse Fourier transform then generalizes (B.25) as

$$f(\mathbf{r}) = \frac{1}{(2\pi)^n}\int_{\mathbb{R}^n}\widehat{f}(\mathbf{K})\,e^{i\mathbf{K}\cdot\mathbf{r}}d^n\mathbf{K} \tag{B.52}$$

This relation is also suitable for generalizing (B.27):

$$\frac{\partial f}{\partial x_j} = \frac{1}{(2\pi)^n}\int_{\mathbb{R}^n}iK_j\widehat{f}(\mathbf{K})\,e^{i\mathbf{K}\cdot\mathbf{r}}d^n\mathbf{K} \tag{B.53}$$

Furthermore, the Poisson summation formula (B.37) is suitable for a generalization within the multidimensional context. To that purpose, a multidimensional Cartesian network with a spatial step R is considered. A demonstration similar to that we have made in dimension 1 then leads to

$$\sum_{\{m_i\}\in\mathbb{Z}^n} f\,(m_1 R, ..., m_n R) = \frac{1}{R^n} \sum_{\{k_i\}\in\mathbb{Z}^n} \widehat{f}\,(k_1 K_R, ..., k_n K_R) \tag{B.54}$$

with

$$K_R \doteqdot \frac{2\pi}{R} \tag{B.55}$$

We will ultimately establish the analytical expressions of the transforms of some usual functions. To that purpose, we consider a dimension $n = 1$, once again adopting the formalism (t, ω). Let us initially define the *Dirac distribution*[3] δ, such as

[3] This is not a function in the usual sense of the term.

$$\forall f, \quad f * \delta = f \tag{B.56}$$

That definition shows that δ is even. It equals zero except at the origin, a point at which it is in principle infinite. The relation (B.10) can be used for immediately deriving its transform:

$$\widehat{\delta}\,(\omega) = 1 \tag{B.57}$$

If the function $\mathbf{1}\,(\omega)$ is defined as contant and equalling 1, we derive, by means of (B.26) and the even nature of δ:

$$\begin{aligned}\widehat{\mathbf{1}}\,(t) &= \widehat{\widehat{\delta}}\,(t) \\ &= 2\pi\delta\,(t)\end{aligned} \tag{B.58}$$

This identity is also written as

$$\int_{-\infty}^{+\infty} e^{-i\omega t} d\omega = 2\pi\delta\,(t) \tag{B.59}$$

The above relation remains valid if the variables t and ω are interchanged. We can derive therefrom the transform of a complex exponential function of frequency ω_0:

$$\begin{aligned}\widehat{e^{i\omega_0 t}}\,(\omega) &= \int_{-\infty}^{+\infty} e^{i(\omega_0-\omega)t} dt \\ &= 2\pi\delta\,(\omega - \omega_0)\end{aligned} \tag{B.60}$$

For $\omega_0 = 0$, we find (B.58) again. This result is crucial in the physical interpretation of the Fourier transform. It can actually be observed that a function oscillating at the frequency ω_0 has a transform exhibiting a peak for that value of the frequency. Through linearity, it can easily be inferred therefrom that the Fourier transform of an arbitrary function highlights its oscillating nature by pointing out the characteristic frequencies.

Furthermore, the *gate function* having a width t_0 as defined by

$$\begin{aligned}F_{t_0}\,(t) &= \frac{1}{t_0} \quad \text{if} \quad |t| \leqslant \frac{t_0}{2} \\ &= 0 \quad \text{if not}\end{aligned} \tag{B.61}$$

may advantageously be considered. Its transform is easily calculated:

$$\begin{aligned}\widehat{F_{t_0}}(\omega) &= \frac{1}{t_0}\int_{-t_0/2}^{t_0/2} e^{-i\omega t}\,dt \\ &= \frac{1}{t_0}\left[-\frac{1}{i\omega}e^{-i\omega t}\right]_{-t_0/2}^{t_0/2} \\ &= \frac{2}{\omega t_0}\sin\frac{\omega t_0}{2}\end{aligned} \tag{B.62}$$

It is a *sine cardinal (or sinc) function*. More generally, the M-ordered *B-spline* function (refer to Monaghan, 1985a, for example) is defined as an M-ordered piecewise even polynomial function, as defined over $\mathbb{R}^+$ by

$$\begin{aligned}f_M(t) \doteqdot \frac{1}{M!}\sum_{k=0}^{M+1}(-1)^k\binom{M+1}{k} \\ \times \max\left(0;\ \frac{t}{t_0}+\frac{M+1}{2}-k\right)^M\end{aligned} \tag{B.63}$$

The particular case $M = 0$ once again yields the function $F_{t_0} = f_0$. It is shown (refer to De Boor, 2001) that

$$\widehat{f_M}(\omega) = \left(\frac{2}{\omega t_0}\sin\frac{\omega t_0}{2}\right)^{M+1} \tag{B.64}$$

which generalizes (B.62). Conversely, if the sine cardinal function is defined by

$$G_{\omega_0}(t) \doteqdot \frac{1}{\omega_0 t}\sin(\omega_0 t) \tag{B.65}$$

(or else $G_{\omega_0}(t) = \widehat{F_{2t}}(\omega_0)$), then (B.26) yields its Fourier transform in the form of a gate function:

$$\begin{aligned}\widehat{G_{\omega_0}}(\omega) &= \frac{\pi}{\omega_0} \quad \text{if} \quad |\omega| \leqslant \omega_0 \\ &= 0 \quad \text{if not}\end{aligned} \tag{B.66}$$

Bibliography

Abramowitz, M., Stegun, I.A. (1972), *Handbook of Mathematical Functions with Formulas, Graphs, and Mathematical Tables*, 10th edition Dover Publications, New York.

Adami, S., Hu, X.Y., Adams, N.A. (2010), *A conservative SPH method for surfactant dynamics*, J. Comput. Phys., **229**:1909–1926.

Agate, G., Guandalini, R. (2010), *The use of 3D SPHERA code to support spillway design and safety evaluation of flood events*, Proc. V^{th} SPHERIC International Workshop, Manchester (UK), pp. 267–272.

Ahrens J.P., Heimbaugh M.S. (1988), *Seawall overtopping model*, Proc. XXIst Int. Conf. Coastal Eng., Malaga (Spain), pp. 795–806.

Aminti P., Franco L. (1988), *Wave overtopping on rubble-mound breakwaters*, Proc. XXIst Int. Conf. Coastal Eng., Malaga (Spain), pp. 770–781.

Arnold, V.I. (1989), *Mathematical Methods of Classical Mechanics*, 2nd edition Springer, New York.

Ataie-Ashtiani B., Shobeyri G. (2008), *Numerical simulation of landslide impulsive waves by incompressible Smoothed Particle Hydrodynamics* , Int. J. Num. Meth. Fluids **56**:209–232.

Balsara, D.S. (1995), *Von Neumann stability analysis of Smoothed Particle Hydrodynamics – Suggestions for optimal algorithms*, J. Comput. Phys., **121**:357–372.

Basdevant, J.-L. (2005), *Principes Variationnels et Dynamique*, Vuibert, Paris, (in French).

Belytschko, T., Xiao, S. (2002), *Stability analysis of particle methods with corrected derivatives*, Comput. Math. Applications **43**:329–350.

Ben Moussa, B. (2006), *On the convergence of SPH method for scalar conservation laws with boundary conditions*, Meth. Appl. Analysis **13**:29–62.

Besicovitch, A.S., Ursell, H.D. (1937), *Sets of Fractional Dimensions*, V, J. London Math. Soc., **32**:142–153.

Boltzmann, L. (1964), *Lectures on Gas Theory*, University of California Press, Berkeley.

Bonet, J., Lok, T.-S.L. (1999), *Variational and momentum preservation aspects of Smoothed Particle Hydrodynamics formulations*, Comput. Meth. Appl. Mech. Eng., **180**:97–115.

Bonet, J., Rodriguez-Paz, M.X. (2005), *Hamiltonian formulation of the variable-h SPH equations*, J. Comput. Phys., **209**:541–558.

Boudenot, J.-C. (1989), *Electromagnétisme et Gravitation Relativistes*, Ellipses, Paris (in French).

Boussinesq, J. (1877), *Essai sur la théorie des eaux courantes*, Mémoires Présentées par Divers Savants à l'Académie des Sciences, Vol. 23 (in French).

Bracewell, R.N. (2000), *The Fourier Transform and its Applications*, 3rd edition McGraw-Hill, Boston.

Bradbury A.P., Allsop N.W.H. (1988), *Hydraulics effects of breakwater crown walls*, Proc. Int. Conf. Breakwaters'88, Eastbourne (UK), pp. 385–396.

Bretherton, F.P., Garret, C.J.R. (1969), *Wavetrains in Inhomogeneous Moving Media.* Proc. Roy. Soc. A, **302**:529–554.

Buhmann, M.D. (2003), *Radial Basis Functions: Theory and Implementations*, Cambridge University Press, Cambridge (UK).

Buvat, C., Violeau, D. (2006), *Lagrangian numerical modelling of boom and oil spill motion for the management of coastal pollution risk*, Proc. VIIth Int. Conf. Hydroinformatics, Nice (France) 4–8 Sept 2006.

Campbell, J.C., Vignjevic, R. (2009), *Modelling extreme wave loading of offshore structures*, Proc. IVth *SPHERIC* International Workshop, Nantes (France), pp. 309–314.

Campbell, J.C., Vignjevic, R., Libersky L. (2000), *A contact algorithm for Smoothed Particle Hydrodynamics*, Comput. Meth. Appl. Mech. Eng., **184**:49–65.

Carter, E. (1994), *The generation and application of random numbers*, Forth Dimensions **16**(1–2).

Chaniotis A.K., Poulikakos D., Koumoutsakos P. (2002), *Remeshed Smoothed Particle Hydrodynamics for the simulation of viscous and heat conducting flows*, J. Comput. Phys., **182**:67–90.

Chen, S., Foias, C., Holm, D.D., Olson, E., Titi, E.S., Wynne, S. (1999), *A connection between the Camassa-Holm equations and turbulent flows in channels and pipes*, Phys. Fluids **11**(8):2343–2353.

Chen, H., Orszag, S.A., Staroselsky, I., Succi, S. (2004), *Expanded analogy between Boltzmann kinetic theory of fluids and turbulence*, J. Fluid Mech. **519**:301–314.

Chern, M.J., Borthwick, A.G.L., Eatock Taylor, R. (2005), *Pseudospectral element model for free surface viscous flows*, Int. J. Num. Meth. Heat & Fluid Flow, **15**(6), 517–554.

Chorin, A.J. (1968), *Numerical Solution of the Navier–Stokes Equations*, Math. Comp. **22**:745–762.

Cleary, P.W. (1996), *New implementation of viscosity. Tests with Couette flows*, report of CSIRO Division of Mathematics and Statistics, Maquarie University, North Ryde (Australia).

Cleary, P.W., Prakash, M. (2004), *Discrete-element modelling and smoothed particle hydrodynamics: potential in the environmental sciences*, Phil. Trans. Roy. Soc. A, **362**(1822):2003–2030.

Colagrossi, A. (2006), *6th benchmark test case: 2D Incompressible flow around a moving square inside a rectangular box*, *SPHERIC* community, http://wiki.manchester.ac.uk/spheric/ [accessed 14 November 2011].

Colagrossi, A., Landrini, M. (2003), *Numerical simulation of interfacial flows by SPH*, J. Comput. Phys., **191**:448–475.

Colagrossi, A., Colicchio, G., Lugni, C., Brocchini, M. (2010), *A study of violent sloshing wave impacts using an improved SPH method*, J. Hydr. Res., **48**:94–104.

Colicchio, G., Greco, M., Faltinsen, O.M. (2006), *Fluid-body interaction on a Cartesian grid: dedicated studies for a CFD validation*, Proc. XXIst International Workshop on Water Waves and Floating Bodies, Loughborough (UK).

Courant, R., Friedrichs, K., Lewy, H. (1967), *On the partial difference equations of mathematical physics*, IBM Journal, March 1967:215–234.

Craft, T.J., Gerasimov, A.V., Iacovides, H., Launder B.E. (2002), *Progress in the generalization of wall-function treatments*, Int. J. Heat Fluid Flow, **23**:148–160.

Crespo, A.J.C., Narayanaswamy, M.S., Gomez-Gesteira, M., Dalrymple, R.A. (2008), *A hybrid Boussinesq-SPH model for coastal wave propagation*, Proc. IIIrd *SPHERIC* International Workshop, Lausanne (Switzerland), pp. 11–16.

Crespo, A.J.C., Domínguez, J.M., Barreiro, A., Gómez-Gesteira, M. (2010), *Development of a Dual CPU-GPU SPH model*, Proc. Vth *SPHERIC* International Workshop, Manchester (UK), pp. 401–407.

Cueille, P.-V. (2005), *Modélisation par Smoothed Particle Hydrodynamics des phénomènes de diffusion présents dans un écoulement*, PhD thesis, *INSA* Toulouse (France).

Cummins, S.J., Rudman, M.J. (1999), *An SPH Projection Method*, J. Comput. Phys., **152**:584–607.

Dalrymple, R.A., Rogers, B.D. (2006), *Numerical modelling of water waves with the SPH method*, Coastal Eng., **53**:141–147.

Daly, B.J., Harlow, F.H. (1970), *Transport equations in turbulence*, Phys. Fluids, **13**:2634–2649.

Danielson, D.A., (2003), *Vectors and Tensors in Engineering and Physics*, 2nd edition Westview, Boulder.

Dean, R., Dalrymple, R.A., (1984), *Water Wave Mechanics for Engineers and Scientists*, Prentice-Hall Inc., Englewood Cliffs, New Jersey.

De Boor, C., (2001), *A Practical Guide to Splines*, 2nd edition Springer, New-York.

Dehnen, W., Aly, H. (2012), *Improving convergence in smoothed particle hydrodynamics simulations without pairing instability*, Mon. Not. R. Astron. Soc., **425**:1068–1082.

De Leffe, M., Le Touzé, D., Alessandrini, B. (2009), *Normal flux method at the boundary for SPH*, Proc. IVth *SPHERIC* International Workshop, Nantes (France), pp. 149–156.

De Leffe, M., Le Touzé, D., Alessandrini, B. (2010), *SPH modelling of shallow-water coastal flows*, J. Hydr. Res. **48**:118–125.

De Padova, D., Mossa, M., Sibilla, S., Torti, E. (2010), *Hydraulic jump simulation by SPH*, Proc. V^{th} SPHERIC International Workshop, Manchester (UK), pp. 50–55.

Di Lisio R., Grenier E., Pulvirenti, M. (1998), *The convergence of the SPH Method*, Comput. Math. Appl. **32**:95–102.

Didier, E., Neves, M.G. (2009), *Coastal flow simulation using SPH: wave overtopping on an impermeable coastal structure*, Proc. IVth SPHERIC International Workshop, Nantes (France), pp. 356–362.

Dominguez, J.M., Crespo, A.J.C., Gomez-Gesteira, M. (2010), *Improvements on SPH neighbour list*, Proc. IVth SPHERIC International Workshop, Nantes (France), pp. 364–370.

Doring, M. (2005), *Développement d'une méthode SPH pour les applications surface libre en hydrodynamique*, PhD thesis, Ecole Centrale Nantes (in French).

Dyka, C.T., Randles, P.W., Ingel, R.P. (1997), *Stress points for tension instability in SPH*, Int. J. Num. Meth. Eng. **40**:2325–2341.

Dynkin, E.B. (1947), *Calculation of the coefficients in the Campbell-Hausdorff formula*, Doklady Akad. Nauk. USSR, **57**:323–326.

Einstein, A., *Investigations on the Theory of Brownian Movement*, Dover, New York, 1956.

Ellero, M., Español, P., Flekkoy, E.G. (2003), *Thermodynamically consistent fluid particle model for viscoelastic flows*, Phys. Rev. E **68**:041504.

Ellero, M., Serrano, M., Español, P. (2007), *Incompressible smoothed particle hydrodynamics*, J. Comput. Phys. **226**:1731–1752.

Español, P. (1995), *Hydrodynamics from dissipative particle dynamics*, Phys. Rev. E **52**(2)1734–1742.

Español, P. (2003), *Statistical Mechanics of Coarse-Graining*, in Lecture notes of SoftSimu2002 – Novel Methods in Soft Matter Simulations , Lecture Notes in Physics, Mikko Karttunen, Ilpo Vattulainen, Ari Lukkarinen editors, Springer-Verlag.

Español, P. (2007), *SPH in the mesoscopic world*, Invited lecture, IInd SPHERIC International Workshop, Madrid (Spain).

Español, P., Öttinger, H.C. (1993), *On the interpretation of random forces derived with projection operators*, Zeitschrift fuer Physik B, **90**:377.

Español, P., Revenga, M. (2003), *Smoothed dissipative particle dynamics*, Phys. Rev. E **67**:026705-1.

Favre, A. (1989), *Formulation of the statistical equations of turbulent flows with variable density, in* Studies in Turbulence, Springer Verlag Heidelberg, pp. 324–341.

Feldman, J., Bonet, J. (2007), *Dynamical refinement and boundary contact forces in SPH with applications in fluid flow problems*, Int. J. Num. Meth. Eng., **72**:295–324.

Ferrand, M., Laurence, D., Rogers, B., Violeau, D. (2010), *Improved time scheme integration approach for dealing with semi-analytical wall boundary conditions in Spartacus2D*, Proc. V^{th} SPHERIC International Workshop, Manchester (UK), pp. 98–105.

Ferrand, M., Rogers, B., Laurence, D., Violeau, D. (2011), *Consistent wall boundary treatment for laminar and turbulent flows in SPH*, Proc. VIth SPHERIC International Workshop, Hamburg (Germany).

Ferrand, M., Laurence, D., Rogers, B., Violeau, D., Kassiotis, C. (2012), *Unified semi-analytical wall boundary conditions for inviscid, laminar or turbulent flows in the meshless SPH method*, Int. J. Num. Meth. Fluids, **71**(4):446–472.

Foias, C., Holm D.D., Titi E.S. (2001), *The Navier–Stokes-alpha model of fluid turbulence*, Physica D **152–153**:509–519.

Fourey, G., Le Touzé, D., Alessandrini, B., Oger, G. (2010), *SPH-FEM coupling to simulate Fluid-Structure Interactions with complex free-surface flows*, Proc. V^{th} SPHERIC International Workshop, Manchester (UK), pp. 369–374.

Frisch, U. (1995), *Turbulence: The Legacy of A. N. Kolmogorov*, Cambridge University Press, Cambridge.

Gardiner, C.W. (1985), *Handbook of Stochastic Methods*, 2nd edition Springer-Verlag, Berlin.

Gaskill, J.D. (1978), *Linear Systems, Fourier Transforms, and Optics*, John Wiley & Sons, New York.

Gatski, T.B., Bonnet, J.-P. (2009), *Compressibility, Turbulence and High-speed Flow*, Elsevier, Oxford.

Gatski, T.B., Speziale, C.G. (1993), *On explicit algebraic stress models for complex turbulent flows*, J. Fluid Mech. **254**:59–78.

Ghia, U., Ghia, K. and Shin, C.A. (1982), *High-Resolutions for incompressible flow using the Navier–Stokes equations and a multigrid method*, J. Comput. Phys., **48**:387–411.

Gingold, R.A., Monaghan, J.J. (1977), *Smoothed particle hydrodynamics: Theory and application to non-spherical stars*, Mon. Not. R. Astron. Soc., **181**:375–389.

Girimaji, S.S. (1995), *Fully-explicit and self-consistent algebraic Reynolds-stress model*. Report ICASE No. 95-82.

Goldstein, H., Poole C., Safko J. (2002), *Classical Mechanics*, 3rd edition Addison Wesley, San Francisco (USA).

Gonzales, L.M., Sanchez, J.M., Macia, F., Souto-Iglesias, A. (2009), *Analysis of WCSPH laminar viscosity models*, Proc. IVth *SPHERIC* International Workshop, Nantes (France), pp. 179–186.

Gotoh H., Shao S., Memita T. (2004), *SPH-LES model for numerical investigation of wave interactions with partially immersed breakwater*, Coastal Eng. J., **46**:39–63.

Gotoh, H., Ikari, H., Memita, T., Sakai, T. (2005), *Lagrangian particle method for simulation of wave overtopping on a vertical seawall*, Coast. Eng. J. **47**:157–181.

Gourlay, M.R. (1996), *Wave set-up on coral reefs. 1. Set-up and wave-generated flow on an idealised two dimensional horizontal reef*, Coastal Eng. **27**:161–193.

Grenier, N., Le Touzé, D., Colagrossi A., Antuono, M. (2009), *A SPH multiphase formulation with a surface tension model applied to oil-water separation*, Proc. IVth *SPHERIC* International Workshop, Nantes (France), pp. 22–29.

Groenenboom, P., Cartwright, B.K. (2010), *Hydrodynamics and fluid-structure interaction by coupled SPH-FE method*, J. Hydr. Res., **48**:61–73.

Guermond, J.L., Oden J.T., Prudhomme S. (2003), *An interpretation of the Navier–Stokes-alpha model as a frame-indifferent Leray regularisation*, Physica D, **177**:23–30.

Guilcher, P.-M., Ducrozet, G., Alessandrini, B., Ferrant, P. (2007), *Water wave propagation using SPH models*, Proc. IInd *SPHERIC* International Workshop, Madrid (Spain), pp.119–122.

Guimet, V., Laurence, D. (2002), *A linearised turbulent production in the $k - \varepsilon$ model for engineering applications*, Proc. Vth International Symposium on Engineering Turbulence Modelling and Measurements, Majorqua (Spain).

Hebsgaard M., Sloth. P., Juhl J. (1998), *Wave overtopping of rubble-mound breakwaters*, Proc. XXVIth Int. Conf. Coastal Eng., Copenhagen (Denmark), pp. 2235–2248.

Hérault, A., Bilotta, G., Dalrymple, R.A. (2010), *SPH on GPU with CUDA*, J. Hydr. Res., **48**:74–79.

Hervouet, J.-M. (2007), *Hydrodynamics of Free Surface Flows*, John Wiley & Sons, Chichester.

Holm, D.D. (1999), *Fluctuation effects on 3D Lagrangian mean and Eulerian mean fluid motion*, Physica D, **133**:215–269.

Hoyas, S., Jimenez, J. (2006), *Scaling of the velocity fluctuations in turbulent channels up to Re_τ* = 2000, Phys. Fluids, **18**:011702.

Hu, X.Y., Adams, N.A. (2006), *A multiphase SPH method for macroscopic and mesoscopic flows*, J. Comput. Phys., **213**:844–861.

Hu, X.Y., Adams, N.A. (2008), *A constant-density approach for incompressible multiphase SPH*, Proc. IIIe SPHERIC International Workshop, Lausanne (Switzerland), pp. 61–66.

Hu, X.Y., Adami, S., Adams, N.A. (2009), *Formulating surface tension with reproducing divergence approximation for multi-phase SPH*, Proc. IVth *SPHERIC* International Workshop, Nantes (France), pp. 38–44.

Issa, R. (2005), *Numerical assessment of the Smoothed Particle Hydrodynamics gridless method for incompressible flows and its extension to turbulent flows*, PhD thesis, University of Manchester (UK).

Issa, R. (2006), *2nd benchmark test case: 3D schematic dam break and evolution of the free surface*, *SPHERIC* community, http://wiki.manchester.ac.uk/spheric/ [accessed 14 November 2011].

Issa, R., Violeau, D. (2009), *Modelling a plunging breaking solitary wave with eddy-viscosity turbulent SPH models*, Comput., Materials and Continua **8**(3):151–164.

Issa, R., Lee, E.-S., Violeau, D., Laurence, D. (2004). *Incompressible separated flows simulations with the Smoothed Particle Hydrodynamics gridless method.* Int. J. Num. Meth. Fluids, **47** (10–11):1101–1106.

Issa, R., Violeau, D., Laurence, D. (2005), *A first attempt to adapt 3D Large Eddy Simulation to the Smoothed Particle Hydrodynamics gridless method*, Proc. Int. Conf. Comput. and Experimental Eng. and Sciences, Ist Symposium on Meshless Methods, Stara-Lesna (Slovakia).

Issa, R., Lee, E.S., Violeau, D., Gariah, A., Stansby, P.K., Laurence, D. (2007), *Wave interactions with coastal structures: quantitative predictions using SPH*, Proc. IInd *SPHERIC* International Workshop, Madrid (Spain), pp.15–18.

Issa, R., Moulinec, C., Latino, D., Violeau, D., Biddiscombe, J., Thibaud, G. (2008), *HPC for Spartacus-3D SPH code and applications to industrial environmental flows*, Proc. IIIe SPHERIC International Workshop, Lausanne (Switzerland), pp. 121–135.

Issa, I., Violeau, D., Lee, E.S., Flament, H. (2009a), *Modelling nonlinear water waves with RANS and LES SPH models, in Advances in Numerical Simulation of Nonlinear Water Waves*, Q.W. Ma ed., Series of Advances in Coastal and Ocean Engineering, Vol. 11, World Scientific Publishing Co., London.

Issa, R., Rougé, D., Benoit, M., Violeau, D., Joly, A. (2009b), *Modelling algae transport in coastal areas with the shallow water equations*, J. Hydro-Environment Res., **3**:215–223.

Jang, Y., Marongiu, J.-C., Parkinson, E. (2008), *Analysis of SPH and mesh based simulations using point based processing tool*, Proc. IIIrd SPHERIC International Workshop, Lausanne (Switerland), pp. 198–202.

Janosi, I.M., Jan, D., Szabo, K.G., Tel, T. (2004), *Turbulent friction reduction in dam-break flows*, Experiments in Fluids, **37**:219–229.

Jiang, S., Oliveira, M.S.A., Sousa, A.C.M. (2006), *SPH simulation of transition to turbulence for planar shear flow subjected to a streamise magnetic field*, J. Comput. Phys., **217**:485–501.

Kajtar, J., Monaghan, J.J. (2008), *SPH simulations of swimming linked bodies*, J. Comput. Phys. **227**:8568–8587.

Kambe, T. (2008), *Variational formulation of ideal fluid flows according to gauge principle*, Fluid dynamics Research, **40**:399–426.

Kassiotis, C., Rogers, B.D., Ferrand, M., Violeau, D., Stansby, P.K., Benoit, M. (2011), *Coupling SPH with a 1–D Boussinesq type wave model*, Proc. VIth SPHERIC International Workshop, Hamburg (Germany).

Kendall, M.G. (1961), *A Course in the Geometry of n Dimensions*, Charles Griffin & Company Limited, London.

Khayyer, A., Gotoh, H., Shao, S.D. (2008), *Corrected incompressible SPH method for accurate water-surface tracking in breaking waves*, Coastal Eng., **55**: 236–250.

Kim, J., Moin, P., Moser, R. (1987), *Turbulence statistics in fully developed channel flow at low Reynolds number*, J. Fluid Mech., **177**:133–166.

Kleefsman, K.M.T., Fekken, G., Veldman, A.E.P., Iwanowski, B., Buchner, B. (2005), *A volume-of-fluid based simulation method for wave impact problems*, J. Comput. Phys., **206**:363–393.

Knapp, C.E. (2000), *An implicit smooth particle hydrodynamics code*, PhD thesis, Los Alamos National Laboratory (USA).

Kolmogorov, A.N. (1962), *A refinement of previous hypotheses concerning the local structure of turbulence in a viscous incompressible fluid at high Reynolds number*, J. Fluid Mech. **13**:82–85.

Kolmogorov, A.N. (1991), *The local structure of turbulence in incompressible viscous fluid for very large Reynolds number*, Proc. R. Soc. London Ser. A, **434**:9–13.

Koshizuka, S., Oka, Y. (1996), *Moving-particle semi-implicit method for fragmentation of compressible fluid*, Nuclear Sc. Eng., **123**:421–434.

Kulasegaram, S., Bonet, J., Lewis, R.W., Profit, M. (2004), *A variational formulation based contact algorithm for rigid boundaries in two-dimensional SPH applications*, Comput. Mech., **33**:316–325.

Landau, L.D. (1944), *On the problem of turbulence*, Doklady Akademii Nauk SSSR **44**:339–342.

Landau, L.D., Lifshitz, E. (1976), *Course of Theoretical Physics. Vol. 1: Mechanics*, 3rd edition Butterworth-Heinemann, Oxford.

Landau, L.D., Lifshitz, E. (1975), *Course of Theoretical Physics. Vol. 2: The Classical Theory of Fields*, 4th edition Butterworth-Heinemann, Oxford.

Landau, L.D., Lifshitz, E. (1980), *Course of Theoretical Physics. Vol. 5: Statistical Physics, Part 1*, 3rd edition Butterworth-Heinemann, Oxford.

Landau, L.D., Lifshitz, E. (1987), *Course of Theoretical Physics. Vol. 6: Fluid Mechanics*, 2nd edition Butterworth-Heinemann, Oxford.

Landau, L.D., Berestetskii, V., Lifshitz, E., Pitaevskii, L. (1982), *Course of Theoretical Physics. Vol. 4: Quantum Electrodynamics*, 2nd edition Butterworth-Heinemann, Oxford.

Landau, L.D., Lifshitz, E., Kosevich, A. (1986), *Course of Theoretical Physics. Vol. 7: Theory of Elasticity*, 3rd edition Butterworth-Heinemann, Oxford.

Laplace, P.-S., (1814), *Théorie analytique des probabilités*, 2nd edition, Courcier, Paris (in French).

Lastiwka, M., Basa, M., Quinlan, N.J. (2008), *Permeable and non-reflecting boundary conditions in SPH*, Int. J. Num. Meth. Fluids **61**(7):709–724.

Launder, B.E. (1990), *Phenomenological modelling: present... and future?*, in *Whither Turbulence? Turbulence at the Crossroads*, J.L. Lumley (ed.), Springer-Verlag, New York.

Launder, B. E., Spalding, D. B. (1974), *The numerical computation of turbulent flows*, Comput. Meth. Applied Mech. Eng. **3**(2):269–289.

Launder B.E., Reece G.J., Rodi W. (1975), *Progress in the development of a Reynolds stress turbulent closure*, J. Fluid Mech. **68**:537–566.

Lee, E.S. (2007), *Truly incompressible approach for computing incompressible flow in SPH and comparisons with the traditional weakly compressible approach*, PhD thesis, University of Manchester (UK).

Lee, E.S., Violeau, D., Benoit, M., Issa, R., Laurence, D., Stansby, P. (2006), *Prediction of wave overtopping on coastal structures by using extended Boussinesq and SPH models*, Proc. XXXth Int. Conf. Coastal Eng., San Diego (USA) pp. 4727–4739.

Lee, E.S., Moulinec, C., Violeau, D., Laurence, D., Stansby, P. (2007a), *2-D flow past a bluff body in a closed channel.* Proc. XXXIInd IAHR Biennial Congress, Venice (Italy) pp. 108–110.

Lee, E.S., Violeau, D., Laurence, D., Stansby, P., Moulinec, C. (2007b), *SPH Benchmark Test Case 6: 2-D Incompressible flow around a moving square inside a rectangular box*, Proc. IInd *SPHERIC* International Workshop, Madrid (Spain), pp. 37–40.

Lee, E.S., Moulinec, C., Xu, R., Violeau, D., Laurence, D., Stansby, P.K. (2008), *Comparisons of weakly compressible and truly incompressible algorithms for the SPH mesh free particle method*, J. Comput. Phys. **227**(18):8417–8436.

Lee, E.S., Violeau, D., Issa, R., Ploix, S., Marc, R. (2009), *Simulating a real dam spillway flow with 3-D SPH*, Proc. IVth *SPHERIC* International Workshop, Nantes (France), pp. 339–345.

Lee, E.S., Violeau, D., Issa, R., Ploix, S. (2010), *Application of weakly compressible and truly incompressible SPH to 3-D water collapse in waterworks*, J. Hydr. Res. **48**:50–60.

Lee, H.Y., Chen, Y.H., You, J.Y., Lin, Y.T. (2000), *Investigations of continuous bed load saltating process*, J. Hydr. Eng. **126** (9):691–700.

Leimkuhler, B., Reich, S. (2005), *Simulating Hamiltonian dynamics*, Cambridge University Press, Cambridge (UK).

Leray, J. (1934), *Sur le mouvement d'un liquide visqueux emplissant l'espace*, Acta Math. **63**:193–248 (in French).

Leroy, A., Violeau, D., Ferrand, M., Kassiotis, C. (2014), *Unified semi-analytical wall boundary conditions for 2-D incompressible SPH*, J. Comput. Phys., **261**:106–129.

Levin, B., Nosov, M. (2009), *Physics of Tsunamis*, Springer, Dordrecht.

Li, Y., Raichlen, F. (2003), *Energy balance model for breaking solitary wave runup*, J. Waterway, Port., Coast. Ocean Eng., **129** (2):47–59.

Lilly, D.K. (1992), *A proposed modification of the Germano subgrid-scale closure method*, Phys. Fluids A **4**:633–635.

Litvinov, S., Hu, X.Y., Adams, N.A. (2008), *Splitting for highly dissipative smoothed particle dynamics*, Proc. IIIrd *SPHERIC* International Workshop, Lausanne (Switzerland), pp. 187–191.

Liu, W.K., Li, S.F., Belytschko, T. (1996), *Moving least square kernel Galerkin method. (II) Fourier analysis*, Comput. Meth. Applied Mech. Eng., **139**:159–193.

Liu, G.R., Liu, M.B. (2003), *Smoothed Particle Hydrodynamics. A Meshfree Particle Method*, World Scientific, Singapore.

Lobovsky, L. (2008), *SPH interaction of fluids and solids*, Proc. IIIrd SPHERIC International Workshop, Lausanne (Switerland), pp. 156–161.

López, D., Marivela, R., Garrote, L. (2009a), *Smoothed Particle Hydrodynamics model applied to hydraulic structures: a hydraulic jump test case*, J. Hydr. Res. **47**: 142–158.

López, D., Marivela, R., Aranda, F. (2009b), *SPH model calibration using data from pressure of the prototype still basin of Villar Del Rey dam, Spain*, Proc. XXXIIIrd IAHR Biennial Congress, Vancouver (Canada) pp. 189–197.

Lorentz, E.N. (1963), *Deterministic non-periodic flow*, J. Atmosph. Sc. **20**:130–141.

Lucy, L.B. (1977), *A numerical approach to the testing of the fission hypothesis*, Astron. J. **82**:1013–1024.

Lumley, J.L. (1978), *Computational modeling of turbulent flows*, Adv. Applied Mech., **18**:123–176.

Mahmood, O., Violeau, D., Kassiotis, C., Ferrand, M. (2011), *Effect of wall boundary treatment in SPH for modelling turbulent flows with inlet/outlet boundary conditions*, Proc. VIth SPHERIC International Workshop, Hamburg (Germany).

Mandelbrot, B.B. (1982), *The Fractal Geometry of Nature*, W.H. Freeman & Co. New York.

Manenti, S., Di Monaco, A., Gallati, M., Sibilla, S., Agate, G., Guandalini, R., Maffio, A. (2009), *Simulating rapid sediment scour by water flow in SPH*, Proc. IVth SPHERIC International Workshop, Nantes (France), pp. 144–148.

Marivela, R., López, D., Lara, Á. Pena, L. (2009), *Applications of the SPH Model to the Design of Fishways*, Proc. XXXIIIrd IAHR Biennial Congress, Vancouver (Canada) 6944–6953.

Marongiu, J.-C. (2007), *Méthode numérique lagrangienne pour la simulation d'écoulements à surface libre - Application aux turbines Pelton*, PhD thesis, Ecole Centrale de Lyon (in French).

Marongiu, J.-C., Lebœuf, F., Caro, J., Parkinson, E. (2009), *Low Mach number numerical schemes for the SPH-ALE method. Application in free surface flows in Pelton turbines*, Proc. IVth SPHERIC International Workshop, Nantes (France), pp. 323–330.

Marongiu, J.-C., Parkinson, E., Lais, S., Leboeuf, F., Leduc, J. (2010), *Application of SPH-ALE method to Pelton hydraulic turbines*, Proc. Vth SPHERIC International Workshop, Manchester (UK), pp. 253–258.

Marrone, S., Colagrossi, A., Le Touzé, D., Graziani, G. (2010), *Fast free-surface detection and level-set function definition in SPH solvers*, J. Comput. Phys., **229**(10):3652–3663.

Marsden, J.E., West, M. (2001), *Discrete mechanics and variational integrators*, Acta Numer. **10**(1):357–514.

Maruzewski, P., Oger, G., Le Touzé, D., Biddiscombe, J. (2010), *High performance computing 3D SPH model: sphere impacting the free-surface of water*, J. Hydr. Res **48**:126–134.

Massel, S.R., Gourlay, M.R. (2000), *On the modelling of wave breaking and set-up on coral reefs*, Coastal Eng. **39**:1–27.

Mayrhofer, A., Gomez-Gesteira, M., Crespo, A.J.C., Rogers, B.D. (2010), *Advanced Pre-Processing for SPHysics*, Proc. V^{th} SPHERIC International Workshop, Manchester (UK), pp. 200–208.

Mayrhofer, A., Rogers, B.D., Violeau, D., Ferrand, M. (2013), *Investigation of wall bounded flows using SPH and the unified semi-analytical wall boundary conditions*, Comput. Phys. Com., **184**:2515–2527.

Mayrhofer, A., Ferrand, M., Kassiotis, C., Violeau, D. (2014), *Unified semi-analytical wall boundary conditions in SPH: analytical extension to 3-D*, Num. Alg., **68**(1): 15–34.

Maurel, B. (2008), *Modélisation par la la méthode SPH de l'impact d'un réservoir rempli de fluide*, PhD thesis, Institut National des Sciences Appliquées de Lyon (in French).

Mellor, G.L., Yamada, T. (1974), *A hierarchy of turbulence closure models for planetary boundary layers*, J. Atmosph. Sc. **31**:1792–1806.

Meriam, J.L. (1966), *Dynamics*, 2nd edition John Wiley, New York.

Minier, J.-P., Peirano, E. (2001), *The pdf approach to turbulent polydispersed two-phase flows*, Physics Reports **352**:1–214.

Monaghan, J.J. (1985a), *Extrapolating B-Splines for interpolation*, J. Comput. Phys. **60**:253–262.

Monaghan, J.J. (1985b), *Particle methods for hydrodynamics*, Comput. Phys. Rep. **3**:71–124.

Monaghan, J.J. (1992), *Smoothed particle hydrodynamics*, Ann. Rev. Astr. Astroph. **30**:543–574.

Monaghan, J.J. (1994), *Simulating free surface flows with SPH*, J. Comput. Phys. **110**:399–406.

Monaghan, J.J. (1998), *An SPH formulation of surface tension*, University of Monash Applied Mathematics Reports and Preprints.

Monaghan, J.J. (1999), *SPH without a tensile instability*, J. Comput. Phys. **159**: 290–311.

Monaghan, J.J. (2002), *SPH compressible turbulence*, Month. Notice Roy. Astr. Soc., **335**(3):843–852.

Monaghan, J.J. (2005), *Smoothed particle hydrodynamics*, Rep. Prog. Phys **68**: 1703–1759.

Monaghan, J.J., Kos, A. (1999), *Solitary Waves on a cretan beach*, J. Waterway, Port, Coastal Ocean Eng., **125**:145–154.

Monaghan, J.J., Kocharyan, A. (1995), *SPH simulation of multi-phase flow*, Comput. Phys. Commun., **87**:225–235.

Monaghan, J.J., Kajtar, J.B. (2009), *SPH boundary forces*, Proc. IVth SPHERIC International Workshop, Nantes (France), pp. 216–219.

Monaghan, J.J., Kos., A., Issa, N. (2003), *Fluid motion generated by impact*, J. Waterway, Port, Coastal Ocean Eng., **129**(6):250–259.

Morris, J.P. (1996a), *A study of the stability properties of Smooth Particle Hydrodynamics*, Publ. Astron. Soc. Aust., **13**:97–102.

Morris, J.P. (1996b), *Analysis of smoothed particle hydrodynamics with applications*, PhD thesis, University of Monash, Melbourne (Australia).

Morris, J.P., Fox, P.J., Zhu, Y. (1997), *Modelling low Reynolds number incompressible flows using SPH*, J. Comput. Phys., **136**:214–226.

Morse, P.M., Feshbach, H. (1953), *Methods of Theoretical Physics*, McGraw-Hill, New-York.

Moser, R.D., Kim, J., Mansour, N.N. (1998), *Direct numerical simulation of turbulent channel flow up to* $Re_t = 590$, Phys. Fluids, **11**(4):943–945.

Moulinec, C., Issa, R., Violeau, D., Marongiu, J.-C., Lebœuf, F. (2007), *Parallel 3-D SPH Simulations over Periodic Hills*, Proc. Int. Conf. Comput. and Experimental Eng. and Sciences, IInd Symposium on Meshless Methods, Patras (Greece).

Moulinec, C., Issa, R., Marongiu, J.-C., Violeau, D. (2008), *Parallel 3-D SPH simulations*, Comput. Model. Eng. Sciences, **25** (3):133–148.

Murnaghan, F.D. (1944), *The Compressibility of Media under Extreme Pressures*, Proc. Nat. Acad. Sc., **30**:244–247.

Narayanaswamy, M., Crespo, A.J.C., Gomez-Gesteira, M., Dalrymple, R.A. (2010), *SPHysics-FUNWAVE hybrid model for coastal wave propagation*, J. Hydr. Res. **48**:85–93.

Nestor, R.M., Quinlan, N.J. (2009), *Incompressible moving boundary flows with the Finite Volume Particle Method*, Proc. IVth *SPHERIC* International Workshop, Nantes (France), pp. 192–199.

Newhouse, S., Ruelle, D., Takens, F. (1978), *Occurrence of strange axiom-A attractors near quasi-periodic flows on* T^m, $m \geqslant 3$, Communications in Mathematical Physics **64**:35–44.

Nezu, I., Nakagawa, H., *Turbulence in open channel. IAHR* Monograph, Balkema, Rotterdam, 1993.

Noether, E. (1918), *Invariante Variationsprobleme*, Nachr. D. Kö nig. Gesellsch. D. Wiss. Zu Göttingen, Math-phys. Klasse, 235–257.

Nwogu, O.G. (1993), *Alternative form of Boussinesq equations for nearshore wave propagation*, J. Waterway, Port, Coastal and Ocean Eng., **119**:618–638.

Obukhov, A.M. (1962), *Some specific features of atmospheric turbulence*, J. Fluid Mech., **13**:77–81.

Oger, G., Doring, M., Alessandrini, B., Ferrant, P. (2005), *Two-dimensional SPH simulations of water wedge entries*, J. Comput. Phys. **213**:803–822.

Oger, G., Doring, M., Alessandrini, B., Ferrant, P. (2007), *An improved SPH method: towards higher order convergence*, J. Comput. Phys., **225**:1472–1492.

Oger, G., Leroy, C., Jacquin, E. (2009), *Specific pre/post treatments for 3-D SPH applications through massive HPC simulations*, Proc. IVth *SPHERIC* International Workshop, Nantes (France), pp. 52–60.

Ottinger, H.C., Grmela, M. (1997), *Dynamics and thermodynamics of complex fluids. II. Illustrations of a general formalism*, Phys. Rev. E **56**:6633–6655.

Orr, W.M.F. (1907a), *The stability or instability of the steady motions of a liquid. Part I*, Proc. Royal Irish Academy, A **27**:9–68.

Orr, W.M.F. (1907b), *The stability or instability of the steady motions of a liquid. Part II*, Proc. Royal Irish Academy, A **27**:69–138.

Orszag, S.A. (1971), *Accurate solution of the Orr–Sommerfeld stability equation*, J. Fluid Mech., **50**:689–703.

Peirano, E., Chibbaro, S., Pozorsky, J., Minier, J.-P. (2006), *Mean-field/PDF numerical approach for polydispersed turbulent two-phase flows*, Progr. Energy Combust. Sc., **32**:315–371.

Pitaevskii, L., Lifshitz, E. (1981), *Course of Theoretical Physics. Vol. 10: Physical Kinetics*, Butterworth-Heinemann, Oxford.

Planck, M. (1899), *Uber Irresversible Strahlungsvorgänge*, Verl. d. Kgl. Akad. d. Wiss., Berlin (in German).

Poincaré, H. (1899), Comptes-Rendus de l'Académie des Sciences **128**:1065–1069 (in French).

Pope, S.B. (1975), *A more general effective-viscosity hypothesis*, J. Fluid Mech. **72**:331–340.

Pope, S.B. (1994), *Lagrangian PDF methods for turbulent flows*, Ann. Rev. Fluid Mech. **26**:23–63.

Pope, S.B. (2000), *Turbulent Flows*. Cambridge University Press, Cambridge.

Pottier, N. (2007), *Physique Statistique Hors d'Equilibre*, EDP Sciences CNRS Editions, Paris (in French).

Prandtl, L. (1925), *Bericht über Untersuchungen zur ausgebildeten Turbulenz*, Z. Angew. Math. Meth., **5**:136–139 (in German).

Quinlan, G.D., Tremaine, S. (1990), *Symmetric multistep methods for the numerical integration of planetary orbits*, Astronom. J. **100**:1694–1700.

Quinlan, N.J., Basa, M., Lastiwka, M. (2006). *Truncation error in mesh-free particle methods*, Int. J. Num. Meth. Eng. **66**:2064–2085.

Rabczuk T., Belytschko T., Xiao S.P. (2004), *Stable particle methods based on Lagrangian kernels*, Comput. Meth. Applied Mech. Eng. **193**:1035–1063.

Randles, P.W., Libersky, L.D. (1996), *SPH: some recent improvements and applications*, Comput. Meth. Applied Mech. Eng. **139**:375–408.

Revell, A., Benhamadouche, S., Craft, T., Laurence, D. (2006), *A stress-strain lag eddy-viscosity model for unsteady mean flow*, Int. J. Heat Fluid Flow **27**: 821–830.

Reynolds, O. (1883), *An experimental investigation of the circumstances which determine whether the motion of water shall be direct or sinuous, and of the law of resistance in parallel channels*, Phis. Trans. Roy. Soc. **174**:935–982.

Reynolds, O. (1894), *On the dynamical theory of incompressible viscous fluids and the determination of the criterion*, Phis. Trans. Roy. Soc., **186** 123–164.

Ricci, G., Levi-Civita, T. (1900), *Méthodes de calcul différentiel absolu et leurs applications*, Mathematische Annalen (Springer) **54**(1–2):125–201 (in French).

Rief, F. (1965), *Fundamentals of Statistical and Thermal Physics*, McGraw-Hill, New York.

Roach, G.F. (1982), *Green's Functions*, Cambridge University Press, Cambridge.

Robinson, M. (2009), *Turbulence and viscous mixing using smoothed particle hydrodynamics*, PhD thesis, University of Monash, Melbourne (Australia).

Rodi., W. (1976), *A new algebraic relation for calculating the Reynolds stresses*, Z. Angew. Math. Mech., **56**:219–221.

Rodi, W. (2000), *Turbulence Models and Their Applications in Hydraulics*. IAHR monograph, Brookfield, Rotterdam.

Rodriguez-Paz, M., Bonet, J. (2005), *A corrected smooth particle hydrodynamics formulation of the shallow-water equations*, Comput. Struct. **17–18**:1396–1410.

Rogers, B.D. (2006), *3rd benchmark test case: 2-D lid-driven cavity flow without gravity*, *SPHERIC* community, http://wiki.manchester.ac.uk/spheric/ [accessed 14 November 2011].

Rogers, B.D., Leduc, J., Marongiu, J.-C., Lebœuf, F. (2009), *Comparison and evaluation of multi-phase and surface tension models*, Proc. IVth *SPHERIC* International Workshop, Nantes (France), pp. 30–37.

Rotta, J.C. (1951), *Statistische theorie nichthomogener turbulenz*, Z. Phys., **129**: 547–572 (in German).

Schwaiger, H.F. (2007), *An implementation of smoothed particle hydrodynamics for large deformation, history dependent geomaterials with application to tectonic deformation*, PhD thesis, University of Washington (USA).

Schwaiger, H.F. (2008), *An implicit corrected SPH formulation for thermal diffusion with linear free-surface boundary condition*, Int. J. Num. Meth. Eng., **75**: 647–671.

Shakibaeinia, A., Jin, Y.C. (2009), *Lagrangian modeling of flow over spillways using the Moving Particle Semi-Implicit method*, Proc. XXXIIIrd *IAHR* Biennial Congress, Vancouver (Canada) 1809–1816.

Shao, S. (2006), *Incompressible SPH simulation of wave breaking and overtopping with turbulence modelling*, Int. J. Numer. Meth. Fluids **50**:597–621.

Shao, S., Lo, E.Y.M. (2003), *Incompressible SPH method for simulating Newtonian and non-Newtonian flows with a free surface*, Adv. Water Resources, **26**:787–800.

Shao, S., Ji, C., Graham, D.I., Reeve, D.E., James, P.W., Chadwick, A.J. (2006), *Simulation of wave overtopping by an incompressible SPH model*, Coastal Eng., **53**:723–735.

Shepard, D. (1968), *A two dimensional function for irregularly spaced data*, Proc. XXIIIrd *ACM* National Conference, New-York (USA), pp. 517–524.

Shkoller, S. (1999), *The geometry and analysis of non-Newtonian fluids and vortex methods*, E-print Math. AP/9908109.

Sigalotti, L.D.G., Lopez, H., Trujillo, L. (2009), *An adaptative SPH method for strong shocks*, J. Comput. Phys. **228**:5888–5907.

Smagorinsky, J. (1963), *General Circulation experiments with the primitive equations I. The basic experiment*, Monthly Weather Review, **91**(3):99–164.

Sommerfeld, A. (1908), *Ein Beitrag zur hydrodynamische Erklärung der turbulenten Flüssigkeitsbewegungen*, Proc. IVth International Congress of Mathematicians, Vol. III, Rome (Italy), pp. 116–124 (in German).

Springel, V., Yoshida, N., White, S.D.M. (2001), *GADGET: A code for collisionless and gasdynamicalal cosmological simulations*, New Astron., **6**:79–117.

Stansby, P.K., Feng., T. (2004), *Surf zone wave overtopping a trapezoidal structure: 1-D modelling and PIV comparison*, Coastal Eng. **51**:483–500.

Stansby, P.K., Chegini, A., Barnes, T.C.D. (1998), *The initial stages of dam-break flow*, J. Fluid Mech. **370**:203–220.

Stein, E., Weiss, G. (1971), *Introduction to Fourier Analysis on Euclidean spaces*, Princeton University Press, Princeton.

Swaters, G.E. (2000), *Introduction to Hamiltonian Fluid Dynamics and Stability Theory*, Chapman & Hall/CRC, Boca Raton.

Swegle, J.W., Hicks, D.L., Attaway, S.W. (1995), *Smoothed particle hydrodynamics stability analysis*, J. Comput. Phys., **116**:123–134.

Tait, P.G. (1888), *Report on some of the physical properties of fresh water and sea water*, Rept. Sci. Results Voy., H.M.S. Challenger, Phys. Chem. **2**:1–76.

Takeda, H., Miyama, S.M., Sekiya, M. (1994), *Numerical simulation of viscous flow by smoothed particle hydrodynamics*, Prog. Theor. Phys., **92**(5):939–960.

Tarrade, L., Texier, A., David, L., Larinier, M. (2008), *Experimental approach to adapt the turbulent flow in the vertical slot fishways to the small fish species*, J. Hydrob. **609**(1):177–188.

Tartakofsky, A.M., Tartakofsky, D.M., Meakin, P. (2008), *Smoothed Particle Hydrodynamics stochastic model for flow and transport in porous media*, Proc. IIIrd *SPHERIC* International Workshop, Lausanne (Switerland), pp. 1–5.

Tartakofsky, A.M., Ferris, K., Meakin, P. (2009), *SPH Van der Waals model for multiphase flows*, Proc. IVth SPHERIC International Workshop, Nantes (France), pp. 45–51.

Temam, R. (1968), *Une méthode d'approximation des solutions des équations de Navier–Stokes*, Bull. Soc. Math. France, **98**:115–152 (in French).

Uribe, J.C., Laurence, D. (2002), *10th Ercoftac (SIG15) / IAHR / QNET-CFD Workshop on Refined Turbulence Modelling*, University of Manchester (UK), http://cfd.mace.manchester.ac.uk/CfdTm/TestCase014.

U.S. Department of the Army, Corps of Engineers (1990), *Hydraulic Design of Spillways*, EM 1110-2-1603.

Vacondio, R., Mignosa, P., Rogers, B.D., Stansby, P.K. (2010), *SPH Shallow Water Equation Solver for real flooding simulation*, Proc. V^{th} SPHERIC International Workshop, Manchester (UK), pp. 106–113.

Valizadeh, A., Shafieefar, M., Monaghan, J.J., Salehi-Neyshaboori, S.A.A. (2008), *Modeling two-phase flows using SPH method*, J. Applied Sci. **8**:3817–3826.

Van der Vorst, H.A. (1992), *Bi-CGSTAB: A fast and smoothly converging variant of Bi-CG for the solution of non-symmetric linear systems*, S.J. Sci. Stat. Comput **13**: 631–644.

Van der Pol, B. (1920), *A theory of the amplitude of free and forced triode vibrations*, Radio Review **1**:701–710, 754–762.

Van der Waals, J.D. (1873), *On the continuity of the gaseous and liquid states*, PhD thesis, University of Leiden (The Netherlands).

Van Liedekerke, P., Tijskens, E., Ramon, H., Ghysels, P., Samaey, G., Roose, D. (2009), *A particle based model to simulate plant cells dynamics*, Proc. IVth *SPHERIC* International Workshop, Nantes (France), pp. 208–215.

Van Rijn, H.C. (1984), *Sediment transport. Part III: Bed forms and alluvial roughness*, J. Hydr. Eng. **110**(12):1733–1754.

Vaschy, A. (1892), *Sur les considérations d'homogénéité en Physique*. Réponse à une Note de M. Clavenad, Comptes Rendus des Séances de l'Académie des Sciences **115**:597–599 (in French).

Vasquez-Quesada, A., Ellero, M., Español, P. (2009), *SPH model for viscoelastic fluids with thermal fluctuations*, Proc. IVth *SPHERIC* International Workshop, Nantes (France), pp. 102–109.

Ven Te Chow (1959), *Open-Channel Hydraulics*, McGraw-Hill, New-York.

Vignjevic, R., Reveles, J., Campbell, J. (2006), *SPH in a total Lagrangian formalism*, Comput. Meth. Eng. Science **14**(3):181–198.

Vila, J.-P. (1999), *On particle weighted methods and SPH*, Math. Model Meth. Appl. Sci. **9**:161–210.

Violeau, D. (2004), *One and two-equations turbulent closures for Smoothed Particle Hydrodynamics*, Proc. 6th Int. Conf. Hydroinformatics, 21–24 June 2004, Singapore, pp. 87–94.

Violeau, D. (2009a), *Explicit algebraic Reynolds stresses and scalar fluxes for density-stratified shear flows*. Phys. Fluids **21**:035103.

Violeau, D. (2009b), *Dissipative forces for Lagrangian models in computational fluid dynamics and application to smoothed-particle hydrodynamics*, Phys. Rev. E. **80**:036705.

Violeau, D., Issa, R. (2005), *Modelling wave overtopping with Smoothed Particle Hydrodynamics*, Proc. XXXIst IAHR Biennial Congress, Seoul (Korea).

Violeau, D., Issa, R. (2007a), *Numerical modelling of complex turbulent free surface flows with the SPH Lagrangian method: an overview*, Int. J. Num. Meth. Fluids **53**(2):277–304.

Violeau, D., Issa, R. (2007b), *Influence of turbulence closure on free-surface flow modelling with SPH*, Proc. XXXIInd *IAHR* Biennial Congress, Venice (Italy).

Violeau, D., Issa, R. (2007c), *SPHERIC benchmark test case number 5: sensitivity analysis to numerical and physical parameters*, Proc. IInd *SPHERIC* International Workshop, Madrid (Spain), pp. 47–50.

Violeau, D., Piccon, S., Chabard, J.-P. (2001), *Two attempts of turbulence modelling in smoothed particle hydrodynamics*, Advances in Fluid Modelling and Turbulence Measurements, World Scientific 2002, pp. 339–346.

Violeau, D., Buvat, C., Abed-Meraïm, K., de Nanteuil, E. (2007). *Numerical modelling of boom and oil spill with SPH*. Coastal Eng., **54**:895–913.

Violeau, D., Issa, R., Benhamadouche, S., Saleh, K., Chorda, J., Maubourguet, M.-M. (2008), *Modelling a fish passage with SPH and Eulerian codes: the influence of turbulent closure*, Proc. IIIrd SPHERIC International Workshop, Lausanne (Switerland), pp. 85–91.

Violeau, D., Leroy, A. (2014), *On the maximum time step in weakly compressible SPH*, J. Comput. Phys., **256**:388–415.

Violeau, D., Leroy, A. (2015), *Optimal time step for Incompressible SPH*, J. Comput Phys., **288**:119–130.

Wallin N, Johansson AV. (2000), *An explicit algebraic Reynolds stress model for incompressible and compressible turbulent flows*, J. Fluid Mech. **403**:89–132.

Warner, J.C., Sherwood, C.R., Arango, H.G., Signel, R.P. (2005), *Performance of four turbulence closure models implemented using a generic length scale method*, Ocean Modelling **8**:81–113.

Welton W.C., Pope S.B. (1997), *PDF model calculations of compressible turbulent flows using smoothed particle hydrodynamics*, J. Comput. Phys., **134**:150–168.

Wendland, H. (1995), *Piecewise polynomial, positive definite and compactly supported radial functions of minimal degree*, Adv. Comput. Math., **4**:389–396.

Wilcox, D.C. (1988), *Re-assessment of the scale-determining equation for advanced turbulence models*, AIAA Journal **26**:1414–1421.

Xu, R., Stansby, P., Laurence, D. (2009), *Accuracy and stability in incompressible SPH (ISPH) based on the projection method and a new approach*, J. Comput. Phys. **228**:6703–6725.

Xu, H., Dao, M.H., Chan, E.S. (2010), *A SPH model with C1 particle consistency*, Proc. Vth *SPHERIC* International Workshop, Manchester (UK), pp.194–200.

Yakhot, V., Orszag, S.A. (1986), *Renormalization group analysis of turbulence. I – Basic theory*, J. Sc. Comput. **1**(1):3–51.

Yap, C.J. (1987), *Turbulent heat and momentum transfer in recirculating and impinging flows*, PhD thesis, University of Manchester (UK).

Yildiz, M., Rook, R.A., Suleman, A. (2009), *SPH with the multiple boundary tangent method*, Int. J. Num. Meth. Eng. **77**(10):1416–1438.

Žagar, D., Džebo, E., Četina, M., Petkovšek, G. (2009), *Simulations of dam break and flow through a steep valley using SPH*, Proc. XXXIIIrd *IAHR* Biennial Congress, Vancouver (Canada) pp. 171–177.

Zwiebach, B. (2009), *A First Course in String Theory*, Cambridge University Press, Cambridge 2nd edition.

Index